Cretaceous-Tertiary Mass Extinctions:
Biotic and Environmental Changes

Cretaceous-Tertiary Mass Extinctions:
Biotic and Environmental Changes

Norman MacLeod and Gerta Keller

W. W. Norton & Company
New York - London

Printed in the United States of America

First Edition

The text of this book is composed in Times Roman
with the display set in Times Bold
Composition and design by Gina Webster
Manufacturing by Courier, Westford

Library of Congress Cataloging-in-Publication Data

Cretaceous-Tertiary mass extinctions: biotic and environmental
changes / Norman MacLeod, Gerta Keller [editors].
p. cm.
Includes bibliographical references and index (pp. 557–75).
1. Extinction (Biology) 2. Cretaceous-Tertiary boundary.
3. Paleoecology. I. MacLeod, Norman, 1953– . II. Keller, Gerta.
QE721.2.E97C74 1995
560'.45—dc20 95–10161

ISBN 0-393-96657-7 (pbk.)

W. W. Norton & Company, Inc., 500 Fifth Avenue, New York, N.Y. 10110
W. W. Norton & Company Ltd., 10 Coptic Street, London WC1A 1PU

1 2 3 4 5 6 7 8 9 0

Foreword

The Cretaceous-Tertiary Cloud Chamber

Niles Eldredge,
*The American Museum of Natural History,
New York, New York*

Extinction is a hot topic these days. The scientific community, along with a good measure of the body politic, is growing increasingly alarmed at the "biodiversity crisis," that is, the mounting loss of species via human alteration of the landscape. This current phase of mass extinction has many implications for the human future, and as we approach the millennium, with attention correspondingly (if artificially) intensified on what lies ahead, extinction looms as a critical phenomenon that all of humanity must confront.

There are purely scientific reasons why extinction has emerged in the past two decades as a subject of serious research. And that work has already paid high dividends. The turning point, the event that focused so much attention on extinction, was the publication of a single paper by a father and son team, Luis and Walter Alvarez, and their colleagues (Alvarez and others, 1980). That paper, as all readers of this book are sure to know already, presented evidence for an extraterrestrial bolide impact at the end of the Cretaceous, and proposed the hypothesis that the collision was the root cause of the mass extinction at the Cretaceous-Tertiary (K-T) boundary. The Alvarez hypothesis garnered extra attention because the K-T event was the one that finally took out those most famous of all extinct organisms—the (nonavian!) dinosaurs.

There truly is little new under the sun, and the more I peruse the nineteenth century writings of the then-young science of paleontology, the more I am persuaded that the phenomena (if not their causal explanations) we attach so much importance to today were well known to our early predecessors. The phenomenon of stasis is the closest to my intellectual home. The relative stability of species once they first appear in the stratigraphic record was well known to nineteeth-century paleontologists. All of the paleontological reviewers of Darwin's

(1859) *On the Origin of Species* (as collected in Hull, 1973) commented on Darwin's omission of this phenomenon from his evolutionary discussions. Stasis only came to be seen as an important phenomenon in evolution in the early 1970s (Eldredge, 1971; see Maynard Smith, 1984).

Much as stasis had to await rediscovery, so too did the realization that major faunal turnovers constitute a truly huge theme in the history of life—a theme that lay dormant until well past the middle of the twentieth century. Norman D. Newell (my mentor at Columbia University and The American Museum of Natural History in the 1960s) was a virtual lone voice in the wilderness, persistently pointing to what he often called "crises" in the history of life (cf. Newell, 1967), while those around him (including his students) vastly preferred to focus on evolution and other matters. Extinction seemed to us the down side, a purely negative aspect in contrast to the positive nature of the evolutionary process. That we now understand that much of evolution is heavily contingent on prior extinction events represents a profound revolution in our grasp of the history of life, and perhaps the most important example of the large dividends that the relatively recent focus on mass extinction has already begun to pay.

But there is more to the genesis of new fields of inquiry in general—and to the renaissance of interest in mass extinction in particular—than the publication of a single paper, however galvanizing its appearance might have been. Paleontology, in its overtly more theoretical and analytical guise as "paleobiology," had been showing signs for some time of breaking into the mainstream of causal explanation typical of conventional science. The problem we have always faced as a science, it seems to me, has been a basically false distinction traditionally drawn between "historical" sciences (which paleontology supposedly epitomizes) and "functional" ("real") science. Historical sciences, it is commony said, deal with the events of history, and must therefore leave their explanation, in true uniformitarian, inductive fashion, to the sciences that deal with entities and processes that can be studied experimentally. This is not an attitude foised on our profession from those outside (although see Maynard Smith [1984] for such an articulation). No less distinguished personages as George Gaylord Simpson (1963) and Ernst Mayr (1982) have written at great length that there is something fundamentally different between "historical" and the other sciences: in my estimation, both esteemed gentlemen were quite wrong.

The great gift, the sine qua non, of historical science is routinely said to be time itself. However, how do we take advantage of the depths of geologic time when we confront the history of life? Traditionally we take events and processes known to occur in the modern world, as detected in the lab and in the field, and extrapolate them over long periods of time. Such a general modus operandi of course confirms abundantly the general supposition that such is all a historical scientist can ever do—apply the causal interactions known to occur in the modern world by extrapolation back to encompass events in the geologic past.

It is true that no paleontologist, surveying the most favorable data set imaginable, could ever have formulated Mendelian genetics, let alone discover that genes are information-bearing segments of DNA, located on chromosomes that are housed, for the most past, within nuclei (in eukaryotes). However, we historian-scientists should never forget that natural selection itself was a discovery made independently by two people (at least), neither of whom had the foggiest notion of the underlying causes of heritable variation. A knowledge of genetics was not a prerequisite to the discovery of natural selection—which tells us something profound about the organization of biological systems. In a striking passage, Dobzhansky (1937) drew a sharp distinction between what he called "physiological genetics" (the statics and dynamics of inheritance) and "population genetics," those processes, such as natural selection and genetic drift, that alter the frequencies of alleles within populations. Population genetics are not in any meaningful sense reducible to the principles of inheritance. As Dobzhansky (1937) said, different rules apply to the genetics of organisms than obtain with the genetics of populations. The systems are hierarchically arranged: organisms are parts of populations.

The real significance of the depths of time in geologic history is not that we can extrapolate back the events and processes observable in the temporal dimensions of human lifespans, but that we are at liberty to search out the behavior of spatio-temporally larger-scale entities that are intrinsically difficult to observe within the confines of laboratories and human lifetimes. Viewed this way, there really is no meaningful distinction between an experimental physicist's cyclotron and the sedimentological and fossil records of the Maastrichtian and Danian. The tracks in the cyclotron's cloud chamber record the evanescent histories of various elementray particles. The physicist is interested not in the details of the history of this or that individual particle, but rather in the characteristically similar behaviors of groups, or classes, of such particles.

In a similar way, we paleobiologists can look beyond the (intrinsically fascinating) details of the histories of individual clades. We can look for patterns in the history of life, just as cyclotron physicists look for patterns in cloud chambers. We see such patterns at characteristically very large spatio-temporal scales, embracing ecosystems and larger-scale ecological entities such as biomes and provinces, as well as species and monophyletic taxa. We have our own particles (as the late T. J. N. Schopf referred to species, at least in a certain context). We can see that species come and go, and that species appear or disappear at characteristically different rates both within different clades and within different ecological systems. Because no students of the Holocene biota have adduced a theory to explain such patterns of "species sorting," it is up to those of us who confront these large-scale entities seen in our K-T version of the cloud chamber to propose such explanations. As long as what we choose to say is thoroughly consistent with well-established microevolutionary and ecological postulates, we are involved in causal science as much as is a cyclotron physicist—or, for that matter, a population geneticist.

It is ironic that perhaps the best model of how we connect patterns in paleontological data to novel causal theory comes from George Gaylord Simpson, the same paleontologist who later stressed the supposed fundamental disparity between historical and nonhistorical sciences. In his *Tempo and Mode in Evolution* (Simpson, 1944), Simpson acknowledged that population size, mutation rates, and other phenomena in the purview of population genetics indeed constitutes what he termed the "determinants" of evolution. However, genetic studies "May reveal what happens to a hundred rats in the course of ten years under fixed and simple conditions, but not what happened to a billion rats in the course of a million years under the fluctuating conditions of earth history" (Simpson, 1944, p. xvii). To Simpson, the actual fossil record of the history of life posed some major barriers to a simple extrapolaion of population genetics results obtained over a few experimental generations.

Simpson (1944) had in mind what we are now accustomed to calling "patterns" in the history of life. In this context, patterns are classes of events—similar events affecting different species, clades, or ecosystems at different times and in different places. Simpson's (1944) *Tempo and Mode in Evolution* is remembered best for its original articulation of quantum evolution, a hypothesis put forth to explain a recurrent phenomenon of evolutionary history. Simpson felt that gradual, linear transformation within lineages in the manner of the original Darwinian formulation constitutes the basically accurate causative explanation of the origin of species and genera. He also felt that gaps in the record between species or closely related genera are attributable to taphonomic artifact; that is, imperfections in the fossil record.

So far, so traditionally Darwinian. However, Simpson also noted that gaps between higher taxa—families, orders, and so forth—could not be attributed so easily to gaps in the fossil record. The earliest whales and bats (both Eocene, then as now) were primitive to some degree with reference to their modern descendants, but nonetheless display the majority of what we now call the "synapomorphies" of their respective clades. Simpson proposed that the average rate of evolution observed between the earliest members of higher taxa, from oldest fossil representatives to modern species, if extrapolated *back* to presumed ancestral states (four-legged, fully terrestrial placental mammals, in the case of both whales and bats) would yield absurd results. Whales and bats would have to have arisen far back in the Mesozoic, perhaps even in the Paleozoic—absurd, simply because their origins would have predated the earliest known dates of appearance of the most primitive known placental mammal.

Simpson had a pattern, one which could not be dismissed as yet another artifact of the fossil record. The record for case after case of the "abrupt" appearance of well-diversified higher taxa, seemed to him to proclaim a mode of exceptionally rapid evolution. Simpson (1944) spent much of his time arguing for the existence of three discrete classes of evolutionary rates, and then examining exceptionally low rates ("living fossils") to present a mirror-image case to shed light on factors governing episodes of exceptionally rapid evolution (Eldredge, 1984). Though couched in the terms of population genetics, Simpson actually adduced a novel combination of "evolutionary determinants," a theory to explain a common (and still important and not fully addressed) pattern in the history of life. Although Simpson (1953), under criticism (cf. Wright, 1945), later retreated from his position, his example should stand as a model to all of us in paleobiology: the fossil record reveals patterns in life's history wholly unanticipated by studies of the modern biota. These patterns often require novel formulations of causal explanation.

The renewed emphasis on extinction in paleobiology is, in my view, an exciting example of the more general trend to take the fossil record more seriously, to examine it for repeated ecological and evolutionary patterns, and to adduce causal theory to explain such patterns. Such theory yields predictions about what might be observed in the record should the theory be true: on the smaller scale of regional ecosystems, the work of Brett and Baird (1955) on patterns of "coordinated stasis" in some 13 successive mid-Paleozoic faunas of the Appalachian basin, and Vrba (1985, 1993) on "turnover pulse" phenomena come to mind. Among other issues, Brett and Baird (1995) raise a chicken-and-egg causal dilemma; that is, is species stasis the cause, or the result, of monotonous recurrence of assemblages within each of these stable mid-Paleozoic faunal assemblages? Vrba (1985, 1993), confronted with evidence of an abrupt faunal turnover in East Africa about 2.5 m.y. ago, proposed an interesting combination of ecological processes (emigration and immigration via habitat tracking) and evolutionary processes (true extinction and speciation caused by habitat transformation and fragmentation) as an overall explanation of the turnover pattern. All such theoretical postulates are consonant with well-entrenched biological theory; however, the phenomena they address have gone essentially undreamed of by biologists, whose gaze has traditionally been fixed strictly on the modern biota. In addition, as I argue elsewhere (Eldredge, 1995, 1996), new causative models, especially species sorting, are required in order to yield a complete explanation of such phenomena.

Thus, the Maastrichtian-Danian fossil record truly can be seen as a large-scale analogue to the experimental physicist's cloud chamber. That mass extinctions are patterns is indisputable; prior to Darwin, such historical events were recognized empirically and served as the basis of the original division of Phanerozoic time. The geologic time scale itself is the best, albeit informal, evidence that mass extinctions (and sundry events of lesser, more regional, scale) have occurred repeatedly throughout the history of life. Some papers in this volume treat the globat K-T event as a single instantiation of the general pattern of global mass extinction. Others, just as reasonably, take the data from a single region as one local pattern to be compared with others throughout the world and from different physical environments, the better to characterize the total global K-T pattern. Each treats the data of the fossil record as a prize at least equal in value to their causal explanation: generalized patterns can only emerge if we are completely clear on what the fossil record really says. The healthy mix of data and theory in these pages is a prime example of the robust condition of our newly revivified discipline of paleobiology. There truly is meaning in the history of life, and we paleobiologists are most fortunate to be charged with discovering this meaning.

REFERENCES CITED

Alvarez, L. W., Alvarez, W., Asaro, F., and Michel, H., 1980, Extraterrestrial cause for the Cretaceous-Tertiary extinction: Science, v. 208, p. 1095–1108.

Brett, C. E., and Baird, C., 1995, Coordinated stasis and evolutionary ecology of Silurian to Middle Devonian faunas in the Appalachian Basin, *in* Anstey, R., and Erwi, D. H., eds., Speciation in the fossil record: New York, Columbia University Press (in press).

Darwin, C., 1859, On the origin of species: London, John Murray, 513 p.

Dobzhansky, T., 1937, Genetics and the origin of species (reprint, 1982): New York, Columbia University Press, 364 p.

Eldredge, N., 1971, The allopatric model and phylogeny in Paleozoic invertebrates: Evolution, v. 25, p. 156–167.
Eldredge, N., 1984, Simpson's inverse: Bradytely and the phenomenon of living fossils, *in* Eldredge, N., and Stanley, S. M., eds., Living fossils: New York, Springer-Verlag, p. 272–277.
Eldredge, N., 1995, Reinventing Darwin. The great debate at the high table of evolutionary theory: New York, John Wiley and Sons (in press).
Eldredge, N., 1996, Hierarchies in macroevolution, *in* Erwin, D. H., Jablonski, D., and Lipps, J., eds., Evolutionary paleobiology: Essays in honor of J. W. Valentine: Chicago, University of Chicago Press (in press).
Hull, D. L. 1973, Darwin and his critics: Cambridge, Harvard University Press, 473 p.
Maynard Smith, J., 1984, Palaeontology at the high table: Nature, v. 309, p. 401–402.
Mayr, E., 1982, The growth of biological thought: Cambridge, Harvard University Press, 974 p.
Newell, N. D., 1967, Revolutions in the history of life, *in* Albritton, C. C., ed., Uniformity and simplicity: A symposium on the principle of the uniformity of nature: Geological Society of America Special Paper 89, p. 63–91.
Simpson, G. G., 1944, Tempo and mode in evolution: New York, Columbia University Press, 237 p.
Simpson, G. G., 1953, The major features of evolution: New York, Columbia University Press, 434 p.
Simpson, G. G., 1963, Historical science, *in* Albritton, C. C., Jr., ed., The fabric of geology: Stanford, California, Freeman Cooper and Co., p. 24–48.
Vrba, E. S., 1985, Environment and evolution: alternative causes of the temporal distribution of evolutionary events: South African Journal of Science, v. 81, p. 229–236.
Vrba, E. S., 1993, Turnover pulses, the Red Queen, and related topics: American Journal of Science, v. 293A, p. 418–452.
Wright, S., 1945, Tempo and mode in evolution: A critical review: Ecology, v. 26, p. 415–419.

Cretaceous-Tertiary Mass Extinctions:
Biotic and Environmental Changes

1

Introduction

Norman MacLeod, *Department of Palaeontology, The Natural History Museum, London, United Kingdom,*
and Gerta Keller, *Department of Geological and Geophysical Sciences, Princeton University, Princeton, New Jersey*

The nature of the Cretaceous-Tertiary (K-T) boundary mass extinction has always been controversial. This event is known to most people by its association with dinosaur extinctions; however, through the years professional paleontological opinion on the K-T boundary extinction event has swung between the extremes of promoting it as the defining moment of the modern biota's evolutionary history and denying its existence as a biological phenomenon. At the same time there has been no shortage of causal mecahnisms proposed to have been the driving force behind aspects of this biotic turnover. These have ranged from the biologically sublime (e.g., competitive interactions with radiating species, evolutionary changes in the plant fauna) to the patently ridiculous (e.g., cigarette smoking, AIDS), including both terrestrial (e.g., volcanism, eustatic sea-level change) and extraterrestrial (e.g., cosmic rays from a proximal supernova) calamities.

The 1980 proposal of the bolide impact/extinction scenario for K-T boundary extinctions, however, was different. This model, which was developed by a team of four scientists from the University of Berkeley led by the Nobel laureate Luis W. Alvarez, rapidly galvinized popular and scientific opinion on the K-T extinction controversy and set off debates among earth scientists and astronomers that continue to this day. Why the impact/extinction scenario had this effect is an interesting question to ponder. As mentioned above, theirs was by no means the first extraterrestrial model that had been proposed to account for an episode of mass extinction. In his 1970 Paleontological Society presidential address, Digby McLaren proposed meteorite impact as the most likely cause of the terminal

Devonian mass extinction. The empirical evidence produced by the Berkely group—in the form of an iridium anomaly in a K-T boundary section near the northern Italian town of the Gubbio—coupled with Luis Alvarez's stature within the United States science establishment meant that their model had to be taken seriously. In addition, the K-T boundary iridium anomalies that were soon recovered from more than 30 sections and deep-sea cores worldwide (Alvarez and others, 1982) provided an initial test of the impact/extinction model's predictive power. Many people in the popular media also first took note of the impact/extinction scenario in a dramatic and largely favorable manner when the term "nuclear winter" (the physical model of which was inspired by the impact/extinction scenario) made its way into current political discussions.

Through the initial burst of excitement and media attention that surrounded the proposal and popularization of the impact/extinction scenario, many, if not most, paleontologists remained skeptical. Whereas a comparatively small number of scientists in this field had ever worked actively on any of the mass extinction events, most were familiar enough with the empirical data available at the time to realize that almost all discussions focusing on the biotic effects of an impact grossly oversimplified the level of empirical support for this model, specified kill mechanisms in only the most general of terms (thus making it difficult to formulate specific biotic predictions of the scenario), or neglected to mention and/or discuss data that seemed inconsistent with the impact/extinction link. For example, the much-discussed familial and generic data compilations assembled by J. J. Sepkoski (and analyzed by Sepkoski and David Raup, 1986; Raup and Sepkoski, 1984, 1986) seemed to suggest that rates of faunal turnover rose dramatically at the K-T boundary, marking this as the third-largest Phanerozoic extinction event (Fig. 1). However, most paleontologists were aware of the fact

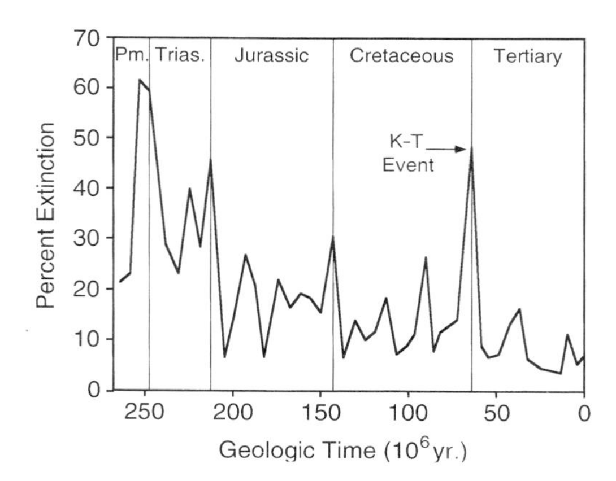

Figure 1. Percent extinction of marine genera from the middle Permian (Pm.) to recent. In this type of summary extinction events appear to be concentrated at stage boundaries, because the stage represents the lowest unit of stratigraphic resolution attempted, and all taxic ranges are extended to the stage boundaries regardless of where the last appearance horizon for individual species was observed. Trias. is Triassic. Redrawn from Raup and Sepkoski (1986).

that Sepkoski's compilations are plagued by serious taxonomic inconsistencies (e.g., inclusion of large numbers of synonymous taxa, paraphyletic and polyphyletic groups, incorrect biostratigraphic ranges; see Patterson and Smith, 1988; Smith and Patterson, 1988; Sepkoski, 1994) and were presented in a manner that artificially extends observed stratigraphic ranges of fossil taxa to the boundaries of extremely coarse units of geologic time (the nearest stage). These factors, especially the latter, bias the Sepkoski compilations in favor of large faunal turnovers at stage boundaries such as that between the Maastrichtian (uppermost Cretaceous stage) and the Paleogene (lowermost tertiary stage) or the Danian (lowermost Tertiary epoch; see Fig. 2).

With respect to providing evidence for or against the impact/extinction sce-

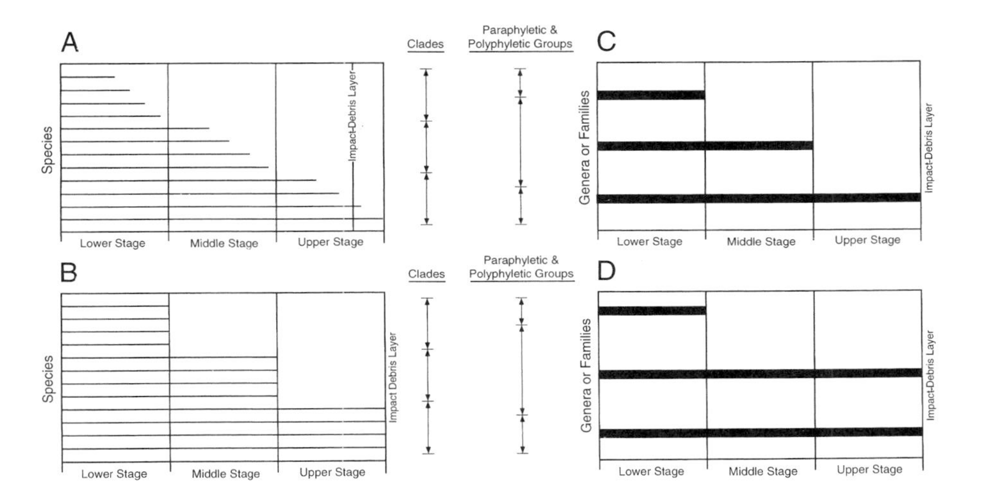

Figure 2. Sources of bias in higher level taxonomic summaries in which observed stratigraphic ranges are extended to the boundaries of higher level stratigraphic units (e.g., stage). A: Hypothetic stratigraphic ranges of 12 species belonging to three different clades undergoing a progressive faunal turnover. Note the presence of an impact-debris layer in the uppermost stage that does not coincide with the last appearance data of any species. B: The effect of coarsening the temporal resolution of the observed data by regarding stages as the basal stratigraphic unit. In this simulation, 33% of the represented taxa now appear to undergo a synchronous extinction coincident with the inferred occurrence of impact debris in the succession. C: The effect of coarsening the taxonomic resolution to higher monophyletic groupings. Note that the general outlines of a progressive faunal turnover are retained, and the interpretation of these patterns within an evolutionary context facilitated. D: The effect of coarsening the taxonomic resolution and misrepresenting the temporal distribution of evolutionary units by organizing species into evolutionarily artificial (paraphyletic and polyphyletic) groupings. Note that this organizational convention will tend to underestimate or overestimate the temporal distribution of phylogenetically meaningful units as well as obscuring their natures. In this simulation, 66% of the represented "taxa" now appear to undergo a synchronous extinction coincident with the inferred occurrence of impact debris in the succession. Unless sufficient attention is paid to the taxonomic consistency and stratigraphic fidelity of paleontological compendia, the patterns represented therein (D.) can bear little relation to the observational data upon which they are ultimately based (A.)

nario, the obvious problem with higher taxonomic level, biostratigraphically generalized compilations such as those of Sepkoski (1982, and more recently Benton, 1993) is that these do not provide a sufficiently faithful representation of the individual paleontological observations upon which they are based. Moreover, these global compilations are referenced to a unique, although seldom-discussed internal biostratigraphic system within which it is difficult, if not impossible, to place specific stratigraphic horizons (e.g., the K-T boundary, carbon-13 shift, iridium anomalies) with any real degree of precision. Thus, associations between species extinctions and physical evidence for impact occurrence within these summaries are at best conjectural, and at worst demonstrably incorrect.

Despite these problems, the Alvarez impact/extinction scenario offered such exciting possibilities for understanding a potentially major factor controlling biotic diversity and evolutionary processes that several new paleontological research efforts targeting the K-T boundary turnover were initiated, while other established programs were refocused to include an examination of various local boundary successions. These more empirically oriented studies are the only direct means by which the actual nature of biotic turnover and environmental variation across the K-T boundary can ever be assessed. Whereas much exciting new evidence that seems to support the recognition of a bolide impact at the close of the Cretaceous has come to light in recent years (e.g., shocked quartz, proposal of the Chicxulub impact structure), these data address only the question of whether such an event took place, not the effect that such an event might or might not have had on the Earth's biota. The Phanerozoic fossil record contains many examples of enormous environmental changes that did not result in mass extinction (e.g., the Pleistocene ice ages, the Paleogene-Neogene transition) as well as instances of mass extinction that are not associated with definitive independent evidence for extraterrestrial forcing (e.g., the Tithonian, Cenomanian-Turonian, Paloecene-Eocene, Eocene-Oligocene extinction events). Because of the long time it takes to collect, analyze, and interpret paleontological data, the fine structure of the K-T biotic turnover event and its relation to the physical evidence for bolide impact remain largely unknown for most fossil groups.

Several of these paleontological research efforts inspired by the impact/extinction scenario are now mature in the sense that the investigators involved have had sufficient time to thoroughly familiarize themselves with the local stratigraphic and paleoenvironmental context of their faunas, sort out taxonomic problems, and construct an impressive database, in some cases ranging into thousands of samples containing millions of individual fossils. As a result, these research projects have produced the most detailed record available for understanding the nature of organismal and evolutionary responses to major environmental upheavals. The significance of the fossil record with respect to examining the process of extinction (even among modern species) cannot be overstated and should not be ignored.

This volume contains a collection of 19 reports from this forefront of paleontological research into the patterns of, and processes responsible for, the K-T mass extinction event. No attempt has been made to provide a comprehensive review of K-T turnover for all major fossil groups. Such a review would be largely anecdotal because, whereas the global K-T fossil records of some groups (e.g., planktonic foraminifera, calcareous nannoplankton, palynomorphs, ammonites) have received intense study, the fossil records of others (e.g., diatoms, brachiopods, bryozoans, echinoderms) are known in only the most general terms. Therefore, instead of putting together a series of review chapters, the levels of detail of which were strikingly uneven, we present here a selection of progress reports from paleontologists who are actively studying the K-T fossil records of various biotic groups. Many of the fossil groups discussed herein have played important roles in the ongoing series of scientific debates surrounding the impact/extinction scenario, whereas others (as yet) have not. Most of the authors included herein take an explicitly stratigraphic approach to the question of evaluating the responses of their groups to environmental changes at the close of the Cretaceous, but others have adopted a less-traditional approach to this question. Without exception, these are innovative, thoroughly documented, and provocative contributions to perhaps the most popular scientific controversy of the late twentieth century. Publication of this volume is sure to spark even more controversy as additional reports from these and other authors fill gaps in our knowledge of the K-T environmental events, and a comprehensive picture of this widely misunderstood biotic turnover begins to be pieced together.

Although intended primarily as a summary of paleontological research on K-T boundary biotic and environmental change, this volume was also assembled and edited to provide students and interested laypersons with a lively and accessible account of the paleontological side of the K-T boundary impact/extinction debates and to serve the researcher as an introduction to the voluminous technical literature on K-T boundary paleontology. At this time the most neglected aspects of the impact/extinction controversy are the lack of sufficiently detailed predictions that identify the types of organisms most at risk by the various proposed killing mechanisms, and our almost complete ignorance of the K-T transition for many benthic marine invertebrate and terrestrial (vertebrate and invertebrate) groups. If this volume serves to stimulate additional research by both paleontologists and specialists in modern representatives of the various survivor lineages, then our efforts will have been well rewarded.

In organizing this book we have indebted ourselves to many individuals and organizations that we would like to acknowledge. First, we thank all the authors who participated in this project. Most of the arguments detailed herein were first presented in a theme session entitled "The Cretaceous-Tertiary Boundary Event: Biotic and Environmental Changes," at the 1993 Annual Meeting of the Geological Society of America in Boston, Massachusetts. We thank the Geological Society of America for providing the forum for that theme session, which served as the catalyst for the creation of this book. Steven Mosberg, our editor at W. W. Norton, and his staff also provided encouragement and invaluable technical assistance in this book's production. Finally, we thank our colleagues on all sides of the geological and paleontological K-T debates with whom we have described, discussed, debated, and disputed the results of our (and their) research into this fascinating period of Earth history. The final version of the K-T boundary story will not be written for many years to come (if ever). Nonetheless, it is satisfying to look back over this and other volumes devoted to K-T topics and realize how much we have accomplished.

REFERENCES CITED

Alvarez, W., Alvarez, L. W., Asaro, F., and Michel, H., 1982, Current status of the impact theory for the terminal Cretaceous extinction, *in* Silver, L. T., and Schultz, P. H., eds., Geological implications of impacts of large asteroids and comets on the Earth: Geological Society of America Special Paper 190, p. 305–315.

Benton, M. J., 1993, The fossil record: London, Chapman & Hall, 845 p.

Patterson, C., and Smith, A. B., 1988, Is periodicity of mass extinctions a taxonomic artefact?: Nature, v. 330, p. 248–251.

Raup, D. M., and Sepkoski, J. J., Jr., 1984, Periodicity of extinctions in the geologic past: Natural Academy of Sciences Proceedings, v. 81, p. 801–805.

Raup, D. M., and Sepkoski, J. J., Jr., 1986, Periodic extinction of families and genera: Science, v. 231, p. 833–836.

Sepkoski, J. J., Jr., 1982, A compendium of fossil marine families: Milwaukee Public Museum Contributions in Biology and Geology, v. 51, p. 1–125.

Sepkoski, J. J., Jr., 1994, What I did with my research career: Or how research on biodiversity yielded data on extinction, *in* Glenn, W., ed., The mass extinction debates: How science works in a crisis: Stanford, California, Stanford University Press, p. 132–144.

Sepkoski, J. J., Jr., and Raup, D. M., 1986, Periodicity in marine extinction events, *in* Elliott, D. K., ed., Dynamics of extinction: New York, Wiley-Interscience, p. 3–36.

Smith, A. B., and Patterson, C., 1988, The influence of taxonomic method on the perception of patterns of evolution: Evolutionary Biology, v. 23, p. 127–216.

2

Palynological Change across the Cretaceous-Tertiary Boundary on Seymour Island, Antarctica: Environmental and Depositional Factors

Rosemary A. Askin and Stephen R. Jacobson,
Byrd Polar Research Center, Ohio State University, Columbus, Ohio

INTRODUCTION

Palynomorphs are the only microfossil group preserved consistently and abundantly throughout the Cretaceous-Tertiary (K-T) transition on Seymour Island, Antarctica. Calcareous microfossils (planktonic foraminifera and calcareous nannofossils), typically used to define marine K-T successions elsewhere in the world, are absent in Seymour Island K-T boundary sections.

On Seymour Island, palynomorphs include (1) marine dinoflagellate cysts (dinocysts), acritarchs, and other algae, including colonial forms; and (2) terrestrially derived spores and pollen from land plants, fungal spores and fruiting bodies, and various fresh– to brackish–water algae. These organic-walled microfossils are useful for interpreting age and environmental conditions at and near the K-T boundary.

We outline factors that could affect palynomorph assemblage composition and K-T palynostratigraphy on Seymour Island, describe the marine and nonmarine palynomorph records, and discuss their relations to environmental changes.

An asteroid impact was suggested by Alvarez and others (1980) as the direct and primary cause of worldwide, terminal Cretaceous extinctions. In testing this global devastation scenario, one might expect evidence of abrupt and probably irreversible changes in the Seymour biotas. Some workers (e.g., McLean, 1985; Officer and Drake, 1985) consider massive volcanism at the end of the Cretaceous a major cause of biotic turnover. Others invoke progressive environmental changes through the late Maastrichtian as agents for a more gradual, longer term biotic turnover (e.g., Keller, 1988; Hallam, 1989; Canudo and others, 1991). In the palynomorph record of Seymour Island, we see evidence of both long-term (through Maastrichtian–early Danian) gradual turnover and more rapid change.

SEYMOUR ISLAND K-T STRATIGRAPHY

On Seymour Island the Maastrichtian to Danian succession comprises shallow-marine siliciclastic sediments of the upper López de Bertodano Formation. There is no obvious lithologic expression of the K-T boundary, which occurs in the middle of a glauconitic interval at the base of Macellari's (1984, 1988) unit 10 (Fig. 1). This glauconite-rich interval crops out for 5.5 km along strike and varies from a discrete glauconite bed one to several meters thick, to a more diffuse interval of glauconitic silty sands. Except for ubiquitous palynomorphs, the glauconite interval is sparsely fossiliferous.

Calcareous microfossils (planktonic foraminifera and calcareous nannofossils) are absent through the K-T transition on Seymour Island. Foraminiferal assemblages are entirely arenaceous in the K-T and basal Danian beds, which Huber

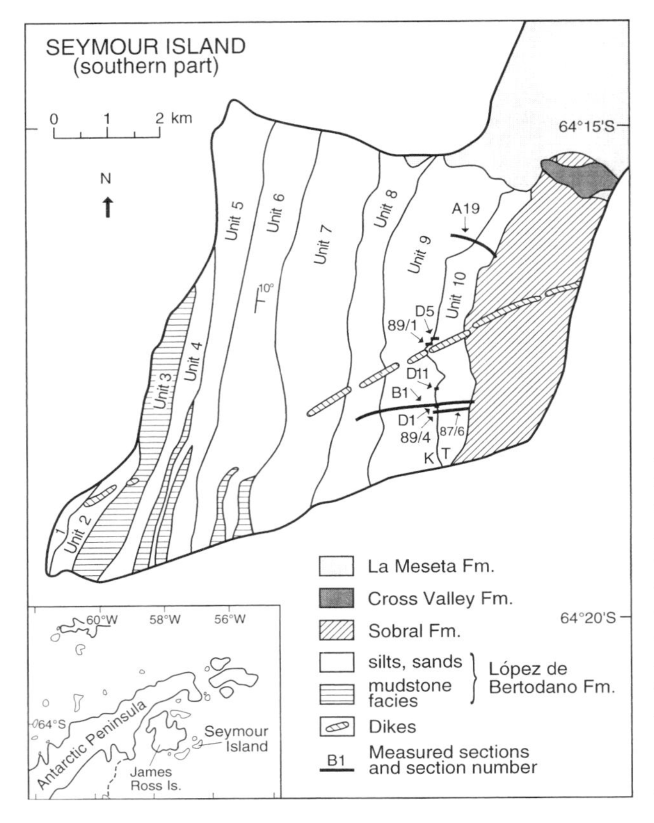

Figure 1. Locality and simplified geology map of southern Seymour Island (López de Bertodano members after Macellari, 1984, 1988). López de Bertodano Formation (Fm.) is upper Campanian to lower Danian; Sobral Formation is Danian; Cross Valley Fm. is Thanetian?; La Meseta Fm. is Eocene.

(1988) interpreted as a carbonate dissolution interval. Siliceous microfossils (silicoflagellates, diatoms) are very sparse, although an increase in relative abundances of diatom resting spores was noted above the boundary (Harwood, 1988). A gradual turnover of the invertebrate fauna (e.g., bivalves, gastropods, ammonites) from lower in the Maastrichtian into the Danian was documented by Zinsmeister and others (1989). The stratigraphically highest occurrence of ammonites is in the lower part of the glauconitic interval (e.g., Fig. 3 of Elliot and others, 1994).

Dinocysts provide the most precise biostratigraphic means of bracketing the K-T boundary. In the Seymour Island Filo Negro section of Askin and others (1993) and Elliot and others (1994), dinocysts demarcate a 20–30 cm K-T transitional interval and provide evidence for an overlying basal Danian flooding event. A single sharp iridium peak (1.6 ng/g, 40x background, Kyte, *in* Askin and others [1993]; Kyte, *in* Elliot and others [1994]) near the base of this dinocyst transitional interval is a likely candidate for the K-T boundary, on the basis of correlation with the main iridium peak at the K-T boundary elsewhere.

Integration of available lithologic and fossil data from Seymour Island K-T sediments (Elliot and others, 1994) suggests deposition in low-energy conditions below wave base, on a wide (tens of kilometers) shallow-marine shelf. These sediments were deposited in the James Ross basin (Elliot, 1988; Del Valle and others, 1992), a back-arc basin flanking the active Antarctic Peninsula magmatic arc. In Maastrichtian-Danian time, the Seymour location was at about the same latitude as it is today (lat 64°S; Lawver and others, 1992). To date it is the most complete siliciclastic, shallow-marine, high southern latitude K-T boundary site. Other southern K-T outcrop sites with marine and nonmarine palynomorph records at comparable paleolatitudes are in New Zealand at about lat 60° to lat 55°S (Lawver and others, 1992).

Several stratigraphic sections were measured and sampled across the K-T boundary on Seymour Island. Figure 1 shows the locations of eight of these sections, including the B1 reference section of Askin (1988b; B1 = section C of Macellari [1988]; Huber [1988]; Zinsmeister and others [1989]); and the 16 m section 89/1 (Filo Negro) discussed by Elliot and others (1994). Section B1 was sampled at 50 cm intervals across the K-T boundary, and section 89/1 at 10 cm intervals. Figure 2 illustrates changes in the Seymour marine palynomorph record over the K-T boundary in the middle 2.5 m of the 89/1 section. Relative abundances of dinocyst taxa are based on counts of more than 300 specimens; thousands of specimens were scanned to obtain diversity data.

FACTORS AFFECTING THE PALYNOMORPH RECORD

Palynomorph records may be affected by the selective preservation of palynomorph taxa, by physical and chemical depositional factors, and by postburial changes, including thermal alteration and surficial (weathering) processes. Factors that could have affected the Seymour K-T palynostratigraphy include preburial transportation and remobilization, bioturbation, reworking, and recent

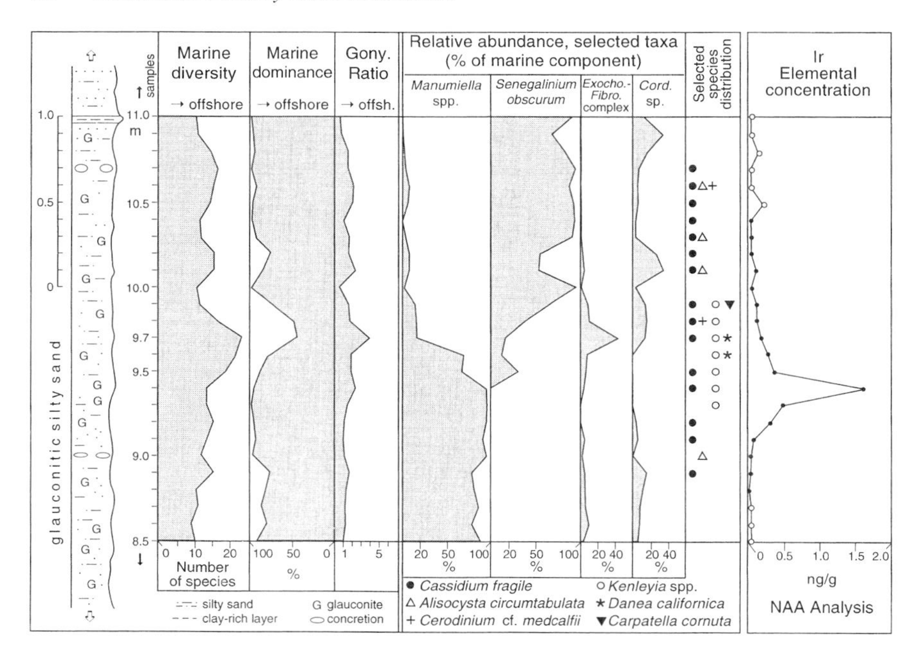

Figure 2. Distribution of selected marine palynomorph taxa and ratios between 8.5 and 11.0 m in Seymour Island section 89/1. Samples at 10 cm spacings are shown to left of graph. Abbreviations: Gony. Ratio = Gonyaulacacean ratio; Exocho-Fibro. *complex* = Exochosphaeridium bifidum–Fibrocysta bipolaris *complex;* Cord. *sp.*= Cordosphaeridium *sp. Iridium data from Kyte (*in *Elliot et al., 1994).*

weathering. However, our data show that these factors do not pose problems for detecting the K-T boundary and interpreting environmental trends.

Preburial Transportation and Remobilization

Palynomorphs have low density, are buoyant, and are carried easily by wind and water. Thus, they are transported easily away from where they originated. They can be remobilized many times prior to burial. Spores and pollen from land plants, and other terrestrial debris, often travel long distances to depositional sites (e.g., Muller, 1959; Cross and others, 1966). While settling through the water column, dinocysts may be transported by onshore and offshore currents and by currents flowing along the shelf and slope (e.g., Wall and others, 1977). Palynomorph transportation and remobilization may complicate interpretations of environment, and sometimes stratigraphic ranges, which is especially relevant to nonmarine taxa carried onto the Seymour marine shelf.

Bioturbation

Bioturbation can move palynomorph specimens downsection, lowering (to the depth of burrows) the first stratigraphic appearances of key taxa. A realistic maximum measure of the downsection mixing by bioturbation on the Seymour K-T shelf (for section 89/1) is 30 cm. This is the depth below the sharp iridium spike where higher than background iridium concentrations can still be detected (Fig. 2; Elliot and others, 1994). We consider 30 cm a maximum because iridium movement is enhanced by other mechanisms, such as diagenetic mobilization and chemical diffusion, that do not affect palynomorphs. Bioturbation can also smear biostratigraphic ranges upsection by bringing deeper sediment and palynomorph grains to burrow tops.

Reworking

Already-buried palynomorphs can be exhumed by erosion and redeposited in younger sediments. Both marine and nonmarine palynomorphs may be subject to reworking on Seymour Island. If undetected, reworking can extend biostratigraphic ranges upsection. Reworking does not present a problem where there are discernible preservational, thermal (color), or age differences between in-place and reworked specimens; moreover, in Antarctica reworked palynomorphs have provided useful geologic insights (e.g., Askin and Elliot, 1982; Truswell, 1983).

Recent Weathering and Freeze-Thaw Effects

Today, the unconsolidated Seymour Island K-T sediments are permeated by permafrost; grains in surface layers are subject to minor displacement by freezing and thawing. Although the physical results are negligible, chemical effects in the waterlogged sediments could encourage biodegradation or oxidation of organic matter (Askin and Jacobson, 1989).

MARINE PALYNOMORPH RECORD

Dinoflagellates produce a selective fossil record. Only a small proportion (~10%) of dinoflagellates living today form fossilizable, acid-resistant, organic-walled cysts during their lifecycle, and dinoflagellates may have changed their ability or preference to produce cysts during their geologic range (e.g., Evitt, 1985). First appearances of dinocyst species could reflect evolutionary or immigratory first appearances, or mark a change in a species' propensity for encystment. Environmental change may initiate or inhibit cyst formation in a species. Furthermore, the encystment capability of dinoflagellates may have increased their chances for survival during environmental crises, depending on the taxon and the nature and length of the crises (e.g., Griffis and Chapman, 1988).

Dinocyst Assemblage Turnover

Throughout the Maastrichtian record, Seymour Island dinocyst assemblages are dominated by *Manumiella* spp. (Askin, 1988a), which are mainly *Manumiella seelandica* in the upper 40 m of the Maastrichtian (note: *M. seelandica* includes *Manumiella druggii* of Askin, 1988a, 1988b; and other reports; Firth, 1987). In the basal few meters of the Danian, the marine component is overwhelmed by *Senegalinium obscurum.* Above this interval, Paleocene assemblages are more varied and are characterized by common *Spinidinium* spp. *Palaeoperidinium pyrophorum* predominates sporadically in the Danian.

The change from an almost monospecific *Manumiella seelandica* dinocyst assemblage at the top of the Maastrichtian to an almost monospecific *Senegalinium obscurum* assemblage occurs relatively abruptly. In the section with the most detailed sampling and analyses (89/1), this change takes place within 60 cm (from 9.4 to 10.0 m; Fig. 2).

The predominant Seymour uppermost Cretaceous species, *Manumiella seelandica*, is distributed worldwide, although associated dinocyst taxa in mid to low latitudes differ substantially from those on Seymour Island. *M. seelandica* may have been opportunistic or well adapted to demanding marginal marine, shallow shelf conditions, when sea level dropped and other environmental conditions changed or deteriorated before the end of the Cretaceous. Possibly restricted conditions for terminal Cretaceous or basal Danian samples containing high relative abundances of *M. seelandica* have been noted previously (e.g., Hultberg, 1986; Firth, 1987, 1993).

Manumiella seelandica was a relatively short lived upper Maastrichtian–lowermost Danian nearshore marine form. In some less-complete sections it is preserved in either Maastrichtian or basal Danian strata. It is reported in K-T sections from, for example, New Zealand (Woodside Creek, Wilson, 1978; Waipara, Wilson, 1987; Helby and others, 1987; Te Hoe River area, Wilson and Moore, 1988), Gippsland basin, southeastern Australia (Stover, 1973; Partridge, 1976; Helby and others, 1987), central California (as *Deflandrea cretacea*; Drugg, 1967), Albany, Georgia (Firth, 1987), Braggs, Alabama (Moshkovitz and Habib, 1993), Stevns Klint, Denmark, especially in the Fish Clay (e.g., Lange, 1969; Wilson, 1971; Hultberg, 1986), and El Kef, Tunisia (Brinkhuis and Leereveld, 1988; Brinkhuis and Zachariasse, 1988). On Seymour Island, *M. seelandica* predominates in uppermost Maastrichtian and K-T transitional beds. Relative abundances of this dinocyst decrease rapidly in basal Danian sediments (Fig. 2). It is unknown whether the dwindling numbers represent a gradual die out on the Seymour shelf, an overwhelming influx of competitive new forms, a facies change, or if Seymour Danian *M. seelandica* specimens are reworked.

The first significant numbers of *Senegalinium obscurum* and the corresponding decline in *Manumiella* spp. occur in sample 9.5 m in section 89/1. This sample is 10 cm above the iridium-marked K-T boundary. This relatively abrupt change in dominant dinocyst species could be the result of the iridium-related K-T event. However, the stratigraphic position of the *S. obscurum* influx is variable. For example, in section B1 (where spacing between samples is wide; i.e., 50 cm), *S. obscurum* first appears higher (Askin, 1988b), 1 m above a small iridium anomaly (0.4 ng/g; Kyte, *in* Zinsmeister and others, 1989) and 0.5 m above a peak in offshore dinocysts (see below).

The variable influx of *Senegalinium obscurum* on Seymour Island, together with the gradual first appearances of new typically Paleocene species starting some distance below the boundary, may reflect gradual driving forces in late Maastrichtian and early Danian time such as rising sea level, facies change, changes in water masses and currents, and/or changes in temperature, water chemistry, or nutrient supply. There is local evidence for Maastrichtian and terminal Cretaceous cooling (e.g., Barrera and others, 1987; Stott and Kennett, 1990). Changes in water chemistry and nutrient availability may be related to global trends or local volcanism.

Pre-Boundary Changes

An important observation for evaluating K-T environmental upheaval is that environmental and biotic changes on the Seymour shelf began significantly before the end of the Cretaceous. Relatively stable shelf conditions lasted from the latest Campanian through most of the Maastrichtian (over 800 m of section; Askin, 1988a), based in part on only minor variations in dinocyst assemblages. Following this period of apparent stability, shown in the range chart of Askin (1988a, Fig. 5), the rate of dinocyst turnover began to increase about 100 m below the top of the Cretaceous. Invertebrate turnover also showed an accelerated rate of change in the late Maastrichtian, including declining ammonite diversity (e.g., Macellari, 1986, 1988; Zinsmeister and others, 1989; Zinsmeister and Feldmann, 1993). The uppermost Maastrichtian *Manumiella seelandica* dinocyst zone (zone 4 of Askin, 1988a) spans the upper 40 m of Cretaceous strata and is equivalent to Macellari's (1986, 1988) *Pachydiscus ultima* and *Zelandites varuna* ammonite zones.

Latest Cretaceous marine biotic turnover may be explained, at least in part, by relative sea-level change (Macellari, 1988). A gradual deepening trend culminated in the middle of unit 9, about 80 m below the K-T boundary. On the basis of lithologic and fossil evidence, these beds represent the deepest or most offshore facies of the Seymour Island López de Bertodano Formation (Macellari, 1988); this late Maastrichtian highstand was followed by a sea-level drop in the latest Maastrichtian (as in the sea-level curve of Haq and others, 1988), and another sea-level rise across the K-T boundary.

Dinocyst First Appearances Across the K-T Boundary

The dinocyst fossil record worldwide shows a gradual pattern of last and first appearances across the K-T boundary. Dinocysts do not show the high degree of extinction that occurs in calcareous microplankton near and at the K-T boundary in other marine localities. Some genus- and species-level dinocyst disappearances occurred in the Maastrichtian, as illustrated in the collated global data of Williams and Bujak (1985) and Northern Hemisphere data of Williams and others (1993). Perhaps more important is the trend of first appearances of new dinocyst species through the Maastrichtian-Danian transition, for example, as documented in Denmark (Hansen, 1977). A similar trend or influx of new forms beginning in the late Maastrichtian and continuing into the Danian occurs in the Seymour Island succession (Askin, 1988a, 1988b).

Distribution of six selected typically Paleocene taxa in part of section 89/1 are shown in Figures 2 and 3. These taxa are characteristic of Paleocene rocks, although some first appear in rocks independently dated as latest Maastrichtian. They are *Cassidium fragile*, *Cerodinium* cf. *medcalfii*, *Alisocysta circumtabulata*, *Kenleyia* spp., *Carpatella cornuta*, and *Danea californica*.

In the Waipara section, South Island, New Zealand (Wilson, 1987), both *Cassidium fragile* and *Cerodinium* cf. *medcalfii* first appear below the K-T

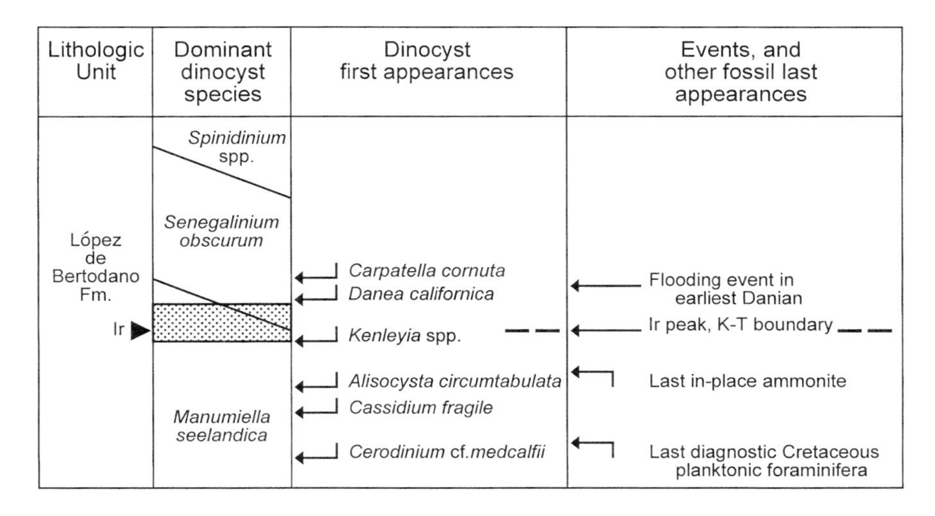

Figure 3. Schematic diagram showing relative order of first and last appearances of important taxa near the iridium-marked K-T boundary on Seymour Island. Stippled area shows K-T transition interval.

boundary. This is also true on Seymour Island. For example, in section B1, *C.* cf. *medcalfii* first occurs 10 m below the boundary in upper unit 9 beds with Maastrichtian ammonites and planktonic foraminifera (in sample B1-88; Askin, 1988b).

Alisocysta circumtabulata is typically Paleocene in southern localities, although in some it first occurs in uppermost Maastrichtian, and Williams and others (1993) showed it ranging throughout the Maastrichtian in the Northern Hemisphere. In the Albany core, Georgia, United States (Firth, 1987), it is very rare in the upper Maastrichtian Providence Sand and lower Danian Clayton Formation. In Seymour section 89/1, its first occurrence is at 9.0 m, 40 cm below the iridium-picked K-T boundary, and in section A19 it occurs at least 5 m below the highest *Manumiella seelandica* zone sample, well below the extent of bioturbated mixing from Danian sediments. *A. circumtabulata* is restricted to the Paleocene and Eocene in New Zealand (Wilson, 1987, 1988), although a form listed as *A.* cf. *circumtabulata* was reported from the uppermost Maastrichtian at Waipara (Wilson, 1987). *A. circumtabulata* first appears in the uppermost Maastrichtian in northwestern Australia (McMinn, 1988) and on the southern Kerguelen Plateau, Ocean Drilling Program (ODP) Site 738 (Tocher, 1991).

Carpatella cornuta also occurs below the K-T boundary at Waipara (Wilson, 1987). *C. cornuta* is very rare on Seymour Island; there is one specimen in sample 9.9 m in section 89/1. In section B1, a specimen of *C.* cf. *cornuta* occurs at least 1.5 m below the K-T boundary (a perforate form in sample B1-106; Askin, in prep.). Outside of the southern high latitudes, Moshkovitz and Habib (1993) noted that the first appearance of *C. cornuta* is an important marker for the early

Danian in Alabama K-T sections; and at El Kef, Tunisia, the probably conspecific *Kenleyia nuda* (Damassa, 1988) occurs only in the Danian (Brinkhuis and Leereveld, 1988; Brinkhuis and Zachariasse, 1988).

On Seymour Island, *Kenleyia* spp. comprise a complex of interrelated forms, most being similar to *K. pachycerata*, and others having characteristics of *K. lophophora* and *K. leptocerata*. *Kenleyia* spp. are closely related to *Carpatella cornuta* (Damassa, 1988). *Kenleyia* spp. were originally described from the lower Paleocene Dartmoor Formation of southeastern Australia (Cookson and Eisenack, 1965); they were also reported from the El Kef Danian section (Brinkhuis and Leereveld, 1988; Brinkhuis and Zachariasse, 1988), and have not been reported below the Danian. Although the name *Kenleyia* is not used in New Zealand, specimens that could be included in this genus occur in the Danian and in the top of the Maastrichtian (G. J. Wilson, 1994, personal commun.). *Kenleyia* spp. may be useful K-T transition and Danian markers. In section 89/1 they occur in every sample from 9.3 to 9.9 m. The base of the *Kenleyia* spp. range at 9.3 m, associated with still-dominant *Manumiella seelandica*, marks the base of the K-T transitional interval.

Danea californica appears at the base of the Danian at El Kef (Brinkhuis and Zachariasse, 1988), the Mussel Creek section at Braggs, Alabama (Moshkovitz and Habib, 1993), and elsewhere (as *D. mutabilis*, e.g., Denmark; Hansen, 1977). With its first appearance immediately above the base of the Danian, and last appearance within the upper Danian, *Danea californica* was shown to be a marker species by L. E. Stover on the Mesozoic-Cenozoic cycle chart of Haq and others (1988). Very rare, poorly preserved specimens occur at 9.6 and 9.7 m in section 89/1, indicating that the K-T boundary occurs below the 9.6 m level.

Inshore-Offshore Trends, Sea Level, and the Marine Environment

A key chronostratigraphic horizon, traceable along strike on Seymour Island, is the marine flooding event near the base of the Danian (Askin and others, 1993; Elliot and others, 1994). The event was detected with dinocyst data, and can be correlated with the earliest Danian flooding elsewhere in the world. Dinocyst evidence for the flooding event includes diversity and dominance trends, and higher frequencies of more offshore dinocyst taxa.

The distribution of modern motile dinoflagellates in the oceans is controlled by temperature, salinity, distance from shoreline, light, and nutrient supply (Taylor, 1987). Marine dinoflagellates prefer neritic habitats, and most species that form organic-walled cysts are neritic. Dinocyst diversity is low close to shore, increases in the offshore direction, and drops off again at the shelf edge (e.g., Wall and others, 1977; Goodman, 1987; Taylor, 1987). Goodman (1979) used the relation of diversity and dominance to identify inshore-offshore trends in Eocene dinocyst floras from Maryland. "Dominance" was defined as the ratio of the summed abundance of the two most-abundant species to the total number of counted dinocysts in the sample. High-diversity–low-dominance assemblages suggest a more offshore paleoenvironmental trend, whereas low-diversity–high-dominance assemblages indicate inshore conditions.

In Seymour section 89/1 (2.5 m of section shown; Fig. 2), a marine diversity peak at 9.7 m suggests the most offshore assemblage. Marine species counts include dinocysts, acritarchs, and other marine algae such as *Palambages*. The 9.7 m sample records the lowest dominance, also indicating the most offshore assemblage. (The dominance curve in Figure 2 is plotted in reverse, so that both more offshore data peaks trend in the same direction.)

A third indication of a more offshore trend for sample 9.7 m (in section 89/1) is a peak in more offshore dinocyst taxa (increases in both number of taxa and their relative abundances). Many studies (e.g., Downie and others, 1971; Harland, 1973) equate increased chorate or gonyaulacacean dinocyst frequencies with more offshore or open-marine environments, and peridiniacean dinocyst predominance with inshore or protected environments such as estuarine or lagoonal, low salinity, or brackish water. Harland (1973) showed this relation in his gonyaulacacean ratio, the ratio of gonyaulacacean to peridiniacean dinocysts, for Campanian assemblages from southern Alberta, Canada. Low values (most <2) for this ratio in the Seymour K-T section are consistent with nearshore conditions, and a small peak at 9.7 m equates with a more offshore assemblage.

An increase in relative abundances of chorate gonyaulacacean taxa is traceable along strike in basal Danian samples on Seymour Island (e.g., 89/1—9.7 m, B1—109). Chorate forms are spine or process-bearing dinocysts. The more offshore chorate taxa are mainly members of an *Exochosphaeridium bifidum–Fibrocysta bipolaris* complex, other species of *Exochosphaeridium* and *Fibrocysta*, and species of *Cordosphaeridium*, including a short-spined form listed in Figure 2 as *Cordosphaeridium* sp. Other taxa include very rare *Danea californica* and species of *Hystrichosphaeridium* and *Oligosphaeridium*, including *H. tubiferum* and *O. complex*.

The more offshore peak at 9.7 m in section 89/1 is interpreted as reflecting the maximum flooding of the K-T to earliest Danian sea-level rise. Coeval flooding is recognized in other K-T sections, including Braggs, Alabama (Donovan and others, 1988; Habib and others, 1992, at the base of unit 6) and El Kef (Brinkhuis and Zachariasse, 1988), near the top of the basal Danian foraminifera P0 zone (MacLeod and Keller, 1991).

The dinocyst diversity and dominance curves, along with dinocyst assemblage compositions, provide paleoenvironmental information for the Seymour Cretaceous-Tertiary shelf. For much of the uppermost Maastrichtian–basal Danian, preserved assemblages are low diversity and high dominance, often nearly monospecific, and dominated by nearer shore indicators—the peridiniacean dinocysts, which include *Manumiella* spp. and *Senegalinium obscurum*. These suggest inshore, restricted environments. Almost monospecific assemblages (e.g., *Manumiella seelandica* and *S. obscurum*–dominated samples on Seymour Island) represent extreme examples of Goodman's (1979) inshore model. Hultberg (1986) suggested low salinity or even brackish conditions for his almost monospecific *M. seelandica* assemblages in the Fish Clay of Stevns Klint. However, Seymour invertebrate evidence (e.g., Zinsmeister and Macellari, 1988; Zinsmeister, 1993, personal commun.) seems incompatible with an inshore or low-salinity depositional environment. Rather, these sediments, which are bioturbated and lack obvious wave-generated current or tidal bedding structures, were interpreted as being deposited mid-shelf, below wave base (Zinsmeister and others, 1989). Elliot and others (1994) concluded that a depositional environment that takes into account the abundant glauconite, lack of primary bedding, deposition below wave base, and the fossil evidence is a low-energy, shallow shelf tens of kilometers from the shoreline. Alternative explanations are that the environment is indeed nearshore and sedimentary structures in these fine-grained sediments have been destroyed by bioturbation, or that deposition was in a large sheltered embayment. The dinoflagellates may have lived in the lower salinity, upper water layers offshore from a river mouth, with the invertebrates in more normal salinity benthic habitats. Stressed conditions other than low salinity could also explain the restricted (nearly monospecific) dinocyst assemblages, such as unusual pH or other chemical conditions (global or local) likely during the Maastrichtian-Danian transition.

DISCUSSION

The Seymour K-T dinocyst record shows (1) a long-term gradual genus- and species-level turnover that accelerated shortly before the end of the Cretaceous, especially with the influx of new forms; (2) a K-T transitional interval that includes a single sharp iridium spike in its lower part; (3) a relatively abrupt replacement of the dominant *Manumiella seelandica* by *Senegalinium obscurum* following the iridium event; this turnover varying in time relative to an earliest Danian flooding event; and (4) a subsequent gradual influx of *Spinidinium* and other Paleocene forms.

Changes in environmental parameters through the Maastrichtian and across the K-T boundary have been widely documented from many locations. From oxygen isotope analyses of benthic foraminifera on Seymour Island, Barrera and others (1987) recorded cooling bottom waters through the middle and late Maastrichtian, followed by warmer conditions some time later in the Danian (from lowermost Sobral Formation specimens). Cooling during the Maastrichtian was also reported from elsewhere in the James Ross basin by Pirrie and Marshall (1990), based on oxygen isotope data from molluscan fossils.

Among the deep-sea records, nearby ODP Leg 113, Maud Rise Sites 689 and 690 are of particular relevance to Seymour Island. Oxygen isotope data (Barrera and Huber, 1990; Stott and Kennett, 1990) indicate a Maastrichtian cooling trend, an abrupt, short-lived, latest Maastrichtian warming event, a marked cooling (mainly in surface waters) at the end of the Cretaceous and across the K-T boundary, and slight warming again (in surface waters) in earliest Danian.

A global carbon isotope excursion at the end of the Cretaceous is also recorded on the Maud Rise. However, unlike lower latitude K-T sequences, with their diminished productivity "Strangelove Ocean" (e.g., Hsü and McKenzie, 1985), the surface to deep water $\delta^{13}C$ gradient was not eliminated in the Antarctic (Stott and Kennett, 1990).

Trends in Maastrichtian dinocyst genus- and species-level turnover included an influx of new forms that accelerated shortly before the end of the Cretaceous. These trends correlate with Maastrichtian eustatic sea-level and temperature changes. They are consistent with other fossil evidence of environmental and biotic change that began in the Maastrichtian with declining marine faunal and floral diversity and increased rates of extinction toward the end of the Cretaceous. On Seymour Island, new dinocysts began to appear during the final phases of the Late Cretaceous environmental crisis, perhaps to replace other phytoplankton.

The iridium spike found on Seymour Island is evidence of fallout from the postulated asteroid impact. It remains unknown, however, if environmental changes that had already begun before the bolide impact (e.g., the K-T cooling event in Antarctic surface waters, eustatic sea-level rise, carbon flux, and productivity-related changes) were responsible for the change from *Manumiella* assemblages to *S. obscurum* assemblages; or if a combination of the deleterious chemical and physical effects of the impact was the final straw that triggered this abrupt dinocyst turnover. Certainly, its effects were less dramatic in the high latitudes, as attested to by the Seymour Island record (e.g., Askin, 1988b; Zinsmeister and others, 1989) and other studies (e.g., Keller, 1993; Keller and others, 1993).

NONMARINE PALYNOMORPH RECORD

Spores and pollen of land plants, with other nonmarine palynomorphs (fungal and algal remains) and plant debris, were transported into the James Ross basin and deposited with the marine fossils. The nonmarine Upper Cretaceous–Paleocene palynomorph record on Seymour Island (Askin, 1990a) shows that the dominant components of this conifer-Proteaceae-*Nothofagidites* palynoflora remained stable across the K-T transition. There was a gradual turnover of angiosperm pollen (Askin, 1990a) and cryptogam spore (Askin, 1990b) species through the late Maastrichtian and early Danian. Several factors may have caused lower spore and pollen species diversity in the Danian. These factors are somewhat ambiguous, but may have included cooler temperatures and local volcanism (Askin, 1988a, 1990a). There is evidence of frequent local volcanism in the Danian of the Antarctic Peninsula (e.g., Macellari, 1988; Elliot and others, 1992), volcanism that probably repeatedly denuded much of the landscape and disrupted local climate patterns and the chemical/nutrient balance.

Independent evidence, summarized by Askin (1992), indicates a temperature change in the Antarctic Peninsula area that probably affected both terrestrial and marine ecosystems. Evidence includes fossil wood from Seymour Island, suggesting cooling through the Maastrichtian into the Paleocene (Francis, 1986; 1991), and oxygen isotope data from the James Ross basin and Maud Rise outlined above.

Transportation, remobilization, and reworking play roles in the distribution of terrestrial palynomorphs in the Seymour marine sediments. Nonmarine palynomorphs were probably remobilized several times during their journeys down rivers and across the shelf before reaching their final resting place. Spore and pollen taxa that disappeared from the Seymour K-T succession are mostly rare forms. It is unclear what these highest occurrences really show, and if they in any way approximate the time of disappearance of the parent plant from the land vegetation, within the fine-scale resolution necessary for documentation of a possible extinction event. Furthermore, land plants are difficult to kill off completely; they may sprout new shoots from broken or burnt stumps, their viable seeds may survive for years, or they may survive in refugia (e.g., Tschudy and Tschudy, 1986). Thus, for nonmarine palynomorphs preserved in Seymour marine sediments, we might not detect evidence of vegetational crisis or abrupt extinctions. Despite these factors, certain Maastrichtian angiosperm pollen species that occurred throughout the southern high latitude region *did* disappear near the end of the Cretaceous from Seymour Island (Askin, 1990a), from New Zealand (e.g., Raine, 1984, 1988), and from southeastern Australia (e.g., Helby and others, 1987). These disappearances suggest a latest Maastrichtian or terminal Cretaceous event or environmental trend that affected the land vegetation throughout this region.

The major part of the land vegetation in southern middle- to high-latitude areas seems to have continued unchanged across the K-T boundary, on the basis of spore and pollen assemblages from Australian and New Zealand marine and nonmarine sections (e.g., Helby and others, 1987; Raine, 1988) and from Seymour Island (Askin, 1990a). Vegetation was dominated by the conifer *Lagarostrobus* (pollen *Phyllocladidites mawsonii*), a conifer that had a wide Late Cretaceous distribution, suggesting a broad ecological tolerance. This form, and many other elements of the flora, evolved in the southern high to middle latitudes, and thrived from the Campanian through the Paleocene throughout this region (e.g., Dettmann, 1989; Askin, 1990a; Dettmann and Jarzen, 1990; Baldoni and Askin, 1993). In the event of a vegetation-destroying crisis, such an adaptive, successful vegetation could regenerate rapidly. Would a break in its pollen record be preserved in the transported and possibly reworked Seymour succession? It is very unlikely.

The most compelling evidence for little vegetational response to any terminal Cretaceous event in the southern high latitudes is from nonmarine sections in New Zealand. These include data from palynomorph (Raine, 1988) and leaf assemblages (Johnson and Greenwood, 1993) which show little change at the K-T boundary. Raine (1988) noted turnovers of some taxa, especially among the angiosperms, but they were less pronounced than the more dramatic events in the New Zealand mid-Cretaceous, at the Paleocene-Eocene boundary, and the late Eocene. Leaf assemblages represent locally derived assemblages and provide a good indication of vegetational change or stability (Johnson and Greenwood, 1993). Leaves are less-easily transported long distances like spores and pollen, and are unlikely to be reworked.

The nonmarine palynomorph assemblages on Seymour Island provide useful paleoclimatic information. The general nature of the vegetation suggests high rainfall (also indicated in climate models; e.g., Parrish and others, 1982), and the abundance of *Sphagnum* moss–type spores (*Stereisporites* spp.) and relatively common remains of water ferns such as *Azolla* suggest that the Cretaceous-Tertiary lowlands adjacent to the depositional basin included wetlands and bogs (Askin, 1990b). The generally waterlogged nature of the Seymour lowlands during the K-T transition and early Danian is also indicated by relatively common *Botryococcus* (fresh- to brackish-water colonial algae). Waterlogged lowlands are consistent with maximum transgression just above the K-T boundary.

CONCLUSIONS

The most detailed palynological results across the K-T boundary on Seymour Island come from section 89/1. Dinocysts define a 20–30 cm K-T transitional interval, which (1) is between the first occurrences of *Kenleyia* spp. and *Danea californica*; (2) includes the start of the *Manumiella seelandica* to *Senegalinium obscurum* turnover; (3) immediately underlies an earliest Danian flooding event; and (4) includes a sharp iridium peak that is correlated with the main iridium peak at the K-T boundary elsewhere. This evidence shows that the dinocyst biostratigraphy works well on Seymour Island, despite some bioturbation and reworking.

Environmental and biotic changes on the Seymour shelf began significantly before the K-T boundary. The marine dinocyst record shows long-term gradual genus- and species-level turnover that accelerated shortly before the end of the Cretaceous, especially with an influx of new forms. The nonmarine spore and pollen record also shows long-term gradual turnover of minor elements, whereas the dominant components of the land vegetation continued across the K-T boundary.

Latest Cretaceous biotic turnover on Seymour Island may be explained, at least in part, by sea-level changes and cooling temperatures. These, and possibly deteriorating environmental conditions, such as water chemistry and nutrient availability, near the end of the Maastrichtian probably initiated changes in the Seymour biotas.

Above the iridium peak, which is presumed fallout from asteroid impact, there is relatively rapid turnover of a *Manumiella seelandica*–dominated dinocyst assemblage to a *Senegalinium obscurum* assemblage. This may be related to sea-level rise, global oceanic trends in chemistry and productivity, deleterious chemical or physical effects of impact, or other factors affecting an already stressed environment, such as local volcanism. A gradual influx of *Spinidinium* and other Paleocene forms followed.

The K-T floral and faunal biotic turnover on Seymour Island probably results from a complex combination of both long-term gradual trends and shorter-term events. Biotic responses to environmental change on the Seymour shelf began before the K-T boundary and appear to be less dramatic than biotic changes recorded at lower latitudes.

ACKNOWLEDGMENTS

We especially appreciate the valuable suggestions made by reviewers David K. Goodman and Graham L. Williams. Askin acknowledges support from National Science Foundation grant DPP 90-191378 and a grant from the Petroleum Research Fund administered by the American Chemical Society.

REFERENCES CITED

Alvarez, L. W., Alvarez, W., Asaro, F., and Michel, H. V., 1980, Extraterrestrial cause for the Cretaceous-Tertiary extinction: Science, v. 208, p. 1095–1108.

Askin, R. A., 1988a, Campanian to Paleocene palynological succession of Seymour and adjacent islands, northeastern Antarctic Peninsula, *in* Feldmann, R. M., and Woodburne, M. O., eds., Geology and paleontology of Seymour Island, Antarctic Peninsula: Geological Society of America Memoir 169, p. 131–153.

Askin, R. A., 1988b, The palynological record across the Cretaceous/Tertiary transition on Seymour Island, Antarctica, *in* Feldmann, R. M., and Woodburne, M.O., eds., Geology and paleontology of Seymour Island, Antarctic Peninsula: Geological Society of America Memoir 169, p. 155–162.

Askin, R. A., 1990a, Campanian to Paleocene spore and pollen assemblages of Seymour Island, Antarctica: Review of Palaeobotany and Palynology, v. 65, p. 105–113.

Askin, R. A., 1990b, Cryptogam spores from the upper Campanian and Maastrichtian of Seymour Island, Antarctica: Micropaleontology, v. 36, p. 141–156.

Askin, R. A., 1992, Late Cretaceous–Early Tertiary Antarctic outcrop evidence for past vegetation and climate, *in* Kennett, J. P., and Warnke, D. A., eds., The Antarctic paleoenvironment: A perspective on global change: American Geophysical Union, Antarctic Research Series, v. 56, p. 61–73.

Askin, R. A., and Elliot, D. H., 1982, Geologic implications of recycled Permian and Triassic palynomorphs in Tertiary rocks of Seymour Island, Antarctic Peninsula: Geology, v. 10, p. 547–551.

Askin, R. A., and Jacobson, S. R., 1989, Total organic carbon content and Rock Eval pyrolysis on outcrop samples across the Cretaceous/Tertiary boundary, Seymour Island, Antarctica: Antarctic Journal of the United States, v. 23, p. 37–39.

Askin, R. A., Elliot, D. H., Kyte, F. T., Jacobson, S. R., and Li, X., 1993, The Cretaceous-Tertiary boundary on Seymour Island, Antarctica: Palynology, iridium and paleoenvironments: Geological Society of America Abstracts with Programs, v. 25, no. 7, p. A295.

Baldoni, A. M., and Askin, R. A., 1993, Palynology of the lower Lefipán Formation (Upper Cretaceous) of Barranca de los Perros, Chubut Province, Argentina. Part II. Angiosperm pollen and discussion: Palynology, v. 17, p. 241–264.

Barrera, E., and Huber, B. T., 1990, Evolution of Antarctic waters during the Maastrichtian: Foraminifer oxygen and carbon isotope ratios, ODP Leg 113, *in* Proceedings of the Ocean Drilling Program, Scientific results, Volume 113: College Station, Texas, Ocean Drilling Program, p. 813–827.

Barrera, E., Huber, B. T., Savin, S. M., and Webb, P. N., 1987, Antarctic marine temperatures: Late Campanian through early Paleocene: Paleoceanography, v. 2, p. 21–47.

Brinkhuis, H., and Leereveld, H., 1988, Dinoflagellate cysts from the Cretaceous/Tertiary boundary sequences of El Kef, northwest Tunisia: Review of Palaeobotany and Palynology, v. 56, p. 5–19.

Brinkhuis, H., and Zachariasse, W. J., 1988, Dinoflagellate cysts, sea level changes and planktonic foraminifers across the Cretaceous-Tertiary boundary at El Haria, northwest Tunisia: Marine Micropaleontology, v. 13, p. 153–191.

Canudo, J. I., Keller, G., and Molina, E., 1991, Cretaceous/Tertiary boundary extinction pattern and faunal turnover at Agost and Caravaca, S.E. Spain: Marine Micropaleontology, v. 17, p. 319–341.

Cookson, I. C., and Eisenack, A., 1965, Microplankton from the Dartmoor Formation, S.W. Victoria: Royal Society of Victoria Proceedings, v. 79, p. 133–137.

Cross, A. T., Thompson, G. G., and Zaitzeff, J. P., 1966, Source and distribution of palynomorphs in bottom sediments, southern part of Gulf of California: Marine Geology, v. 4, p. 467–524.

Damassa, S. P., 1988, *Carpatella cornuta* Grigorovich 1969 (Dinophyceae)—A member of the Aptiana-Ventriosum complex: Palynology, v. 12, p. 167–177.
Del Valle, R. A., Elliot, D. H., and MacDonald, D. I. M., 1992, Sedimentary basins on the east flank of the Antarctic Peninsula: Proposed nomenclature: Antarctic Science, v. 4, p. 477–478.
Dettmann, M. E., 1989, Antarctica: Cretaceous cradle of austral temperate rainforests?, *in* Crame, J. A., ed., Origins and evolution of the Antarctic biota: Geological Society of London Special Publication 147, p. 89–105.
Dettmann, M. E., and Jarzen, D. M., 1990, The Antarctic/Australian rift valley: Late Cretaceous cradle of northeastern Australasian relicts?: Review of Palaeobotany and Palynology, v. 65, p. 131–144.
Donovan, A. D., Baum, G. R., Blechschmidt, G. L., Loutit, T. S., Pflum, C. E., and Vail, P. R., 1988, Sequence stratigraphic setting of the Cretaceous-Tertiary boundary in central Alabama, *in* Wilgus, C. K., Hastings, B. S., Kendall, C. G., Posamentier, H. W., Ros, C., and Van Wagoner, J. C., eds., Sea-level changes: An integrated approach: Society of Economic Paleontologists and Mineralogists Special Publication 42, p. 299–307.
Downie, C., Hussain, M. A., and Williams, G. L., 1971, Dinoflagellate cysts and acritarch associations in the Paleogene of southeast England: Geoscience and Man, v. 3, p. 29–35.
Drugg, W. S., 1967, Palynology of the upper Moreno Formation (Late Cretaceous–Paleocene), Escarpado Canyon, California: Palaeontographica, v. 120B, p. 1–71.
Elliot, D. H., 1988, Tectonic setting and evolution of the James Ross Island Basin, northern Antarctic Peninsula, *in* Feldmann, R. M., and Woodburne, M. O., eds., Geology and paleontology of Seymour Island, Antarctic Peninsula: Geological Society of America Memoir 169, p. 541–555.
Elliot, D. H., Hoffman, S. M., and Rieske, D. E., 1992, Provenance of Paleocene strata, Seymour Island, *in* Yoshida, Y., and others, eds., Recent progress in Antarctic earth science: Tokyo, TERRAPUB, p. 347–355.
Elliot, D. H., Askin, R. A., Kyte, F. T., and Zinsmeister, W. J., 1994, Iridium and dinocysts at the Cretaceous-Tertiary boundary on Seymour Island, Antarctica: Implications for the K-T event: Geology, v. 22, p. 675–678.
Evitt, W. R., 1985, Sporopollenin dinoflagellate cysts: Their morphology and interpretation: Houston, Texas, American Association of Stratigraphic Palynologists Foundation, 333 p.
Firth, J. V., 1987, Dinoflagellate biostratigraphy of the Maastrichtian to Danian interval in the U.S. Geological Survey Albany core, Georgia, U.S.A.: Palynology, v. 11, p. 199–216.
Firth, J. V., 1993, Dinoflagellate assemblages and sea-level fluctuations in the Maastrichtian of southwest Georgia: Review of Palaeobotany and Palynology, v. 79, p. 179–204.
Francis, J. E., 1986, Growth rings in Cretaceous and Tertiary wood from Antarctica and their palaeoclimatic implications: Palaeontology, v. 29, p. 665–684.
Francis, J. E., 1991, Palaeoclimatic significance of Cretaceous–early Tertiary fossil forests of the Antarctic Peninsula, *in* Thomson, M. R. A., Crame, J. A., and Thomson, J. W., eds., Geological evolution of Antarctica: London, Cambridge University Press, p. 623–627.
Goodman, D. K., 1979, Dinoflagellate "communities" from the lower Eocene Nanjemoy Formation of Maryland, U.S.A.: Palynology, v. 3, p. 169–190.
Goodman, D. K., 1987, Dinoflagellate cysts in ancient and modern sediments, *in* Taylor, F. J. R., ed., The biology of dinoflagellates (Botanical Monograph 21): Oxford, Blackwell Scientific Publications, p. 649–722.

Griffis, K., and Chapman, D. J., 1988, Survival of phytoplankton under prolonged darkness: Implications for the Cretaceous-Tertiary boundary darkness hypothesis: Palaeogeography, Palaeoclimatology, Palaeoecology, v. 67, p. 305–314.

Habib, D., Moshkovitz, S., and Kramer, C., 1992, Dinoflagellate and calcareous nannofossil response to sea-level change in Cretaceous-Tertiary boundary sections: Geology, v. 20, p. 165–168.

Hallam, A., 1989, The case for sea-level change as a dominant causal factor in mass extinction of marine invertebrates: Royal Society of London Philosophical Transactions, ser. B, v. 325, p. 437–455.

Hansen, J. M., 1977, Dinoflagellate stratigraphy and echinoid distribution in upper Maastrichtian and Danian deposits from Denmark: Geological Society of Denmark Bulletin, v. 26, p. 1–26.

Haq, B. U., Hardenbol, J., and Vail, P. R., 1988, Mesozoic and Cenozoic chronostratigraphy and eustatic cycles of sea-level change, *in* Wilgus, C. K., Hastings, B. S., Kendall, C. G., Posamentier, H. W., Ros, C., and Van Wagoner, J. C., eds., Sea-level changes: An integrated approach: Society of Economic Paleontologists and Mineralogists Special Publication 42, p. 71–108.

Harland, R., 1973, Dinoflagellate cysts and acritarchs from the Bearpaw Formation (upper Campanian) of southern Alberta, Canada: Palaeontology, v. 16, p. 665–706.

Harwood, D. M., 1988, Upper Cretaceous and lower Paleocene diatom and silicoflagellate biostratigraphy of Seymour Island, eastern Antarctic Peninsula, *in* Feldmann, R. M., and Woodburne, M. O., eds., Geology and paleontology of Seymour Island, Antarctic Peninsula: Geological Society of America Memoir 169, p. 55–129.

Helby, R. J., Morgan, R., and Partridge, A. D., 1987, A palynological zonation of the Australian Mesozoic, *in* Jell, P. A., ed., Association of Australasian Paleontologists Memoir 4, p. 1–93.

Hsü, K. J., and McKenzie, J. A., 1985, A "Strangelove" Ocean in the earliest Tertiary, *in* Sunquist, E. T., and Broecker, W. S., eds., The carbon cycle and atmospheric CO_2: Natural variations Archean to present: American Geophysical Union Geophysical Monograph 32, p. 487–492.

Huber, B. T., 1988, Upper Campanian–Paloecene foraminifera from the James Ross Island region, Antarctic Peninsula, *in* Feldmann, R. M., and Woodburne, M. O., eds., Geology and paleontology of Seymour Island, Antarctic Peninsula: Geological Society of America Memoir 169, p. 163–252.

Hultberg, S. U., 1986, Danian dinoflagellate zonation, the C-T boundary, and the stratigraphical position of the Fish Clay in southern Scandinavia: Journal of Micropalaeontology, v. 5, p. 37–47.

Johnson, K. R., and Greenwood, D., 1993, High-latitude deciduous forests and the Cretaceous-Tertiary boundary in New Zealand: Geological Society of America Abstracts with Programs, v. 25, no. 5, A295.

Keller, G., 1988, Extinction, survivorship and evolution of planktic foraminifers across the Cretaceous/Tertiary boundary at El Kef, Tunisia: Marine Micropaleontology, v. 13, p. 239–263.

Keller, G., 1993, The Cretaceous-Tertiary boundary transition in the Antarctic Ocean and its global implications: Marine Micropaleontology, v. 21, p. 1–45.

Keller, G., Barrera, E., Schmitz, B., and Mattson, E., 1993, Gradual mass extinction, species survivorship, and long-term environmental changes across the Cretaceous-Tertiary boundary in high latitudes: Geological Society of America Bulletin, v. 105, p. 979–997.

Lange, D., 1969, Mikroplankton aus dem Fischton von Stevns-Klint auf Seeland: Beiträge zur Meereskunde, v. 24–25, p. 110–121.

Lawver, L. A., Gahagan, L. M., and Coffin, M. F., 1992, The development of paleoseaways around Antarctica, *in* Kennett, J. P., and Warnke, D. A., eds., The Antarctic paleoenvironment: A perspective on global change: American Geophysical Union Antarctic Research Series, v. 56, p. 7–30.

Macellari, C. E., 1984, Late Cretaceous stratigraphy, sedimentology and macropaleontology of Seymour Island, Antarctic Peninsula [Ph.D. thesis]: Columbus, Ohio State University, 599 p.

Macellari, C. E., 1986, Late Campanian–Maastrichtian ammonite fauna from Seymour Island (Antarctic Peninsula): Journal of Paleontology Memoir 18, 55 p.

Macellari, C. E., 1988, Stratigraphy, sedimentology, and paleoecology of Upper Cretaceous/Paleocene shelf-deltaic sediments of Seymour Island, *in* Feldmann, R. M., and Woodburne, M. O., eds., Geology and paleontology of Seymour Island, Antarctic Peninsula: Geological Society of America Memoir 169, p. 25–53.

MacLeod, N., and Keller, G., 1991, How complete are Cretaceous/Tertiary boundary sections? A chronostratigraphic estimate based on graphic correlation: Geological Society of America Bulletin, v. 103, p. 1439–1457.

McLean, D. M., 1985, Deccan Traps mantle degassing in the terminal Cretaceous marine extinctions: Cretaceous Research, v. 6, p. 235–259.

McMinn, A., 1988, Outline of a Late Cretaceous dinoflagellate zonation of northwestern Australia: Alcheringa, v. 12, p. 137–156.

Moshkovitz, S., and Habib, D., 1993, Calcareous nannofossil and dinoflagellate stratigraphy of the Cretaceous-Tertiary boundary, Alabama and Georgia: Micropaleontology, v. 39, p. 167–191.

Muller, J., 1959, Palynology of Recent Orinoco Delta and shelf sediments: Micropaleontology, v. 5, p. 1–32.

Officer, C. B., and Drake, C. L., 1985, Terminal Cretaceous environmental events: Science, v. 227, p. 1161–1167.

Parrish, J. T., Ziegler, A. M., and Scotese, C. R., 1982, Rainfall patterns and the distribution of coals and evaporites in the Mesozoic and Cenozoic: Palaeogeography, Palaeoclimatology, Palaeoecology, v. 40, p. 67–101.

Partridge, A. D., 1976, The geological expression of eustacy in the early Tertiary of the Gippsland Basin: APEA Journal, v. 16, p. 73–79.

Pirrie, D., and Marshall, J. D., 1990, High-paleolatitude Late Cretaceous paleotemperatures: New data from James Ross Island, Antarctica: Geology, v. 18, p. 31–34.

Raine, J. I., 1984, Outline of a palynological zonation of Cretaceous to Paleogene terrestrial sediments in West Coast region, South Island, New Zealand: New Zealand Geological Survey Report 109, 82 p.

Raine, J. I., 1988, The Cretaceous/Cainozoic boundary in New Zealand terrestrial sequences: International Palynological Congress, 7th, Brisbane, Australia, Abstracts, p. 137.

Stott, L. D., and Kennett, J. P., 1990, The paleoceanographic and paleoclimatic signature of the Cretaceous/Paleogene boundary in the Antarctic: Stable isotope results from ODP Leg 113, *in* Proceedings, Ocean Drilling Program, Scientific results, Volume 113: College Station, Texas, Ocean Drilling Program, p. 829–848.

Stover, L. E., 1973, Paleocene and Eocene species of *Deflandrea* (Dinophyceae) in Victorian coastal and offshore basins, Australia: Geological Society of Australia Special Publication 4, p. 167–188.

Taylor, F. J. R., 1987, Ecology of dinoflagellates: A. General and marine ecosystems, *in* Taylor, F. J. R., ed., The biology of dinoflagellates (Botanical Monograph 21): Oxford, Blackwell Scientific Publications, p. 399–502.

Tocher, B. A., 1991, Late Cretaceous dinoflagellate cysts from the southern Kerguelen Plateau, Site 738, *in* Proceedings, Ocean Drilling Program, Scientific results, Volume 119: College Station, Texas, Ocean Drilling Program, p. 631–633.

Truswell, E. M., 1983, Recycled Cretaceous and Tertiary pollen and spores in Antarctic marine sediments: A catalogue: Palaeontographica, v. 186B, p. 121–174.

Tschudy, R. H., and Tschudy, B. D., 1986, Extinction and survival of plant life following the Cretaceous/Tertiary boundary event, Western Interior, North America: Geology, v. 14, p. 667–670.

Wall, D., Dale, B., Lohmann, G. P., and Smith, W. K., 1977, The environmental and climatic distribution of dinofla gellate cysts in modern marine sediments from regions in the North and South Atlantic oceans and adjacent seas: Marine Micropaleontology, v. 2, p. 121–200.

Williams, G. L., and Bujak, J. P., 1985, Mesozoic and Cenozoic dinoflagellates, *in* Bolli, H. M., Saunders, J. B., and Perch-Nielsen, K., eds., Plankton stratigraphy: London, Cambridge University Press, p. 847–964.

Williams, G. L., Stover, L. E., and Kidson, E. J., 1993, Morphology and stratigraphic ranges of selected Mesozoic-Cenozoic dinoflagellate taxa in the Northern Hemisphere: Geological Survey of Canada Paper 92-10, 137 p.

Wilson, G. J., 1971, Observations on European Late Cretaceous dinoflagellate cysts, *in* Farinacci, A., ed., Proceedings of the II Planktonic Conferences, Rome 1970: Rome, Edizioni Technoscienza, p. 1259–1275.

Wilson, G. J., 1978, The dinoflagellate species *Isabelia druggii* (Stover) and *I. seelandica* (Lange): Their association in the Teurian of Woodside Creek, Marlborough, New Zealand: New Zealand Journal of Geology and Geophysics, v. 21, p. 75–80.

Wilson, G. J., 1987, Dinoflagellate biostratigraphy of the Cretaceous-Tertiary boundary, mid-Waipara River section, North Canterbury, New Zealand: New Zealand Geological Survey Record, v. 20, p. 8–15.

Wilson, G. J., 1988, Paleocene and Eocene dinoflagellate cysts from Waipawa, Hawkes Bay, New Zealand: New Zealand Geological Survey Paleontological Bulletin, v. 57, p. 1–96

Wilson, G. J., and Moore, P. R., 1988, Cretaceous-Tertiary boundary in the Te Hoe River area, western Hawkes Bay: New Zealand Geological Survey Record, v. 35, p. 34–37.

Zinsmeister, W. J., and Feldmann, R. M., 1993, Late Cretaceous faunal changes in the high southern latitudes: A harbinger of impending global biotic catastrophe?: Geological Society of America Abstracts with Programs, v. 25, no. 7, p. A295.

Zinsmeister, W. J., and Macellari, C. E., 1988, Bivalvia (Mollusca) from Seymour Island, Antarctic Peninsula, *in* Feldmann, R. M., and Woodburne, M. O., eds., Geology and paleontology of Seymour Island, Antarctic Peninsula: Geological Society of America Memoir 169, p. 253–284.

Zinsmeister, W. J., Feldmann, R. M., Woodburne, M. O., and Elliot, D. H., 1989, Latest Cretaceous/earliest Tertiary transition on Seymour Island, Antarctica: Journal of Paleontology, v. 63, p. 731–738.

3

Calcareous Nannofossils at the Cretaceous-Tertiary Boundary

S. Gartner, *Department of Oceanography, Texas A&M University, College Station, Texas*

INTRODUCTION

The very great turnover of calcareous nannoplankton at the Cretaceous-Tertiary (K-T) boundary was first documented by Bramlette and Martini (1964). Yet 30 years later, after much study, documentation, and debate, we seem to be no closer to understanding how and why nearly all of the very successful Cretaceous calcareous phytoplankton disappeared. There is somewhat better agreement on the general outline of how the early Cenozoic calcareous phytoplankton evolved, but examination of the early Cenozoic repopulation of the photic water column of the oceans, has yielded only vague clues about the nature of preceding events. The most difficult puzzle is precisely how and why the many Cretaceous species became extinct. Some read and interpret the evidence to indicate that these extinctions were abrupt, essentially instantaneous, when viewed in the perspective of geologic time. A large body impact is a suitable grand scenario for this interpretation. However, the lack of experience with such phenomena by paleontologists and geologists leaves many details unanswered. Others see these extinctions as more gradual, with at least some of the dominant Cretaceous phytoplankton species surviving for a time into the Cenozoic. They favor a sequence of adverse changes in the environment extending across a spectrum from near catastrophic (e.g., large body impact, extensive volcanic outpouring, tectonic rearrangement of continents and oceans) to the geologically mundane (e.g., cooling climate, sea-level drop, and regression). This review considers the evidence bearing on the turnover of the calcareous nannoplankton at the K-T boundary, that has been gathered by a number of investigators, and tries to determine whether the available data can in any way limit, or even circumscribe, the range of causal mechanisms.

Events in Earth history are best understood when viewed within a geographical framework. Climate changes, sea-level changes, tectonic rearrangement—all have significant impact on the biota, and all have an important geographical context. Consequently, the record of calcareous nannoplankton turnover at the K-T transition has been examined by various investigators in many different geographical settings. Close attention has been given to latitudinal differences in the pattern of nannoplankton extinction and recovery, especially in the Southern Hemisphere, where coring of the boundary interval in deep-sea sediments has yielded several remarkable records.

The abruptness of biotic change at the K-T transition was for a long time taken as clear evidence that the boundary was marked by a discontinuity in the historical record. Consequently, boundary sections recovered during the initial years of the Deep Sea Drilling Project (DSDP) were often assumed to contain hiatuses of varying magnitudes. Similar reasoning was used to explain turnover of the biota in boundary sections exposed on land.

A survey of the literature indicates that, as far as the nannofossil record is concerned, nearly all of the K-T boundary transitions fit into one of three general categories. The first is the pelagic carbonate record, which consists of late Maastrichtian pelagic ooze or chalk overlain by basal Paleocene pelagic ooze or chalk. The lithologic change in this type of K-T boundary section may be so subtle that the boundary can be identified with certainty only from the microfossil turnover. Sometimes only a thin layer of clay, a few millimeters to a few centimeters, can be identified to mark the position of the boundary. Occasionally the boundary corresponds to a solution surface or submarine hardground. The vicinity of the boundary may also contain geochemical anomalies. At most sites the clay at the boundary represents a period of calcium carbonate starvation in the pelagic sediment supply. At some sites a volcanic ash has been identified. An impressive number of sections of this type have been recovered by drilling in all oceans and at all latitudes, especially in the Southern Hemisphere (South Atlantic—sites 356, 524, 527, 528; Kerguellen Plateau—sites 738, 750; Broken Ridge—site 752; Ninety East Ridge—site 216; Exmouth Plateau—site 761; Maud Rise—site 690; North Atlantic—site 384; and North and equatorial Pacific—sites 47, 577). On land the exposures near Gubbio (Contessa, Bottaccione), Italy, are of this type, as are some sections in northern Jylland in Denmark (Fig. 1).

A second type of transition is found in calcareous hemipelagic, chalky to marly sediments, and also exhibits a distinctive boundary clay that is usually a few centimeters to tens of centimeters thick. However, in these instances the boundary clay may be nearly or completely barren of fossils. This type of transition is found throughout the Tethyan area (e.g., Spain, Tunisia, Israel, Kazakhstan). Some of the boundary sections of northern Europe also belong to this group. These hemipelagic sections, owing to their high sedimentation rate, have yielded some of the most detailed lithostratigraphic and geochemical records of the transition.

The third type of transition, which is very different, is the Gulf of Mexico type boundary. This type of boundary is characteristically developed in chalky or marly hemipelagic sediments (mudstones). It is significant that the Gulf of

Mexico type boundary exhibits a distinct disconformity with a scoured surface (although often there seems to be very little section missing) overlain by tens of centimeters to several meters of sediments that indicate clearly high-energy transport of some sort. These high-energy sediments may include, from the bottom up, any of the following: a breccia; rip-up clast; single or multiple pebbly, locally spherule-bearing, coarse, sometimes graded, sandy beds; the last may grade upward into or is followed by fine, well-sorted rippled sand or sandstone, that in turn may gradually or abruptly give way to basal Paleocene marls. Locally a tabular, well-cemented calcareous mudstone or chalk up to 10 cm thick may be present immediately beneath the coarse sand or just above the rippled sandstone.

The marls immediately above the high-energy sediment packett are much like those in the Maastrichtian, although often they become sandier upward within the outcrop or exposure. The Brazos sections in Texas and the Braggs section of Alabama are of this general type, as are also sections in Mexico, Haiti, and possibly deep-water Gulf of Mexico sections recovered by the DSDP. The interpretation of these complex deposits at or immediately adjacent to the K-T boundary is the subject of considerable controversy. In the Brazos River sections the calcareous nannofossil succession has been documented in very closely spaced samples (Gartner and Jiang, 1985; Jiang and Gartner, 1986). At the best known of these sections, usually referred to as Brazos 1, the boundary has been placed at between 15 and 20 cm above the top of the high energy deposit. The boundary is designated as the level at which the dominant Maastrichtian species start to decline numerically (following along a decay curve as if by redeposition), and

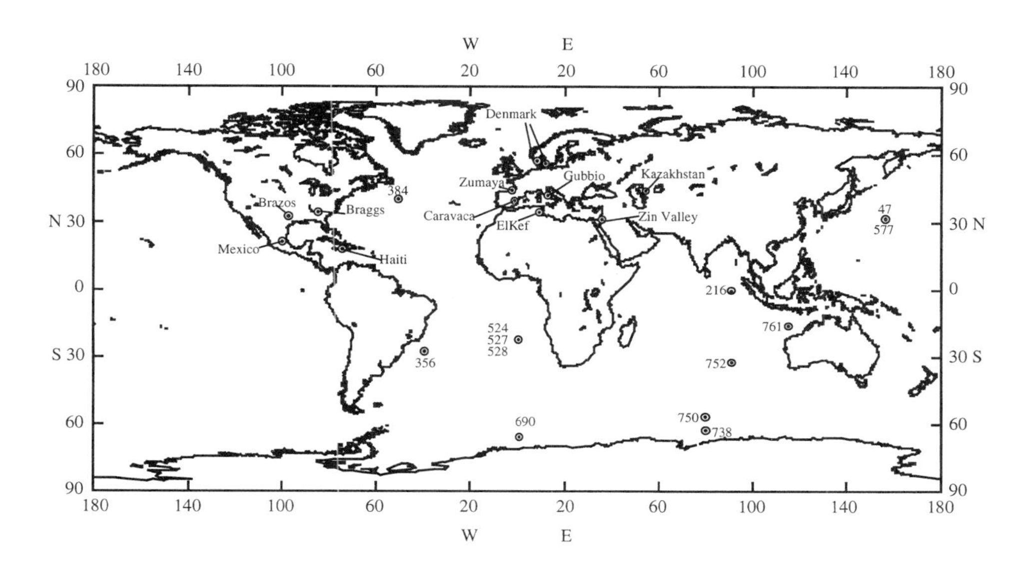

Figure 1. Locations of various Cretaceous-Tertiary boundary sites mentioned in text. Numbers refer to Deep Sea Drilling Project and Ocean Drilling Program sites. Outcrop sections may consist of multiple exposures.

which corresponds precisely to the increase of persistent species, in this case *Thoracosphaera imperforata* +spp. Unfortunately, calcareous nannofossils contribute little to resolving questions about the nature of the high-energy deposits.

Despite sharp differences of sediment types and sedimentation styles at the boundary, the calcareous nannofossil succession generally is very similar in essential features at these various boundary sites (Figs. 2 and 3). The several stages of the succession are as follows: The youngest clearly undisturbed and uncontaminated sediments, which normally occur only a few centimeters below the boundary, contain an unremarkable late Maastrichtian assemblage. In the first two boundary types (those lacking high-energy sediments) this normal Maastrichtian assemblage may be overlain by an interval of nearly or entirely barren clay, often referred to as the boundary clay. When nannofossils are present in this clay, they are usually the most solution-resistant Upper Cretaceous species, a residue of the normal late Maastrichtian assemblage. The third (Gulf of Mexico) boundary type appears to be very complex, but the complexity is related to lithology, which in turn is attributable to the high-energy sediments that are regionally so prominent. These appear to have no real bearing on the paleontological record, although they may cause further confusion because of the sorting and redeposition of Maastrichtian fossils.

PRELUDE: THE LATE CRETACEOUS (LATE MAASTRICHTIAN) RECORD OF CALCAREOUS NANNOFOSSILS

The K-T transition was geologically extraordinary. There may not be another event like it in all of Earth history. The sheer magnitude of plankton extinctions, indeed, all biotic extinctions and attendant side effects, renders suspect for the

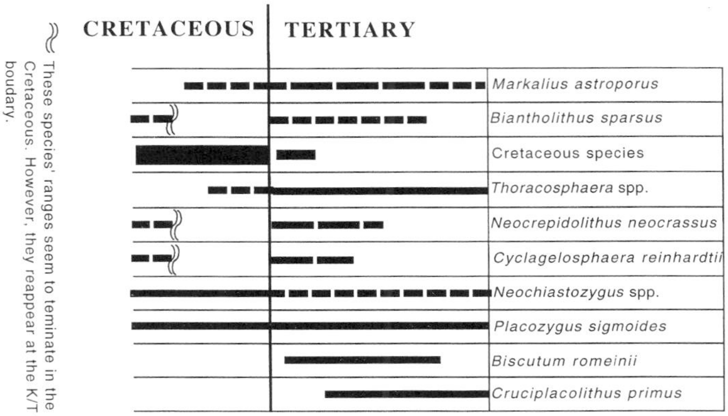

Figure 2. Synthetic summary representation of the ranges and approximate abundance relations of important nannofossils in the vicinity of the Cretaceous-Tertiary boundary.

purpose of interpretation the record of the extinctions as well as the period immediately following it. For example, a pH change of the oceans has been invoked both as cause (Worsley, 1974), and effect (McLean, 1978) of the extinction of calcite-secreting phytoplankton, and there seems to be no way to distinguish which interpretation is correct. One line of evidence that may not be compromised by whatever caused the extinctions is the record of uppermost Maastrichtian events immediately preceding the extinctions. Whether the extinctions were brought on gradually or abruptly, the record of preceding events should be intact. In this regard most students of calcareous nannofossils are in reasonable agreement. There is little or nothing in the coccolith fossil record that would suggest or foreshadow impending massive extinctions. Extensive documentation prepared by Ehrendorfer (1993a, 1993b), confirms general observations by other investigators (see e.g., Percival and Fischer, 1977; Thierstein, 1981; Jiang and Gartner, 1986), that the late Maastrichtian calcareous nannoplankton community was essentially stable to the very end of its existence.

Eshet and others (1992) concluded otherwise: they noted an abundance of *Micula decussata* immediately below the boundary in the Zin Valley (Hor Hahar) section in Israel and, pointing to similarities in some other sections for which checklists have been prepared, they suggested the occurrence of a monospecific phenomenon attributable to harsh conditions and a drop in sea level related to the end of the Maastrichtian. However, monospecific assemblages are most often associated with the unusual success of a particular member of the assemblage (i.e., blooms), but are not known to lead to and, hence, to predict impending extinctions. The *Braarudosphaera* chalks of the South Atlantic Tertiary, and the

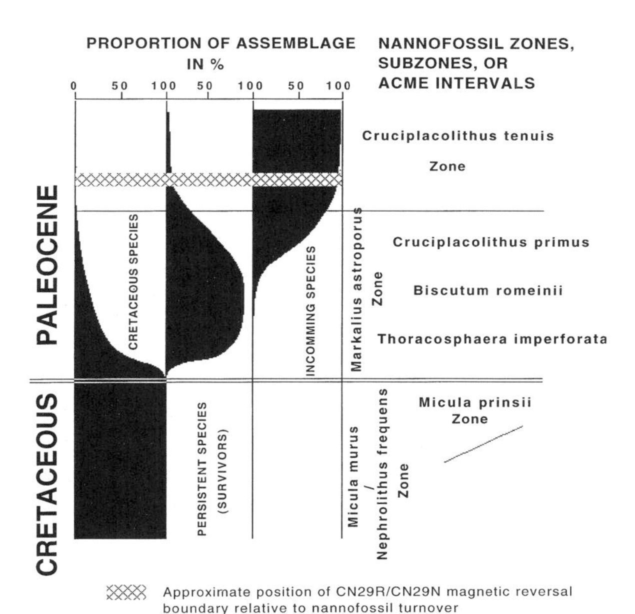

Figure 3. Diagrammatic representation of the succession of Cretaceous species, persistent species, and incoming species at and immediately above the Cretaceous-Tertiary boundary within a general framework of nannofossil zones.

very different *Braarudosphaera-Thoracosphaera* (or other) species dominated earliest Danian marls and chalks in a number of K-T boundary sections, blooms of *Emiliania huxleyi* and of *Umbilicosphaera sibogae* in modern oceans, all reflect success rather than presaging a decline. Seasonal or episodic blooms of the modern coccolithophores *Emiliania huxleyi* and *Umbilicosphaera sibogae*, laminated sediment layers of *Emiliania huxleyi* in pre-Holocene Black Sea sediments (Bukry, 1974), and the remarkable monospecific coccolith laminae (of several coccolith species) in the Lower Cretaceous Munk Marl of the North Sea (Thomsen, 1989) all seem to reflect short-term success. These are not associated with any extinctions.

It has also been noted that in some sections (Texas, Tunisia) there is a marked increase in the latest Maastrichtian of the numbers of a relatively diminutive and otherwise altogether innocuous species, *Prediscosphaera quadripunctata* (Jiang and Gartner, 1986). A very similar (possibly identical) form, *Prediscosphaera stoverii*, dominates to an even greater degree the latest Maastrichtian assemblages at the austral high-latitude boundary section of Ocean Drilling Program (ODP) Site 690C (Pospichal and Wise, 1990). The increase in abundance of this species is not clearly understood, and its simultaneous occurrence in low and high latitudes may suggest a decrease in the intensity of latitudinal gradient. However, there is no external evidence that would link this not very spectacular success to the extinctions that followed. The species that are conspicuous in the late Maastrichtian are so because they seemed to prosper, not decline. It is important that none of them would be considered a disaster form or an opportunistic species. Consequently, there is no good justification for inferring unusual stress of the photic water column during the latest Maastrichtian.

THE CRETACEOUS-TERTIARY BOUNDARY: WHERE EXACTLY IS IT?

Prior to about 1980 nannofossil biostratigraphers had no difficulty identifying the K-T boundary. The turnover of the assemblage was an unmistakable marker. Our search for the cause of this turnover has, however, led to ever-closer sampling and scrutiny of the interval containing the boundary (a sample spacing of centimeters or millimeters is not uncommon), and what previously was obvious now requires strict definition and explanation.

The boundary may be defined from above or from below (Fig. 2). Hay and Mohler (1967) fixed it as a conventional biostratigraphic boundary by defining the *Markalius astroporus* zone as the lowermost Paleocene nannofossil zone, its lower limit being marked by the lowest occurrence of *Markalius astroporus*. This was an unfortunate designation because *Markalius astroporus* was already known to occur in the Maastrichtian (Bramlette and Martini, 1964). Subsequently, the first occurrence of *Biantholithus sparsus* was suggested as an alternative (Perch-Nielsen, 1979). Strictly speaking, the second alternative is also flawed because *Biantholithus sparsus* was described by Caratini (1960) from Turonian-Cenomanian sediments under the name *Cyclodiscolithus radialis*. Where *Biantholithus sparsus* occurs in the lowermost Tertiary in middle- and low-latitude sections, it is always very rare and notably sporadic (Jiang and Gartner,

1986; Percival and Fischer, 1977; Perch-Nielsen, 1979, 1981; Eshet and others, 1992), an undesirable attribute for an important biostratigraphic marker. Moreover, *Biantholithus sparsus'* appearance in the vicinity of the K-T boundary is not consistent with other criteria variously cited as characterizing the boundary. In Southern Hemisphere high-latitude pelagic sections, both of the above two marker species are equally unreliable because of their extremely rare occurrences in nannofossil assemblages (e.g., Pospichal and Wise, 1990; Pospichal, 1991; Wei and Pospichal, 1991; Pospichal and Bralower, 1992).

Extinction of Cretaceous species (i.e., defining the boundary from below) is a more robust criterion and is applied widely as a practical biostratigraphic tool for determining the position of the K-T boundary. However, this criterion is troublesome where precision is essential because of pervasive redeposition (tens of centimeters to meters) above the boundary, complicated further by burrowing across the boundary interval. The two combined can ultimately lead to a very subjective determination of the precise position of the boundary, a position that may not agree with placement of the boundary based on other physical or chemical evidence (e.g., a boundary clay or trace-element spike).

An added complexity is that redeposition may result in sorting, so that the most solution-resistant and/or mechanically robust specimens (species) may increase in proportion in the redeposited assemblage, a pattern that can also be interpreted as survival of a particular solution-resistant species.

The only alternative that remains is to identify the level at which unequivocally earliest Paleocene age species are first encountered. This approach probably is the most objective but has drawbacks. The lowest occurrence of Cenozoic species does not coincide precisely with a physical discontinuity that is almost invariably present, albeit often very subtle (e.g., lithologic change, color change, barren interval, dissolution surface, submarine hardground, scoured surface). The first occurrence of truly Paleocene species does not necessarily coincide with a boundary identified by chemists, petrologists, volcanologists, tectonicists, astronomers, and others.

The conventional explanation for this lack of correspondence is that a finite period of time was required to allow for evolution of a new nannoflora, presumably from survivor species, nearly all of which have left a very cryptic record of their own. *Neocrepidolithus neocrassus*, *Cyclagellosphaera reinhardtii*, *Neochiastozygus modestus*, and *Placozygus sigmoides* all trace their ancestry back into the Mesozoic, but it is a broken record. The first two, for example, are prominent in Jurassic and Lower Cretaceous sediments, but are virtually unknown in Upper Cretaceous sediments. *Biscutum romeinii* and *Biscutum parvulum* are two very small Danian species that are, in ultrastructure and optical pattern of structural elements, similar to Maastrichtian representatives of the genus *Biscutum*, but the inferred relation may be an artifice; the Maastrichtian and Danian forms are really not very similar. Electron microscope photographs of the two Danian species from El Kef show these two to have a cribrate center (Perch-Nielsen, 1981), a feature not found in Maastrichtian representatives of the genus and a characteristic absent in the one survivor of the genus, *Biscutum castrorum* (Pospichal and Wise, 1990). Qualitatively, there seems to be closer similarity between *Crepidolithus crassus*, a Late Jurassic–Early Cretaceous species, and *Neocrepidolithus neocras-*

sus, a Danian species, than between the two Danian species of *Biscutum* and their putative Maastrichtian and/or Danian relatives.

Ignoring for now the problem of *Thoracosphaera*, the first truly Cenozoic species is probably *Biscutum romeinii*. However, that would be a poor choice for marking the boundary for two reasons. One is that this species is so small (<2 μm) that identifying it with certainty can only be done with an electron microscope, except when present in overwhelming abundance and when seen in the context of the nannoflora succession immediately above the K-T boundary. The second reason is that *Biscutum romeinii* seems not to be present in high-latitude sections, particularly those of the Southern Hemisphere. In view of the above objections, a better choice would be the slightly younger first occurrence of the more cosmopolitan *Cruciplacolithus primus*. However, both of these markers first appear above the boundary and designating either as a boundary marker for practical reasons would only cause further confusion.

Some nannofossil students now simply ignore the problem of where precisely to place the boundary, recognizing that the variety and richness of available biotic evidence can be a curse when compared to the apparent tidiness of iridium-enriched clay, tektites, shocked quartz, soot, buckminsterfullerines (buckyballs), or other unwholesome exotica. In any event, pinpointing the boundary to within millimeters is not always possible, and may not be all that important. What is important is the events that transpired near the boundary, and so far as the calcareous nannoplankton are concerned, all of the obvious and important events seem to occur from the boundary up.

THE EARLIEST TERTIARY: NANNOPLANKTON ATTENUATION AND RECOVERY

Early detailed studies of the calcareous nannofossil turnover across the K-T boundary for the most part had a built-in bias toward extracting paleoecological data, emphasizing broad geographic pattern rather than details of temporal succession (e.g., Worsley, 1974; Thierstein, 1981). When taken in the context of the thinking that prevailed at the time, this clearly reflects a predisposition to seek the cause for the biotic turnover in a gradual environmental deterioration compatible with a uniformitarian model such as a major regression, probably accompanied by a cooling climate spreading outward from polar areas; nutrient depletion of the surface oceans; or critical changes in ocean chemistry (e.g., Bramlette, 1965; Worsley, 1974; Percival and Fischer, 1977; Romein, 1977). However, commencing with the study of Percival and Fischer (1977), and continuing with that of Romein (1977, 1979), Perch-Nielsen (1979, 1981), Monechi (1979), and others, documentation of the nannofossil succession of the K-T transition interval has been presented in such detail as to reveal the intrinsic similarities of the record. The boundary sections examined in the above-listed studies, except the two sections near Zumaya on the Bay of Biscay coast studied by Percival and Fisher (1977), are strictly within the Tethyan realm. The Gubbio sections of Monechi (1979) are deep-water pelagic limestone in which calcareous nannofossils are not well preserved. The three remaining sections, Barranco del

Gredero (Caravaca), Spain (Romein, 1977), Zin Valley (Nahal Avdat), Israel (Romein, 1979), and El Kef, Tunisia (Perch-Nielsen, 1979, 1981), are similar marly hemipelagic sections and may be thought of as representative of the nannofossil succession in a subtropical hemipelagic setting.

Declining Species

The first indication of a change in the nannofossil assemblage may be marked by a lithologic discontinuity, such as a clayey layer, that superficially resembles the marls above and below except for a reduced carbonate content. At this level the number and quality of Cretaceous specimens start to diminish, almost imperceptibly at first but then more noticeably, over a stratigraphic interval of tens of centimeters to as much as several meters (Fig. 3; also see Romein, 1977, 1979; Perch-Nielsen, 1979, 1981; Eshet and others, 1992). The same pattern of declining abundance can be seen in other hemipelagic sections, among them Zumaya (Percival and Fischer, 1977) and the Texas Brazos River and Littig sections (Gartner and Jiang, 1985; Jiang and Gartner, 1986). Where sample spacing is sufficiently close to resolve the pattern, this attenuation of the Cretaceous assemblage has also been documented in truly oceanic pelagic sections. Among these are DSDP Sites 356 and 384 (Thierstein, 1981) and 577 (Monechi, 1985); the austral high-latitude ODP Sites 690C (Pospichal and Wise, 1990), 750 (Ehrendorfer and Aubry, 1992), and 752 (Pospichal, 1991); and the Southern Hemisphere mid-latitude ODP Site 761 (Pospichal and Bralower, 1992). There are probably other sections in which the decline can be seen, but in many studies the objectives did not require the sample spacing needed to document clearly the numerical decline and ultimate disappearance of Cretaceous specimens. How this decline should be interpreted is not yet resolved. Nevertheless, the decline of the Cretaceous assemblage is easily recognized.

Persistent Species

There is an increase in the persistent species (Figs. 3 and 4) concomitant with the decline of Cretaceous nannofossils. The term persistent species, introduced by Percival and Fischer (1977), refers to those species (sometimes genera) that have a clearly documented record extending from the Maastrichtian or earlier into the Danian (see Perch-Nielsen and others, 1982; Jiang and Gartner, 1986). Included among these are species of *Thoracosphaera*, *Braarudosphaera bigelowi*, and possibly *Braarudosphaera discula*, *Micrantholithus* sp., *Biscutum castrorum* (=?*Markalius panis*), *Cyclagelosphaera reinhardtii*, and possibly other species of *Cyclagelosphaera*, *Markalius astroporus*, *Neocrepidolithus*, *Neochiastozygus*, *Placozygus*, notably *P. sigmoides*, and holococcoliths variously assigned to *Ottavianus*, *Octolithus*, *Tetralithus*, and *Lanternithus*. Several of the above species may be common to abundant—10% to 50%—of the assemblage. Also included among the persisting species, but seldom exceeding 1% of the assemblage, are *Goniolithus fluckigeri*, *Scapholithus fossilis*, species of *Lapideacassis* and *Scampanella*, and *Biantholithus sparsus*.

All of the above are probably genuine survivors. However, whether a species is a survivor or merely redeposited is not always easy to determine. When a species attains substantially greater abundance above the boundary than within

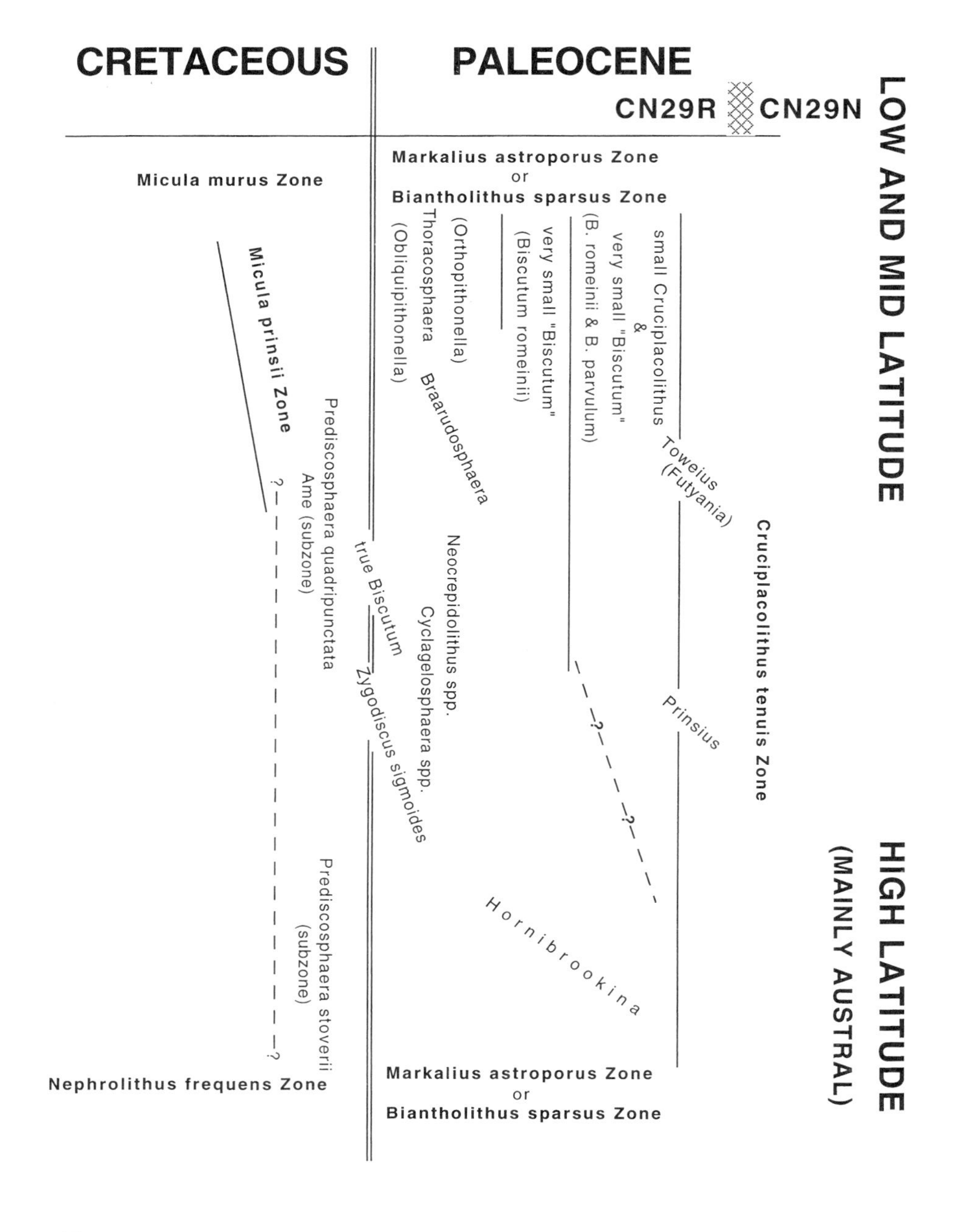

Figure 4. Generalized latitudinal distribution of major coccolith groups near the Cretaceous-Tertiary boundary. Names in bold face type are zonal markers. Some genera are represented by several very small species that may be ecophenotypes.

the late Maastrichtian, it could very well be a survivor. If a species persists into the Danian to well above the interval of redeposition of Cretaceous species, it is probably a survivor. If a species occurs sporadically, is rare, or occurs in the same proportion as in Maastrichtian assemblages, it could have been redeposited. It follows that, for rare species, survivorship above the K-T boundary is virtually impossible to prove or disprove.

Prominent among those cited as persistent species, that is, survivors, are several species of *Thoracosphaera*, which are not coccolithophores but calcareous dinoflagellates. Nearly all authors recognize that several distinct species seem to be involved, but because of a lack of uniform nomenclature it is unclear whether there are regional differences in the species composition of this largely artificial group. This is a problem well worth investigating further. A *Thoracosphaera*-dominated interval occurs in all of the Tethyan sections; Barranco del Gredero, El Kef, Zin Valley, Gubbio; at the Atlantic coast Zumaya sections; in the Texas and Alabama sections; at DSDP Sites 384 and 577 (but seemingly less-intensely developed in DSDP Site 356); and at ODP Sites 690C, 738, 750, 752, and 761. In some sections—Zumaya, the Brazos sections in Texas, Gubbio, DSDP Site 577—there appear to be successive blooms of at least two different species of *Thoracosphaera*. The same pattern may be present at other locations, but the frequent fragmentation of *Thoracosphaera* specimens combined with their ambiguous character has often left them incompletely documented.

Fütterer (1990) used electron microscopy to study closely the various calcareous dinoflagellate species that occur near the boundary at ODP Sites 689 and 690 (*Thoracosphaera* of coccolithophore nomenclature, *Orthopithonella, Obliquipithonella,* and *Centosphaera* of Fütterer; *Pithonella* and calcispherers in older references). Fütterer reported a total turnover of calcareous dinoflagellates at the boundary, with no survivors. The genus *Obliquipithonella* occurs only above the K-T boundary, and *Obliquipithonella operculata* (=*Thoracosphaera operculata*) and *Orthopithonella minuta* may be abundant (bloom?). *Thoracosphaera operculata* has been found to be abundant at other K-T boundary sites, and has been reported to occur in latest Maastrichtian sediments by several investigators. Most documentation of this species was done with light microscopes. Consequently, Cretaceous occurrences of *Obliquipithonella* (i.e., *Thoracosphaera operculata*) may be based on fragments that could have originated from Maastrichtian species of *Orthopithonella*, the optical pattern of which may closely resemble that of *Thoracosphaera operculata*.

If the stratigraphic relation found by Fütterer is true elsewhere, it implies that the first species to be successful after the extinction event were not survivors, but had newly evolved. This is difficult to reconcile with the notion of rapid recovery in an environment temporarily hostile even to true survivors of the Cretaceous calcareous nannoflora; if sufficient time passed to allow for the evolution of *Obliquipithonella* from a Cretaceous ancestor, then why did no new coccolithophore species evolve during that same period? It would be more reasonable to expect that *Thoracosphaera operculata* may actually occur in Maastrichtian sediments elsewhere. It is possible that Fütterer's data have an ecological bias in that during the Maastrichtian some species of *Thoracosphaera* may have been excluded from the open pelagic environment in high latitudes

(ODP Sites 689 and 690) because they were unable to compete successfully with the better-adapted coccolithophores. Alternatively, the very condensed section at these high-latitude sites may be incomplete, although that may not be detectable within the resolution limit of the calcareous nannofossil zonation.

In some boundary sections another problematic nannofossil, *Braarudosphaera bigelowi* (and sometimes *Braarudosphaera discula*, which may be the same species), becomes dominant in the assemblage (Figs. 4 and 5). At Zumaya, Barranco del Gredero, Braggs, and Brazos, *Braarudosphaera* specimens occur in great abundance just above the level where *Thoracosphaera* first dominate, and may in turn be followed by another interval of *Thoracosphaera* dominance. At

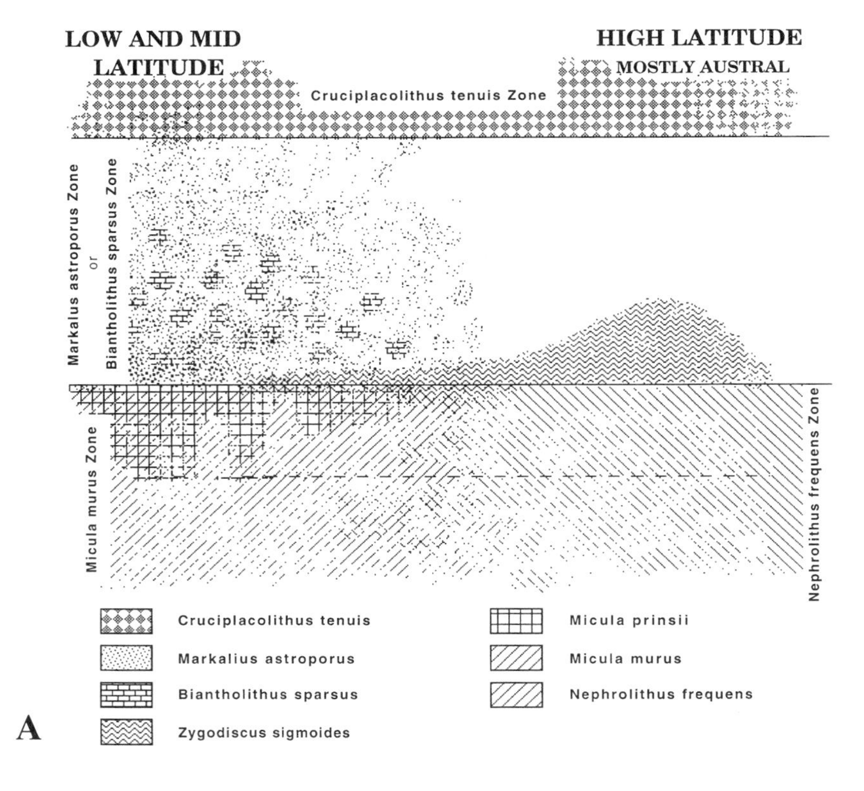

Figure 5. Generalized latitudinal distribution of major coccolith groups near the Cretaceous-Tertiary boundary. The patterns (see legend) indicate the approximate latitudinal and temporal range of a particular group. Each group is prominent at one of the sites discussed, except Micula murus, *which is normally a minor species in the assemblage. A:* Cruciplacolithus tenuis, Markalius astroporus, Biantholithus sparsus, Zaygodiscus sigmoides, Micula prinsii, Micula murus, *and* Nephrolithus frequens*; B:* Thoracosphaera *spp.,* Braarudosphaera bigelowi, Prediscosphaera quadripunctata–Prediscosphaera stoveri, Cruciplacolithus primus, *and* Prinsius *spp.; C:* Biscutum parvulum *and* Biscutum romeinii *(very small* Biscutum*),* Hornibrookina, *and true* Biscutum.

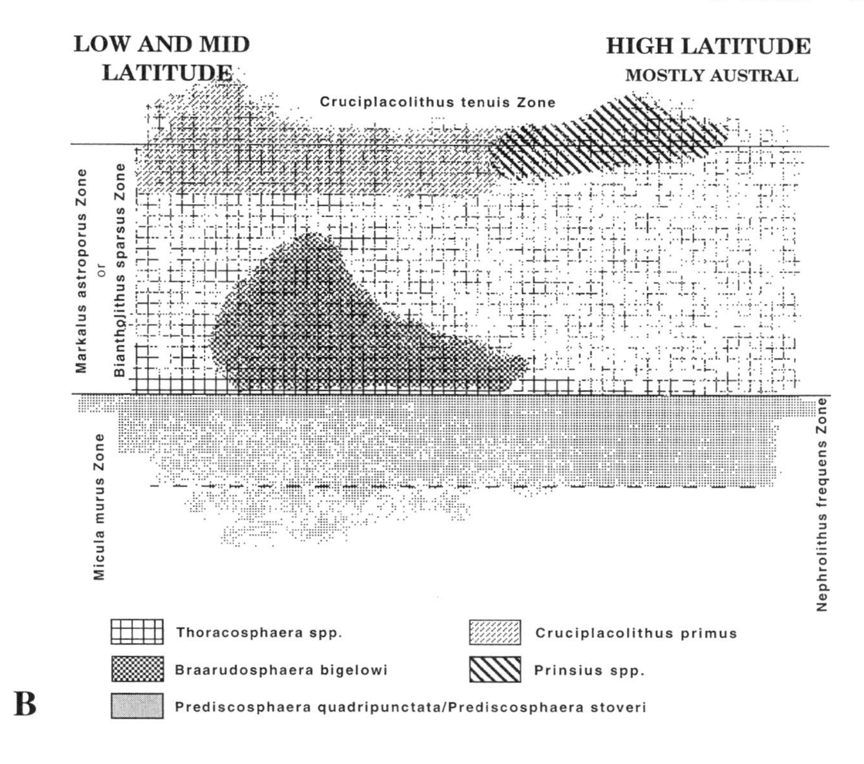

LOW AND MID LATITUDE
HIGH LATITUDE
MOSTLY AUSTRAL
Cruciplacolithus tenuis Zone
Markalus astroporus Zone
or
Biantholithus sparsus Zone
Micula murus Zone
Nephrolithus frequens Zone
Thoracosphaera spp.
Braarudosphaera bigelowi
Prediscosphaera quadripunctata/Prediscosphaera stoveri
Cruciplacolithus primus
Prinsius spp.
B

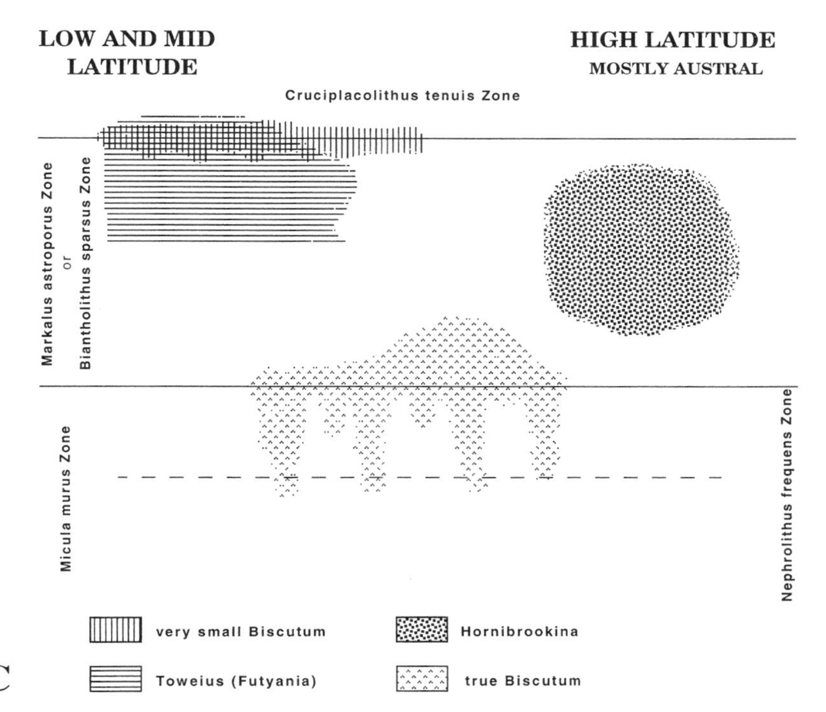

LOW AND MID LATITUDE
HIGH LATITUDE
MOSTLY AUSTRAL
Cruciplacolithus tenuis Zone
Markalus astroporus Zone
or
Biantholithus sparsus Zone
Micula murus Zone
Nephrolithus frequens Zone
very small Biscutum
Toweius (Futyania)
Hornibrookina
true Biscutum
C

Gubbio, abundant *Braarudosphaera* specimens precede abundant *Thoracosphaera*, and at DSDP Site 384 they more or less coincide. At the Southern Hemisphere ODP Sites 690C, 738, 752, and 761, which are generally higher latitude but are also more pelagic, *Braarudosphaera* is exceedingly rare or not present. The record at the Zin Valley sections in Israel and at El Kef in Tunisia is also curious. At the Zin Valley sections, *Braarudosphaera* are not present, possibly because of a hiatus in this greatly condensed section; at the El Kef section, the species first appears ~0.5 m above the boundary (Jiang and Gartner, 1986) and is present in significant numbers (few to common) only ~1.3 to 1.5 m above the level of the first abundant *Thoracosphaera*. As a first-order approximation this is much higher (and probably later) than expected, based on other sections (Barranco del Gredero, Brazos), even after the greater sedimentation rate of the expanded section at El Kef is taken into account.

It is possible that where abundant *Braarudosphaera* specimens precede or coincide with abundant *Thoracosphaera*, the initial *Thoracosphaera* dominance interval may be missing because of a hiatus. However, an alternative inference can be made, namely that the *Braarudosphaera* dominance interval is not synchronous, as implied by Thierstein (1981). Hence, the bloom that produced the dominance interval was not synchronous oceanwide, and probably did not occur as widely as the *Thoracosphaera* bloom. The second alternative would also explain discrepancies in the ranges of *Braarudosphaera bigelowi* and *Biscutum romeinii* in the several low-latitude hemipelagic sections. A third possibility is that the *Thoracosphaera* dominance interval represents a cosmopolitan event, whereas the bloom of *Braarudosphaera* was limited environmentally, even while the ocean was largely depopulated of coccolithophores.

Cyclagellosphaera reinhardtii has an abundance peak at the lowermost peak abundance level of *Thoracosphaera* at Zumaya (recorded as *Markalius reinhardtii*), at Braggs, and at the equivalent level at ODP Site 761 (i.e., immediately above the level where the Cretaceous assemblage starts to decline). In most sections, however, this species occurs as a minor element of the assemblage, or it may be absent altogether. Species of *Cyclagellosphaera* apparently bloomed only locally or regionally.

The two problematic forms (genera), *Thoracosphaera* and *Braarudosphaera*, both have extant species that probably are not very different from the earliest Cenozoic representatives (although the modern *Thoracosphaera heimii* is a vegetative cell). Other than this the two may not have much in common. The former occurs primarily in the open ocean, whereas the latter is able to survive in marginal marine environments: indeed, *Braarudosphaera* may thrive there. However, it is also known to occur in normal marine environments, and there may be a danger of overinterpreting the environmental significance of this genus. *Thoracosphaera* are often recorded as fragments and may include unrelated objects, some perhaps not dinoflagellates. *Cyclagellosphaera reinhardtii* (and other species of *Cyclagellosphaera* cited at this level) do not suggest abnormal environments. It is reasonable to conclude, therefore, that the nannofossil assemblages that seem to have had the greatest success following the decline of the Cretaceous assemblage need not indicate a stressed or otherwise abnormal environment at the time when they flourished. It is possible that they simply occupied what was vacated by a previously very successful phytoplankton community. The species that prospered in the earliest Danian apparently did so by default.

The question that occurs at this point is what is it in the nature of *Thoracosphaera, Braarudosphaera, Cyclagellosphaera*, and the other species that prospered early in the Paleocene that favored them, while the many previously more successful species were obliterated from all of the oceans? The answer to this has not yet been found.

The terms dominant, dominance interval, and bloom used in description of the calcareous nannoplankton recovery following the decline or extinction of Cretaceous coccolithophores may be somewhat deceptive, but they are used for lack of better terms, although peak abundance may come close in some cases. In closely spaced samples in pelagic as well as hemipelagic sections, abundances of dominant species (as well as subordinate species) may fluctuate by three-fold or four-fold or more in adjacent samples separated by 10 cm or less. Thus, within the dominance interval or peak abundance interval of a species, that species may not be particularly abundant in any randomly picked sample. Taking into account sediment mixing, it is reasonable to conclude that the phytoplankton that produced the record must have varied greatly in abundance over a time span as short as or shorter than that recorded in the sediments. Because even the best hemipelagic record can resolve events only on the time scale of millennia, it is not possible to determine the length of putative blooms, which actually need not last longer than a season. Whether the dominance intervals can be equated with blooms also cannot be answered unequivocally. In the absence of any other species, even a relatively low abundance *Thoracosphaera* in the water column would appear as a dominance interval in the sediment and might be interpreted incorrectly as a bloom.

Incoming Species

A number of additional species flourished in their turn, all of them incoming species. Virtually all of these species are very small (<3 μm, some <2 μm) and difficult to identify with the light microscope. Consequently, it is also difficult to make a valid comparison from one investigator to another. At Barranco del Gredero, Romein (1977) noted successive abundance of *Biscutum* sp. (=*Biscutum romeinii*), *Tetralithus multiplus* (=*Octolithus multiplus*), and *Toweius petalosus*. At El Kef the succession is similar but perhaps more complex and includes (in order) *Biscutum romeinii, Biscutum parvulum* (plus *Lanternithus duocavus*), and *Cruciplacolithus primus* (plus *Toweius petalosus* plus *Prinsius* aff. *P. dimorphosus*) (Perch-Nielsen, 1981). In the Brazos sections distinct dominance intervals can be identified with *Biscutum romeinii*, followed by *Cruciplacolithus primus*, then *Biscutum parvulum* plus *Toweius petalosus* (Jiang and Gartner, 1986), very much like the succession at Barranco del Gredero, although the relative positions of peak abundances may be in different order. A *Cruciplacolithus primus* dominance interval also was identified at pelagic DSDP Sites 356 and 384, but not a *Biscutum romeinii* dominance interval that precedes it in well-developed hemipelagic sections (Thierstein, 1981). At the Pacific DSDP Site 577, which was in the tropics at the end of the Cretaceous, initial dominance in the Paleocene is by *Biscutum romeinii,* followed by *Cyclagel-*

losphaera reinhardtii plus several species of *Thoracosphaera*, then *Cruciplacolithus primus,* and finally *Placozygus sigmoides* (Monechi, 1985). Oceanic pelagic sediments sometimes lack small coccoliths. The significance of this may have to do with the greater susceptibility to dissolution of very small specimens such as those of *Biscutum romeinii*; it may indicate that small coccoliths were winnowed during sedimentation and deposited elsewhere; or it may simply mean that the very small specimens were not recorded in the studies of these sites. However, it might also mean that no small coccoliths were produced in the overlying water column.

The succession of incoming species at the mid-latitude Southern Hemisphere ODP Site 761 is different from sections considered above in that the small species of *Biscutum* so prominent elsewhere, as well as the abundance interval of *Cruciplacolithus primus*, are not reported from this site. Prominent to dominant are *Cruciplacolithus tenuis, Prinsius dimorphosus*, and *Coccolithus pelagicus*, species suggestive of a significantly younger age; thus, Pospichal and Bralower (1992) inferred a hiatus between the interval of predominantly surviving species and the interval of predominantly incoming species.

The several austral high-latitude sites have some similarity, but they also differ. At ODP Site 690 (Maud Rise), *Hornibrookina edwardsii* is the first incoming species to dominate, followed by *Cruciplacolithus primus* plus *Cruciplacolithus tenuis* (Pospichal and Wise, 1990). At ODP Site 738, which is at about the same latitude, *Hornibrookina* attains (a lesser) maximum abundance just before *Cruciplacolithus primus* plus *Cruciplacolithus tenuis* become dominant. This last event also coincides with a peak abundance of *Prinsius dimorphosus* (Wei and Pospichal, 1991), a very small species that may suggest an environment signalled elsewhere by the very small *Biscutum romeinii* or *Biscutum parvulum*, although the *Prinsius dimorphosus* peak abundance seems to come later. That, in turn, is followed by peak abundances of *Coccolithus pelagicus*, then *Chiasmolithus danicus* plus *Chiasmolithus bidens*, and then *Prinsius bisulcus* plus *Prinsius martinii*. This succession extends into the Paleocene higher than most that have been documented in detail, into the interval of evolving but established Danian nannoflora. However, there may be significant hiatuses in this section, within the interval characterized by survivors. The corresponding succession at Site 750 (Kerguellen Plateau) has a very short interval of *Hornibrookina* dominance that yields to *Cruciplacolithus primus*, which agrees with the inferred hiatus for this level (Ehrendorfer and Aubry, 1992).

The above survey suggests that the order of the succession of dominance or unusual (peak) abundance of incoming species appears to some degree to be site specific, differing from one ocean area to another, although there are some regional similarities. The succession of dominant species is essentially identical at Brazos 1 and Brazos 3, two localities less than two km apart. (It may be significant that the data from these two sites were recorded by the same investigators.) The succession at the Braggs section appears to be different, even though that section was deposited in a similar setting (Thierstein, 1981; Moshkovitz and Habib, 1993). It may be important that the sample spacing for the Braggs section studies is not uniformly close, and thus it is not possible to resolve the details of assemblage succession documented elsewhere.

CALCAREOUS NANNOPLANKTON AND THE NATURE OF THE EXTINCTION EVENT—GRADUAL, STEPWISE, OR ABRUPT

Few students of calcareous nannofossils will dispute that the extinction of the Cretaceous coccolith-producing phytoplankton was catastrophic. Whether there was one massive extinction (i.e., all extinctions occurred within as short a time span as one to 100 years, or even one to 100 hours), or whether the extinctions were gradual (i.e., occurring within a stratigraphically resolvable, 10^3 to 10^5 year, time frame) is still disputed. The evidence indicates that the extinction episode was too short in duration to be clearly resolvable in any of the available stratigraphic records of calcareous nannofossils. Most of the Cretaceous species were eliminated in an initial massive extinction episode, but a few survived; the "persistent species" of Percival and Fischer (1977) and Jiang and Gartner (1986); and the "survivors" of Pospichal (1991), and Pospichal and Bralower (1992). Only the species that are regularly more abundant above the boundary than below can be included unequivocally among the survivors, as is suggested by the analysis of Pospichal and Bralower (1992). Others might have survived, but that cannot be proved.

None of the "persistent species" or "survivors" are numerically important elements of the Maastrichtian assemblage; however, they are among the first species to be successful following the extinction event. Assuming that *Thoracosphaera* are survivors, then the several species of this genus that occur as dominant elements in the earliest Tertiary, along with *Braarudosphaera bigelowi* and *Cyclagelosphaera reinhardtii,* constitute the initial bloom episodes. Their sudden success was short lived, however, and with the advent of the incoming species—*Biscutum romeinii, Biscutum parvulum, Hornibrookina* spp., and *Cruciplacolithus primus*—they declined abruptly. After the appearance of *Cruciplacolithus tenuis*, which marks the establishment of a stable Danian calcareous nannoflora, persistent species are no longer prominent, although they are still present as rare constituents. This fading from the record may suggest a second episode of extinction, one that affected only the hapless survivors and, hence, might be attributable to their inability to compete with the newly evolving phytoplankton. Apparently these persistent species inhabited an environment unaffected by the cause of the extinctions, or were adapted to conditions lethal to other calcareous phytoplankton species. But when the post-Cretaceous oceans were restored to a state resembling the Late Cretaceous oceans, the persistent species, the hardy survivors, again retreated to unpreserved niches or perished. One of these persistent species, *Braarudosphaera bigelowi,* has survived to the present, whereas others such as *Neochiastozygus, Placozygus, Scapholithus*, and the holococcoliths, have left closely related descendents. Most have disappeared.

As far as the occurrence of Cretaceous, specifically late Maastrichtian, nannofossils above the boundary is concerned, it is clear that extensive redeposition took place at sites that were located in hemipelagic, shelf-type settings as well as in deep-sea pelagic settings. The abundance of Cretaceous species declines and approaches zero usually within tens of centimeters. Often anomalously large numbers of Cretaceous species are found well above the level of the onset of decline of Cretaceous species. Usually these can be traced to burrow filling, attesting to the prevalence of bioturbation and displacement of Maastrichtian nannofossils into younger sediments (e.g., Pospichal and others, 1990). Burrowing can account for the often jagged nature of the decline curve for Cretaceous species immediately above the boundary.

An increase of species relative abundances in the potentially redeposited assemblage from in-place Cretaceous abundances might be taken to indicate that the species persisted beyond the end of the Cretaceous. Such is the case with many of the "persistent species" that bloom during the earliest Paleocene. This pattern may also be found for *Micula decussata*. However, the latter is also the most solution-resistant Cretaceous species and, where its abundance increases above the boundary, it could be due entirely to selective dissolution of less-robust Cretaceous species during redeposition.

Perch-Nielsen and others (1982) argued for survival of some Cretaceous nannoplankton species, noting that carbon isotope values of earliest Paleocene sediments (whole rock) require an unrealistically negative isotopic composition for the early Paleocene ocean, if it is assumed that all Cretaceous species in these sediments were redeposited, and thus contributed an exclusively Cretaceous ocean isotope signal in proportion to their abundance as redeposited elements in the postboundary sediments. However, similar isotopic analyses made on samples in which everything but the coccolith size fraction was eliminated yielded results that do not require unrealistic negative isotopic composition for the earliest Paleocene ocean (Alcala-Herrera and others, 1990). Consequently, survival of common Maastrichtian coccolithophores into the earliest Paleocene is not required to explain carbon isotope ratios of the coccolith size fraction.

What then does the nannofossil record tell about the extinctions that mark the K-T boundary? Several scenarios have been developed that make planktonic algae, particularly coccolithophores, central players causally linked to changes in Earth's atmosphere-hydrosphere-biosphere that ultimately precipitated the catastrophic extinctions (Bramlette, 1965; Tappan, 1968; McLean, 1978). From the current perspective, it seems far more likely that coccolithophores were not active players. Rather, they were victims, as were all the other successful Cretaceous organisms that disappeared. The impact of a large asteroid seems to provide an entirely adequate mechanism for the debacle—as do a number of other models involving short term earthly or unearthly cataclysms.

CONCLUSION

1. There are no changes in the late Maastrichtian calcareous nannofossil assemblage that anticipate or predict the impending extinctions. The coccolithophores were not in decline and they do not appear to have been unusually stressed.

2. The calcareous nannofossil succession in the various settings from which presumed continuous records across the K-T boundary have been recovered (mainly open ocean pelagic and hemipelagic shelf settings) do not require a major change of sea level in association with the boundary. The hemipelagic sections of the greater Gulf of Mexico–northern Caribbean area mark the boundary by a sharp lithologic break and high-energy sediments, whereas the same general setting in

the Tethyan area (Spain, Tunisia, Israel) and in northern and central Europe, as well as many oceanic pelagic boundary sections, are marked by only a minor lithologic change; that is, a reduction in carbonate sedimentation. Yet in all areas the latest Maastrichtian sediments preserved are essentially of the same age.

3. The succession of earliest Tertiary nannofossil assemblages is similar, although by no means identical in the various settings. The earliest Tertiary is dominated by an assemblage of persistent species, which are true survivors, followed shortly by an interval dominated by incoming species.

4. The assemblage of persistent species is dominated initially by the calcareous dinoflagellate *Thoracosphaera*. Other persistent species such as *Zygodiscus sigmoides*, species of *Cyclagellosphaera*, and the problematic coccolithophore *Braarudosphaera* may be abundant, rare, or absent, and their order of occurrence is not necessarily the same in different locations.

5. The assemblages of persistent species vary significantly with latitude, but may also vary at locations from similar latitudes.

6. The assemblages of incoming species in subtropical pelagic and hemipelagic locations (i.e., the Tethys) are markedly different from assemblages in temperate- to high-latitude (pelagic) locations.

7. During the recovery phase (characterized by blooms of persistent species followed by blooms of incoming species), the phytoplankton appear to have been very different in different ocean areas, which is in sharp contrast to a relatively more uniform coccolithophore community during the latest Maastrichtian.

8. Persistent species were, for the most part, extremely rare during the Maastrichtian and they became very rare again just as soon as a stable incoming Danian assemblage became established. At the species level, this fading into the background could be interpreted as a second extinction, although several of the persistent species left descendent species that remain extant.

9. Other than the persistent species, no important Maastrichtian coccolith species seems to have survived the extinction marking the end of the Cretaceous.

REFERENCES

Alcala-Herrera, J., Gartner, S., and Grossman, E., 1990, Single or multiple extinctions at the Cretaceous/Tertiary boundary: The South Atlantic carbon isotope record: Geological Society of America Abstracts with Programs, v. 22, no. 7, p. 279.

Bramlette, M. N., 1965, Massive extinctions of biota at the end of Mesozoic time: Science, v. 148, p. 1696–1699.

Bramlette, M. N., and Martini E, 1964, The great change of calcareous nannoplankton fossils between the Maastrichtian and Danian: Micropaleontology, v. 10, p. 291–322.

Bukry, D., 1974, Coccoliths as paleosalinity indicators—Evidence from the Black Sea, *in* Degens, E. T., and Ross, D. A., eds., The Black Sea—Geology, chemistry, and biology: American Association of Petroleum Geologists Memoir 20, p. 353–363.

Caratini, C., 1960, Contribution a l'étude des coccolithes du Cénomanien Supérieur et du Turonien de la région de Rouen [Ph.D. thesis]: Algeria, Université D'Alger, 73 p.

Ehrendorfer, T. W., 1993a, High resolution investigation of calcareous nannoplankton associations during the last 500 ka of the Cretaceous (Abstracts, INA Conference, 5th, Salamanca, 1993): International Nannoplankton Association Newsletter, v. 15, p. 59.

Ehrendorfer, T. W., 1993b, Late Cretaceous (Maestrichtian) calcareous nannoplankton biogeography [Ph.D. thesis]: Woods Hole, Massachusetts, Woods Hole Oceanographic Institute–Massachusetts Institute of Technology (WHOI-93-15), 288 p.

Ehrendorfer, T. W., and Aubry, M. P., 1992, Calcareous nannoplankton changes across the Cretaceous/Paleocene boundary in the southern Indian Ocean (Site 750), *in* Proceedings of the Ocean Drilling Program, Scientific results, Volume 120: College Station, Texas, Ocean Drilling Program, p. 451–470.

Eshet, Y., Moshkovitz, S., Habib, D., Benjamini, C., and Magaritz, M., 1992, Calcareous nannofossil and dinoflagellate stratigraphy across the Cretaceous/Tertiary boundary at Hor Hahar, Israel: Marine Micropaleontology, v. 18, p. 199–228.

Fütterer, D. K., 1990, Distribution of calcareous dinoflagellates at the Cretaceous-Tertiary boundary of Queen Maude Rise, eastern Weddell Sea, Antarctica (ODP Leg 113), *in* Proceedings of the Ocean Drilling Program, Scientific results, Volume 113: College Station, Texas, Ocean Drilling Program, p. 533–548.

Gartner, S., and Jiang, M. J., 1985, The Cretaceous/Tertiary boundary in east central Texas: Gulf Coast Association of Geological Societies Transactions, v. 35, p. 373–380.

Hay, W. W., and Mohler, H. P., 1967, Calcareous nannoplankton from early Tertiary rocks at Pont Labau, France, and Paleocene–early Eocene correlations: Journal of Paleontology, v. 41, p. 1504–1541.

Jiang, M. J., and Gartner, S., 1986, Calcareous nannofossil succession across the Cretaceous/Tertiary boundary in east central Texas: Micropaleontology, v. 32, p. 232–255.

McLean, D. M., 1978, A terminal Mesozoic "Greenhouse": Lessons from the past: Science, v. 201, p. 401–406.

Monechi, S., 1979, Variations in nannofossil assemblage at the Cretaceous/Tertiary boundary in the Bottaccione section (Gubbio, Italy), *in* Christensen, W. K., and Birkelund, T., eds., Proceedings, Cretaceous-Tertiary Boundary Events Symposium, Volume 2: Copenhagen, University of Copenhagen, p. 164–167.

Monechi, S., 1985, Campanian to Pleistocene calcareous nannofossil stratigraphy from the northwestern Pacific Ocean, Deep Sea Drilling Project Leg 86, *in* Initial reports of the Deep Sea Drilling Project, Volume 86: Washington, D.C., U.S. Government Printing Office, p. 301–336.

Moshkovitz, S., and Habib, D., 1993, Calcareous nannofossil and dinoflagellate stratigraphy of the Cretaceous-Tertiary boundary, Alabama and Georgia: Micropaleontology, v. 39, p. 167–191.

Perch-Nielsen, K., 1979, Calcareous nannofossils at the Cretaceous/Tertiary boundary in Tunisia, *in* Christensen, W. K., and Birkelund, T., eds., Proceedings, Cretaceous-Tertiary Boundary Events Symposium, Volume 2: Copenhagen, University of Copenhagen, p. 238–243.

Perch-Nielsen, K., 1981, Nouvelles observations sur les nannofossiles calcaires à la limite Crétacé-Tertiaire près de El Kef (Tunisie): Cahiers de Micropaléontologie, v. 3, p. 25–36

Perch-Nielsen, K., McKenzie, J., and He, Q., 1982, Biostratigraphy and isotope stratigraphy and the "catastrophic" extinction of calcareous nannoplankton at the Cretaceous/Tertiary boundary, *in* Silver, L. T., and Schultz, P. H., eds., Geological implications of impacts of large asteroids and comets on the Earth: Geological Society of America Special Paper 190, p. 353–371.

Percival, S. F., and Fischer, A. G., 1977, Changes in calcareous nannoplankton in the Cretaceous-Tertiary biotic crisis at Zumaya, Spain: Evolutionary Theory, v. 2, p. 1–35.

Pospichal, J. J., 1991, Calcareous nannofossils across the Cretaceous/Tertiary boundary at Site 752, Eastern Indian Ocean, *in* Proceedings of the Ocean Drilling Program, Scientific results, Volume 121: College Station, Texas, Ocean Drilling Program, p. 297–355.

Pospichal, J. J., and Bralower, T. J., 1992, Calcareous nannofossils across the

Cretaceous/Tertiary boundary at Site 761, Northwest Australian Margin, *in* Proceedings of the Ocean Drilling Program, Scientific results, Volume 122: College Station, Texas, Ocean Drilling Program, p. 735–752.

Pospichal, J. J., and Wise, S. W., Jr., 1990, Calcareous nannofossils across the K/T boundary, ODP Hole 690C, Maud Rise, Weddell Sea, *in* Proceedings of the Ocean Drilling Program, Scientific results, Volume 113: College Station, Texas, Ocean Drilling Program, p. 515–532.

Pospichal, J. J., Wise, S. W., Asaro, F., and Hamilton, N., 1990, The effects of bioturbation across a biostratigraphically complete, high southern latitude K/T boundary, *in* Sharpton, V., and Ward, P., eds., Proceedings of the Conference on Global Catastrophes in Earth History: An interdisciplinary conference on impacts, volcanism, and mass mortality: Geological Society of America Special Paper 247, p. 497–508.

Romein, A. J. T., 1977, Calcareous nannofossils from the Cretacous/Tertiary boundary interval in the Barranca del Gredero (Caravaca, Province Murcia, S. E. Spain), Part I and II: Koninklijke Nederlandse Akademie van Wettenschappen Proceedings, ser. B, v. 80, p. 256–279.

Romein, A. J. T., 1979, Calcareous nannofossils from the Cretaceous/Tertiary boundary interval in the Nahal Avdat section, the Negev, Israel, *in* Christensen, W. K., and Birkelund, T., eds., Proceedings, Cretaceous-Tertiary Boundary Events Symposium, Volume 2: Copenhagen, University of Copenhagen, p. 202–206.

Tappan, H., 1968, Primary production, isotopes, extinctions and the atmosphere: Palaeogeography, Palaeoclimatology, Palaeoecology, v. 4, p. 187–210.

Thierstein, H. R., 1981, Late Cretaceous nannoplankton and the change at the Cretaceous/Tertiary boundary, *in* Warme, J. E., Douglas, R. G., and Winterer, E. L., eds., The Deep Sea Drilling Project: A decade of progress: Society of Economic Paleontologists and Mineralogists Special Publication 32, p. 355–394.

Thomsen, E., 1989, Seasonal variability in the production of Lower Cretaceous calcareous nannoplankton: Geology, v. 17, p. 715–717.

Wei, W., and Pospichal, J. J., 1991, Danian calcareous nannofossil succession at Site 738 in the southern Indian Ocean, *in* Proceedings of the Ocean Drilling Program, Scientific results, Volume 119: College Station, Texas, Ocean Drilling Program, p. 495–512.

Worsley, T., 1974, The Cretaceous/Tertiary boundary event in the ocean, *in* Hay, W. W., ed., Studies in paleooceanography: Society of Economic Paleontologists and Mineralogists Special Publication 20, p. 94–125.

4

The Cretaceous-Tertiary Mass Extinction in Planktonic Foraminifera: Biotic Constraints for Catastrophe Theories

Gerta Keller, *Department of Geological and Geophysical Sciences, Princeton University, Princeton, New Jersey*

INTRODUCTION

At various times in Earth's history rare events have caused the extinction of a major part of the Earth's biota followed by the evolution and radiation of new organismal groups better adapted to a changing environment. Such times of mega-mass extinctions are known from the end of the Ordovician, end of the Devonian, Permian-Triassic, and Cretaceous-Tertiary (K-T) boundaries. Many more less-extensive mass-extinction events have also occurred through time. These were generally restricted to specific organismal groups inhabiting particular environments. Hypotheses regarding the causal agents for both types of mass extinction events have included extraterrestrial bolide impacts (Alvarez and others, 1980; Alvarez, 1987; McLaren and Goodfellow, 1990; Raup, 1990), volcanism (McLean, 1985; Officer, 1990; Courtillot and others, 1988), CO_2 poisoning (McLean, 1985), O_2 suffocation (Landis and others, 1993; Hengst and others, this volume), reproductive failure (McLean, 1991, 1994), sea-level fluctuations, and climate change (Hallam, 1989). Whereas each of these scenarios provides plausible explanations for mass extinctions, factual supporting evidence has remained elusive, incomplete, or even contradictory. This is not surprising, because primary evidence of causal factors is often difficult to find in the preserved stratigraphic or fossil record, and in many cases may no longer exist due to postdepositional processes (e.g., erosion, redeposition, solution, reprecipitation).

Of the currently proposed hypotheses, only bolide impact calls for an instantaneous catastrophic event which would leave a single horizon in the stratigraphic record that should coincide with the mass extinction, even though some organisms may have survived the event itself for some time. All other hypotheses require an extended period for the buildup of lethal effects and therefore predict that the mass extinction should have occurred gradually over an extended period of time. Because no single stratigraphic horizon can be identified, these alternative hypotheses are difficult, if not impossible, to test in the stratigraphic record. In contrast, the impact hypothesis should, at least in theory, be easy to test, although in practice commonly recognized impact signatures may have been eroded, geochemically altered beyond recognition, or may have multiple origins not related to extraterrestrial sources. Such concerns were raised in a number of publications, including Carter and others (1990), Officer (1990), and Officer and Carter (1991) on shocked quartz; Jéhanno and others (1992) and Lyons and Officer (1992) on glassy spherules from Haiti; Sawlowicz (1993) and Wang and others (1993) on iridium; Meyerhoff and others (1994) on Chicxulub as an impact crater; and Savrda (1993), Stinnesbeck and others (1993, this volume), Keller and others (1993a, 1994a) and Keller and Stinnesbeck (this volume) on so-called mega-tsunami deposits in Alabama, the Gulf of Mexico, and northeastern Mexico.

Despite these doubts, there is an impressive dataset that suggests that a large bolide struck Earth at the K-T boundary 65 m.y. ago. Here I examine, not whether or where a bolide impact occurred, but what effect it would have had on life on Earth. It is generally speculated that an impactor larger than 5 km in size would cause biological extinctions (McLaren and Goodfellow, 1990; Raup, 1990; Jansa, 1993). The K-T boundary impactor is believed to have been one of the largest in Earth history with a size of at least 10 km (e.g., Alvarez and others, 1980; Sharpton and others, 1992). In theory, biological extinctions occurred worldwide, were geologically instantaneous, and indiscriminately affected nearly all organismal groups (Alvarez, 1987; Raup, 1990; Sepkoski, 1990). In practice, this theory is not supported by the current very extensive biological global database. That the biological predictions are not supported by evidence in the fossil record does not necessarily negate the impact hypothesis, but it does raise questions regarding the proposed biological consequences of sudden catastrophes of large magnitude, the nature of the kill mechanism(s), and the possible long-term evolutionary effects.

The current global database accumulated for all organismal groups in general, and planktonic foraminifera in particular, provides a comprehensive history of the nature and tempo of evolution and extinctions across the K-T boundary. This database details the precise stratigraphic horizons of extinct species, including (1) whether they occur at the impact horizon (e.g., Ir anomaly, shocked minerals, microspherules, Ni-rich spinels), below, or above; (2) whether the extinctions are catastrophic (e.g., concentrated at the impact horizon) or gradual (extending above or below); (3) whether the extinctions are random (affecting all organismal groups in varied environments) or selective (affecting only certain organismal groups in specific environments); (4) whether survivors existed; and (5) what environments favored survivorship and evolution. This database provides necessary biological and environmental constraints for the impact-catastrophe hypothesis and permits the separation of long-term environmental effects (e.g., climate, sea level, volcanism, sea-floor spreading) from short-term effects of a bolide impact.

Why choose planktonic foraminifera as test case for evaluating the K-T catastrophe hypothesis? There are many reasons why this is the best fossil group for this test. Planktonic foraminifera are the only microfossil group for which nearly total species extinction has been claimed (Smit, 1982, 1990; Premoli-Silva and McNulty, 1984; Liu and Olsson, 1992; Olsson and Liu, 1993). No other microfossil group shows a similar dramatic biotic response to the K-T boundary event. Planktonic foraminifera are ubiquitous in all marine sediments in shallow to deep water and high to low latitudes, and thus permit evaluation of the biotic response on a global basis. The largest global database exists for this microfossil group, and includes more than 45 K-T boundary sections. Planktonic foraminifera are single-celled organisms that are highly sensitive to temperature and chemical variations in the oceans; this is reflected in their extinction, evolution, and relative abundance patterns. Because planktonic foraminifera are very abundant in marine sediments, evaluation of environmental effects of the K-T boundary event is not restricted to species extinctions as in most fossil groups, but permits quantitative studies to examine the timing and nature of these effects upon each species population.

This study examines the mass-extinction record in planktonic foraminifera across the K-T boundary transition in the deep sea and epicontinental seas as well as low and high latitudes in order to determine the biotic effects in each environment and establish biotic constraints for K-T boundary catastrophe hypotheses. To date, I and my collaborators have studied more than 45 K-T boundary sections spanning a host of biomes ranging from nearshore shallow marine to epicontinental seas and open oceans. The geographic locations of the best and most complete of these sections are illustrated in Figure 1 and their depositional depths are

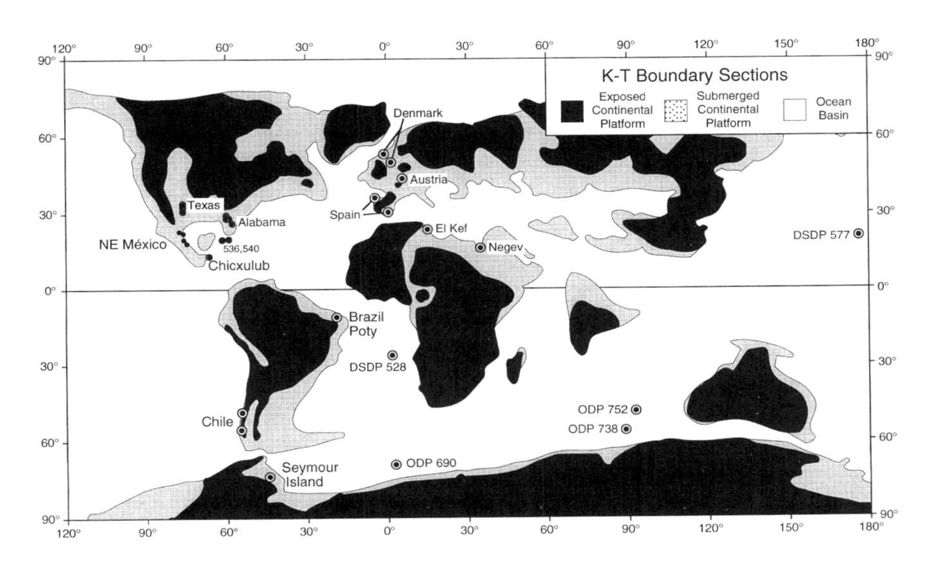

Figure 1. Location map of the most complete K-T boundary sections used in this study. Stippled pattern marks submerged shallow continental shelf areas. Note that dots mark localities, many of which represent between 4 and 10 sections (e.g., Spain, Texas, Mexico, Negev). DSDP is Deep Sea Drilling Project, ODP is Ocean Drilling Program.

shown in Figure 2. These K-T boundary sequences form the basis for a comprehensive global evaluation of the biotic effects of the K-T boundary event upon planktonic foraminifera. Previous globally based studies have indicated that the K-T boundary event did not cause a global mass extinction in planktonic foraminifera (Keller, 1993; Keller and others, 1993b) as commonly claimed, nor are global mass extinctions demonstrated in invertebrates (Stinnesbeck, 1986, this volume; Zinsmeister and Macellari, 1983; Zinsmeister and others, 1989), palynofloras (Brinkhuis and Zachariasse, 1988; Askin, 1988, 1992; Sweet and others, 1992; Méon, 1990; Askin and Jacobson, this volume), radiolarians (Hollis, 1993, this volume; Hollis and Rodgers, in prep.), ostracodes (Donce and others, 1985; Peypouquet and others, 1986), or benthic foraminifera (Thomas, 1989, 1990; Keller, 1988b, 1992; Nomura, 1991). In nannofossils, the biotic response is obscured by reworking (Jiang and Gartner, 1986; Pospichal, 1991; Gartner and others, 1994, this volume). In planktonic foraminifera, the major adverse biotic effects and mass extinction appear to be restricted to the low-latitude open marine environment and epicontinental seas. No mass extinction occurred in shallow, near-shore environments of either low or high latitudes. Examination of the biotic effects in these environments provides constraints for causal hypotheses of the K-T boundary event.

CRITERIA FOR PLACEMENT OF THE K-T BOUNDARY

Evaluating the timing and magnitude of biotic extinctions across the K-T boundary worldwide necessitates the placement of the boundary at the same strati-

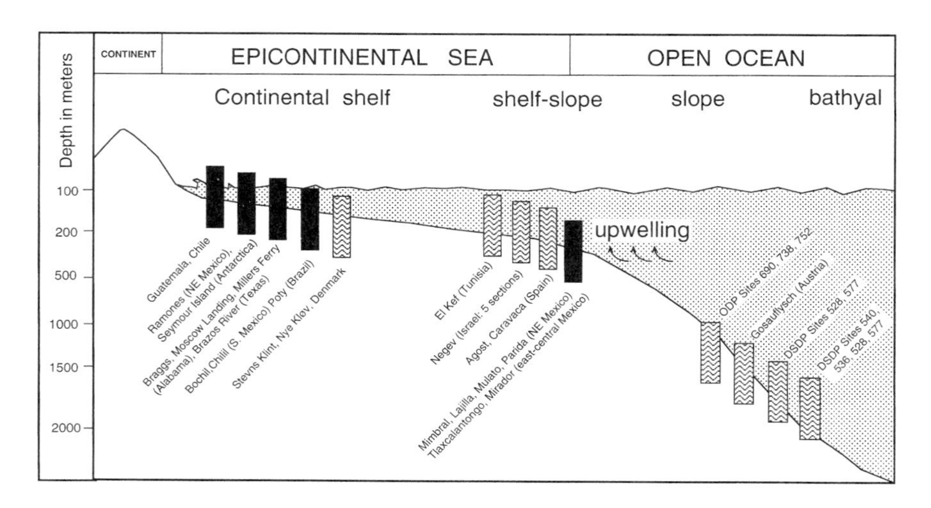

Figure 2. Depositional depths of K-T boundary sections span from shallow continental shelf (middle neritic) to the upwelling area of the shelf-slope break and to the deep open ocean. Note that deep-sea sections contain very condensed sedimentation across the K-T transition due to reduced sedimentation, dissolution, and sometimes erosion. ODP is Ocean Drilling Program, DSDP is Deep Sea Drilling Program.

graphic horizon. Fortunately, the K-T boundary is identified easily on the basis of the following lithological, geochemical, and paleontological criteria, which are exemplified in the El Kef stratotype section as illustrated in Figure 3. There is a lithologic break from chalk or marl deposition of the Cretaceous to a thin layer of dark, organic-rich clay with very little carbonate, and $CaCO_3$ poor clay, known as the boundary clay. At the El Kef stratotype, this clay layer is 55 cm

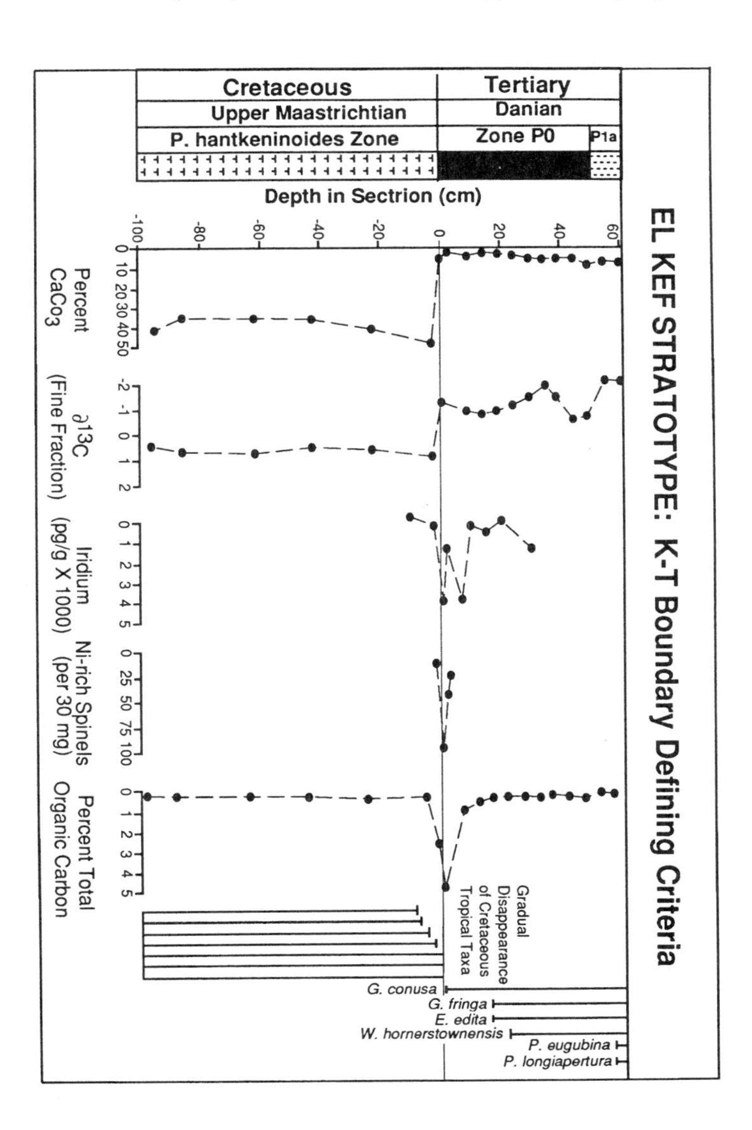

Figure 3. K-T boundary defining criteria at the El Kef stratotype section. $CaCO_3$, total organic carbon, and $\delta^{13}C$ data are from Keller and Lindinger (1989), iridium and Ni-rich spinel data are from Robin and others (1991), planktonic fora-miniferal data are from Keller and others (1995) and Ben Abdelkader (1992).

thick and represents the most expanded boundary clay observed to date in any K-T boundary sections. More frequently, the boundary clay is only 4 to 6 cm thick (e.g., Agost, Caravaca; Smit, 1990; Canudo and others, 1991; Stevns Klint, Nye Kløv; Schmitz and others, 1992; Mimbral; Keller and others, 1993b, 1994b). A 2 to 3 mm oxidized red layer is also typically present at the base of the boundary clay in most complete K-T boundary sections. Maximum iridium concentrations generally peak in the red layer and boundary clay, although they may tail several tens of centimeters above the boundary clay. Ni-rich spinels are frequently present in the red layer or base of the boundary clay, as also observed at El Kef (Robin and others, 1991). A negative $\delta^{13}C$ shift in marine plankton is observed in low and middle latitudes (fine fraction or planktonic foraminifera). The first appearance of Tertiary microfossils occurs at the base or within a few centimeters of the boundary clay, red layer, iridium anomaly, and Ni-rich spinels. At El Kef, Ben Abdelkader (1992) reported the first appearance datum (FAD) of Tertiary species (*Globoconusa conusa*) at the base of the boundary clay. In addition, *Eoglobigerina fringa, E. edita*, and *Woodringina hornerstownensis* first appear in the lower 15–20 cm of the boundary clay. Finally, there is a gradual disappearance of Cretaceous tropical taxa at or below the K-T boundary (Fig. 3).

Most of these criteria are present in all the best and most complete K-T boundary sequences (e.g., Agost, Caravaca, Nye Kløv, Brazos River, Mimbral, Ocean Drilling Program [ODP] Site 738C). The coincidence of these multiple lithological, geochemical, and paleontological criteria is unique in the geologic record and virtually ensures that the stratigraphic placement of the boundary is uniform and coeval in marine sequences across latitudes. Any of these criteria used in isolation, however, diminish the stratigraphic resolution of the K-T boundary. Moreover, the coincidence of impact and paleontological markers permits the evaluation of biotic effects as a direct consequence of the impact event.

Some workers have suggested that the sudden extinction of all but one or at most three planktonic foraminiferal species should define the K-T boundary (Smit, 1982, 1990; Berggren and Miller, 1988; Olsson and Liu, 1993; Liu and Olsson, 1992; Zachos and others, 1992). This is not practical, however, because such a sudden extinction horizon has only been observed in sections with a K-T boundary hiatus or extremely condensed interval (MacLeod and Keller, 1991a, 1991b; Keller and others, 1993a, 1993b). In temporally continuous low-latitude sections such as El Kef, Brazos, Mimbral, Agost, and Caravaca, only specialized tropical to subtropical forms disappeared, whereas most cosmopolitan taxa survived (Keller, 1988a, 1989; Keller and others, 1994b; Canudo and others, 1991). Moreover, in high-latitude sections such as Nye Kløv and ODP Site 738C, where no tropical taxa are present, there are no significant species extinctions at the K-T boundary (Keller, 1993; Keller and others, 1993b). Equally impractical is the suggestion that the first abundance increase of *Guembelitria* spp. defines the K-T boundary (Smit, 1982; Liu and Olsson, 1992; Olsson and Liu, 1993), because species abundance peaks are not unique events, but rather reflect favorable local environmental conditions (see MacLeod, 1995a, 1995b, for an in-depth discussion of these problems).

BIOTIC EFFECTS IN LOW LATITUDES

The planktonic foraminiferal mass extinction has been described as (1) a sudden catastrophic mass extinction with all or nearly all species extinct at the K-T boundary (Smit, 1982, 1990; Premoli-Silva and McNulty, 1984; Liu and Olsson, 1992; Olsson and Liu, 1993; Peryt and others, 1993), or (2) gradual, occurring over an extended time period with approximately one-third of the species surviving well into the Tertiary (Keller, 1988a, 1989, 1993; Canudo and others, 1991; Keller and Benjamini, 1991; Keller and others, 1993a, 1993b, 1994a, 1994b). These apparently contradictory interpretations actually reflect differences in faunal turnover between low vs. high latitudes and deep-sea vs. continental shelf sections, and, in some cases, different author's interpretations.

On the basis of continental shelf to upper slope sections from El Kef in Tunisia and Agost and Caravaca in Spain, Smit (1982, p. 339, and 1990) concluded that, "The mass extinction event at the Cretaceous/Tertiary boundary exterminated all but one species of planktonic foraminifera (*Guembelitria cretacea*)." This conclusion was not corroborated by Keller (1988a) for the El Kef section or by Canudo and others (1991) for the Agost and Caravaca sections. In all three sections, these authors found that about one-third of the species extended into the Danian, approximately one-half of the species disappeared near the K-T boundary, and the remainder disappeared earlier. These two extinction patterns were recently examined further at the El Kef stratotype in a blind sample analysis by four investigators. Results show that all four investigators reported between 2% and 21% of the tropical-subtropical Cretaceous species disappeared below the K-T boundary, between 35% and 64% disappeared at or near the K-T boundary, and between 33% and 46% of the Cretaceous species are present in lower Danian sediments. None of the four blind test results confirm the extinction pattern described and illustrated by Smit (1982, 1990) showing all but one species extinct precisely at the K-T boundary. Smit's extinction pattern, in fact, reflects his interpretation that all but one species were exterminated by the bolide impact and, therefore, any Cretaceous species present above the K-T boundary (except *G. cretacea*) must be reworked. The same view is held by Liu and Olsson (1992) and Olsson and Liu (1993), who, however, considered two more species (*Hedbergella monmouthensis* and *H. holmdelensis*, both of which were first described by Olsson) as K-T survivors, because they believe them to be the ancestors of all Tertiary species. It is inexplicable that they consider all other Cretaceous species present in earliest Tertiary sediments as reworked (for further discussion see MacLeod, 1995a, 1995b).

Foraminiferal workers have long believed that nearly all Cretaceous species were exterminated by the K-T boundary event because in low-latitude deep-sea sections Cretaceous species frequently terminate precisely at the K-T boundary (Premoli-Silva and McNulty, 1984; D'Hondt and Keller, 1991; Keller and others, 1993b). This observation, coupled with the now known to be erroneous belief that deep-sea sections always represent the most complete and continuous sedimentation records, led to the belief that any Cretaceous species found in early Tertiary sediments must be reworked. High-resolution studies of expanded continental shelf sequences that have high sediment accumulation rates have demonstrated that nearly all deep-sea sections, which were deposited at depths below 1000 m, are very condensed across the K-T boundary, and that the basal Tertiary zone P0 and part of zone P1a are missing (MacLeod and Keller, 1991a, 1991b). To date, only two deep-sea sections are known that have zone P0 present (ODP Site 738, Keller, 1993; Flyschgosau section of Austria, Peryt and others, 1993). Analysis of K-T boundary sections worldwide indicates that condensed or missing intervals in the deep sea are correlated with the sea-level transgression across the K-T boundary and into the earliest Tertiary (Donovan and others, 1988; Haq and others, 1987; Brinkhuis and Zachariasse, 1988; Brinkhuis and Leereveld, 1988; Schmitz and others, 1992; Keller and others, 1993b; Keller and Stinnesbeck, this volume). During sea-level transgressions, the locus of sedimentation shifts from shelf margins to the continental shelves themselves, leading to sediment starvation in proximal deep sea areas. As a result, sediment accumulation rates increase on continental shelves and decrease in the deep sea. The effects of this process on the fossil record of calcareous organisms (e.g., foraminifera, calcareous nannoplankon) may be particularly severe because increased dissolution due to decreased carbonate sedimentation further reduces accumulation rates for these particles (see MacLeod and Keller, 1991a, 1991b). During sea-level regressions, this process is reversed. The mass-extinction horizon in which all Cretaceous species are extinct that is commonly observed in the deep sea is thus an artifact of a highly condensed or incomplete stratigraphic record, where species extinctions appear to be instantaneous because of very low sediment accumulation rates, nondeposition, or erosion (MacLeod and Keller, 1991a, 1991b; Keller and others, 1993b).

Although recognizing the highly condensed K-T boundary interval in deep-sea sections, Olsson and Liu (1993) nevertheless argued that this represents a continuous depositional record where the missing biozones are obscured by bioturbation. If bioturbation is the culprit for the missing biozones, then there must be evidence of mixed sediments, a disrupted sequence of first and last appearances of species, and smeared-out abundance peaks of species. Because in most deep-sea sections there is little evidence of bioturbation or sediment mixing, this interpretation is untenable as a general model.

The low rates of sediment accumulation in the deep-sea preclude an accurate assessment of the nature and timing of the mass-extinction event (D'Hondt and Keller, 1991; Keller and others, 1993a). The true mass-extinction pattern can only be observed in the expanded and biostratigraphically complete records of shallower continental shelf to upper slope regions such as El Kef in Tunisia, Agost and Caravaca in Spain, Stevns Klint and Nye Kløv in Denmark, Brazos River in Texas, and Mimbral, Lajilla, Mulato, and other sections in northeastern Mexico. Examination of these and many other K-T boundary sections reveals that the effect of the K-T boundary event upon planktonic foraminifera was very

Figure 4. Stratigraphic ranges of planktonic foraminiferal species across the K-T boundary at El Kef with scanning electron microscope illustrations of characteristic extinct, surviving, and evolving species shown at their relative sizes. Note the early disappearance of all tropical large species, the survival of smaller, cosmopolitan species, and the evolution of small, unornamented Tertiary species (modified from Keller, 1988a).

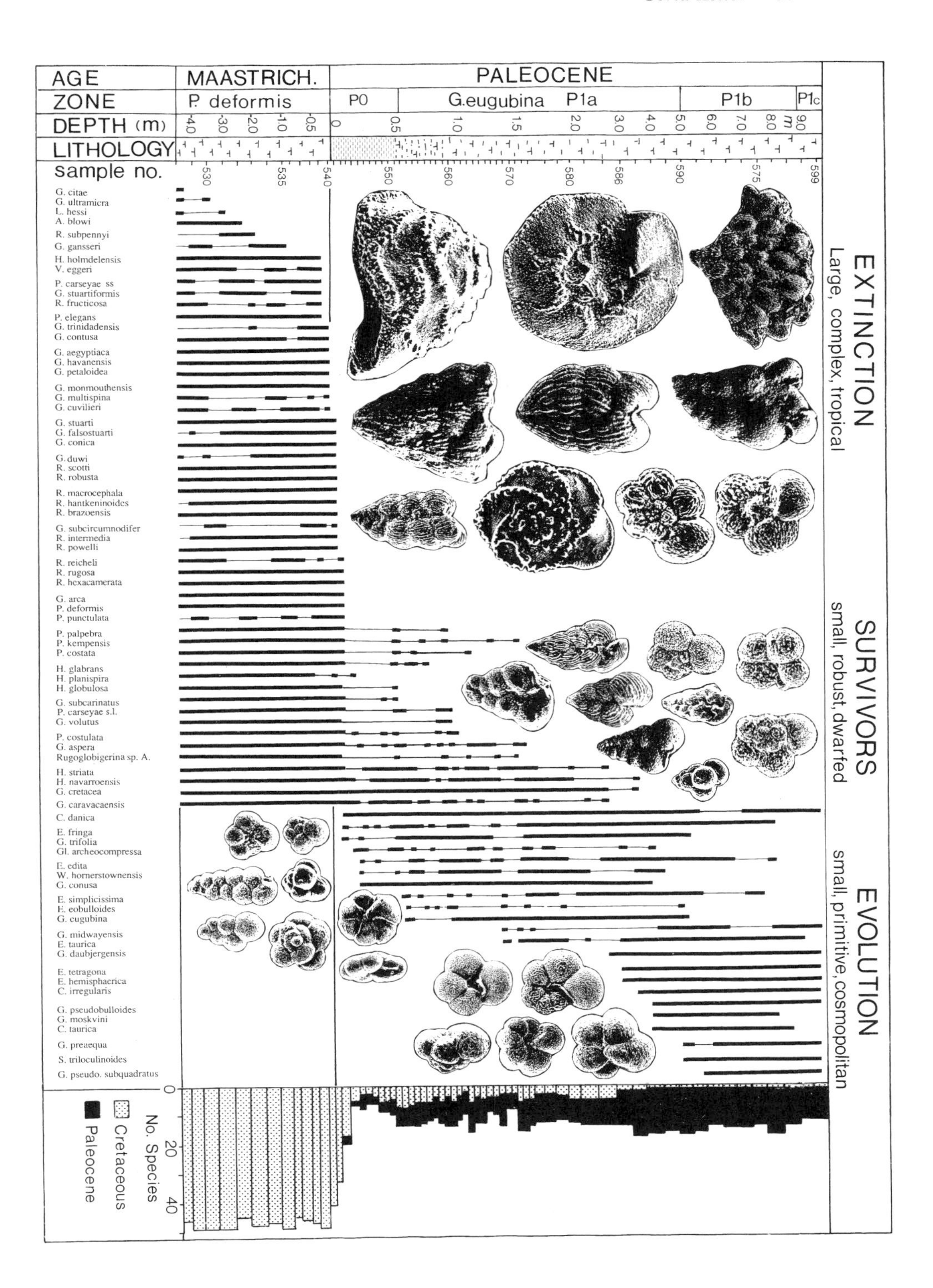

AGE
MAASTRICH.
PALEOCENE
ZONE
P. deformis
P0
G.eugubina P1a
P1b
P1c
DEPTH (m)
-4.0 -3.0 -2.0 -1.0 -0.5 0 0.5 1.0 1.5 2.0 3.0 4.0 5.0 6.0 7.0 8.0 9.0
LITHOLOGY
sample no.
530 535 540 550 560 570 580 586 590 575 599
G. citae
G. ultramicra
L. hessi
A. blowi
R. subpennyi
G. gansseri
H. holmdelensis
V. eggeri
P. carseyae ss
G. stuartiformis
R. fructicosa
P. elegans
G. trinidadensis
G. contusa
G. aegyptiaca
G. havanensis
G. petaloidea
G. monmouthensis
G. multispina
G. cuvilieri
G. stuarti
G. falsostuarti
G. conica
G. duwi
R. scotti
R. robusta
R. macrocephala
R. hantkeninoides
R. brazoensis
G. subcircumnodifer
R. intermedia
R. powelli
R. reicheli
R. rugosa
R. hexacamerata
G. arca
P. deformis
P. punctulata
P. palpebra
P. kempensis
P. costata
H. glabrans
H. planispira
H. globulosa
G. subcarinatus
P. carseyae s.l.
G. volutus
P. costulata
G. aspera
Rugoglobigerina sp. A.
H. striata
H. navarroensis
G. cretacea
G. caravacaensis
C. danica
E. fringa
G. trifolia
Gl. archeocompressa
E. edita
W. hornerstownensis
G. conusa
E. simplicissima
E. eobulloides
G. eugubina
G. midwayensis
E. taurica
G. daubjergensis
E. tetragona
E. hemisphaerica
C. irregularis
G. pseudobulloides
G. moskvini
C. taurica
G. praeequa
S. triloculinoides
G. pseudo. subquadratus
EXTINCTION
Large, complex, tropical
SURVIVORS
small, robust, dwarfed
EVOLUTION
small, primitive, cosmopolitan
0
20
40
No. Species
Cretaceous
Paleocene

different in low and high latitudes, and a mass extinction was apparently restricted to low latitudes in open ocean and epicontinental seas (Keller and others, 1993b, 1994a, 1994b; Keller, 1993; Peryt and others, 1993).

The low-latitude extinction pattern is best documented in sections deposited in continental shelf to upper slope settings such as the El Kef K-T boundary stratotype (Brinkhuis and Zachariasse, 1988; Keller, 1988a; Keller and others, in press), Agost and Caravaca (Canudo and others, 1991), and Mimbral and Lajilla in northeastern Mexico (Keller and others, 1994a, 1994b). Figure 4 illustrates the species ranges and extinction pattern of the El Kef stratotype originally published by Keller (1988a). Characteristic morphologies of extinct, surviving, and evolving species are illustrated via scanning electron microscopy with the same magnification to indicate size differences. This figure illustrates that the K-T boundary event resulted in the disappearance of large, complex, tropical and subtropical species, the survivorship of small, robust, generalist species that are frequently dwarfed in early Tertiary sediments, and the evolution of very small, unornamented species of primitive morphologies and generally short life ranges. In the original study (Keller, 1988a), nearly two-thirds of the Cretaceous species disappeared between 25 cm below to 4 cm above the K-T boundary, some species disappeared below this interval, and the remaining one-half to one-third extended into the early Tertiary and are considered survivors (Fig. 4).

It is possible that some pre-K-T boundary species extinctions are due to sampling effects (e.g., very rare occurrences and larger 25 cm sample spacing). To investigate this further, the El Kef stratotype section plus one nearby K-T boundary outcrop (El Kef II of Keller and others, 1995) were resampled at 5 cm intervals between 50 cm below and 50 cm above the K-T boundary, and at 2 cm intervals across the boundary itself. Results of these new analyses are shown in Figures 5 and 6.

In the new El Kef stratotype analysis, of 55 Cretaceous species identified, 20% (11 species) gradually disappear between 10 and 50 cm below the K-T boundary (Fig. 5) (between 2% and 21% are reported disappearing in this interval by the four blind sample test participants; Ginsburg and others, in prep.; MacLeod and Keller, in prep.). Between 5 cm below and the first centimeter of the boundary clay, 33% (18 species) disappeared. Another 16% (9 species) disappeared 10 cm above the K-T boundary (Fig. 5). Thus, 53% of species disappear between 10 cm below and 10 cm above the K-T boundary (between 33% and 64% were reported by the blind test investigators). Above this interval 31% (17 species) are present in zones P0, P1a and beyond (Fig. 6; Keller and others, in prep.), and a similar percentage (33%–46%) is reported by the blind test investigators. Many of the latter species occur only sporadically in zone P0, which spans the 50-cm-thick boundary clay at El Kef, but reappear above the boundary clay in zone P1a (Fig. 4). A similar pattern of species extinctions was reported by the four blind test participants (Ginsburg and others, in prep.; MacLeod and Keller, in prep.). The expanded K-T boundary sequence of the El Kef stratotype thus records clearly a gradual or extended pattern of species extinctions across the K-T boundary with significantly accelerated extinctions coinciding with the boundary event. A similar mass-extinction pattern is observed at Agost and Caravaca in Spain (Canudo and others, 1991), Mimbral, Mulato, and Lajilla in northeastern

Contents

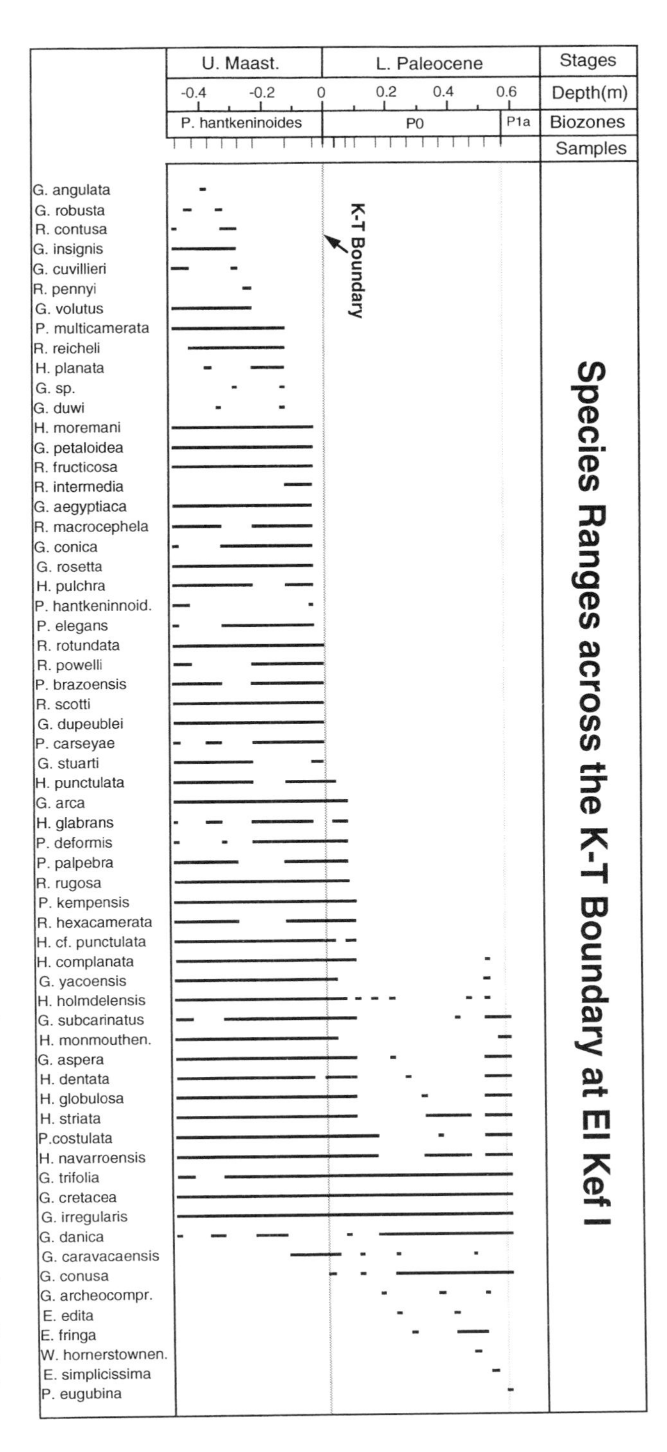

Figure 5. Biostratigraphic ranges of species across the K-T boundary at the El Kef stratotype based on 5 cm sample spacing between 50 cm below and 50 cm above the K-T boundary and closer 2 cm sample spacing at the boundary (from Keller and others, 1995). Note the gradual and extended species disappearances across the K-T boundary.

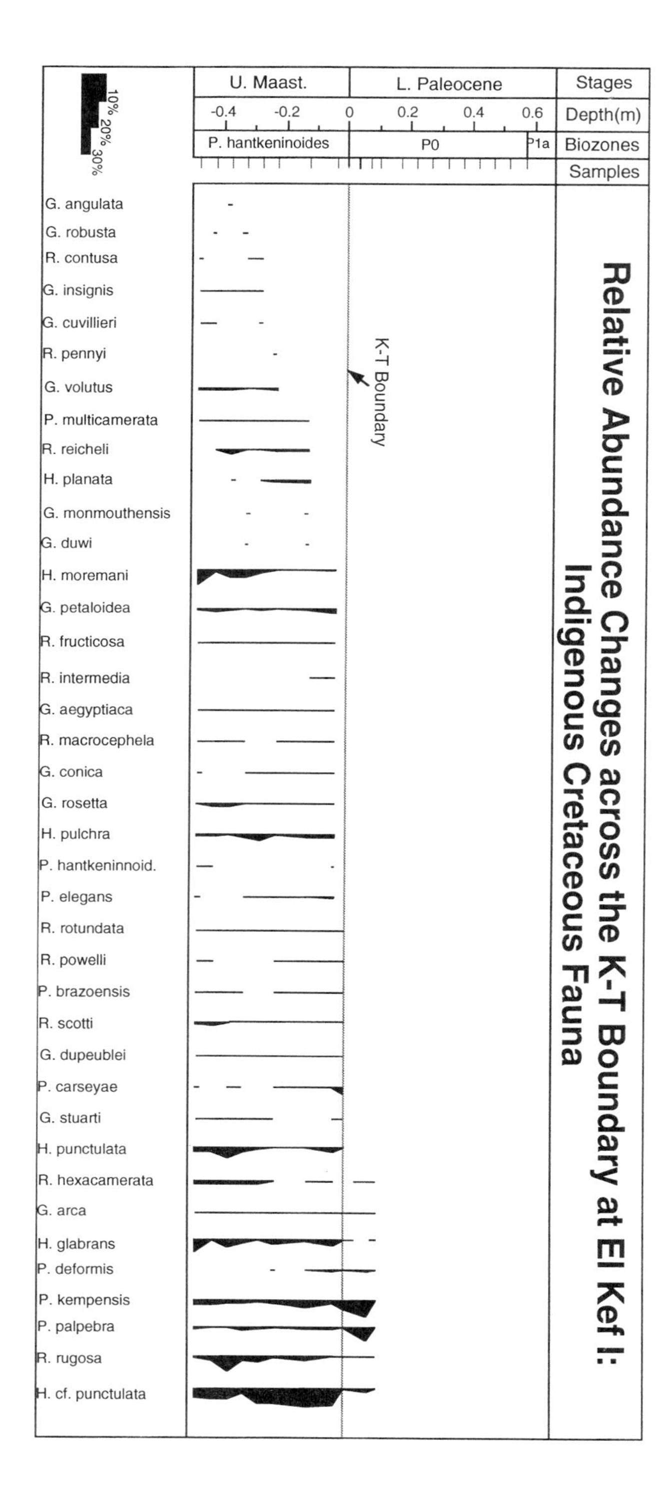

Figure 6. A: Relative abundance changes of the two-thirds of the Cretaceous species disappearing between the 50 cm below to 10 cm above the K-T boundary. Note that all of these taxa are relatively rare and their combined relative abundance is less than 20% of the total fauna.

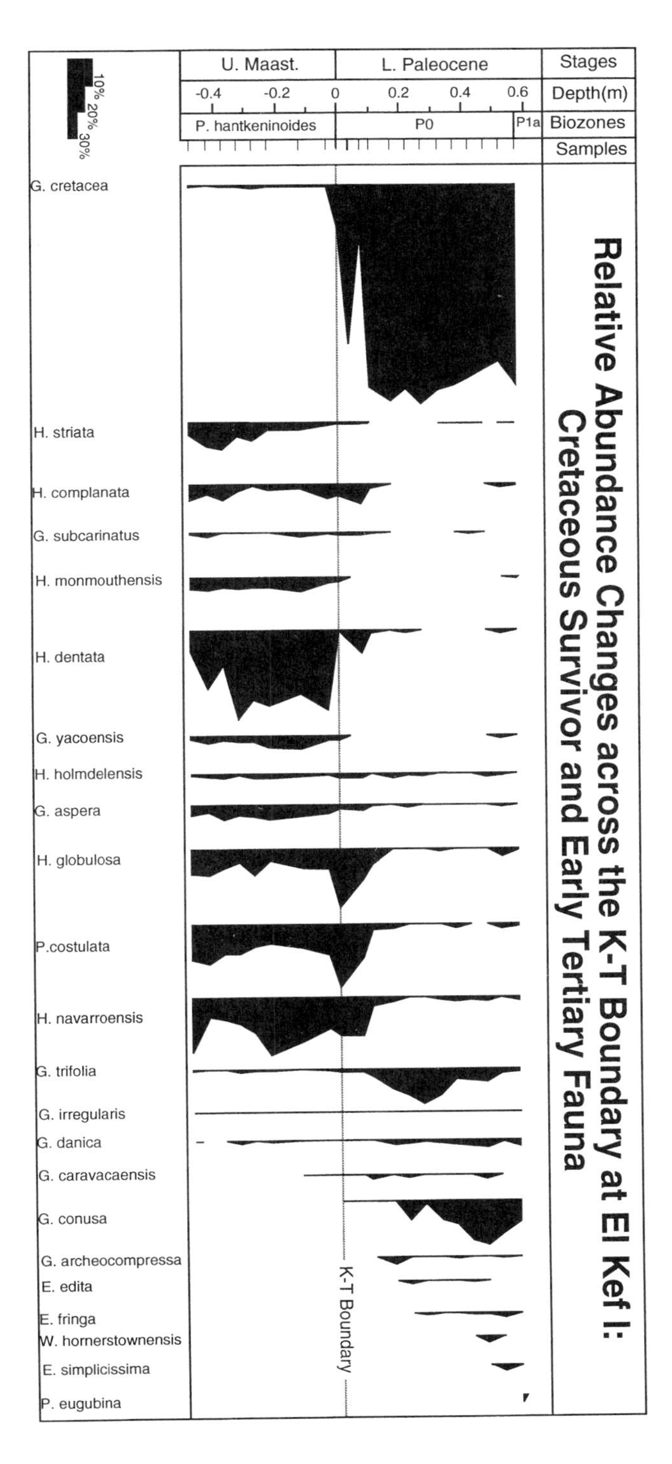

Figure 6. B: Relative abundance changes of the one-third of the Cretaceous species extending well into the Tertiary and the first evolving Tertiary species. Note that all of this species group dominates the late Maastrichtian ocean, but at El Kef declines in the post-K-T environment when the triserial low-oxygen-tolerant guembelitrids dominate.

Mexico (Keller and others, 1994a, 1994b; Lopez-Oliva and Keller, 1994, 1995), and Millers Ferry, Alabama (Olsson and Liu, 1993; Liu and Olsson, 1992).

SELECTIVE NATURE OF KILL MECHANISM

The K-T boundary mass extinction was neither random nor instantaneous, but affected specific groups of species inhabiting particular environments. By determining which environments were most severely affected, potential kill mechanisms can be isolated and biotic constraints for K-T mass-extinction hypotheses can be established. The selective nature of the K-T mass extinction can be determined by the morphology affected (large, small, ornamented, plain), the relative abundances of extinct species (rare, common, abundant), the biogeographic distribution (tropical-subtropical, temperate, polar), and the depth habitat (surface or deep dwellers). A list of genera grouped by their common distributions in tropical-subtropical, cosmopolitan, and polar environments is shown in Table 1. Note that early Danian taxa are generally of cosmopolitan distributions with the pos-

TABLE 1
List of genera grouped by their common distribution in tropical, subtropical, cosmopolitan, or polar environments

Tropical-Subtropical	Cosmopolitan	High Latitudes (endemic)
Late Maastrichtian	**Late Maastrichtian**	**Late Maastrichtian**
Archeoglobigerina	*Abathomphalus*	*Chiloguemb. waiparaensis*
Gansserina	*Globigerinelloides*	
Gublerina	*Globotruncanella**	
Globotruncana	*Guembelitria*	
Globotruncanita	*Hedbergella*	
Globotruncanella	*Heterohelix*	
Pseudotextularia	*Rugoglobigerina (rugosa)**	
Pseudoguembelina	*Pseudotextularia**	
Plummerita	*Pseudoguembelina**	
Racemieguembelina	*Shackoina*	
Rugotruncana		
Rugoglobigerina	(*taxa which are relatively rare	
Ventrilabella	in middle to high latitudes)	
Early Tertiary-Danian	**Early Tertiary-Danian**	**Early Tertiary-Danian**
	Chiloguembelina	*Murciglobigerina chascanona*
	Eoglobigerina	*M. aquiensis*
	Globoconusa	*Igorina spiralis*
	Globanomalina	*Chiloguemb. waiparaensis*
	Parvularugoglobigerina	
	Subbotina	
	Morozovella	
	Woodringina	

sible exception of *Murciglobigerina* and *Igorina*, which appear to have evolved first in southern high latitudes and later migrated to lower latitudes (Keller, 1993).

The morphologies of some extinct and surviving species at El Kef are illustrated in Figure 4, and the relative abundances are shown in Figures 5 and 6. With few exceptions (5 species at El Kef), all of the 53% (29 species) species extinct at or below the K-T boundary are complex, large, ornamented morphologies of the genera *Globotruncana, Globotruncanita, Rosita, Rugoglobigerina, Planoglobulina, Pseudotextularia, Ventrilabella*, and *Racemieguembelina*, which are characteristic of tropical-subtropical environments. Each of these extinct species is rare; relative abundances are between 0.1% and 1% (Fig. 5). The combined relative abundance of this disappearing tropical-subtropical species group averages less than 8% of the total fauna at El Kef (Fig. 5) as well as at Agost, Caravaca, Mimbral, and Lajilla (Canudo and others, 1991; Keller and others, 1994a, 1994b; Lopez-Oliva and Keller, 1995). Therefore, although 53% of the species disappeared, the effect on the total population was less than 8%. This demonstrates that using species (or genera) extinctions alone is a misleading indicator of the severity of a mass-extinction event. However, it is equally important that it demonstrates that highly specialized morphologies were ill-suited for the environmental changes across the K-T boundary, and in fact were the first to disappear. This is not unexpected. Throughout the fossil record, overspecialization and adaptation to particular environments have lead to extinction when environments change.

The group of species (16% or 9 species) that disappeared at El Kef within the first 10 to 20 cm of the Tertiary is composed of primarily mid-sized taxa of the genus *Pseudoguembelina* (Fig. 5). These species vary in relative abundances between less than 1% and 6% and have a combined relative abundance of ~12% (Fig. 5). Even if we combine this group with the group extinct at or below the K-T boundary, the combined relative abundance of 69% of the extinct species amounts to less than 20% of the total population. The same pattern of species extinctions has been documented at El Mimbral, La Lajilla, and El Mulato in northeastern Mexico (Keller and others, 1994a, 1994b; Lopez-Oliva and Keller, 1995). This means that survival favored species populations with large numbers of individuals and smaller size.

Biogeographic distribution patterns indicate that, with the exception of some pseudoguembelinids, all of the extinct species (globotruncanids, rugoglobigerinids, planoglobulinids, pseudotextularids, racemiguembelinids) are generally restricted to tropical-subtropical open oceanic environments and are rarely found in shallow epicontinental seas (except rugoglobigerinids) or cool temperate environments, as also observed by Sliter (1972a, 1972b), Douglas (1972), and Leckie (1987). This supports the observation that the K-T boundary event preferentially killed off specialized tropical and subtropical taxa that were more sensitive to changes in temperature, salinity, oxygen, nutrients, and watermass stratification than cosmopolitan generalists. The apparent preference of most of these species to open oceanic environments also indicates deeper depth habitats, as suggested by Sliter (1972a) and Leckie (1987).

Most planktonic foraminifera live within the upper 400 m of the water column and, within this interval, they can be grouped into surface, intermediate, or deep dwellers on the basis of their oxygen and carbon isotopic ranking (Douglas and Savin, 1978; Boersma and Shackleton, 1981; Stott and Kennett, 1989, 1990; Keller, 1985; Keller and Perch-Nielsen von Salis, 1995; Keller and others, 1993b; Lu and Keller, in press). Species living in warm surface waters have the lightest $\delta^{18}O$ and heaviest $\delta^{13}C$ values, whereas species living in deeper, cooler waters have successively heavier $\delta^{18}O$ and lighter $\delta^{13}C$ values.

Stable isotopic ranking of Cretaceous species indicates that all deeper-dwelling species living at or below thermoclinal depths and some surface dwellers disappeared at or near the K-T boundary. Only surface dwellers survived (Fig. 7). This indicates that the kill mechanism most severely affected subsurface habitats (>200 m depth), probably by disrupting the water-mass stratification, and had only a secondary effect on surface waters. Because deeper-water dwellers are generally rare or absent from shallow epicontinental seas (e.g., Brazos River, Nye Kløv, Stevns Klint), the absence of a mass extinction at the K-T boundary observed in these regions seems to be related primarily to water depth, and possibly to differential environmental effects (e.g., salinity, oxygen, temperature, and pH variations) between open ocean and epicontinental seas (Schmitz and others, 1992; Keller, 1989; Keller and others, 1993b).

In addition to the total extinction of deeper-water dwellers, about one-half of the surface dwellers also disappeared; the most significant of these is the iso-

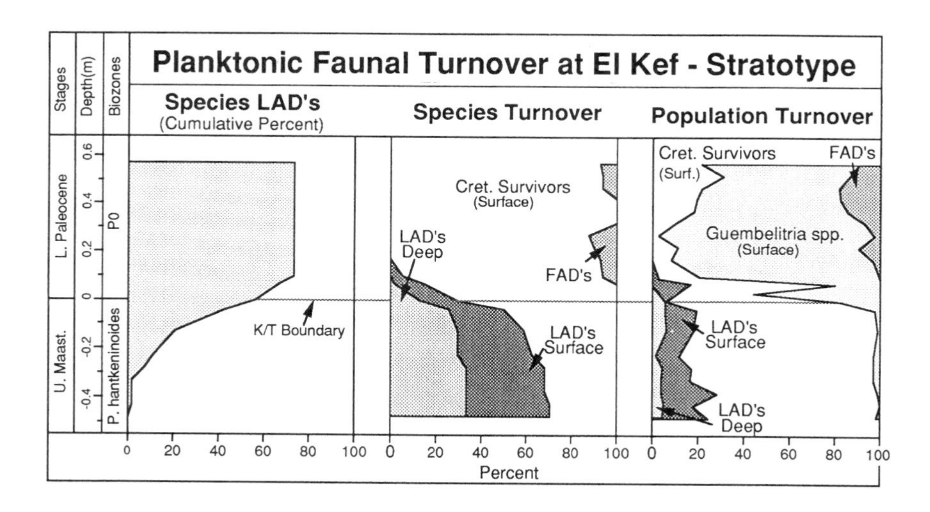

Figure 7. Planktonic foraminiferal turnover at El Kef in terms of species richness, species turnover, and population turnover. Note that, although two-thirds of the Cretaceous species disappear by K-T boundary time, their combined relative abundance is less than 20% of the total population. The dominant Cretaceous species (>80% of the population) extend well into the Tertiary. All deep dwellers and half of the surface dwellers die out, but only surface dwellers survive into the Tertiary. Dominance of the guembelitrids in the post-K-T environment suggests expansion of the oxygen minimum zone. LAD is last appearance datum, FAD is first appearance datum.

topically lightest and hence near-surface-living *Rugoglobigerina* group (Keller, 1988a; Keller and others, 1993b, 1995). Their disappearance suggests that surface waters were also disrupted severely by the K-T boundary event and that species tolerance to changes in nutrients, oxygen, salinity, temperature, and pH may have ultimately determined the nature of their survivorship. It is unknown whether rugoglobigerinids were particularly sensitive to any of these parameters, or whether their demise may have been caused by the demise of symbionts dependent on photosynthesis.

SELECTIVE NATURE OF SURVIVORSHIP

Cretaceous species survivorship across the K-T boundary in low latitudes is not random, but highly selective. Only small species with simple morphologies and little ornamentation survived (heterohelicids, pseudoguembelinids, guembelitrids, hedbergellids, globigerinellids), and all of these are isotopically light or surface dwellers (Figs. 6 and 7). Moreover, all survivor taxa are relatively abundant in the late Maastrichtian and, as a group, dominate the foraminiferal assemblages in open ocean and epicontinental seas from low to high latitudes (Fig. 6). The global biogeographic distribution and high relative abundance of this group indicate that survivor species were ecologic generalists able to tolerate wide-ranging temperature, salinity, oxygen, and nutrient conditions. This tolerance to variable environmental conditions apparently enabled them to survive the K-T boundary event. The adverse biotic effects of the K-T event upon this group of survivors include dwarfing, decrease in relative abundance, and increase in number of species, often beginning below the K-T boundary, but generally accelerating above the boundary. The eventual demise of this fauna 200 to 300 ka after the K-T boundary seems to be related to the evolution and ecologic competition of the new Tertiary faunas, which were better adapted to prevailing environmental conditions (Keller, 1988a, 1989, 1993; Keller and Benjamini, 1991; Canudo and others, 1991; D'Hondt and Keller, 1991; Keller and others, 1993a, 1994a, 1994b).

Guembelitria spp. (*G. cretacea, G. trifolia, G. irregularis, G. danica*), the only group that thrived after the K-T boundary event, are very small (<80 μm) triserial morphotypes that are present in low numbers in late Maastrichtian assemblages in both deep-sea and outer continental shelf regions, but common to abundant in shallow epicontinental seas (e.g., Brazos River, Nye Kløv, Stevns Klint; Keller, 1989; Keller and others, 1993b; Schmitz and others, 1992). After the K-T boundary event, this group proliferates to dominate assemblages in both open ocean and epicontinental seas (Bang, 1979; Brinkhuis and Zachariasse, 1988; Smit, 1982, 1990; Smit and Romein, 1985; Keller, 1988a, 1989, 1993; D'Hondt and Keller, 1991; Keller and Benjamini, 1991; Keller and others, 1993a, 1994a, 1994b; Schmitz and others, 1992; Liu and Olsson, 1992; Peryt and others, 1993).

In all sections examined thus far, the *Guembelitria* proliferation begins in the boundary clay immediately following the K-T boundary event; however, maximum abundance peaks and duration of dominance vary, probably due to differing local environmental conditions (Keller and others, 1995; Figs. 6 and 7). The significance is not that abundance distributions vary, but that the *Guembelitria* group thrives after the K-T boundary event in environmental conditions that were apparently adverse for all other Cretaceous foraminifera, and that, in contrast to the Maastrichtian, this group thrives in both open ocean and epicontinental seas during the earliest Tertiary. This suggests that major differences between open ocean and epicontinental seas were largely eliminated in the post-K-T environment, possibly as a direct consequence of the K-T boundary event. It also indicates that *Guembelitria* spp. are ecologic opportunists able to take advantage of new environmental conditions.

Clues to the environmental preference or tolerance of *Guembelitria* species are found in their association with sedimentologic, geochemical, and other faunal parameters (e.g., stable isotopes, clay mineralogy, ostracodes, benthic foraminifera), as well as by analogy with living triserial morphotypes and their ecologic distribution. D'Hondt and Keller (1991) argued that the extant triserial species *Gallitellia vivans*, which lives in upwelling regions of the Indian Ocean, the semienclosed (low oxygen?) basin of the Persian Gulf (Kroon and Nederbragt, 1990), and shallow seas with significant riverine input (low salinity; Wang and others, 1985), suggests that these small triserial morphotypes are adapted to unstable environments and, by implication, to upwelling, low-oxygen, and low-salinity environments. By analogy, the morphologically similar Cretaceous *Guembelitria* spp. may have thrived in similar environments. For example, *Guembelitria* spp. thrived in upwelling regions (ODP Site 738, Deep Sea Drilling Project [DSDP] Sites 528, 577), cool-water regions of the Southern Hemisphere (McGowran and Beecroft, 1985), and in shallow epicontinental seas that often have significant riverine input (El Kef, Nye Kløv, Brazos River, Negev).

A low-salinity environment has been suggested for Brazos River and Nye Kløv (Barrera and Keller, 1990; Keller and others, 1993b), and low-oxygen conditions appear to be associated with El Kef and Negev sections and the high-latitude ODP Site 738 (Keller, 1988a; Brinkhuis and Zachariasse, 1988; Keller and Benjamini, 1991; Magaritz and others, 1992; Almogi-Labin and Bein, 1993). At El Kef, the oxygen minimum zone appears to have temporarily invaded shallow continental shelves, as suggested by Rohling and others (1991) and Keller and others (1995). Both benthic foraminifera and ostracodes indicate that low-oxygen conditions began coincident with organic-rich black-clay deposition at the K-T boundary at El Kef (Peypouquet and others, 1986; Donce and others, 1985; Keller, 1988b, 1992). In benthic foraminifera, this resulted in the expansion and dominance (>80%) of low-oxygen-tolerant species, disappearance of some species, and decreased relative abundance of Cretaceous survivors intolerant of lower oxygen conditions, as illustrated in Figure 8. A similar faunal turnover is apparent in planktonic foraminifera (Fig. 7); the Cretaceous survivor group is dominated (>80%) by *Guembelitria* spp. Similar to the benthic low-oxygen group, other survivors decreased in relative abundance and many Cretaceous species disappeared. These analogous benthic and planktonic foraminiferal turnovers suggest that *Guembelitria* spp. were low-oxygen tolerant and that a temporary expansion of the oxygen minimum zone into shallow waters occurred in the earliest Tertiary.

Evidence for an expanded oxygen minimum zone is also present at Site 738,

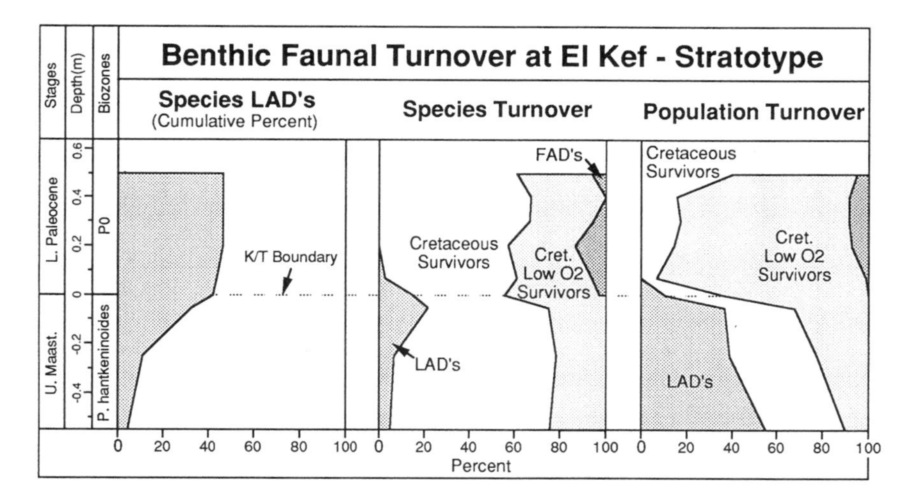

Figure 8. Benthic foraminiferal turnover at El Kef based on species richness, species turnover, and population turnover. Note that low-oxygen-tolerant species increase and dominate the post-K-T benthic environment, similar to the rise to dominance of the guembelitrids in the planktonic foraminifera (from Keller, 1988b, 1992). LAD is last appearance datum.

where the biserial low-oxygen-tolerant Cretaceous species *Chiloguembelina waiparaensis* begins to dominate (>80%) the planktonic foraminiferal assemblage across the K-T boundary and well into the early Tertiary (zones P0, P1a, P1b). In southern high latitudes, expansion of the oxygen minimum zone coincides with increased upwelling (Keller, 1993; Barrera and Keller, 1994), as observed also in the thick siliceous K-T boundary deposits of New Zealand (Hollis, 1993; Hollis and Rodgers, in prep.). However, if there was increased upwelling in low latitudes, it did not lead to increased surface productivity, as indicated by the 2% to 3% drop in surface $\delta^{13}C$ values observed at El Kef and other low-latitude sections (Keller and Lindinger, 1989; Zachos and Arthur, 1986).

A variety of evidence thus indicates that post-K-T environmental conditions substantially reduced or eliminated the open ocean–epicontinental sea differences that led to the extinction of deeper-dwelling and specialized surface-dwelling planktonic foraminifera and to preferential survivorship of species tolerant of variable conditions in temperature, salinity, oxygen, and nutrients. The rising dominance of low-oxygen-tolerant triserial and biserial planktonic foraminifera (*Guembelitria, Heterohelix globulosa, C. waiparaensis*), coincident with the dominance of the benthic low-oxygen-tolerant faunas, indicates an expanding oxygen minimum zone into shallow waters (see also Rohling and others, 1991). Palynofloral evidence from high latitudes suggests a period of increased rainfall, humidity, and riverine runoff beginning at or below the K-T boundary (Askin, 1988, 1990, 1992; Askin and Jacobson, this volume), and increased fresh-water influx and lower salinities during this time in the Brazos River (Keller, 1989; Barrera and Keller, 1994) and Nye Kløv regions (Keller and others, 1993b) have also been suggested.

BIOTIC EFFECTS IN HIGH LATITUDES

The K-T boundary transition in southern high latitudes (Antarctica, New Zealand) has been reported to exhibit no major species extinctions and no major faunal and floral changes among invertebrates (Zinsmeister and others, 1989; Zinsmeister and Macellari, 1983; Zinsmeister and Feldmann, this volume; Stinnesbeck, 1986, this volume), plants and palynomorphs (Askin, 1988, 1992; Johnson, 1992; Askin and Jacobson, this volume), and radiolarians (Hollis and Rodgers, in prep.; Hollis, 1993, this volume). Instead, faunal and floral changes are relatively minor and very gradual, suggesting slowly changing environmental conditions including increased precipitation, a rising sea level, and increased upwelling (Hollis, this volume). Generally, planktonic foraminiferal preservation in Antarctic sediments is too poor to yield biostratigraphic or faunal information. Relatively well preserved planktonic foraminiferal faunas, however, have been recovered across the K-T boundary transition from Antarctic Ocean ODP Sites 690, 738, 750, and 752 (Stott and Kennett, 1989, 1990; Huber, 1991; Zachos and others, 1992; Van Eijden and Smit, 1991). Restudy of these sections based on higher resolution sample spacing and quantitative faunal analysis revealed, however, that a hiatus is present in Sites 690, 750, and probably 752, and the basal Tertiary zones P0, P1a, and possibly P1b are missing (Keller, 1993). At ODP Site 738, the K-T boundary transition was recovered near the base of a 15-cm-thick laminated interval marked by a major iridium anomaly (Schmitz and others, 1991; Thierstein and others, 1991) and the presence of the basal Tertiary zones P0 and P1a, although short intrazonal hiatuses are present at the P0-P1a and P1a-P1b boundaries (Keller, 1993; MacLeod and Keller, 1994). In his shipboard study of Site 738, Huber (1991) did not recognize or find most Danian species, and in a critique of the Keller study, he claimed that they were either wrongly identified or not present (see Huber and others, 1995, and reply by Keller and MacLeod, 1995). Huber's K-T studies have concentrated on Antarctica (James Ross Island), where planktonic foraminifera are virtually absent across the K-T boundary and no early Danian faunal record has been recovered to date (Huber, 1988). As a result, Danian faunas were not identified in the Antarctic region prior to the Keller (1993) study, and Huber believed them to be absent. Thus, the discrepancy between the Huber (1991) and Keller (1993) studies has little to do with species identification, but rather with discovering the first early Danian fauna in the region. Huber's failure to discover this fauna in Site 738 appears to have been a result of sample processing. He examined only the size fraction >64 μm (Huber, 1991; Keller and MacLeod, 1995), whereas the Danian fauna resides in the smaller 38–63 μm size fraction.

The absence of a mass extinction at the K-T boundary has also been reported from the northern high latitudes (Alaska, Denmark) for invertebrates (Marincovitch, 1993; this volume), ostracodes (Browers, 1993; this volume), palynofloras (Sweet and others, 1992) and planktonic foraminifera (Keller and others, 1993b). In Denmark, the Stevns Klint section has a hiatus with at least zone P1a missing (Schmitz and others, 1992), but the Nye Kløv section is biostratigraphically complete; all biozones are present, although there is a short intrazonal hiatus at the P1a-P1b boundary (Keller and others, 1993b).

K-T BOUNDARY FAUNAL CHANGE

No mass extinction occurred in planktonic foraminifera at Nye Kløv; in fact, only one species (the surface dweller *Rugoglobigerina rugosa*) disappeared at the K-T boundary (Fig. 9). All other species survived well into the Tertiary and disappeared 200–300 k.y. above the K-T boundary (Keller and others, 1993b). Despite this difference between high and low latitudes, the sequence of newly evolving Tertiary species remains the same. Possible explanations for the different response to the K-T boundary event in high and low latitudes include different faunal compositions (tropical-subtropical vs. cool temperate), different envi-

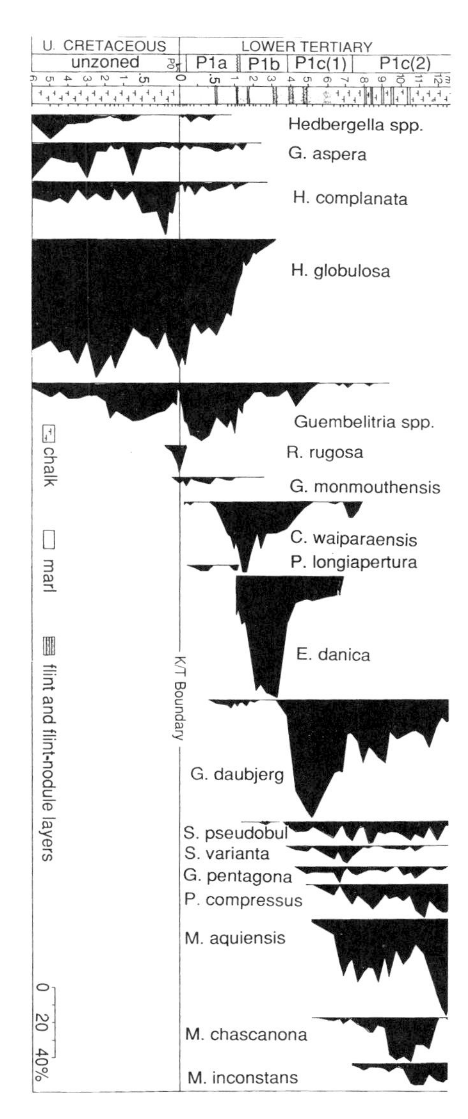

Figure 9. Biostratigraphic ranges and relative abundances of planktonic foraminifera from Nye Kløv, Denmark. Note the pre-K-T boundary decline in H. complanata, *the brief appearance of abundant* R. rugosa, *the continued abundance of* H. globulosa, *and absence of significant species extinctions, all suggesting long-term environmental changes rather than a single catastrophic event at K-T boundary time.*

ronmental effects of the K-T event across latitudes and in shallow- vs. deep-water environments, and a change in the source of bottom waters and in the water-mass stratification.

The faunal turnover pattern observed at Nye Kløv is strikingly similar to that originally observed at Brazos River, Texas (Keller, 1989), which has remained anomalous within low latitudes. The faunal similarities between the two regions appear to be due primarily to similar depositional environments in shallow epicontinental seas (middle to inner neritic depths, ~100 m; Fig. 2), including low salinity and a well-developed oxygen minimum zone (Keller, 1989; Keller and others, 1993a). At Brazos River, the shallow middle neritic environment excludes all deeper-dwelling species that in low latitudes account for 50% of all the species extinctions (Fig. 7). Many of these taxa are present, although extremely rare, prior to the latest Maastrichtian sea-level regression, which culminates in a hiatus at the base of the clastic-event deposit just below the K-T boundary (Keller, 1989; Keller and Stinnesbeck, this volume). At the high-latitude Nye Kløv section the deeper-dwelling tropical fauna is absent because of the shallow depth and cooler temperatures. Shallow continental shelf sections in low latitudes can thus mirror the low-diversity faunal assemblages and faunal turnovers of high latitudes as a result of ecologic exclusion of tropical-subtropical deep-water dwellers and surface dwellers intolerant to temperature, oxygen, and salinity fluctuations.

The major differences in biotic responses between high and low latitudes, however, seem to be independent of water depth, as indicated by the absence of a mass extinction in the deep-water (~1000 m) ODP Site 738. Figure 10 shows all species present across the K-T boundary at Site 738 in their relative abundances. Of these, 15 species are rare or only sporadically present (single lines in Fig. 10); all of these are commonly found in low-latitude assemblages where they tend to extend up to the K-T boundary (Fig. 4). Their rare presence in high latitudes marks the limits of their biogeographic distribution, and it cannot be assumed that they were present at K-T boundary time. Most large, complex tropical-subtropical taxa that become extinct at the K-T boundary are conspicuously absent in southern (as well as northern) high latitudes. Similar to those in low latitudes, however, species dominating the faunal assemblage are the small, generalist, and cosmopolitan heterohelicids, hedbergellids, and globigerinellids, all of which survived well into the Tertiary. Species evolving include the same group of taxa observed in low latitudes plus some high-latitude endemic species (discussed below).

It is obvious from Figures 5, 6, 9, and 10 that the major difference between high- and low-latitude assemblages is the near absence of tropical-subtropical taxa in latest Maastrichtian sediments (see Table 1). These are the taxa that comprise the one-half to two-thirds of the species assemblage disappearing at or below the K-T boundary in low latitudes. The absence of this environmentally sensitive group in high latitudes is largely responsible for the absence of a mass extinction in high latitudes.

What are the biotic effects of the K-T boundary event in high latitudes? Figure 10 shows that at Site 738 no major faunal abundance changes or extinctions coincide with the iridium anomaly that marks the K-T boundary event. Instead, faunal

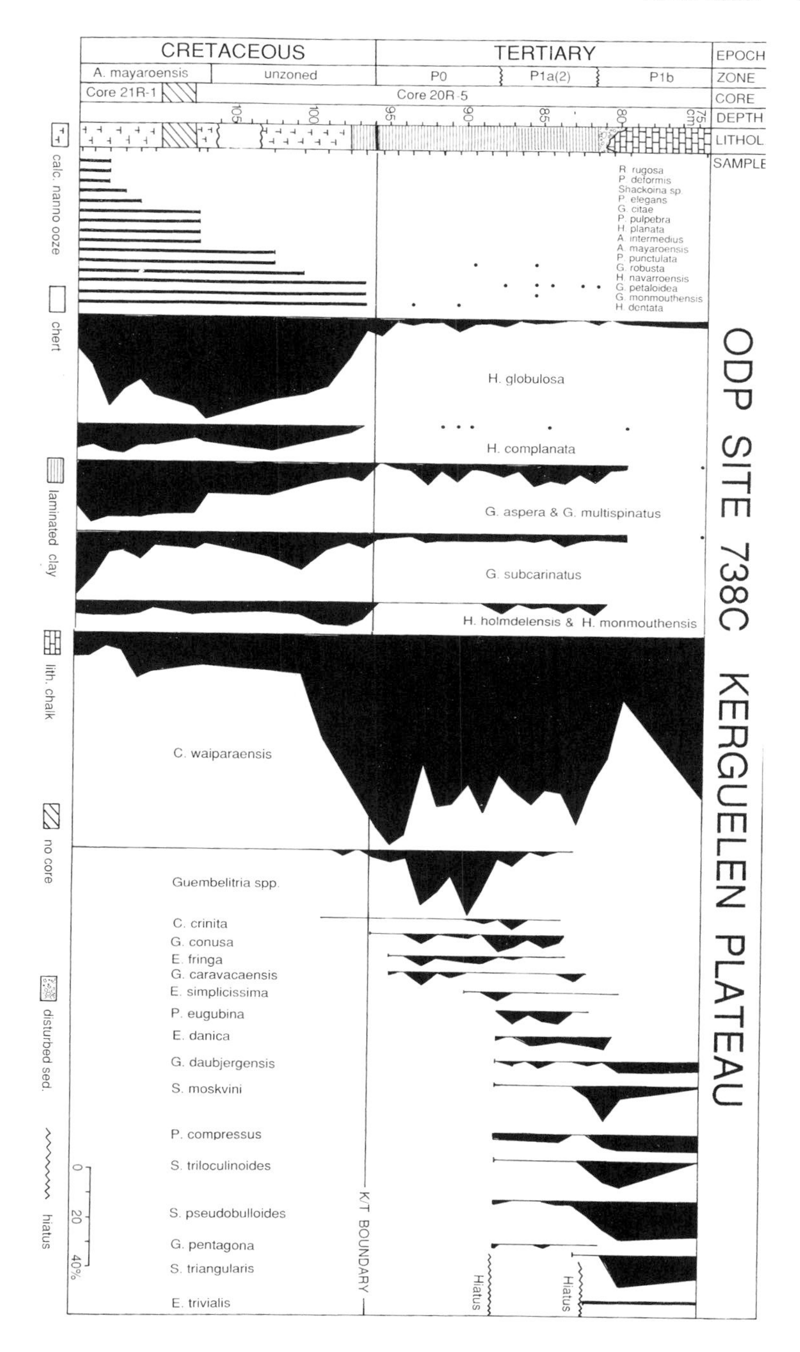

Figure 10. Biostratigraphic ranges and relative abundances of planktonic foraminifera from Antarctic Ocean Ocean Drilling Program (ODP) Site 738. Note the pre-K-T boundary decline in species abundances and concurrent rise in the low-oxygen-tolerant species C. waiparaensis, *indicating expansion of the oxygen minimum zone.*

abundance changes are gradual, beginning 10 cm below the K-T boundary. Because sediments are laminated and undisturbed from 4 cm below to 15 cm above the K-T boundary, this gradual faunal change cannot be attributed to bioturbation, but rather indicates that environmental changes began before the end of the Cretaceous in southern high latitudes. In the northern high-latitude Nye Kløv section, hedbergellids, globigerinellids, and some heterohelicids also begin their terminal decline in abundance below the K-T boundary, whereas *H. globulosa* shows no change until well into the Tertiary (Fig. 9, top of zone P1a; Keller and others, 1993b). This indicates that at Nye Kløv, environmental changes also began before the K-T boundary and continued well into the Tertiary. Prevailing evidence thus indicates that in neither of the two high-latitude sections can specific biotic effects be attributed to a catastrophic K-T boundary event marked by the iridium anomaly. Although the absence of K-T boundary-specific biotic effects in planktonic foraminifera must still be confirmed in additional high-latitude sections, current evidence from many fossil groups indicates that, rather than a single catastrophic event, long-term environmental changes beginning long before the end of the Cretaceous were responsible for the observed biotic changes.

NATURE OF ENVIRONMENTAL CHANGES IN HIGH LATITUDES

What was the nature of K-T boundary environmental changes? At Stevns Klint and Nye Kløv, stable isotopic, sedimentologic, benthic, and planktonic foraminiferal analyses indicate a major sea-level regression during the latest Maastrichtian, followed by a rising sea level beginning below and continuing across the K-T boundary clay (Schmitz and others, 1992; Keller and others, 1993b; Keller and Stinnesbeck, this volume). The pre-K-T boundary sea-level rise is associated with an influx of the warm-water surface-dweller *R. rugosa* (previously absent in this region), and suggests climatic warming. Stott and Kennett (1989, 1990) observed high-latitude warming just preceding the K-T boundary in the Weddell Sea ODP Site 690. This episode of warming began during the last 100 000 yr of the Maastrichtian and accelerated during the last 50 k.y. before the K-T boundary (Pardo and others, this volume; Keller and Stinnesbeck, this volume). Other than this brief warming, planktonic foraminifera indicate no major changes in the Danish basin, where late Maastrichtian dominance of *H. globulosa* continues across the K-T boundary and well into the Tertiary (~200–300 k.y.). This suggests the presence of a well-developed oxygen minimum zone in the latest Maastrichtian to early Tertiary Danish basin (Keller and others, 1993a). The only notable difference across the K-T boundary is an expansion of the oxygen minimum zone coincident with climatic warming, and a rising sea level, as suggested by the increase in *Guembelitria* spp. (Keller and others, 1993b).

A similar pattern of environmental change is present in Site 738, where *Chiloguembelina waiparaensis*, a biserial low-oxygen-tolerant species, dominates (>80%) from just below the K-T boundary through the earliest Tertiary (zones P0, P1a), indicating the establishment of a well-developed oxygen minimum zone (Keller, 1993; Barrera and Keller, 1994). Similar to northern high latitudes and low latitudes, increased abundance of *Guembelitria* spp. in the earliest Tertiary zones P0 and P1a at Site 738 signals the further expansion of the oxygen minimum zone in the post-K-T environment.

PALEOECOLOGICAL EVOLUTION OF THE TERTIARY FAUNAS

Unlike the high- vs. low-latitude differences in extinction and eventual disappearance of the Cretaceous faunas, the evolution of the early Tertiary faunas is very similar across latitudes. Evolution of the first Tertiary species begins immediately after the K-T boundary event with the appearance of very small, unornamented, trochospiral and biserial morphologies (*Globoconusa conusa, Eoglobigerina fringa, E. simplicissima, E. edita, Woodringina hornerstownensis, W. claytonensis*). In the basal zone P0, these species are very rare and Cretaceous survivor taxa dominate. In the succeeding zone P1a (50 to 230 k.y. above the K-T boundary), the first significant evolutionary diversification occurs with the appearance of a dozen new species, including *Parvularugoglobigerina eugubina* and *P. longiapertura* (Keller, 1988a, 1989, 1993; Canudo and others, 1991; Keller and others, 1993b, 1994a, 1994b). Few new species are added in zones P1b or P1c in low latitudes. In high latitudes, more species evolve and later radiate into lower latitudes. For example, *C. waiparaensis* was originally believed to be a Tertiary species, but has now been found to appear first during the late Maastrichtian, after which it migrates to low latitudes during the early Tertiary (Keller, 1993). *Murciglobigerina aquiensis*, *M. chascanona*, and *Igorina spiralis* also first appear in zone P1c and migrate to lower latitudes during the late Paleocene, along with *Eoglobigerina danica* and *Globoconusa extensa* (Keller, 1993; Keller and others, 1993b). These data strongly suggest that, during the earliest Tertiary, high-latitude regions temporarily acted as centers of origin and dispersal for planktonic foraminifera. Similar high-latitude endemism has been observed in reptiles (Molnar, 1989), terrestrial mammals (Case, 1989), marine mammals (Fordyce, 1989; Walting and Thurston, 1989), crustaceans (Feldmann and Tschudy, 1989), noncrustaceans (Zinsmeister and Feldmann, 1984), and plants (Dettmann, 1989; Askin, 1989).

Early Tertiary planktonic foraminifera indicate that the post-K-T boundary environment was similar across latitudes. From the boreal seas of Denmark to the Tethys and Antarctic Ocean, the world oceans were populated by an impoverished planktonic foraminiferal fauna consisting of Cretaceous survivors and an evolving Tertiary fauna tolerant of wide-ranging temperature, oxygen, salinity, and nutrient conditions. However, stable isotopic data, as well as high-latitude endemism and increased species richness, indicate that the biotic effects of the post-K-T boundary event were significantly more severe in low latitudes. The major difference appears to have been in decreased surface productivity in low latitudes, as suggested by the 2‰ to 3‰ drop in $\delta^{13}C$ values of plankton of low- to middle-latitude sections (Zachos and Arthur, 1986; Zachos and others, 1992; Keller and Lindinger, 1989; Barrera and Keller, 1994). In contrast, in Nye Kløv the $\delta^{13}C$ shift is a negligible 0.5‰ and in the Antarctic Ocean Site 738 it is absent (Keller, 1993; Barrera and Keller, 1994). Carbon isotopic data thus indicate that, whereas surface productivity decreased significantly in lower latitudes for ~300 k.y., it remained stable (Site 738) or only slightly decreased (Nye Kløv) in high latitudes.

DISCUSSION AND CONCLUSIONS

The current global database indicates that the biotic effects of the K-T boundary event upon planktonic foraminifera are more complex than originally assumed and predicted by the bolide impact theory. The nature and magnitude of biotic effects vary between low and high latitudes, epicontinental seas, and open ocean environments.

The mass extinction in planktonic foraminifera is limited to low-latitude open ocean and epicontinental seas. In deep-sea sections at depths below 1000 m, low sediment accumulation, dissolution, and erosion result in very condensed K-T boundary sections that juxtapose species extinctions at the K-T boundary horizon, giving the erroneous impression of a sudden catastrophe (e.g., Gubbio, Flyschgosau, DSDP Sites 577, 528, 690, 750, 752; MacLeod and Keller, 1991a, 1991b; Keller, 1993; Peryt and others, 1993). Higher sediment accumulation rates in continental shelf and upper slope environments reveal a gradual and extended extinction pattern beginning below and extending well above the K-T boundary (e.g., El Kef, Agost, Caravaca, Negev, Mimbral, Lajilla, Mulato, Brazos River, Millers Ferry). The extinction of Cretaceous species was selective, not random, and was morphology specific in that large, complex, ornate taxa were eliminated prior to small, less-complex taxa; extinction was also temperature and/or biogeography specific, because all deep tropical-subtropical taxa were eliminated, but not cosmopolitan species. In addition, extinction was depth-habitat specific, because all deep dwellers and approximately one-half of the surface dwellers were eliminated, and species specific in that only rare already endangered species were eliminated. Cretaceous survivor species include all taxa that are common or dominant in the late Maastrichtian and often have a cosmopolitan distribution. They are small, have simple morphologies, and are ecologic generalists, able to tolerate fluctuations in temperature, salinity, oxygen, and nutrients. The most dominant of these are biserial and triserial morphotypes tolerant of low-oxygen conditions; dominance of these taxa reflects the expansion of the oxygen minimum zone in the post-K-T environment.

On the basis of extinction patterns from shelf to upper slope environments (e.g., El Kef, Agost, Caravaca, Brazos River, Negev sections, Mimbral, Lajilla, Mulato), the nature and magnitude of the K-T boundary mass extinction can be assessed. In these sections approximately two-thirds of all Cretaceous species are extinct by K-T boundary time, and one-third survive for ~200 to 300 ka into the early Tertiary. Although two-thirds of all extinct species represents a major mass extinction resulting in a severe reduction in species diversity, the effect on the overall planktonic foraminiferal population (number of individuals disappearing) is significantly less severe because only rare species disappear. Their combined relative abundance is less than 20% of the planktonic foraminiferal population. These K-T extinct species are generally specialized morphotypes adapted to specific environments (depth, temperature, salinity, nutrients, symbionts), and their low abundance in late Maastrichtian sediments suggests that their habitats were already stressed prior to the K-T boundary event. The extinction of these rare, already endangered species could be due to relatively minor ecologic variations, and is therefore a poor measure of the magnitude of the K-T boundary event.

Relative abundance changes of survivor species provide a better indicator of the magnitude of the K-T boundary biotic effects. Survivor species are ecologic generalists (heterohelicids, hedbergellids, globigerinellids) that began to dominate the planktonic foraminiferal assemblage during the late Maastrichtian climatic cooling (Barrera and Keller, 1994; Pardo and others, this volume). One group of survivors (guembelitrids) proliferates after the K-T boundary event. With the exception of *Guembelitria* spp., all survivors decline terminally in relative abundance in the early Danian. The onset of this decline seems variable among species and may occur below, at, or above the K-T boundary (Keller, 1988a; Canudo and others, 1991; Keller and others, 1994a, 1994b). These data indicate that all Cretaceous planktonic foraminiferal species were affected by the K-T boundary event, and that, by inference, all of their marine habitats were strongly altered. However, the timing of the species extinctions and faunal abundance changes also shows that environmental changes began well before and continued well after the K-T boundary. Therefore, the mass-extinction pattern and environmental changes cannot have been caused by a single catastrophic event, such as a bolide impact, at the K-T boundary. A bolide impact may have accelerated the demise of already declining tropical-subtropical Cretaceous faunas in an environment stressed by climatic fluctuations (e.g., cooling followed by warming). However, it is very unlikely that these long-term climatic fluctuations alone could have caused the observed extinction pattern. A catastrophic event (whether impact or volcanism) or the attainment of threshold conditions for marine plankton as a result of long-term deteriorating environmental conditions must be considered as possible causes.

No mass extinction occurred in high-latitude planktonic foraminifera, and nearly all species present in the uppermost Cretaceous survived well into the Danian. The planktonic foraminiferal record across the K-T boundary transition in high latitudes is currently known from Denmark (Stevns Klint and Nye Kløv, Hultberg and Malmgren, 1987; Schmitz and others, 1992; Keller and others, 1993b) and the Antarctic Ocean (ODP Sites 690, 738, 750, 752, Stott and Kennett, 1989, 1990; Huber, 1991; Keller, 1993; Zachos and others, 1992). Unfortunately, only two of these sections (Nye Kløv and ODP Site 738) are biostratigraphically complete; all other sections have hiatuses with at least the early Tertiary biozones P0 and P1a missing (Keller, 1993; Keller and others, 1993b). Although the high-latitude planktonic foraminiferal record needs to be investigated in other complete K-T sections, the absence of a K-T mass extinction in these habitats is corroborated by other microfossil and macrofossil groups, including radiolaria (Hollis, 1993, this volume; Hollis and Rodgers, in prep.), diatoms (Harwood, 1988), palynofloras (Askin, 1988, 1992; Askin and Jacobson, this volume), plants (Johnson, 1992), and invertebrates (Zinsmeister and others, 1989; Zinsmeister and Feldmann, this volume; Stinnesbeck, 1986).

The fossil records of Nye Kløv and ODP Site 738 demonstrate that there is no K-T mass extinction in high-latitude planktonic foraminiferal faunas and that nearly all Cretaceous species survived well into the Danian (Keller, 1993; Keller and others, 1993b). Because in low latitudes only large specialized tropical-subtropical taxa became extinct, the absence of a mass extinction in high latitudes may simply reflect the absence of this specialized low-latitude fauna. If that were the case, then the biotic effects of the K-T boundary event could have been the same across latitudes. Faunal and stable isotopic evidence indicates that this was not the case, and that the K-T biotic and environmental effects were most severe in low latitudes and strongly diminished into high latitudes. This is suggested by the dwarfing of species and relative abundance changes of planktonic foraminifera, which for most species show the onset of the terminal decline in abundance below the K-T boundary. This indicates that some environmental changes preceded the K-T boundary. For example, the high-latitude endemism and evolution of Tertiary planktonic foraminifera prior to their appearance in low latitudes suggest that more favorable environmental conditions prevailed in high latitudes. In addition, there is an absence of a $\delta^{13}C$ shift in the surface waters of the Antarctic Ocean Site 738 and near-absence (0.5‰) at Nye Kløv, Denmark, as compared to the 2‰ to 3‰ $\delta^{13}C$ shift observed in all low-latitude sections (Keller, 1993; Keller and others, 1993b; Keller and Lindinger, 1989; Zachos and Arthur, 1986). Because the negative $\delta^{13}C$ shift in surface waters of low latitudes is generally interpreted as a major drop in surface productivity, its absence in high-latitude sections indicates that surface productivity remained nearly stable or increased slightly in high latitudes (Keller and others, 1993b; Keller, 1993). A similar conclusion was reached by Stott and Kennett (1989, 1990), who reported a reduced $\delta^{13}C$ drop at the K-T boundary at Site 690. In contrast, Zachos and others (1992) reported a significant $\delta^{13}C$ shift at Site 750 in the Antarctic Ocean. Biostratigraphic and isotopic comparisons of this section with Site 738 revealed, however, that the $\delta^{13}C$ shift reported by Zachos and others (1992) as occurring at the K-T boundary occurs in zone P1b-P1c, or ~300 ka after the boundary event (Barrera and Keller, 1994).

The current database thus indicates that long-term environmental changes as a result of climate cooling followed by short-term warming, sea-level fluctuations, ocean anoxia, and volcanism that characterized the latest Maastrichtian to earliest Tertiary were primarily responsible for the observed long-term trends in planktonic foraminiferal turnovers. Superimposed upon these long-term faunal trends is a short-term catastrophic event (impact or volcanism) at the K-T boundary, with its attendant biotic effects which were largely limited to low latitudes (no mass extinction in high latitudes). The observed mass extinction of rare tropical species at or near the K-T boundary may have been accelerated by this catastrophe. Whatever the nature of this catastrophe, or the long-term environmental changes, they resulted in the disruption of the well-stratified Maastrichtian ocean (elimination of deeper-water dwellers), expansion of the oxygen minimum zone, and a drop in surface productivity in low latitudes. The biotic effects of these environmental changes upon the planktonic foraminifera were the most severe in low-latitude tropical faunas and diminished greatly into high latitudes. These data provide some constraints for K-T catastrophe theories, and also suggest that current impact theories overestimate the global biotic effects of the proposed K-T impact.

ACKNOWLEDGMENTS

I am grateful for reviews, discussions, and suggestions from G. Jenkins, N. MacLeod, D. McLean, and S. Gartner. This research was supported by National Science Foundation grant OCE-9021338, the Petroleum Research Fund (ACS) PRF# 26780-AC8, and National Geographic Society grant no. 4620-91.

REFERENCES CITED

Almogi-Labin, A., and Bein, A., 1993, Late Cretaceous upwelling system along the southern Tethys margin (Israel): Interrelationship between productivity, bottom water environments, and organic matter preservation: Paleoceanography, v. 8, p. 671–690.

Alvarez, L. W., 1987, Mass extinctions caused by large bolide impacts: Physics Today, July, p. 24–33.

Alvarez, L. W., Alvarez, W., Asaro, F., and Michel, H., 1980, Extraterrestrial cause for the Cretaceous-Tertiary extinction: Science, v. 208, p. 1095–1108.

Askin, R. A., 1988, The palynological record across the Cretaceous/Tertiary transition on Seymour Island, Antarctica, *in* Feldmann, R. M., and Woodburne, M.O., eds., Geology and paleontology of Seymour Island, Antarctic Peninsula: Geological Society of America Memoir 169, p. 155–162.

Askin, R. A., 1989, Endemism and heterochroneity in the Late Cretaceous (Campanian) to Paleocene palynoflorals of Seymour Island, Antarctica: Implications for origins, dispersal and paleoclimates of southern floras, *in* Crame, J. A., ed., Origins and evolution of Antarctic biota: Geological Society of London Special Publication 47, p. 107–120.

Askin, R. A., 1990, Campanian to Paleocene spore and pollen assemblages of Seymour Island, Antarctica: Review of Paleobotany and Palynology, v. 65, p. 105–113.

Askin, R. A., 1992, Preliminary palynology and stratigraphic interpretations from a new Cretaceous/Tertiary boundary section from Seymour Island: Antarctic Journal of the United States, v. 25, p. 42–44.

Bang, I., 1979, Foraminifera from the type section of the *eugubina* Zone compared with those from Cretaceous-Tertiary boundary localities in Jylland, Denmark, *in* Birkelund, T., and Bromley, R. G., eds., Cretaceous-Tertiary boundary events 1, The Maastrichtian and Danian of Denmark: Copenhagen, University of Copenhagen, p. 127–130.

Barrera, E. and Keller, G., 1990, Stable isotope evidence for gradual environmental changes and species survivorship across the Cretaceous-Tertiary boundary: Paleoceanography, v. 5, p. 867–890.

Barrera, E. and Keller, G., 1994, Productivity across the Cretaceous/Tertiary boundary in high latitudes: Geological Society of America Bulletin, v. 106, p.1254–1266.

Ben Abdelkader, O., 1992, Planktonic foraminifera content of El Kef Cretaceous-Tertiary (K/T) boundary type section (Tunisia) [abs.]: International Workshop on Cretaceous-Tertiary Transitions (El Kef Section) Part I: Tunis, Geological Survey of Tunisia, p. 9.

Berggren, W. A., and Miller, K. G., 1988, Paleogene tropical planktic foraminiferal biostratigraphy and magnetobiochronology: Micropaleontology, v. 34, p. 362–380.

Boersma, A., and Shackleton, N. J., 1981, Oxygen and carbon isotopic variations in planktonic foraminiferal depth habitats: Late Cretaceous to Paleocene, Central Pacific DSDP Sites 463 and 465, Leg 65, *in* Initial reports of the Deep Sea Drilling Project, Volume 65: Washington, D.C., U.S. Government Printing Office, p. 513–526.

Brinkhuis, H., and Leereveld, H., 1988, Dinoflagellate cysts from the Cretaceous/Tertiary boundary sequence of El Kef, N.W. Tunisia: Review of Paleobotany and Palynology, v. 56, p. 5–19.

Brinkhuis, H., and Zachariasse, W. J., 1988, Dinoflagellate cysts, sea level changes and planktonic foraminifera across the Cretaceous-Tertiary boundary at El Haria, northwest Tunisia: Marine Micropaleontology, v. 13, p. 153–190.

Brouwers, E., 1993, Earliest origins of temperate nonmarine ostracode taxa: Evolutionary development and survival through the K/T mass extinction event: Geological Society of America Abstracts with Programs, v. 25, no. 6, p. A430.

Canudo, I., Keller, G., and Molina, E., 1991, K/T boundary extinction pattern and faunal turnover at Agost and Caravaca, SE Spain: Marine Micropaleontology, v. 17, p. 319–341.

Carter, N. L., Officer, C. B., and Drake, C. L., 1990, Dynamic deformation of quartz and feldspar: Clues to causes of some natural crises: Tectonophysics, v. 171, p. 77–118.

Case, J. A., 1989, Antarctica: The effect of high latitude heterochroneity on the origin of the Australian marsupials, *in* Crame, J. A., ed., Origins and evolution of the Antarctic biota: Geological Society of London Special Publication 47, p. 217–226.

Courtillot, V., Féraud, G., Maluski, H., Vandamme, D., Moreau, M. G., and Besse, J., 1988, Deccan flood basalts and the Cretaceous/Tertiary boundary: Nature, v. 333, p. 843–846.

Dettmann, E. M., 1989, Antarctica: Cretaceous cradle of austral temperate rain forests?, *in* Crame, J. A., ed., Origins and evolution of the Antarctic biota: Geological Society of London Special Publication 47, p. 89–106.

D'Hondt, S., and Keller, G., 1991, Some patterns of planktic foraminiferal assemblage turnover at the Cretaceous-Tertiary boundary: Marine Micropaleontology, v. 17, p. 77–118.

Donce, P., Jardine, S., Legoux, O., Masure, E., and Meon, H., 1985, Les évènements à la limite Crétacé-Tertiaire: au Kef (Tunisie septentrionale), l'analyse palynoplanctologique montre qu'un changement climatique est décelable à la base du Danian: Tunis, Actes du Premier Congrès National des Sciences de la Terre, p. 161–169.

Donovan, A. D., Baum, G. R., Blechschmidt, G. L., Loutit, T. S., Pflum, C. E., and Vail, P. R., 1988, Sequence stratigraphic setting of the Cretaceous/Tertiary boundary in central Alabama, *in* Wilgus, C. K., Hastings, B. S., Kendall, C. G., Posamentier, H. W., Ros, C., and Van Wagoner, J. C., eds., Sea-level changes: An integrated approach: Society of Economic Paleontologists and Mineralogists Special Publication 42, p. 300–307.

Douglas, R. G., 1972, Paleozoogeography of Late Cretaceous planktonic foraminifera in North America: Journal of Foraminiferal Research, v. 2, p. 14–34.

Douglas, R. G., and Savin, S. M., 1978, Oxygen isotopic evidence for the depth stratification of Tertiary and Cretaceous planktic foraminifera: Marine Micropaleontology, v. 3, p. 175–196.

Feldmann, R. M., and Tschudy, D. M., 1989, Evolutionary patterns in macrurous decapod crustaceans from Cretaceous to early Cenozoic rocks of the James Ross Island region, Antarctica, *in* Crame, J. A., ed., Origins and evolution of the Antarctic biota: Geological Society of London Special Publication 47, p. 183–196.

Fordyce, R. E., 1989, Origins and evolution of Antarctic marine mammals, *in* Crame, J. A., ed., Origins and evolution of the Antarctic biota: Geological Society of London Special Publication 47, p. 253–268.

Gartner, S., Alcal, J., and Grossman, E., 1994, Coccolithophore extinction at the K/T boundary: Gradual or abrupt?, *in* New developments regarding the K/T event and other catastrophes in Earth history: Houston, Texas, Lunar and Planetary Institute Contribution 825, p. 40.

Hallam, A., 1989, The case for sea-level change as a dominant causal factor in mass extinction of marine invertebrates: Royal Society of London Philosophical Transactions, ser. B, v. 325, p. 437–455.

Hansen, T. A., Farrand, R., Montgomery, H., Billman, H., and Blechschmidt, G., 1987, Sedimentology and extinction patterns across the Cretaceous-Tertiary boundary interval in east Texas: Cretaceous Research, v. 8, p. 229–252.

Haq, B. U., Hardenbol, J., and Vail, P. R., 1987, Chronology of fluctuating sea levels since the Triassic: Science, v. 235, p. 1156–1166.

Harwood, D. M., 1988, Upper Cretaceous and lower Paleocene diatom and silicoflagellate biostratigraphy of Seymour Island, eastern Antarctic Peninsula, *in* Feldmann, R. M., and Woodburne, M. O., eds., Geology and paleontology of Seymour Island, Antarctic Peninsula: Geological Society of America Memoir 169, p. 55–129.

Hollis, C. J., 1993, Latest Cretaceous to late Paleocene radiolarian biostratigraphy: A new zonation from the New Zealand region: Marine Micropaleontology, v. 21, p. 295–327.

Huber, B. T., 1988, Upper Campanian–Paleocene foraminifera from the James Ross Island region, Antarctic Peninsula, *in* Feldmann, R. M., and Woodburne, M. O., eds., Geology and paleontology of Seymour Island, Antarctic Peninsula: Geological Society of America Memoir 169, p. 163–251.

Huber, B. T., 1991, Maastrichtian planktonic foraminifer biostratigraphy and the Cretaceous-Tertiary boundary at ODP Hole 738C (Kerguelen Plateau, southern Indian Ocean), *in* Proceedings of the Ocean Drilling Program, Scientific results, Volume 119: College Station, Texas, Ocean Drilling Program, p. 451–466.

Huber, B. T., Liu, C., Olsson, R. K., and Berggren, W. A., 1995, The Cretaceous-Tertiary boundary transition in the Antarctic Ocean and its global implications: Comment: Marine Micropaleontology, v. 24, p. 91–99.

Hultberg, S. U., and Malmgren, B. A., 1987, Quantitative biostratigraphy based on late Maastrichtian dinoflagellates and planktonic foraminifera from southern Scandinavia: Cretaceous Research, v. 8, p. 211–228.

Jansa, L. F., 1993, Cometary impacts into the ocean: Their recognition and the threshold constraints for biological extinctions: Palaeogeography, Palaeoclimatology, Palaeoecology, v. 104, p. 271–286.

Jéhanno, C., Bocket, D., Froget, L., Lambert, B., Robin, E., Rocchia, R., and Turpin, L., 1992, The Cretaceous-Tertiary boundary at Beloc, Haiti: No evidence for an impact in the Caribbean area: Earth and Planetary Science Letters, v. 109, p. 229–241.

Jiang, M. J., and Gartner, S., 1986, Calcareous nannofossil succession across the Cretaceous/Tertiary boundary in east-central Texas: Micropaleontology, v. 32, p. 232–255.

Johnson, K. R., 1992, High latitude deciduous vegetation and muted-floral response to the Cretaceous/Tertiary boundary event in New Zealand: Geological Society of America Abstracts with Programs, v. 24, no. 4, p. A333.

Keller, G., 1985, Depth stratification of planktonic foraminifers in the Miocene Ocean, *in* Kennet, J. P., ed., The Miocene ocean: Geological Society of America Memoir 163, p. 177–196.

Keller, G., 1988a, Extinction, survivorship and evolution of planktic foraminifera across the Cretaceous/Tertiary boundary at El Kef, Tunisia: Marine Micropaleontology, v. 13, p. 239–263.

Keller, G., 1988b, Biotic turnover in benthic foraminifera across the Cretaceous/Tertiary boundary at El Kef, Tunisia: Palaeogeography, Palaeoclimatology, Palaeoecology, v. 66, p. 153–171.

Keller, G., 1989, Extended Cretaceous/Tertiary boundary extinctions and delayed population changes in planktonic foraminifera from Brazos River, Texas: Paleoceanography, v. 4, p. 287–332.

Keller, G., 1992, Paleoecologic response of Tethyan benthic foraminifera to the Cretaceous Tertiary boundary transition, *in* Takayanagi, Y., and Saito, T., eds., Studies in benthic foraminifera: Sendai, 1990: Tokyo, Tokyo University Press, p. 77–91.

Keller, G., 1993, The Cretaceous-Tertiary boundary transition in the Antarctic Ocean and its global implications: Marine Micropaleontology, v. 21, p. 1–45.

Keller, G., and Benjamini, C., 1991, Paleoenvironment of the eastern Tethys in the early Paleocene: Palaios, v. 6, p. 439–464.

Keller, G., and Lindinger, M., 1989, Stable isotope, TOC and $CaCO_3$ records across the Cretaceous/Tertiary boundary at El Kef, Tunisia: Palaeogeography, Palaeoclimatology, Palaeoecology, v. 73, p. 243–265.

Keller, G., and MacLeod, N., 1995, The Cretaceous-Tertiary boundary transition in the Antarctic Ocean and its global implications: Reply: Marine Micropaleontology, v. 24, p. 101–108.

Keller, G., MacLeod, N., Lyons, J. B., and Officer, C. B., 1993a, Is there evidence for Cretaceous/Tertiary boundary-age deep-water deposits in the Caribbean and Gulf of Mexico?: Geology, v. 21, p. 776–780.

Keller, G., Barrera, E., Schmitz, B., and Mattson, E., 1993b, Long-term oceanic instability but no mass extinction or major $\delta^{13}C$ shift in planktic foraminifera across the K/T boundary in northern high latitudes: Evidence from Nye Kløv, Denmark: Geological Society of America Bulletin, v. 105, p. 979–997.

Keller, G., Stinnesbeck, W., Adatte, T., MacLeod, N. and Lowe, D. R., 1994a, Field guide to Cretaceous-Tertiary boundary sections in northeastern Mexico: Houston, Texas, Lunar and Planetary Institute Contribution 827, 110 p.

Keller, G., Stinnesbeck, W., and Lopez-Oliva, J. G., 1994b, Age, deposition and biotic effects of the Cretaceous/Tertiary boundary event at Mimbral NE Mexico: Palaios, v. 9, p. 144–157.

Keller, G., and Perch-Nielsen von Salis, K., 1995, K/T mass extinction: Effect of global change on calcareous nannoplankton, *in* Stanley, S., ed., The effects of post-global change on life: Washington, D.C., National Academy of Sciences/National Research Council.

Keller, G., Li, L., and MacLeod, N., 1995, The Cretaceous/Tertiary Boundary Stratotype section at El Kef, Tunisia: How catastrophic was the mass extinction?: Palaeogeography, Palaeoclimatology, Palaeoecology.

Kroon, D., and Nederbragt, A. J., 1990, Ecology and paleoecology of triserial planktic foraminifera: Marine Micropaleontology, v. 16, p. 25–38.

Landis, G. P., Rigby, J. K., Jr., Sloan, R. E., and Hengst, R., 1993, Pele hypothesis: A unified model for ancient atmosphere and biotic crisis: Geological Society of America Abstracts with Programs, v. 25, no. 7, p. A362.

Leckie, R. M., 1987, Paleoecology of mid-Cretaceous planktonic foraminifera: A comparison of open ocean and epicontinental sea assemblages: Micropaleontology, v. 33, p. 164–176.

Liu, G., and Olsson, R. K., 1992, Evolutionary radiation of microperforate planktonic foraminifera following the K/T mass extinction event: Journal of Foraminiferal Research, v. 22, p. 328–346.

Lopez-Oliva, J. G., and Keller, G., 1994, Biotic effects of the K/T boundary event in northeastern Mexico, *in* New developments regarding the K/T event and other catastrophes in Earth history: Houston, Texas, Lunar and Planetary Institute Contribution 825, p. 72–73.

Lopez-Oliva, J. G., and Keller, G., 1995, Age and stratigraphy of near-K/T boundary clastic deposits in NE Mexico, *in* Ryder, G., Gartner, S., and Fastovsky, D., eds., New developments regarding the Cretaceous-Tertiary boundary event and other catastrophes in Earth history: Geological Society of America Special Paper (in press).

Lu, G., and Keller, G., in press, Stable isotope paleobiology and paleoecology of late Paleocene and early Eocene planktic foraminifera, DSDP Site 577: Palaios.

Lyons, J. B., and Officer, C. B., 1992, Mineralogy and petrology of the Haiti Cretaceous/Tertiary section: Earth and Planetary Science Letters, v. 109, p. 205–224.

MacLeod, N., 1995, Cretaceous/Tertiary (K/T) biogeography of planktic foraminifera: Historical Biology (in press).

MacLeod, N., 1995, Graphic correlation of high latitude Cretaceous/Tertiary (K/T) boundary sequences at Nye Klov (Denmark), ODP Site 690 (Weddell Sea), and ODP

Site 738 (Kerguelen Plateau): Comparison with El Kef (Tunisia) Boundary stratotype: Modern Geology (in press).
MacLeod, N., and Keller, G., 1991a, Hiatus distribution and mass extinction at the Cretaceous/Tertiary boundary: Geology, v. 19, p. 497–501.
MacLeod, N., and Keller, G., 1991b, How complete are Cretaceous/Tertiary boundary sections? A chronostratigraphic estimate based on graphic correlation: Geological Society of America Bulletin, v. 103, p. 1439–1457.
MacLeod, N., and Keller, G., 1994, Mass extinction and planktic foraminiferal survivorship across the Cretaceous/Tertiary boundary: A biogeographic test: Paleobiology, v. 20, p. 143–177.
MacLeod, N., and Keller, G., 1995, An interpretation of the El Kef planktic foraminiferal blind test results. Marine Micropaleontology (in press).
Margaritz, M., Benjamini, C., Keller, G., and Moskovitz, S., 1992, Early diagenetic isotopic signal at the Cretaceous/Tertiary boundary, Israel: Palaeogeography, Palaeoclimatology, Palaeoecology, v. 91, p. 291–304.
Marincovich, L., Jr., 1993, Delayed extinction of Mesozoic marine mollusks in the Paleocene Arctic ocean basin: Geologic Society of America Abstracts with Programs, v. 25, no. 6, p. A362.
McGowran, B., and Beecroft, A., 1985, *Guembelitria* in the early Tertiary of southern Australia and its paleoceanographic significance: South Australia Department of Mines and Energy Special Publication 6, p. 247–261.
McLaren, D. J., and Goodfellow, W. D., 1990, Geological and biological consequences of giant impact: Annual Review of Earth and Planetary Sciences, v. 18, p. 123–171.
McLean, D. M., 1985, Deccan traps mantle degassing in the terminal Cretaceous marine extinctions: Cretaceous Research, v. 6, p. 235–259.
McLean, D. M., 1991, A climate change mammalian population collapse mechanism, *in* Kainlauri, E., and others, eds., Energy and environment: Atlanta, Georgia, ASHRAE, p. 93–100.
McLean, D. M., 1994, Proposed law of nature linking impacts, plume volcanism, and Milankovitch cycles to terrestrial vertebrate mass extinctions via greenhouse-embryo death coupling? *in* New developments regarding the K/T event and other catastrophes in Earth history: Houston, Texas, Lunar and Planetary Institute Contribution 825, p. 82–83.
Méon, H., 1990, Palynologic studies of the Cretaceous/Tertiary boundary interval at El Kef outcrop, northwest Tunisia: Paleogeographic implication: Review of Paleobotany and Palynology, v. 65, p. 85–94.
Meyerhoff, A. A., Lyons, J. B., and Officer, C. B., 1994, Chicxulub structure: A volcanic sequence of Late Cretaceous age: Geology, v. 22, p. 3–4.
Molnar, R. E., 1989, Terrestrial tetrapods in Cretaceous Antarctica, *in* Crame, J. A., ed., Origins and evolution of Antarctic biota: Geological Society of London Special Publication 47, p. 131–140.
Nomura, R., 1991, Paleoceanography of upper Maastrichtian to Eocene benthic foraminiferal assemblages at Sites 752, 753 and 754, Eastern Indian Ocean, *in* Proceedings of the Ocean Drilling Program, Scientific results, Volume 121: College Station, Texas, Ocean Drilling Program, p. 3–29.
Officer, C. B., 1990, Extinctions, iridium and shocked minerals associated with the Cretaceous/Tertiary transition: Journal of Geological Education, v. 38, p. 402–425.
Officer, C. B., and Carter, N. L., 1991, A review of the structure, petrology and dynamic deformation characteristics of some enigmatic terrestrial structures: Earth Science Reviews, v. 30, p. 1–49.
Olsson, R. K., and Liu, G., 1993, Controversies on the placement of the Cretaceous-Paleogene boundary and the K/T mass extinction of planktonic foraminifera: Palaios, v. 8, p. 127–139.
Peryt, D., Lahodynsky, R., Rocchia, R., and Boclet, D., 1993, The Cretaceous/Paleogene

boundary and planktonic foraminifera in the Flyschgosau (Eastern Alps, Austria): Palaeogeography, Palaeoclimatology, Palaeoecology, v. 104, p. 239–252.

Peypouquet, J. P., Grousset, F., and Mourguiart, P., 1986, Paleoceanography of the Mesogean Sea based on ostracods of the northern Tunisian continental shelf between the Late Cretaceous and early Paleogene: Geologishe Rundschau, v. 75, p. 159–174.

Pospichal, J. J., 1991, Calcareous nannofossils across the Cretaceous/Tertiary boundary, ODP Leg 121, Site 752, Eastern Indian Ocean, *in* Proceedings of the Ocean Drilling Program, Scientific results, Volume 121: College Station, Texas, Ocean Drilling Program, p. 395–414.

Premoli-Silva, S., and McNulty, G. L., 1984, Planktonic foraminifers and calpionellids from Gulf of Mexico sites, Deep Sea Drilling Project Leg 77, *in* Initial reports of the Deep Sea Drilling Project, Volume 77: Washington, D.C., U.S. Government Printing Office, p. 547–584.

Raup, D. M., 1990, Impact as a general cause of extinction; a feasibility test, *in* Sharpton, V. L., and Ward, P., eds., Global catastrophes in Earth history: An interdisciplinary conference on impacts, volcanism, and mass mortality: Geological Society of America Special Paper 247, p. 27–32.

Robin, E., Boclet, D., Bonté, P., Froget, L., Jéhanno, C., and Rocchia, R., 1991, The stratigraphic distribution of Ni-rich spinels in Cretaceous-Tertiary boundary rocks at El Kef (Tunisia), Caravaca (Spain) and Hole 761C (Leg 122): Earth and Planetary Science Letters, v. 107, p. 715–721.

Rohling, E. J., Zachariasse, W. J., and Brinkhuis, H., 1991, A terrestrial scenario for the Cretaceous-Tertiary boundary collapse of the marine pelagic ecosystem: Terra Nova, v. 3, p. 41–48.

Savrda, C. E., 1993, Ichnosedimentologic evidence for a noncatastrophic origin of Cretaceous-Tertiary boundary sands in Alabama: Geology, v. 21, p. 1075–1078.

Sawlowicz, Z., 1993, Iridium and other platinum-group elements as geochemical markers in sedimentary environments: Palaeogeography, Palaeoclimatology, Palaeoecology, v. 104, p. 253–270.

Schmitz, B., Asaro, F., Michel, H. V., Thierstein, H. R., and Huber, B. T., 1991, Element stratigraphy across the Cretaceous/Tertiary boundary in ODP Hole 738C, *in* Proceedings of the Ocean Drilling Program: Scientific results, Volume 119: College Station, Texas, Ocean Drilling Program, p. 719–730.

Schmitz, B., Keller, G., and Stenvall, O., 1992, Stable isotope and foraminiferal changes across the Cretaceous/Tertiary boundary at Stevns Klint, Denmark: Arguments for long-term oceanic instability before and after bolide impact: Palaeogeography, Palaeoclimatology, Palaeoecology, v. 96, p. 233–260.

Sepkoski, J. J., Jr., 1990, The taxonomic structure of periodic extinctions, *in* Sharpton, V. L., and Ward, P., eds., Global catastrophes in Earth history: An interdisciplinary conference on impacts, volcanism, and mass mortality: Geological Society of America Special Paper 247, p. 33–44.

Sharpton, V. L., Dalrymple, G. B., Marin, L. E., Ryder, G., Schuraytz, B. C., and Urrutia-Fucugauchi, J., 1992, New links between the Chicxulub impact structure and the Cretaceous/Tertiary boundary: Science, v. 359, p. 819–821.

Sliter, W. V., 1972a, Upper Cretaceous planktonic foraminiferal zoogeography and ecology—Eastern Pacific margin: Palaeogeography, Palaeoclimatology, Palaeoecology, v. 12, p. 15–31.

Sliter, W. V., 1972b, Cretaceous foraminifers—Depth habits and their origin: Nature, v. 239, p. 514–515.

Smit, J., 1982, Extinction and evolution of planktonic foraminifera after a major impact at the Cretaceous/Tertiary boundary, *in* Silver, L. T., and Schultz, P. H., eds., Geological implications of impacts of large asteroids and comets on the Earth: Geological Society of America Special Paper 190, p. 329–352.

Smit, J., 1990, Meteorite impact, extinctions and the Cretaceous-Tertiary boundary: Geologie en Mijnbouw, v. 69, p. 187–204.

Smit, J., and Romein, A. J. T., 1985, A sequence of events across the Cretaceous-Tertiary boundary: Earth and Planetary Science Letters, v. 74, p. 155–170.

Stinnesbeck, W., 1986, Zu den Faunistischen und Paloko—logischen verhältnissen in der Quiriquina Formation (Maastrichtium) Zentral-Chiles: Palaeontographica, abt. A, v. 194, p. 1–237.

Stinnesbeck, W., Barbarin, J. M., Keller, G., Lopez-Oliva, J. G., Pivnik, D. A., Lyons, J. B., Officer, C. B., Adatte, T., Graup, G., Rocchia, R., and Robin, E., 1993, Deposition of channel deposits near the Cretaceous-Tertiary boundary in northeastern Mexico: Catastrophic or "normal" sedimentary deposits?: Geology, v. 21, p. 797–900.

Stott, L. D., and Kennett, J. P., 1989, New constraints on early Tertiary paleoproductivity from carbon isotopes in foraminifers: Nature, v. 342, p. 526–529.

Stott, L. D., and Kennett, J. P., 1990, The paleoceanographic and paleoclimatic signature of the Cretaceous/Paleogene boundary in the Antarctic: Stable isotopic results from ODP Leg 113, *in* Proceedings of the Ocean Drilling Program, Scientific results, Volume 113: College Station, Texas, Ocean Drilling Program, p. 829–848.

Sweet, A. R., Braman, D. R., and Lerbekmo, J. R., 1992, Palynofloral response to K/T boundary events; a transitory interruption within a dynamic system, *in* Silver, L. T., and Schultz, P. H., eds., Geological implications of impacts of large asteroids and comets on the Earth: Geological Society of America Special Paper 190, p. 457–470.

Thierstein, H. R., Asaro, F., Ehrmann, W. U., Huber, B., Michel, H., Sakai, H., and Schmitz, B., 1991, The Cretaceous/Tertiary boundary at Site 738, South Kerguelen Plateau, *in* Proceedings of the Ocean Drilling Program, Scientific results, Volume 119: College Station, Texas, Ocean Drilling Program, p. 849–868.

Thomas, E., 1989, Development of Cenozoic deep-sea benthic foraminiferal faunas in Antarctic waters, *in* Crame, J. A., ed., Origins and evolution of the Antarctic biota: Geological Society of London Special Publication 47, p. 283–296.

Thomas, E., 1990, Late Cretaceous through Neogene deep-sea benthic foraminifers (Uland Rise, Weddell Sea, Antarctica), *in* Proceedings of the Ocean Drilling Program, Scientific results, Volume 113: College Station, Texas, Ocean Drilling Program, p. 571–594.

Van Eijden, A. J. M., and Smit, J., 1991, Eastern Indian Ocean Cretaceous and Paleogene quantitative biostratigraphy, *in* Proceedings of the Ocean Drilling Program, Scientific results, Volume 121: College Station, Texas, Ocean Drilling Program, p. 77–124.

Walting, L., and Thurston, H. M., 1989, Antarctica as an evolutionary incubator: Evidence for the cladistic biogeography of the Amphipod Family Iphimediidae, *in* Crame, J. A., ed., Origins and evolution of Antarctic biota: Geological Society of London Special Publication 47, p. 297–313.

Wang, P., Min, Q., and Bian, Y., 1985, Distribution of foraminifera and ostracods in bottom sediments of the northwestern part of the South Huanghai (Yellow) Sea and its geological significance, *in* Wang, P., ed., Marine micropaleontology of China: Bejing, China Ocean Press and Berlin, Springer, p. 93–115.

Wang, K., Attrep, M., Jr., and Orth, C. J., 1993, Global iridium anomaly, mass extinction, and redox change at the Devonian Carboniferous boundary: Geology, v. 21, p. 1071–1074.

Zachos, J. C., and Arthur, M. A., 1986, Paleoceanography of the Cretaceous-Tertiary boundary event: Inferences from stable isotopic and other data: Paleoceanography, v. 1, p. 5–26.

Zachos, J. C., Berggren, W. A., Aubry, M. P., and Mackenen, A., 1992, Chemostratigraphy of the Cretaceous/Paleocene boundary at Site 750, southern Kerguelen Plateau, *in* Proceedings of the Ocean Drilling Program, Scientific results, Volume 119: College Station, Texas, Ocean Drilling Program, p. 961–977.

Zinsmeister, W. J., and Feldmann, R. M., 1984, Cenozoic high latitude heterochroneity of Southern Hemisphere marine faunas: Science, v. 224, p. 281–283.
Zinsmeister, W. J., and Macellari, C. E., 1983, Changes in the macrofossil faunas at the end of the Cretaceous on Seymour Island, Antarctic Peninsula: Antarctic Journal of the United States, v. 18, p. 68–69.
Zinsmeister, W. J., Feldmann, R. M., Woodburne, M. O., and Elliot, D. H., 1989, Latest Cretaceous/earliest Tertiary transition on Seymour Island, Antarctica: Journal of Paleontology, v. 63, p. 731–738.

5

Nature of the Cretaceous-Tertiary Planktonic Foraminiferal Record: Stratigraphic Confidence Intervals, Signor-Lipps Effect, and Patterns of Survivorship

Norman MacLeod, *Department of Palaeontology, The Natural History Museum, London, United Kingdom*

INTRODUCTION

The 1980 proposal of the Alvarez impact-extinction scenario is often described as a singular development in the history of extinction studies (see Gould, 1984; Raup, 1986, 1991; Glenn, 1994a) because it is claimed to represent the first testable hypothesis that attempts to explain the cause of mass extinctions. Many previous mass-extinction models have been proposed and rightly criticized for making no detailed predictions with respect to patterns that should be found in the stratigraphic record (Gould, 1984; Raup, 1986, 1991; Glenn, 1994a). Others were originally advanced to explain the demise of particular organismal groups (e.g., dinosaurs, ammonites, trilobites) but foundered as general theories when it was recognized that other groups, regarded as equally sensitive to particular environmental changes, survived (see Benton, 1990).

For example, most paleontologists found it difficult to understand how electromagnetic radiation from a proximal supernova (Schindewolf, 1962) could doom nonavian dinosaurs, but allow birds and amphibians to weather the event with little or no obvious change in diversity. Similarly, it was difficult to see what ecological factor ammonites and planktonic foraminifera could share that would make them more susceptible to extinctions at the close of the Cretaceous than a host of other marine clades (e.g., benthic foraminifera, fish, *Nautilus*, sharks, and rays). Difficulties in identifying a particular ecological theme among clades that underwent differentially high extinction rates at mass-extinction horizons led some paleontologists to suggest that the such events are not biological phenomena at all, but rather the result of sampling effects (e.g., outcrop area effect, monograph effect; see Raup, 1976a, 1976b). Thus, the indeterminacy problem, coupled with abundant and compelling counterexamples for proposed mechanisms, relegated mass-extinction studies to a highly speculative backwater of mainstream paleontology for most of this century.

The perception that mass-extinction studies lacked the necessary rigor to be taken seriously by the scientific community was shattered in 1980 by publication of the impact-extinction scenario (Alvarez and others, 1980). This approach to developing an explanation for mass extinctions in general, and the Cretaceous-Tertiary (K-T) mass extinction in particular, was noteworthy because it appealed to observational evidence (an iridium anomaly) to establish the presence and estimate the size of the proposed extinction mechanism (bolide impact) and relate this mechanism to the extinctions themselves. No previous mass-extinction causal hypothesis had exhibited this degree of specificity.

Whereas most of the original Alvarez and others (1980) paper was given over to a lengthy discussion of the Ir anomaly and its interpretation as evidence for the occurrence of an impact at the K-T boundary, that paper was frustratingly vague on the crucial second part of the hypothesis test, namely the specification of extinction mechanisms and the stratigraphic comparison of the Ir anomaly to actual species extinctions. For example, Alvarez and others (1980, p. 1105) envisioned the environmental effects of the impact in only the most general terms, to wit, "An asteroid struck the earth, formed an impact crater, and some of the dust-sized material ejected from the crater reached the stratosphere and was spread around the globe. This dust effectively prevented sunlight from reaching the surface for a period of several years... Loss of sunlight suppressed photosynthesis, and as a result most food chains collapsed and the extinctions resulted." No predictions were made with respect to the types of organisms most likely to be affected by these events and no paleontological data were presented to corroborate this mass extinction scenario. Indeed, organisms were barely mentioned at all.

Alvarez and others (1982) revised their estimate of stratospheric dust residence times to a few months and expanded the range of conceivable killing mechanisms to include acid rain and the release of noxious chemicals. However, no discussion was provided as to which organisms were predicted to be most severely affected and which organisms were predicted to have survived. As in the previous 1980 paper the only biotic prediction that was made is that extinctions attributable to bolide impact should have taken place suddenly. Thus, the 1982 paper cited Smit's (1981) estimate of planktonic foraminiferal extinctions at the Caravaca boundary section (southern Spain) to have been completed within "less than 50 years" (Smit, 1981, p. 309) of the proposed impact.

Throughout the 1980s to the present the impact-extinction controversy has raged around these two separate but related issues: the physical evidence for large body impact at the K-T boundary, and the biotic evidence required to confirm a link between impact debris and species extinctions. However, because of the difference in the way the original scenario treated these issues, its testability is usually presented as residing entirely with the physical evidence for impact occurrence. To many, but by no means all, paleontologists, the idea that a mass-extinction hypothesis could be developed and tested without extensive reference to the fossil record seemed curious, to say the least. However, the lack of sufficiently high resolution biostratigraphic data, along with appeals to widely appreciated but (then) little described sampling and preservational problems inherent in paleontological data (e.g., Signor-Lipps effect, confidence estimates on stratigraphic ranges), introduced a sense of uncertainty among earth scientists, including many paleontologists, with respect to the degree to which paleontological data could be expected to support the impact-extinction scenario.

Since 1980 a large number of paleontological studies, most of which were inspired by the impact-extinction controversy, have shown that the fossil records of many organismal groups fail to exhibit the expected catastrophic faunal turnovers coincident with horizons containing impact debris (see Perch-Nielsen and others, 1982 [coccolithophorids]; Askin, 1990, 1992 [dinoflagellates]; Hollis, 1993 [radiolaria]; Keller, 1988a, 1989, 1993; Brinkhuis and Zachariasse, 1988; Canudo and others, 1991; Keller and Benjamini, 1991; MacLeod and Keller 1991, 1994 [planktonic foraminifera]; Keller, 1988b, 1992; Kaiho, 1988; Thomas, 1990; Widmark and Malmgren, 1992 [benthic foraminifera]; Donce and others, 1982, 1985; Peypouquet and others, 1986; Brouwers and De Deckker, 1993 [ostracodes]; Surlyk, 1990 [brachiopods]; Labandiera, 1992 [insect and insect plant associations]; Sloan and others, 1986 [dinosaurs]; Archibald and Clemens, 1984 [microvertebrates]; Lerbekmo and others, 1987; Sweet and Braman, 1992; Sweet and others, 1990; McIver and others, 1991; Méon, 1990 [pollen]; Knobloch and others, 1993; Golovneva, 1993 [plant body fossils]). In addition, it is now widely accepted that rudistid and inoceramid bivalves, the extinctions of which were described in many of the early impact-extinction papers as being sudden and coincident with the K-T boundary, actually disappear from the fossil record well below the K-T boundary, prior to any trace of impact debris (e.g., Johnson and Kauffman, 1993; Kauffman, 1984 respectively). Nevertheless, Alvarez and Asaro (1990, p. 83) criticized Courtillot's (1990) volcanic eruption scenario for K-T extinctions by arguing that it cannot be regarded as "a serious suspect in a killing that evidently took place over .001 Myr or less." Alvarez and others (1994, p. 3) also cited studies by Ward and others (1991) and Sheehan and others (1991) as evidence that the terminal Cretaceous extinctions of ammonites and dinosaurs were "indistinguishable from a sudden extinction" and therefore supportive of their pro-impact interpretation, despite the fact that neither of these 1991 reports demonstrated an unambiguous link between the disappearances of the fossil groups they studied and either the K-T boundary or impact debris.

Owing to recent data from the proposed Chicxulub impact structure, along with studies of coeval sediments from the Carribean and Gulf of Mexico, many would argue that a causal link between a K-T impact and the K-T mass extinction has been proven beyond reasonable doubt. However, this argument is grounded on the assumption that the establishment of impact occurrence is sufficient to demonstrate a link between that occurrence and species extinctions. In order to establish causation, the processes by which impact-driven extinctions were brought about must be specified and the fossil record examined to determine whether the groups predicted to have been differentially susceptible to impact-induced environmental effects exhibit differentially high extinction rates. In the absence of such predictions any biotic pattern perceived to be odd or unusual occurring at the K-T boundary or in the lowermost Tertiary interval could be attributed to the impact event regardless of its true cause. Moreover, in the absence of such predictions, any biotic pattern that does not conform to the impact-extinction scenario (e.g., survival of frogs, salamanders, birds, crocodiles, nautiloids) can be reconciled with the scenario through ad hoc means (e.g., birds survived in caves [see Clemens *in* Glenn, 1994b] or mammals survived because they ate seeds [Alvarez and others, 1980]).

Ad hoc explanations such as these risk obscuring the very patterns paleontologists wish to interpret. If all extinctions near the K-T boundary are attributed to an impact event regardless of their true cause, it is unlikely that we will ever be able to say anything more about the biotic effects of sudden, large-scale environmental change. In other words, if there is no way to tell if or when a particular hypothesis concerning impact-extinction linkage is wrong, there will be no way to determine whether it is right. Proponents of the impact-extinction scenario should be interested in making the relevant biotic predictions and testing their models against the data of the fossil record, if for no other reason than to demonstrate their case. Until this has been done the much discussed "testability" of the impact-extinction scenario will remain little more than a half-realized rhetorical device.

Although few specific biotic predictions have been formally made by proponents of the impact-extinction scenario, it should be possible to identify at least some of the organismal groups that should have been differentially affected by the proposed environmental effects of a large bolide impact (e.g., a marked reduction in sunlight reaching Earth's surface, a sharp temperature decrease, global wildfire, acid rain). Planktonic foraminifera must be within the group of predicted victims. These organisms live in the uppermost portion of the marine water column, many modern (and presumably many ancient) species being most commonly found within particular depth zones. Ecologically, planktonic foraminifera comprise a mixed group of herbivores, carnivores, and scavengers. Many species seem to require the presence of symbiotic zooxanthellae in order to maintain normal life processes. Their tests (= shells) are composed of calcium carbonate, which is highly susceptible to attack by corrosive fluids and which must be secreted through a complex biochemical process that is sensitive to a large number of environmental variables. These organisms are distributed worldwide and possess an exceedingly fine grained stratigraphic record. (Note that planktonic foraminifera are the standard index fossils for the Upper Cretaceous–lower Paleocene interval.) Variations in the species richness, relative abundance, and the isotopic composition of planktonic foraminiferal tests have been used in high-resolution studies of environmental change throughout the past 100 m.y. If an unambiguous signal of biotic change induced by large body impact can be read in the fossil record, it should be read easily in the fossil record of planktonic foraminifera.

Maastrichtian planktonic foraminiferal extinctions precipitated by environmental changes wrought by a large body impact should coincide stratigraphically with impact debris, or should occur at a horizon that can be correlated on the basis of independent evidence to such a debris-bearing horizon. If a single impactor is accepted (as with the Chicxulub scenario) this horizon should be demonstrably coeval with other such horizons scattered throughout the affected area. In addition, the impact-extinction scenario also predicts that the uniqueness of this event should be preserved in distinctions between the disappearing and surviving taxa (Gould, 1984; Jablonski, 1986a, 1986b, 1986c). Specifically, patterns of survivorship across the K-T boundary should not resemble patterns of planktonic foraminiferal survivorship during other turnover events regarded to be driven by "normal" terrestrially based processes (e.g., the Eocene-Oligocene transition, Neogene background extinctions).

The set of null hypotheses for the impact-extinction scenario in these areas are as follows. (1) Local species disappearances should be distributed throughout an extended transition K-T interval (representing a period of time in excess of the duration of the inferred killing mechanisms) within all local sections preserving a complete record of K-T events. (2) Disappearances of species should occur diachronously between the various local sections (reflecting either the operation of different causal factors or the spatially heterogeneous expression of a single causal mechanism). (3) Distinctions between disappearing and surviving taxa should mirror morphologic, taxonomic, and/or ecologic patterns recorded from other nonimpact driven events. In a formal hypothesis test it is these statements that must be disproven prior to the provisional acceptance of their alternatives. If these statements cannot be rejected on objective grounds, the predictions of the impact-extinction scenario must be considered uncorroborated and the scenario must be accorded a corresponding probability of being correct.

This contribution tests these biotic predictions of the impact-extinction scenario against the K-T planktonic foraminiferal record from four well-known K-T boundary successions representing a large suite of marine environments. In addition to reviewing the biostratigraphy of each section and core, series of tests are made to assess the reliability of the apparent faunal turnover record. These tests are required because of complications in the interpretation of stratigraphic data that arise from a large number of factors, including sampling problems (e.g., bias in the estimation of stratigraphic ranges due to fluctuating sample sizes, species-specific differences in susceptibility to diagenesis) and the possibility of reworking. Once these tests have been made and the appropriate corrections instituted (where possible), the tempo and mode of the K-T planktonic foraminiferal turnover may be summarized by constructing a synthetic model of the turnover pattern from these local records and calibrating that model against available radiometrically "dated" horizons with the uppermost Maastrichtian and lowermost Danian interval.

MATERIALS AND METHODS

This report is based on samples collected from Nye Kløv, Denmark (Keller and others, 1993), Agost, southern Spain (Canudo and others, 1991), Brazos core, Brazos River, Texas (Keller, 1989), and Ocean Drilling Program (ODP) Site 738, Kerguelen Plateau (Keller, 1993). The distribution of these sections and cores plotted on a paleogeographic reconstruction of continental positions at 65 Ma is shown in Figure 1 and a summary of their depositional environments is shown in Figure 2. The exact sample positions were provided in the biostratigraphic literature cited above.

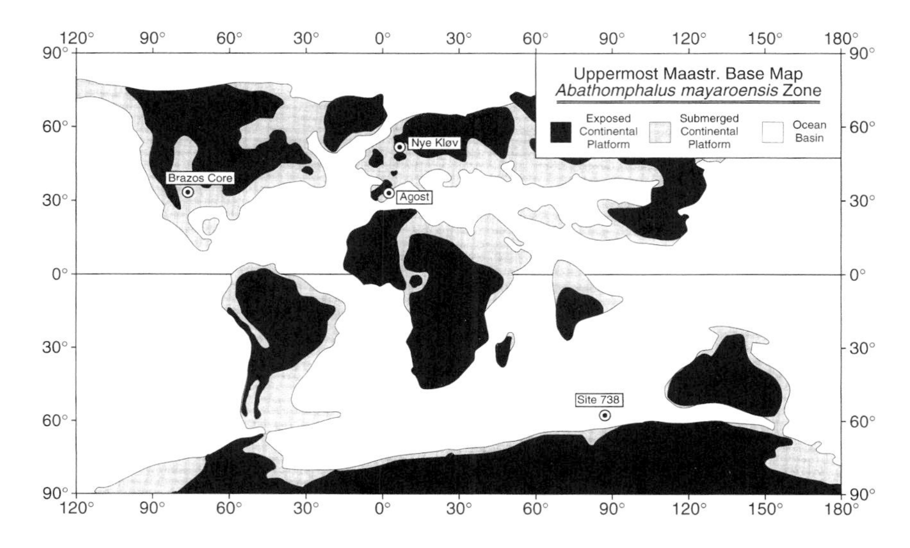

Figure 1. Upper Maastrichtian (Maastr.) paleogeographic base map showing location of the four study sections and/or cores (redrawn from MacLeod and Keller, 1994).

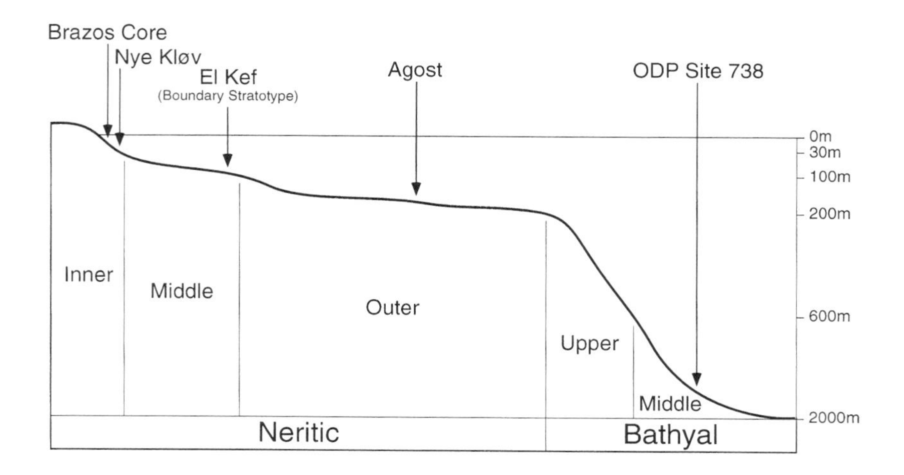

Figure 2. Depositional environments represented by Nye Kløv, Agost, Brazos core, Ocean Drilling Program (ODP) Site 738 and the Cretaceous-Tertiary boundary stratotype (El Kef, Tunisia). These environments were inferred on the basis of associated benthic foraminiferal faunas (redrawn from MacLeod and Keller, 1994).

All foraminiferal samples were processed according to standard micropaleontological techniques (see Keller, 1986) that included the recovery of all particulate matter ≥ 63 μm in diameter for all boundary sections and ≥ 38 μm in diameter for the boundary clay interval of ODP Site 738. Recovered planktonic foraminiferal faunas were randomly split into 200–400 specimen aliquots that were identified in their entirety to the species level (see Buzas, 1990, for a justification of this procedure). Identified sample aliquots were mounted on paper slides for reference and are currently in the planktonic foraminiferal collections of Princeton University.

Consistent species concepts were applied across the entire dataset. Whereas the Danian taxonomy employed herein differs from some recent revisions (e.g., Olsson and others, 1992), the empirical data presented in those contributions primarily relate to changes in the generic assignment of particular species. These revisions have little significance in terms of species-level data summaries or biostratigraphic correlations. Several alternative biostratigraphic zonations have been proposed for the uppermost Maastrichtian and lowermost Danian interval, and were summarized in numerous publications, including Keller (1988a), Canudo and others (1991), Keller and Benjamini (1991), Ben Abdelkader and others (1992), and Keller (1993). Throughout this report the zonation of Keller (1993) is employed.

Although most mass-extinction studies to date have relied on raw biostratigraphic data, it is widely appreciated that, given the variable nature of paleontological samples, it cannot be assumed that the observed last appearance datum (LAD) provides an accurate estimate of the true LAD of a species (Signor and Lipps, 1982). This concept, known as the Signor-Lipps effect, simply restates the truism that fossil data are complex and often imperfect records of the past. Many quantitative techniques are available to test aspects of local fossil records to determine whether they have been biased in particular directions. Herein stratigraphic confidence limits and tests associated with the Signor-Lipps effect are used to examine the planktonic foraminiferal fossil records of the four study sections and/or cores.

Signor-Lipps Effect

The Signor-Lipps effect, a version of which was first described by Shaw (1964), states that because of vagaries in the processes of sampling, fossil preservation, and/or ecologic occurrence pattern, it cannot be assumed that the observed LAD (or first appearance datum [FAD]) provides an accurate estimate of the true LAD or FAD (Signor and Lipps, 1982). Hence Signor and Lipps' (1982, p. 291) conclusion that "The recorded ranges of fossils, especially of uncommon taxa or taxa in habitats not represented by a continuous record, may be inadequate to test either gradual or catastrophic [extinction] hypotheses." Note that this statement is inconsistent with some authors' blanket invocation of the Signor-Lipps effect to suggest that gradual extinction patterns support catastrophic extinction mechanisms (see Kerr, 1994; Smit, 1994). The Signor-Lipps effect represents an attempt to set limits on the interpretability of biostratigraphic data that are contingent upon either the assumption or the demonstration of significant differences between observed and true LADs. If such a discrepancy is assumed (as is most often the case), then the Signor-Lipps effect negates any possibility of making an objective interpretation of K-T paleontological data. However, if evidence supporting a suspicion that a Signor-Lipps effect is operative within a particular dataset is collected and presented prior to the interpretation of such data, the Signor-Lipps effect is transformed from a statement of futility into a testable hypothesis.

It is fortunate that the Signor-Lipps effect makes several predictions of patterns that should be observed in any fauna whose biostratigraphic record has been systematically biased to favor progressive, rather than abrupt, turnover. First, a Signor-Lipps effect may be produced by the selective removal of species susceptible to either syndepositional or postdepositional diagenesis. Such effects are known to be problematic in assemblages of calcareous microfossils (such as planktonic foraminifera), where it is not unusual to encounter intervals of moderate to severe dissolution corresponding to fluctuations in physical oceanographic parameters (e.g., depths of the marine lysocline and calcite compensation depth [CCD]; see Berger, 1968, 1970, 1979). This source of bias may be examined by independently assessing the relative dissolution susceptibilities for various species and then determining if there is a systematic bias in the ratio of dissolution-resistant to dissolution-susceptible species as the K-T boundary is approached.

Second, a Signor-Lipps effect can be generated in faunas whose total abundance, or species whose relative abundance, is either already low or declining as they approach an arbitrarily defined stratigraphic horizon (Fig. 3A). This well-known sampling problem has received extensive treatment in the ecological and paleoecological literature (Raup, 1972, 1976b; Signor, 1978; Koch and Morgan, 1988). The predicted pattern of decreasing abundance with decreasing distance from the horizon can be quantitatively tested using a wide variety of statistical techniques (see Sokal and Rohlf, 1981) or modeled using the method of Koch and Morgan (1988). If such a relation is confirmed, the biased species richness patterns present in the raw data may, in principal, be corrected using rarefaction (Maguran, 1988), after which deterministic interpretation may proceed.

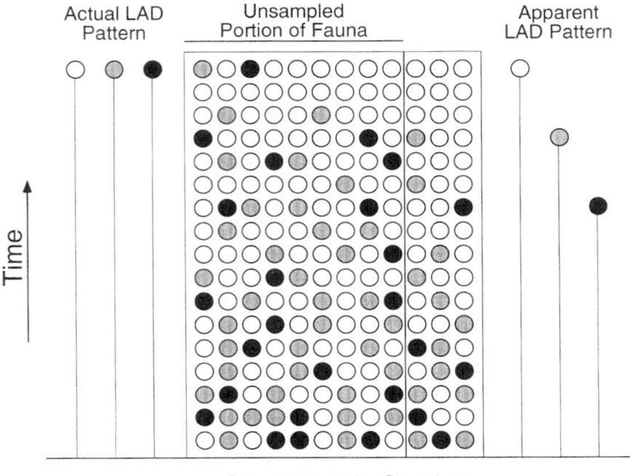

Figure 3A. Potential types of stratigraphic range bias in the fossil record: The Signor-Lipps effect, which occurs when the relatively low abundance taxa are present in the interval, but fail to be recognized due to the collection of finite-sized samples. Low abundances may be produced by a wide range of biotic and sampling factors (see text for discussion).

Third, a Signor-Lipps effect may be produced if the local environment shifts from a set of conditions within the tolerance range of a particular species to a set of conditions outside that range. Such habitat shifts were interpreted by Signor and Lipps (1982) to be responsible for the patchy distribution of species occurrences through a stratigraphic sequence. Patchy occurrence distributions often represent an intrinsic paleoecological attribute of individual species that may play an important role in controlling their extinction susceptibility (see Hanski, 1982, Hanski and Gilpin, 1991).

There are at least two methods by which the testing of this environmental migration hypothesis may be approached. The most direct involves the quantitative comparison of sediment analysis and/or geochemical results with the distribution of particular taxa in an effort to determine whether the presence of a particular species is correlated with the presence or absence of some physical factor. Alternatively, biostratigraphic data can be used to identify assemblages of species sharing common occurrence or relative abundance patterns throughout the sampled interval. Once putative ecologic guilds have been identified, it can be determined whether they disappear in a structured manner as the boundary is approached. If any of these tests yield statistically significant patterns consistent with predictions of the Signor-Lipps effect, that model can be advocated plausibly to account for the observed data. However, if such tests fail to yield the predicted patterns, further appeals to the Signor-Lipps effect as being responsible for systematically biased patterns of LADs lose credibility.

Although these tests provide important data and should be undertaken in many different paleontological contexts, it is important to remember that the demonstration of a Signor-Lipps effect does not necessarily provide support for a catastrophic extinction pattern. Rather, such an effect, if supported by empirical data, can only suggest that the apparent pattern of first and last occurrences might not be indicative of the true pattern. In addition, the impact-extinction scenario does not seek to account for the disappearance of species whose abundances are already so low as to challenge the ability of experienced biostratigraphers to find them. Such species are inherently at risk. Instead, this scenario predicts that numerically dominant uppermost Maastrichtian planktonic foraminiferal species were catastrophically eliminated by a single very short term environmental perturbation arising as a direct result of bolide impact. Because the disappearance of very rare taxa is inherently ambiguous with respect to the testing of this model (see Signor and Lipps, 1982), appeals to a Signor-Lipps effect alone (e.g., Kerr, 1994; Smit, 1994) cannot provide the type of data necessary to discriminate between alternative explanatory scenarios. Moreover, if evidence supporting a Signor-Lipps bias in these data is found, the implied range extensions must be recognized to have potentially affected all species in the fauna, not just those LADs observed to fall below the K-T boundary.

Stratigraphic Confidence Limits

The calculation of stratigraphic confidence limits has been suggested as one approach to testing hypotheses of faunal turnover across the K-T boundary (Jablonski, 1994). Starting with the same assumption as the Signor-Lipps effect, this method attempts to use observed data to infer the hypothetical limits of species occurrence within a local section or core, subject to various assumptions. Unfortunately, there are both practical and theoretical difficulties involved in applying stratigraphic confidence limits to the K-T boundary problem.

Like the Signor-Lipps effect, the concept of stratigraphic confidence intervals can be traced to Shaw (1964), who used the principles of binomial probability to test whether the absence of comparatively rare species from a sample just above its LAD could be the result of failing to obtain adequate sample sizes. From this relation Shaw (1964) tabulated the probabilities of finding low abundance species in variously sized samples. Paul (1982) applied Shaw's strategy to the analysis of gaps between fossil occurrences, which he theorized should form an exponential distribution. Provided it is reasonable to assume that local preservation patterns are random and that sediment accumulation rates are uniform throughout the sampled interval, Paul (1982) estimated the 95% and 99% confidence intervals to be (respectively) four and seven times the median size of the gaps between species occurrence horizons. Paul (1982) also provided quantitative strategies for determining whether observed gap distributions are sufficiently similar to an exponential distribution to apply his methods.

The techniques currently used for inferring stratigraphic confidence intervals were developed by Strauss and Sadler (1989). These require that sections or cores be sampled continuously for species occurrences and that those occurrences be randomly distributed throughout the sampled interval. Marshall (1944a) developed a variation of the Strauss and Sadler method that enables confidence intervals to be estimated for all species with non-random occurrence patterns. However, this alternative method still requires continuous sampling. Micropaleontological samples are typically taken over discrete stratigraphic intervals, transported to a laboratory, and processed to separate the fauna from the surrounding inorganic sediment. This produces a time-averaged, discrete sample of species occurrence patterns that violates the continuous-sampling assumption of the Strauss and Sadler (1989) and Marshall (1994a) methods.

It is currently unknown to what degree the results of these confidence interval estimation techniques are robust to deviations from continuous sampling. Many statistical tests (e.g., t-tests, ANOVA) are routinely applied to paleontological data that violate the basic assumptions of these tests (e.g., normality) and it may be possible to estimate reliable stratigraphic confidence intervals for some types of discreetly sampled data. In addition, there is no test available for determining whether a section that is simply scanned for fossil occurrences had truly been continuously sampled.

Unlike traditional descriptive statistics that estimate population parameters lying within the limits of the observed data (e.g., means, standard deviations, confidence intervals on the mean), stratigraphic confidence limits attempt to describe population parameters that, by definition, lie outside their sampled universe. Therefore, these techniques are extraordinarily sensitive to a wide range of stratigraphic assumptions. Moreover, even if the issue of assumption testing is set aside, the question of how to interpret the results of a stratigraphic confidence intervals with respect to K-T boundary turnover patterns remains.

Stratigraphic confidence limits represent intervals within which, to a stated degree of certainty, one expects the true LAD (or FAD) of a taxon to reside. The determination of such intervals does not provide for the inductive selection of certain horizons that might fall within the confidence limit (e.g., the K-T boundary) as being more likely candidates for the level of true LADs (or FADs) than other included horizons (Fig. 3B). In order to estimate the probabilities associated with particular geometries of LADs (or FADs) within the confidence band, those geometries must be compared with all possible LAD (or FAD) geometries that could be constructed within the specified confidence band.

Graphic Correlation

Once biostratigraphic data have been assessed for bias and (where possible) corrected, a composite model of species occurrence patterns can be constructed using graphic correlation (Fig. 4). Graphic correlation and compositing of these biostratigraphic data into a lower Danian composite standard (LD-CS) were carried out according to the methods described in Miller (1977) and Edwards (1984) with the procedural revisions for estimating final correlation models proposed by MacLeod (1994). Positions for the line of correlation (LOC) were estimated using the qualitative method of Miller (1977; see also MacLeod and Sadler, 1995), which distinguishes between the chronostratigraphic implications of FADs and LADs, and composited using multiple comparison cycles until a stable CS was achieved. FADs were judged to be more chronologically reliable than

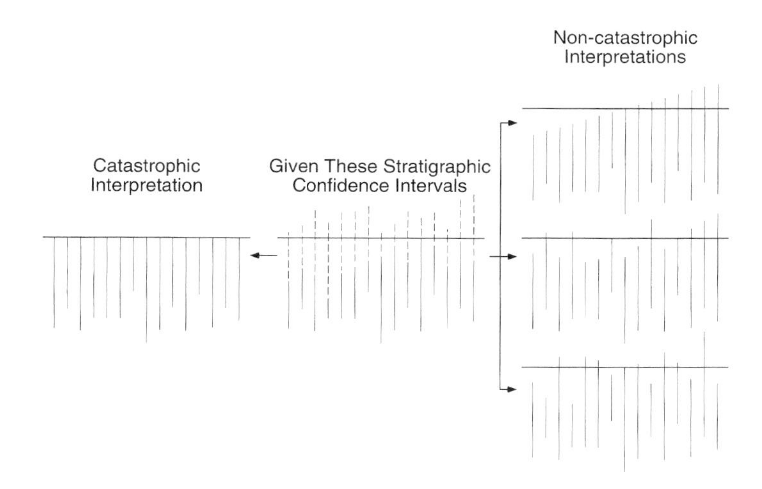

Figure 3B. Potential types of stratigraphic range bias in the fossil record: Stratigraphic confidence intervals (dashed lines), which arise as a result of patchy occurrence distributions within the stratigraphic range of fossil species. Both the Signor-Lipps effect and stratigraphic confidence intervals are statements of uncertainty with respect to the deterministic interpretation of raw biostratigraphic data. Because this uncertainty covers a very wide range of possible geometries for last appearance datums (LADs), their recognition in a particular section or core cannot be invoked as evidence supporting a particular faunal turnover model.

LADs due to the potential for upward reworking that some authors believe to be a serious problem for the interpretation of lowermost Danian microfossil biostratigraphy (e.g., Liu and Olsson, 1992; Olsson and Liu, 1993). Initial CS estimation was achieved via designation of the El Kef boundary stratotype as the standard reference section (SRS) and compositing data from the study sections and/or cores into this sequence. (Note: see MacLeod [1995a] for a parallel analysis of the El Kef stratotype section with respect to Signor-Lipps effect, strati-

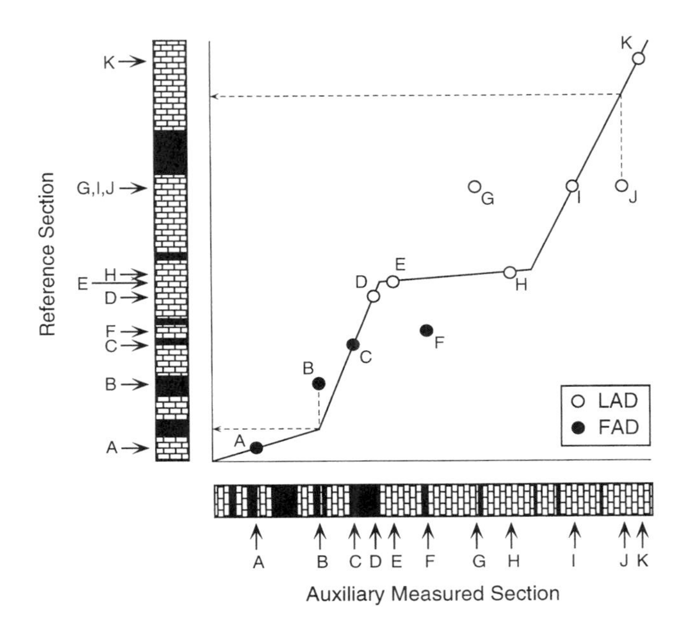

Figure 4. An example of the graphic correlation of biostratigraphic data. Given two stratigraphic sections, each of which contain 11 biostratigraphic datums (filled circles = first appearance datums [FADs], open circles = last appearance datums [LADs]), the correlation between these sections can be expressed as that line which maps the position of datums in one section onto their corresponding positions in the other. FADs and LADs located on or near the line of correlation (LOC) (datums A, C, D, E, H, I, and K) represent events that occur at approximately the same temporal horizon in each section, while those located at some distance from the LOC (datums B, F, G, and J) are indicative of intersection temporal diachrony. If the datum sequence on the abscissa of such a plot is designated as the "reference" section, FADs occurring below the LOC (datum J) and FADs occurring above the LOC (datum B) represent datums that occupy (respectively) lower and higher horizons in the auxiliary section (ordinate) than in the reference section (abscissa). Because accurate chronostratigraphic correlation requires that only global highest and lowest occurrences be employed in relating depositional sequences to one another, erroneous (= nonmaximal) occurrences of these datums in the reference section may be corrected by using the LOC to infer their appropriate positions (dashed lines). Once this is done, the reference section becomes a composite estimate of the actual datum sequence, or a composite sequence (CS). By performing several rounds of comparison between the CS and a set of auxiliary measured sections, a stable estimate of the global datum sequence of datums can be achieved. Redrawn from MacLeod and Keller (1991).

graphic confidence intervals, and tests for reworked faunas.) As in previous graphic studies of K-T boundary correlation (MacLeod and Keller, 1991; MacLeod, 1995a, 1995b, 1995c) LOC placement during compositing utilized each section's "best-case" correlation model (i.e., most-complete sequence; sediment accumulation rate as constant as possible; simplest LOC geometry). Employment of this convention produces a conservative estimate of the true event sequence.

RESULTS

Observed Biostratigraphic Patterns

There is little support for a catastrophic planktonic foraminiferal mass extinction coinciding with the K-T boundary to be found among the raw biostratigraphic data recovered from these four boundary successions (Fig. 5). In each case the overwhelming majority of Cretaceous taxa record their last observed occurrences either below or above the independently recognized local K-T boundary horizons. (Note that in all instances the location of the K-T boundary is based on multiple criteria, including the first appearance of Tertiary planktonic foraminiferal species, an Ir anomaly, a zone enriched in Fe-Ni spinels, and the initial stages of a $\delta^{13}C$ excursion.) No LADs of Cretaceous taxa were found to coincide with the local K-T boundaries within the two high-latitude successions (Nye Kløv and ODP Site 738), whereas faunas recovered from the middle-latitude sections of Agost and Brazos Core record extinction intensities of only 25.49% and 6.52% of their respective Cretaceous assemblages.

The interval over which the K-T faunal transition takes place in these successions ranges from 2 to 20 m. Because each of these sections and cores contains a sharp iridium anomaly and no obvious evidence of sediment disturbance on this lithostratigraphic scale, there is little reason to suspect that local sediment mixing can be responsible for the extended interval over which the biotic transition appears to take place. These extended intervals of K-T faunal turnover have now been observed in high-resolution planktonic foraminiferal biostratigraphic studies from more than 30 different K-T boundary successions (see MacLeod and Keller, 1991, 1994, for reviews). In virtually all cases the observational evidence suggests that last appearances of Cretaceous taxa are spread over intervals, many up to several meters thick, that straddle the boundary horizon. Typically, few LADs coincide with the boundary itself. Exceptions to this pattern are known (see MacLeod and Keller, 1991), but in all cases these are most parsimoniously interpreted to indicate the presence of a hiatus coinciding with the K-T boundary. Moreover, this pattern in the observational data has been confirmed by an independent "blind test" carried out using the boundary stratotype succession (see Ginsburg and others, in prep.; McLeod and Keller, in prep.) and recorded for other groups of calcareous plankton (Pospichal, 1994).

Because the pattern of observed planktonic foraminiferal last appearances in biostratigraphically complete successions has now been established beyond reasonable doubt, the relevant questions become (1) how reliable are the LADs of Cretaceous taxa that fall below the boundary? and (2) what is the most appropri-

Depth in Section (cm)
-100.0 -50.0 0.0 50.0 100.0 150.0
Agost
Ventilabrella multicamerata
Globotruncanita pettersi
Globotruncana rosetta
Shackoina sp.
Globotruncanella pschadae
Globotruncanita conica
Gublerina cuvillieri
Rugoglobigerina milamensis
Globigerinelloides rosebudensis
Heterohelix americana
Planoglobulina brazoensis
Plummerita hantkenoides
Rugoglobigerina pennyi
Globotruncana aegyptiaca
Abathomphalus mayaroensis
Globotruncana dupeublei
Globotruncana esnehensis
Globotruncanita contusa
Globotruncanita stuarti
Pseudoguembelina sp.
Pseudotextularia deformis
Pseudotextularia elegans
Racemiguembelina fructicosa
Racemiguembelina intermedia
Racemiguembelina powelli
Rugoglobigerina scotti
Rugo. macrocephala
Heterohelix striata
Heterohelix complanata
Globotruncana spp.
Hedbergella caravacaensis
Pseudoguembelina costata
Pseudoguembelina palpebra
Globotruncanella minuta s.l.
Globotruncanella petaloidea
Abathomphalus intermedius
Globotruncana arca
Hedbergella monmouthensis
Pseudo. kempensis
Rugo. hexacamerata
Rugoglobigerina rugosa
Globigerinell. subcarinatus
Hexterohelix pseudotessera
Globigerinelloides aspera
Globigerinell. yaucoensis
Hedbergella holmdelensis
Heterohelix globulosa
Parvularugo. eugubina
Pseudo. costulata
Heterohelix navarroensis
Globoconusa conusa
Globastica fodina
Subbotina minutula
Eo. edita
P. longiapertura
G. fringa
P. eugubina-G. taurica
Eoglobigerina edita
Guembelitria cretacea
Guembelitria trifolia
W. claytonensis
E. fringa
Guembelitria danica
P.compressus
C. midwayensis
S. pseudobulloides
Guembelitria irregularis
W. hornerstownensis
C. morsei
G. daubjergensis
Globanomalina pentagona
Murciglobigerina chasconona
Subbotina triloculinoides
Eoglobigerina trivialis
Subbotina varianta
Morozovella inconstans
Morozovella pseudoinconstans

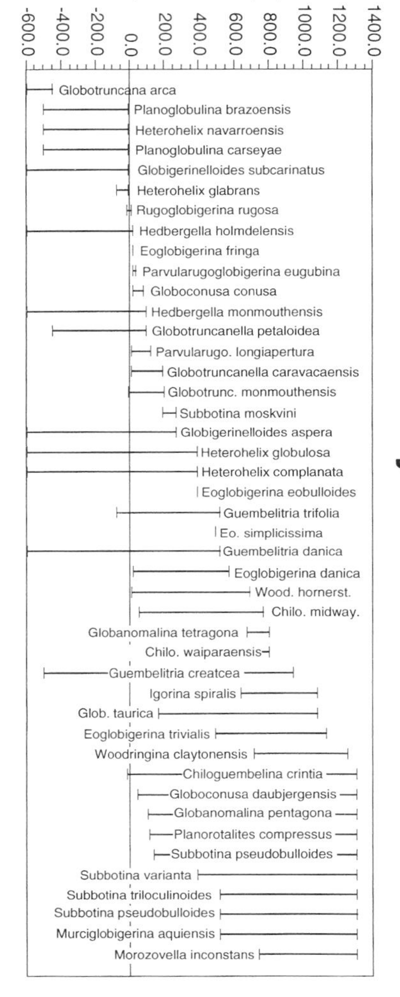

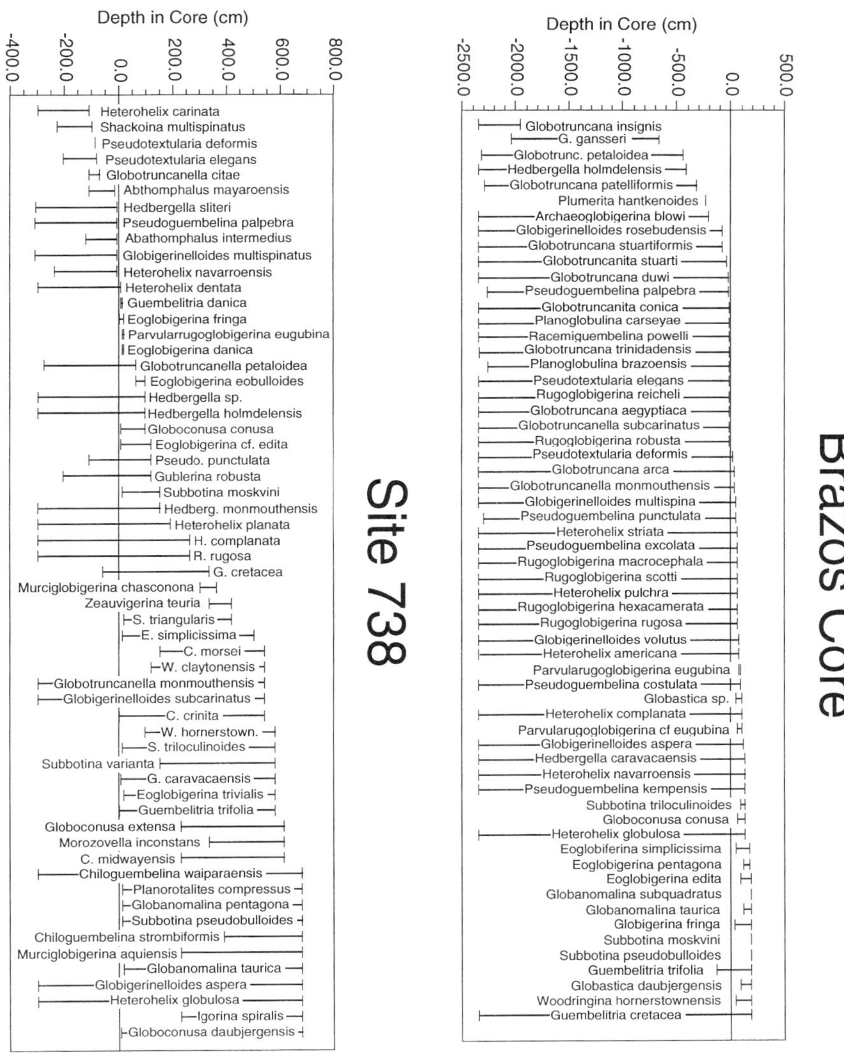

Figure 5. Observed biostratigraphic ranges of Cretaceous-Tertiary planktonic foraminiferal faunas from Nye Kløv (data from Schmitz and others, 1992), Agost (data from Canudo and others, 1991), Brazos core (data from Keller, 1989), and Ocean Drilling Program Site 738 (data from Keller, 1993).

ate interpretation for the Danian occurrences of "Cretaceous" species? Surely it is inappropriate to pick and choose among these observed stratigraphic ranges such that those fitting the predictions of a particular causal model are "believed" while all others are discounted. However, it is useless to pretend that sampling patterns, preservational factors, and relative abundance–sample size interactions are not present in these data or cannot influence their interpretation.

For example, Figure 6A shows that the rank-ordered total relative abundance distributions for all four faunas exhibit a sharp decline, most species exhibiting values of 0.01 or below. It is interesting that the pattern of relative abundance distribution for the Nye Kløv fauna is strongly reminiscent of a geometric series, whereas those of the other three faunas correspond to log-normal models. Following Maguran (1988), the Nye Kløv pattern is consistent with a fauna exposed to harsh conditions (Nye Kløv represents an inner neritic high-latitude environment) within which niche preemption predominates. Similarly, the log-normal pattern is the most common encountered in ecological field studies and is usually interpreted to reflect the existence of random variation in a large number of independent factors that together control the abundance distribution of individual species. Looking at these curves purely on the phenomenological

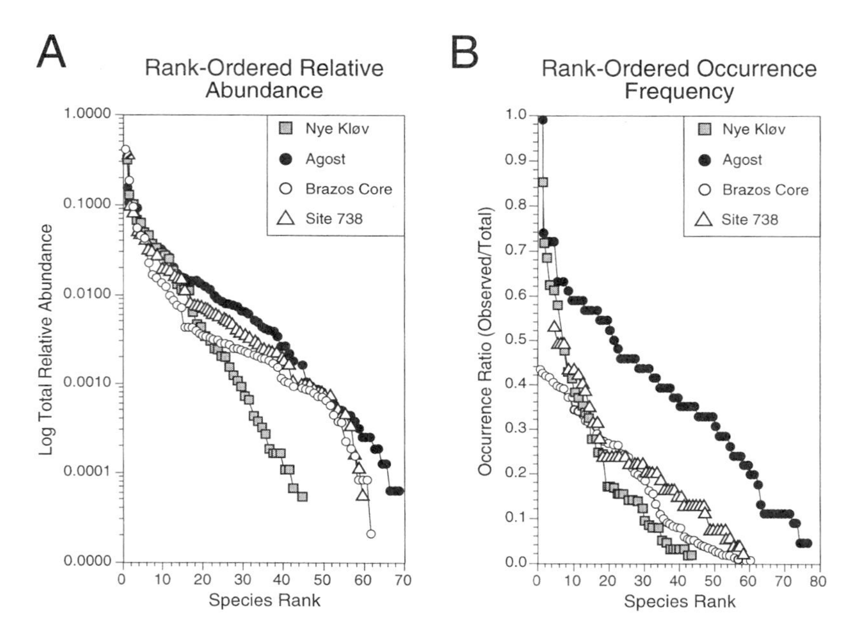

Figure 6. Rank-ordered plots of species-specific relative abundance and occurrence frequency in each of the four study sections and/or cores. As is typical for living animal communities, a few species exhibit large relative abundances and high occurrence frequencies, while the majority of the fauna exists at low relative abundances and occurs infrequently. This leads to several potential sampling problems that must be taken into consideration when formulating paleobiologic interpretations of biostratigraphic data. See text for additional discussion.

level, it seems clear that relatively rare species (which, after all, compose the majority of these faunas) may not appear in samples unless a sufficiently large number of specimens are examined.

The occurrence patterns of species composing these faunas also exhibit distributions (Fig. 6B) that may conspire with sampling patterns to affect our perception of the fossil record. Here, the two middle-latitude sections are composed of faunas that exhibit a more or less constantly declining, rank-ordered occurrence pattern while the high-latitude sequences exhibit a more striking exponential or log-decay pattern. Under these conditions we might expect the high-latitude records to be more seriously influenced by fluctuations in sample spacing than the middle-latitude successions.

Do the sample sizes and sample spacings used to infer the K-T biotic record in these four successions contain contrasts that may be problematic with respect to the literal interpretation of these observations? There is ample cause for concern. Figure 7 shows that sample sizes in all four successions exhibit marked fluctuations in the vicinity of the boundary horizon. Although these fluctuations do not (with the exception of ODP Site 738) approach the magnitude of those that Koch (1991) observed in his demonstration of a Signor-Lipps effect at the Braggs, Alabama, K-T boundary section, they do occupy a substantial percentage of the sample size range and occur in the context of patterns, the effects of which on the observed record are difficult to predict without the aid of statistical models. Much the same can be said of the stratigraphic sampling patterns shown in Figure 8. Whereas the interval closest to the boundary horizon (± 20 cm) was almost always sampled continuously, intervals at some distance from the boundary were sampled at a variety of resolutions reflecting the idiosyncracies of outcrop exposure and field time. Nevertheless, these variations can affect the observed fossil record and, to the maximum extent possible, their influence should be evaluated before any interpretations are made.

Stratigraphic Confidence Intervals

Determination of stratigraphic confidence intervals investigates the extent to which patterns of species occurrences and sample spacing interact to alter patterns in the fossil record. Provided that the gaps between individual occurrences of a species can be assumed to exhibit an exponential distribution and not be serially correlated (fairly robust assumptions given the character of most biostratigraphic data), the interval overlying the last observed appearance within which failure to find a species may result from the presence of a gap between occurrences (rather than genuine absence) can, in principal, be determined. Additional requirements for accurate confidence interval determination include a knowledge of the entire local species range (not just its range within an arbitrary sample interval) and a robust chronostratigraphic model. The latter is a particularly important, although widely neglected, condition because the geometry of the confidence interval calculation assumes that a unit distance within the observed stratigraphic range be equivalent to a unit distance outside of that range. This means that confidence intervals based on the lithostratigraphic separation between first and last appearances must either assume a constant sediment accumulation rate throughout the observed and predicted range (with no included hiatuses), or have some

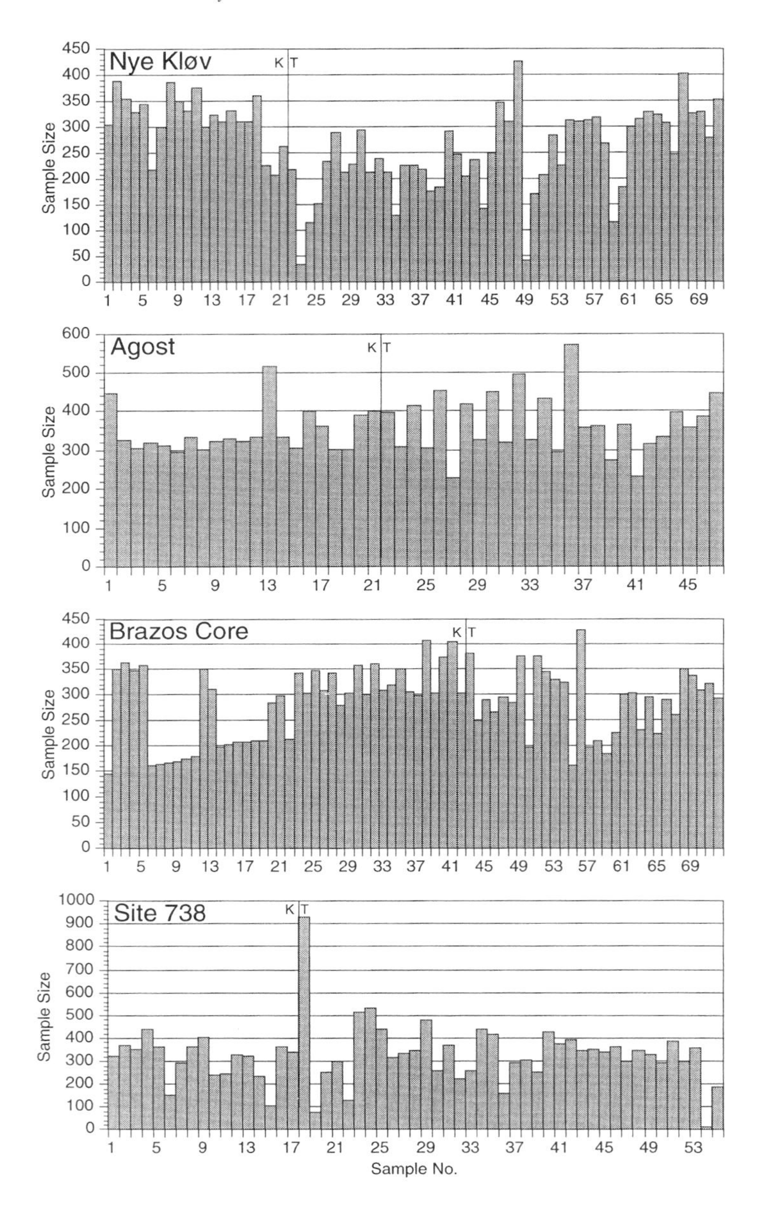

Figure 7. Sample sizes for the Cretaceous-Tertiary (K-T) planktonic foraminiferal data used in this study. Note that there is no systematic trend toward larger or smaller samples in the uppermost Maastrichtian or lowermost Danian intervals. Observed fluctuations in sample size represent, for the most part, fluctuations in the ratio of fossils to sediment particles or the intensive analysis of particular intervals (e.g., the first Danian sample at Ocean Drilling Program Site 738). Interactions between sample size, relative abundance, and occurrence frequency (see Fig. 6) can lead to the misrepresentation of biotic turnover patterns. However, the lack of systematic sample size bias around the K-T boundary horizon in these four sections and cores should minimize the effects of such bias.

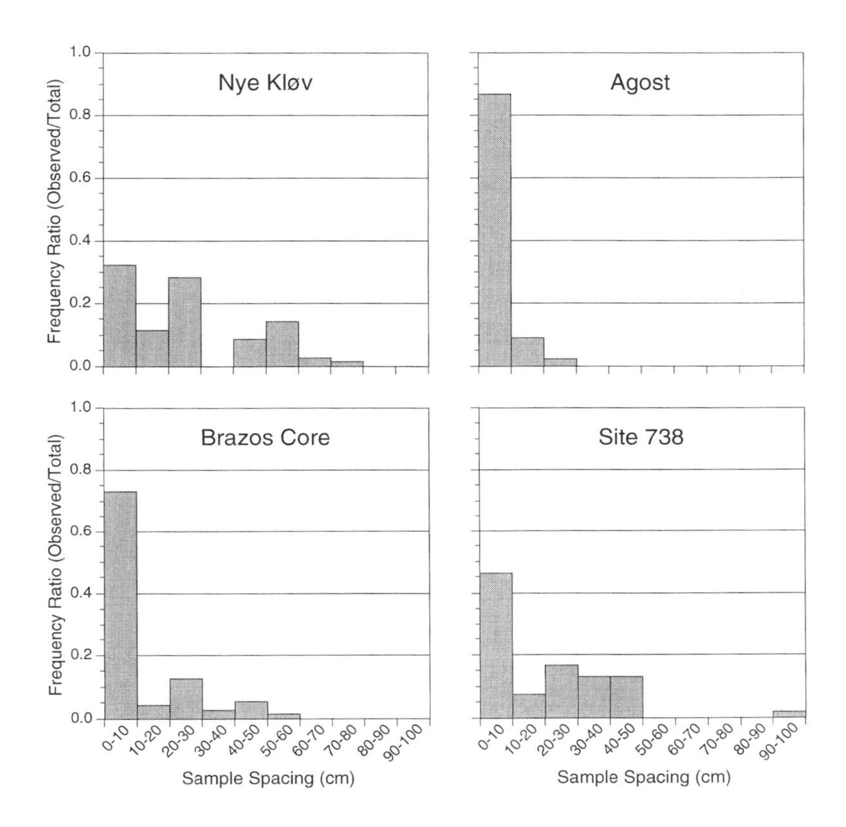

Figure 8. Sample spacings for the four study sections and/or cores. Samples present within the first spacing category (0–10 cm) predominantly represent continuously sampled intervals. Nevertheless, the wide variation in sampling patterns among the different sequences can lead to additional sources of bias. Note, however, that all of these distributions are skewed toward the higher-resolution end of the sampling spectrum, indicating that they represent the imposition of time and outcrop availability on a sampling pattern designed to recover the maximum amount of resolution. Lack of systematic differences among these four sampling patterns should reduce sample-spacing bias to a minimum.

stratigraphically reasonable means to recognize and estimate the duration of hiatuses and convert lithostratigraphic distance into time.

Unfortunately very few fossiliferous sequences, and no K-T boundary sequences, are known to these levels of resolution. MacLeod and Keller (1991) provided high-resolution chronostratigraphic models for the Agost and Brazos Core sections, and MacLeod (1995b, 1995c) formulated similar models for Nye Kløv and ODP Site 738. However, these models covered only the lowermost Danian intervals of those sections and core in detail. From the sediment accumulation rate fluctuations and hiatuses documented in those two studies, it seems highly unlikely that the upper Maastrichtian records of these same successions are entirely free from such complications. Moreover, none of the sampled intervals within these successions is known to contain the entire local stratigraphic range of any Maastrichtian species. The sampling pattern used to obtain these data also represents a complex mixture of continuous samples (= occurrence gaps probably represent real gaps in species distributions) and discontinuous samples (= occurrence gaps overestimated due to the presence of unsampled intervals). These factors severely limit the degree to which accurate stratigraphic confidence limits could be determined from these data.

Despite such problems, a simple experiment can be performed that illustrates the ambiguity of this proposed approach to the analysis of faunal turnover patterns. Let us assume that for each of the four successions species disappearing below the K-T boundary (Fig. 9) all possess 95% stratigraphic confidence intervals that include the boundary horizon. Would we then be able to say anything definite about the true pattern of faunal turnover? We could not reject the possibility that all of these species were synchronously eliminated at the boundary horizon. However, we would also not be able to reject any of a large number of alternative turnover geometries, each of which are consistent with such data. Even if we arbitrarily exclude the uppermost Maastrichtian intervals that were sampled, there exist from 2.13×10^8 (ODP Site 738) to 4.48×10^3 (Nye Kløv) different possible LAD arrangements for these data (Table 1). Because stratigraphic confidence intervals are measures of uncertainty, none of the alternative turnover geometries can be excluded from consideration in any test of K-T extinction scenarios. All are equally likely to represent the "true" pattern. Table 1 also lists the

TABLE 1

Number of alternative last appearance datum (LAD) configurations and probabilities associated with catastrophic extinction models based on hypothetical stratigraphic confidence intervals (see text)

	Catastrophic Extinction: Single Horizon		Catastrophic Extinction: K-T-5cm	
Section/Core	**No. of LAD config.**	**Probability**	**No. of LAD config.**	**Probability**
Nye Kløv	4.4800×10^3	2.23×10^{-4}	4.4720×10^3	2.24×10^{-4}
Agost	1.2672×10^7	7.89×10^{-8}	1.2671×10^7	7.89×10^{-6}
Brazos Core	2.0808×10^6	4.81×10^{-7}	2.0798×10^6	4.81×10^{-7}
Site 738	2.1289×10^8	4.70×10^{-9}	2.0971×10^8	4.77×10^{-9}

Figure 9. Hypothetical stratigraphic confidence intervals (dashed lines) for Cretaceous species whose observed last appearance datums (LADs) fall below the Cretaceous-Tertiary (K-T) boundary horizon. Tick marks along the abscissa represent the mid-point position of each sampled interval. Shaded zone represents 5 cm interval below K-T boundary. Even if these confidence intervals are accepted they cannot be used to infer the "true" geometry of LADs. See text for discussion.

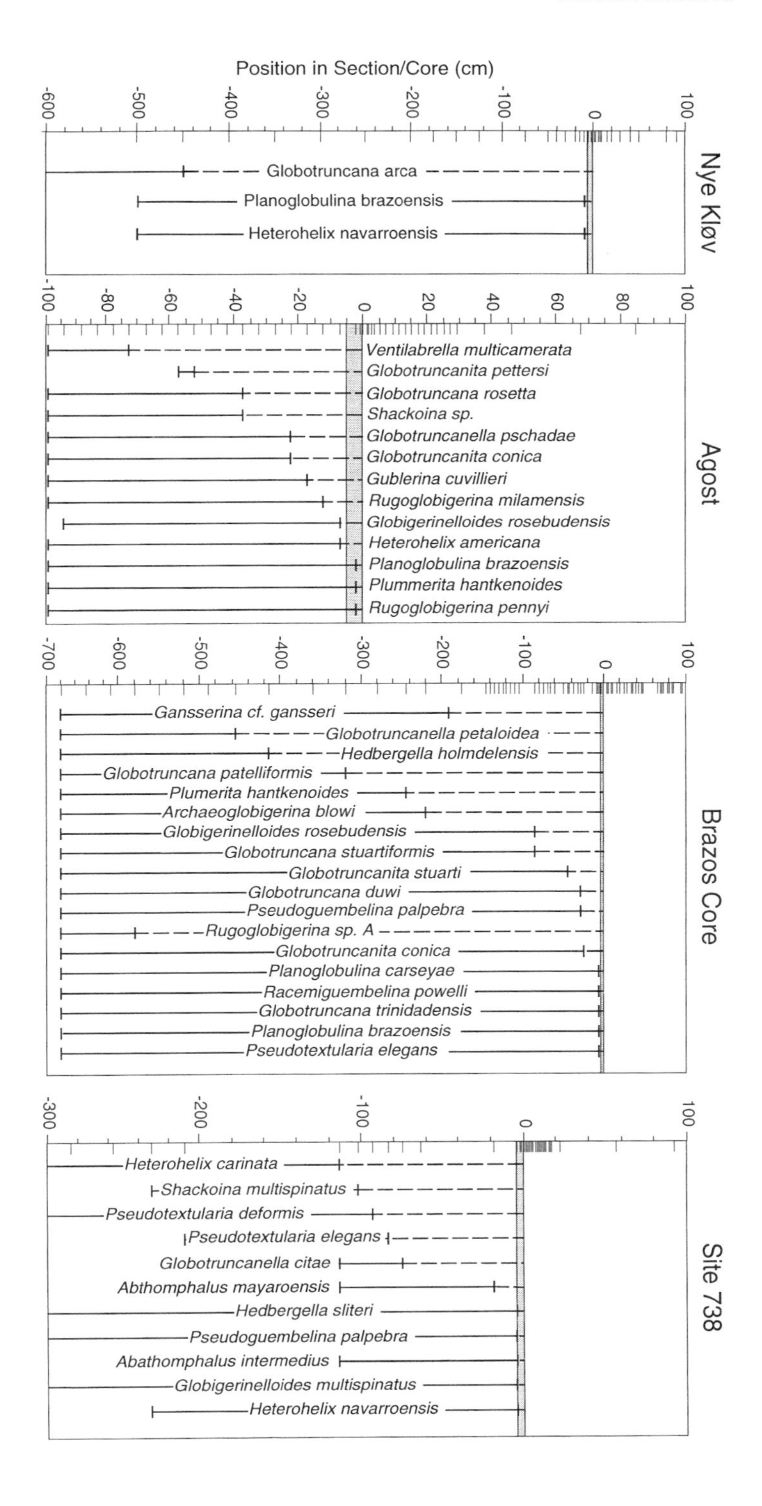
Position in Section/Core (cm)
Nye Kløv
Globotruncana arca
Planoglobulina brazoensis
Heterohelix navarroensis
Agost
Ventilabrella multicamerata
Globotruncanita pettersi
Globotruncana rosetta
Shackoina sp.
Globotruncanella pschadae
Globotruncanita conica
Gublerina cuvillieri
Rugoglobigerina milamensis
Globigerinelloides rosebudensis
Heterohelix americana
Planoglobulina brazoensis
Plummerita hantkenoides
Rugoglobigerina pennyi
Brazos Core
Gansserina cf. gansseri
Globotruncanella petaloidea
Hedbergella holmdelensis
Globotruncana patelliformis
Plumerita hantkenoides
Archaeoglobigerina blowi
Globigerinelloides rosebudensis
Globotruncana stuartiformis
Globotruncanita stuarti
Globotruncana duwi
Pseudoguembelina palpebra
Rugoglobigerina sp. A
Globotruncanita conica
Planoglobulina carseyae
Racemiguembelina powelli
Globotruncana trinidadensis
Planoglobulina brazoensis
Pseudotextularia elegans
Site 738
Heterohelix carinata
Shackoina multispinatus
Pseudotextularia deformis
Pseudotextularia elegans
Globotruncanella citae
Abthomphalus mayaroensis
Hedbergella sliteri
Pseudoguembelina palpebra
Abathomphalus intermedius
Globigerinelloides multispinatus
Heterohelix navarroensis

probabilities associated with the synchronous disappearance of these species at a single horizon and their diachronous disappearance within 5 cm of the boundary horizon for each of the successions. With the exception of the diachronous (5 cm) model for Nye Kløv, these probabilities are all extremely low.

Springer (1990) and Marshall (1994b) developed confidence-interval–based methods for testing the hypothesis of whether data such as these are compatible with a coextinction model and identifying the stratigraphic interval over which such a model may be valid. While both of these authors applied their methods to the Macellari (1986) upper Maastrichtian ammonite data in an attempt to test the mass-extinction model for this group in the Seymour Island section, neither was able to demonstrate a unique association between an inferred coextinction (mass extinction) horizon and the K-T boundary. Moreover, both approaches assume coextinction (which is, in fact, the very hypothesis we would like to test) and fail to take into consideration the large number of independent extinction and combined independent-coextinction models that are also consistent with the Macellari (1986) data and their associated stratigraphic confidence intervals.

Tests for the Signor-Lipps Effect

Given the nature of the present planktonic foraminiferal K-T biostratigraphic database, analyses of stratigraphic confidence intervals cannot provide more than a crude test of various turnover models. A practical alternative to the confidence interval approach lies in developing a series of explicit null and alternative hypothesis based on reasonable interactions between relative abundance, sample size, occurrence pattern, and sampling density that predict the presence of systematic bias in Maastrichtian LADs, and then evaluating statistically the predictions of those hypotheses. (Note that because of the failure of most paleontological data to conform to assumptions of normality, statistical testing of these data should be conducted by nonparametric methods whenever possible.) While this approach does not seek to "correct" biostratigraphic observations per se, it does enable much more sophisticated testing of those observations for bias. If systematic bias is revealed by these tests, it may be able to be removed, normalized, or otherwise compensated for. However, if systematic bias is not detected one may regard the observed data as reflecting biologically meaningful patterns.

Differential Preservation Potential

One source of potentially important bias is differential preservation potential. Within this model Maastrichtian LADs may occur below the boundary horizon, rather than at the horizon itself, because these species represent morphotypes prone to selective in situ destruction by diagenetic processes. As is typical for temporally complete K-T boundary successions, Nye Kløv, Agost, Brazos core, and ODP Site 738 are all characterized by a boundary clay layer that most likely represents a condensed interval resulting from eustatic sea-level rise (see Haq and others, 1987; Haq, 1991; Donovan and others, 1988, 1990; MacLeod and Keller, 1991). The presence of this boundary clay may have served as a barrier to the postdepositional migration of fluids through the sediment, resulting in the concentration of these fluids in the uppermost Maastrichtian marls.

In order to test this aspect of the Signor-Lipps model, the dissolution susceptibility scales of Douglas (1971), along with the modifications to this scale proposed by Malmgren (1987), for Upper Cretaceous planktonic foraminifera were used to classify the Nye Kløv, Agost, Brazos core, and Site 738 faunas into dissolution susceptible, resistant, and unaffected categories (see MacLeod and Keller, 1994). Once these groupings were made, each of the four faunas was subdivided further into forms observed to disappear below the K-T boundary "Cretaceous" species observed to extend at least to the boundary horizon itself. If the dissolution-based Signor-Lipps model is correct, the indigenous Maastrichtian fauna as a whole should be differentially composed of dissolution-susceptible forms relative to the cohort of taxa that reach the K-T boundary. Results of this analysis are shown as histograms in Figure 10.

Contrary to predictions of the alternative hypothesis, the indigenous Cretaceous faunas of all four study successions are differentially composed of either unaffected or dissolution-resistant taxa, whereas those species whose

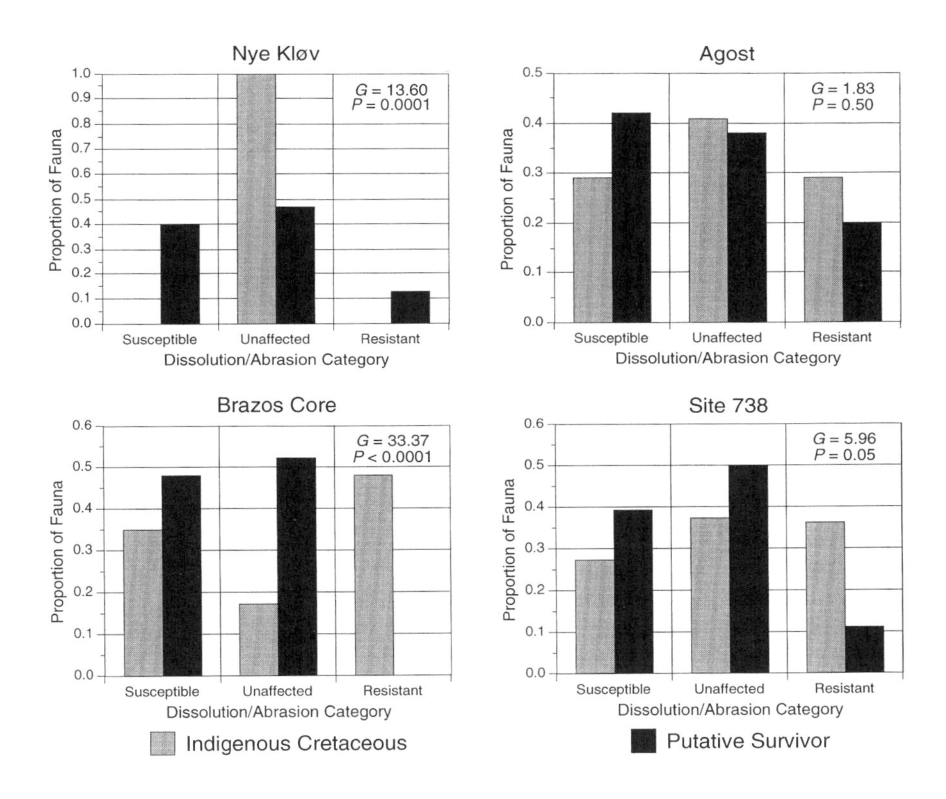

Figure 10. Proportionate frequency histograms for three dissolution and/or abrasion categories within the indigenous Cretaceous and putative survivor cohorts for each of the study sections and cores. G *statistics and accompanying probability estimates (*P*-values) indicate that the indigenous Cretaceous and putative survivor faunas at Nye Kløv and Brazos core represent samples drawn from populations whose faunas had a significantly different composition of dissolution-resistant and dissolution-susceptible species. Note that in all four sections the putative survivor fauna is differentially composed of dissolution-susceptible species.*

observed LADs reach or overlie the local K-T boundary contain a high proportion of dissolution-susceptible morphotypes. Thus, if differential preservation potential were responsible for producing a systematic bias with respect to the distributions of LADs throughout these sequences, we would expect to find precisely the opposite patterns that are observed. Statistical comparison of these distributions, via the likelihood ratio test (see Zar, 1974; Sokal and Rohlf, 1981), confirms this interpretation at the 95% confidence level for two out of the four analyses (Fig. 10). The associated confidence level for the Canudo and others (1991) Agost data is 0.50, indicating that there is a one in two chance of these two sample distributions being drawn from the same underlying population. Accordingly, the null hypothesis of no distinction between Cretaceous taxa exhibiting Maastrichtian LADs and those exhibiting K-T boundary or Danian LADs cannot be rejected for this sequence. With respect to the other three successions, however, there seems to be no reason to suspect that a dissolution-based Signor-Lipps effect has artificially biased their fossil records.

Relative Abundance–Sample Size Interactions

As noted above, species exhibiting low abundances can be underrepresented across a biostratigraphic dataset if large contrasts in sample size exist. Within the four study sections and/or cores every effort was made to keep the sample size as constant as possible. Variations in sample size do exist, however, and even though there does not appear to be a systematic decline in sample size in the uppermost Maastrichtian portions of these sequences, the potential for bias due to relative abundance–sample size interactions should be evaluated.

Koch and Morgan (1988, p. 126) proposed a probability-based procedure designed to carry out such an analysis: this test assumes that the true range of each species extends throughout the entire sampled interval, and employs observed sample sizes along with the pooled relative abundance of each species to determine the "average or expected number of species that will [appear to] range across any two specified [stratigraphic] levels." Koch and Morgan (1988, p. 126) recommended their method "to aid the investigator in sorting out real differences from sample size effects." Substantial agreement between expected and observed species range distributions constitutes evidence that observed biostratigraphic patterns have been influenced systematically by relative abundance–sample size interactions. Failure to detect significant levels of correspondence between expected and observed range distributions indicates a lack of systematic bias with respect to this factor.

Results of the Koch and Morgan test for the uppermost Maastrichtian intervals of Nye Kløv, Agost, Brazos core, and Site 738 are shown in Figures 11 and 12. The observed number of species ranges at each locality decreases as the boundary horizon is approached, suggesting either the operation of a Signor-Lipps effect or the onset of species extinctions within the uppermost Maastrichtian interval—well below any evidence for impact debris within these successions. Sample size fluctuates markedly in the intervals leading up to the boundary horizon. This leads to a predicted species range distribution containing a modest Signor-Lipps effect at Agost, Brazos core, and Site 738. The question that needs to be asked of these results is whether the observed patterns of species range dis-

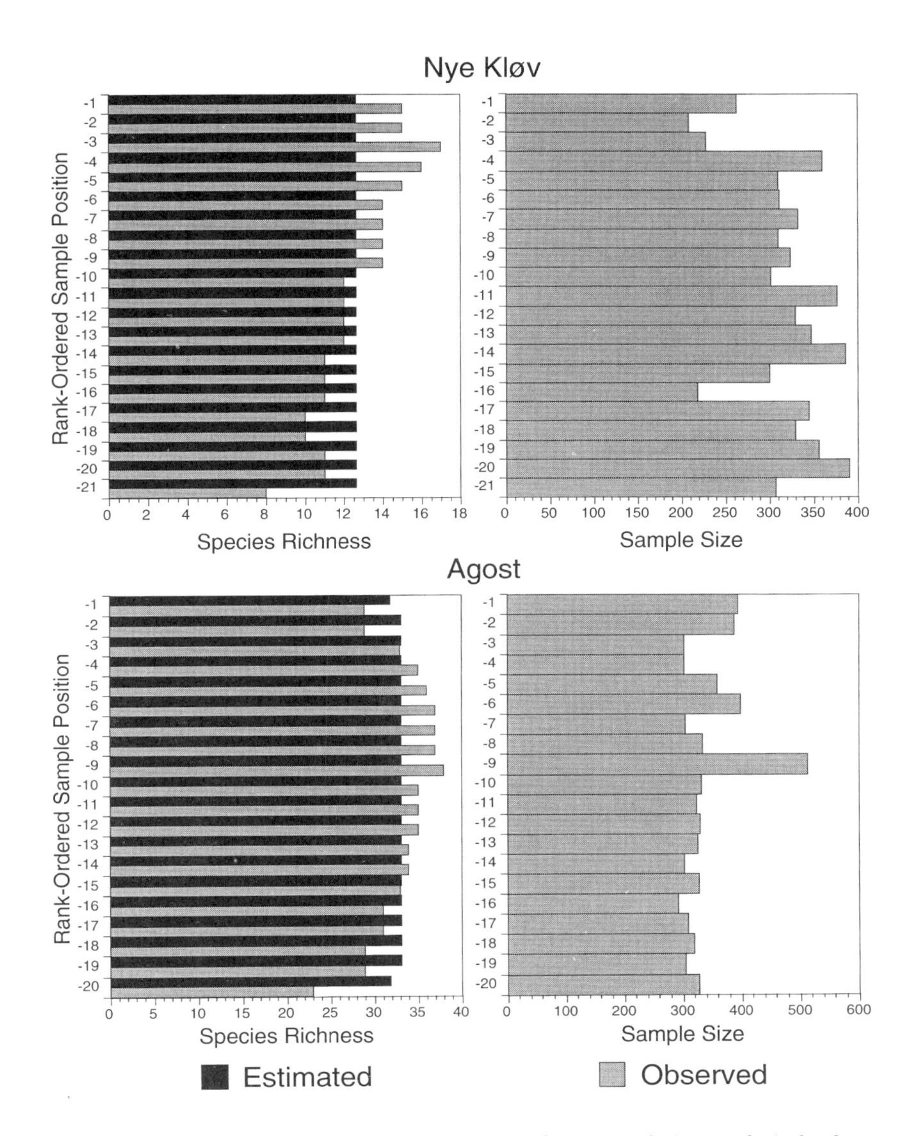

Figure 11. Results of a Koch and Morgan species richness simulation analysis for faunas composing the uppermost Cretaceous interval at Nye Kløv and Agost. Estimated species richness values represent the distribution of sample ranges expected (given existing relative abundance and sample size interactions) if all species ranged throughout the sampled interval. A Kolmogorov-Smirnov analysis of these results (see Table 2) shows that the observed distribution of species ranges at Nye Kløv are significantly (95%, two-sided) different from expectations in this model that the observed data cannot be regarded as conforming to the expectations of a measurable Signor-Lipps effect. Similar results for the Agost data, however, support the recognition of a Signor-Lipps effect within this interval. See text for discussion.

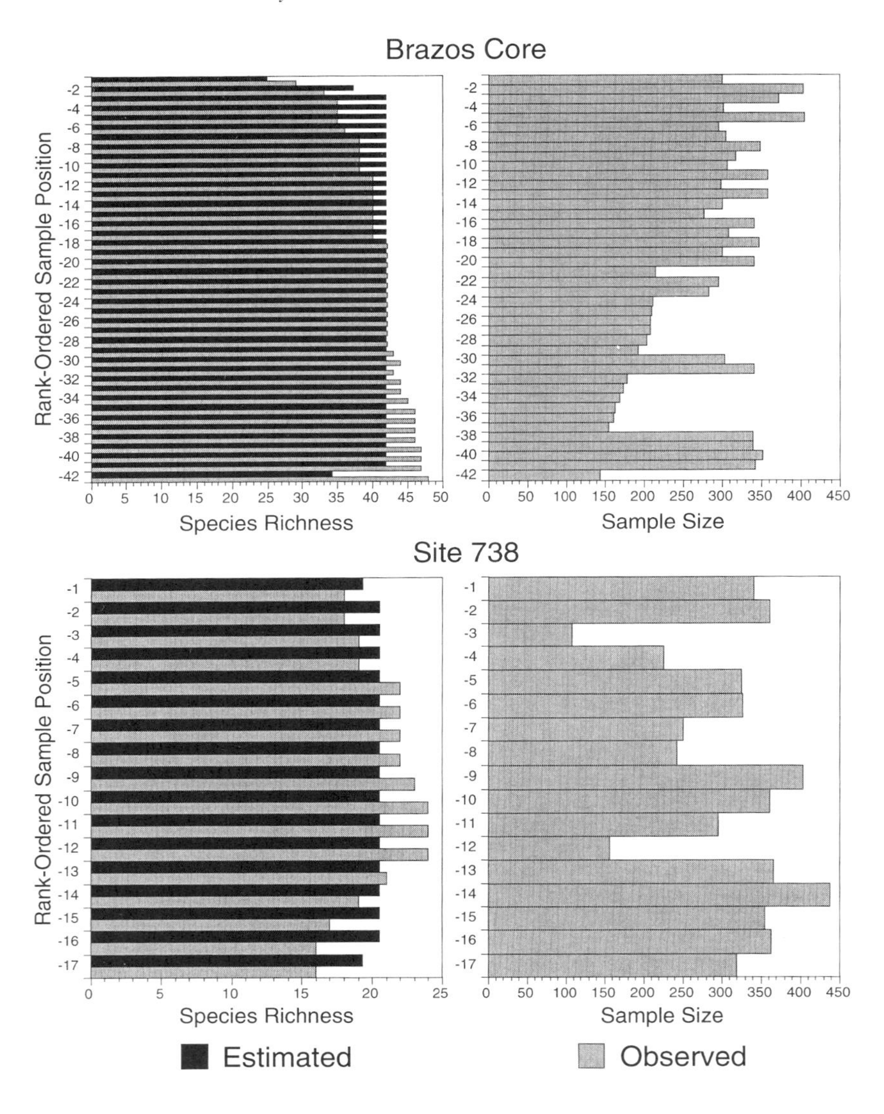

Figure 12. Results of a Koch and Morgan species richness simulation analysis for faunas composing the uppermost Cretaceous interval at Brazos core and Ocean Drilling Program (ODP) Site 738. Estimated species richness values represent the distribution of sample ranges expected (given existing relative abundance and sample size interactions) if all species ranged throughout the sampled interval. A Kolmogorov-Smirnov analysis of these results (see Table 2) shows that the observed distributions of species ranges at Brazos core are significantly (95%, two-sided) different from expectations in this model that the observed data cannot be regarded as conforming to the expectations of a measurable Signor-Lipps effect. Similar results for the Ocean Drilling Program Site 738 data, however, support the recognition of a Signor-Lipps effect within this interval. See text for discussion.

tribution differs significantly from the predicted pattern that assumes that all species range throughout the entire interval.

It is interesting that in the two previous applications of the Koch and Morgan procedure (Koch and Morgan, 1988; Koch, 1991), no attempt was made to evaluate statistically the correspondence between observed and predicted range distributions. This evaluation can be made by applying the two-sample Kolmogorov-Smirnov test (Zar, 1974; Sokal and Rohlf, 1981) to the species range data shown in Figures 11 and 12. Results of these tests (Table 2) show that the null hypothesis of no difference between predicted and observed range distributions for the Maastrichtian sample interval can be rejected for Nye Kløv and Brazos core at well above a 95% confidence level. The same test, however, fails to reject the null hypothesis for the Agost data of Canudo and others (1991) and Keller's (1993) ODP Site 738 data. This suggests that in these sequences a Signor-Lipps effect may have been developed as a result of relative abundance–sample size interactions.

The lack of statistically significant deviations from the Koch and Morgan predictions for the Agost and Site 738 data is interesting, but insufficient to demonstrate a definite association between the boundary horizon and the upper limit of species range co-occurrence. If species co-occurrences actually terminate at the K-T boundary, the zone of statistically significant correspondence between the Koch and Morgan predictions and the observed ranges of Cretaceous species should not encompass the lowermost Danian interval. (Note that this prediction assumes that these Danian occurrences are the result of survivorship rather than reworking, an interpretation that has been discussed for these data in numerous publications [e.g., Keller, 1989, 1993; Canudo and others, 1991; MacLeod and Keller, 1991, 1994; Schmitz and others, 1992; MacLeod, 1994, 1995a, 1995b, 1995c.) Results from a repetition of the Koch and Morgan–Kolmogorov-Smirnov test for Agost and Site 738 using "Cretaceous" species range data from the entire

TABLE 2

Results of Kolmogorov-Smirnov two-sample tests for differences between expected and observed species richness values based on the Koch and Morgan (1988) expected range simulation

Model	**D**	**Critical Value (95%, Two-Sided)**
Nye Kløv (Up. Maas.)	0.077	0.075
Agost (Up. Maas.)	0.029	0.048
Brazos Core (Up. Maas.)	0.033	0.029
Site 738 (Up. Maas.)	0.037	0.066
Agost (Up. Maas. + Zn. P0)	0.078	0.044
Site 738 (Up. Maas. + Zn. P0)	0.036	0.055

Note: Up. Maas. is Upper Maastrichtian, Zn. is zone, D is value of the Kolmogorov-Smirnov statistic.

Maastrichtian and lowermost Danian zone P0 intervals are shown in Figure 13. These results show that the Koch and Morgan predictions remain valid well into the lower Danian at Site 738 (thereby suggesting that the Signor-Lipps effect in this succession is not restricted to the K-T boundary), but are rejected for the Agost data. Taking the results obtained from all four sequences into consideration, the following inferences can be supported. First, fluctuations in the number of species ranges throughout the upper Maastrichtian intervals at Brazos core and Nye Kløv appear to be real and not the product of a Signor-Lipps effect based on relative abundance–sample size interactions. Second, fluctuations in the number of species ranges in both the upper Maastrichtian and Maastrichtian + lowermost Danian intervals at Site 738 are consistent with the predicted effects of relative abundance–sample size interactions. This is consistent with recognition of a Signor-Lipps effect operating within the Site 738 data. However, this effect does not appear to be linked to the K-T boundary horizon. Third, that there are fluctuations in the number of species ranges in the upper Maastrichtian, but not the Maastrichtian + lowermost Danian interval at Agost, is also consistent with a Signor-Lipps effect based upon relative abundance–sample size interactions. In this instance the Signor-Lipps effect does seem to be linked to the K-T boundary, suggesting that the LADs of Cretaceous taxa occurring within the upper Maastrichtian interval may be effectively regarded as synchronous with LADs occurring at the boundary horizon itself.

Coordinated Response to Habitat Shift

A final test of a potential systematic Signor-Lipps effect deals with the possibility of environmental changes excluding a component of the Cretaceous fauna from particular areas prior to their ultimate extinction. With the exception of the basal Danian clay layer that marks the boundary in these successions, no other lithostratigraphic contrast between uppermost Maastrichtian and lowermost Danian sediments has been reported. The lack of a lithostratigraphic change covarying with the faunal disappearance pattern offers little support for the habitat shift scenario. Nevertheless, systematic variation of many different types of important biotic factors would not be expected to leave a detailed lithostratigraphic record. A more powerful test of this hypothesis lies in the examination of species-specific occurrence patterns or relative abundance logs.

If different cohorts of planktonic foraminiferal species with contrasting environmental tolerances exist within the upper Maastrichtian planktonic foraminiferal faunas, these cohorts should exhibit a similar within-group response to environmental fluctuations in terms of either coordinated presence-absence, or relative abundance patterns. Accordingly, the predicted existence of these patterns can be used to identify groups of species with similar inferred environmental tolerances. This biofacies approach can be used to test the proposition that those forms that disappear below the boundary horizon appear to do so as a result of their differential response to subtle environmental changes not recorded in the physical or chemical lithostratigraphy.

For the four datasets under consideration, coordinated presence-absence patterns were used to identify species with similar response patterns to environmental fluctuation. Relative abundance can also be used and would perhaps be a more

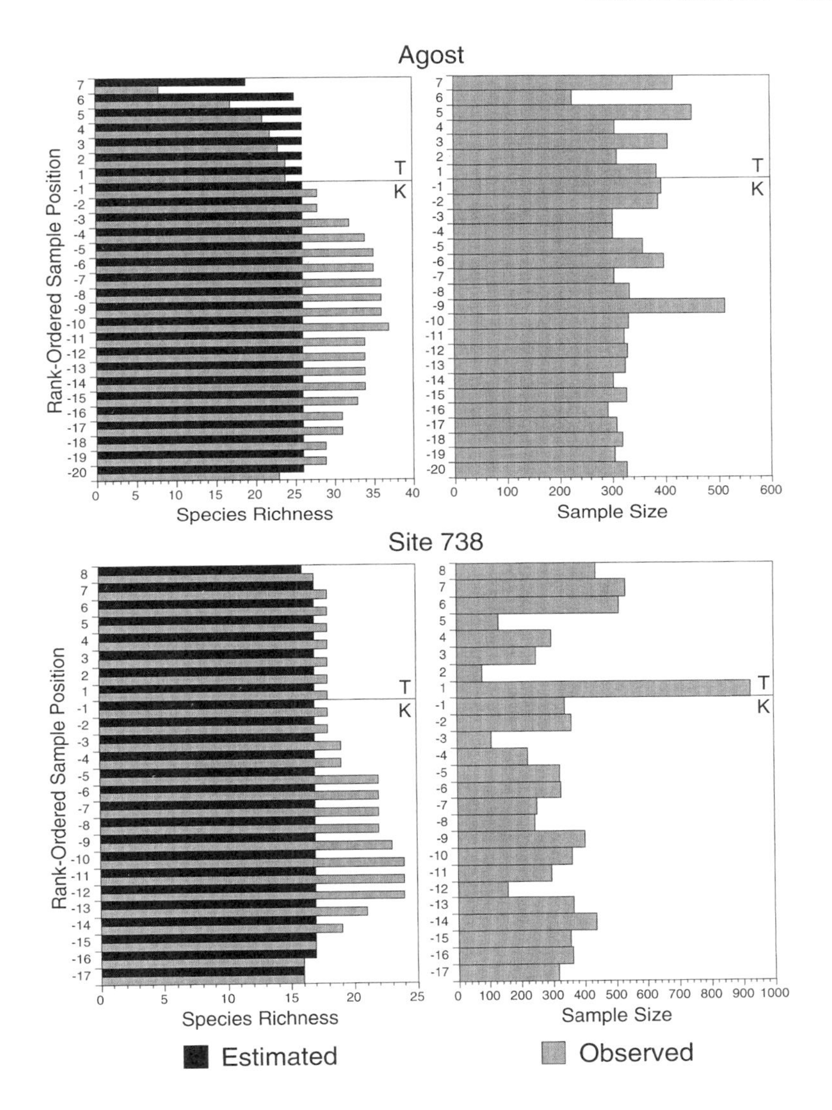

Figure 13. Results of an additional Koch and Morgan species richness simulation analysis for faunas composing the uppermost Cretaceous and lowermost Danian (zone P0) intervals at Agost and Ocean Drilling Program (ODP) Site 738. Estimated species richness values represent the distribution of sample ranges expected (given existing relative abundance and sample size interactions) if all species ranged throughout the entire Cretaceous and Danian sampled interval. A Kolmogorov-Smirnov analysis of these results (see Table 2) shows that the observed distribution of species ranges at Agost are significantly (95%, two-sided) different from expectations in this model that the observed data cannot be regarded as conforming to the expectations of a measurable Signor-Lipps effect. Similar results for the ODP Site 738 data, however, support the recognition of a Signor-Lipps effect within this entire interval. These results indicate that the Signor-Lipps effect recognized at Agost seems to be associated with the Cretaceous-Tertiary (K-T) boundary horizon, while that observed at ODP Site 738 extends throughout the lowermost Danian interval as well. See text for discussion.

sensitive indicator of ecologic similarity. However, the published data for these successions do not permit recalculation to species-specific relative abundance. This difficulty, along with the very patchy distribution of many species, makes a relative abundance approach unworkable for these successions. However, many appropriate analytic techniques are available for use with presence-absence data.

Interspecific similarities based on presence-absence observations for the upper Maastrichtian interval were quantified using the Euclidean distance metric with pairwise deletion of missing values (samples above the observed LAD) and summarized via multidimensional scaling (MDS). This approach is preferable to cluster analysis in that the interspecific similarities are not expected to be hierarchically organized and MDS represents a unified methodology with a small number of variant forms (as opposed to very large number of different clustering algorithms), thus enhancing the reproducibility of the results. In all instances a two-dimensional MDS solution accounted for more than 85% of the observed variation.

Once MDS coordinates were obtained, the data were subdivided into species disappearing prior to the K-T boundary and those whose observed biostratigraphic ranges reach the boundary or extend into the lower Danian. If this subdivision reflects a coordinated response to an environmental shift, a pronounced clustering between these two groups within the two-dimensional MDS space should be evident. Results are shown in Figure 14.

These plots fail to reproduce the patterns predicted by the coordinated response to habitat shift model. In all cases there is substantial overlap between the two faunas, suggesting broad overlap in occurrence patterns and, by inference, species-specific responses to environmental change. This suggests that in many cases, species whose stratigraphic ranges extend to the local K-T boundary exhibit upper Maastrichtian occurrence patterns that are indistinguishable from species disappearing from the record below the boundary horizon. The presence of a marked Signor-Lipps effect due to habitat shift is not corroborated by these data: however, even if it were, this type of Signor-Lipps effect is inherently indeterminate with respect to mass-extinction models because it only reveals that a group of taxa have tracked environmental change out of the local area. This test cannot address the comparatively more important question of whether these species survived a catastrophic environmental change later, that is, farther up in the section or core.

Tests for Survivorship

Throughout the previous discussion of the Signor-Lipps effect, the occurrence of traditional "Cretaceous" species in sediments overlying the K-T boundary in each of the four sections and cores has been ignored and these taxa have been treated as though their stratigraphic ranges extended to the boundary horizon and no higher. However, observational evidence for the consistent appearance of these species in the lowermost Danian sediments from all temporally complete, and many incomplete, sections and cores is overwhelming (see MacLeod and Keller, 1994, for a review). In several sections these anomalous faunas have been observed to rise many meters into the Danian (e.g., Liu and Olsson, 1992). Although the interpretation of these occurrences differs among various planktonic foraminiferal biostratigraphers, their existence can no longer be denied.

Two alternative interpretations are available to account for these occurrences.

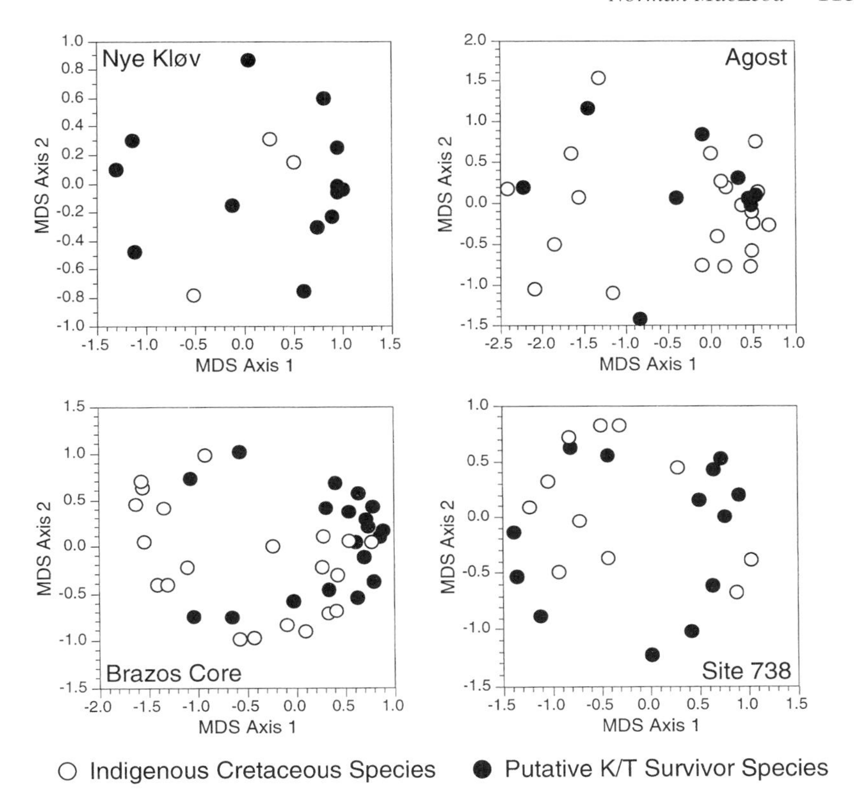

Figure 14. Metric multidimensional scaling results for an occurrence pattern analysis of species occurring within the uppermost Maastrichtian intervals of the four study sections and cores. The lack of a clear separation between those species restricted to Cretaceous strata and those that go on to occur in Danian strata suggests that two faunas were responding to upper Maastrichtian environmental changes in much the same way. These results offer no empirical support for the invocation of a Signor-Lipps effect to due habitat shift to account for the pre-boundary disappearance of Cretaceous species

Either they represent the reworking of indigenous Cretaceous species from lower horizons in each local sequence or they represent the survivorship of substantial faunas for as much as 200,000 years into the Tertiary (see MacLeod and Keller, 1991, 1994; MacLeod, 1995a, 1995b, 1995c). The problem is how to devise tests for these alternatives that can be applied to the data of the fossil record. It is insufficient to assume that one interpretation or the other is correct on the basis of extra-paleontological data (e.g., identification of geochemical anomalies) because, while these data may be sufficient to identify the presence of an environmental event in the stratigraphic record, they fail to address the fundamental question of the nature of biotic response to that event. Similarly, it is inadequate to demand that others provide reasons why a particular interpretation, for which there is no positive evidence, should not be accepted because this line of reasoning cannot separate uncorroborated from untestable hypotheses. As always, the burden of proof must fall on those who seek to advance deterministic explanations for any natural phenomenon.

It is interesting to note that in routine planktonic foraminiferal biostratigraphy, indeed in the biostratigraphy of any fossil group, the consistent occurrence of well-preserved specimens stratigraphically above a particular horizon is universally regarded as sufficient evidence for the actual existence of those organisms at that time and in that locality. If such were not the case, biostratigraphy would be impossible. Nevertheless, the paradigm of an instantaneous K-T planktonic foraminiferal mass extinction is so deeply ingrained in the technical and popular literature that the adequacy of this widely used principle has been questioned when applied to the interpretation of these anomalous Danian faunas. Fortunately, other tests are available (see MacLeod, 1994).

Comparative Analysis of Species-Specific Stable Isotopic Logs

Barrera and Keller (1990) found that Danian occurrences of the formerly Cretaceous species *Heterohelix globulosa* exhibited different $\delta^{18}O$ and $\delta^{13}C$ isotopic ratios than underlying Cretaceous occurrences of the same species in the Brazos core. Olsson and Liu (1993) questioned the original interpretation of these data, and speculated that the Danian deviations in isotopic ratios could have been induced by diagenesis or by nonequilibrium isotopic fractionation during this species' ontogeny. (Note that Danian populations exhibit smaller test sizes than their Cretaceous counterparts, which Olsson and Liu [1993] interpreted to represent a collection of juveniles; see MacLeod and Keller, 1990.) However, MacLeod and Keller (1994) showed that the isotopic patterns predicted by these alternative hypotheses were not present in the Brazos core *H. globulosa* data. Since the original 1991 study many more survivor species have been recognized on the basis of species-specific K-T stable isotopic logs from Nye Kløv (Keller and others, 1993) and ODP Site 738 (Barrera and Huber, 1994). This method remains the most powerful and unambiguous indicator of K-T survivorship.

Preservational State

The state of test preservation is often used by planktonic foraminiferal micropaleontologists to infer the presence of reworked specimens in their samples. Obviously worn or abraded tests with holes in the test walls, along with specimens exhibiting discolorations of the test wall relative to the majority of specimens in the sample, are often regarded as reworked. Results detailing dissolution and/or abrasion susceptibility contrasts between indigenous Cretaceous and putative survivor populations for the four study sections (see also MacLeod and Keller, 1994) are pertinent to this test. In all sections and cores thus far investigated the putative survivor fauna has been made up of a greater proportion of species susceptible to dissolution and/or abrasion than the underlying indigenous Cretaceous faunas. Thus, "Cretaceous" species commonly found above the K-T boundary would be more likely to record the effects of reworking than underlying indigenous Cretaceous forms.

The K-T planktonic foraminiferal literature contains many references to anomalous occurrences of Cretaceous species (not faunas) in Danian sediments that are anecdotally accounted for via reference to degraded preservational state.

Nevertheless, prior to Keller (1988a, 1989), no explicit comparative data on the state of preservation within the same species from Maastrichtian and Danian samples were made available. Keller's studies show fully comparable states of preservation for specimens representing these two faunas at El Kef and Brazos core. Although some faunas (e.g., Brazos core) contain subpopulations of differently colored specimens, the preservational state exhibited by "Cretaceous" species occurring in Danian strata is often remarkably high and always fully comparable with underlying indigenous Cretaceous occurrences. This observation was confirmed by Olsson and Liu (1993), who studied similar faunas from the Miller's Ferry K-T boundary section.

Relative Abundance/Population Ratio

Olsson and Liu (1993, p. 136) explained their relative abundance/population ratio test for mass-extinction survivorship as follows: "If one species survived [the K-T boundary] its population size in the lower Paleocene would be composed of both the reworked and indigenous surviving fractions while that of the extinct species consists only of the reworked fraction. As a result, the relative abundance of the survivor taxon to that of the extinct species would significantly increase after the extinction of the other species." The logic of the relative abundance/population ratio test is outlined diagrammatically in Figure 15.

Applying this test to the most abundant species in the four study sections and/or cores (Fig. 16), it can be shown that for these fossil records the predicted patterns are not observed for either accepted survivor species (*Guembelitria cre-*

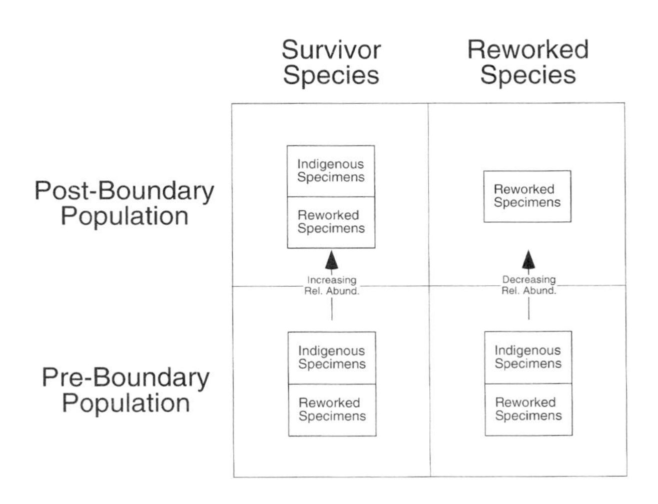

Figure 15. Conceptual basis for the Olsson and Liu (1993) relative abundance/population ratio test for Cretaceous-Tertiary survivorship. Owing to the catastrophic elimination of the indigenous faunal component, surviving species should undergo a dramatic relative abundance increase and reworked species should undergo a similarly dramatic relative abundance decrease coincident with the mass-extinction horizon. See text for discussion.

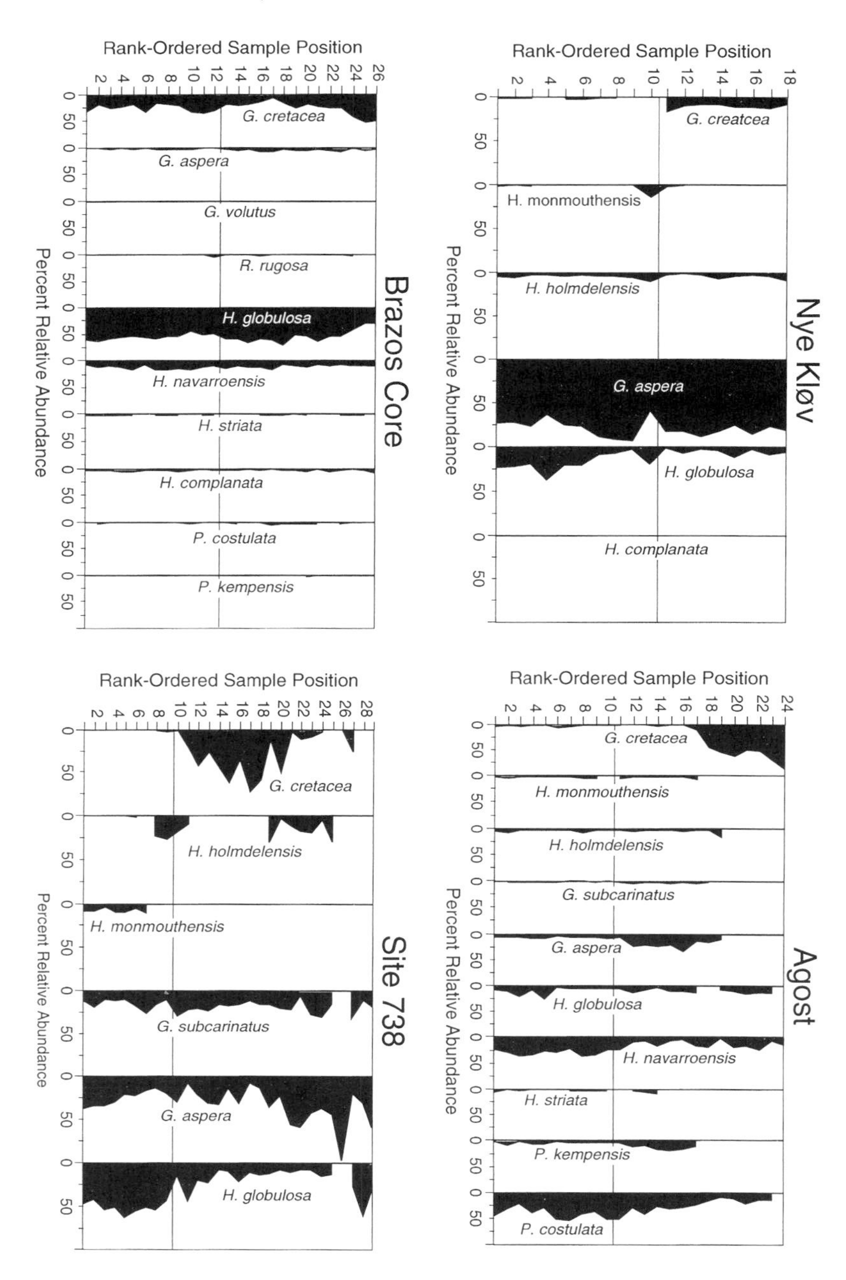

Figure 16. Results of a relative abundance-population ratio for the most abundant species at each of the four study sections and cores. These data show that the patterns predicted by Olsson and Liu (1993) are not observed for either widely recognized survivor species (e.g., G. cretacea, H. holmdelensis, H. monmouthensis*) or for more controversial survivor species. Accordingly, the power of this test for identifying unambiguously Cretaceous-Tertiary survivors must be regarded as very low.*

tacea, Hedbergella holmdelensis, H. monmouthensis) or for putative survivor species. In all instances the relative abundance record of *G. cretacea, H. holmdelensis,* and *H. monmouthensis* fail to exhibit the sharp rise in relative abundance predicted for survivor taxa. Similarly, the putative survivors *Globigerinelloides aspera, Heterohelix globulosa*, and *H. complanata* fail to exhibit the sharp decreases in relative abundance above the boundary horizon predicted by Olsson and Liu (1993). These relative abundance data are most parsimoniously interpreted as reflecting the long-term operation of environmental processes that are driving a progressive turnover in taxic dominance patterns among the most abundant morphotypes while leaving the less-abundant taxa more or less unaffected. Whereas this turnover interval appears to straddle the boundary horizon in each section, most taxa do not exhibit an anomalous excursion in their relative abundance logs at the boundary horizon itself. In addition, these data show clearly that relative abundance fluctuations for these species are diachronous among the four study sequences, and therefore cannot be used to support any interpretation involving globally coordinated biotic response to a single event.

Estimating and Scaling a Global Turnover Sequence

With the results detailed above (Table 3), we can now make an objective decision as to which of these four fossil records contain evidence of systematic bias resulting from a wide range of potential sources and which can be used confidently to construct and temporally scale a global K-T sequence of events. As shown above, stratigraphic confidence intervals are largely irrelevant to arguments arising from the observation of Cretaceous taxa disappearing from local fossil records below and above the boundary horizon. Despite the fact that the temporal resolution of these data is unmatched for any other paleontologic dataset across any major stratigraphic boundary, the partial fossil ranges recorded therein are unable to provide sufficient constraints on the estimated LAD to preclude a very large number of alternative interpretations. If the biostratigraphic records of these species were sampled at comparable levels of detail throughout their entire range, the number of occurrences would most likely result in the estimated confidence

TABLE 3
Summary of Analytic Results

Source of Potential Bias	Nye Kløv	Agost	Brazos Core	Site 738
Signor-Lipps Effect				
Preservation Potential	No	Yes	No	No
Rel. Abundance/Sample Size	No	Yes	No	Yes
Habitat Shift	No	No	No	No
Reworking Above Boundary				
Stable Isotopes	No (some species)	?	No (some species)	?
Preservational State	No	No	No	No
Rel. Abundance/Pop. Ratio	No	No	No	No

interval being practically indistinguishable from the observed LAD. Of course, this would not be true for many invertebrate taxa, the fossil records of which are typically much more discontinuous than that of planktonic foraminifera. Nevertheless, it is interesting to note that, in terms of the uncertainty introduced by gaps in the distribution of microfossil species, the standard micropaleontologic practice of regarding species' FAD and LAD horizons as robust indicators of deterministically interpretable biotic patterns seems fully justified.

With respect to results of the Signor-Lipps and reworking and/or survivorship tests described above (Table 3), there seems ample reason to suspect that the biostratigraphic record of Agost (Canudo and others, 1991) and ODP Site 738 (Keller, 1993) may have been systematically biased by either diagenetic factors and/or relative abundance–sample size interactions. At least the null hypothesis that the observed biotic patterns are inconsistent with these interpretations cannot be rejected at the standard level of statistical confidence. This does not mean that the fossil records of these two sequences are consistent with a catastrophic faunal turnover at the K-T boundary. As pointed out by Signor and Lipps (1982), recognition of a Signor-Lipps effect simply means that the LAD pattern cannot be interpreted with confidence. Moreover, these two sequences contain abundant faunas of "Cretaceous" species within their lowermost Danian intervals that fail to conform to various predictions of the reworking hypothesis. However, in order to provide the most conservative estimate of K-T biotic turnover possible, the entire planktonic foraminiferal fossil record from these two sequences will be eliminated from consideration in estimating a global turnover sequence. This leaves the observed fossil records from Nye Kløv, Brazos core, and El Kef (which was found to be unbiased by either the Signor-Lipps effect or reworking in a previous study; MacLeod, 1995c) available to form the basis of a global turnover estimate.

Graphic correlation diagrams for Brazos core and Nye Kløv plotted against a K-T composite sequence (K-T–CS) based on new El Kef planktonic foraminiferal data (Keller and others, 1995) representing a synthesis of FAD and LAD positions within all three successions are presented in Figure 17. Correlation models for these two sections indicate that the K-T boundary resides within an interval of condensed sediment accumulation relative to the lowermost and uppermost portions of the study interval. Although the interval contains the K-T boundary, there is no evidence in these data that a boundary event caused the drop in sediment accumulation in these two inner neritic localities. This drop in the relative sediment accumulation rate in nearshore settings is consistent, however, with the expected effects of the well-documented eustatic sea-level rise that began in the uppermost Maastrichtian and continued throughout the first 100 ka of the Danian (see Haq and others, 1987; Brinkhuis and Zachariasse, 1988, 1990; Haq, 1991).

Figure 18 illustrates the estimated overall pattern of faunal turnover based on the planktonic foraminiferal records of El Kef, Brazos core, and Nye Kløv. As can be seen readily from these data, extinction rates for this organismal group underwent a dramatic increase in the uppermost Maastrichtian—well below the first evidence for impact debris in any of these three successions. Once it had begun, the faunal turnover continued at a fairly continuous rate throughout the 2 m of composite section bracketing the boundary horizon. The boundary horizon itself is located nearer to the beginning of the turnover interval, but not at its beginning.

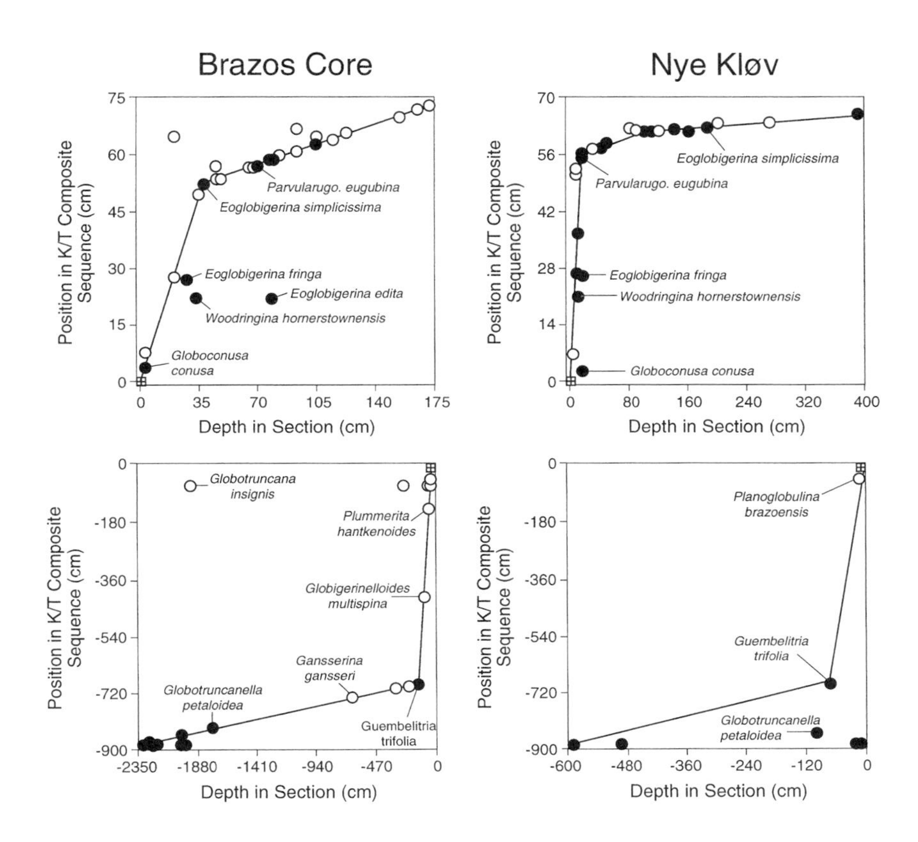

Figure 17. Graphic correlation plots for the Cretaceous-Tertiary (K-T) intervals of Brazos core and Nye Kløv against a composite sequence based on these two successions and the El Kef boundary stratotype sequence. Note that the boundary stratotype served as the composite standard. The lowermost Danian portion of these sequences is shown in the upper row of plots and the uppermost Maastrichtian interval is shown in the lower row of plots. The position of the K-T boundary horizon is show at level 0 along both axes of all plots and is represented within the coordinate system by a square box containing a cross. Relative to the composite sequence, both Brazos core and Nye Kløv exhibit condensed relative sediment accumulation rates within the uppermost Maastrichtian and lowermost Danian intervals. This interval of condensed sediment accumulation is the expected result of a eustatic sea-level rise that took place at this time (see text). This interval is also coincident with a pronounced $\delta^{13}C$ excursion and with the planktonic foraminiferal faunal turnover. It does not coincide, however, with the first occurrence of impact debris in any of these successions. These data suggest that environmental changes associated with a rapid eustatic sea-level rise and accompanying volcano-tectonic activity were both underway and driving the early stages of a planktonic foraminiferal faunal turnover hundreds of thousands of years before the occurrence of any impact debris.

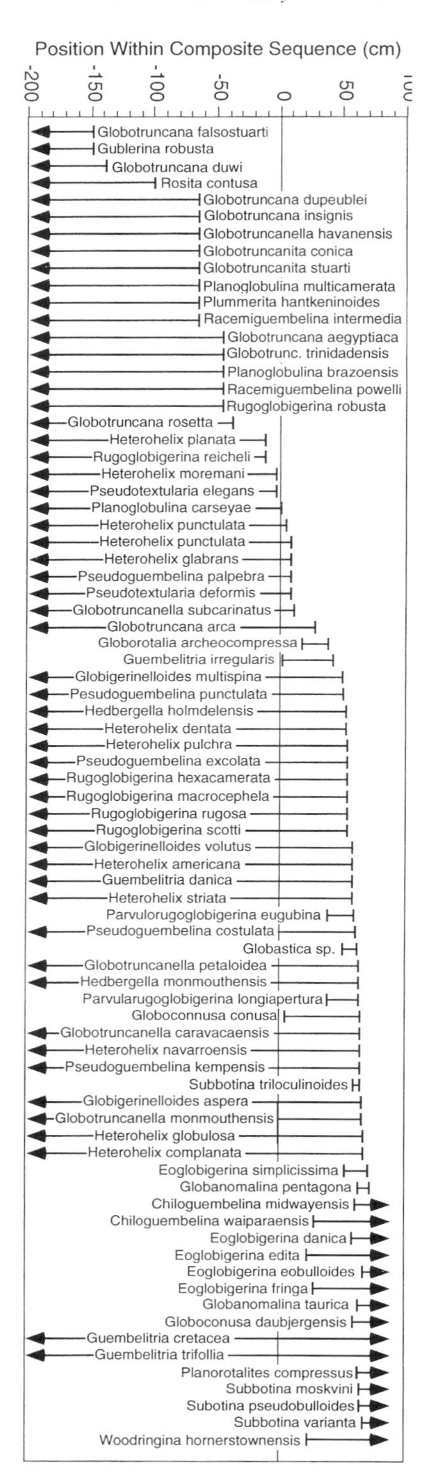

Figure 18. A composite sequence of 86 planktonic foraminiferal datums constructed by summarizing the biostratigraphic information present in the El Kef, Nye Kløv, and Brazos core sequences, none of which have been found to contain a significant Signor-Lipps effect or to provide data supporting the recognition of substantial reworking of Cretaceous faunas into lowermost Danian sediments. As with most vertebrate and invertebrate data, the Cretaceous-Tertiary planktonic foraminiferal turnover appears to have taken place over an extended stratigraphic interval, and not to have been concentrated at horizons bearing impact debris. Whereas this pattern of faunal turnover is consistent with a wide variety of causal models, it is not consistent with published predictions of the Alvarez impact-extinction scenario.

Of the 54 species occurring in the uppermost meter of Cretaceous sediments in this summary, 35 species (representing 65% of the fauna) are found in the lowermost Danian sediments above the boundary horizon. Nevertheless, within the space of this 2 m interval, which encompasses the uppermost *P. deformis* zone (Maastrichtian), all of zone P0 (Danian), and the lowermost portion of zone P1a (Danian), virtually all of the well-developed and diverse uppermost Cretaceous planktonic foraminiferal faunas are replaced by fully Tertiary species of distinct morphological character and phylogenetic ancestry. Given current chronostratigraphic estimates for the duration of these biozones (see Harland and others, 1989; MacLeod and Keller, 1991; Cande and Kent, 1992), this interval represents no less than 500 ka and probably no more than 1.5 m.y.

Morphologic Patters of Extinction and Survivorship

The most far-reaching implications of the impact-extinction scenario lies in the realm of evolutionary theory. If, as Gould (1984, 1985), Raup (1986, 1991), Jablonski (1986a, 1986b, 1986c, 1989, 1994) and others claim, the five or so widely recognized mass-extinction events represent fundamental discontinuities in the fabric of natural selection, then mass extinction can be seen as constituting a unique evolutionary process. However, recognition of such a hierarchy of selection levels must be based upon the predictions that are confirmed through empirical observation.

Jablonski (1986a, 1986b, 1986c) proposed an ecological distinction between the nature of selection pressures characterizing background and mass-extinction events in K-T molluscs by comparing observed survivorship levels in different life-history guilds as inferred from morphologic evidence. He concluded that because mollusc groups with planktotrophic larvae (large protoconch) were differentially resistant to extinction during background events but suffered the same rate of extinction susceptibility as nonplanktotrophic groups (small protoconch) during the K-T mass extinction, the nature of selection pressures during these two types of biotic crises was profoundly different.

Although planktonic foraminifera cannot be separated into different life-history guilds in the same way as molluscs, we can use the concept behind Jablonski's test to ask whether there is a morphologic signal in the pattern of extinction and survivorship through the K-T faunal turnover. If so, Jablonski's model predicts that the nature of species sorting during a mass extinction should differ from the type of species sorting characteristic of background extinction events.

Stanley and others (1988) surveyed the Neogene history of the globigerinid and globorotalid planktonic foraminiferal clades and concluded that during this interval members of the globigerinid clade were significantly less prone to extinction than members of the globorotalid clade. There are several marked ecological differences between these two clades, and Stanley and others (1988) concluded that the shorter duration of globorotalid species was the result of its characteristic occupation of small, patchy, deeper-water habitats at the base of the thermocline, a scavenging lifestyle, and/or unstable population sizes. Because the Neogene interval investigated by Stanley and others (1988) contains no recognized mass-extinction events, this pattern can be regarded as typical of planktonic foraminiferal background extinctions.

Figure 19 shows the relative frequency of heterohelicid (= tests with serially

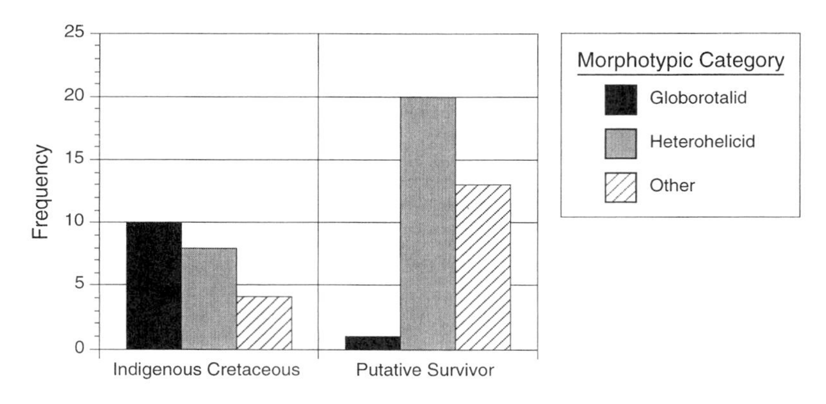

Figure 19. Analysis of the morphotypic composition of indigenous Cretaceous and putative survivor populations within the composite sequence shown in Figure 18. The Cretaceous-Tertiary (K-T) biotic event seems to have affected globorotalid species early (prior to the K-T boundary horizon), and heterohelicid, globigeriniform, and planispiral morphologies much later (during the lowermost Danian). This morphotypic structuring of a major planktonic foraminiferal turnover is consistent with the patterns observed by Stanley and others (1988) for Neogene background extinction events. See text for discussion.

arranged chambers), globorotalid (= flattened trochospiral tests with peripheral keels), and "other" (= trochospiral and planispiral globigerinacid) morphotypes for indigenous Cretaceous and putative survivor species in the composite sequence (see Fig. 18). These data confirm that at each site the relative percentage of generalized globorotalid morphotypes declined dramatically while the relative proportion of heterohelicid morphotypes rose. An X^2 analysis of the 2 x 2 contingency table for globigerinid and globorotalid morphotypes within these two occurrence categories yields a value of 28.33, which is highly significant ($P << 0.0001$). A similar X^2 analysis that considers the data from all three categories results in a value of 19.34, which is also significant. Thus, patterns of taxonomic and presumed ecologic variation in planktonic foraminiferal extinction susceptibility in a composite sequence constructed out of the most reliable data currently available are consistent with patterns documented from a Neogene interval characterized by background extinction intensities. This characteristic extinction susceptibility pattern is also consistent with similar data summaries for the global K-T fauna (MacLeod and Keller, 1994; MacLeod, 1995c).

DISCUSSION

No student of the fossil record can avoid the conclusion that a profound faunal turnover occurred in some (but certainly not all) organismal groups at the close of the Mesozoic. Had this event not taken place, it is difficult to imagine what the biotic world might look like today. However, if any one of an astronomically large number of events had not taken place in the 65 m.y. since the K-T event, it is difficult to imagine what the world might look like today. This is simply a restatement of the historically contingent nature of evolutionary processes.

No student of the K-T biotic record could also fail to be impressed by the speed at which this transition apparently took place. Compared to the Eocene-Oligocene (E-O) turnover (see Prothero, 1994) or the Permian-Triassic (P-Tr) event (see Erwin, 1993), the K-T event was completed on a time scale that was, even by the most conservative estimates, several orders of magnitude shorter. However, due to the small number of comparably sized events (the E-O event was much smaller and the P-Tr event was much larger than the K-T event), it is currently unknown whether the K-T event represents an anomalously short time interval for major environmental change. Much confusion has been generated by those who claim that most planktonic foraminifera became extinct abruptly at the end of the Cretaceous when they fail to explain that by the word "abruptly" they are referring to the lithostratigraphic distance over which the turnover takes place. Seen from the vantage point of the entire Cretaceous interval, encompassing many hundreds of meters of section in some localities, that these terminal Cretaceous extinctions occur over several meters appears abrupt. Nevertheless, these extinctions occur over a measurable stratigraphic interval, the temporal significance of which can be estimated by means independent of the impact-extinction scenario. When this analysis is carried out, it is clear that the K-T planktonic foraminiferal turnover event did not take place in hours, days, or even hundreds of years. Rather, it extended over an interval that predates and postdates any proposed impact debris horizon by many hundreds of thousands of years.

Given these data, and armed with the recently gained appreciation for the role that extraterrestrial impacts may have played in Earth history, it is very tempting to link the evidence for bolide impact at the K-T boundary and the existence of this faunal turnover in a cause and effect couplet. Not only would this linkage seem to account for one of the long-standing mysteries of the fossil record (the cause of the dinosaur extinction in particular and mass extinctions in general), but it is often argued that acceptance of a link between impacts and extinctions would help break the uniformitarian mind set of modern geology, and provide another, albeit very dramatic, example of the punctuated mode of evolutionary change (e.g., Gould, 1984, 1985; Raup, 1986, 1989a, 1991; Hsü, 1983, 1989; Marvin, 1990). Moreover, the proposed link between impacts and extinctions is in perfect accord with the reductionist tradition of the physical sciences (see Mayr, 1982), and its sensationalistic aspects have captured enormous and largely favorable attention from the scientific and popular media. It is therefore fair to ask why this proposed linkage remains controversial, especially among paleontologists.

The simple answer to this question is that, after the philosophical rhetoric is set aside, the proposal of a causal link between impacts and the K-T mass extinction remains empirically based on the related, but very different, question of whether a bolide impact occurred at or near the K-T boundary. The answer to this question appears to be yes, although disagreements remain as to the number, size, composition, and site of the impacting body or bodies (see Silver and Schultz, 1982; Sharpton and Ward, 1990; Ryder and others, in prep.). Many studies have been published that appear to show abrupt ecologic changes, faunal turnovers, and/or catastrophic extinction patterns for one or another organismal group at horizons interpreted to correspond to the K-T boundary. However, when these studies are examined in detail they are almost invariably compromised by several serious interpretational problems, including failure to demonstrate that the study section or core contains a temporally complete record across the K-T interval (e.g., Ward, 1990; Khunt and Kaminski, 1993), failure to observed last appearances of extinct taxa coinciding with the boundary horizon (e.g., Ward, 1990; Ward and others, 1991; Sheehan and others, 1991), and failure to report the observation of Cretaceous species in lowermost Danian sediments (e.g., Smit, 1981, 1982, 1990; Smit and others, 1988; see Brinkhuis and Zachariasse, 1988; Keller, 1988a; Canudo and others, 1991, for alternative biostratigraphies). In the absence of such data, the interpretation of these patterns must be regarded as ambiguous (see below). It is more important, though, that there now exists an extensive paleontological literature (see Introduction) demonstrating both the disappearance of many taxa previously thought to extend to the boundary in Upper Cretaceous sediment *and* the occurrence of many Cretaceous taxa previously thought to disappear at the boundary in lower Danian sediments. With respect to testing the predictions of the impact-extinction scenario, the significance of these observations cannot be overestimated.

The disappearance of Cretaceous species prior to the boundary horizon might be explained by the Signor-Lipps effect, and their occurrence in lowermost Danian strata might be explained by reworking (e.g., Kerr, 1994). However, the blanket invocation of these factors to dismiss any and all paleontological data deemed inconsistent with the impact-extinction scenario without any corroborating evidence, in effect dismisses the entire fossil record from having any role to play in the testing of a scenario that purports to account for one of that record's most outstanding features. The positions of LADs and FADs for fossil taxa—especially microfossils—have a well-established and well-deserved record of being able to be located accurately and interpreted deterministically in a wide variety of stratigraphic and paleobiological contexts. How reasonable is it to suppose that these data cannot be relied upon when it comes to testing the impact-extinction scenario?

Perhaps one of the most ironic features of this controversy is that even when the impact-extinction scenario is accepted, very little is explained with respect to the biotic structure of the K-T event or of mass extinctions in general. The primary biotic predictions of the original Alvarez and others (1980) study was that the extinction event should be geologically instantaneous (less than 50 years in duration; see also Alvarez and others, 1982, 1994; Alvarez and Asaro, 1990) and should involve a large number of ecologically distinct clades that were not already undergoing a diversity decline in the uppermost Maastrichtian. Biotic predictions for marine species offered in the Alvarez and others (1980, p. 1106) report are as follows:

> A temporary absence of sunlight would effectively shut off photosynthesis and thus attack food chains at their origins. ... The food chain in the open ocean is based on microscopic floating plants such as the coccolith-producing algae which show a nearly complete extinction. The animals at successively higher levels in the

> food chain were also very strongly affected, with nearly total extinction of the [planktic] foraminifera and complete disappearance of the belemnites, ammonites, and marine reptiles.

It goes on to note,

> The situation among shallow marine bottom-dwelling invertebrates is less clear; some groups became extinct and others survived. A possible base for a temporary food chain in this environment is nutrients originating from decaying land plants and animals brought by rivers to the shallow marine waters.

The global LAD of belemnites, ammonites, and marine reptiles (along with those of rudistid and inoceramid bivalves) have not been observed to coincide with the K-T boundary in any complete marine succession investigated thus far. As shown by the data presented above, the patterns of LADs for Cretaceous planktonic foraminiferal species in the Brazos River and Nye Kløv sequences do not support an interpretation of catastrophic extinction at a horizon coinciding with the proposed emplacement of impact debris in the sediments. In addition, there is no evidence that the local fossil records of these sections have been biased by either a Signor-Lipps effect or by the reworking of Cretaceous species into lowermost Danian sediments. The observed coccolith record is essentially the same, many Cretaceous species being found in strata that overlie the boundary and impact-debris horizon (see Gartner, this volume). In addition, there is no hint of the predicted grading of extinction intensity through the different trophic levels of the uppermost Cretaceous marine planktonic or benthic fauna.

A large negative excursion in $\delta^{13}C$ values coinciding with the K-T boundary in deep-sea cores is often cited as evidence for a catastrophic collapse of planktonic marine productivity at the K-T boundary (see Hsü and others, 1982; Zachos and Arthur, 1986; Zachos and others, 1989; Hsü and McKenzie, 1990; Hsü, 1994). However, similar stable isotopic analyses of more complete neritic marine successions (e.g., Barrera and Keller, 1990; Schmitz and others, 1992; MacLeod and Keller, 1994) have shown that this $\delta^{13}C$ isotopic excursion begins in the uppermost Maastrichtian—well below any proposed evidence for impact debris—and continues throughout the lowermost Danian interval. This excursion coincides with the planktonic foraminiferal turnover documented above and exhibits a close match to the K-T eustatic sea-level excursion. Although the K-T impact horizon is embedded within the $\delta^{13}C$ isotopic excursion, it does not occupy a position within the excursion pattern that would support a cause and effect interpretation.

With respect to the neritic marine benthos, it is apparent that the authors of the original impact scenario were disturbed by their failure to find an extinction pattern for these organisms that would match their scenario. In fact, the evidence for an extended period of faunal turnover accompanied by a progressive change in the paleoecology of faunas through the K-T interval is, if anything, clearer for benthic marine invertebrates than it is for planktonic marine invertebrates (see Kauffman, 1984). In addition, that Cretaceous (or indeed modern) marine invertebrate faunas are capable of switching to nutrient sources derived from decaying plant and animal matter remains undemonstrated.

Implicitly acknowledging the existence of biotic evidence that contradicts predictions of the original impact-extinction model, Hsü (1989), Hsü and McKenzie 1990), and Ryder (1994) have argued that even though the K-T faunal turnover was dominantly driven by impact-induced environmental changes, the characteristic short-term signature of this event might be smeared out by (unspecified) processes over such a long temporal interval as to appear indistinguishable from a progressive faunal turnover. Alternatively, Hut and others (1987) argued that mass-extinction events are spread out over stratigraphically measurable intervals because they are driven by multiple impacts (= comet showers). Both of these arguments constitute slight variations on the original Alvarez and others (1980) impact-extinction model. However, in order to be useful in explaining the K-T biotic event, these variants must make predictions for the fossil record that can in principle be, and ultimately are, corroborated by paleontological studies. Neither of these variations accomplish this task.

These variations on the theme of the original impact-extinction scenario deny the possibility of being able to separate impact-driven mass extinctions from non-impact-driven mass extinctions. The comet-shower variant makes no testable biotic predictions and fails to provide any independent support (other than the existence of a progressive extinction event) for the presence of multiple impact horizons within the K-T interval. In addition, many predictions of dire events following bolide impact (e.g., acid rain, "nuclear" winter, global wildfires) have been proposed without any corresponding evaluation of the K-T biotic record to determine whether there is any evidence for their existence in either the lithostratigraphic or biostratigraphic records (see Archibald, this volume, for an initial attempt).

If the scientific community is mystified and frustrated by paleontologist's failure to accept the impact-extinction scenario as the most plausible explanation for mass extinctions, the paleontological community is no less frustrated by seeing its data either ignored or misrepresented. The paleontological community has largely accepted the evidence for some type of bolide impact at or near the end of the Cretaceous. However, further progress on the question establishing a causal link between this impact and organismal extinctions is unlikely to be made until the physical-science community (hopefully in conjunction with paleontologists and biologists) develops realistic biotic predictions based on their various physical scenarios and accepts the role that paleontological data must play in testing those predictions.

SUMMARY

Despite 15 years of intense research and often acrimonious debate surrounding the Alvarez and others (1980) impact-extinction scenario, little has been resolved with respect to establishing a causal link between bolide impacts and mass extinctions and identifying the mechanisms involved in establishing that link. References to the testability of the impact-extinction scenario refer almost exclusively to testing the hypothesis that a bolide impact occurred near the end of the

Cretaceous. Tests of this question, while interesting, cannot establish a compelling level of support for recognizing a casual link between that impact (should such be proven to have occurred) and extinctions, mass or otherwise. Gould (reported *in* Glenn, 1994c, p. 266) summed up the position of most impact-extinction supporters when he noted that, while the possibility of impact debris at the K-T boundary being unrelated to the K-T faunal turnover exists, "My first-class prejudice is not to accept coincidence on that scale."

Rather than allowing such questions to remain forever in the realm of educated guessing, hypothesis testing exists as a means whereby the degree of support for particular interpretations can be judged more objectively. One of the most consistent predictions of the impact-extinction scenario since its inception is that planktonic foraminifera should undergo a dramatic and instantaneous mass extinction resulting from the proximate effects of a bolide impact over a time interval that would appear instantaneous in the stratigraphic record (Alvarez and others, 1980). In order to test this prediction, the K-T planktonic foraminiferal records from Nye Kløv (Denmark), Agost (southern Spain), Brazos core (Texas), and ODP Site 738 (Kerguelen Plateau) were examined, and particular attention was paid to factors that might bias their observed fossil records in favor of progressive, rather than catastrophic, turnover.

In each of these sections and/or cores the observed record of turnover through the K-T interval exhibited Cretaceous species' LADs occurring at horizons below, at, and above the local K-T boundary. In addition, this boundary was located on the basis of the presence of an impact-debris layer (e.g., Ir anomalies, Ni-rich spinels, shocked quartz), which is assumed, for the purposes of this study, to be chronostratigraphically correlative with a single impact event.

First, stratigraphic confidence intervals for the cohort of Cretaceous taxa that disappear from each succession below the K-T boundary horizon could not be empirically determined owing to violation of the distributional assumptions on which these statistical inferences are based. Nevertheless, working on the assumption that all Cretaceous taxa occurring within the last meter of Maastrichtian sediments possess confidence intervals that extend (at least) to the boundary horizon, it is possible to estimate the number of alternative turnover patterns that would be consistent with such data. Although on the basis of this scenario a catastrophic mass extinction of all taxa at the K-T boundary horizon cannot be rejected, this represents but one of a very large number (10^3 to 10^8 in the sections and cores examined herein) of alternative turnover geometries, none of which can be rejected on an a priori basis. These results suggest that the probabilities of a single catastrophic extinction horizon, or a very rapid but diachronous extinction within the last 5 cm of the Cretaceous section, being the "true" interpretation of these data are low enough to be judged unlikely.

The potential for the Cretaceous portions of these four fossil records to have been biased by a Signor-Lipps effect due to differential preservation, relative abundance–sample size interactions, and habitat shift were evaluated via the nonparametric statistical analysis of dissolution and/or abrasion frequency histograms, Koch and Morgan (1988) modeling with two-sample Kolmogorov-Smirnov analysis, and multidimensional scaling analysis, respectively. Results of these tests suggest that the faunal records of the Canudo and others (1991) Agost data may have been biased by diagenesis and the record of this section and Keller (1993) ODP Site 738 data may have been biased by relative abundance–sample size interactions. There was no evidence for substantial bias by coordinated responses to small-scale environmental changes by taxa of similar environmental tolerance as measured by comparing species-specific occurrence patterns.

The potential for bias due to reworking of Cretaceous faunas into lowermost Danian strata was evaluated via analysis of species-specific stable isotope logs, relative preservational state, and the relative abundance/population ratio test (Olsson and Liu, 1993). Results indicate that there is no definable difference between Cretaceous and Danian occurrences of the most abundant species in terms of preservational state or relative abundance/population ratio, but that there are measurable differences between the stable isotopic signatures of many species as they cross the K-T boundary that cannot be attributed to diagenetic or life history effects. All of these lines of evidence suggest that the Danian occurrence of "Cretaceous" species represents the survivorship of these populations into the lowermost Danian, and cannot be objectively recognized to be the result of reworking.

Graphic correlation was then used to combine the biostratigraphic data from the two unbiased successions (Nye Kløv and Brazos core) with comparable data from the K-T boundary stratotype succession at El Kef, Tunisia (previously found to be free of Signor-Lipps effect and reworking bias; see MacLeod, 1995a) to estimate and scale a global turnover sequence. This summary confirms the original pattern of planktonic foraminiferal extinctions occurring prior to, at, and above the proposed K-T boundary impact-debris layer. Moreover, this interval of faunal turnover appears to coincide with a rise in eustatic sea level and a progressive excursion in $\delta^{13}C$ isotopic values, both of which are widely interpreted as proxies for long-term environmental change in other parts of the geologic record. The K-T iridium anomaly horizons are embedded within this interval of rapid environmental and faunal change. However, none of the data presented herein provide any justification for a belief that the observed faunal change was in any way related to the proposed emplacement of impact debris within these successions.

The taxonomic and evolutionary structure of species sorting throughout the K-T turnover event was examined by comparing the number of globorotaliform to globigeriniform and planispiral species in cohorts of indigenous and putative survivor population in the four study sections and/or cores. Stanley and others (1988) identified consistently greater susceptibility to Neogene background extinction events by globorotaliform species, and suggested that this reflected macroevolutionary distinctions characteristic to the globorotaliform clade. Examination of the K-T record shows that globorotaliform morphotypes are differentially prone to extinction, whereas globigeriniform and planispiral species, for the most part, compose the putative survivor fauna. These data suggest that selection pressures during the K-T event were qualitatively similar to (albeit greater in magnitude than) the types of selection pressures characteristic of Neogene background extinction events.

Throughout this study the results have shown consistent discrepancies between the biotic predictions of the impact-extinction scenario and the empirical structure of the K-T planktonic foraminiferal fossil record. Proximal effects of a bolide impact at or near the K-T boundary may have contributed to the observed faunal turnover at least in part. However, because the biotic turnover had been underway for several hundreds of thousands of years prior to any evidence for impact debris in any K-T boundary succession, scenarios proposing that all organismal extinctions lying stratigraphically above (and in some instances those that lie below) the proposed impact-debris layer be causally linked to the impact itself require a "leap of faith" that is not only uncharacteristic of the scientific enterprise, but is also uninformative with respect to any deep biotic understanding of the event itself. Further progress on the issue of demonstrating a causal link between impacts and extinctions must await the development of realistic biotic predictions based on the various physical scenarios and acceptance of the role that paleontological data must play in testing those predictions.

ACKNOWLEDGMENTS

I would like to thank Gerta Keller for encouraging my interest in the K-T boundary and the many colleagues (on both sides of the K-T ideological divide) with whom I have discussed various aspects of K-T research over the years. The discussion of applicability of the Strauss and Sadler method for estimating stratigraphic confidence intervals in the face of variations in sampling continuity and assumptions pertaining to the existence of species across sampling gaps presented herein benefited greatly from conversations on these topics with Charles Marshall. This paper represents a contribution from the Global Change & The Biosphere Program of the Natural History Museum, London.

REFERENCES CITED

Alvarez, L. W., Alvarez, F., Asaro, F., and Michel, H. V., 1980, Extraterrestrial cause for the Cretaceous-Tertiary extinction: Science, v. 208, p. 1095–108.

Alvarez, W., and Asaro, F., 1990, An extraterrestrial impact: Scientific American, v. 263, p. 78–84.

Alvarez, W., Alvarez, L. W., Asaro, F., and Michel, H., 1982, Current status of the impact theory for the terminal Cretaceous extinction, *in* Silver, L. T., and Schultz, P. H., eds., Geological implications of impacts of large asteroids and comets on the Earth: Geological Society of America Special Paper 190, p. 305–315.

Alvarez, W., Asaro, F., Claeys, P., Grajales-N., J. M., Montanari, A., and Smit, J., 1994, Developments in the K/T impact theory since Snowbird II, *in* New developments regarding the K/T event and other catastrophes in Earth history: Houston, Texas, Lunar and Planetary Institute Contribution 825, p. 3–5.

Archibald, J. D., and Clemens, W. A., 1984, Mammal evolution near the Cretaceous-Tertiary boundary, *in* Berggren, W. A., and Vancouvering, J. A., eds., Catastrophes and Earth history: The new Uniformitarianism: Princeton, New Jersey, Princeton University Press, p. 229–371.

Askin, R. A., 1990, The palynological record across the Cretaceous/Tertiary transition on Seymour Island, Antarctica: Geology and paleontology of Seymour Island, Antarctic Peninsula, *in* Feldmann, R. M., and Woodburne, M. O., eds., Geology and paleontology of Seymour Island, Antarctic Peninsula: Geological Society of America Memoir 169, p. 131–156.

Askin, R. A., 1992, Preliminary palynology and stratigraphic implications from a new Cretaceous-Tertiary boundary section from Seymour Island: Antarctic Journal of the United States, v. 25, p. 42–44.

Barrera, E., and Huber, B. T., 1994, Stable carbon isotopic evidence for Cretaceous planktic species survivorship and reworking, *in* New developments regarding the K/T event and other catastrophes in Earth history: Houston, Texas, Lunar and Planetary Institute Contribution 825, p. 8–9.

Barrera, E., and Keller, G., 1990, Foraminiferal stable isotope evidence for gradual decrease of marine productivity and Cretaceous species survivorship in the earliest Danian: Paleoceanography, v. 5, p. 867–870.

Ben Abdelkader, O., Haj Ali, N. B., Ben Salem, H., and Razgallah, S., 1992, International workshop on Cretaceous-Tertiary transitions (El Kef Section, Part II, Field trip guidebook): Tunisia, IUGS/GSGP/ATEIG/ Tunisian Geological Survey, 25 p.

Benton, M. J., 1990, Scientific methodologies in collision: The history of the study of the extinction of dinosaurs: Evolutionary Biology, v. 24, p. 371–400.

Berger, W. H., 1968, Planktonic foraminifera: Selective solution and paleoclimatic interpretation: Deep-Sea Research, v. 15, p. 31–43.

Berger, W. H., 1970, Planktic foraminifera: Selective solution and the lysocline: Marine Geology, v. 8, p. 111–138.

Berger, W. H., 1979, Preservation of foraminifera, *in* Lipps, J. H., and Berger, W. H., eds., Foraminiferal ecology and paleoecology: Houston, Texas, Society of Economic Paleontologists and Mineralogists Short Course no. 6, p. 105–155.

Berggren, W. A., Kent, D. V., and Flynn, J. J., 1985, Jurassic to Paleogene: Part 2, Paleogene geochronology and chronostratigraphy, *in* Snelling, N. J., eds., The chronology of the geologic record: Geological Society of London Memoir 10, p. 141–195.

Brinkhuis, H., and Zachariasse, W. J., 1988, Dinoflagellate cysts, sea level changes and planktonic foraminifers across the Cretaceous-Tertiary boundary at El Haria, northwest Tunisia: Marine Micropaleontology, v. 13, p. 153–191.

Brouwers, E. M., and De Deckker, P., 1993, Late Maastrichtian and Danian ostracode faunas from northern Alaska: Reconstructions of environment and paleogeography: Palaios, v. 8, p. 140–154.

Buzas, M. A., 1990, Another look at confidence limits for species proportions: Journal of Paleontology, v. 64, p. 842–843.

Cande, S. C., and Kent, D. V., 1992, A new geomagnetic polarity time scale for the Late Cretaceous and Cenozoic: Journal of Geophysical Research, v. 97, p. 13,917–13,951.

Canudo, J. J., Keller, G., and Molina, E., 1991, Cretaceous/Tertiary boundary extinction pattern and faunal turnover at Agost and Caravaca, S. E. Spain: Marine Micropaleontology, v. 17, p. 319–341.

Courtillot, V. E., 1990, A volcanic eruption: Scientific American, v. 263, p. 85–92.

Donce, P., Colin, J. P., Damotte, R., Oertli, H. J., Peypouquet, J.-P., and Said, R., 1982, Les ostracodes du Campanien terminal à l'Eocene inférieur de la coupe du Kef, Tunisie Nord-orientale: Bulletin des Centres de Recherches Exploration-Production Elf-Aquataine, v. 6, p. 307–335.

Donce, P., Jardine, S., Legoux, O., Masure, E., and Méon, H., 1985, Les évènements à la limite Crétacé-Tertiaire: au Kef (Tunisie septentrioale), l'analyse palynoplactogique montre qu'un changement climatique est décelable à la base du Danian: Tunis, Tunisia, Actes du Premier Congrès National des Sciences de la Terre, September 1, 1981, p. 161–169.

Donovan, A. D., Baum, G. D., Blechschmidt, G. L., Loutit, T. S., Pflum, C. E., and Vail, P. R., 1988, Sequence stratigraphic setting of the Cretaceous-Tertiary boundary in central Alabama, *in* Wilgus, C. K., Hastings, B. S., Kendall, C. G., Posamentier, H. W., Ros, C., and Van Wagoner, J. C., eds., Sea-level changes: An integrated approach: Society of Economic Paleontologists and Mineralogists Special Publication 42, p. 299–307.

Donovan, A. D., Loutit, T. S., and Greenlee, S. M., 1990, Looking at the forest as well as the trees: The sequence stratigraphic setting of the K/T boundary in the southern United States: Geological Society of America Abstracts with Programs, v. 22, no. 7, p. A279.

Douglas, R. G., 1971, Cretaceous foraminifera from the northwestern Pacific Ocean: Leg 6, Deep Sea Drilling Project, *in* Initial reports of the Deep Sea Drilling Project, Volume 6: Washington, D.C., U.S. Government Printing Office, p. 1027–1053.

Edwards, L. E., 1984, Insights on why graphic correlation (Shaw's method) works: Journal of Geology, v. 92, p. 583–597.

Erwin, D. H., 1993, The Great Paleozoic Crisis: Life & Death in the Permian: New York, New York, Columbia University Press, 327 p.

Glenn, W., 1994a, What the impact/volcanism/mass-extinction debates are about, *in* Glenn, W., ed., The Mass-extinction debates: How science works in a crisis: Stanford, California, Stanford University Press, p. 7–38.

Glenn, W., 1994b, On the mass-extinction debates: An interview with William A. Clemens, *in* Glenn, W., ed., The Mass-extinction debates: How science works in a crisis: Stanford, California, Stanford University Press p. 237-252.

Glenn, W., 1994c, On the mass-extinction debates: An interview with Stephen Jay Gould, *in* Glenn, W., ed., The Mass-extinction debates: How science works in a crisis: Stanford, California, Stanford University Press, p. 253–267.

Golovneva, L. B., 1994, The flora of the Maastrichtian-Danian deposits of the Koryak Upland, NE Russia: Cretaceous Research, v. 15, p. 89–100.

Gould, S. J., 1984, The cosmic dance of Siva: Natural History, v. 93, p. 8.

Gould, S. J., 1985, The paradox of the first tier: An agenda for paleobiology: Paleobiology, v. 11, p. 2–12.

Hanski, I., 1982, Dynamics of regional distribution: The core and satellite species hypothesis: Okios, v. 31, p. 210–211.

Hanski, I., and Gilpin, M., 1991, Metapopulation dynamics: Brief history and conceptual domain: Linnaean Society Biological Journal, v. 42, p. 3–16.

Haq, B., 1991, Sequence stratigraphy, sea-level change and significance for the deep sea, *in* Mac Donald, D. I. M., ed., Sedimentation, tectones, and eustacy: International Association of Sedimentologists Special Publication, p. 3–39.

Haq, B., Hardenbol, J., and Vail, P. R., 1987, Chronology and fluctuating sea levels since the Triassic: Science, v. 235, p. 1156–1166.

Harland, W. B., Armstrong, R. L., Cox, A. V., Craig, L. E., Smith, A. G., and Smith, D. G., 1989, A geologic time scale 1989: London, Cambridge University Press, 263 p.

Hollis, C. J., 1993, Latest Cretaceous to late Paleocene radiolarian biostratigraphy: A new zonation from the New Zealand region: Marine Micropaleontology, v. 21, p. 295–327.

Hsü, K. J., 1983, Actualistic catastrophism: Address of the retiring president of the International Association of Sedimentologists: Sedimentology, v. 30, p. 3–9.

Hsü, K. J., 1989, Catastrophic extinctions and the inevitability of the improbable: Geological Society of London Journal, v. 146, p. 749–754.

Hsü, K. J., 1994, Uniformitariansim vs. catastrophism in the extinction debate, *in* Glenn, W., ed., The Mass-extinction debates: How science works in a crisis: Stanford, California, Stanford University Press, p. 217–229.

Hsü. K. J., and McKenzie, J. A., 1990, Carbon-isotope anomalies at era boundaries: Global catastrophes and their ultimate cause, *in* Sharpton, V. L., and Ward, P. D., eds., Global catastrophes and Earth history: An interdisciplinary conference on impacts, volcanism, and mass mortality: Geological Society of America Special Paper 247, p. 61–70.

Hsü, K. J., and 19 others, 1982, Mass mortality and its environmental consequences: Science, v. 216, p. 249–256.

Hut, P., Alvarez, W., Elder, W. P., Hansen, T., Kauffman, E. G., Keller, G., Shoemaker, E. M., and Weissman, P. R., 1987, Comet showers as a cause of mass extinctions: Nature, v. 329, p. 118–126.

Jablonski, D., 1986a, Causes and consequences of mass extinctions: Implications for macroevolution, *in* Elliott, D. K., eds., Dynamics of extinction: New York, New York, Wiley, p. 183–229.

Jablonski, D., 1986b, Background and mass extinctions: The alteration of macroevolutionary regimes: Science, v. 231, p. 129–133.

Jablonski, D., 1986c, Evolutionary consequences of mass extinctions, *in* Raup, D. M., and Jablonski, D., eds., Patterns and processes in the history of life: Berlin, Germany, Springer-Verlag, p. 313–329.

Jablonski, D., 1989, The biology of mass extinction: A palaeontological view: Royal Society of London Philosophical Transactions, ser. B, v. 325, p. 357–368.

Jablonski, D., 1994, Mass extinctions: Persistent problems and new directions, *in* New developments regarding the K/T event and other catastrophes in Earth history: Houston, Texas, Lunar and Planetary Institute Contribution 825, p. 56–57.

Johnson, C. C., and Kauffman, E. G., 1993, Rates and patterns of Maastrichtian reef extinction associated with environmental changes in the Carribean province: Geological Society of America Abstracts with Programs, v. 25, no. 7, p. A363.

Kaiho, K., 1988, Uppermost Cretaceous to Paleogene bathyal benthic foraminiferal biostratigraphy of Japan and New Zealand: Latest Paleocene–middle Eocene benthic species turnover: Revue de Paléibiologie, v. 2, p. 553–559.

Kauffman, E. G., 1984, The fabric of Cretaceous extinctions, *in* Berggren, W. A., and Van Couvering, J. A., eds., Catastrophes and Earth history: The new Uniformitarianism: Princeton, New Jersey, Princeton University Press, p. 151–246.

Keller, G., 1986, Stepwise mass extinctions and impact events: Late Eocene to early Oligocene: Marine Micropaleontology, v. 10, p. 267–293.

Keller, G., 1988a, Extinction, survivorship and evolution of planktic foraminifera across the Cretaceous/Tertiary boundary at El Kef, Tunisia: Marine Micropaleontology, v. 13, p. 239–263.

Keller, G., 1988b, Biotic turnover in benthic foraminifera across the Cretaceous-Tertiary boundary at El Kef, Tunisia: Palaeogeography, Palaeoclimatology, Palaeoecology, v. 66, p. 153–171.

Keller, G., 1989, Extended Cretaceous/Tertiary boundary extinctions and delayed population change in planktonic foraminiferal faunas from Brazos River, Texas: Paleoceanography, v. 4, p. 287–332.

Keller, G., 1992, Paleoecologic response of Tethyan benthic foraminifera to the Cretaceous/Tertiary boundary transition, *in* Takayanagi, Y,. and Saito, T., eds., Studies in benthic foraminifera: Tokyo, Japan, Tokai University Press, p. 77–91.

Keller, G., 1993, The Cretaceous-Tertiary boundary transition in the Antarctic Ocean and its global implications: Marine Micropaleontology, v. 21, p. 1–46.

Keller, G., and Benjamini, C., 1991, Paleoenvironment of the eastern Tethys in the early Paleocene: Palaios, v. 6, p. 439–464.

Keller, G., Barrera, E., Schmitz, B., and Mattson, E., 1993, Gradual mass extinction, species survivorship, and long-term environmental changes across the Cretaceous-Tertiary boundary in high latitudes: Geological Society of America Bulletin, v. 105, p. 979–997.

Keller, G., Li, L., and MacLeod, N., 1995, The Cretaceous/Tertiary boundary stratotype section at El Kef, Tunisia: How catastrophic was the mass extinction?: Palaeogeography, Palaeoclimatology, Palaeoecology (in press).

Kerr, R. A., 1994, Testing an ancient impact's punch: Science, v. 263, p. 1371–1372.

Khunt, W., and Kaminski, M. A., 1993, Changes in the community structure of deep water aggultinated foraminifers across the K/T boundary in the Basque Basin (Northern Spain): Revista Espanola Micropaleontología, v. 25, p. 57–92.

Koch, C. F., 1991, Species extinction across the Cretaceous-Tertiary boundary: Observed patterns versus predicted sampling effects, stepwise or otherwise: Historical Biology, v. 5, p. 355–361.

Koch, C. F., and Morgan, J. P., 1988, On the expected distribution of species ranges: Paleobiology, v. 14, p. 126–138.

Knoboch, E., Kvacek, Z., Buzek, C., Mai, D. H., and Batten, D. J., 1993, Evolutionary significance of floristic changes in the Northern Hemisphere during the Late Cretaceous and Palaeogene, with particular reference to central Europe: Review of Palaeobotany and Palynology, v. 78, p. 41–54.

Labandiera, C. C., 1992, Diets, diversity, and disparity: Determining the effect of the terminal Cretaceous extinction on insect evolution, *in* North American Paleontological Convention, 5th, Abstracts with Programs: Paleontological Society Special Publication, v. 6, p. 174.

Lerbekmo, J. F., Sweet, A. R., and St. Louis, R. M., 1987, The relationship between the iridium anomaly and palynological floral events at three Cretaceous-Tertiary boundary localities in western Canada: Geological Society of America Bulletin, v. 99, p. 325–330.

Liu, C., and Olsson, R. K., 1992, Evolutionary radiation of microperforate planktonic foraminifera following the K/T mass extinction: Journal of Foraminiferal Research, v. 22, p. 328–346.

Macellari, C. E., 1986, Late Campanian–Maastrichtian ammonites from Seymour Island, Antarctic Penninsula: Journal of Paleontology, v. 60, p. 1–55.

MacLeod, N., 1994, An evaluation of criteria that may be used to identify species surviving a mass extinction, *in* New developments regarding the K/T event and other catastrophes in Earth history: Houston, Texas, Lunar and Planetary Institute Contribution 825, p. 75–77.

MacLeod, N., 1995a, Graphic correlation of high latitude Cretaceous-Tertiary boundary sequences at Nye Kløv (Denmark), ODP Site 690 (Weddell Sea), and ODP Site 738 (Kerguelen Plateau): Comparison with the El Kef (Tunisia) boundary stratotype: Modern Geology, v. 19 (in press).

MacLeod, N., 1995b, Graphic correlation of new Cretaceous/Tertiary (K/T) boundary sections, *in* Mann, K. O., and Lane, H. R., eds., Graphic correlation: Society for Sedimentary Geology Special Publication 53 (in press).

MacLeod, N., 1995c, Abrupt or progressive extinction among planktic foraminifera across the El kef Cretaceous-Tertiary (K/T) boundary: A reality check, *in* Ryder, G., Fastovsky, D., and Gartner, S., eds., New developments regarding the K/T event and other catastrophes in Earth history: Geological Society of America Special Paper (in press).

MacLeod, N., and Keller, G., 1990, Foraminiferal phenotypic response to environmental changes across the Cretaceous-Tertiary boundary: Geological Society of America Abstracts with Programs, v. 22, no. 7, p. A106.

MacLeod, N., and Keller, G., 1991, How complete are Cretaceous/Tertiary boundary sections? A chronostratigraphic estimate based on graphic correlation: Geological Society of America Bulletin, v. 103, p. 1439–1457.

MacLeod, N., and Keller, G., 1994, Comparative biogeographic analysis of planktic foraminiferal survivorship across the Cretaceous/Tertiary (K/T) boundary: Paleobiology, v. 20, p. 143–177.

MacLeod, N., and Sadler, P., 1995, Estimating the line of correlation, *in* Mann, K., and Lane, H. R., eds., Graphic correlation: Society for Sedimentary Geology, Special Publication (in press).

Maguran, A. E., 1988, Ecological diversity and its measurement: Princeton, New Jersey, Princeton University Press, 179 p.

Malmgren, B. A., 1987, Differential dissolution of Upper Cretaceous planktonic foraminifera from a temperate region of the South Atlantic Ocean: Marine Micropalontology, v. 11, p. 251–271.

Marshall, C. R., 1994a, Confidence intervals on stratigraphic ranges: Partial relaxation of the assumption of randomly distributed fossil horizons: Paleobiology, v. 20, p. 459–469.

Marshall, C. R., 1994b, Using the ammonite fossil record to predict the positions of the K-T boundary Ir anomaly on Seymour Island, Antarctica: Geological Society of America Abstracts with Programs, v. 26, no. 7, p. A394.

Marvin, U. B., 1990, Impact and its revolutionary implications for geology, *in* Sharpton, V. L., and Ward, P. D., eds., Global catastrophes in Earth history: An interdisciplinary conference on impacts, volcanism, and mass mortality: Geological Society of America Special Paper 247, p. 147–154.

Mayr, E., 1982, The growth of biological thought: Diversity, evolution, and inheritance: Cambridge, Massachusetts, Harvard University Press, 974 p.

McIver, E. E., Sweet, A. R., and Basinger, J. F., 1991, Sixty-five-million-year-old flowers bearing pollen of the extinct triprojectate complex—A Cretaceous-Tertiary boundary survivor: Review of Paleobotany and Palynology, v. 70, p. 77–88.

Méon, H., 1990, Palynologic studies of the Cretaceous/Tertiary boundary interval at El Kef outcrop, northwestern Tunisia: Review of Paleobotany and Palynology, v. 65, p. 85–94.

Miller, F. X., 1977, The graphic correlation method in biostratigraphy, *in* Kauffman, E. G., and Hazel, J. E., eds., Concepts and methods of biostratigraphy: Stroudsburg, Pennsylvania, Dowden, Hutchinson & Ross p. 165–186.

Olsson, R. K., and Liu, C., 1993, Controversies on the placement of Cretaceous-Paleogene boundary and the K/P mass extinction of planktonic foraminifera: Palaios, v. 8, p. 127–139.

Olsson, R. K., Hemleben, C., Berggren, W. A., and Liu, C., 1992, Wall texture classification of planktonic foraminifera genera in the lower Danian: Journal of Foraminiferal Research, v. 22, p. 195–213.

Paul, C. R. C., 1982, The adequacy of the fossil record, *in* Joysey, K. A. and Friday, A. E., eds., Problems of phylogenetic reconstruction: New York, New York, Academic Press, p. 75–117.

Perch-Nielsen, K., McKenzie, J., and He, Q., 1982, Biostratigraphy and isotope stratigraphy and the catastrophic extinction of calcareous nannoplankton at the Cretaceous/Tertiary boundary, *in* Silver, L. T., and Schultz, P. H., eds., Geological implications of impacts of large asteroids and comets on the Earth: Geological Society of America Special Paper 190, p. 353–371.

Peypouquet, J.-P., Grousset, F., and Mourguiart, P., 1986, Paleoceanography of the Mesogean Sea based on ostracodes of the northern Tunisian continental shelf between the Late Cretaceous and early Paleogene: Geologsch Rundschdau, v. 75, p. 159–174.

Pospichal, J. J., 1994, Calcareous nannofossils at the K-T boundary, El Kef: No evidence for stepwise, gradual, or sequential extinctions: Geology, v. 22, p. 99–102.

Prothero, D. R., 1994, The Eocene-Oligocene transition: Paradise lost: New York, New York, Columbia University Press, 291 p.

Raup, D. M., 1972, Taxonomic diversity during the Phanerozoic: Science, v. 177, p. 1065–1071.

Raup, D. M., 1976a, Species diversity in the Phanerozoic: A tabulation: Paleobiology, v. 2, p. 279–288.

Raup, D. M., 1976b, Species diversity in the Phanerozoic: An interpretation: Paleobiology, v. 2, p. 289–297.

Raup, D. M., 1986, The Nemesis affair: A story of the death of the dinosaurs and the ways of science: New York, W. W. Norton, 220 p.

Raup, D. M., 1989, The case for extraterrestrial causes of extinction: Royal Society of London Philosophical Transactions, ser. B, v. 325, p. 421–435.

Raup, D. M., 1991, Extinction: Bad genes or bad luck: New York, W. W. Norton, 210 p.

Ryder, G., 1994, K/T boundary: Historical context, counter-revolutions, and strawmen, *in*

New developments regarding the K/T event and other catastrophes in Earth history: Houston, Texas, Lunar and Planetary Institute Contribution 825, p. 101–103.

Schindewolf, O., 1962, Neokatastophismus?: Deutschen Geologischen Geselleschaft zeitschrift, v. 114, p. 430–445.

Schmitz, B., Keller, G., and Stenvall, O., 1992, Stable isotope and foraminiferal changes across the Cretaceous-Tertiary boundary at Stvens Klint, Denmark: Arguments for long-term oceanic instability before and after bolide-impact event: Palaeoceanography, Palaeoclimatology, Palaeogeography, v. 96, p. 233–260.

Shaw, A., 1964, Time in stratigraphy: New York, McGraw-Hill, 365 p.

Sharpton, V. L., and Ward, P. D., eds., 1990, Global catastrophes in Earth history: An interdisciplinary conference on impacts, volcanism, and mass mortality: Geological Society of America Special Paper 247, 631 p.

Sheehan, P. M., Fastovsky, D. E., Hoffmann, R. G., Berghaus, C. B., and Gabriel, D., 1991, Sudden extinction of the dinosaurs: Latest Cretaceous, upper Great Plains, U.S.A.: Science, v. 254, p. 835–839.

Signor, P. W., 1978, Species richness in the Phanerozoic: A reflection of labor by systematists?: Paleobiology, v. 7, p. 36–53.

Signor, P. W., and Lipps, J. H., 1982, Sampling bias, gradual extinction patterns and catastrophes in the fossil record, *in* Silver, L. T. and Schultz, P. H., eds., Geological implications of impacts of large asteroids and comets on the Earth: Geological Society of America Special Paper 190, p. 291–296.

Silver, L. T., and Schultz, P. H., eds., 1982, Geological implications of impacts of large asteroids and comets on the Earth: Geological Society of America Special Paper 190, 500 p.

Sloan, R. E., Rigby, J. K., Van Valen, L. M., and Gabriel, D., 1986, Gradual dinosaur extinction and simultaneous ungulate radiation in the Hell Creek Formation: Science, v. 232, p. 629–633.

Smit, J., 1981, Synthesis of stratigraphical, micropaleontological and geochemical evidence from the K-T boundary: Indication for cometary impact: Houston, Texas, Lunar and Planetary Institute, Snowbird Abstracts, p. 52.

Smit, J., 1982, Extinction and evolution of planktonic foraminifera after a major impact at the Cretaceous/Tertiary boundary, *in* Silver, L. T., and Schultz, P. H., eds., Geological implications of impacts of large asteroids and comets on the Earth: Geological Society of America Special Paper 190, p. 329–352.

Smit, J., 1990, Meteorite impact, extinctions and the Cretaceous-Tertiary boundary: Geologie en Mijnbouw, v. 69, p. 187–204.

Smit, J., 1994, Blind tests and muddy waters: Nature, v. 368, p. 809–810.

Smit, J., Groot, H., de Jonge, R., and Smit, P., 1988, Impact and extinction signatures in complete Cretaceous Tertiary (KT) boundary sections [abs.]: Global Catastrophes in Earth History: An Interdisciplinary Conference on Impacts, Volcanism, and Mass Mortality: Houston, Texas, Lunar and Planetary Institute, p. 182–183.

Smith, A. B., 1994, Systematics and the fossil record: Documenting evolutionary patterns: London, Blackwell, 223 p.

Springer, M. S., 1990, The effect of random range truncations on patterns of evolution in the fossil record: Paleogeology, v. 16, p. 512–520.

Sokal, R. R., and Rohlf, J. F., 1981, Biometry: San Francisco, W. H. Freeman and Co., 859 p.

Stanley, S. M., Wetmore, K. L., and Kennett, J. P., 1988, Macroevolutionary differences between two major clades of Neogene planktonic foraminifera: Paleobiology, v. 14, p. 235–249.

Strauss, D., and Sadler, P. M., 1989, Classical confidence intervals and bayesian probability estimates for ends of local taxon ranges: Mathematical Geology, v. 21, p. 411–427.

Surlyk, F., 1990, Mass extinction events, Section 2.13.6 Cretaceous-Tertiary (Marine), *in* Briggs, D. E. G., and Crowther, P. R., eds., Palaeobiology: A synthesis: Oxford, United Kingdom, Blackwell Publishers, p. 198–203.

Sweet, A. R., and Braman, D. R., 1992, The K-T boundary and contiguous strata in western Canada: Interactions between paleoenvironments and palynological assemblages: Cretaceous Research, v. 13, p. 31–79.

Sweet, A. R., Braman, D. R., and Lerbekmo, J. F., 1990, Palynofloral response to K/T boundary events; a transitory interruption within a dynamic system, *in* Sharpton, V. L., and Ward, P. D., eds., Global catastrophes in Earth history: An interdisciplinary conference on impacts, volcanism, and mass mortality: Geological Society of America Special Paper 247, p. 457–469.

Thomas, E., 1990, Late Cretaceous-Early Eocene mass extinction in the deep sea, *in* Sharpton, V. L., and Ward, P. D., eds., Global catastrophes in Earth history: An interdisciplinary conference on impacts, volcanism, and mass mortality: Geological Society of America Special Paper 247, p. 481–495.

Ward, P. D., 1990, A review of Maastrichtian ammonite ranges, *in* Sharpton, V. L., and Ward, P. D., eds., Global catastrophes in Earth history: An interdisciplinary conference on impacts, volcanism, and mass mortality: Geological Society of America Special Paper 247, p. 519–530.

Ward, P. D., Kennedy, W. J., MacLeod, K. G., and Mount, J. F., 1991, Ammonite and inoceramid bivalve extinction patterns in Cretaceous/Tertiary boundary sections of the Biscay region (southwestern France, northern Spain): Geology, v. 19, p. 1181–1184.

Widmark, J. G., and Malmgren, B. A., 1992, Benthic foraminiferal changes across the Cretaceous-Tertiary boundary in the deep sea; DSDP Sites 525, 527, and 465: Journal of Foraminiferal Research, v. 22, p. 81–113.

Zachos, J. C., and Arthur, M. A., 1986, Paleoceanography of the Cretaceous/Tertiary boundary event from stable isotopic and other data: Paleoceanography, v. 1, p. 5–26.

Zachos, J. C., Arthur, M. A., and Dean, W. E., 1989, Geochemical evidence for suppression of pelagic marine productivity at the Cretaceous/Tertiary boundary: Nature, v. 337, p. 61–64.

Zar, J. H., 1974, Biostatistical analysis: Englewood Cliffs, New Jersey, Prentice Hall, 620 p.

6

Latest Maastrichtian and Cretaceous-Tertiary Boundary Foraminiferal Turnover and Environmental Changes at Agost, Spain

Alfonso Pardo, Nieves Ortiz, and Gerta Keller,
Department of Geological and Geophysical Sciences, Princeton University, Princeton, New Jersey

INTRODUCTION

After more than 12 years of controversy and a multitude of publications, the Cretaceous-Tertiary boundary (K-T) mass extinction is still one of the most controversial topics in earth sciences. There is no doubt that planktonic foraminifera underwent a major mass extinction at this time, at least in low to middle latitudes. In fact, they suffered the most dramatic mass extinction among all coeval marine organisms (Smit, 1982, 1990; Brinkhuis and Zachariasse, 1988; Keller, 1988a, 1989a, 1989b, this volume; Canudo and others, 1991; Liu and Olsson, 1992; Olsson and Liu, 1993; Peryt and others, 1993; Keller and others, 1994). However, the present controversy concerns not whether a mass extinction occurred, but the nature and tempo of this mass extinction. Did all but one or at most three Cretaceous planktonic foraminiferal species suddenly go extinct at the K-T boundary, as suggested by Smit (1982, 1990) and Olsson and Liu (1993), or was the mass extinction more gradual, beginning below and extending well above the K-T boundary, as suggested by Keller (1988a, 1989a, 1989b)? If nearly all species went extinct at the K-T boundary, then the causal agent is likely a bolide impact. But, if species extinctions were more gradual and selective, then long-term environmental changes must have caused at least some extinctions, and the bolide impact accelerated the demise of already stressed faunas.

There is compelling evidence that the mass extinction was selective, eliminating large, ornate tropical-subtropical species and favoring the survival of smaller cosmopolitan taxa tolerant of environmental fluctuations (Keller, 1988a, 1989a,

1989b, this volume). In high latitudes, where tropical-subtropical taxa are absent, no mass extinction is observed and nearly all species survived well into the Danian (Keller, 1993; Keller and others, 1993). The observed pattern of the mass extinction in low latitudes is typically on the order of 10% of the species disappearing below the K-T boundary, 50%–60% disappearing at or near the KT boundary, and about 20%–30% ranging into the early Tertiary (Keller, 1988a, this volume; Keller and others, 1994).

This low-latitude mass-extinction pattern was recently confirmed by a blind sample test of the El Kef stratotype section (MacLeod and Keller, in prep.). However, is this observed mass-extinction pattern real or an artifact of the sedimentary record? Are the species disappearances below the K-T boundary artifacts of sample size and the rarity of taxa (the Signor-Lipps effect)? Are the Cretaceous species continuously present above the K-T boundary a result of reworking of older sediments into Tertiary strata, or are they survivor taxa? The question of reworking has been addressed in several publications based on stable isotopic data and global biogeographic distribution patterns (MacLeod and Keller, 1994; MacLeod, 1995).

Late Maastrichtian and early Paleocene stable isotopic signals of planktonic foraminifera vary by approximately 2‰–3‰ (Zachos and Arthur, 1986; Zachos and others, 1989; Keller and Lindinger, 1989; Barrera and Keller, 1990). Therefore, if Cretaceous species present above the K-T boundary are survivors, they should have Paleocene isotopic signals, and if they are reworked, they should have Maastrichtian isotopic signals. To date, isotopic measurements of about a dozen Cretaceous species from early Tertiary strata reveal that they are Cretaceous survivors (Barrera and Keller, 1990, 1994; Keller and others, 1993; Keller, 1993). Furthermore, a global biogeographic analysis of putative Cretaceous survivor species in early Paleocene sediments yields a consistent pattern of species survivorship across latitudes that cannot be reconciled statistically with random sediment reworking (MacLeod and Keller, 1994; MacLeod, in prep.).

The reality of the pattern of species disappearing below the K-T boundary is difficult to assess. Most of these species are so rare, or only intermittently present, that the last specimens may not be found in a small sample (e.g., Signor-Lipps effect). These species may have lived up to the K-T boundary, or even into the early Paleocene, but because of their rarity this is impossible to determine. Even if a single isolated specimen can be found, there is no way of assessing whether it is in situ or reworked. Furthermore, there is no statistical support for interpreting these species as having lived right up to the K-T boundary, where they went extinct due to the bolide impact, as suggested by some workers (Smit, 1982, 1990; Olsson and Liu, 1993). They are more likely to have disappeared at random intervals (MacLeod, this volume).

The question of increased species extinctions prior to the K-T boundary cannot be addressed by the currently available mass-extinction database that generally examines only a very short (50 to 100 cm) interval below the K-T boundary. To shed some light on this problem, a longer interval must be investigated. This will allow examination of environmental changes preceding the K-T boundary and their specific effects on Cretaceous taxa. This report details such a study for both planktonic and benthic foraminifera at the relatively continuous Agost section in Spain. We have chosen the Agost section to examine pre-K-T boundary environmental changes because of its apparently continuous sedimentation record, good preservation of benthic and planktonic foraminifera, and good paleomagnetic control. Our analysis integrates the planktonic foraminiferal work of Canudo and others (1991) for the K-T transition interval and continues the benthic foraminiferal analysis into the Danian. We are thus able to evaluate the K-T boundary mass extinction as well as long-term environmental changes beginning some 0.5 m.y. prior to the K-T boundary.

METHODOLOGY

The Agost section is located 600 m from the center of Agost on the west side of the Agost-Castalla road (Fig. 1). Sediments consist of gray to cream marls interlayered with marly limestone beds in the lower 6 m of the exposed Maastrichtian interval. The K-T boundary is marked by a 6.5-cm-thick layer of black-green clay with a basal 2-mm-thick red oxidized layer. The boundary clay is composed of almost pure smectite (Groot and others, 1989) and contains a few 50–70 μm sized spherules composed of smectite. Gray marls of the early Paleocene (Danian) overlie the black-green clay.

We trenched the section to remove surface contamination and obtain fresh unweathered bedrock. Samples were then collected at 20 cm intervals through the 11.5 m of uppermost Maastrichtian and at 2 to 5 cm intervals across the K-T boundary beginning 65 cm below to 50 cm above the boundary. A total of 58 samples were analyzed for this study, including 33 samples in the 11.55 m of upper Maastrichtian strata and 25 samples in the 1.3 m of early Danian strata above the K-T boundary. The Danian samples were originally studied by Canudo and others (1991) and have been integrated into this study.

Samples were disaggregated in water and a weak detergent with H_2O_2 added to remove organic matter. Each sample was cleaned with ultrasonic agitation dur-

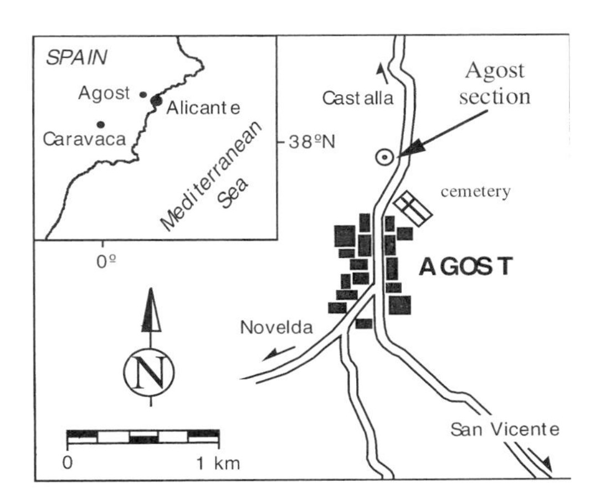

Figure 1. Location map of Agost.

ing 10 to 15 s intervals and washed through a 63 μm screen. The procedure was repeated until a clean foraminiferal residue was recovered. The final residue was dried in an oven at 50 °C. Planktonic and benthic foraminifera are well preserved, although original calcite shells are recrystallized.

Population counts for both planktonic and benthic foraminifera are based on representative random splits (using an Otto microsplitter) of 300 to 400 specimens from the 63 μm and 105 μm size fractions, respectively. Planktonic and benthic foraminifera were picked from each sample and mounted on separate microslides for a permanent record and identified. The remaining sample was searched for rare species. Relative abundance data for planktonic foraminifera are listed in Tables 1, 2, and 3.

BIOSTRATIGRAPHY

The Agost section is stratigraphically similar to the Caravaca section of southern Spain (Canudo and others, 1991) and the El Kef stratotype section in Tunisia (Keller, 1988a). At these sections the biozonation of Keller (1988a) was used and modified by Canudo and others (1991). This biozonation was later updated in Keller (1993), as shown in Figure 2, in comparison with other commonly used biozonations. In this study we follow the biozonation of Keller (1988a, 1993) for

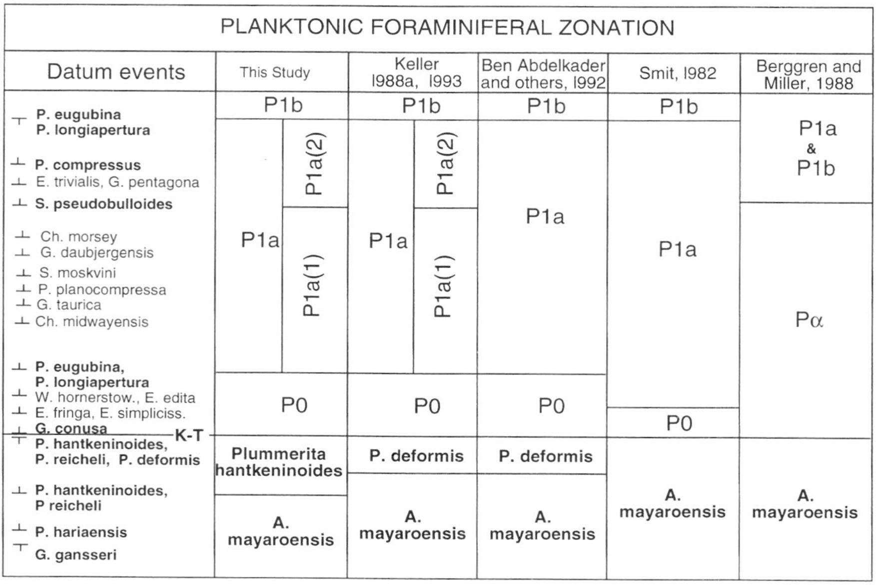

PLANKTONIC FORAMINIFERAL ZONATION

Datum events	This Study		Keller 1988a, 1993		Ben Abdelkader and others, 1992	Smit, 1982	Berggren and Miller, 1988
⊤ **P. eugubina** **P. longiapertura**	P1b		P1b		P1b	P1b	P1a & P1b
⊥ **P. compressus** ⊥ E. trivialis, G. pentagona ⊥ **S. pseudobulloides**	P1a	P1a(2)	P1a	P1a(2)	P1a	P1a	P1a & P1b
⊥ Ch. morsey ⊥ G. daubjergensis ⊥ S. moskvini ⊥ P. planocompressa ⊥ G. taurica ⊥ Ch. midwayensis	P1a	P1a(1)	P1a	P1a(1)	P1a	P1a	Pα
⊥ **P. eugubina,** **P. longiapertura** ⊥ W. hornerstow., E. edita ⊥ E. fringa, E. simpliciss. ⊥ **G. conusa**	P0		P0		P0	P1a / P0	Pα
K-T ⊤ **P. hantkeninoides,** **P. reicheli, P. deformis**	**Plummerita hantkeninoides**		**P. deformis**		**P. deformis**	**A. mayaroensis**	**A. mayaroensis**
⊥ **P. hantkeninoides,** **P reicheli** ⊥ **P. hariaensis** ⊤ **G. gansseri**	**A. mayaroensis**		**A. mayaroensis**		**A. mayaroensis**	**A. mayaroensis**	**A. mayaroensis**

Figure 2. Planktonic foraminiferal biozonations and datum events of this study compared with other commonly used biozonation for the Cretaceous-Tertiary transition. Zonal marker species are in bold. A new biozone, Plummerita hantkeninoides, *has been defined to mark the last 170–200 ka of the Maastrichtian.*

TABLE 1A

Relative percent abundances of planktonic foraminifera in cm below the K-T boundary (latest Maastrichtian upper part of C29R) of Agost. x = rare species, < 1%

SPECIES	0.0–1.0	1.0–3.0	5.0–10	15–20	30–35	45–50	60–65	100–105	140–145	180–185	220–225
Abathomphalus mayaroensis	x				x		x		x		
Achaeoglobigerina cretacea	2	1	1		1	1	1	1	1		
Gansserina gansseri											
Globigerinelloides aspera	5	4	x	2	1	2	4	2	5	1	3
G. subcarinatus	x	2	x	1	x	1	2	2	x	3	4
G. volutus			x					1			
G. yaucoensis	4	3	7	5	3	4	4	3	4	3	8
Globotruncana aegyptiaca						x	x	x	x	x	
G. angulata	1	2					1				
G. arca	2	1	3	1	1	x	2	x	2	3	1
G. dupeublei	x	1	x	x	x	1	2	x		x	x
G. esnehensis	5	2		2	2	2	3	2	2	1	x
G. falsostuarti											
G. insignis	2	1	x	1	1	x			x	x	
G. orientalis	x	1									
G. rosetta	2	3	x		1	2	1	1	2	1	1
Globotruncanella petaloidea	1	1	1	x	1	1		1	2	2	
G. monmouthensis	1	1	x	x		2		x	x		1
Globotruncanita conica	1		x	x			x	x	x	x	
G. pettersi			1						x	x	x
G. stuarti	1	x	1	1	2	1	3	1	1	x	1
G. stuartiformis	x	x	1	1	x	x	x	x	1	x	1
Gublerina acuta									x		
G. cuvillieri										x	
Guembelitria cretacea		x			1	1	x		x	x	
G. trifolia						x					
Hedbergella holmdelensis	2	2	1	1	2	1	2	2	1		3
H. monmouthensis	3	5	2	2	5	3	1	3	1		1
Heterohelix complanata	2		3		1	1	x		1	1	x
H. dentata	5	12	13	8	9	11	8	7	6	7	10
H. glabrans	4		4	7	7	7	4	4	6	5	6
H. globulosa	15	15	15	17	17	16	13	19	16	16	13
H. moremani	1		2	2	2	3	2	1	2	2	2
H. navarroensis	4	10	11	10	8	8	8	18	9	9	12
H. planata			2	8	2	3	2	2	2	2	1
H. pulchra	1	x	1	1	2	0	x				x
H. punctulata	3	1	1				1		1	x	
H. cf. puntulata		1	x	1	x						x
H. striata		2	1	1	2	2	3	3	1	1	1
Planoglobulina brazoensis	1	1	x	x	x	1		x	x	x	x
P. carseyae	x			1	x		x		1	2	
P. cf. carseyae		1			x						1
P. multicamerata	1		x			x	x			x	x
Plummerita hantkeninoides		x	x	1		1	x	x			
Pseudoguembelina costulata	11	9	10	11	13	13	16	13	15	17	12
P. hariaensis						x	x	x			
P. kempensis	5	5	2	8	5	1	3	3	2	4	7
P. palpebra	5	4	6	4	3	3	5	5	5	3	4
Pseudotextularia deformis	1	2	1	x	x	x	1	1	2	1	
P. elegans	1	1	3	2	1	1	2	1	1	8	2
Racemiguembelina fructicosa	1	x	1	1	x	1	1	x	x	x	1
R. intermedia	1										1
R. powelli	1	x	1		1		x	x	x		
Rosita contusa									x		x
R. patelliformis								x			x
Rugoglobigerina hexacamerata	1		1	x	1	2	2	1	2	3	3
R. macrocephala			1				1		x	x	1
R. milamensis								x			
R. pennyi		x	x			1					
R. rugosa	2	2	1	2	2	2	2	1	2	1	1
R. scotti	2	x		1	1	x	1	1	1	3	x
Shakoina cretacea	x		x	x	1	x	x		x	x	
Sigalia decoratissima									x		
S. deflaensis					x			x			
juveniles/no identification			1		2	1		2		x	
TOTAL # COUNTED	345	329	317	315	323	322	326	314	326	325	318

TABLE 1B
Relative percent abundances of planktonic foraminifera in cm below the K-T boundary (latest Maastrichtian lower C29R) of Agost. x = rare species, < 1%

SPECIES	260–255	300–305	340–345	380–385	420–425	460–465	500–505	540–545	580–585	620–625
Abathomphalus mayaroensis							x	x		x
Achaeoglobigerina cretacea	1	x	1	x	x					
Gansserina gansseri										
Globigerinelloides aspera	2	2	3	2	2	1	2	2	x	2
G. subcarinatus	1	3	1	2	1	x	1	1	3	x
G. volutus	x	1	x	1		1	x			1
G. yaucoensis	3	4	1	4	4	3	4	2	6	6
Globotruncana aegyptiaca		1			x			1		
G. angulata	x							x	x	1
G. arca	1	x	1	1	1	x	2	x	x	x
G. dupeublei	x	x	x	x	x	x	x	1	1	3
G. esnehensis	1	2	2	1		2	1		x	
G. falsostuarti						x	x	1		x
G. insignis	x		x							
G. orientalis								x	x	1
G. rosetta	1	2	1	2	1	2	1	x	3	
Globotruncanella petaloidea	1	1		4	2	2	1	1	x	1
G. monmouthensis	2	2	1	1	x		1	x	x	1
Globotruncanita conica	x	x		x		x	x			
G. pettersi		x		x				1	1	x
G. stuarti	1	1	1	x	1	1	1		x	x
G. stuartiformis	x	x	x	x	x	x	x			
Gublerina acuta	1									
G. cuvillieri								x		
Guembelitria cretacea				x	x					
G. trifolia					2					
Hedbergella holmdelensis	1	1	2	2	1	1	1	2	2	1
H. monmouthensis	2	3	3	1			3	5	1	1
Heterohelix complanata						1	1	1	1	1
H. dentata	10	1	9	7	6	6	9	6	6	9
H. glabrans	5	6	8	7	5	5	5	8	5	7
H. globulosa	17	16	16	20	20	21	13	16	18	15
H. moremani	2	2	2	3	3	6	3	3	1	1
H. navarroensis	6	12	9	7	7	6	11	9	11	5
H. planata	2	1	2	1	1	3	1	2	1	x
H. pulchra	x			x			x			
H. punctulata	2	3	1	1	1	2	2	1	1	2
H. cf. puntulata										
H. striata	2	2	3	1	1	1			1	2
Planoglobulina brazoensis		x	1	1	x	1	x	x	x	x
P. carseyae	x		1	x	x	1		2	1	1
P. cf. carseyae										
P. multicamerata				1			x			
Plummerita hantkeninoides	x	x	x							
Pseudoguembelina costulata	16	12	15	14	20	14	20	20	23	28
P. hariaensis						x	x			x
P. kempensis	4	7	4	4	5	4	4	2	4	2
P. palpebra	6	4	4	5	5	5	5	5	4	5
Pseudotextularia deformis	x	1	1	2	1	x	x	x		x
P. elegans	1	2	1	x	1	1	1	1		2
Racemiguembelina fructicosa	1	2	x	x	1	2	1	1	x	x
R. intermedia										
R. powelli	1				1	2				x
Rosita contusa										
R. patelliformis					x	x				
Rugoglobigerina hexacamerata	1	1	1	4	2	3	2	x		
R. macrocephala	1	1						2	2	3
R. milamensis									x	
R. pennyi			1	x		x		x		
R. rugosa	4	3	3	1	3	2	3	1	2	1
R. scotti	x	3	2	1	2	2	1	2	1	
Shakoina cretacea	x	x						x		
Sigalia decoratissima										
S. deflaensis			x	x	x			1		x
juveniles/no identification			1	x	x		1		2	1
TOTAL # COUNTED	326	305	315	307	307	311	320	314	306	315

TABLE 1C

Relative percent abundances of planktonic foraminifera in cm below the K-T boundary (late Maastrichtian lower C29R) of Agost. x = rare species, < 1%

SPECIES	660–665	700–705	750–755	790–795	840–845	880–885	920–925	960–965	1000–1005	1040–1045	1110–1115
Abathomphalus mayaroensis	x	x	x		x	x					x
Achaeoglobigerina cretacea		x	x								
Gansserina gansseri							x	3	1	1	1
Globigerinelloides aspera	1	3	2	2	3	1	3	2	4	x	4
G. subcarinatus	1	1	1	1	1	1	x	2	1	2	x
G. volutus		1	x					x		1	x
G. yaucoensis	2	6	4	3	3	3	2	3	3	3	2
Globotruncana aegyptiaca			1		x	x	1			x	
G. angulata		1	x	x		x	x		x		2
G. arca	1	2	1	1	x	1	1	x	1	1	2
G. dupeublei	x	x	x	1	1	1	x	2	x		x
G. esnehensis	1	2	2	1	1	1	1	2	2	x	x
G. falsostuarti	x		x		x	x	x	x	1	x	x
G. insignis	1	x	x	x	x	1	x	x	x	1	x
G. orientalis		1									
G. rosetta	x	1	1	x	1	2	2	2	1	2	1
Globotruncanella petaloidea	1	1	1	2	1	1	1	1	1	2	1
G. monmouthensis	1	1	1	1	2		1	2	2	1	x
Globotruncanita conica	x	1		x	x	x	x	x	x		x
G. pettersi										x	x
G. stuarti	2	1	1	1	x	1	2	x	1	x	x
G. stuartiformis	1	x			x	2		1		1	x
Gublerina acuta				1	x		x	x	x		
G. cuvillieri	1			1	x	1	x				
Guembelitria cretacea				x							x
G. trifolia											
Hedbergella holmdelensis	2	1	1	2	x	1	2	2	2	3	1
H. monmouthensis	4	4	4	3	1	3	3	3	2	6	2
Heterohelix complanata	1	2	1			1				1	1
H. dentata	7	7	7	11	8	7	6	7	6	8	5
H. glabrans	7	8	9	7	7	6	6	7	10	9	12
H. globulosa	14	13	12	10	17	20	17	15	15	11	9
H. moremani	7	2	2	3	3	3	5	2	2	1	4
H. navarroensis	4	8	7	8	8	8	10	9	13	11	11
H. planata	2	1	1	6	3	2	5	3	5	3	3
H. pulchra						1	x	2			
H. punctulata	1	2	1	1	1	2	3			x	3
H. cf. puntulata											
H. striata	1	1	2	2	1		1		x		1
Planoglobulina brazoensis	1	x	1	1		x	x	x	x	x	
P. carseyae	x		1	1	2	2	3		2	1	3
P. cf. carseyae											
P. multicamerata		x		x		x	x	x		x	x
Plummerita hantkeninoides											
Pseudoguembelina costulata	20	20	21	21	20	20	12	17	17	22	14
P. hariaensis											
P. kempensis	1	2	4	2	4	3	1	1	2	1	x
P. palpebra	6	4	3	1	3	1	4	3	2	2	4
Pseudotextularia deformis	x	1	x	1	x	x	3	1		x	1
P. elegans	2		1	2	2	1	2	3	1	4	3
Racemiguembelina fructicosa	x	x	x	x	x	x	x	x	1	x	x
R. intermedia					x	x	x				
R. powelli	1	1	1	1	2	x	x	1	1		x
Rosita contusa	x	x	x	x	x	x	x	x	x	x	x
R. patelliformis				x	x		x	x	x	x	x
Rugoglobigerina hexacamerata	1	1	x		1	x	1	1	1	2	1
R. macrocephala		x	1		x					x	x
R. milamensis					x						
R. pennyi				x							
R. rugosa	1	2	2	x	1	1		1	1	1	1
R. scotti	1	1	1	1	2	1	1	1	1	1	x
Shakoina cretacea					x				x	x	1
Sigalia decoratissima											
S. deflaensis									x		x
juveniles/no identification	1			1	1		x	3			1
TOTAL # COUNTED	322	308	321	321	314	336	316	317	321	310	297

TABLE 2

Relative percent abundances of Cretaceous planktonic foraminifera in cm above the K-T boundary at Agost. x = rare species, < 1%

SPECIES	0.00	1.00	1.0–2.0	2.0–3.0	3.0–4.0	4.0–6.0	6.0–8.0	8.0–10.0	10.0–12.0
G. aspera	15	13	17	14	21	4	4	1	
G. subcarinatus	2	1	2	3	1	1	1		
G. yaucoensis	7	5	6	6	7	8	2	1	
G. petaloidea	4	x	3	x	x	2			
H. holmdelensis	3	3	2	3	4	1	1	2	
H. monmouthensis	2	2	1	1	1	2			
H. glabrans	3	2	3	1	1	2	1		
H. globulosa	8	5	3	7	7	3	x	1	1
H. navarroensis	6	4	12	5	3	4	5	x	2
H. striata	2	3	5						
P. costulata	18	23	18	20	19	7	3	1	1
P. kempensis	8	6	22	13	10	3			
P. palpebra	2	8	1	2	4				
R. hexacamerata	1	3	6	2	4				
R. rugosa	2	2	1	3	3				
TOTAL # COUNTED	331	415	308	452	230	518	327	450	320

SPECIES	12.0–14.0	14.0–16.0	16.0–18.0	18.0–22.0
G. aspera				
G. subcarinatus				
G. yaucoensis				
G. petaloidea				
H. holmdelensis				
H. monmouthensis				
H. glabrans				
H. globulosa	2	1	1	
H. navarroensis	2	1	x	x
H. striata				
P. costulata	2	1	1	
P. kempensis				
P. palpebra				
R. hexacamerata				
R. rugosa				0
TOTAL # COUNTED	494	328	433	296

TABLE 3

Relative percent abundances of early Paleocene (Danian) planktonic foraminifera in cm above the K-T boundary at Agost. x = rare species, < 1%

SPECIES	0	1	1–2	2–3	3–4	4–6	6–8	8–10	10–12	12–13	14–16	16–18
G. creatacea + G. trifolia		2	6	1	2	31	11	9	13	8	5	6
Ch. midwayensis											3	3
Ch. morsey												
Ch. danica						5	5	1	2	3	2	1
E. edita												
E. cf. edita						1	3		x	2	2	5
E. fringa						x	3	x	2	2	6	7
E. cf. fringa										1	4	6
E. trivialis												
G. aff. planocompressa										1	2	
G. daubjergensis												
G. conusa			1	x	x	18	21	12	11	10	8	12
P. eugubina + P. taurica								2	12	9	20	17
P. longiapertura						5	40	67	57	55	40	36
S. pseudobulloides												
S. triloculinoides												
W. claytonensis			x		x	1	1	1	1		1	1
W. hornerstownensis												
TOTAL # COUNTED	331	415	308	452	230	518	327	450	320	494	328	433

the Danian and propose the biozone, *Plummerita hantkeninoides*, to mark the latest Maastrichtian (Fig. 2).

Abathomphalus mayaroensis zone

The *Abathomphalus mayaroensis* zone is defined by the total range of *A. mayaroensis*. This taxon is generally rare and often absent in uppermost Maastrichtian strata (Masters, 1984, 1993; Keller, 1988a; Canudo and others, 1991; Olsson and Liu, 1993). At El Kef, Keller (1988a) did not find *A. mayaroensis* in the uppermost 4 m of Maastrichtian sediments and therefore proposed the *Pseudotextularia deformis* zone to characterize the interval from the last occurrence of *A. mayaroensis* to the K-T boundary. Similarly, Masters (1984) proposed the *Plummerita reicheli* zone to mark this interval. However, the last appearance of *A. mayaroensis* is diachronous and therefore a poor stratigraphic marker. For example, because *A. mayaroensis* lives in thermoclinal or deeper waters, this species is generally absent in shallow neritic environments such as the Brazos River section in Texas (United States), Stevns Klint and Nye Kløv sections in Denmark, the El Kef section in Tunisia (Hultberg and Malmgren, 1986; Keller, 1989a; Schmitz and others, 1992; Keller and others, 1993), and the Poty section of Brazil (Stinnesbeck, 1989; Stinnesbeck and Keller, this volume). Thus, the last appearance datum (LAD) of *A. mayaroensis* is not a reliable stratigraphic marker for the K-T boundary or for the base of the *P. deformis* and *P. reicheli* zones. In addition, the *A. mayaroensis* zone has a long stratigraphic range (~1.2 m.y.), which does not permit high-resolution biostratigraphy or evaluation of the completeness of the sedimentary record during the late Maastrichtian. Increased biostratigraphic resolution is therefore needed for the late Maastrichtian. This is not easy because very few new species appear in the upper part of the late Maastrichtian and last appearances of species are suspect due to reworking and the Signor-Lipps effect.

TABLE 3 (CONTINUED)

18–20	20–22	22–24	24–26	26–28	28–30	35–40	45–47	65–70	82–86	98–102	114–118	130–134
3	3	2	3	5	1	1	4	4	33	33	46	44
4	6	7	5	5	7	7	2	4	2	3	4	3
1	x	2	1	3	5	x	4	3	2	2	4	2
x	1	3	1	1	x	3	1	1		1	1	
								x	x	1	1	x
5	5	4	9	8	5	9	2	1				
8	7	4	9	6	4	15	3	2	1	1	2	1
10	7	9	6	4	7	7	5	1				
								x	1	1	x	2
1			1	2	2	6	2	1		1	1	1
	1	1	2				5	2	4	2	7	1
2	2	7	1	14	16	9	9					
41	18	18	19	15	12	18	37	20	4			
19	17	10	10	13	8	1	2	1				
								x	1	1	x	
								1		x	x	
4	41	29	31	18	25	22	24	53	54	50	33	44
	1	2	3	3	5	2	1	1	1	3	2	1
296	572	359	364	275	366	233	318	335	398	359	387	448

We have searched different late Maastrichtian sedimentary sequences for potential new index fossils, which ideally should be (1) geographically widespread to permit correlation between distant outcrops, (2) morphologically distinct and easy to identify, (3) morphologically constant during their defining range, (4) consistently present, (5) short lived, and (6) dissolution resistant. Examination of upper Maastrichtian sequences at El Kef in Tunisia (Keller, 1988a), La Lajilla, El Mulato, La Parida in northeastern Mexico (Lopez-Oliva and Keller, 1996), and the Poty section in Brazil (Stinnesbeck, 1989; Stinnesbeck and Keller, this volume) reveals that *Plummerita hantkeninoides* is a morphologically distinct, short ranging, and easily identifiable species that characterizes uppermost Maastrichtian strata and provides an ideal zonal marker species (Plate 1).

Plummerita hantkeninoides was originally described by Brönnimann (1952) along with several other morphotypes and was placed in the genus *Rugoglobigerina* with two subgenera, *Rugoglobigerina* and *Plummerella* (e.g., *Rugoglobigerina* [*Rugoglobigerina*] *reicheli reicheli*, *Rugoglobigerina* [*Plummerella*] *hantkeninoides costata*, *Rugoglobigerina* [*Plummerella*] *hantkeninoides inflata*). Masters (1984, 1993) documented the occurrence of these morphotypes in the Caribbean, Middle East, and eastern Europe and proposed the new *P. reicheli* zone to characterize their range at the top of the Maastrichtian. In a recent taxonomic revision, Masters (1993) proposed that all of these morphotypes (*P. hantkeninoides*, *P. costata*, *P. inflata*, and *P. reicheli*) are, in fact, different ontogenetic stages of a single species, of which *P. reicheli* and *P. hantkeninoides* are the adult and neanic forms, respectively. For this reason, Masters (1993) proposed that the name of *P. reicheli* should be the preferred and only name for this group. We agree in principle. However, we also believe that further study of these morphotypes is necessary before the two forms are synonomized for the following reasons. If *P. reicheli* is the adult form, then its neanic (juve-

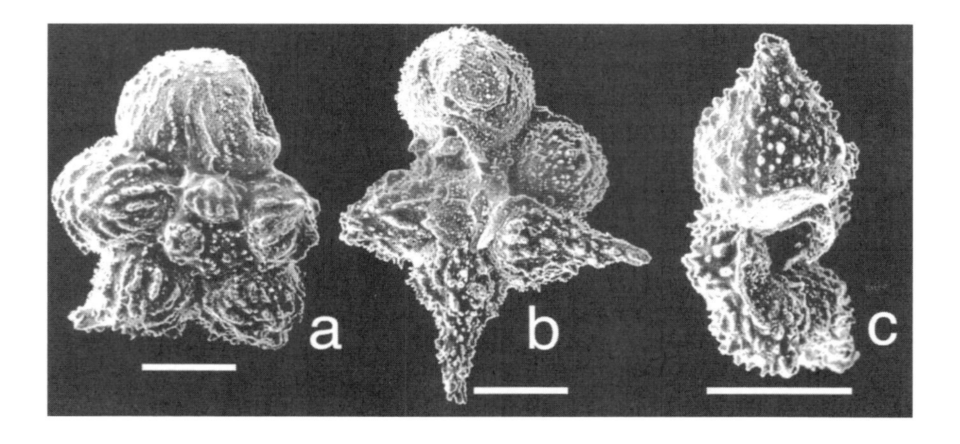

Plate 1. a: Plumerita reicheli *(Brönnimann); b, c:* Plumerita hantkeninoides *(Brönnimann). Scale bar = 100 μm. Specimens from top 5 m of Maastrichtian at El Kef, Tunisia.*

nile) chambers should have *hantkeninoides*-like spines. We have broken off adult chambers from many *reicheli* forms and found no vestiges of spines on its juvenile chambers. Moreover, early chambers of the *hantkeninoides* form also seem to lack spines. Our test thus provides no supporting evidence of a *hantkeninoides* like juvenile stage in *reicheli* forms and the absence of spines in the early chambers of the *hantkeninoides* forms suggests that acquisition of spines may be a later development. Although our test does not disprove the Masters (1993) hypothesis, it suggests the need for further study and until such time, we prefer to keep the two species names.

Plummerita reicheli is not as easy to identify as the distinct *P. hantkeninoides* morphotype with its long spines. This is illustrated in Plate 1, where the *P. reicheli* morphotype (Plate 1a) shows only vestigial remains of the characteristic long spines of the neanic stage shown in Plate 1b and Plate 1c as the *P. hantkeninoides* morphotype. The difficulty in identifying the *P. reicheli* morphotype is also evident in studies of different sections where *P. hantkeninoides* has generally been identified, but the less-distinct morphotype *P. reicheli* has not been recorded (Keller, 1988a; Lopez-Oliva and Keller, 1996; Stinnesbeck, 1989; Stinnesbeck and Keller, this volume; Canudo and others, 1991). As a biostratigraphic marker, the spinose *P. hantkeninoides* is therefore preferable over the less-distinctive *P. reicheli*.

J. Smit (1994, personal commun.) also suggested a potential new biozone marker, *Pseudoguembelina hariaensis* Nederbragt, to mark the uppermost Maastrichtian based on the total range of this new species (see Plate 2). Nederbragt (1990) reported *P. hariaensis* from the top 15 m of the Maastrichtian strata at the El Kef section. We have also observed it in the latest Maastrichtian interval at Agost as well as in the Mexico K-T boundary sections. However, we found *P. hariaenesis* to be quite similar to *P. palpebra*, from which it is distinguished by its multiserial terminal whorl, whereas *P. palpebra* remains biserial

Plate 2. a: Pseudoguembelina hariaensis *Nederbragt, 1990, side view of paratype; b, c: edge and side views of holotype. Scale bar = 20 µm. (See Nederbragt, 1990, from which SEM photomicrographs were taken.)*

in its gerontic stage. Unfortunately, the multiserial chambers of adult *P. hariaensis* are fragile and rarely preserved. We find the neanic stage of *P. hariaensis*, including adult specimens with the final chambers broken or missing, indistinguishable from *P. palpebra*. As a result we were not confident in identifying the first appearance of this new taxon. However, if these identification problems can be solved in the future, the short-ranging *P. hariaensis* could constitute another excellent latest Maastrichtian index taxa.

Plummerita hantkeninoides zone

This new zone spans the interval from the first appearance datum (FAD) of *P. hantkeninoides* to the LAD, which coincides with the mass extinction of large tropical-subtropical taxa at the K-T boundary. The first Tertiary species, including *Globoconusa conusa*, *Eoglobigerina fringa*, *E. edita*, *E. simplicissima*, and *Woodringina hornerstownensis*, appear in the first few centimeters of Tertiary sediments (Keller, 1988a; Ben Abdelkader and others, 1992). On the basis of the Agost and El Kef sections, the *P. hantkeninoides* zone corresponds to the upper three-fourths of the *Micula prinsii* nannofossil biozone (Fig. 3, Pospichal, 1993, written commun.). At the El Kef stratotype, the total range of *P. hantkeninoides* spans the top 6 m of the Maastrichtian, whereas at Agost it spans the top 3.45 m. Occurrence of *P. hantkeninoides* in uppermost Maastrichtian sediments is well documented in sections from Tunisia (Keller, 1988a), Mexico (Lopez-Oliva and Keller, 1996), and Brazil (Stinnesbeck, 1989; Stinnesbeck and Keller, this volume), as well as from Trinidad, Cuba, Haiti, Hungary, Nigeria, Egypt, Jordan, Iraq, and Pakistan (Masters, 1984, 1993).

On the basis of magnetostratigraphic data from Agost (Groot and others, 1989), the age of the *P. hantkeninoides* zone in this section can be estimated. The C30N-C29R boundary was identified by Groot and others (1989) 6 m below the K-T boundary between limestone layers 3 and 4, whereas in our field work we measured this interval at 7.5 m below the K-T boundary. The age of the portion of C29R below the K-T boundary is estimated at 350 Ka (Berggren and others, 1985; Herbert and D'Hondt, 1990). Thus the average sedimentation rate (assuming continuous sediment accumulation), according to Groot and others (1989), is 1.7 cm/Ka, whereas according to our measured section the rate is 2.1 cm/ka. On the basis of these sedimentation rates, the total range (3.45 m) of the *P. hantkeninoides* zone spans the uppermost 170 to 200 k.y. of the Maastrichtian at Agost (Fig. 3).

K-T Boundary

The definition of the K-T boundary as the mass extinction of all but one or at most three species (Smit, 1982, 1990; Liu and Olsson, 1992; Olsson and Liu, 1993) is a poor criterion because in all K-T complete sections approximately one-third of the Cretaceous taxa range well into the Tertiary (Keller, 1988a, 1989a, 1989b, 1993; Canudo and others, 1991; MacLeod, 1995). For this reason, the FADs of Tertiary species *Globoconusa conusa*, *Eoglobigerina fringa*, *E. edita*, *Woodringina hornerstownensis*, which appear within the first few centimeters of the base of the K-T boundary clay at the El Kef stratotype, are the most reliable planktonic foraminiferal markers for the K-T boundary.

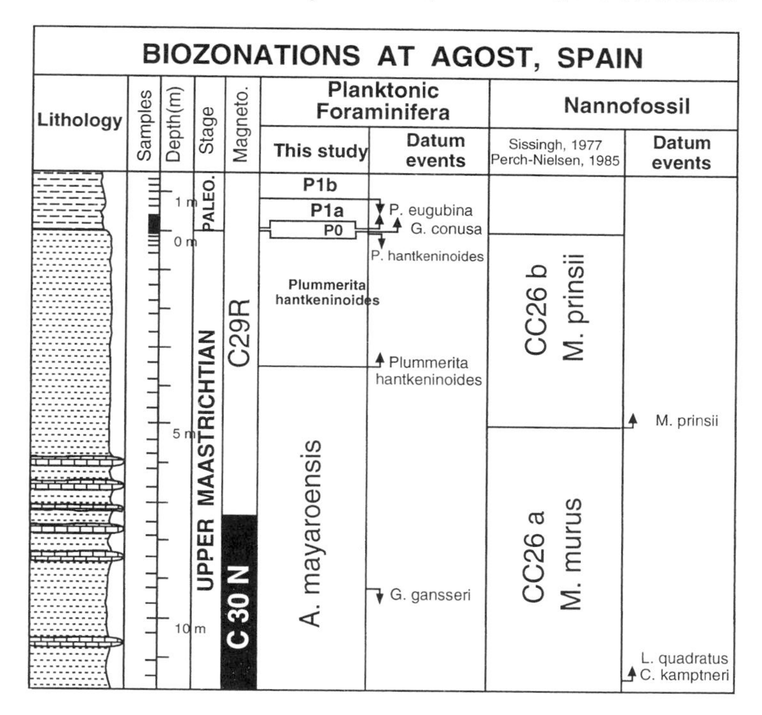

Figure 3. Planktonic foraminiferal and nannofossil biozonations at the Agost section in Spain. Note that the new Plummerita hantkeninoides *biozone spans the last 170–200 ka of the Maastrichtian.*

The K-T boundary is usually marked by a major lithologic change from gray carbonate-rich marls of the latest Maastrichtian to dark clay in the earliest Danian. In complete K-T sections a thin oxidized red layer is typically present at the base of the boundary clay that contains Ni-rich spinels, anomalously high iridium concentrations, and other elements of the platinum group (Robin and others, 1991). The clay layer in low latitudes is also marked by a drop in $CaCO_3$ to near zero and a negative excursion in $\delta^{13}C$ (Keller, 1988a; Keller and Lindinger, 1989; Barrera and Keller, 1990; Zachos and Arthur, 1986; Zachos and others, 1989). In the Agost section, the red layer shows a drop in $CaCO_3$ to less than 1%, a negative shift in $\delta^{13}C$, and an enrichment of iridium (Smit and ten Kate, 1982; Smit, 1990; Canudo and others, 1991).

Zone P0

At Agost, the first occurrences of Tertiary species, including *Globoconusa conusa*, *Eoglobigerina fringa*, *E. edita*, *E. simplicissima*, and *Woodringina claytonensis*, mark the base of this zone and the K-T boundary. The top of zone P0 is defined by the first appearance of *Parvularugoglobigerina eugubina* and/or *P. longiapertura* (Keller, 1988a, 1989a, 1993; Canudo and others, 1991). On the basis of inferred Milankovitch cyclicity in marine sediments across the K-T boundary at

Deep Sea Drilling Project (DSDP) Site 528, Herbert and D'Hondt (1990) and MacLeod and Keller (1991a, 1991b) estimated that zone P0 spans 40 to 50 ka.

Zone P1a

This zone is defined as the interval from the FAD of *P. eugubina* or *P. longiapertura* to the LAD of these taxa (Keller, 1989a, 1993; Canudo and others, 1991). In DSDP Site 577 (equatorial Pacific) and in the Brazos River sections in Texas, the top of this zone corresponds to the boundary between magnetochrons 29R and 29N (Keller, 1989a; D'Hondt and Keller, 1991). On the basis of the magnetostratigraphic data of Agost by Groot and others (1989), the top of the *P. eugubina* zone in this section is located near the top of magnetochron 29R, similar to DSDP Site 528 in the South Atlantic (D'Hondt and Keller, 1991).

Zone P1a spans a minimum of 180 k.y. (Berggren and others, 1985; Herbert and D'Hondt, 1990) and a maximum of 240 k.y. (MacLeod and Keller, 1991a, 1991b; Keller, 1993). At Agost zone P1a is characterized by the first post-K-T boundary evolutionary diversification, including *Chiloguembelina midwayensis*, *C. morsei*, *Globoconusa conusa*, *G. daubjergensis*, and *Woodringina hornerstownensis*. Keller (1993) divided zone P1a into two subzones, based on the FAD

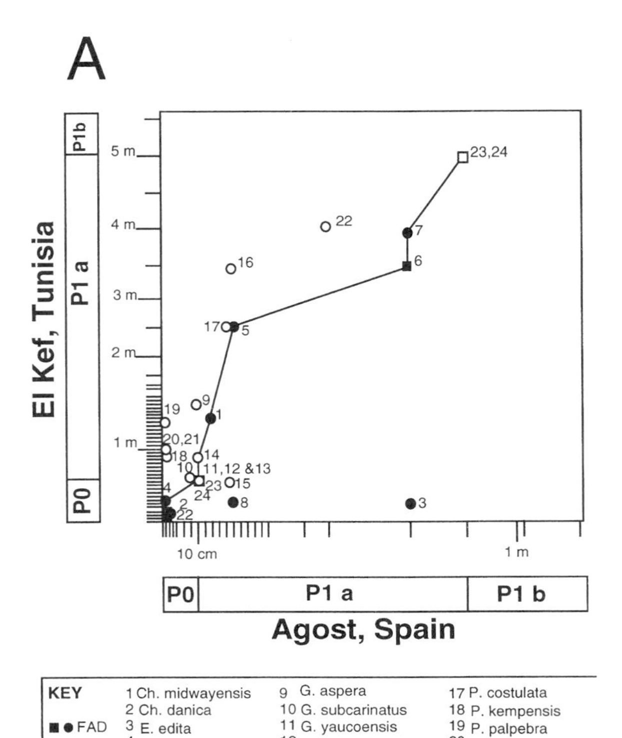

Figure 4. A: Graphic correlation of the lower Danian of El Kef and Agost showing condensed sedimentation or short hiatuses at the P0/P1a and near the zone P1a/P1b zone boundaries at Agost.

of *Subbotina pseudobulloides*. Subzone P1a (1) spans the interval from the FAD of *P. eugubina* and/or *P. longiapertura* to the FAD of *S. pseudobulloides*. Subzone P1a (2) spans the interval from the FAD of *S. pseudobulloides* to the LAD of *P. eugubina* and/or *P. longiapertura*.

Zone P1b

This zone spans the interval from the last occurrence of *P. eugubina* or *P. longiapertura* to the first appearance of *Subbotina varianta*. The studied interval at Agost does not encompass the top of zone P1b.

IS THERE A HIATUS AT OR NEAR THE K-T BOUNDARY AT AGOST?

We cannot demonstrate whether the K-T section at Agost is complete or contains a hiatus at or near the K-T boundary solely on the basis of biozones. However, graphic correlation of the Agost section with the El Kef stratotype section (Fig. 4, A and B) can provide some clues regarding sedimentation rates and stratigraphic continuity in the lower Danian at Agost. Figure 4 shows the graphic cor-

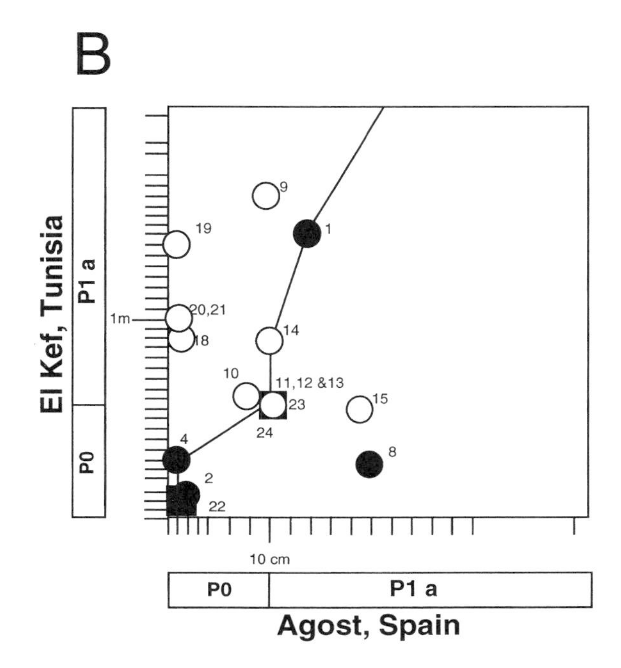

Figure 4. B: Details of the zone P0/P1a hiatus. Tick marks on axis mark sample locations. White circles are last appearance datums (LADs), open squares are LADs of zonal index species. Black circles are first appearance datums (FADs), black squares are FADs of zonal index species.

relation of the lower Danian strata (note the different scale in the Tunisian and Spanish sections). FADs are plotted as filled circles and LADs are plotted as open circles. Zone-determining taxa are plotted as black (FADs) and white (LADs) squares.

Graphic correlation of Agost with El Kef reveals highly condensed sediment accumulation in the first 25 cm at Agost that corresponds to an interval of more than 150 cm at El Kef. Within this condensed interval there appears to be a hiatus at Agost, as indicated by the FADs of *Parvularugoglobigerina eugubina* and *P. longiapertura* and simultaneous LADs of *Globigerinelloides yaucoensis*, *Hedbergella holmdelensis*, *H. monmouthensis,* and *Heterohelix glabrans*, 10 cm above the K-T boundary. Figure 5 illustrates that this hiatus is marked by the very high abundances of *P. longiapertura* and *P. eugubina* at their initial appearances and the sudden increase in *Globoconusa conusa* populations. Because evolutionary first appearances of species are characteristically marked by only a few specimens, such high abundances of *P. longiapertura* and *P. eugubina* indicate that the lower part of their range (lower part of zone P1a) is missing. The succeeding interval up to the FAD of *G. daubjergensis* appears to be very condensed, as suggested by graphic correlation (Fig. 4, A and B) as well as the near-simultaneous evolutionary first appearances of 7 species (Fig. 5) within a 10 cm interval that corresponds to an interval spanning more than 2 m at El Kef. Graphic correlation also suggests the presence of a second hiatus that corresponds to the second evolutionary pulse of early Danian species, including *Subbotina pseudobulloides*, *S. triloculinoides*, and *Eoglobigerina trivialis* (Fig. 4A). Figure 5 shows that this hiatus corresponds to the sudden disappearance of dominant *G. conusa* and a second increase in the *W. claytonensis* population. In general, the ratio of sediment accumulation rates between Agost and El Kef is 1:5 during the earliest Danian.

At Agost the low sediment accumulation rates are due primarily to two hiatuses at the P0-P1a boundary and within the upper part of zone P1a. We cannot evaluate how complete sediment accumulation is in the uppermost Maastrichtian at Agost and El Kef, because there are insufficient data available for a reliable estimate. However, the first appearance of *P. hantkeninoides* at 3.45 m at Agost and at 6 m at El Kef below the K-T boundary suggests either a condensed interval and/or hiatus at Agost or one-half the sediment accumulation rate of El Kef. Stratigraphic and paleomagnetic control at Agost suggests that if a hiatus is present, the missing interval must be relatively short (<100 k.y.).

Planktonic Foraminiferal Turnover

Latest Maastrichtian

There have been 63 Maastrichtian species identified in the Agost section (Fig. 6). Two of these species, *Gansserina gansseri* and *Globotruncana falsostuarti*, disappeared 9 m and 5 m, respectively, below the boundary. In the top 2 m of the Maastrichtian six additional species (9.3%) disappeared (*Gublerina acuta*, *G. cuvillieri*, *Rosita patelliformis*, *Rosita contusa*, *Rugoglobigerina milamensis*, and *Sigalia decoratissima*). Figure 6 shows that these taxa became rarer and were only sporadically present in the latest Maastrichtian sediments prior to their extinction. Their disappearances from the fossil record prior to the K-T bound-

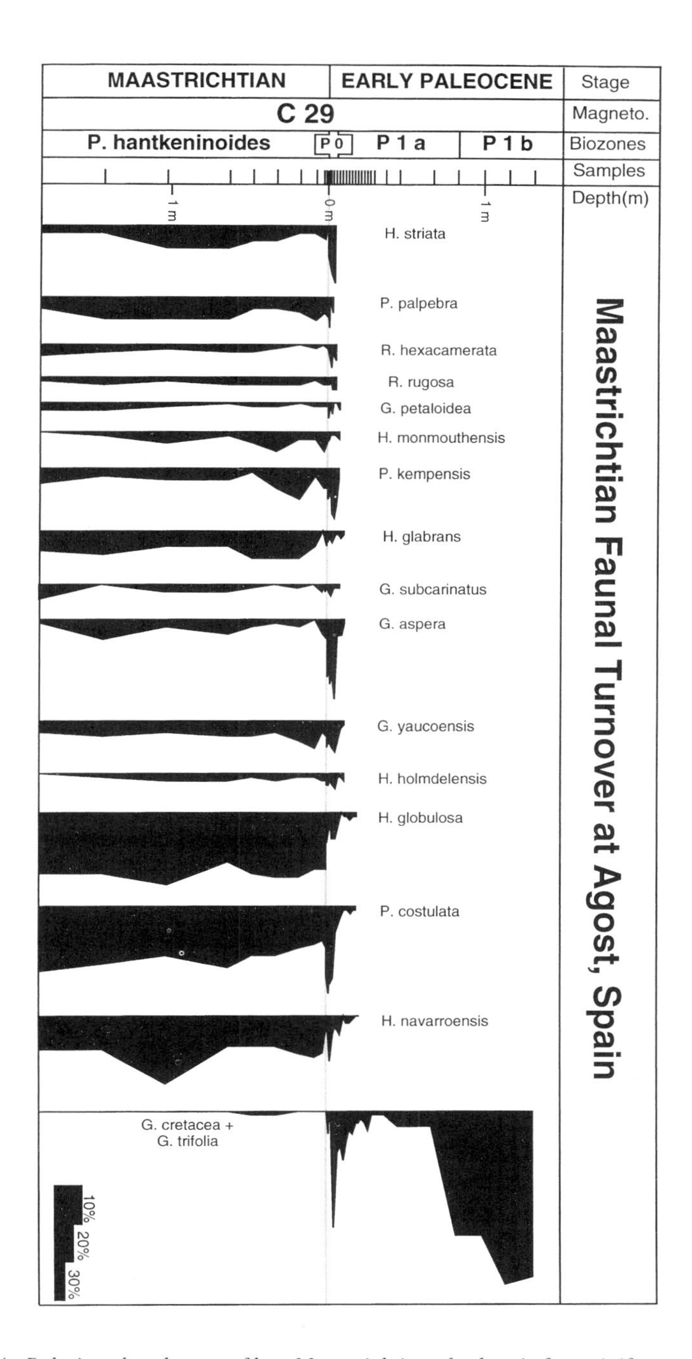

Figure 5. A: Relative abundances of late Maastrichtian planktonic foraminifera across the K-T boundary and within P. Hantkeninoides *zone. Note increased abundance of Maastrichtian species after the K-T boundary and their abrupt termination at the P0/P1a hiatus.*

ary seem to be related to pre-K-T boundary environmental changes. In addition, nearly all other large tropical-subtropical species declined in abundance or were only sporadically present during the last 300 k.y. of the Maastrichtian.

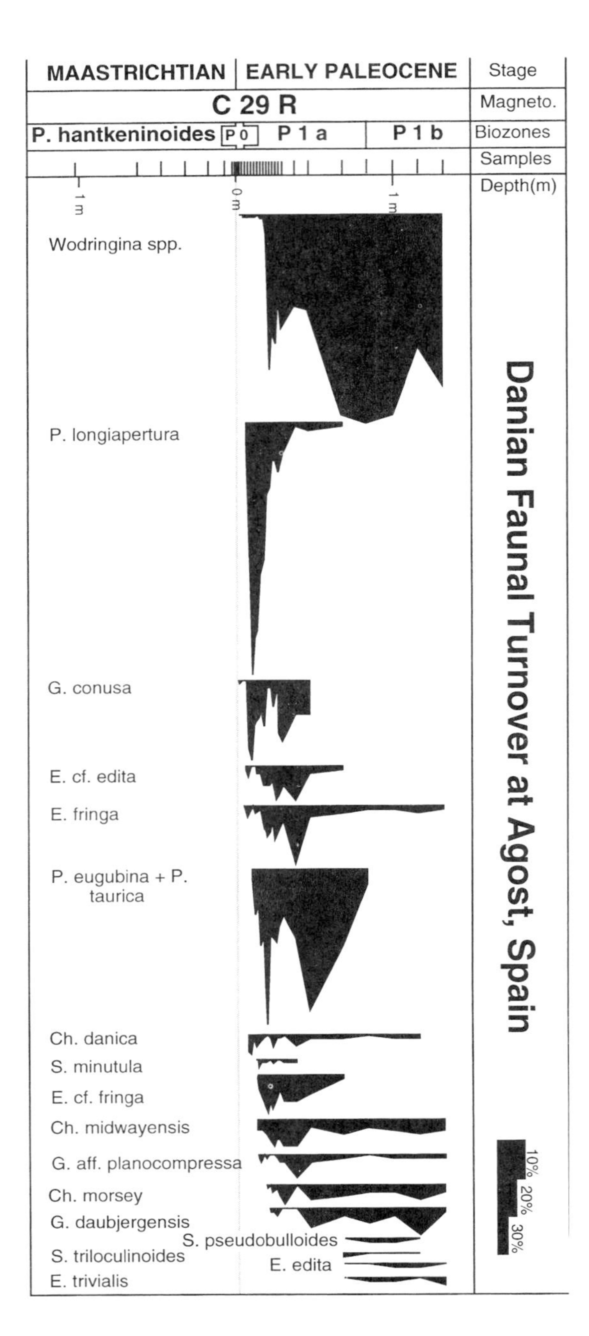

Figure 5. B: Relative abundances of evolving Danian species above the K-T boundary. Note that the sudden appearance of abundant P. eugubina *and* P. longiapertura *marks a hiatus at the zone P0/P1a boundary. A second hiatus near the zone P1a/P1b boundary is marked by the simulatneous first appearances of four species and the sudden disappearance of* G. conusa.

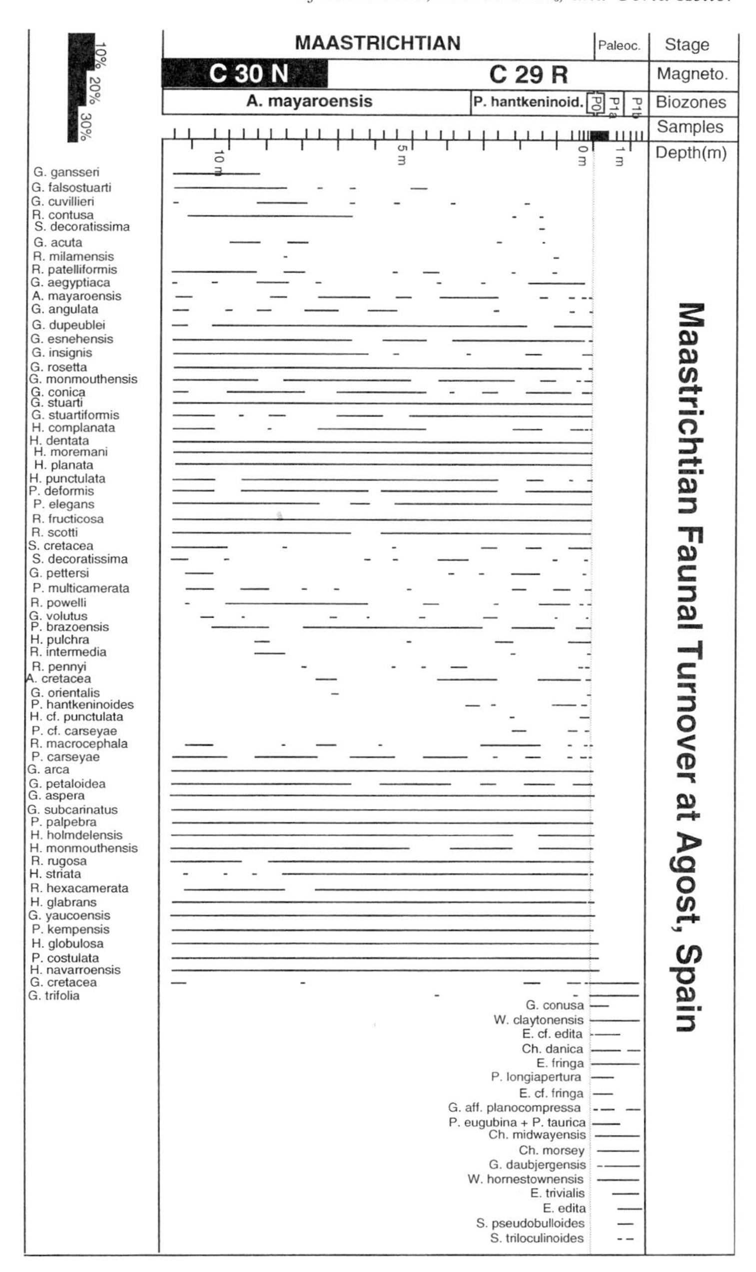

Figure 6. Planktonic foraminiferal species ranges during the latest Maastrichtian and earliest Paleocene at Agost, Spain. Note the rare and sporadic occurrences and disappearances of eight species (12%) during the pre-Cretaceous-Tertiary (K-T) environmental changes, and the very rare and sporadic occurrence of most large tropical species during this time. Of these, 57% disappeared at the K-T boundary; they included all large, complex, and ornate tropical-subtropical taxa.

What was the biotic effect of the pre-K-T boundary environmental changes upon the dominant species group, which consisted largely of ecologic generalists? An estimate can be obtained from the relative abundances of the most common species, as shown in Figure 7. Species that were common to abundant during the latest Maastrichtian do not show a consistent trend of either increasing or decreasing abundances as they approached the K-T boundary. Thus, the relative abundance data of common taxa do not indicate major environmental or sea-level changes at Agost during the latest Maastrichtian. In contrast, benthic foraminiferal data show pronounced environmental changes in the 300–400 ka below the K-T boundary, as discussed below. The absence of any major biotic effects in the ecologic generalist group is probably due to their tolerance of environmental changes.

K-T Boundary and Early Danian

The mass extinction centered at the K-T boundary at Agost and all other low- to middle-latitude sections (Brinkhuis and Zachariasse, 1988; Keller, 1988a, 1989a, this volume; Keller and Benjamini, 1991; Canudo and others, 1991; Olsson and Liu, 1993; Peryt and others, 1993; Keller and others, 1994). At this time a sudden biotic crisis, possibly induced by a low-latitude bolide impact, was superimposed upon the long-term stresses caused by climatic and sea-level changes. At Agost, 36 species (57%) of the planktonic foraminifera disappeared, including all tropical-subtropical, large, ornate and complex morphotypes (e.g., *Globotruncana*, *Globotruncanita*, *Planoglobulina*, *Pseudotextularia*, *Racemiguembelina*, *Rosita*). The combined relative abundance of this group, however, totals only 26% of the population. These taxa were already stressed and in decline during the latest Maastrichtian climatic and sea-level changes.

There are 20 species (32%) that cross the Agost K-T boundary and range well into the early Danian. Their combined relative abundance totals 74% of the population. These higher ranging species were generally small, with little surface ornamentation and simple morphologies. They were ecologic generalists able to survive in variable environmental conditions and they thrived during the climatic and sea-level changes of the last 300 k.y. of the Maastrichtian. However, at the K-T boundary many of these species declined in abundance, providing strong support for a sudden crisis (e.g., bolide impact, volcanism) beyond the effects of normal climatic and sea-level changes. (For a discussion of survivor taxa, see Keller and others, 1993; MacLeod and Keller, 1994; MacLeod, 1995.) Thus, the K-T boundary bolide impact coincided with a time of greenhouse warming (Stott and Kennett, 1990; Johnson, 1992; Keller and others, 1993; Barrera, 1994; Barrera and Keller, 1994; Keller and Stinnesbeck, this volume), rising sea level, an expanded oxygen minimum zone (Keller and others, 1993), and already highly stressed tropical-subtropical faunas. It appears to have caused or accelerated the extinction of this stressed tropical-subtropical fauna and the decline of the ecologic generalists at Agost and in low latitudes in general (Brinkhuis and Zachariasse, 1988; Keller, 1988a, 1989a; D'Hondt and Keller, 1991; Canudo and others, 1991; Olsson and Liu, 1993; Peryt and others, 1993; Keller and others, 1994).

These data show that, based on relative species abundances, the biotic effects of the K-T boundary mass extinction seem significantly less severe than when

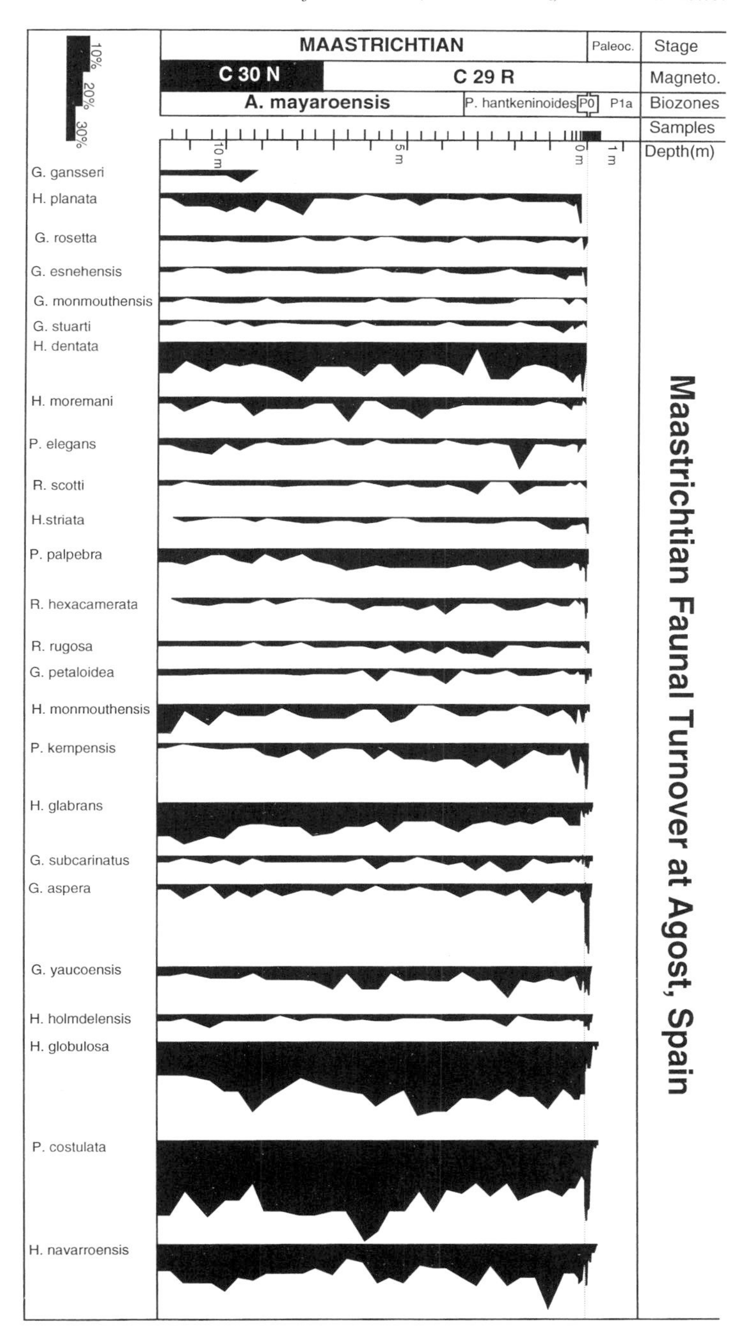

Figure 7. Relative abundances of common and dominant planktonic foraminiferal species during the latest Maastrichtian. Note that the dominant species were ecologic generalists able to tolerate environmental fluctuations. They show no consistent trends in either decreasing or increasing abundance during the latest Maastrichtian.

based solely upon species census data (26% as compared to 57% respectively). Extinct taxa are generally those that were not able to maintain stable populations during the late Maastrichtian environmental changes, which included cooling and lower sea levels followed by warming and a rising sea level during the last 50-100 k.y. of the Maastrichtian. Species that disappeared at the K-T boundary were already endangered and even small environmental stresses could have lead to their extinction.

Benthic Foraminiferal Turnover

Benthic foraminiferal assemblages are prime indicators for environmental changes and sea-level fluctuations based on upper depth limits of species, environmentally controlled morphologic traits, the preservation of assemblages, and the nature of associated biota and sediments (Douglas, 1979; Murray, 1991; Jackson, 1994; Buzas and Culver, 1994). On the basis of these indicators, major environmental and sea-level changes can be observed during the latest Maastrichtian and across the K-T boundary at Agost. Figures 8 and 9 illustrate these environmental changes by grouping species according to their upper depth limits and environmental associations (e.g., middle shelf, outer shelf, upper bathyal; Mallory, 1959; Sliter and Baker, 1972; Aubert and Berggren, 1976; Ingle, 1980; von Morkhoven and others, 1986; Keller, 1988b, 1992).

Latest Maastrichtian

Benthic foraminifera at Agost show a strong faunal turnover indicating a sea-level lowstand and rise during the latest Maastrichtian beginning near paleomagnetic boundary C30N-C29R, ~9 m below the K-T boundary. Figure 8 shows that deeper water species, *Nuttallides trumpyi* and *Cibicidoides sandidgei*, decreased from 25% to 10% in relative abundance beginning 9 m below the K-T boundary, suggesting decreasing water depth. About 6 m below the K-T boundary, shallow-water species with upper depth limits in the middle shelf, such as *Stensioina beccariiformis* forma *parvula*, *Anomalinoides danicus*, and *Globorotalites conicus*, increased rapidly and dominated the latest Maastrichtian assemblages. Outer shelf species also increased at this time and continued through the latest Maastrichtian. Characteristic outer shelf species such as *Anomalinoides acuta* and *Tritaxia globulifera* first appeared, whereas other species, such as *Buliminella carseyae*, *Globorotalites tappanae*, *Karrierella conversa*, *Anomalinoides welleri*, and *Eouvigerina maqfiensis*, became common to abundant. This benthic foraminiferal assemblage change suggests that shallowing of about 100 m occurred from uppermost bathyal to outer neritic depths during this sea-level lowstand. All of the common middle to outer shelf species that dominated the faunal assemblage after the sea-level lowstand were either infaunal, living within the top few centimeters of surface sediments, or low-oxygen epifaunal, living on the sediment surface (see Table 4). The progressive increase of outer shelf infaunal and low-oxygen epifaunal species, and some special traits such as the increased size of chambers in *B. carseyae*, *Coryphostoma midwayensis*, *C. incrassata*, and *A. acuta*, suggest decreased oxygen levels and expansion of the oxygen minimum zone after the sea-level lowstand.

On the basis of benthic foraminiferal assemblage changes and the paleomagnetic record at Agost, the sea-level lowstand reached a maximum ~300 ka before the K-T boundary. This sea-level lowstand coincided in both magnitude and timing with the eustatic sea-level lowstand recognized by Haq and others (1987). A sea-level drop of similar magnitude has been identified based on benthic foraminifera from many sections, including Denmark (Hultberg and Malmgren, 1986; Schmitz and others, 1992), Israel (Keller, 1992), Mexico (Keller and oth-

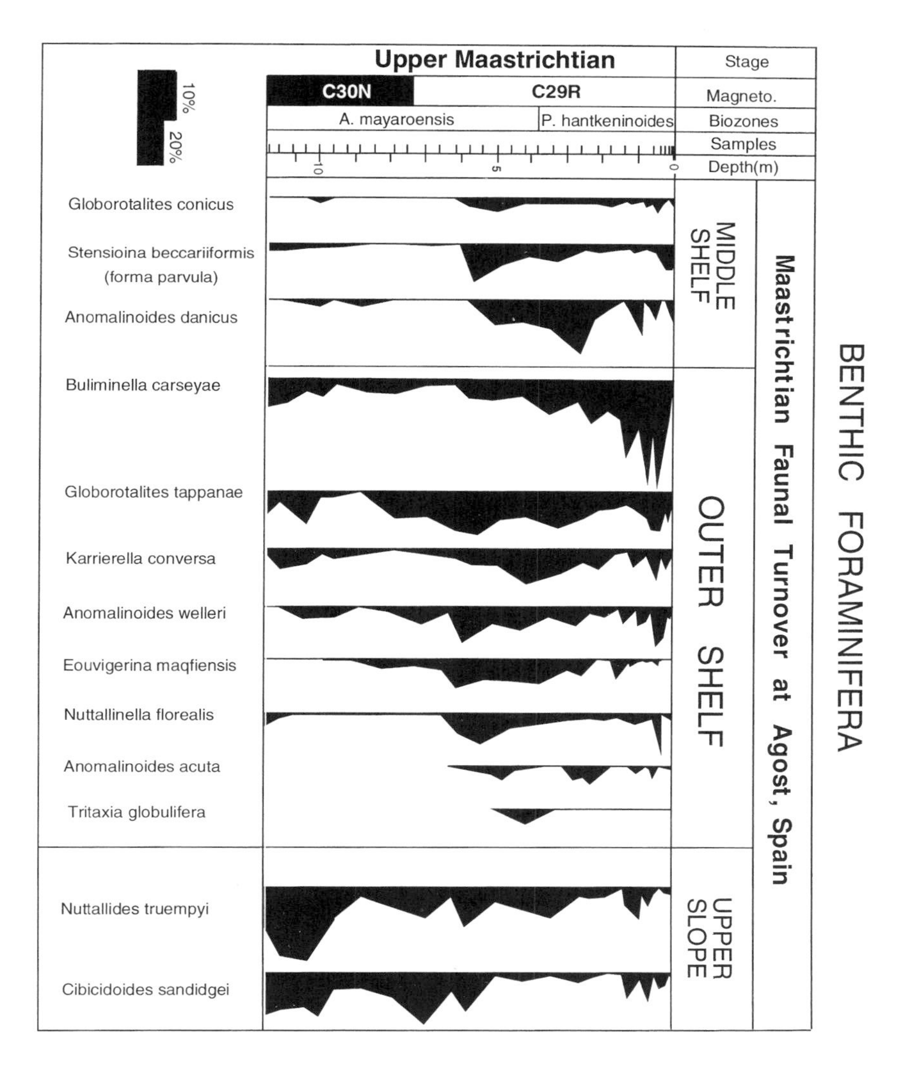

Figure 8. Dominant benthic foraminifera during the latest Maastrichtian showing increasing abundances of middle shelf and outer shelf species and decreasing abundances of upper bathyal species near the C30N/C29R boundary. This faunal turnover indicates a sea-level shallowing from upper bathyal to outer shelf depths, or a drop of about 100 meters.

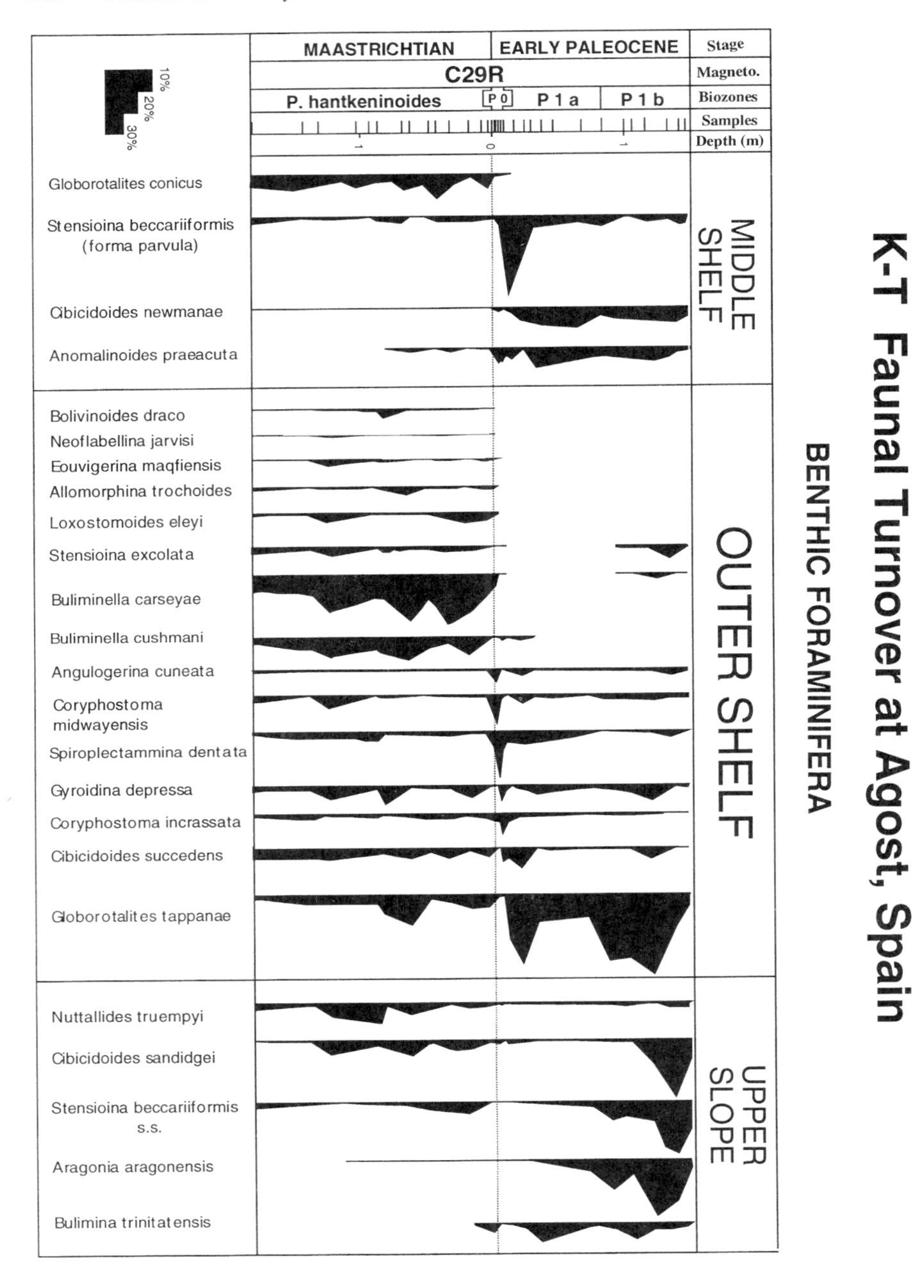

Figure 9. Benthic foraminiferal turnover across the Cretaceous-Tertiary (K-T) boundary at Agost, Spain. Note that at the K-T boundary and in zone P0 to the lower part of zone P1a, the benthic faunas are dominated by low oxygen tolerant infaunal and epifaunal species (see Table 4). During the upper part of zones P1a and in P1b increased abundance of upper bathyal species indicates a return to a deeper water environment.

TABLE 4
Benthic foraminiferal species, habitat, paleoecology, and upper depth limit at Agost. See Figures 8 and 9 for species abundances

SPECIES	HABITAT	PALEOECOLOGY	UPPER DEPTH LIMIT
Allomorphina trochoides	Infauna		Outer shelf
Angulogerina cuneata	Infauna	Low 02	Outer shelf
Anomalinoides acuta	Epifauna	Low 02	Outer shelf
Anomalinoides danicus	Epifauna	Low 02	Middle shelf
Anomalinoides praecuta	Epifauna	Low 02	Middle shelf
Anomalinoides welleri	Epifauna		Outer shelf
Aragonia aragonensis	Infauna		Upper bathyal
Bolivinoides delicatulus	Infauna	Low 02	Outer shelf
Bolivinoides draco	Infauna	Low 02	Outer shelf
Bulimina trinitatensis	Infauna	Low 02	Upper bathyal
Buliminella carseyae	Infauna	Low 02	Outer shelf
Buliminella cushmani	Infauna	Low 02	Outer shelf
Cibicidoides newmanae	Epifauna	Low 02	Middle shelf
Cibicidoides succedens	Epifauna	Low 02	Outer shelf
Cibicidoides sandidgei	Epifauna	Low 02	Upper bathyal
Coryphostoma incrassata	Infauna	Low 02	Outer shelf
Coryphostoma midwayensis	Infauna	Low 02	Outer shelf
Eouvigerina maqfiensis	Infauna		Outer shelf
Globorotalites conicus	Epifauna		Middle shelf
Globorotalites tappanea	Epifauna	Low 02	Outer shelf
Gyroidina depressa	Epifauna		Outer shelf
Gyroidinoides globosus	Epifauna		Upper bathyal
Loxostomoides eleyi	Infauna		Outer shelf
Neoflabellina jarvisi	Infauna		Outer shelf
Nuttallides truempyi	Epifauna		Upper bathyal
Nuttalinella florealis	Epifauna		Outer shelf
Spiroplectammina dentata	Infauna		Outer shelf
Stensioina beccariiformis	Epifauna		Middle/Outer shelf
Stensioina excolata	Epifauna		Outer shelf

ers, 1994), and Brazil (Stinnesbeck and Keller, this volume), and is discussed in Keller and Stinnesbeck (this volume). Near the top of the Maastrichtian (uppermost part of *P. hantkeninoides* zone), about 1.5 m below the K-T boundary, a sea-level rise is indicated by the increased abundance of bathyal species and decreased abundance of middle shelf species (Fig. 8). This sea-level rise appears to be associated with an expansion of the oxygen minimum zone, as indicated by the increase and dominance of the low-oxygen-tolerant outer shelf species *Buliminella carseyae*, and correlates with the greenhouse warming observed in southern high latitudes (Stott and Kennett, 1990; Barrera, 1994).

K-T Boundary and Early Danian

The second major benthic foraminiferal turnover occurred across the K-T boundary, as shown in Figure 9. This faunal turnover affected 42% of the Cretaceous fauna. Species richness, which began to decline in the latest Maastrichtian, rapidly decreased in the lower part of zone P0 and remained low but variable through most of the Danian (Keller, 1992). The decrease in species richness is the result of the temporary disappearance of some taxa and the extinction of others, especially in the outer shelf environment (e.g., *Loxostomoides eleyi*, *Stensioina excolata*, *Eouvigerina maqfiensis*, *Globorotalites conicus*, *Allomorphina trochoides*, *Bolivinoides draco*, and *Neoflabellina jarvisi,* among others). Some of these taxa have been observed to extend into the early Danian in other sections (Keller, 1992) and so may represent local disappearances at Agost. There appears to be no major mass extinction in benthic foraminifera associated with the K-T boundary, although a major faunal turnover is apparent and continued into the Danian. At the K-T boundary faunal turnover, most species declined in dominance or temporarily disappeared and/or emigrated during the K-T boundary crisis, only to reappear when favorable conditions returned in the later Danian planktonic foraminiferal zones P1b–P1c (Keller, 1988b, 1992). The benthic foraminiferal trend at Agost suggests a general sea-level rise from middle-outer neritic to bathyal depth by zone P1b and low-oxygen conditions throughout the early Danian. This is indicated by the progressive increase in abundance of bathyal species and the dominance of infaunal and low-oxygen epifaunal species between zones P0 and P1b, as discussed below.

After the K-T boundary crisis, benthic foraminiferal assemblages show peak abundances in low-oxygen-tolerant infaunal taxa in the first 5 cm of the early Danian (*Angulogerina cuneata*, *C. midwayensis*), which suggests strongly anoxic bottom waters (Fig. 9). Thereafter, *Spiroplectammina dentanta* and other aggultinated species that thrived in low $CaCO_3$ environments, such as zone P0, reached dominance. Near the P0-P1 boundary the shallow-water morphotype of *S. beccariiformis* (forma *parvula*) reached peak abundance (33%), and common bathyal species, such as *Nuttallides truempyi, Cibicidoides sandidgei*, and *Bulimina trinitatensis*, were nearly absent. This faunal change suggests a short-term sea-level drop near the P0-P1 boundary.

A second short-term sea-level lowstand is indicated in the upper part of zone P1a at a short hiatus near the P1a-P1b boundary. In the lower part of zone P1a species richness increased along with open marine trochospiral species (*G. tappanae*, *C. succedens*, and *B. trinitatensis*). In the middle to upper part of zone P1a, species richness decreased and some middle neritic species, such as *Anomalinoides praeacuta* and *Cibicidoides newmanae*, became more abundant, whereas open marine species, such as *G. tappanae*, declined. This suggests a sea-level regression with a maximum lowstand near the P1a-P1b zone boundary. Sea level rose again in the lower part of zone P1b, as indicated by the increased relative abundance of outer neritic and bathyal species (*A. aragonensis*, *S. beccariiformis s.s.*). Upper bathyal depths, similar to the late Maastrichtian, were reached by zone P1c (Keller, 1992).

The sea-level fluctuations indicated by benthic foraminiferal assemblages are similar to those observed at El Kef in Tunisia, the Negev of Israel, Stevns Klint and Nye Kløv in Denmark, Brazos River in Texas, and Mimbral in northeastern Mexico (Brinkhuis and Zachariasse, 1988; Keller, 1988b, 1989a, 1989b, 1992; Schmitz and others, 1992; Keller and others, 1993, 1994), and hence suggest eustatic sea-level changes. Benthic foraminifera seem to have responded primarily to these sea-level fluctuations and the associated environmental changes in temperature, oxygen, and nutrient conditions. Although the K-T boundary event significantly increased the environmental stress, it is difficult to separate the specific short-term K-T boundary stress from long-term environmental changes. It seems, however, that the K-T boundary event greatly accelerated ongoing environmental changes in benthic foraminifera.

DISCUSSION

Numerous studies have documented the K-T boundary mass extinction and early Danian recovery in the marine environment, but few have focused on pre-K-T boundary faunal, climatic, and environmental changes. Consequently, we have many data on the structure, mode, and tempo of the K-T boundary mass extinction, but very few on the environmental changes that preceded this event. Studies of the late Maastrichtian are generally restricted to biostratigraphic summaries, which for planktonic foraminifera yield little time control because one zone (*A. mayaroensis*) spans over 1 m.y. In our study we have improved this biostratigraphic control by adding the new *Plummerita hantkeninoides* zone, which marks the last 170–200 k.y. of the Maastrichtian. This biozone permits improved biostratigraphic evaluation of the completeness of the stratigraphic record just below the K-T boundary.

Quantitative faunal studies that yield environmental information are rare and usually hampered by the lack of time control. Nevertheless, they have shown major climatic and sea-level changes (Hultberg and Malmgren, 1986; Keller, 1989a, 1989b; Schmitz and others, 1992; Keller and others, 1993). These studies show that a major sea-level lowstand occurred during the late Maastrichtian between 300 and 400 k.y. below the K-T boundary (see also Haq and others, 1987). In the Agost and Danish sections a sea-level rise is observed beginning below the K-T boundary and continuing through the early Danian. The pre-K-T boundary sea-level rise was accompanied by climatic warming and marked by the incursion of warm-water surface dwellers that disappeared just above the K-T boundary (e.g., *Rugoglobigerina*, Schmitz and others, 1992; Keller and others, 1993) and low-oxygen-tolerant bottom dwellers. In numerous K-T boundary and early Paleocene sections, benthic foraminifera, ostracodes, and dinoflagellates indicate a generally rising sea level in the early Paleocene zones P0 to P1c, interrupted by two short sea-level lowstands at the P0-P1a and P1a-P1b boundaries (Peypouquet and others, 1986; Brinkhuis and Zachariasse, 1988; Brinkhuis and Leereveld, 1988; Keller, 1988b, 1989a, 1989b, 1993; MacLeod and Keller, 1991a, 1991b; Schmitz and others, 1992; Keller and others, 1993, 1994; Habib and others, 1992; Savrda, 1993; Donce and others, 1985, 1994).

Our study of Agost, where paleomagnetic control is available (Groot and others, 1989), confirms these sea-level changes and provides time control for the late Maastrichtian to early Paleocene sea-level fluctuations, as shown in Figure 10. The late Maastrichtian sea-level lowstand reached a maximum near the base of C29R about 300 k.y. below the K-T boundary, assuming that no hiatus is present at Agost. Benthic foraminifera indicate a change from an upper bathyal to outer shelf environment, suggesting a sea-level drop of ~100 m. About 100 k.y. below the K-T boundary, sea level began to rise: it accelerated in the last 50 k.y. below the boundary and was accompanied by an expanded oxygen minimum zone. This rapid sea-level rise seems to correlate with the greenhouse warming observed in Ocean Drilling Program (ODP) Site 690 (Stott and Kennett, 1990; Barrera, 1994), the incursion of warm-water low-latitude species into high latitudes (Schmitz and others, 1992; Keller and others, 1993), and the climatic warming observed in low latitude plants (Johnson, 1992).

Benthic foraminifera at Agost indicate the same generally rising seas during the early Danian, interrupted by short-term lowstands at the P0-P1a and P1a-P1b zonal boundaries, as at El Kef, the Negev, Denmark, Texas, Mexico, and Site 738 in the southern Indian Ocean (Peypouquet and others, 1986; Brinkhuis and Zachariasse, 1988; Brinkhuis and Leereveld, 1988; Keller, 1988b, 1989a, 1993; Keller and Benjamini, 1991; Keller and others, 1993, 1994). The global nature of these sea-level changes indicates that they are controlled eustatically, rather than by local tectonic events. At Agost, as elsewhere, early Danian benthic faunas are dominated by infaunal and low-oxygen epifaunal taxa reflecting a low-oxygen environment (see also Keller, 1988b, 1992). This is also supported by low-oxygen-tolerant ostracodes (Peypouquet and others, 1986; Donce and others, 1985, 1994). Plants indicate that climate cooled at this time, and there was increased rainfall (Johnson, 1992; Méon, 1990). Stable isotopic data are inconclusive due to diagenetic alteration of foraminiferal tests in most sections, although prevailing data suggest a generally cooler climate in the early Paleocene.

Benthic foraminiferal assemblages thus indicate that major environmental changes beginning ~300 k.y. before the K-T boundary were related to climatic cooling and a major sea-level drop. About 50–100 k.y. before the K-T boundary, greenhouse warming, a rapidly rising sea level, and expanded oxygen minimum zone created highly stressful conditions for marine organisms. How did planktonic foraminifera respond to these pre-K-T environmental changes? In the rela-

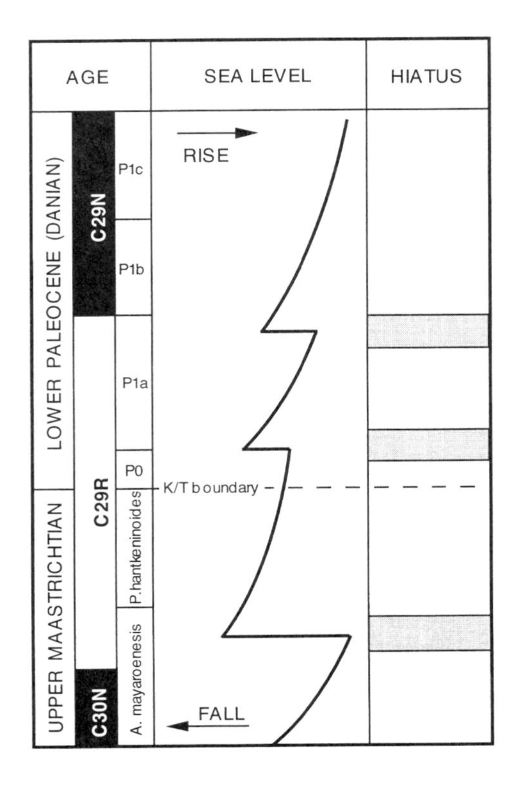

Figure 10. Sea-level changes across the Cretaceous-Tertiary (K-T) transition based on benthic foraminiferal analysis at Agost as well as K-T boundary sections globally. Paleomagnetic time scale from Agost by Groot and others (1989).

tively deep water upper bathyal to outer shelf environment of Agost, planktonic foraminifera show no dramatic changes. *Gansserina gansseri* and *Globotruncana falsostuarti* disappeared during the cooling trend (Fig. 6). Another six species disappeared during the greenhouse warming (*Gublerina acuta, G. cuvilieri, Rosita contusa, R. patelliformis, Sigalia decoratissima, Rugoglobigerina milamensis*). These taxa are more consistently present in the upper part of C30N prior to the sea-level lowstand, but are extremely rare thereafter. Therefore, it cannot be determined whether their last occurrence during the greenhouse warming signals their extinction, or whether they are too rare to be found in the sediments (e.g., Signor-Lipps effect). These data indicate clearly, however, that increasing environmental stress during the late Maastrichtian caused their decline and eventual extinction (Fig. 6). More than one-half of the remaining taxa are also rare and only sporadically present during the last 300 k.y. of the Maastrichtian. All of these are large, complex, ornate tropical and subtropical taxa intolerant of environmental changes. It is largely this group, which has a combined relative abundance of 26% of the total population, that became extinct at the K-T boundary. The dominant, smaller, and less-ornate cosmopolitan taxa show little change in their relative abundances until the K-T boundary event (Fig. 7). At that time they declined in abundance in zones P0 and P1a and gradually disappeared. Their eventual extinctions in the Danian appear to be related to the evolution and rise to dominance of the early Danian fauna.

CONCLUSIONS

Benthic foraminifera indicate a major sea-level regression ~300 k.y. prior to the K-T boundary, followed by greenhouse warming, expansion of the oxygen minimum zone, and a rapid sea-level rise beginning 50–100 k.y. prior to the K-T boundary. These latest Maastrichtian environmental changes led to highly stressful conditions for benthic foraminifera and particularly for tropical-subtropical planktonic foraminifera. A significant number of planktonic foraminiferal species (eight species, or 12%) disappeared during this time and only one species evolved (*P. hantkeninoides*). The rare and sporadic presence of these species after the sea-level lowstand suggests that their decline and eventual disappearance below the K-T boundary was related to increasing environmental stress. Most other large tropical-subtropical species also suffered, as suggested by their decreased abundance during the last 300 k.y. of the Maastrichtian.

A major mass extinction occurred in planktonic foraminifera, centered at the K-T boundary with 57% of the species extinct at a combined relative abundance of 26%. The extinct faunas included all large tropical-subtropical species that suffered most severely during the latest Maastrichtian environmental changes. About 32% of the species, with a combined relative abundance 74%, extended into the early Danian. These species were ecologic generalists able to tolerate the late Maastrichtian environmental changes. Their decline beginning at the K-T boundary in low latitudes (Agost, El Kef) strongly supports a major biotic crisis, possibly induced by the bolide impact, that exceeded the biotic stresses of the late Maastrichtian climate and sea-level changes.

ACKNOWLEDGMENTS

This study was supported by travel grant no. 4620-91 from the National Geographic Society, National Science Foundation grant OCE-9021338, and a post-doctoral fellowship (EX92-17215112) from the Spanish Ministry of Education and Science. We gratefully acknowledge discussions with B. Masters, N. MacLeod, and J. Smit. James Pospichal provided nannofossil data for the Agost section.

REFERENCES CITED

Aubert, J., and Berggren, W. A., 1976, Paleocene benthic foraminiferal biostratigraphy and paleoecology of Tunisia: Bulletin du Centre de Recherches de Pau-SNPA, v. 10, p. 379–469.

Barrera, E., 1994, Global environmental changes preceding the Cretaceous-Tertiary boundary: Early-late Maastrichtian transition: Geology, v. 22, p. 877–800.

Barrera, E., and Keller, G., 1990, Foraminiferal stable isotope evidence for gradual decrease of marine productivity and Cretaceous species survivorship in earliest Danian: Paleoceanography, v. 5, p. 867–870.

Barrera, E., and Keller, G., 1994, Productivity across the Cretaceous/Tertiary boundary in high latitudes: Geological Society of America Bulletin, v. 106, p. 1254–1266.

Ben Abdelkader, O. B., Haj Ali, N. B, Salem, H. B., and Razgallah, S., 1992, Field trip guide book part II: Workshop on Cretaceous-Tertiary transitions at El Kef: Tunis, IUGS/GSGP/ATEIG, Geological Survey of Tunisia, p. 4.

Berggren, W. A., and Miller, K. G., 1988, Paleogene tropical planktic foraminiferal biostratigraphy and magnetobiochronology: Micropaleontology, v. 34, p. 362–380.

Berggren, W. A., Kent, D. V., Flynn, J. J., and Van Couvering, J. A., 1985, Cenozoic geochronology: Geological Society of America Bulletin, v. 96, p. 1407–1418.

Brinkhuis, H., and Leereveld, H., 1988, Dinoflagellate cysts from the Cretaceous/Tertiary boundary sequence of El Kef, N. W. Tunisia: Review of Paleobotany and Palynology, v. 56, p. 5–19.

Brinkhuis, H., and Zachariasse, W. J., 1988, Dinoflagellate cysts, sea level changes and planktonic foraminifera across the Cretaceous-Tertiary boundary at El Haria, northwest Tunisia: Marine Micropaleontology, v. 13, p. 153–190.

Brönnimann, P., 1952, Globigerinidae from the Upper Cretaceous (Cenomanian-Maastrichtian) of Trinidad. B.W.I.: Bulletins of American Paleontology, v. 34, p. 1–71.

Buzas, M. A., and Culver, S. J., 1994, Species poll and dynamics of marine paleocommunities: Science, v. 264, p. 143–144.

Canudo, J. I., Keller, G., and Molina, E., 1991, Cretaceous/Tertiary extinction pattern and faunal turnover at Agost and Caravaca: S.E. Spain: Marine Micropaleontology, v. 17, p. 319–341.

D'Hondt, S., and Keller, G., 1991, Some patterns of planktic foraminiferal assemblage turnover at Cretaceous-Tertiary boundary: Marine Micropaleontology, v. 17, p. 77–118.

Donce, P., Jardine, S., Legoux, O., Masure, E., and Méon, H., 1985, Les évènements à la limite Crétacé-Tertiaire au Kef (Tunisie septentrionale), l'analyse palynoplanctologique montre qu'un changement climatique est décelable à la base du Danian: Tunis, Actes du Premier Congrès National des Sciences de la Terre, p. 161–169.

Donce, P., Méon, H., Rocchia, R., Robin, E., and Froget, L., 1994, Biological changes at the K/T stratotype of El Kef (Tunisia) [abs.], *in* New developments regarding the K/T event and other catastrophes in Earth history: Houston, Texas, Lunar and Planetary Institute Contribution 825, p. 30–31.

Douglas, R. G., 1979, Benthic foraminiferal ecology and paleoecology: A review of concepts and methods, *in* Lipps, J. H., ed., Foraminiferal ecology and paleoecology: Tulsa, Oklahoma, Society of Economic Paleontologists and Mineralogists Short Course Notes no. 6, p. 21–53.

Groot, J. J., de Jonge, R. B. G., Langereis, C. G., ten Kate, W. G. H. Z., and Smit, J., 1989, Magnetostratigraphy of the Cretaceous-Tertiary boundary at Agost (Spain): Earth and Planetary Science Letters, v. 94, p. 385–397.

Habib, D., Moshkowitz, S., and Kramer, C., 1992, Dinoflagellate and calcareous nannofossil response to sea-level change in Cretaceous-Tertiary boundary sections: Geology, v. 20, p. 165–168.

Haq, B. U., Hardenbol, J., and Vail, P. R., 1987, Chronology of fluctuating sea levels since the Triassic: Science, v. 235, p. 1156–1166.

Herbert, T. D., and D'Hondt, S., 1990, Environmental dynamics across the Cretaceous-Tertiary extinction horizon measured by 21 thousand year climate cycles in sediments: Earth and Planetary Science Letters, v. 99, p. 263–275.

Hultberg, S. U., and Malmgren, B. A., 1986, Dinoflagellate and planktonic foraminiferal paleobathymetrical indices in the boreal uppermost Cretaceous: Micropaleontology, v. 32, p. 316–323.

Ingle, J. C., 1980, Cenozoic paleobathymetry and depositional history of selected sequences within the southern California continental borderland: Cushman Laboratory Foraminiferal Research Special Publication 19, p. 163–195.

Jackson, J. B., 1994, Community unity?: Science, v. 264, p. 1412–1413.

Johnson, K. R., 1992, Leaf-fossil evidence for extensive floral extinction at the Cretaceous-Tertiary boundary, North Dakota, USA: Cretaceous Research, v. 13, p. 91–117.

Keller, G., 1988a, Extinction, survivorship and evolution of planktic foraminifers across the Cretaceous/Tertiary boundary at El Kef, Tunisia: Marine Micropaleontology, v. 13, p. 239–263.

Keller, G., 1988b, Biotic turnover in benthic foraminifera across the Cretaceous/Tertiary boundary at El Kef, Tunisia: Palaeoceanography, Palaeoclimatology, Palaeoecology, v. 66, p. 153–171.

Keller, G., 1989a, Extended period of extinctions across the Cretaceous/Tertiary boundary in Planktonic foraminifera of continental shelf sections: Implications for impact and volcanism theories: Geological Society of America Bulletin, v. 101, p. 1408–1419.

Keller, G., 1989b, Extended Cretaceous/Tertiary boundary extinctions and delayed population change in planktonic foraminiferal faunas from Brazos River, Texas: Paleoceanography, v. 4, p. 287–332.

Keller, G., 1992, Paleoecologic response of Tethyan benthic foraminifera to the Cretaceous-Tertiary boundary transition, *in* Takagayanagi, Y., and Saito, T., eds., Studies in benthic foraminifera: Tokyo, Tokai University Press, p. 77–91.

Keller, G., 1993, The Cretaceous Tertiary boundary transition in the Antarctic Ocean and its global implications: Marine Micropaleontology, v. 21, p. 1–45.

Keller, G., and Benjamini, C., 1991, Paleoenvironment of the eastern Tethys in the early Paleocene: Palaios, v. 6, p. 439–464.

Keller, G., and Lindinger, M., 1989, Stable isotope, TOC and $CaCO_3$ record across the Cretaceous-Tertiary boundary at El Kef, Tunisia: Palaeogeography, Palaeoclimatology, Palaeoecology, v. 73, p. 243–265.

Keller, G., Barrera, E., Schmitz, B., and Mattson, E., 1993, Gradual mass extinction, species survivorship and long-term environmental changes across the Cretaceous-Tertiary boundary in high latitudes: Geological Society of America Bulletin, v. 105, p. 979–997.

Keller, G., Stinnesbeck, W., and Lopez-Oliva, J. G., 1994, Age, deposition and biotic effects of the Cretaceous/Tertiary boundary event at Mimbral NE Mexico: Palaios, v. 9, p. 144–157.

Liu, G., and Olsson, R. K., 1992, Evolutionary radiation of microperforate planktic foraminifera following the K/T mass extinction event: Journal of Formainiferal Research, v. 22, p. 328–346.

Lopez-Oliva, G. J., and Keller, G., 1996, Age and stratigraphy of near-K/T boundary clastic deposits in NE Mexico, *in* Ryder, G., Gartner, S., and Fastovsky, D., eds., New developments regarding the Cretaceous-Tertiary boundary event and other catastrophes in Earth history: Geological Society of America Special Paper (in press).

MacLeod, N., 1995, Cretaceous/Tertiary (K/T) biogeography of planktic foraminifera: Historical Biology (in press).

MacLeod, N., and Keller, G., 1991a, Hiatus distribution and mass extinctions at the Cretaceous/Tertiary boundary: Geology, v. 19, p. 497–501.

MacLeod, N., and Keller, G., 1991b, How complete are the K/T boundary sections?: Geological Society of America Bulletin, v. 103, p. 1439–1457.

MacLeod, N., and Keller, G., 1994, Comparative biogeographic analysis of planktic foraminiferal survivorship across the Cretaceous/Tertiary (K/T) boundary: Paleobiology, v. 20, p. 143–177.

Mallory, V. S., 1959, Lower Tertiary biostratigraphy of the California Coast Ranges: Tulsa, Oklahoma, American Association of Petroleum Geologists, 416 p.

Masters, B. A., 1984, Comparison of planktonic foraminifers at the Cretaceous-Tertiary boundary from the El Haria shale (Tunisia) and the Esna shale (Egypt), *in* Proceedings of the 7th Exploration Seminar, March, 1984: Cairo, Egyptian General Petroleum Corporation, p. 310–324.

Masters, B. A., 1993, Re-evaluation of the species and subspecies of the genus *Plummerita* Brönimann and a new species of *Rugoglobigerina* Brönimann (Foraminiferida): Journal of Foraminiferal Research, v. 23, no. 4, p. 267–274.

Méon, H., 1990, Palynologic studies of the Cretaceous/Tertiary boundary interval at El Kef outcrop, northwest Tunisia: Paleogeographic implications: Review of Paleobotany and Palynology, v. 65, p. 85–94.

Murray, J. W., 1991, Ecology and paleoecology of benthic foraminifera: Essex, Longman Scientific and Technical Publishers, 397 p.

Nederbragt, A. J., 1990, Biostratigraphy and paleoceanographic potential of the Cretaceous planktic foraminifera Heterolicidae: Amsterdam, Centrale Huisdrukkerij Vrije Univesiteit, 203 p.

Olsson, R. K., and Liu, G., 1993, Controversies on the placement of the Cretaceous-Tertiary events and K/T mass extinction of planktonic foraminifera: Palaios, v. 8, p. 127–139.

Perch-Nielsen, K., 1985, Mesozoic culcareous nannofossils, *in* Bolli, H. M., Saunders, J. B., and Perch-Nielsen, K., eds., Plankton stratigraphy: London, Cambridge University Press, p. 329–426.

Peryt, D., Lahodynsky, R., Rocchia, R., and Boclet, D., 1993, The Cretaceous/Paleogene boundary and planktonic foraminifera in the Flyschgosau (Eastern Alps, Austria): Palaeogeography, Palaeoclimatology, Palaeoecology, v. 104, p. 239–252.

Peypouquet, J. P., Grousset, F., and Mourgniart, P., 1986, Paleoceanography of the Mesogean Sea based on ostracods of the northern Tunisian continental shelf between the Late Cretaceous and early Paleogene: Geologische Rundschau, v. 75, p. 159–174.

Robin, E., Boclet, D., Bonté, P., Froguet, L., Jéhanno, C., and Rocchia, R., 1991, The stratigraphic distribution of Ni-rich spinels in Cretaceous-Tertiary boundary rocks at El Kef (Tunisia), Caravaca (Spain) and Hole 761C (Leg 122): Earth and Planetary Science Letters, v. 107, p. 715–721.

Savrda, C. E., 1993, Ichnosedimentologic evidence for a noncatastrophic origin of Cretaceous-Tertiary boundary sands in Alabama: Geology, v. 21, p. 1075–1078.

Schmitz, B., Keller, G., and Stenwall, O., 1992, Stable isotope and foraminiferal changes across the Cretaceous-Tertiary boundary at Stevns Klint, Denmark: Arguments for long-term oceanic instability before and after bolide impact: Palaeogeography, Palaeoclimatology, Palaeoecology, v. 96, p. 233–260.

Sissingh, W., 1977, Biostratigraphy of Cretaceous calcareous nannoplankton: Geologie en Mijnbouw, v. 56, p. 37–65.

Sliter, W. V., and Baker, R. A., 1972, Cretaceous bathymetric distribution of benthic foraminifers: Journal of Foraminiferal Research, v. 2, p. 167–183.

Smit, J., 1982, Extinction and evolution of planktonic foraminifera after a major impact at the Cretaceous-Tertiary boundary, *in* Silver, L. T., and Schultz, P. H., eds., Geological implications of impacts of large asteroids and comets on the Earth: Geological Society of America Special Paper 190, p. 329–352.

Smit, J., 1990, Meteorite impact, extinctions and the Cretaceous-Tertiary boundary: Geologie en Mijnbouw, v. 69, p. 187–204.

Smit, J., and ten Kate, W. G. H. Z., 1982, Trace element patterns at the Cretaceous-Tertiary boundary—Consequence of a large impact: Cretaceous Research, v. 3, p. 307–332.

Stinnesbeck, W., 1989, Fauna y microflora en el limite Cretácio-Terciario en el estado de Pernambuco, Noreste de Brasil: Contribucionas a los Simposios sobre el Cretacio de America Latina, Parte A: Buenos Aires, Argentina, Eventos y Registro Sedimentario, p. 215–230.

Stott, L. D., and Kennett, J. P., 1990, The paleoceanographic and paleoclimatic signature of the Cretaceous/Paleogene boundary in the Antarctic, *in* Proceedings of the Ocean Drilling Program, Scientific results, Volume 113: College Station, Texas, Ocean Drilling Program, p. 829–848.

von Morkhoven, F. P. C. M., Berggren, W. A., and Edwards, A. S., 1986, Cenozoic cosmopolitan deep-water benthic foraminifera: Bulletin des Centres de Recherches Exploration-Production Elf-Aquitaine, Memoire 11, p. 1–421.

Zachos, J. C., and Arthur, M. A., 1986, Paleoceanography of the Cretaceous/Tertiary boundary event from stable isotopic and other data: Paleoceanography, v. 1, p. 5–26.

Zachos, J. C., Arthur, M. A., and Dean, W. E., 1989, Geochemical evidence for suppression of pelagic marine productivity at the Cretaceous/Tertiary boundary: Nature, v. 337, p. 61–67.

7

Radiolarian Faunal Change through the Cretaceous-Tertiary Transition of Eastern Marlborough, New Zealand

Christopher J. Hollis,
Institute of Geological and Nuclear Sciences,
Lower Hutt, New Zealand

INTRODUCTION

Radiolarians are sarcodine protozoans with shells of opaline silica. They are closely related to foraminifera, but are exclusively marine and planktonic. The group ranges from Cambrian to recent and occurs in all marine facies, although it is most common in pelagic sediments of the open ocean. Studies of fossil radiolarians have increased dramatically over the past two decades as the stratigraphic resolution of deep-sea drilling, combined with the scanning electron microscope and the hydrofluoric acid method of retrieving radiolarians from siliceous rocks, provided solutions to many of the taxonomic and phylogenetic problems that led earlier workers (e.g., Campbell, 1954) to despair of finding any biostratigraphic utility for the group. Radiolarians are now indispensable dating tools for Paleozoic, Mesozoic, and Cenozoic deep-sea sediments; they have a major role in the paleogeographic fingerprinting of displaced terranes, and are being used increasingly as an environmental proxy in paleoceanographic studies.

Despite the present high profile of the group, and in striking contrast to information available for their foraminiferal cousins, the global effects of events at the Cretaceous-Tertiary (K-T) boundary on radiolarian populations remain poorly known. Limited data have led to wide variations in estimates of their levels of K-T extinction. These estimates have ranged from 7% extinction (Emiliani and others, 1981) to 85% (Thierstein, 1982). This extreme range is not due to lack of interest, but rather to the absence of radiolarians from the vast majority of identified K-T boundary sections. Not only radiolarians but all siliceous microfossils are exceedingly rare in latest Cretaceous and early Paleocene sediments. It is only within the past six years that any reports of faunal or floral changes across well-delineated K-T boundary horizons have become available for radiolarians (Hollis, 1988, 1991, 1993), diatoms, or silicoflagellates (Harwood, 1988).

In modern oceans, siliceous plankton are only abundant in areas of high surface productivity (Kennett, 1982; Casey, 1993). Such areas occur where wind and current systems cause surface waters to diverge, resulting in the upwelling of nutrient-rich deeper waters. Biosiliceous oozes accumulate in east-west–trending belts beneath polar and equatorial divergent currents and beneath zones of coastal upwelling at continental margins. Similar oceanographic regimes are believed to have led to accumulation of past biosiliceous deposits (Ramsay, 1973; Hein and Parrish, 1987), although the intensity and location of circulation systems have undoubtedly changed in response to climatic changes and plate movements. Consequently, the location and extent of biosiliceous deposits in the past, as well as their biotic character, can be used to infer oceanographic phenomena. Obvious examples are extensive accumulations of biosiliceous sediments in Late Cretaceous, middle Eocene, and early Miocene oceans that imply periods of high global oceanic productivity (Leinen, 1979; Hein and Parrish, 1987; Blueford, 1989). Conversely, diminishment of siliceous belts indicates a decline in oceanic productivity. Perhaps the most extreme of these declines, since the appearance of diatoms in the mid-Cretaceous, spans the K-T boundary. The extreme scarcity of radiolarians and other siliceous microfossils through the K-T transition supports isotopic evidence for a gross reduction in oceanic productivity at the end of the Cretaceous (Hsü and McKenzie, 1985; Zachos and Arthur, 1986; Zachos and others, 1989).

A few high-latitude areas continued to sustain siliceous biotas through the K-T transition; notably in Seymour Island, Antarctica, where diatoms and silicoflagellates occur throughout a thick fine clastic K-T boundary sequence (Harwood, 1988), and in the pelagic succession of eastern Marlborough, on the northeastern edge of New Zealand's South Island (Fig. 1). An abundance of radiolarians in the much-studied Woodside Creek K-T boundary section in coastal eastern Marlborough was first noted 15 years ago (Strong, 1977). More recent studies (Hollis, 1991, 1993; Strong and others, 1995) have confirmed the richness of the Woodside radiolarian faunas and produced similar faunas from another four well-delineated K-T boundary sections in Marlborough: Wharanui Point, Chancet Rocks, Flaxbourne River, and Mead Stream (Fig. 1). Whereas chertification of these sediments has destroyed or rendered unidentifiable many diatom and silicoflagellate tests, the more robust tests of radiolarians remain intact and retrievable by standard extraction methods (Sanfilippo and Riedel, 1985).

Two attributes make these Marlborough faunas extremely valuable for improving understanding of environmental changes through the K-T transition. First, species originating in the Cretaceous form a significant component of early Paleocene assemblages. Therefore it is possible to apply knowledge of their ecological tolerances in the Cretaceous to assess the predictions for changes in abundance and local disappearances in the Paleocene made by alternative causal scenarios. This is impossible for groups that underwent mass extinction at the boundary, where arguments for evolutionary changes in the Paleocene are bounded by ad hoc ecological assessments of new Tertiary elements alone. Second, the five Marlborough sections represent ranges of depositional environments that allow for comparisons of patterns of faunal change in order to distinguish changes that are synchronous and common to all sections, and thereby relating to a regional event, from localized changes, which may relate to a particular environmental setting within the region.

LITHOLOGIES AND CORRELATION

Lithostratigraphy

Four of the five sections examined are located within a 15 km stretch of coastline between Woodside Creek and Wharanui Point in the south and Chancet Rocks in the north (Fig. 1). The fifth section is 15 km inland in the southern branch of Mead Stream, a tributary of the Clarence River (Fig. 1). The Late Cretaceous and Paleocene interval examined lies within Mead Hill Formation, the lower of two formations composing the Muzzle Group (Reay, 1993).

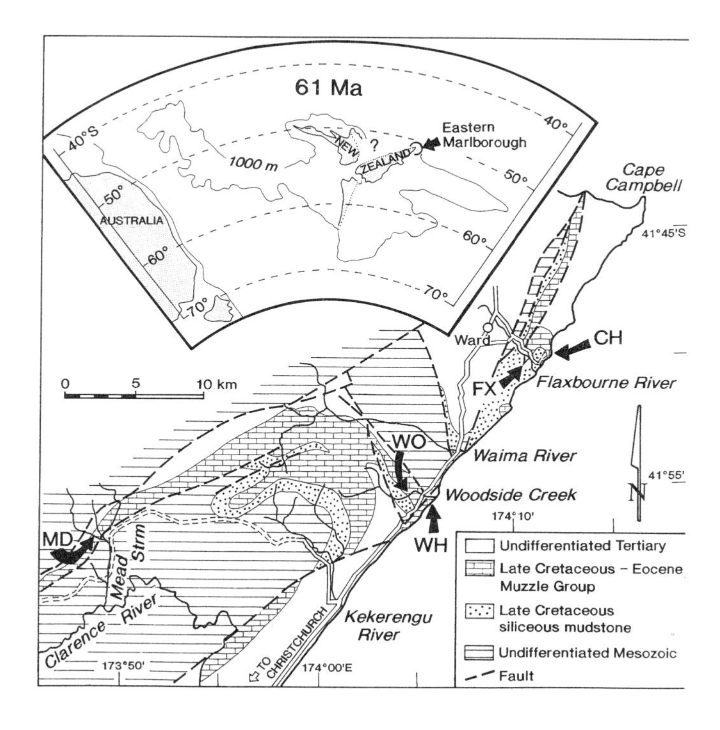

Figure 1. Simplified geologic map of eastern Marlborough, northern South Island, New Zealand (after Lensen, 1962), showing location of Cretaceous-Tertiary boundary sections examined at Mead Stream (MD), Woodside Creek (WO), Wharanui Point (WH), Flaxbourne River (FX) and Chancet Rocks (CH). Inset: Location of Marlborough on Paleocene paleogeographic map of New Zealand (after Weissel and others, 1977)

Lithologies range from fine-grained limestone to bedded chert (Fig. 2); there is only a minor terrigenous component. They are interpreted as open ocean, biogenic sediments. Coastal sections consist of a similar succession of pelagic lithofacies; i.e., Late Cretaceous pale siliceous limestone (10–60 m thick), becoming less siliceous in the upper 5 to 10 m; earliest Paleocene dark gray, clay-rich, calcareous chert (0.2–2 m), with a thin iridium-rich "boundary" clay at the base; pale calcareous chert (11–16 m), becoming less siliceous in the upper part; and limestone (25+ m). The calcareous chert in these sections has been previously labeled "siliceous limestone" (e.g., Strong, 1977; Hollis, 1993) based on field assessment. It is identified herein as calcareous chert because whole-rock geochemical analyses indicate a silica content of >50% (see below). The inland section at Mead Stream has a similar but even more siliceous succession, including a 23-m-thick chert unit in the basal Paleocene (Fig. 2). Other than a sharp lithologic change at the K-T boundary, contacts between all units are gradational.

The K-T boundary is well defined at all sections. It is placed at a distinct lithologic change that separates thick-bedded, moderately siliceous limestone that contains typical Late Cretaceous foraminiferal faunas, from thin-bedded calcareous chert that contains either early Paleocene foraminifera (Woodside Creek, Chancet Rocks—Strong, 1977, 1981, 1984a) or dwarfed Cretaceous foraminifera (Flaxbourne River and Mead Stream—Strong and others, 1987, 1995). Radiolarian

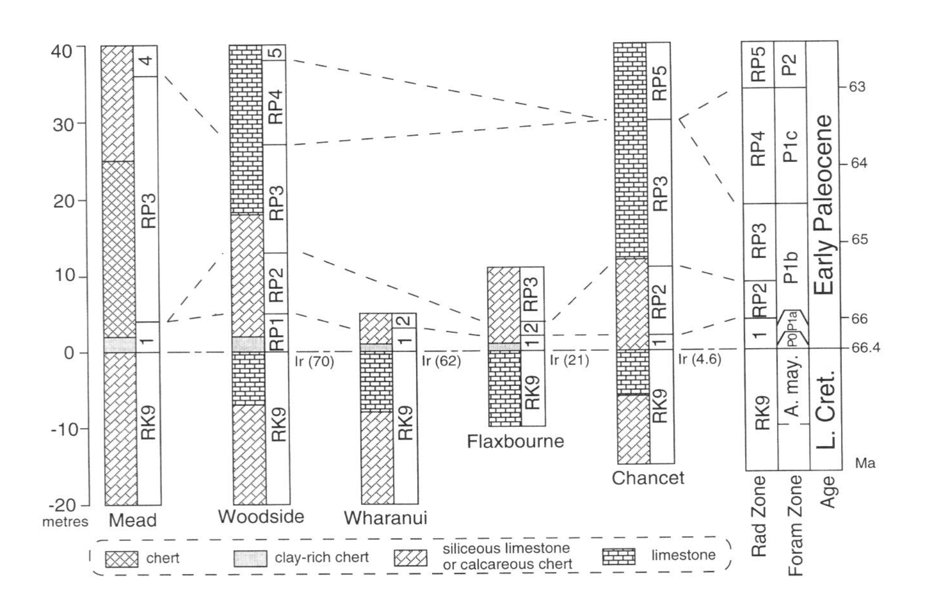

Figure 2. Lithologies and radiolarian correlation of sections (Strong and others, 1991; Hollis, 1993). Time scale after Berggren and others (1985). Basis of correlation with planktonic foraminiferal zones P0, P1a, and P1b discussed in text. Abundance of iridium (ng/g) in the Cretaceous-Tertiary boundary clay, or basal Paleocene, is shown for each of the coastal sections (Strong and others, 1988).

biostratigraphy (Hollis, 1993) supports lithologic and foraminiferal evidence for the boundary location at these four sections, and corroborates the lithologically located boundary at Wharanui Point (Strong and others, 1988). Cretaceous and Tertiary lithologies in the coastal sections are separated by a thin clay layer that bears the characteristic geochemical signature of K-T boundary clays (Alvarez and others, 1980; Brooks and others, 1986; Strong and others, 1987, 1988; Wolbach and others, 1988). A weak iridium anomaly of 0.8 ng/g has also been reported from the inferred boundary clay at Mead Stream (Strong and others, 1995).

Radiolarian Biostratigraphy

Hollis (1993) introduced the first Late Cretaceous–late Paleocene radiolarian zonation for the New Zealand region based on the faunal succession in the four coastal sections and Lord Howe Rise Deep Sea Drilling Project (DSDP) Site 208 (Figs. 1 and 2). The six new zones established are interval zones based on the first appearances of the nominate species: *Lithomelissa*? *hoplites* (RK9, Late Cretaceous); *Amphisphaera aotea* (RP1, earliest Paleocene); *A. kina* (RP2); *Stichomitra granulata* (RP3); *Buryella foremanae* (RP4); and *Buryella tetradica* (RP5, early-late to late-early Paleocene). Integrated biostratigraphic study of radiolarians, foraminifera, and dinoflagellates in the inland Mead section (Strong and others, 1995), where zone RP5 is succeeded by the late Paleocene *Bekoma campechensis* zone of Nishimura (1987), corroborates the relative ages of these zones. This new zonation has substantially improved overall age control for the Marlborough sections and has shown that at least 40 m of early Paleocene sediments of Mead Hill Formation overlie the K-T boundary at Woodside Creek and Chancet Rocks. Lack of age control led previous workers to correlate the upper part of these sections to the latest Paleocene to Eocene Amuri Limestone (e.g., Strong, 1977), the upper formation of the Muzzle Group (Reay, 1993). Woodside strata are particularly rich in radiolarians and, because the faunal succession is well resolved, the section serves as a reference section for five of the new zones (RK9 through RP4). DSDP Site 208 is the reference section of zone RP5.

Graphic Correlation

In order to compare patterns of faunal change between sections it is important to have a biostratigraphic framework that allows the comparison of equivalent time intervals. The established zonation permits such analysis at a broad and qualitative level. In this report the graphic correlation method (Shaw, 1964; Edwards, 1984; MacLeod and Keller, 1991) is used to refine correlation of the early Paleocene by establishing a composite standard reference section (CSRS) consisting of 40 microfossil events (Figs. 3 and 4). The radiolarian zonal stratotype, Woodside Creek, is used as the standard reference section (SRS) because it contains the greatest number of events and event horizons (22 events at 14 horizons). The primary events used for correlation in the earlier zonation and in the present scheme are first appearances of Paleocene radiolarians. These are integrated with datums from planktonic foraminifera and, at Site 208, calcareous nannofossils. These microfossil events were discussed by Hollis (1993) and provide an independent biostratigraphic framework for assessing the Tertiary ranges of the many Cretaceous survivors; last appearances of radiolarians are used only as a final

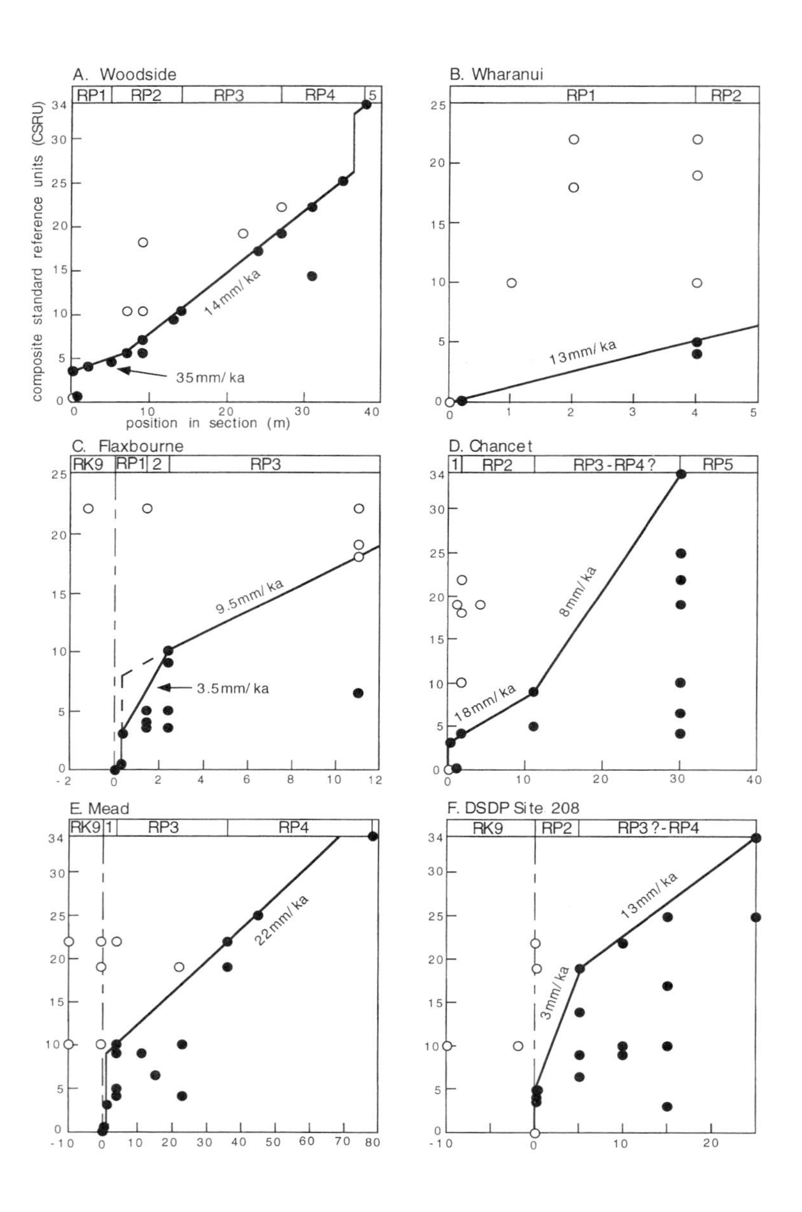

Figure 3. Graphic correlation of (A) Woodside, (B) Wharanui, (C) Flaxbourne, (D) Chancet, (E) Mead sections, and (F) Deep Sea Drilling Project (DSDP) Site 208. Woodside Creek is the standard reference for the composite standard reference section. Filled circles are first appearance datums (FADs); open circles are last appearance datums (LADs). Solid lines of correlation (LOCs) are best-case models (used in Figs. 6 and 9) and dashed LOC at Flaxbourne River is a worst-case model. Where possible LOCs are drawn to intersect FADs which are the primary events for correlation. Vertical LOCs mark the stratigraphic location (x axis) and temporal duration (y axis) of hiatuses.

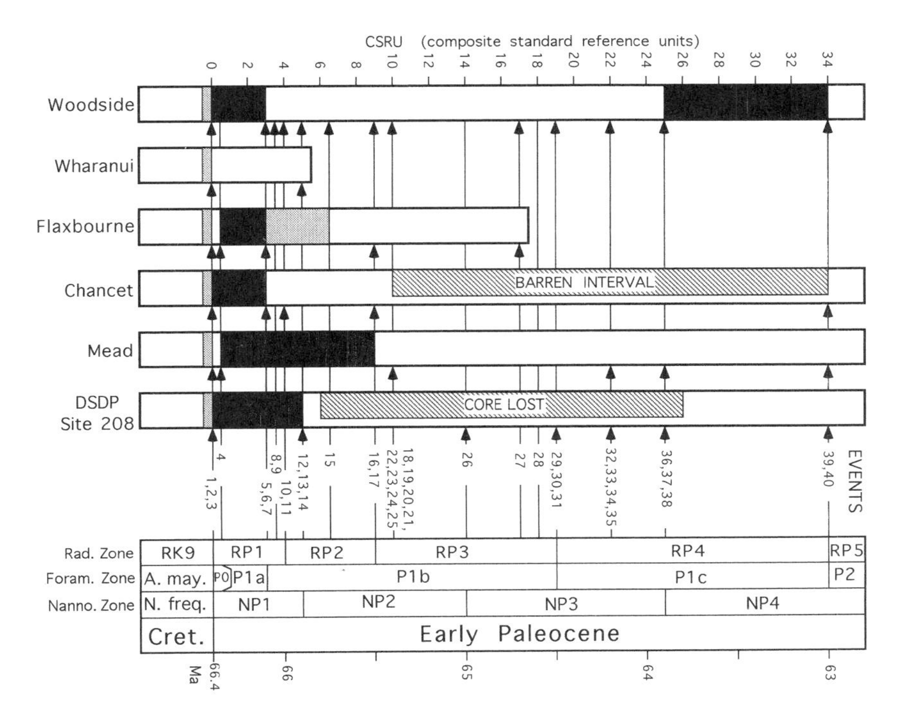

Figure 4. Estimated extent and temporal distribution of hiatuses within five Marlborough sections and Deep Sea Drilling Project (DSDP) Site 208. White = interval present under the best- and worst-case graphic correlation models; stippled = interval present under best-case model but absent under worst-case model; black = interval absent under best- and worst-case models. Arrows indicate event horizons identified at individual sections. Events identified within "core lost" interval at Site 208 are based on core catcher samples (see Hollis, 1993). Preliminary geochronological calibration uses a direct linear relation between composite standard reference units (CSRU) and the time scale of Berggren and others (1985). FAD is first appearance datum, LAD is last appearance datum. The 40 events used for correlation are as follows: 1: LAD large Cretaceous foraminifera; 2: FAD dwarf Cretaceous planktonic foraminifera; 3: FAD Amphisphaera aotea*; 4: FAD* Eoglobigerina *cf.* fringa*; 5–7: FADs* Globoconusa daubjergensis, Subbotina triloculinoides, Subbotina pseudobulloides*; 8–9: FADs* Lithelius *n.sp. A*, Saturnalis kennetti*; 10–11: FADs* Amphisphaera kina, A. goruna*; 12–14: FADs* Stichomitra wero, Clathrocycloma *n.sp. A*, Cruciplacolithus tenuis*; 15: FAD Spyrida gen. et spp. undet.; 16–17: FADs* Stichomitra granulata, Dictyophimus*? sp. A; 18–20: FADs* Amphisphaera radiosa, Stylosphaera *spp. gp. B, Actinomminae gen. gp. B; 21–25: LADs* Amphisphaera aotea, Patulibracchium *spp.*,* Lithomelissa*? *hoplites, Lophophaena*?* polycyrtis, Dictyomitra *sp. A; 26: FAD* Chiasmolithus danicus*; 27: FAD* Spongodiscus communis*; 28: LAD* Amphipyndax *n.sp. A; 29: FAD* Buryella foremanae*; 30–31: LADs* Stylosphaera pusilla, Siphocampe altamontensis*; 32: FAD* Buryella dumitricai*; 33–35: LADs* Prunobrachium kennetti, Orbiculiforma renillaeformis, Stichomitra compsa*; 36–38: FADs* Prinsius martinii, Amphisphaera *spp. gp. D*, Clathrocycloma *spp. gp. B; 39–40: FADs* Morozovella uncinata, Buryella tetradica.

means of control. For expediency, and to allow comparison with earlier age estimates (Hollis, 1993; Strong and others, 1995), composite standard reference units (CSRU) have a linear relation to the geochronometric units in the time scale of Berggren and others (1985; see Fig. 4). This does not imply that they will continue to do so as more biostratigraphic data becomes available for the microfossil succession. No events are available to improve stratigraphic resolution in the Cretaceous of these sections. Although the late Maastrichtian *Abathomphalus mayaroensis* foraminiferal zone, or its New Zealand equivalent, has been identified at all sections (Strong, 1977, 1984a; Strong and others, 1987, 1988, 1995), a sharp lithologic break at the boundary raises the possibility that the very latest Cretaceous strata are not preserved.

Graphic correlation also offers an assessment of temporal completeness (MacLeod and Keller, 1991). Despite a relatively small number of sections and datums, and considerable variation in the quality and quantity of biostratigraphic data, the method provides a useful, if preliminary, estimate of the extent and duration of early Paleocene hiatuses in the Marlborough sections (Figs. 3 and 4). Standard foraminiferal zonations for the K-T boundary interval (e.g., Smit, 1982; Smit and Romein, 1985; Keller, 1988; MacLeod and Keller, 1991) have not been recognized in the New Zealand region. The Flaxbourne and Mead sections appear to contain the most biostratigraphically complete K-T interval in the region (Strong and others, 1987, 1995). These two sections exhibit a similar sequence of foraminiferal events that is comparable to standard zonations. An interval containing dwarfed Cretaceous planktonic foraminifera and considered to be equivalent to the earliest Paleocene *Guembelitria cretacea* zone (zone P0 of Smit, 1982; see also MacLeod and others, this volume) occurs directly above the lithologically located K-T boundary in both sections. It is followed by an interval, containing small Tertiary planktonic species (including *Eoglobigerina* cf. *fringa*), that is provisionally correlated to the *Parvularugoglobigerina eugubina* zone (zone P1a; usage of Berggren and others, 1985). The succeeding *Subbotina pseudobulloides* zone (zone P1b) is recognized by the first appearance of the nominate species at Mead Stream (Strong and others, 1995) and of *Globoconusa daubjergensis* at Flaxbourne River (Strong and others, 1987). Although further study may confirm these provisional correlations, graphic correlation indicates temporal incompleteness at both sections (Figs. 3 and 4). A probable early Paleocene hiatus within radiolarian zones RP1–RP2 is comparable to a hiatus that MacLeod and Keller (1991) inferred for the Flaxbourne section based on Strong's published data (lower part of planktonic foraminiferal zone P1b).

The presence of an iridium-rich boundary clay in the sections is not proof of stratigraphic completeness unless the clay is identified conclusively as a primary deposit derived from a geologically instantaneous event such as the proposed K-T bolide impact (Alvarez and others, 1980). Interpretations of the boundary layer at Woodside Creek as primary fallout from an impact (Alvarez and others, 1980; Bohor and others, 1987; Wolbach and others, 1988; Hollander and others, 1993) fail to take into consideration biostratigraphic evidence indicating that the earliest Paleocene is missing. The boundary clay at Woodside Creek contains the highest iridium concentration of all the Marlborough sections and one of the highest in the world (70 ng/g—Strong and others, 1988). It also contains abundant "impact-shocked" mineral grains (Bohor and others, 1987) and "soot" particles (Wolbach and others, 1988), the latter being used as primary evidence for global wildfires inferred to have been sparked by the impact. However, Strong (1977) reported the foraminiferal species *Globoconusa daubjergensis* and *Morozovella triloculinoides* (events 4 and 5 in Fig. 4) from this same 10-mm-thick boundary clay. Both species occur consistently over the succeeding 2 m and recent reexamination of these critical faunas confirms the identifications (C. P. Strong, 1993, personal commun.). In standard zonations (Smit, 1982; Smit and Romein, 1985; Toumarkine and Luterbacher, 1985) the first appearance datums (FADs) of these two species are at the base of planktonic foraminiferal zone P1b, about 300 ka above the K-T boundary (66.1 Ma—Berggren and others, 1985). Although some studies have reported earlier occurrences of these species in upper P1a (66.3 Ma—MacLeod and Keller, 1991), there is no evidence for such early occurrences in those New Zealand sections in which an interval considered to be equivalent to zone P1a is preserved (Waipara River—Strong, 1984b; Flaxbourne River—Strong and others, 1987; Mead Stream—Strong and others, 1995). The only way to conclude that the earliest Paleocene is preserved at Woodside Creek is to assume that the foraminifera occur near the top of the mudstone and that there is a hiatus within the mudstone separating the zone P1b markers from the geochemical anomalies. Even if these species occur in zone P1a, an extraordinarily low rate of sediment accumulation (no more than 0.1 mm/Ka after compaction) is required for an intact earliest Paleocene interval. The average rate for the section is 14 mm/ka. Although it is possible that both a thin boundary layer is preserved and the two species occur earlier than their established datums, for practical purposes the Woodside section is considered to lack the first 300 ka of the Paleocene.

The earliest Paleocene is also considered to be missing at Chancet Rocks, although there is a possibility that the basal unit is either condensed or a hiatus occurs between the lithologically determined boundary and the appearance of P1b foraminifera, *Subbotina pseudobulloides* and *S. triloculinoides*, 25 cm higher. Wharanui Point is considered to be intact in the absence of evidence to the contrary. Foraminiferal data are lacking for the earliest Paleocene, but radiolarian zone RP1 is relatively thick (> 2 m). Note, however, zone RP1 is at least 3 m thick at Woodside Creek, but foraminiferal evidence indicates that the earliest Paleocene is missing.

RADIOLARIAN SURVIVAL ACROSS THE K-T BOUNDARY

Radiolarian Occurrence

Taxonomic study of the radiolarian assemblages was carried out using standard methods of optical and scanning electron microscopy. Faunal counts were made on total faunas (60–300 individuals) or splits of sample residues (300–1000 individuals). As is the case in most radiolarian studies, faunal diversity was too great and taxonomic distinctions too poorly known for all individuals to be assigned to species. Of the 65 species or species groups identified, 45 were encountered in the Cretaceous (Plates 1 and 2). The ranges of 44 of these species at Woodside

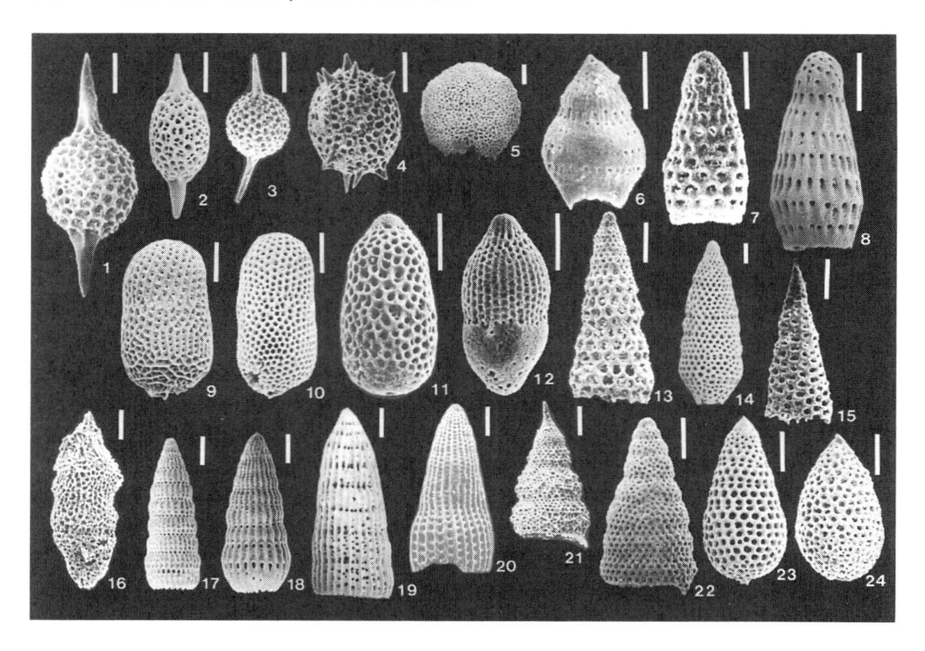

Plate 1. Established Cretaceous radiolarian species occurring in the Paleocene of Marlborough. All specimens are from Woodside Creek and are deposited in the Geology Department, University of Aukland. Notations refer to radiolarian zone, distance below (-) or above (+) the K-T boundary, New Zealand fossil record file number, and Aukland University curation number. Scale bar = 50 µm. 1: Amphisphaera privus *(Foreman). Paleocene, RP1, +2 m, P30/f368, R66. 2:* Protoxiphotractus perplexus *Pessagno. Paleocene, RP4, +31 m, P30/f375, R67. 3:* Stylosphaera pusilla *Campbell and Clark. Paleocene, RP1, +2 m, P30/f368, R68. 4:* Lithomespilus coronatus *Squinabol. Paleocene, RP2, +9 m, P30/f371, R69. 5:* Orbiculiforma renillaeformis *(Campbell and Clark). Paleocene. RP4, +27 m, P30/f374, R70. 6:* Theocampe vanderhoofi *Campbell and Clark. Paleocene, RP1, +2 m, P30/f368, R71. 7:* Siphocampe altamontensis *(Campbell and Clark). Cretaceous, RK9, -30 m, P30/f363, R72. 8:* Siphocampe *sp. A (*Theocampe *sp. A of Empson-Morin, 1984). Paleocene, RP4, +31 M, P30/f375, R73. 9:* Theocapsomma erdnussa *(Empson-Morin). Paleocene, RP4, +31 m, P30/f375, R74. 10:* Theocapsomma? comys *Foreman. Paleocene, RP4, +31 m, P30/f375, R75. 11:* Theocapsomma amphora *(Campbell and Clark). Paleocene, RP4, +27 m, P30/f374, R76. 12:* Myllocercion acineton *Foreman. Paleocene, RP1, +2 m P30/f368, R77. 13:* Amphipternis alamedaensis *(Campbell and Clark). Cretaceous, RK9, -18 m, P30/f364, R78. 14:* Amphipyndax stocki *(Campbell and Clark). Paleocene, RP4, +31 m, P30/f375, R79. 15:* Cornutella californica *Campbell and Clark. Paleocene, RP1, +2 m, P30/f368, R80. 16:* Lithomelissa? hoplites *Foreman. Paleocene, RP3, +14 m, P30/f372, R81. 17:* Dictyomitra andersoni *(Campbell and Clark). Paleocene, RP4, +31 m, P30/f375, R82. 18:* Dictyomitra multicostata *Zittel. Paleocene, RP4, +31 m, P30/f375, R83. 19:* Archaeodictyomitra lamellicostata *(Foreman). Paleocene, RP4, +31 m, P30/f375, R84. 20:* Mita regina *(Campbell and Clark). Paleocene, RP3, +14 m, P30/f372, R85. 21:* Stichomitra asymbatos *Foreman. Paleocene, RP3, +14 m, P30/f372, R86. 22:* Stichomitra compsa *Foreman. Cretaceous, RK9, -30 m P30/f363, R87. 23:* Stichomitra livermorensis *(Campbell and Clark). Paleocene, RP4, +31 m, P30/f375, R88. 24:* Stichomitra campi *(Campbell and Clark). Paleocene, RP1, +2 m, P30/f368, R89.*

Creek are shown in Figure 5, supplemented with indications of the highest range at other sections (excluding isolated occurrences). The remaining species, *Lophophaena? polycyrtis*, occurs in one Cretaceous sample at Site 208 and two early Paleocene samples at Mead Stream. In the following discussion it should be noted that these 45 species probably account for no more than 50% of the total number of Cretaceous species in the sections, and the proportion of Tertiary-restricted taxa allocated to species is substantially less (~30%). Abundances of

Plate 2. Cretaceous-Paleocene radiolarian species of the New Zealand region. All specimens are from Woodside Creek and are deposited in the Geology Department, University of Aukland. Scale bar = 50 μm. 1: Protoxiphotractus *sp. A. Paleocene, RP4, +31 m, P30/f375, R90. 2:* Stylosphaera *sp. A. Paleocene, RP4, +27 m, P30/f374, R91. 3:* Phaseliforma subcarinata *Pessagno. Paleocene, RP4, +27 m, P30/f374, R92. 4:* Phaseliforma *sp. A. Paleocene, RP4, +31 m, P30/f375, R93. 5:* Orbiculiforma *sp. A. Paleocene, RP4, +31 m, P30/f375, R94. 6:* Spongotrochus *sp. Paleocene, RP1, +2 m, P30/f368, R95. 7:* Prunobracchium kennetti *Pessagno. Cretaceous, RK9, -18 m, P30/f364, R96. 8:* Botryostrobus*? sp. Paleocene, RP1, +2 m, P30/f368, R97. 9:* Lophophaena*? sp. Paleocene, RP1, +2 m, P30/f368, R98. 10:* Lithomelissa *cf.* heros *Campbell and Clark. Paleocene, RP1, +2 m, P30/f368, R99. 11:* Myllocercion *sp. A. Paleocene, RP4, +31 m, P30/f375, R100. 12:* Mita *sp. A. Paleocene, RP4, +31 m, P30/f375, R101. 13:* Dictyomitra *sp. A. Cretaceous, RK9, -30 m, P30/f363, R102. 14:* Amphipyndax *sp. A. Paleocene, RP1, +2 m, P30/f368, R103. 15:* Eusyringium*? sp. Paleocene, RP4, +31 m, P30/f375, R104. 16:* Lithocampe *sp. A.* (sensu *Dumitrica, 1973). Paleocene, RP1, +2 m, P30/f368, R105. 17:* Lithocampe *sp. B. Paleocene, RP4, +31 m, P30/f375, R106. 18:* Stichomitra *sp. A* (sensu *Dimitrica, 1973). Paleocene, RP4, +27 m, P30/f374, R107. 19:* Neosciadiocapsa jenkinsi *Pessagno. Cretaceous, RK9, -18 m, P30/f364, R108.*

Cretaceous species are combined with those of well-established Cretaceous genera in the plots shown in Figure 6. These taxa provide a minimum value for the abundance of Cretaceous taxa, because many poorly differentiated groups are excluded, particularly within spumellarian families (actinommids, litheliids,

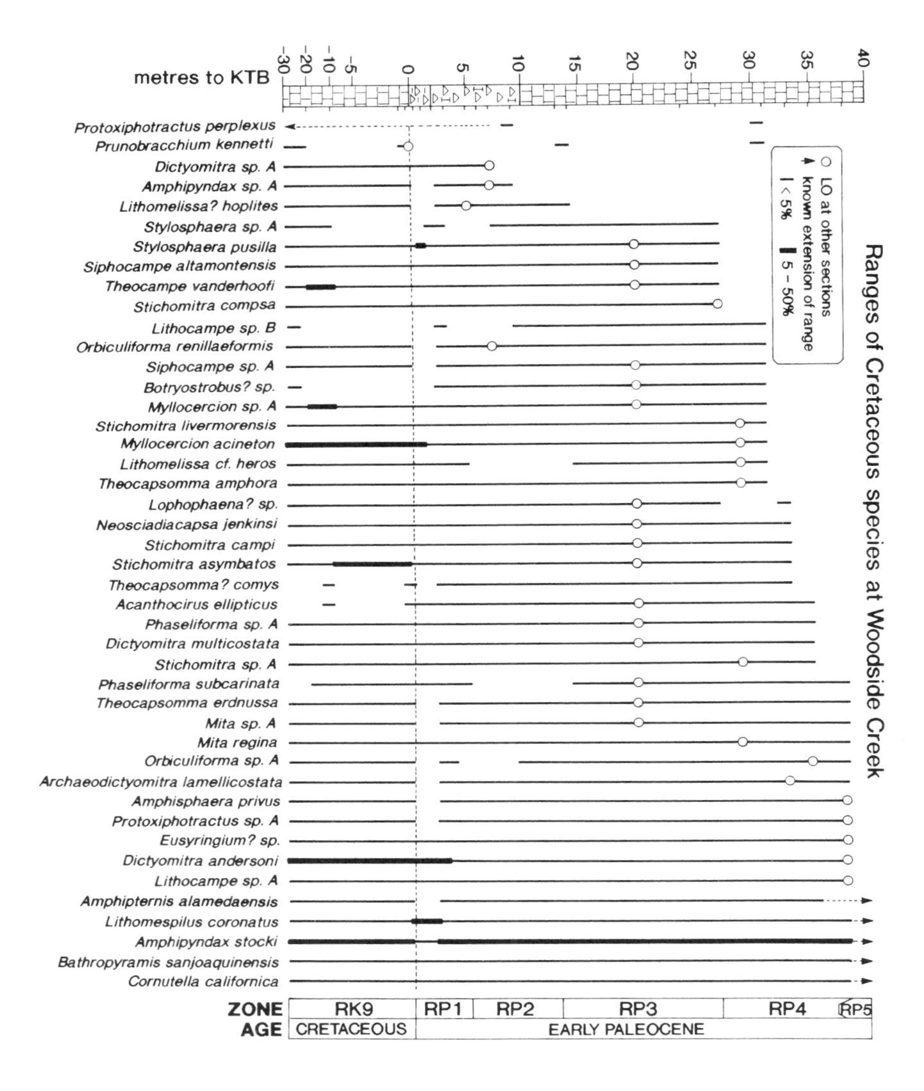

Figure 5. Stratigraphic ranges of species identified in Cretaceous at Woodside Creek. A break in a range represents absence from two or more adjacent samples. Thin lines = less than 5% of total fauna; thick lines = 5%–50%. Open circles represent the last consistent occurrence at other Marlborough sections or Deep Sea Drilling Project (DSDP) Site 208 (Hollis, 1991, 1993); arrow = known extension of range (Petrushevskaya and Kozlova, 1972; Foreman, 1973; Sanfilippo and Riedel, 1973; Strong and others, 1995).

spongurids and spongodiscids). The "counting group" approach of Westberg and Riedel (1978) was used to assign individuals to 94 taxonomic groupings ranging from species to suborder (i.e., Spyrida). Discussion of faunal changes across the K-T boundary focuses on survivorship patterns at the level of species and genus, because most higher level taxonomic groupings are still highly artificial in Radiolaria.

Faunal richness and diversity are greatest at Woodside and Wharanui sections (averaging 50 and 45 taxa per sample, respectively; Fisher α diversity indices of 10–15). The possible ecological significance of apparent lower diversity in other sections (~30 taxa per sample; Fisher α indices 6–10) is uncertain. Differences are partly an artifact of better preservation of Woodside faunas, more time spent studying them, and close similarities between Woodside and Wharanui assemblages. Therefore more taxa are lumped at the other sections. Cretaceous samples frequently yield abundant radiolarians but taxic richness is only moderate (average of 55 taxa at Woodside Creek, 25 at Flaxbourne River), partly due to relatively poor preservation. Tests are altered to microcrystalline quartz and frequently infilled with a similar matrix in the Late Cretaceous and earliest Paleocene.

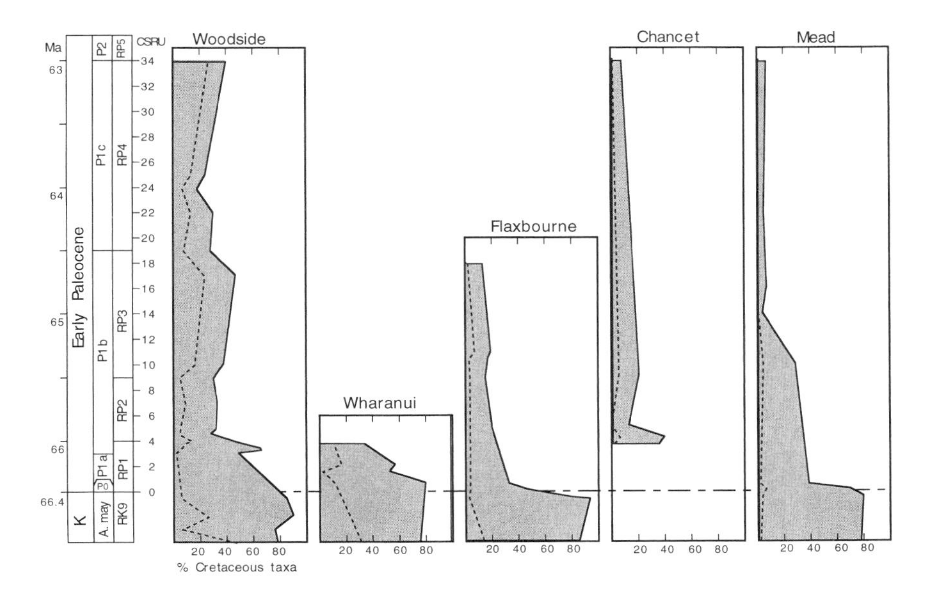

Figure 6. Variations with age in relative abundance of Cretaceous survivors through the early Paleocene of five Marlborough sections. Dashed line indicates the abundance of a single Cretaceous species, Amphipyndax stocki. *Samples are related to composite standard reference units (CSRU) by the best-case graphic correlation model (Fig. 3) and assuming a Cretaceous rate of sediment accumulation (compacted) equal to the average Paleocene rate.*

Above zone RP1 preservation improves. Tests are composed of cryptocrystalline quartz and infilling material is easily removed by standard cleaning methods (Sanfilippo and Riedel, 1985). Improved preservation is considered to be the primary reason for an increase in richness of the Paleocene fauna (average of 65 at Woodside Creek, 35 at Flaxbourne River). This increase in richness is not accompanied by a substantial increase in Fisher α index values at any of the sections.

Cretaceous assemblages lack many low-latitude species (e.g., *Amphipyndax tylotus*), and high-latitude taxa, such as *Prunobrachium*, are rare (Empson-Morin, 1984). They most closely resemble mid-latitude faunas with a slightly stronger high-latitude aspect than Site 208 to the north, where *Prunobrachium* is absent, and less than DSDP Site 275 to the south, where *Prunobrachium* and other robust spongy species are common (Fig. 1). The Marlborough assemblages are similar to those of the Late Cretaceous of California (Campbell and Clark, 1944; Foreman, 1968) with 22 species in common.

K-T Survival

Initial study of the radiolarian succession at Woodside Creek (Hollis, 1988) showed that a surprisingly high number (30) of Cretaceous species crossed the K-T boundary. This conflicted with the common perception that radiolarians underwent substantial extinctions at the end of the Cretaceous (Thierstein, 1982; Sanfilippo and Riedel, 1985; Sanfilippo and others, 1985; Ward, 1990; see Casey, 1993, for a contrasting view). This earlier conclusion was based on marked differences between known latest Cretaceous and late Paleocene faunas and a single corehole through the K-T transition at Cima Hill, California (Foreman, 1968). Foreman inferred an extinction horizon, because of the 77 species she recorded from the latest Cretaceous of California, only 5 persisted into the Paleocene. However, not only is the Cima Hill Cretaceous relatively impoverished (only 38 species), but the boundary is poorly constrained by foraminifera and dinoflagellates (Drugg, 1967) to a 3 m transitional interval. This interval is of possible Paleocene age and contains 25 of the 38 species reported from the Cima Hill Cretaceous. Moreover, Foreman did not document the entire fauna, but restricted her attention to mainly nassellarians. For example, in her reexamination of one fauna, only three species are listed of the 34 spumellarians first reported by Campbell and Clark (1944). Nor did Foreman describe the Paleocene faunas or consider the possibility that the rapid faunal turnover was due to local environmental changes. The Woodside findings did not conflict with reports of numerous Cretaceous species in the early Paleocene at DSDP Site 208 (Dumitrica, 1973), but reports of early Paleocene faunas from other areas were either lacking or inconclusive (e.g., Kozlova, 1984).

Evidence for significant radiolarian survival strengthened after further sampling of Woodside Creek and other known K-T sections in Marlborough, and a reexamination of the Site 208 faunas, was followed by more detailed study of both Cretaceous and Paleocene faunas. None of the 45 species now identified from the Cretaceous of Marlborough disappeared at the K-T boundary (Fig. 5). Three Cretaceous species are restricted to isolated earliest Paleocene occurrences, but 38 species range into mid-early Paleocene (RP3–RP4). Other than the three sporadically occurring species, there is little to suggest that reworking could account for these Paleocene occurrences. They occur with new Paleocene forms in the same state of preservation (improving upsection), and have a high degree of consistency up to their last appearance. Excluding uncertain identifications, 25 of these species have been reported from the Paleocene or Eocene of California (Foreman, 1968), the central Atlantic (Petrushevskaya and Kozlova, 1972; Foreman, 1973; Sanfilippo and Riedel, 1973), Site 208 (Dumitrica, 1973), Siberia (Kozlova, 1984), or Hokkaido (Iwata and Tajika, 1986, 1992). In addition, 12 of the species also occur in possibly Paleocene strata in the Cima Hill core (Foreman, 1968). The many Cretaceous taxa that remain unassigned to species imply that further study will reveal more survivors along with some species that disappear at the boundary.

Survivorship Trends

Despite high continuity in faunal composition across the K-T boundary in eastern Marlborough, the Paleocene ranges of these "Cretaceous survivors" and their relative abundance show significant differences between sections. Most ranges extend higher into the Paleocene at Woodside Creek than they do at other sections (Fig. 5). Relative abundance of Cretaceous survivors (Fig. 6) is also higher in the Woodside and Wharanui Paleocene. Most Cretaceous survivors decrease in abundance across the boundary in all the sections, although two Cretaceous spumellarian species (*Lithomespilus coronatus* and *Stylosphaera*? *pusilla*) experience a short-lived increase. At Woodside Creek, overall abundance of Cretaceous survivors decreases by 50% during the earliest Paleocene (RP1–RP2) but stabilizes at 30%–40% of the total faunas for the remainder of the early Paleocene (RP3–RP5). At the Flaxbourne, Chancet, and Mead sections, Cretaceous survivors decline to 30%–40% of the total faunas during the earliest Paleocene (RP1–RP2), and then decline rapidly to 5%–10% in the remaining Paleocene (RP3–RP5). A similar rapid decline in Cretaceous survivors occurs at DSDP Site 208 (Hollis, 1993). High relative abundance of Cretaceous survivors in the upper Woodside section is partly due to the sustained high abundance (10%–30%) of a single species, *Amphipyndax stocki*, which dominates Cretaceous assemblages at Woodside and Wharanui sections, but is comparatively rare at other sections.

Before considering these patterns in more detail, some brief comments on higher level changes in radiolarians through the Late Cretaceous and early Tertiary are warranted. The spumellarian families dominant in the Marlborough Cretaceous—actinommids, litheliids, spongurids, and spongodiscids—in general increase in abundance and diversity in the Paleocene. Rare families survive into the early Paleocene but are largely gone or extremely rare by late Paleocene—prunobrachiids, patulibracchiids, phaseliformids, and orbiculiformids. Among the nassellarians, many Cretaceous elements within long-ranging families decline rapidly in the early Paleocene. The artostrobiids (*Siphocampe*, *Theocampe*) and cryptocephalic groups (*Myllocercion*, *Theocapsomma*) do not appear to show significant diversification again until the Eocene or Oligocene, and it is likely that many of these later Tertiary species derive from different ancestral stock. Some Cretaceous theoperids (*Stichomitra asymbatos*, *S. compsa*) are rapidly replaced by new species (e.g., *S. wero*, *S. granulata*, and *Buryella* spp.) while others remain at Cretaceous levels of abundance through the early Paleocene (*Stichomitra livermorensis*, *Eusyringium*? sp., *Lithocampe* sp. A). The amphipyndacids and archaeodictyomitrids remain abundant in the early Paleocene at Woodside Creek and persist into the late Paleocene at Mead Stream; *Amphipyndax stocki* and *Amphipternis alamedaensis* both appear to range into the Eocene (Foreman, 1973; Takemura, 1992; Strong and others, 1995). The two Cretaceous plectopyramidiid species (*Bathropyramis sanjoaquinensis*, *Cornutella californica*) appear to thrive in the early Paleocene and may well be the forerunners of many late Paleocene and Eocene species. The earliest Paleocene sees the first appearance of several important Tertiary groups within the actinommids (e.g., the *Amphisphaera aotea/goruna* lineage), theoperids (*Stichomitra wero/Buryella* lineage, *Clathrocycloma*, *Dictyophimus*?) and the first members of the suborder Spyrida (Hollis, 1993).

RADIOLARIAN FAUNAL ANALYSIS

Paleoenvironment

Because radiolarians are planktonic, their fossil assemblages are inevitably thanatocoenoses (skeletal remains that have been brought together after death by current transport or settling through the water column). In principle, with sufficient sampling in sediments from various known depths and environments tied to depth distributions established from collections of living plankton in the water column, it would be possible to subdivide fossil assemblages into their original biotopes (groups of species from a distinct environment). In practice, however, such analysis is fraught with difficulties, because there is still a great lack of data on living radiolarian distributions and the bulk of paleoecological study has been restricted to localized investigations. The problem is extreme in Paleogene and older sediments where many of the radiolarians have no known living analogues and very few paleoecological studies have been made. Nevertheless, the few studies that are available (e.g., Empson-Morin, 1984) can be tied to established distributions and local paleoenvironmental data from other sources to provide insights into the pattern of radiolarian faunal change through the K-T transition of Marlborough.

Foraminifera indicate that Woodside Creek had a fully oceanic outermost-shelf to uppermost-slope setting (uppermost bathyal) in the Late Cretaceous (Strong, 1977) and that the Flaxbourne, Chancet, and Mead sections were situated in a deeper mid-bathyal setting on the upper to middle slope (Strong, 1984a; Strong and others, 1987, 1995). Differences between radiolarian assemblages at Woodside Creek and these deeper sections, as outlined below, support these depth relations. Radiolarians also indicate equivalent depths for the Woodside and Wharanui sections. Although all the sediments are essentially pelagic ooze, a slightly higher sand and clay content at Woodside Creek adds another line of support for these depth relations.

Diatom abundance and silica content provide information on relative nutrient levels. Latest Cretaceous strata are highly calcareous and contain very few diatoms at Flaxbourne River, are moderately siliceous with common diatoms at Woodside Creek and Wharanui Point, and are highly siliceous with abundant diatoms at Mead Stream (Hollis, 1991; Strong and others, 1995). This suggests varying influence of upwelling along the Late Cretaceous Marlborough coast: little upwelling in the area of the Flaxbourne and Chancet sections, moderate upwelling in the area of the Woodside and Wharanui sections, and strong upwelling at Mead Stream. However, present geographic relations of these sections may bear little resemblance to their Cretaceous-Paleogene setting. Intense deformation since inception (25 Ma) of the present transform plate boundary directly west of Marlborough Muzzle Group outcrops may have caused substantial lateral displacements along with some 45° of counterclockwise rotation of the Marlborough block (Fig. 1; Prebble, 1980).

Faunal Associations and Biotopes

Four types of radiolarian thanatocoenosis (associations) are recognized in the Marlborough assemblages on the basis of the dominant faunal elements: *Amphipyndax*, *Theocampe/Stichomitra asymbatos*, *Amphisphaera aotea/kina*, and *Stichomitra wero/granulata* (Fig. 7). The first two are primarily nassellarian-dominated Cretaceous assemblages and reflect differences in paleodepth between the sections. The remaining two are spumellarian dominated and spumellarian rich, respectively, and are intimately connected to environmental changes through the K-T transition.

1. The *Amphipyndax* association occurs in the Cretaceous of the uppermost bathyal Woodside and Wharanui sections. It is characterized by a dominance of *Amphipyndax stocki* (usually >20% of the total fauna) and common *Dictyomitra*

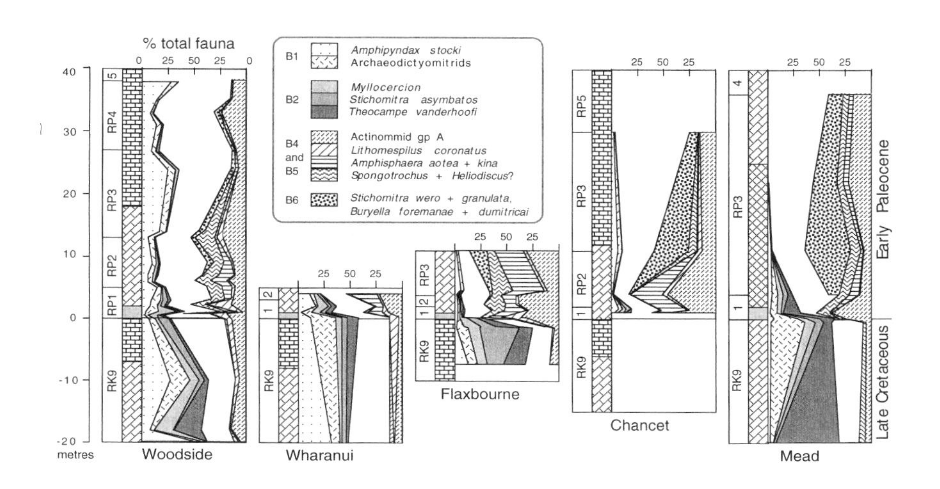

Figure 7. Variations with stratigraphic position in relative abundance of main faunal elements in the latest Cretaceous and early Paleocene of five Marlborough sections. Dominant Cretaceous biotopes (B1 and B2) are plotted from the left axis; dominant Paleocene biotopes (B4, B5, and B6) are plotted from the right axis. Minor biotope B3 is not shown.

andersoni, *Stichomitra asymbatos*, *Myllocercion acineton*, *M.* aff. *echtus*, and *Theocampe vanderhoofi*. Equivalent associations also occur in the upper Paleocene at Woodside Creek (upper RP3–RP4), but differ in that they contain more actinommids (e.g., *Amphisphaera goruna*) and litheliids, and very rare *Myllocercion* spp., *S. asymbatos*, and *T. vanderhoofi*. Despite the scarcity of these latter Cretaceous species, other Cretaceous survivors occur consistently and in moderate abundance (1%–2%) in these early Paleocene assemblages; notably, *Amphisphaera privus*, *Protoxiphotractus* sp. A, the *Cromyodruppa concentrica* group, *Phaseliforma* spp., *Theocapsomma erdnussa*, *Archaeodictyomitra lamellicostata*, *Lithocampe* sp. A, *Eusyringium*? sp. A, *Bathropyramis sanjoaquinensis*, and *Cornutella californica* (Fig. 5). *Cromyodruppa concentrica* group is the most common litheliid group in the Cretaceous and Paleocene, but it is not included in abundance estimates of Cretaceous survivors because it is not known which of its many forms may be restricted to the Cretaceous or to the Paleocene.

2. The *Theocampe/Stichomitra asymbatos* association occurs in the Cretaceous of the mid-bathyal Flaxbourne and Mead sections. It is characterized by the codominant *Theocampe vanderhoofi* and *Stichomitra asymbatos* groups, common *Myllocercion acineton*, *Dictyomitra multicostata*, *D. andersoni*, and comparatively rare *A. stocki* (usually < 5%). Cretaceous Mead faunas also contain common *Prunobrachium* and *Spongotrochus* spp. An intermediate association codominated by *Stichomitra asymbatos* and *Amphipyndax stocki* occurs in the mid-bathyal Cretaceous at Site 208 (Hollis, 1993). This association appears very similar to late Maastrichtian faunas of California (Foreman, 1968), where *Dictyomitra* spp. and *Theocampe vanderhoofi* appear to be the dominant nassellarians.

3. The *Amphisphaera aotea/kina* association occurs in the earliest Paleocene of all sections. In RP1 it is characterized by a codominance of *A. aotea* and the Cretaceous survivor *Lithomespilus coronatus*, and common *Stylosphaera pusilla* (Figs. 7 and 8). In RP2 *A. aotea* is replaced by abundant *A. kina*, while *L. coronatus* and *S. pusilla* decline markedly in abundance and spinose discoidal spumellarians (*Heliodiscus*? and *Spongotrochus* spp.) become common (Fig. 8). A similar association in RP2 at Site 208 includes common *A. kina* and *Amphipyndax stocki*, but is dominated by litheliids and porodiscids and includes very few *Heliodiscus*? or *Spongotrochus* spp. (Hollis, 1993).

4. The *Stichomitra wero/granulata* association occurs above the earliest Paleocene at all mid-bathyal sections. It is characterized by abundant *S. wero*, *S. granulata*, *Buryella* spp. (in RP4–RP5), and numerous undifferentiated spumellarians. *Amphisphaera kina* and *Amphipyndax stocki* are common in the lower samples. Cretaceous survivors *Bathropyramis sanjoaquinensis*, *Cornutella californica*, and *Amphipternis alamedaensis* occur consistently in moderate abundance (1%–2%) although are very rare in the Cretaceous at Flaxbourne River and Mead Stream. This association also occurs in RP4–RP5 at Site 208 (Hollis, 1993).

Within these associations there are six recurring groups of taxa that can be related to distinct environments (i.e., six biotopes).

B1. Uppermost-bathypelagic biotope

This assemblage consists of *Amphipyndax stocki*, *Dictyomitra andersoni*, and

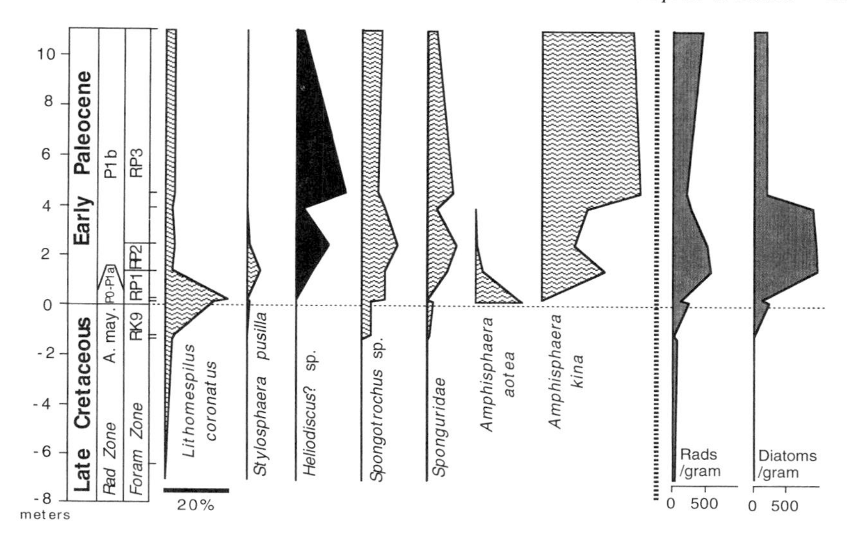

Figure 8. Variations with stratigraphic depth in relative abundance of common early Paleocene species and genera, and in radiolarian and diatom abundance, at Flaxbourne River. Boundary between planktonic foraminiferal zones P0 and P1a is at 0.3 m.

probably *Amphisphaera privus*, *Protoxiphotractus* sp. A, *Phaseliforma* spp., *Archaeodictyomitra lamellicostata*, *Eusyringium*? sp. A, *Lithocampe* sp. A, and *Theocapsomma erdnussa*. All these species are most abundant at Woodside Creek and Wharanui Point (Hollis, 1991, 1993). *A. stocki* and *D. andersoni* are ecological generalists, common at all depths and latitudes in the Late Cretaceous and dominant in many neritic assemblages in low latitudes (Empson-Morin, 1984). They probably did not inhabit deep waters, but their great abundance at shallow depths would result in significant numbers of their robust tests in all assemblages. Most elements within this biotope appear to have a tolerance of or preference for cool waters. Taxa common in the southern Atlantic and Weddell Sea (Ling and Lazarus, 1990; Ling, 1991) include *A. stocki*, *Dictyomitra* spp., *Archaeodictyomitra* spp., *Theocapsomma* spp., *Lithocampe* spp., *Protoxiphotractus perplexus*, and litheliids. This biotope includes most of those Cretaceous survivors that range higher into the Paleocene, occur more consistently, and are more abundant at Woodside Creek than at the other sections (Fig. 5).

B2. Mid-bathypelagic biotope

This assemblage consists of *Theocampe vanderhoofi*, *Stichomitra asymbatos* group, and probably *Dictyomitra multicostata*, *Myllocercion acineton*, and *Stichomitra compsa*. The first two species are much more abundant in the mid-bathyal sections, and the others are slightly less abundant in the shallower sections (Hollis, 1993). Members of the genus *Theocampe* and *Stichomitra asymbatos* are most common in mid-bathyal to abyssal depths at low latitudes

(Empson-Morin, 1984). Other studies indicate both taxa are extremely rare or absent south of 40°–50° latitude (Foreman, 1968; Pessagno, 1975; Ling and Lazarus, 1990; Ling, 1991). This biotope is well represented in the Late Cretaceous siliceous mudstones of California (Foreman, 1968), where coastal upwelling, apparently within a neritic setting (Drugg, 1967), may have introduced an abundance of many of the same species dominant in the deeper Mead section. Species of this biotope, together with *Patulibracchium* spp., *Lithomelissa*? *hoplites*, *Lophophaena* sp. A, and *Theocapsomma amphora*, rapidly decline in abundance in the earliest Paleocene (Fig. 7).

B3. Deep cool-water biotope

This assemblage consists of *Bathropyramis sanjoaquinensis*, *Cornutella californica*, and probably *Amphipternis alamedaensis*. It is curious that these species are very rare in the Cretaceous of the mid-bathyal sections, but are slightly more abundant in the Woodside and Wharanui Cretaceous and ubiquitous in low numbers (1%–2%) in the Paleocene of all sections, including Site 208 (RP2–RP5; Hollis, 1993). In present-day oceans the plectopyramidiins, *Bathropyramis* and *Cornutella*, are deep-dwelling genera (lower bathyal to abyssal) that may be brought into shallower settings by upwelling of cool deep waters (Casey, 1993).

B4. "Opportunist" biotope

This assemblage consists of *Amphisphaera aotea*, undifferentiated small actinommids (within Actinomminae group A), and Cretaceous survivors *Lithomespilus coronatus* and *Stylosphaera pusilla*. *A. aotea* and other small actinommids dominate earliest Paleocene assemblages at all sections, and the other two species increase in abundance at Woodside Creek, Wharanui Point, and Flaxbourne River (Fig. 8). All show a rapid decline above RP1, although *L. coronatus* occurs sporadically into the late Paleocene (Strong and others, 1995) and *A. aotea* gives rise to *A. kina* in RP2 (Hollis, 1993). This biotope is labeled here simply as "opportunist" to signify that the cause of the increase in these actinommids remains unknown; however, it appears to be a short-lived (400 ka) opportunistic response to the environmental changes that lead to the dominance of biotope 5. Interpretation of this and the following two biotopes is hampered by the absence of early Paleocene assemblages from low latitudes.

B5. Shallow cool-water biotope

This assemblage consists of *Amphisphaera kina*, undifferentiated actinommids (also within Actinomminae group A), and the spinose discoidal spumellarians grouped into the genera *Spongotrochus* and *Heliodiscus*?. These species dominate early Paleocene assemblages of all sections (RP2–lower RP3; Figs. 7 and 8). *A. kina* is also common in the spumellarian-dominated basal Paleocene assemblages at Site 208 (Hollis, 1993). An accompanying maximum in diatom abundance over the same interval in the Marlborough sections (Figs. 8 and 9), although not at Site 208, suggests that these radiolarians, and those of biotope B4, are near-surface dwellers that benefited directly from an early Paleocene bloom in microfloras (see discussion below). The cool-water interpretation of this biotope is based largely on the abundance of *Spongotrochus* and other spined discoidal spumellarians.

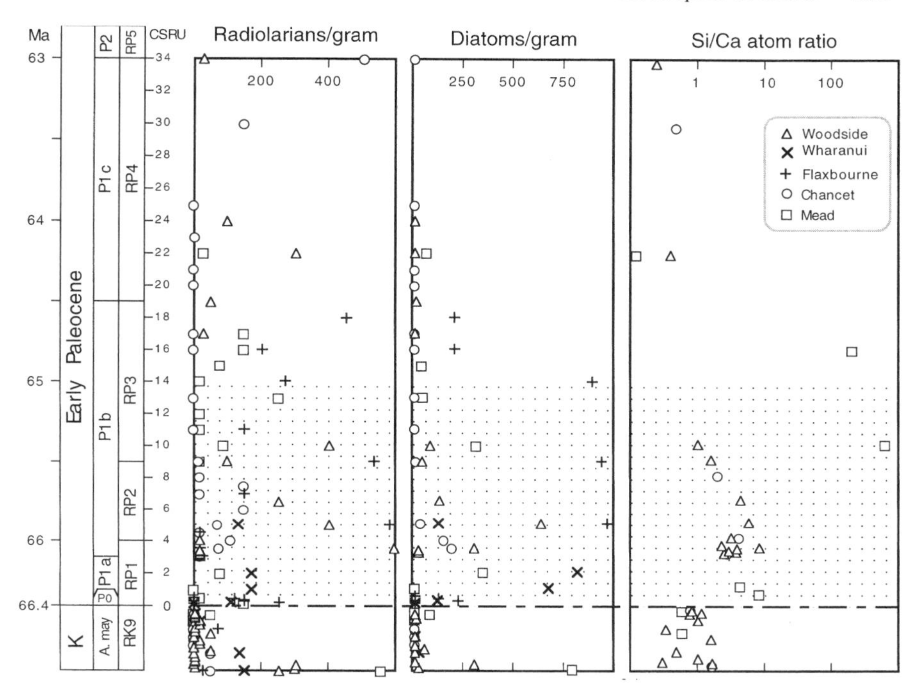

Figure 9. Variations with age of radiolarian and diatom abundance (numbers/gram of sediment) in five Cretaceous-Tertiary boundary sections, and of Si/Ca atom ratios in three of these sections. Samples are related to composite standard reference units (CSRU) by the best-case graphic correlation model (Fig. 3) and assuming a Cretaceous rate of sediment accumulation equal to the average Paleocene rate.

Spongotrochus glacialis is a present-day cool-water species that is common in shallow to intermediate depths in polar regions (Riedel, 1958; Petrushevskaya, 1971) and flourishes in areas where cool nutrient-rich deep waters upwell in lower latitudes (Casey, 1993). This biotope also resembles a present-day shallow-water subpolar association described by Casey (1989) as including abundant discoidal spumellarians ("cenodiscids" [phacodiscids], spongodiscids, spongotrochins) as well as litheliids and lophophaenins (plagoniids). Many of the most common nassellarians of modern polar areas are plagoniids, but unequivocal members of the family do not appear until the Eocene.

It has been suggested that high-latitude Late Cretaceous assemblages are characterized by robust spongy spumellarians and, in particular, the genus *Prunobrachium* (Pessagno, 1975; Empson-Morin, 1984). Because this genus occurs in Marlborough, it would be expected to be a significant component of this early Paleocene biotope. Instead, *Prunobrachium* is common in the Cretaceous at Mead Stream, where it disappears at the K-T boundary, but occurs very rarely in the Cretaceous elsewhere in Marlborough. It occurs sporadically in the early Paleocene at Woodside Creek. The only other occurrences of the genus are in high-latitude Cretaceous faunas from Siberia (Kozlova and Gorbovetz, 1966) and Campbell Plateau DSDP Site 275 (Pessagno, 1975). Although *Prunobrachium* is undoubtedly a high-latitude and probably cool-water genus, it is possible that *Prunobrachium*, along with other large spongodiscids that are abundant in these localities, represents a distinct biotope not well represented in Marlborough. Such forms are typically symbionts inhabiting shallow depths in present-day oceans (Matsuoka and Anderson, 1992; Casey, 1993). Empson-Morin (1984) argued for a setting significantly shallower than present water depth for Site 275, based on the abundance of spumellarians and support from other studies (Wilson, 1975; Kennett and others, 1975). Although Empson-Morin's suggestion of a bathyal setting conflicts with the bathyal nassellarian-dominated assemblages of Marlborough, these relative depth relations may be valid. It is possible that Site 275 was situated in a neritic setting in the Late Cretaceous. Although not mentioned in the Pessagno's (1975) original report, it is significant that Site 275 faunas also contain numerous actinommids similar to the undifferentiated actinommids in this biotope (my personal observation, 1994). Such actinommids are not particularly common in the Late Cretaceous of the southern Atlantic (Ling and Lazarus, 1990; Ling, 1991), where biotope B1 elements are well represented but *Prunobrachium* is absent.

B6. High-latitude mid-bathypelagic biotope

This assemblage consists of *Stichomitra wero*, *S. granulata*, *Buryella foremanae*, and *B. dumitricai*. These species dominate assemblages in the mid-early Paleocene (upper RP3 and RP4) of the mid-bathyal sections and Site 208 (Hollis, 1993); individual species compose 40% of the total fauna in some samples (Fig. 7). This biotope is also evident over the same interval at Woodside Creek, but the species never attain a combined abundance of more than 6% of the total fauna. These species appear to constitute an evolutionary lineage culminating in *Buryella tetradica*, which appears in abundance directly above a sudden decline in the abundance of members of this biotope at Site 208. The high-latitude interpretation of this biotope is based on the great abundance of the ancestral species, *S. wero*, in Marlborough and Hokkaido (Iwata and Tajika, 1986, 1992; my personal observation, 1993), the comparatively low abundance of *S. wero* at Site 208, and the absence of all members of the biotope from late Paleocene assemblages in low latitudes and the North Atlantic (Foreman, 1973; Riedel and Sanfilippo, 1978; Nishimura, 1987, 1992). At Mead Stream all of these species range into the late Paleocene, and *S. granulata* continues into the early Eocene (Strong and others, 1995).

The RP5 index, *Buryella tetradica*, is not considered a member of this biotope. It is abundant in all sections from early–late Paleocene, including the uppermost bathyal setting represented by Woodside Creek. However, the species probably derived from the *B. foremana/dumitricai* complex in high latitudes. It is very abundant in Marlborough and Site 208, where it first appears at the base of foraminiferal zone P2 (Hollis, 1993), and in northern Japan, (personal observation, 1993), but is rare in low latitudes, the earliest recorded appearance being in the early-late Paleocene (planktonic foraminiferal zone P4 in Foreman, 1973; Sanfilippo and Riedel, 1978). This led Foreman (1973) and Sanfilippo and others (1985) to suggest that the species evolved from *B. pentadica*. Site 208 records indicate the reverse is the case; the latter species appears as an evolutionary offshoot of *B. tetradica* in upper RP5 (Hollis, 1993). *B. tetradica* is abundant in the late Paleocene of the northwest Atlantic (Nishimura, 1992), but earlier occurrences from the region are unknown.

SYNTHESIS

In the absence of Cretaceous-Tertiary transitional radiolarian faunas from other areas, the environmental assignments of these biotopes are admittedly highly inferential and any synthesis must be necessarily speculative. Nevertheless, such inferences and speculations are as vital for unraveling the complex way radiolarians respond to K-T events as were the speculations (Alvarez and others, 1980) that led to the development of the bolide impact theory.

Events at the K-T boundary caused no mass extinction of radiolarians in the Marlborough sections. Progressive faunal changes over the first three million years of the Tertiary (RP1–RP4), however, were dramatic and altered irreversibly the structure and composition of radiolarian populations. The first substantial change was the sudden burgeoning of spumellarians directly above the boundary: although they rarely compose more than 10% of the total faunas in the Late Cretaceous, spumellarians dominate earliest Paleocene assemblages and maintain high abundance throughout the Paleocene and into the Eocene at Mead Stream (Strong and others, 1995). The interpretation of biotopes B4 and B5 is critical to understanding the cause of this rapid transition from nassellarian to spumellarian dominance. These two biotopes are clearly closely related and appear to represent a progressive response to a regional event. The *Amphisphaera aotea/kina* association occurs in the basal Paleocene of all sections; a similar association occurs in the basal Paleocene at Site 208. It does not seem to be directly related to K-T sea-level changes, because the same association occurs at the shallowest and deepest sections. Neither does it seem to be related to an intensification of the conditions that promoted high siliceous productivity in the Cretaceous at Mead Stream, where Cretaceous faunas are dominated by nassellarians. Both biotopes, however, are intimately related to an increase of biosiliceous productivity.

In all sections, the burgeoning of spumellarians at the K-T boundary is accompanied by an increase in total radiolarian abundance and a marked increase in diatom abundance (Fig. 8). Siliceous microfossil abundance is influenced strongly by taphonomic factors, which include patchy distribution in the original sediment; variations in silica solubility in interstitial fluids at time of deposition; localized variations in rates of recrystallization under diagenesis; buffering from dissolution by the presence of clay (Blome and Reed, 1992); and protection from mechanical breakage and dissolution by the formation of nodules. Consequently, most intervals within a sedimentary sequence will contain a range of abundances reflecting a continuum from weak to strong degrees of dissolution or destruction of siliceous remains. This does not imply that changes in microfossil abundance bear no relation to original depositional fluxes, but rather that localized variations are not as important as general trends or overall changes in the frequency of microfossil-rich samples. Certainly, the frequency of samples appears to be a meaningful guide to original siliceous microfossil abundance in the Marlborough sections. Radiolarian- and diatom-rich samples are most frequent over the same interval in which sediments are most siliceous (RP1–RP3; Figs. 2 and 9). In three sections in which whole-rock mineral and chemical compositions have been determined (Hollis and others, 1995)—Woodside Creek, Chancet Rocks and Mead Stream—the Si/Ca atom ratio increases from 1:1 in the Cretaceous to ratios exceeding 10:1 in the earliest Paleocene and to more than 100:1 in the highly siliceous Mead section. Estimates of quantitative changes in mineral composition indicate that total quartz increases by 10%–20% across the boundary and by 50%–70% a few meters above the boundary. A similar correlation between abundance of radiolarian and diatoms and percentage of opaline silica was observed in late Paleogene and Neogene strata in the North Atlantic (Lazarus and Pallant, 1989). Opal was also found to correlate with total organic carbon (TOC), indicating to Lazarus and Pallant that covariation between siliceous microfossil abundance, opal, and TOC concentration reflects surface-water productivity.

All silica in the Marlborough sections is in the form of quartz, but only a small proportion (<10% of total quartz) is considered to be of nonorganic or terrigenous origin. No correlation was found between total silica content and the amount of terrigenous or detrital material. The terrigenous content was estimated from the sand content of acid-leached residues, thin-section examination, mineral compositions determined by quantitative X-ray diffraction (i.e., clay and feldspar proportions), and elemental compositions determined by X-ray fluorescence (i.e., Al and Mg proportions). Other than a thin clay-rich interval directly above the K-T boundary (Fig. 2) and below the level of maximum silica, the sediments are essentially biogenic and bimineralic (calcite and quartz). The silica is thought to have been derived primarily from diatoms, which because of their small size are more prone to dissolution than radiolarians. Those diatoms upon which abundance estimates are based are larger robust forms that survived the rigors of diagenesis, hydrofluoric acid treatment, and sieving through a 63 μm mesh. Thus, the cited abundances of diatoms, and to a lesser extent radiolarians, are undoubtedly great underestimates of original numbers.

High silica content in the basal Paleocene is not due solely to reduced rates of carbonate accumulation. Biostratigraphic correlation of the sections (Fig. 3) indicates that earliest Paleocene rates of sediment accumulation were similar to those determined for the later Paleocene, and substantially greater at Woodside Creek and Chancet Rocks, based on foraminiferal evidence for basal Paleocene hiatuses. The silica-rich interval in the basal Paleocene at Mead Stream is 25 m thick, yet only spans ~1 m.y. (allowing for a 0.8 m.y. hiatus in the lower part), indicating a compacted sedimentation rate similar to that determined for the remaining early Paleocene (22 mm/k.y.) and to the average rate for the entire 650-m-thick Late Cretaceous to middle Eocene section (25 mm/k.y. in the Cretaceous, 20 mm/k.y. in the Tertiary; Strong and others, 1991, 1995).

For these reasons, correlated increases in radiolarian and diatom abundance and silica content are inferred to be the result of a marked increase in surface productivity in the earliest Paleocene of Marlborough: a peak in productivity within the first 0.5 m.y. of the Tertiary was followed by a gradual decline to Cretaceous levels over the following 0.5–1 m.y. (Fig. 9). This interval of high surface productivity is the same interval (RP1 to lower RP3) in which biotopes B4 and B5 are dominant (Fig. 7), indicating that these spumellarians benefited directly from increased surface productivity. In the mid-latitude location of the Marlborough sections, enhanced upwelling is the most likely cause for increased surface productivity. This is supported by the abundance of *Spongotrochus* in biotope B5, and a significant increase of biotope B3 (e.g., *Bathropyramis*) in the early Paleocene of the mid-bathyal sections. However, this regional upwelling regime appears markedly different from the type of localized upwelling inferred for the Mead Cretaceous, where members of biotope B3 are absent, although *Spongotrochus* is relatively common, and the species of biotope B2 are dominant. Biotope B2 includes the species most adversely effected by K-T events. At Mead Stream, Flaxbourne River, and Site 208, these species survive into the earliest Paleocene but in greatly reduced numbers. Near the top of the silica-rich interval the nassellarians that fill this vacated niche are the members of biotope 6 (e.g., *Stichomitra wero*); species so far known only from high latitudes (Fig. 7).

Such a shift in upwelling regimes suggests changes in ocean circulation in the Marlborough region through the K-T transition. Over this interval the region was near the eastern boundary of the two dominant surface current systems inferred for the southern Pacific (Kennett, 1982); a warm anticyclonic gyre to the north and a cool cyclonic gyre to the south (Fig. 10). Intensification of either current and its associated wind systems may have enhanced wind-driven upwelling and

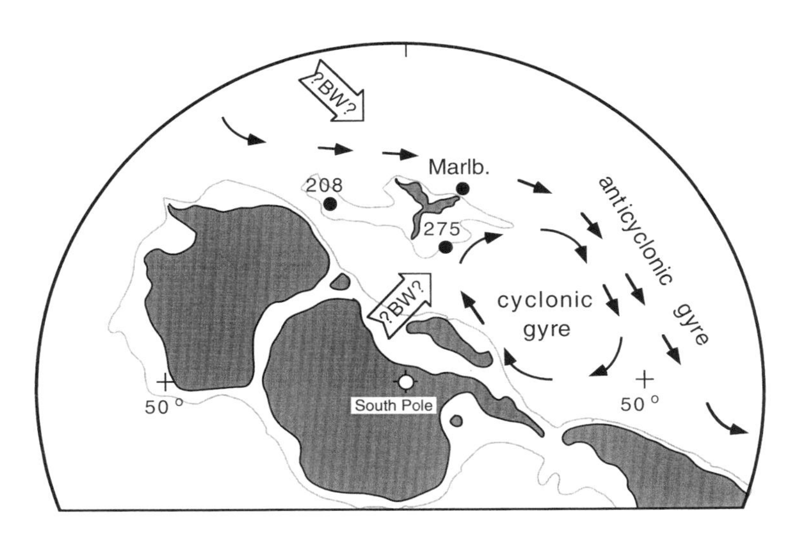

Figure 10. Inferred surface current circulation and postulated bottom-water (BW) currents in the southern Pacific at the Cretaceous-Tertiary transition (66.4 Ma). Reconstruction and surface currents after Kennett (1982). Marlb. is Marlborough.

affected the composition of radiolarian faunas. In the mild climates of the Late Cretaceous it is likely that the northern system was dominant (Kennett, 1982), and Mead upwelling was caused by anticyclonic winds. There is ample evidence from oxygen isotopes, calcareous plankton, and land plants for pronounced cooling in middle to high latitudes during the latest Cretaceous and earliest Paleocene (Haq and others, 1977; Zachos and others, 1989) and particularly in the southern oceans (Askin, 1988; Stott and Kennett, 1990; Francis, 1991; Keller, 1993). Whatever the cause of this cooling, the outcome would have been intensification of the southern cyclonic system at the expense of the northern system. A very plausible result would be development of the new and more widespread upwelling regime seen in the early Paleocene of Marlborough.

These changes in surface or shallow circulation may have led directly to the rapid decline in the mid-bathypelagic Cretaceous elements of biotope B2, which appear to have preferred warm waters, by providing conditions more favorable for the domination of newly evolved high-latitude species (biotope B6). However, the contrast between the pronounced decline experienced by these Cretaceous survivors and the minimal effects on shallow-dwelling species raises the possibility that climatic deterioration was coupled with changes in deep-water circulation. It has been suggested that shallow low-latitude seas may have supplied the bulk of nutrient-rich bottom waters, in the form of dense saline plumes, to the world's oceans during much of the Cretaceous (Brass and others, 1982; Zachos and Arthur, 1986) and in the Tertiary (Woodruff and Savin, 1989; Kennett and Barker, 1990). Zachos and Arthur (1986) argued that high-latitude sources of bottom water, such as in modern oceans, may have become increasingly important as sea levels fell in the latest Cretaceous (Haq and others, 1987) and sources for low-latitude deep waters diminished. Zachos and Arthur (1986) also explained the severity of reductions in surface productivity and calcareous plankton extinctions in low latitudes as due to unstable water-mass stratification caused by the interaction of waning low-latitude bottom waters and new bottom waters from high latitudes.

This provides an appealing refinement to the Marlborough scenario. Late Cretaceous cyclonic gyres may have been strong enough to restrict entry of low-latitude shallow elements (e.g., *Amphipyndax tylotus*), while their deeper-water counterparts (biotope B2) ranged into the region within the warm bottom waters that upwelled in the Mead area. In addition to intensifying the cyclonic surface system, climatic deterioration at the K-T transition may have led to the strengthening of bottom currents originating in Antarctica, which supplied both the cool nutrient-rich upwelling waters and the new group of deep-water high-latitude elements (biotopes B3 and B6) that rapidly replaced the Cretaceous species in mid-bathyal settings from Marlborough to the Lord Howe Rise (Site 208). This model also explains why many shallow-water Cretaceous species tolerant of cool conditions (biotope B1) continued to thrive through the early Paleocene and returned to dominate uppermost-bathyal assemblages at Woodside Creek after the earliest Paleocene spumellarian bloom (biotopes B4 and B5). It seems that upwelling weakened in the mid-early Paleocene (upper RP3), but cool conditions were maintained for a further two million years (top of RP4). Faunas from the later Paleocene in Marlborough and Site 208, and the Eocene of Mead Stream

(Strong and others, 1995), lack many elements common in low latitudes, but high-latitude affinities may have been no stronger than in the comparatively warm Late Cretaceous.

CONCLUSIONS

Ongoing studies of K-T transitional sequences that contain Radiolaria in Japan and the southern Indian Ocean have failed to produce anything like the rich assemblages of Marlborough, either in the latest Cretaceous or the early Paleocene. A major gap in the paleoecological analysis presented here is corroborating evidence for the great abundance of Cretaceous survivors found in the upper Woodside section. There is no evidence in Marlborough, or elsewhere, of how these survivors fared in the uppermost bathyal setting in the late Paleocene. The rarity of these species in late Paleocene sequences certainly indicates that the Woodside environment was a final refuge from the environmental changes that caused rapid faunal turnover in the early Paleocene in other settings.

The primary findings of this study are as follows.

1. Radiolarians show no indications of undergoing mass extinction at the K-T boundary in eastern Marlborough. Although not all Cretaceous taxa have been allocated to species, all of the 45 species that have been differentiated occur in the Tertiary and 38 of these species range well into the early Paleocene.

2. Environmental changes across the K-T boundary resulted in a burgeoning of small actinommids and spinose discoidal spumellarians, and a pronounced increase in diatom and radiolarian productivity. Enhanced upwelling is the most likely cause for continued abundance of these spumellarians and high siliceous productivity over the first one to two million years of the Tertiary. Upwelling was probably promoted by climatic cooling, which would have intensified cyclonic circulation.

3. Within this earliest Paleocene interval, uppermost-bathypelagic Cretaceous survivors undergo only a slight decline in abundance and recover to nearly Cretaceous abundances later in the early Paleocene before a presumed rapid decline in the late Paleocene. The dominant species in this group appear to be ecological generalists able to withstand climatic changes during the K-T transition. In contrast, mid-bathypelagic Cretaceous nassellarians decline rapidly in abundance in the earliest Paleocene. These Cretaceous species may have been more specialized and, in the absence of the ameliorating influence of anticyclonic circulation, gave up their niche to a new group of deep-water nassellarians that probably originated in high latitudes. Long-term changes in bottom-water circulation may have accentuated changes in deeper environments and upwelling settings. Relatively warm bottom currents may have declined during long-term Late Cretaceous regression, which reduced the extent of shallow low-latitude source areas, although localized upwelling appears to have continued in the area represented by the Mead section during the latest Cretaceous. Earliest Paleocene cooling may have strengthened high-latitude bottom currents, and introduced a new regional upwelling regime. Invasion of these cool bottom waters along the eastern margin of New Zealand may have hastened the demise of many deep-

dwelling Cretaceous elements. The high-latitude nassellarians that replaced them remained abundant into at least the late-early Paleocene, indicating that cool conditions were maintained for two millions years after upwelling ceased.

ACKNOWLEDGMENTS

This essay has benefited from the technical assistance and valuable criticism of a great many people. Robin Parker, Ritchie Sims, John Wilmshurst, Sue Courtney, and Michael Laffey carried out the geochemical analyses, and Louise Cotterall prepared Figures 1 and 5 (all University of Auckland). Kerry Rodgers and Peter Ballance (University of Auckland), C. Percy Strong (Institute of Geological and Nuclear Sciences, New Zealand), Donna Hull (University of Texas at Dallas), Jim Kennett (University of California at Santa Barbara), Dave Lazarus (ETH-Zentrum, Zurich), Norm MacLeod (Natural History Museum), and Annika Sanfilippo (Scripps Institution of Oceanography) provided valuable reviews of the manuscript. Fruitful discussions were had with Gerta Keller (Princeton University), Toyosaburo Sakai, and Yoshiaki Aita (Utsunomiya University). The article was prepared while Hollis was recipient of a Japan Society for the Promotion of Science postdoctoral fellowship (based at Utsunomiya University).

REFERENCES CITED

Alvarez, L. W., Alvarez, W., Azaro, F., and Michel, H. V., 1980, Extraterrestrial cause for the Cretaceous-Tertiary extinction: Experimental results and theoretical interpretation: Science, v. 208, p. 1095–1108.

Askin, R. A., 1988, Campanian to Paleocene palynological succession of Seymour and adjacent islands, northeastern Antarctic Peninsula, *in* Feldmann, R. M., and Woodburne, M. O., eds., Geology and paleontology of Seymour Island, Antarctic Peninsula: Geological Society of America Memoir 169, p. 131–153.

Berggren, W. A, Kent, D. V., Flynn, J. J., and van Couvering, J. A., 1985, Cenozoic geochronology: Geological Society of America Bulletin, v. 96, p. 1407–1418.

Blome, C. D., and Reed, K. M., 1992, Acid processing of pre-Tertiary radiolarian cherts and its impact on faunal content and biozonal correlation: Geology, v. 21, p. 177–180.

Blueford, J. R., 1989, Radiolarian evidence: Late Cretaceous through Eocene ocean circulation patterns, *in* Hein, J. R., and Obradovich, J., eds., Siliceous deposits of the Tethys and Pacific regions: New York, Springer-Verlag, p. 19–29.

Bohor, B. F., Modreski, P. J., and Foord, E. E., 1987, Shocked quartz in the Cretaceous-Tertiary boundary clays: Evidence for a global distribution: Science, v. 236, p. 705–709.

Brass, G. W. Southam, J. R., and Peterson, W. H., 1982, Warm saline bottom water in the ancient ocean: Nature, v. 296, p. 620–623.

Brooks, R. R., Strong, C. P., Lee, J., Orth, C. J., Gilmore, J. S., Ryan, D. E., and Holzbecher, J., 1986, Stratigraphic occurrences of iridium anomalies at four Cretaceous/Tertiary boundary sites in New Zealand: Geology, v. 14, p. 727–729.

Cambell, A. S., 1954, Radiolaria, *in* Moore, R. C., ed., Treatise on invertebrate paleontology, Part D, Protista 3: Geological Society of America (and University of Kansas Press), p. D11–D195.

Campbell, A. S., and Clark, B. L., 1944, Radiolaria from Upper Cretaceous of middle California: Geological Society of America Special Paper 57, 61 p.

Casey, R. E., 1989, Model of modern polycystine radiolarian shallow-water zoogeography: Palaeogeography, Palaeoclimatology, Palaeoecology, v. 74, p. 15–22.

Casey, R. E., 1993, Radiolaria, *in* Lipps, J. H., ed., Fossil prokaryotes and protists: Boston, Blackwell Scientific, p. 249–284.

Drugg, W. S., 1967, Palynology of the upper Moreno Formation (Late Cretaceous–Paleocene), Escarpado Canyon, California: Palaeontographica, Abt. B, Bd. 120, p. 1–71.

Dumitrica, P., 1973, Paleocene radiolaria, DSDP Leg 21, Initial reports of the Deep Sea Drilling Project, Volume 21: Washington, D.C., U.S. Government Printing Office, p. 787–817.

Edwards, L. E., 1984, Insights on why graphic correlation (Shaw's method) works: Journal of Geology, v. 92, p. 583–597.

Emiliani, C., Kraus, E. B., and Shoemaker, E. M., 1981, Sudden death at the end of the Cretaceous: Earth and Planetary Science Letters, v. 55, p. 317–334.

Empson-Morin, K. M., 1984, Depth and latitude distribution of Radiolaria in Campanian (Late Cretaceous) tropical and subtropical oceans: Micropaleontology, v. 30, p. 87–115.

Foreman, H. P., 1968, Upper Maestrichtian Radiolaria of California: London, Palaentological Association, Special Papers in Palaeontology, no. 3, 82 p.

Foreman, H. P., 1973, Radiolaria from Leg 10 with systematics and ranges for the Families Amphipyndacidae, Artostrobiidae and Theoperidae, *in* Initial reports of the Deep Sea Drilling Project, Volume 10: Washington, D.C., U.S. Government Printing Office, p. 407–474.

Francis, J. E., 1991, Paleoclimatic significance of Cretaceous–early Tertiary fossil forests of the Antarctic Peninsula, *in* Thomson, M. R. A., Crame, J. A., and Thomson, J. W., eds., Geological evolution of Antarctica: London, Cambridge University Press, p. 623–627.

Haq, B. U., Premoli-Silva, I., and Lohmann, G. P., 1977, Calcareous plankton paleobiogeographic evidence for major climatic fluctuations in the early Cenozoic Atlantic Ocean: Journal of Geophysical Research, v. 82, p. 3861–3876.

Haq, B. U., Hardenbol, J., and Vail, P. R., 1987, Chronology of fluctuating sea levels since the Triassic: Science, v. 235, p. 1156–1167.

Harwood, D. M., 1988, Upper Cretaceous and lower Paleocene diatom and silicoflagellate biostratigraphy of Seymour Island, eastern Antarctic Peninsula, *in* Feldmann, R. M., and Woodburne, M. O., eds., Geology and paleontology of Seymour Island, Antarctic Peninsula: Geological Society of America Memoir 169, p. 55–129.

Hein, J. R., and Parrish, J. T., 1987, Distribution of siliceous deposits in space and time, *in* Hein, J. R., ed., Siliceous sedimentary rock-hosted ores and petroleum: New York, Van Nostrand Reinhold, p. 10–57.

Hollander, D. J., McKenzie, J. A., and Hsü, K. J., 1993, Carbon isotope evidence for unusual plankton blooms and fluctuations of surface water CO_2 in "Strangelove Ocean" after terminal Cretaceous event: Palaeogeography, Palaeoclimatology, and Palaeoecology, v. 104, p. 229–237.

Hollis, C. J., 1988, Uppermost Cretaceous and lower Paleocene Radiolaria from Woodside Creek, NE Marlborough, New Zealand [abs.]: Geological Society of New Zealand Miscellaneous Publication 41a, p. 85.

Hollis, C. J., 1991, Latest Cretaceous to late Paleocene Radiolaria from Marlborough (New Zealand) and DSDP Site 208 [Ph.D thesis]: Auckland, New Zealand, University of Auckland, 308 p.

Hollis, C. J., 1993, Latest Cretaceous to late Paleocene radiolarian biostratigraphy: A new zonation from the New Zealand region: Marine Micropaleontology, v. 21, p. 295–327.

Hollis, C. J., Rodgers, K. A., and Parker, R. J., 1995, Siliceous plankton bloom in the earliest Tertiary of Marlborough, New Zealand: Geology (in press).

Hsü, K. J., and McKenzie, J. A., 1985, A "Strangelove" ocean in the earliest Tertiary, *in* Sundquist, E. T. and Broecker, W. S., eds., The carbon cycle and atmospheric CO_2: Natural variations archean to present: American Geophysical Union Geophysical Monograph 32, p. 487–492.

Iwata, K., and Tajika, J., 1986, Late Cretaceous Radiolarians of the Yubetsu Group, Tokoro Belt, Northeast Hokkaido: Hokkaido University Journal of the Faculty of Science, ser. 4, v. 21, p. 619–644.

Iwata, K., and, Tajika, J., 1992, Early Paleogene radiolarians from green and red mudstones in the Yubetsu Group and reconsideration of the age of their sedimentation: Hokkaido Geological Survey Report, no. 63, p. 23–31.

Keller, G., 1988, Extinction, survivorship and evolution of planktic foraminifera across the Cretaceous/Tertiary boundary at El Kef, Tunisia: Marine Micropaleontology, v. 13, 239–263.

Keller, G., 1993, The Cretaceous-Tertiary boundary transition in the Antarctic Ocean and its global implications: Marine Micropaleontology, v. 21, p. 1–45.

Kennett, J. P., 1982, Marine geology: Englewood Cliffs, New Jersey, Prentice-Hall, 811 p.

Kennett, J. P., and Barker, P. F., 1990, Latest Cretaceous to Cenozoic climate and oceanographic developments in the Weddell Sea, Antarctica: An ocean-drilling perspective, *in* Proceedings of the Ocean Drilling Program, Scientific results, Volume 113: College Station, Texas, Ocean Drilling Program, p. 937–960.

Kennett, J. P., and others, 1975, Cenozoic paleoceanography in the southwest Pacific Ocean, Antarctic glaciation and the development of the Circum-Antarctic Current, *in* Initial reports of the Deep Sea Drilling Project, Volume 29: Washington, D.C., U.S. Government Printing Office, p. 1155–1169.

Kozlova, G. E., 1984, Zonal subdivision of the boreal Paleogene by radiolarians, *in* Petrushevskaya, M. G., and Stepanjants, S. D., eds., Morphology, ecology and evolution of radiolarians: Material for the fourth symposium of European radiolarists, Leningrad (EURORAD IV): Leningrad, Nauka, p. 196–220 [in Russian].

Kozlova, G. E., and Gorbovetz, A. N., 1966, Radiolyarii verkhnemelovykh i verkhne eot senovykh otlozhenii Zapadno-Sibirskoi Nizmennosti. Trudy Vsesoyuznogo Neftyanogo Nauchno-Issledovatellskogo Geologorazvedochnogo Instituta (VNIGRI), v. 248, 1–159.

Lazarus, D., and Pallant, A., 1989, Oligocene and Neogene radiolarians from the Labrador Sea, ODP Leg 105, *in* Proceedings of the Ocean Drilling Program, Scientific results, Volume 105: College Station, Texas, Ocean Drilling Program, p. 349–380.

Leinen, M., 1979, Biogenic silica accumulation in the central equatorial Pacific and its implications for Cenozoic paleoceanography: Summary: Geological Society of America Bulletin, v. 90, p. 801–803.

Lensen, G.J., 1962, Geological map of New Zealand, Sheet 16, Kaikoura (first edition): Wellington, New Zealand, Department of Scientific and Industrial Research, scale 1:250,000.

Ling, H. Y., 1991, Cretaceous (Maestrichtian) radiolarians: Leg 114, *in* Proceedings of the Ocean Drilling Program, Scientific results, Volume 114: College Station, Texas, Ocean Drilling Program, p. 317–324.

Ling, H. Y., and Lazarus, D. B., 1990, Cretaceous Radiolaria from the Weddell Sea: Leg 113, *in* Proceedings of the Ocean Drilling Program, Scientific results, Volume 113: College Station, Texas, Ocean Drilling Program, p. 353–363.

MacLeod, N., and Keller, G., 1991, How complete are K/T boundary sections?: Geological Society of America Bulletin, v. 103, p. 1439–1457.

Matsuoka, A., and Anderson, O. E., 1992, Experimental and observational studies of radiolarian physiological ecology: 5. Temperature and salinity tolerance of *Dictyocoryne truncatum*: Marine Micropaleontology, v. 19, p. 299–313.

Nishimura, A., 1987, Cenozoic Radiolaria in the Western North Atlantic, Site 603, Leg 93 of the Deep Sea Drilling Project, *in* Initial reports of the Deep Sea Drilling Project, Volume 93: Washington, D.C., U.S. Government Printing Office, p. 713–737.

Nishimura, A., 1992, Paleocene radiolarian biostratigraphy in the northwest Atlantic at Site 384, Leg 43 of the Deep Sea Drilling Project: Micropaleontology, v. 38, p. 317–362.

Pessagno, E. A., 1975, Upper Cretaceous Radiolaria from DSDP Site 275, *in* Initial reports of the Deep Sea Drilling Project, Volume 29: Washington, D.C., U.S. Government Printing Office, p. 1011–1029.

Petrushevskaya, M. G., 1971, Radiolaria in the plankton and Recent sediments from the Indian Ocean and Antarctic, *in* Funnell, B. M., and Riedel, W. R., eds., The micropalaeontology of the oceans: London, Cambridge University Press, p. 319–329.

Petrushevskaya, M. G., and Kozlova, G. E., 1972, Radiolaria: Leg 14, Deep Sea Drilling Project, *in* Initial reports of the Deep Sea Drilling Project, Volume 14: Washington, D.C., U.S. Government Printing Office, p. 495–648.

Prebble, W. M., 1980, Late Cainozoic sedimentation and tectonics of the East Coast deformed belt in Marlborough, New Zealand, *in* Ballance, P. F., and Reading, H. G., eds., Sedimentation in oblique-slip mobile zones: International Association of Sedimentologists Special Publication 4, p. 217–228.

Ramsay, A. T. S., 1973, A history of organic siliceous sediments in oceans, *in* Hughes, N. F., ed., Organisms and continents through time: London, Palaeontological Association, Special Papers in Palaeontology, no. 12, p. 199–234.

Reay, M. B., 1993, Geology of the middle part of the Clarence valley, Marlborough, New Zealand: New Zealand Institute of Geological and Nuclear Sciences Geological Map 10, scale 1:50,000.

Riedel, W. R., 1958, Radiolaria in Antarctic sediments: British, Australian, and New Zealand Antarctic Research Expedition Report, ser. B, v. 7, p. 217–255.

Riedel, W. R., and Sanfilippo, A., 1978, Stratigraphy and evolution of tropical Cenozoic radiolarians: Micropaleontology, v. 24, p. 61–96.

Sanfilippo, A., and Riedel, W. R., 1973, Cenozoic Radiolaria (exclusive of theoperids, artostrobiids and amphipyndacids) from the Gulf of Mexico, DSDP Leg 10, *in* Initial reports of the Deep Sea Drilling Project, Volume 10: Washington, D.C., U.S. Government Printing Office, p. 475–611.

Sanfilippo, A., and Riedel, W. R., 1985, Cretaceous Radiolaria, *in* Bolli, H. M., Saunders, J. B., and Perch-Nielsen, K., eds., Plankton stratigraphy: London, Cambridge University Press, p. 573–630.

Sanfilippo, A., Westberg-Smith, M. J., and Riedel, W. R., 1985, Cenozoic Radiolaria, *in* Bolli, H. M., Saunders, J. B., and Perch-Nielsen, K., eds., Plankton stratigraphy: London, Cambridge University Press, p. 631–712.

Shaw, A. B., 1964, Time in stratigraphy: New York, McGraw-Hill, 365 p.

Smit, J., 1982, Extinction and evolution of planktonic foraminifera after a major impact at the Cretaceous/Tertiary boundary, *in* Silver, L. T., and Schultz, P. H., eds., Geological implications of impacts of large asteroids and comets on the Earth: Geological Society of America Special Paper 190, p. 329–352.

Smit, J., and Romein, A. J. T., 1985, A sequence of events across the Cretaceous-Tertiary boundary: Earth and Planetary Science Letters, v. 74, p. 155–170.

Stott, L. D., and Kennett, J. P., 1990, The paleoceanographic and paleoclimatic signature of the Cretaceous/Paleogene boundary in the Antarctic: Stable isotopic results from ODP Leg 113, *in* Proceedings of the Ocean Drilling Program, Scientific results, Volume 113: College Station, Texas, Ocean Drilling Program, p. 829–846.

Strong, C. P., 1977, Cretaceous-Tertiary boundary at Woodside Creek, north-eastern Marlborough: New Zealand Journal of Geology and Geophysics, v. 20, p. 687–696.

Strong, C. P., 1981, Cretaceous-Tertiary boundary at Woodside Creek—Revisited: Geological Society of New Zealand Newsletter, v. 51, p. 4–5.

Strong, C. P., 1984a, Cretaceous-Tertiary boundary at Chancet Rocks Scientific Reserve, Northeast Marlborough: New Zealand Geological Survey Record 3, p. 47–51.

Strong, C. P., 1984b, Cretaceous-Tertiary boundary, mid-Waipara River section, north Canterbury, New Zealand (Note): New Zealand Journal of Geology and Geophysics, v. 27, p. 231–234.

Strong, C. P, Brooks, R. R., Wilson, S. M., Reeves, R. D., Orth, C. J., Mao, X.-Y., Quintana, L. R., and Anders, E., 1987, A new Cretaceous-Tertiary boundary site at Flaxbourne River, New Zealand: Biostratigraphy and geochemistry: Geochimica et Cosmochimica Acta, v. 51, p. 2769–2777.

Strong, C. P., Brooks, R. R., Orth, C. J., and Mao, X.-Y., 1988, An iridium-rich calcareous claystone (Cretaceous-Tertiary Boundary) from Wharanui, Marlborough, New Zealand: New Zealand Journal of Geology and Geophysics, v. 31, p. 191–195.

Strong, C. P., Hollis, C. J. and Wilson, G. J., 1995, Foraminiferal, radiolarian and dinoflagellate biostratigraphy of Late Cretaceous to middle Eocene pelagic sediments (Muzzle Group), Mead Stream, Marlborough, New Zealand: New Zealand Journal of Geology and Geophysics, v. 38, no. 2, p. 165–206.

Takemura, A., 1992, Radiolarian Paleogene biostratigraphy in the southern Indian Ocean, Leg 120, *in* Proceedings of the Ocean Drilling Program, Scientific results, Volume 120: College Station, Texas, Ocean Drilling Program, p. 735–756.

Thierstein, H. R., 1982, Terminal Cretaceous plankton extinctions: A critical assessment, *in* Silver, L. T., and Schultz, P. H., eds., Geological implications of impacts of large asteroids and comets on the Earth: Geological Society of America Special Paper 190, p. 385–399.

Toumarkine, M., and Luterbacher, H., 1985, Paleocene and Eocene planktic foraminifera, *in* Bolli, H. M., Saunders, J. B., and Perch-Nielsen, K., eds., Plankton stratigraphy: London, Cambridge University Press, p. 87–154.

Ward, P. D., 1990, The Cretaceous/Tertiary extinctions in the marine realm; a 1990 perspective, *in* Sharpton, V. L., and Ward, P. D., eds., Global Catastrophes in Earth history: An interdisciplinary conference on impacts, volcanism, and mass mortality: Geological Society of America Special Paper 247, p. 425–432.

Weissel, J. K., Hayes, D. E., and Herron, E. M., 1977, Plate tectonic synthesis: The displacements between Australia, New Zealand, and Antarctica since the Late Cretaceous: Marine Geology, v. 25, p. 231–277.

Westberg, M. J., and Riedel, W. R., 1978, Accuracy of radiolarian correlation in the Pacific Miocene: Micropaleontology, v. 24, p. 1–23.

Wilson, G. J., 1975, Palynology of deep sea cores from DSDP Site 275, southeast Campbell Plateau, *in* Initial reports of the Deep Sea Drilling Project, Volume 29: Washington, D.C., U.S. Government Printing Office, p. 1031–1035.

Wolbach, W. S., Gilmour, I., Anders, E., Orth, C. J., and, Brooks, R. R., 1988, Global fire at the Cretaceous-Tertiary boundary: Nature, v. 334, p. 665–669.

Woodruff, F., and Savin, S. M., 1989, Miocene deepwater oceanography: Paleoceanography, v. 4, p. 87–140.

Zachos, J. C., and Arthur, M. A., 1986, Paleoceanography of the Cretaceous-Tertiary boundary event. Inferences from stable isotope and other data: Paleoceanography, v. 1, p. 5–26.

Zachos, J. C., Arthur, M. A., and Dean, W. E., 1989, Geochemical evidence for suppression of pelagic marine productivity at the Cretaceous-Tertiary boundary: Nature, v. 337, p. 61–64.

8

Earliest Origins of Northern Hemisphere Temperate Nonmarine Ostracode Taxa: Evolutionary Development and Survival through the Cretaceous-Tertiary Boundary Mass-Extinction Event

Elisabeth M. Brouwers, *U.S. Geological Survey, Denver, Colorado* and Patrick De Deckker, *Australian National University, Canberra, Australia*

INTRODUCTION

Outcrops of nonmarine and marine fossiliferous sedimentary rocks of middle to Late Cretaceous and early Tertiary age are well exposed in the northern Alaska coastal plain (Brosge and Whittington, 1966) by spectacular bluff exposures along the Colville River. This study focuses on rocks considered to be Maastrichtian and Danian in age that are ~50 km south of the Arctic Ocean (Marincovich and others, 1985, 1990; Brouwers and others, 1987; Brouwers and DeDeckker, 1993). These Maastrichtian sediments are entirely nonmarine and contain terrestrial vertebrates (primarily hadrosaurs; Clemens and Nelms, 1994) and aquatic invertebrates (gastropods, pelecypods, and ostracodes) along with sparse plant megafossils and root traces. Paleocene sedimentary rocks in this outcrop belt include nonmarine, marginal-marine, and shallow-marine deposits that contain abundant and diverse invertebrate faunas.

Late Cretaceous and Tertiary terrestrial and marine assemblages from polar regions are characterized by the dominance of endemic taxa that had adapted to cold air and water temperatures, seasonal fluctuations of solar insolation, and the geographic isolation characteristic of high-latitude environments (Zinsmeister and Feldmann, 1984; Marincovich and others, 1985; Spicer, 1987; Brouwers and others, 1987). Fossil faunas from high latitudes such as northern Alaska, the Canadian Arctic, and Antarctica represent the only temperate habitats that existed during the Late Cretaceous and earliest Tertiary. Estimates of paleotemperatures based on plant megafossils and floral diversity suggest a mild-temperate to cold-temperate climate for northern Alaska during the latest Cretaceous (Spicer and Parrish, 1986; Spicer, 1987).

Several of the nonmarine ostracode genera discussed herein are believed to record their oldest geologic occurrence worldwide, suggesting that these genera may have evolved in northern Alaska during the Late Cretaceous. As world climate cooled throughout the Tertiary, these "arctic" genera evidently migrated southward and constitute some of the more typical Cenozoic and present-day Nearctic genera of the middle latitudes.

GEOGRAPHIC AND GEOLOGIC SETTING

The study area is located at lat 70°N on the northern Alaska coastal plain (the North Slope) along the Colville River, which drains northward from the Brooks Range (Fig. 1). The Colville River defines the eastern boundary of the National Petroleum Reserve of Alaska (NPRA). Extensive exposures of middle to Upper Cretaceous and lower Tertiary sedimentary rocks form prominent river bluffs that range from 10 to 50 m high.

Sedimentary rocks discussed herein belong to the Kogosukruk Tongue of the Prince Creek Formation and to the Sentinel Hill Member of the Schrader Bluff Formation (Brosge and Whittington, 1966). The Prince Creek Formation is the uppermost unit of the Colville Group, and consists primarily of nonmarine sandstones and siltstones with interbedded marine tongues. The Prince Creek

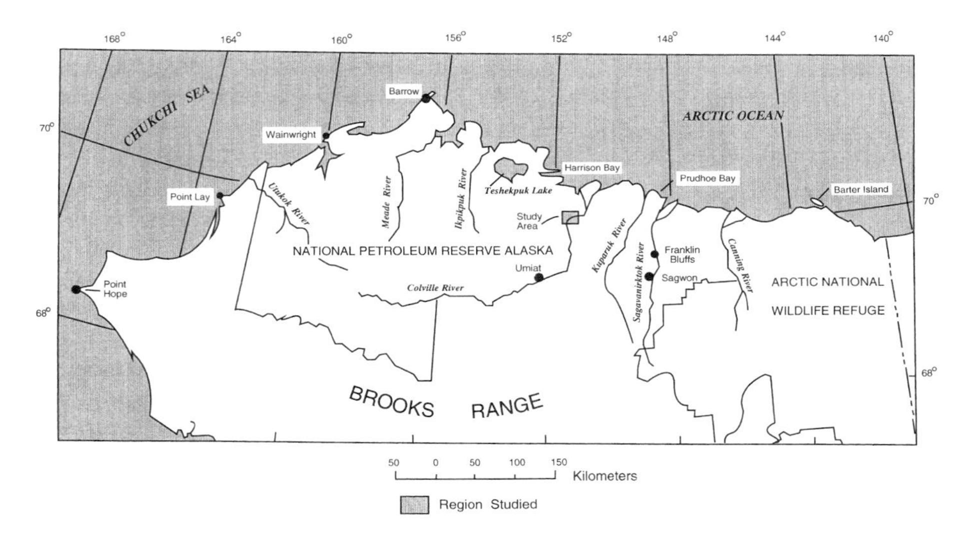

Figure 1. Map showing study area along the northern Colville River.

Formation has been subdivided into the lower Tuluvak Tongue and the upper Kogosukruk Tongue (Brosge and Whittington, 1966). The Sentinel Hill Member of the Schrader Bluff Formation interfingers with the Prince Creek Formation in this area. Observed lithotypes include claystones, siltstones, sandstones, coal, carbonaceous mudstones, and tephras. A prominent unconformity separates the upper Maastrichtian and Danian rocks (dipping ~1°–5° ENE) from the overlying sands and silts of the Pliocene and Pleistocene Gubik Formation (Brosge and Whittington, 1966; Carter and Galloway, 1985).

In 1986 25 stratigraphic sections were measured, beginning at the lowest known horizon containing dinosaur bones and ending at the first fully marine section (the Ocean Point beds; Fig. 2; Phillips, 1988). Dinosaur skeletal remains have been found subsequently in stratigraphically older rocks upriver (W. A. Clemens, 1992, personal commun.; R. Gangloff, 1994, personal commun.). The total sequence is ~180 m thick, of which ~25 m are covered.

Most of the sequence (the lowermost 135 m) consists of flood-plain sediments, including various overbank and channel deposits. Siltstones and claystones dominate in the flood-plain deposits, and sandstone is dominant in the fluvial distributary deposits. Lateral facies changes are pronounced and fairly abrupt. Organic-rich beds (believed to represent paleosols) are developed at the top of fining-up overbank cycles; these highly visible and traceable beds were used to correlate between the measured sections in the lowermost 135 m (Phillips, 1988).

The upper part of the sequence (the uppermost 45 m), referred to by Marincovich and others (1983) and subsequently by numerous workers as the Ocean Point beds, consists of nonmarine, marginal-marine, and shallow-marine sandstones and siltstones. These marginal-marine and shallow-marine deposits

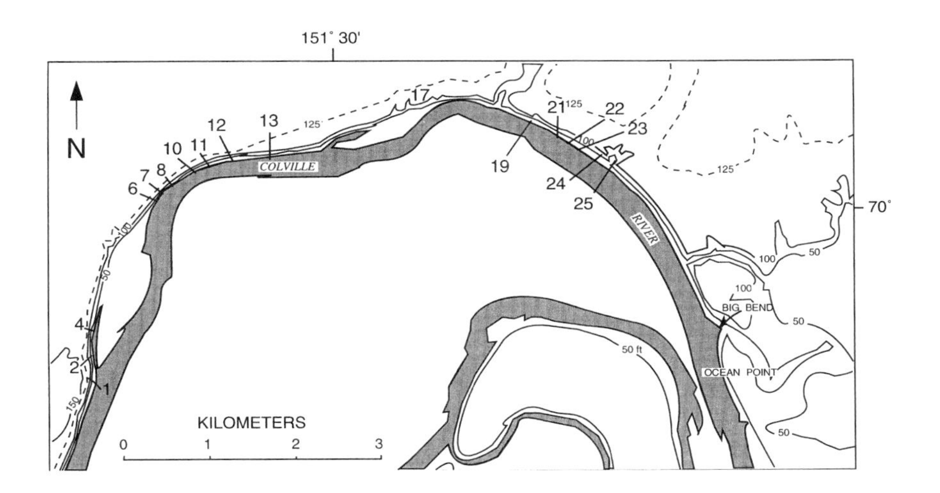

Figure 2. Map showing the 17 sections (of the total 25 measured sections) that contained microfossils.

have few facies changes and consequently have more lateral continuity, allowing beds of a number of different lithologies to be used in correlating between the measured sections. The nature of the change from the lower nonmarine rocks to the upper marginal-marine and shallow-marine rocks is not known because of the presence of a fault of unknown displacement.

TECHNIQUES

A total of 160 paleontological samples were collected from the 180-m-thick sequence (see Phillips, 1988, for the exact stratigraphic position of the samples in each of the measured sections). Sampling was conducted for calcareous and arenaceous microfossils (ostracodes, charophytes, and foraminifers) and for palynomorph microfossils (spores, pollen, and dinoflagellates). All samples were processed for calcareous and arenaceous microfossils, and selected samples were processed for palynomorphs.

Microfossil samples were processed in the Denver Calcareous Microfossil Laboratory, which uses specially designed procedures to recover delicate nonmarine microfossils. For each sample, 500 g (dry weight) of raw sediment was processed using sodium bicarbonate, sodium hexametaphosphate (Calgon), and a freeze-thaw cycle to disaggregate the claystones and siltstones. Samples were washed on a 230 mesh (63 μm) sieve. All residues were examined under a binocular microscope; 53 residues contained ostracodes and 23 contained charophytes. All of the ostracode and charophyte specimens 180 μm or larger (80 mesh sieve) were extracted, identified, and counted.

Four distinct ostracode assemblages can be recognized through the sequence (Fig. 3): (1) a lower nonmarine assemblage, (2) a marginal-marine assemblage, (3) an upper nonmarine assemblage, and (4) a shallow-marine assemblage. These four assemblages are characterized by particular taxonomic composition and species abundance. The four distinct faunas are believed to represent both environmental and temporal changes through the sequence. Arguments for the age assignments of the different facies are presented below, followed by interpretations of the paleoenvironment and paleogeography.

AGE EVIDENCE

The lowermost nonmarine sedimentary rocks (the basal 135 m) have been dated by biostratigraphy and by isotopic dating of the tephras (McKee and others, 1989; Conrad and others, 1990). An abundant, low-diversity arenaceous foraminifer assemblage from a single horizon in the lower part of the sequence, associated with the main dinosaur horizon, indicates a late Campanian to early Maastrichtian age (Sliter, *in* Brouwers and others, 1987). The foraminifer assemblage includes *Verneuilinoides fischeri*, *Trochammina albertensis*, *T. diagonis*, *Reophax texanus*, and *Ammomarginulina sp.*, and correlates with well-dated late Campanian to early Maastrichtian sequences in Alberta, Saskatchewan, and Manitoba. Fossil palynomorphs from the lowermost nonmarine rocks indicate an early Maastrichtian age, on the basis of correlation with dated sections in Arctic Canada and the North American Western Interior (Frederiksen and others, 1988; Frederiksen, 1989, 1991).

A fission-track age of 50.9 ± 7.7 Ma (Eocene) was obtained on a thick, water-lain tephra that crops out stratigraphically below the measured sections (Carter and others, 1977). This isotopic date provides an extreme minimum age, obviously representing alteration, because the tephra occurs well below dinosaur-bearing horizons. Preliminary $^{40}Ar/^{39}Ar$ analyses for a tephra horizon near the base of the measured section and for a tephra from the middle part of the nonmarine section indicate an age range between 68 and 71 Ma (late Maastrichtian; McKee and others, 1989; Conrad and others, 1990).

The lower nonmarine ostracode assemblage includes four species and three

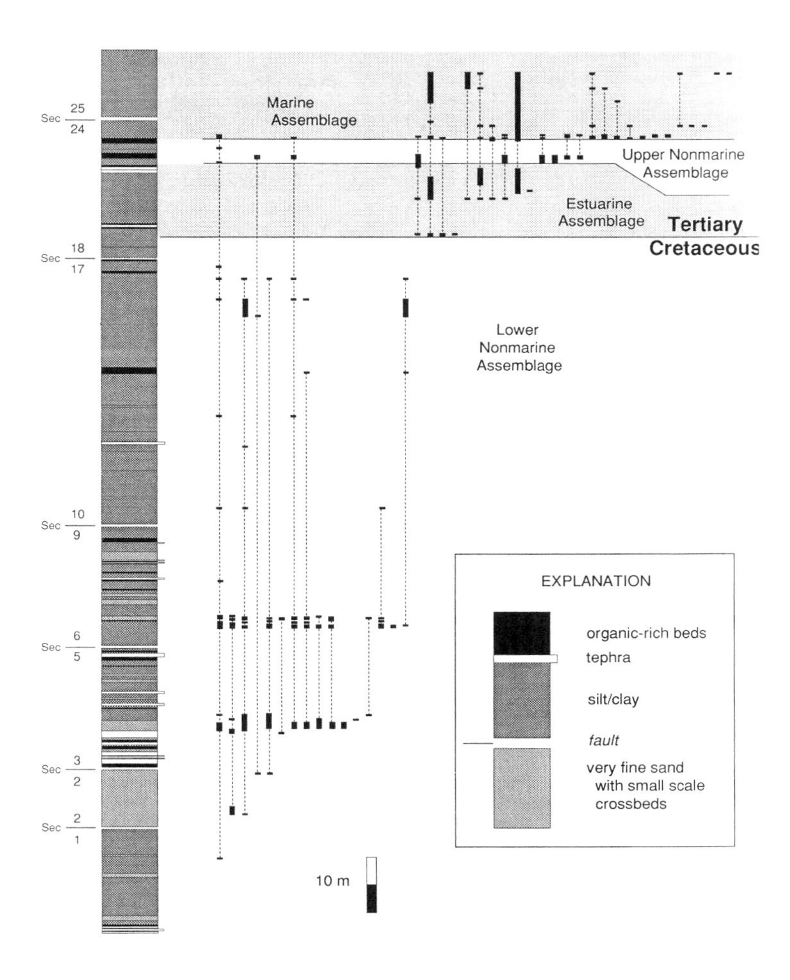

Figure 3. Plot of the stratigraphic ranges of ostracode species versus the total measured section. See Brouwers and De Deckker (1993) for more detailed information.

genera that belong to the Cyprideinae; the genera include *Mongolocypris* and *Cypridea*, which became extinct near the end of the Cretaceous (Szczechura, 1979; Hou, 1983). These two genera do not occur in the upper nonmarine assemblage (which is environmentally comparable to the lower assemblage). *Mongolocypris* is believed to extend well into the latest Cretaceous, and consequently its absence in the upper part of the section implies a post-Maastrichtian age (Figs. 4 and 5). In addition, the occurrence in the lower section of *Ziziphocypris*, a genus well represented in Upper Cretaceous beds of China but not found in any Tertiary deposits (Li and others, 1988; Zhao Yuhong, 1990, written commun.), confirms further the Cretaceous age for the lower beds.

The marginal-marine and shallow-marine sedimentary rocks (the uppermost 45 m of the sequence) have been dated by biostratigraphic analysis because no material suitable for isotopic analysis was recovered. Interpretations of the approximate age for these rocks differ between fossil groups, suggested ages ranging from early Maastrichtian to early Eocene. On the basis of correlation with lower latitude faunas from the Western Interior and northern Europe, the calcareous benthic foraminifer assemblage indicates a Campanian (MacBeth and Schmidt, 1973) or early Maastrichtian (McDougall, 1987) age. The palynomorph spectra is interpreted as middle Maastrichtian in age (Frederiksen and others, 1988; Frederiksen, 1991), on the basis of comparison with coeval Western Interior floras from Canada and the northern United States.

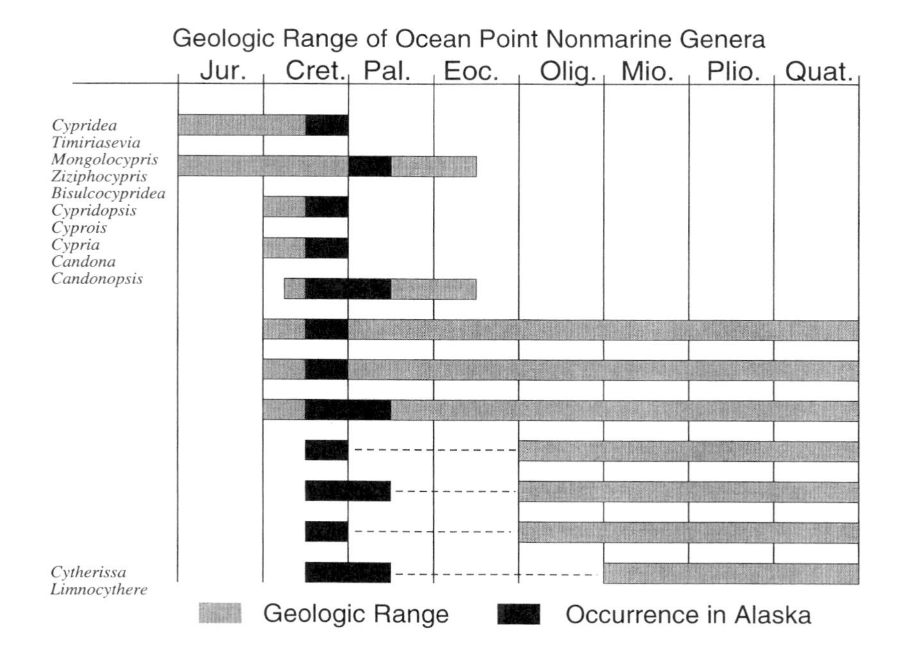

Figure 4. Known geologic ranges of nonmarine genera occurring at Ocean Point.

Ostracode and mollusk assemblages in the shallow-marine Ocean Point beds have been interpreted as Paleocene (post-Danian) to early Eocene in age (Marincovich and others, 1985; Marincovich, 1993), on the basis of generic-level comparisons with the Western Interior of North America (especially the Cannonball Formation of North Dakota) and northwestern Europe. The presence of the ostracode taxon *Paracyprideis similis* in the upper part of the section is a strong indication of Paleocene age, because this species is only known from middle and upper Paleocene beds in northern Europe (Triebel, 1941; M. C. Keen, 1991, personal commun.) and in the Paleocene Eureka Sound Formation on Ellesmere Island, Arctic Canada (Brouwers, 1991, unpublished data). Paleontological documentation of a Danian age for the Eureka Sound Formation was provided by Marincovich and Zinsmeister (1991). Reevaluation of the shallow-marine ostracode fauna (Brouwers and De Deckker, 1993) indicates a Danian age; two pectinid bivalves in the marine mollusk faunas are interpreted to represent an age of latest Maastrichtian or Danian (Waller and Marincovich, 1992).

Species composition of the lower and upper nonmarine ostracode assemblages indicate similar paleoenvironments; genus- and species-level differences between the two faunas are therefore interpreted to represent differences in age. The upper nonmarine suite contains three ostracode genera that imply a Late Cretaceous or Paleogene age: (1) the oldest occurrence of *Cytherissa*, previously known from Paleogene rocks of eastern Mongolia (Danielopol and others,

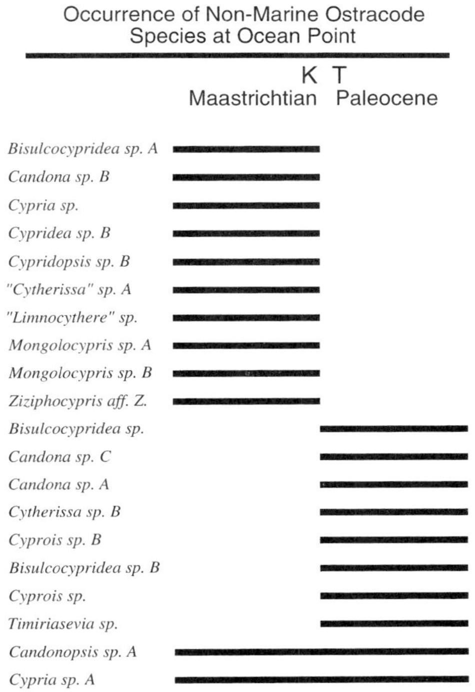

Occurrence of Non-Marine Ostracode
Species at Ocean Point

K T
Maastrichtian Paleocene

Bisulcocypridea sp. A
Candona sp. B
Cypria sp.
Cypridea sp. B
Cypridopsis sp. B
"Cytherissa" sp. A
"Limnocythere" sp.
Mongolocypris sp. A
Mongolocypris sp. B
Ziziphocypris aff. Z.
Bisulcocypridea sp.
Candona sp. C
Candona sp. A
Cytherissa sp. B
Cyprois sp. B
Bisulcocypridea sp. B
Cyprois sp.
Timiriasevia sp.
Candonopsis sp. A
Cypria sp. A

Figure 5. Occurrence of ostracode species in the lower (Maastrichtian) and upper (Paleocene) nonmarine assemblages at Ocean Point.

1990b); (2) the presence of *Cyprois*, previously known from Paleocene rocks (Swain, 1949); and (3) the occurrence of *Candona*, previously documented from Oligocene and younger rocks. *Mongolocypris* and *Cypridea*, two characteristic Cretaceous genera, do not occur above the lower nonmarine assemblage. The lack of dinosaur skeletal remains in the upper nonmarine sedimentary rocks, together with the conspicuous absence of Cretaceous marine mollusks such as inoceramids (which have been recovered from shallow-marine sediments of Campanian age downsection along the Colville River) in the marginal-marine and shallow-marine sedimentary rocks, is consistent with a Tertiary age for the uppermost 45 m of the sequence.

The sequence studied here does not include the actual Cretaceous-Tertiary boundary, which is related to two physical factors. First, the lower part of the measured section represents fluvial-deltaic environments, which by nature are episodic in sedimentation. Second, a normal fault of unknown displacement occurs at the change from lower nonmarine to marginal-marine sediments; if a boundary horizon did exist, it has been subsequently faulted and eroded.

The age suggested for the lowermost 135 m of the sequence, based on biostratigraphy, ranges from late Campanian to late Maastrichtian. $^{40}Ar/^{39}Ar$ isotopic dates of 68–71 Ma (late Maastrichtian) provide a calibration for the various biostratigraphic zonations. The late Maastrichtian age agrees well with the ages implied by the dinosaur fauna (J. Horner, 1989, personal commun.), with the nonmarine ostracode fauna, and with the charophyte flora (Feist and Brouwers, 1991).

The lack of agreement for the age of the uppermost 45 m of the sequence is difficult to resolve. Palynomorph and benthic foraminifer assemblages can be correlated to numerous dated sequences from lower latitudes of the Western Interior of North America. These correlations presume that no significant latitudinal differences exist and that the first and last appearances of taxa are geologically simultaneous throughout their geographic distribution, which may not be the case (Hickey and others, 1983). With the exception of the tephras associated with the dinosaur horizons, biostratigraphic zonations in Alaska have not been calibrated to isotopic dates. The Ocean Point sequence provides an excellent example of a Cretaceous-Tertiary assemblage, in which there are elements of the fauna in both the latest Cretaceous and earliest Tertiary that cannot be separated temporally.

The Paleocene age suggested by the ostracode assemblages for the uppermost 45 m is based on (1) the absence of *Mongolocypris*, *Cypridea,* and related Cyprideinae genera that are restricted worldwide to the Mesozoic; (2) correlation with nonmarine Cretaceous and Tertiary ostracode assemblages in Mongolia and the Western Interior of North America; and (3) correlation with Danian marine ostracode assemblages in the Canadian Arctic and northern Europe. The absence of dinosaurs, inoceramids, and ammonites in the uppermost 45 m provides additional, although not definitive, support for a Tertiary age.

For the purposes of this paper, which focuses on the ostracode assemblages, the age of the lower nonmarine sedimentary rocks is considered to be late Maastrichtian and the age of the marginal-marine, upper nonmarine, and shallow-marine sedimentary rocks (the Ocean Point beds) is considered to be Paleocene, probably Danian.

MAASTRICHTIAN SEQUENCE

Sedimentary Rocks

The lower 135 m represent nonmarine flood-plain environments that include sands deposited by major and minor fluvial distributaries and extensive organic-rich silts, punctuated by tephra deposits (Phillips, 1990). Overbank sediments are most abundant and include crevasse splays, overbank sheet floods, and vertical accretion deposits. Section 4 was measured near the base (Fig. 6) and typifies the

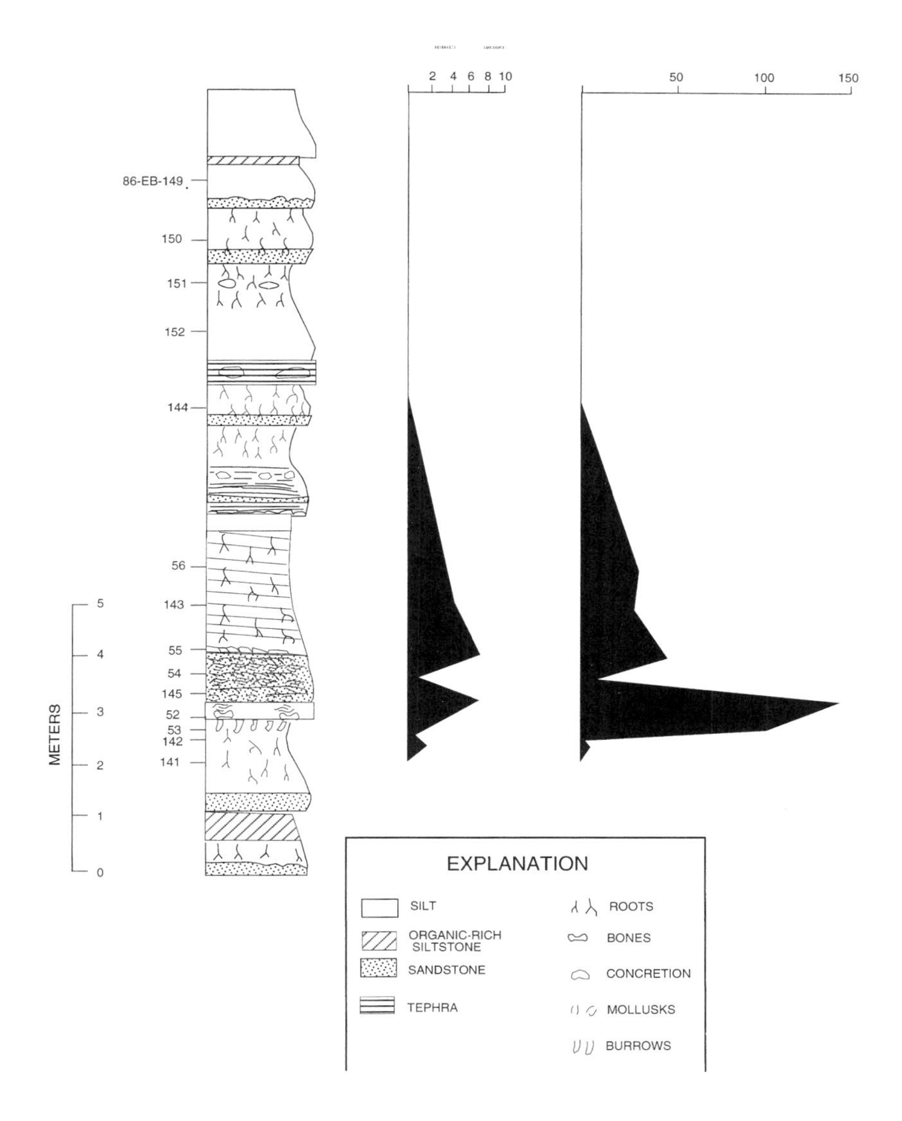

Figure 6. Measured section 4, near the base of the sequence, plotted against ostracode species diversity (number of species) and abundance (number of valves).

overbank strata, incorporating several fining-up cycles and a minor fluvial distributary. Fossils are found in the small channel near the base of the section (samples 52, 54, and 145) and include carbonized wood, fresh-water mollusks, dinosaur skeletal elements, ostracodes, and charophytes.

Ostracodes

Ostracodes occur in discrete horizons throughout the nonmarine sequence (Fig. 3), most commonly in the overbank siltstones. Although distribution is generally sparse and widely scattered, the Colville River sequence contains more ostracode-bearing sedimentary rocks than most Cretaceous marginal fluvial sections elsewhere in North America (for example, the Western Interior of North America; Fouch and others, 1987; R. M. Forester, 1990, personal commun.; E. M. Brouwers, 1981, unpublished data). Ostracodes are rare to absent in the tephras, sandstones, and carbonaceous siltstones of the sequence; obviously these sedimentary rocks are not conducive to the preservation of calcareous material.

A plot of species richness through the sequence (Fig. 7) illustrates that diversity is greatest in mollusk-rich beds that probably represent lag deposits. The high diversity in these horizons is a consequence of accumulation and deposition from several marginal fluvial environments (sedimentologic evidence indicates that these represent temporary ponds, oxbows, and streams). The ostracode assemblage in these lag deposits represents several habitats mixed together. Ostracode abundances (number of valves) parallel the richness trends and show high values in the lag deposits. The presence of fragile nonmarine ostracode valves indicates that these faunas are not reworked, otherwise they would be fragmented or show signs of abrasion and wear (De Deckker, 1986).

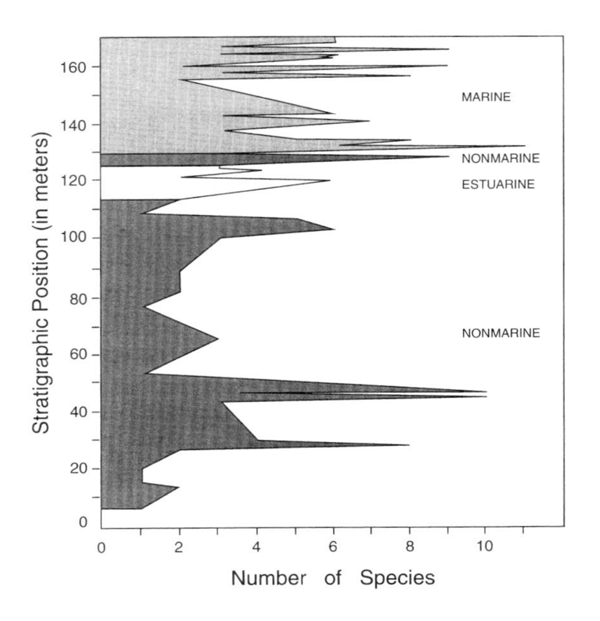

Figure 7. Plot of ostracode species diversity (number of species) through the entire sequence.

The basal 135 m are characterized by the lower nonmarine ostracode assemblage, which dominates the Colville River sequence (Fig. 3). Specimens in this assemblage are exceedingly well preserved; the nonmarine species identified include *Candona sp. B*, *"Cytherissa" sp. A*, *Cypridea sp. A*, *Cypridea sp. B*, *Cypridea sp. C*, *Cypria sp. A*, *Cypridopsis sp. A*, *Cypridopsis sp. B*, *Bisulcocypridea sp. A*, *Ziziphocypris sp.*, *"Limnocythere" sp.*, *Mongolocypris sp. A*, *Mongolocypris sp. B*, and *Candonopsis sp.*

Morphologic similarity between Cretaceous Arctic genera and Tertiary middle-latitude temperate to subarctic genera suggests that the middle-latitude genera originated in high-latitude environments during the Cretaceous and migrated southward (Keen, 1972, 1977) to lower latitudes as global climate cooled through the Tertiary and strong seasonality in climate became prevalent. For example, *Cytherissa* today lives in deep, cold-water lakes in the Northern Hemisphere and has diversified, especially in cold-temperate, ecologically stable environments such as Lake Baikal (Bronstein, 1947; Danielopol and others, 1990a, 1990b). The presence of *Cytherissa* in the Colville River sequence suggests cold water and well-oxygenated conditions (Delorme and Zoltai, 1984; Geiger, 1990), because no warm-water hypoxic *Cytherissa* species are known.

The assemblage composition and the associated sedimentary rocks suggest that most of the ostracodes lived in marginal fluvial habitats (e.g., oxbows, ponds, marshes, and small streams) adjacent to the large distributaries. The abundance of root traces (Phillips, 1988, 1990) and the high organic content of the siltstones (Brouwers and others, 1987) suggest that lush aquatic and terrestrial vegetation existed on the delta, which would have created a number of microhabitats and supported an abundant and diverse microfauna. The presence of taxa such as *Timiriasevia* and *Cytherissa*, being cytherids, implies that the water bodies must have been permanent, because known representatives of this group of ostracodes cannot withstand desiccation. This assumption is made by comparison with living representatives belonging to the same phylogenetic groups.

Depositional Environment

A simplified reconstruction of the late Maastrichtian environment represented in the study area based on the fossil assemblages and sedimentary features is shown in Figure 8. The Colville River sequence was part of a large fluvial-deltaic complex that bordered the paleo-Arctic Ocean, which was located northeast of the delta during the Late Cretaceous. The region was dominated by a low-gradient, low-relief coastal plain that was infrequently inundated by marginal-marine waters, as implied by the presence of occasional arenaceous benthic foraminifers. The water table was probably high, as evidenced by the lack of mudcracks, the abundance of aquatic invertebrates and plants, the presence of numerous organic-rich paleosols (Phillips, 1990), and the occurrence of nonmarine ostracode taxa that require permanent water. The fluvial-deltaic complex was mostly low ground, just above water level, but locally included some high ground. Several major rivers and many small distributary streams meandered across the coastal plain, carrying large volumes of overbank silts (Phillips, 1990).

The rivers and streams had extensive vegetation along their banks, evidenced by the presence of many root traces in the overbank silts. On the basis of a low-

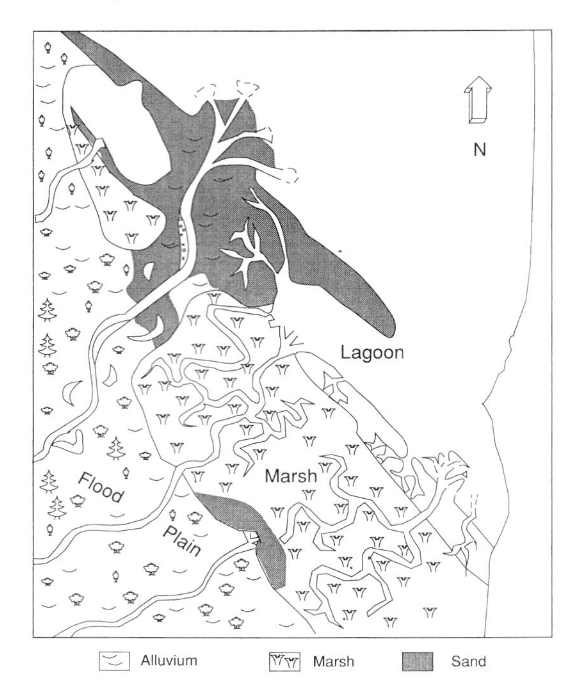

Figure 8. Reconstruction of deltaic environment of the Late Cretaceous and early Tertiary in northern Alaska. The paleo-Arctic Ocean was located northeast of the delta. Low ground is covered by ferns, horsetails, angiosperm shrubs, and aquatic plants. Deciduous gymnosperms occupy local high ground and form forests in the upland regions to the southwest. Ostracodes occur in the flood plains, marshes, small streams, ponds, bay facies, and shallow-marine habitats.

diversity plant megafossil assemblage, Spicer (1987) and Spicer and Parrish (1986, 1990) speculated that the coastal plain was dominated by a deciduous herbaceous ground cover that consisted mainly of ferns, *Equisetites* (horsetails), aquatic and subaquatic wetland plants, and a low-diversity angiosperm understory. The presence of abundant gymnosperm needles in the overbank siltstones and paleosols indicates that a few deciduous conifers probably existed on local high ground within or near the delta complex. Most of the trees probably occurred in upland forests, not on the delta proper. A rich palynomorph flora implies that a diverse forest occurred in the more distant upland regions (Ager, *in* Brouwers and others, 1987). The northern high-latitude forests consisted mainly of deciduous gymnosperms (forms related to *Metasequoia*), which formed a temperate forest system that was dominant from Alaska to southern Alberta (Spicer, 1987, 1989; Wolfe and Upchurch, 1987). This forest type corresponds approximately to the high-latitude coastal mixed conifer forests of North America (Wolfe, 1979).

Interpretations based on paleobotanical data suggest that northern Alaska underwent progressive cooling through the Late Cretaceous, probably coincident with the rapid northward movement of the North American plate (Hillhouse and Grommé, 1982). By Maastrichtian time the climate of the North Slope was probably temperate, comparable in air temperature to the modern Pacific Northwest, with estimated mean annual air temperatures of 2–8 °C (Spicer and Parrish, 1986; Spicer, 1987). Parrish and Spicer (1988) attributed the marked decrease of flora and the increase in richness of temperate floral taxa from the Cenomanian to the Maastrichtian primarily to climatic deterioration, which included cooling and increased seasonality. Alaskan North Slope Cenomanian and Coniacian plant megafossil floras are characterized by conifers and diverse angiosperms; by the end of the Cretaceous, North Slope vegetation was depauperate, consisting predominantly of conifers. The cooling trend evidenced by the flora is paralleled by sparse vertebrate faunas: Albian and Cenomanian Alaskan faunas include turtles as well as dinosaurs (Parrish and others, 1987), and the associated plant megafossils suggest mean annual temperatures of 10 °C (Parrish and Spicer, 1988). By Campanian and Maastrichtian time, amphibians and nondinosaurian reptiles had disappeared, leaving a low-diversity dinosaur assemblage consisting of *Edmontosaurus*, *Pachyrhinosaurus*, *Troodon*, and a tyrannosaurid (Clemens and Nelms, 1994). Associated plant megafossils suggest mean annual temperatures of 2–8 °C and an increase in seasonality (Spicer and Parrish, 1990). The cool temperatures at northern high latitudes is in marked contrast to the subtropical and tropical climates that dominated the middle and low latitudes during the Late Cretaceous (Wolfe and Upchurch, 1987).

Paleogeography

Estimates of paleolatitude for the Colville River area during the Late Cretaceous, based on paleomagnetic data and tectonic reconstructions, range from 70° to 85° N (Witte and others, 1987; Smith and others, 1981). Movements of the North Slope during the Tertiary are thought to have been minor, without significant latitudinal shifts (Jones, 1983; Sweeney, 1983). The North Slope therefore must have undergone profound seasonality of temperature as a consequence of several months of winter dusk or darkness. Low levels of winter insolation are evidenced by the flora, which consists entirely of deciduous gymnosperms and angiosperms (Spicer, 1987). Forest vegetation consists of conifers, a low-diversity angiosperm understory, low-diversity ground cover of ferns and horsetails, and aquatic plants.

Reconstructions of paleogeography based on tectonics and faunal provinces show that the Late Cretaceous Arctic Ocean had narrow, shallow connections with the Western Interior seaway of North America and with the Tethys seaway through Turgai Strait (Fujita and Newberry, 1983; Vinogradov and others, 1968; Beznosov and others, 1978). The Arctic Ocean was connected with the Western Interior seaway continuously from the Cenomanian through the early Maastrichtian (Williams and Stelck, 1975), but this seaway was closed by the end of the early Maastrichtian due to eustatic sea-level fall and tectonism related to uplift of the Rocky Mountain system (Williams and Stelck, 1975; Vail and others, 1977; Matthews, 1984).

Late Cretaceous nonmarine ostracodes of the Colville River sequence show taxonomic affinities with coeval faunas both of the Western Interior (Fouch and others, 1987) and northern China and Mongolia (Hao and others, 1982; Szczechura and Blaszyk, 1970). Western Interior nonmarine faunal and floral assemblages indicate subtropical temperatures, probably warmer than existed in northern Alaska, so the number of genera in common between the two regions is not great (Peck, 1951a; Fouch and others, 1987; Wolfe and Upchurch, 1986). In contrast, the Alaskan ostracode and charophyte genera show strong affinities with Late Cretaceous assemblages from Mongolia and northern China (Szczechura and Blaszyk, 1970; Szczechura, 1979; Karczewska and Ziembinska-Tworzydlo, 1970; Hou, 1983; Zhao Yuhong, 1989, personal commun.). However, Alaskan Nearctic taxa such as *Cyprois* and *Cypria* are missing from the Asian faunas. The close affinity between nonmarine Cretaceous ostracode assemblages from Alaska and China is believed to be a function of similar climatic conditions, presumably consisting of comparable temperature regimes. This in turn suggests the existence of a distinctive nonmarine temperate province at high northern latitudes that extended from Alaska into northern China.

DANIAN SEQUENCE

The uppermost 45 m of the Danian sequence consists of nonmarine, marginal-marine, and shallow-marine sedimentary rocks (Fig. 9). A fault of unknown dis-

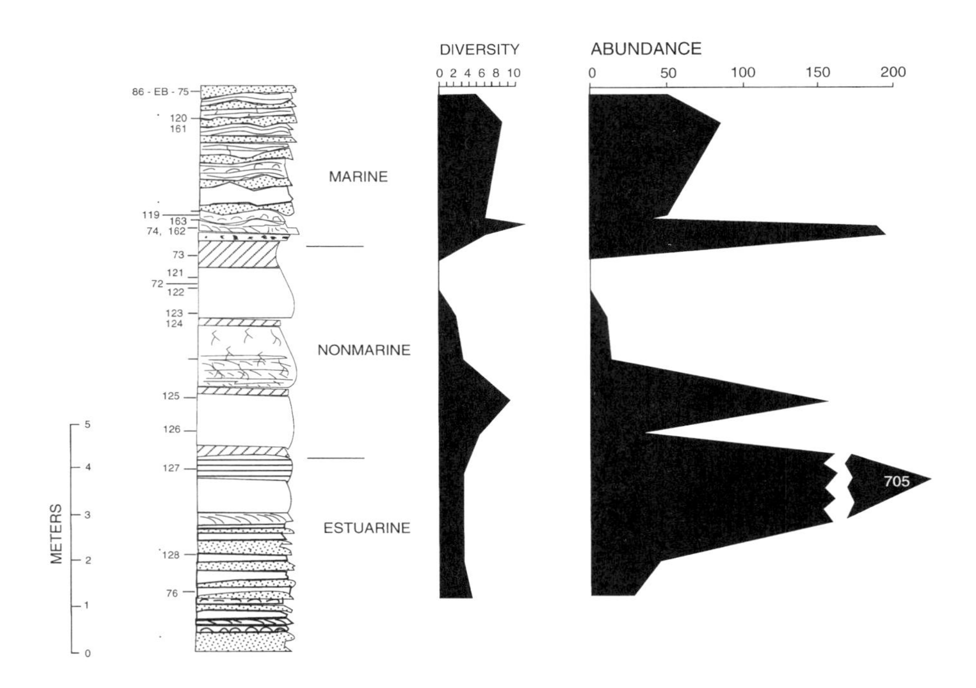

Figure 9. Measured section 24, near the top of the sequence, plotted against ostracode species diversity (number of species) and abundance (number of valves). Locality illustrates three distinct ostracode assemblages. Diversity and abundance peak at base of marine beds identifies a lag deposit with reworked nonmarine and in situ marine species.

placement separates the lower 135 m from marginal-marine deposits that immediately overlie the nonmarine sediments. Sedimentary environments do not change markedly from the Maastrichtian to the Danian; the Maastrichtian consists of flood-plain deposits and the Danian consists of flood-plain, marginal-marine, and shallow-marine deposits.

Three distinct ostracode faunas occur in the upper part of the sequence: a marginal-marine assemblage, a nonmarine assemblage, and a shallow-marine assemblage (Fig. 3). These faunas are most logically discussed in two parts—(1) the nonmarine assemblage and (2) the marginal-marine and shallow-marine assemblages.

Upper Nonmarine Assemblage

Sedimentary Rocks

Nonmarine sedimentary rocks occur only in measured sections 23 and 24, near the top of the sequence (Fig. 9). These consist of several fining-upward overbank cycles, each ending with an organic-rich paleosol. Overbank deposits include siltstones with rhizoliths and fine sandstones. The depositional environment of these nonmarine sediments is interpreted to be a combination of muddy marshes, minor distributary streams, and overbank deposits (Phillips, 1990).

Ostracodes

Ostracodes occur throughout most of the Danian sequence but vary in species richness and abundance. The pronounced peak in species diversity and abundance in sample 125 (Fig. 9) is probably a consequence of deposition from several adjacent environments. Unlike the lower nonmarine Maastrichtian horizons, where high ostracode species richness and abundance correspond to concentrations of mollusks, the upper nonmarine peak of diversity and abundance does not relate to a shell bed.

The upper nonmarine assemblage contains taxa indicative of marginal fluvial habitats similar to those habitats inferred for the lower nonmarine assemblage. Genera such as *Candona*, *Cyprois*, *Bisulcocypridea*, *Timiriasevia*, and *Candonopsis* inhabit oxbows, ponds, small streams, lakes, and other environments marginal to a fluvial system. There is no evidence for substantial water depth on the delta (e.g., >10 m) on the basis of sedimentological evidence. The presence of *Cytherissa*, however, suggests either a shallow perennial water body or the littoral zone of a lake (Bronstein, 1947); a deeper water body would probably have been stratified in the summer (De Deckker and Forester, 1988) and hence would have favored species tolerant of seasonality.

The lower (Maastrichtian) and upper (Danian) nonmarine ostracode assemblages differ in genus and species composition (Figs. 3 and 5). In the entire sequence 20 nonmarine species occur, of which only 2 species occur in both assemblages; 10 are exclusive to the lower assemblage, and 8 species occur only in the upper assemblage (Fig. 5). Because the environment is inferred to be comparable for both assemblages, the change in composition is believed to be temporal.

Stratigraphic distribution of ostracodes through the succession indicates that the nonmarine Cretaceous marker genera *Mongolocypris* and *Cypridea* disappear at the top of the lower nonmarine assemblage (Fig. 5). The only cyprideinid genus that occurs in the upper nonmarine assemblage is *Bisulcocypridea*, which is the sole post-Cretaceous survivor of the subfamily in North America (Fig. 10). *Bisulcocypridea* persists in Tertiary sediments of the Western Interior of North America, and is represented by two described species: *B. arvadensis* ranges from Late Cretaceous to early Paleogene and *B. nyensis* ranges from Paleocene to early Eocene (Swain, 1949; Peck, 1951b). *Bisulcocypridea* underwent morphologic changes from the Late Cretaceous to its extinction, including a gradual loss of the prominent *Cypridea* "beak," a reduction of the bisulcate nature of the carapace, the development of nodes and large pitted ornament, and an increase in size. In the Western Interior these changes are correlated to a well-dated sequence in the Uinta basin (Fouch and others, 1987). *Bisulcocypridea* sp. B, occurring in the uppermost nonmarine assemblage in Alaska, shows a reduction of beak development and complex ornament that correlates with Paleocene forms from the Uinta basin.

The upper nonmarine assemblage includes the first appearance of several taxa that are believed to represent the earliest examples of some characteristic Tertiary Nearctic genera such as *Candona*, *Candonopsis*, *Limnocythere*, and *Cytherissa*, which occur in Tertiary fossil and modern temperate to arctic climates (Fig. 5). For example, the oldest occurrence of *Cyprois* previously documented was from Paleocene and Eocene strata in Nevada and Utah (Fig. 11); the oldest record of *Limnocythere* was from Paleocene strata in Montana, Nevada, and Utah (Fig. 12). These two genera are ubiquitous in Canada and the United States through the Neogene and today, and occur in temperate climates.

Nonmarine ostracodes respond with great sensitivity to the ambient physicochemical environment, to parameters such as water chemistry, temperature, and nutrients, which in turn are controlled by geological, hydrological, botanical, and climatic factors (Delorme, 1969). Because nonmarine aquatic systems are subject to a greater degree of variability and instability, many ostracode taxa have responded by developing eggs capable of withstanding desiccation and freezing

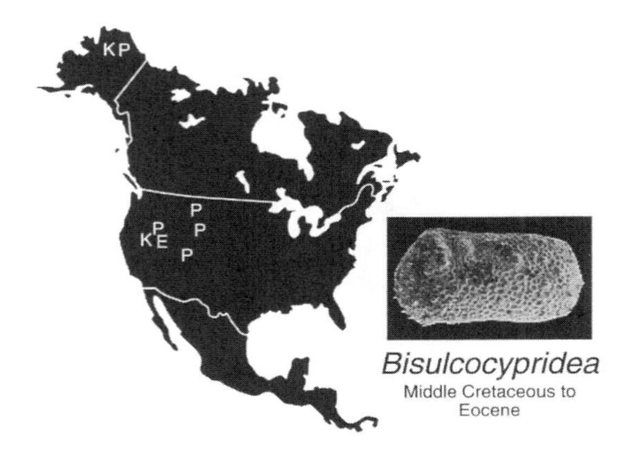

Figure 10. Map showing the oldest known occurrences of Bisulcocypridea *in North America;* Bisulcocypridea *is the only Cyprideinae genus to survive into the Tertiary. Two species occur in the Western Interior, one that ranges from Late Cretaceous (K) to early Paleogene and one that ranges from Paleocene to early Eocene (P, E).*

after developing to the blastula stage (Bronstein, 1947). These adaptations permit rapid, widespread passive dispersal of the eggs by such vectors as wind, birds and flying reptiles, and animals. The resistant, double-walled eggs are capable of surviving extreme environmental conditions for a period of years to decades.

There can be little doubt that the unusually cold climate of the Late Cretaceous in Alaska favored taxa that had these evolutionary adaptations and thus forced related changes to the physiology of the ostracodes, resulting in an endemic group of genera at northern high latitudes. Consequently, with new physiological adaptations for living in cold seasonal climatic conditions, these genera were able to migrate successfully southward during the early Tertiary, when cooling climates in the mid latitudes became prevalent. Keller and others (1993) noted that the northern high latitudes acted as centers of origin and dispersal for temperate planktonic foraminifers that subsequently migrated to lower latitudes.

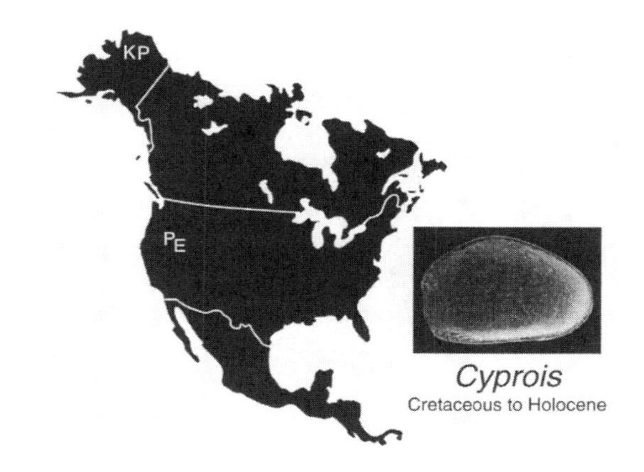

Figure 11. Map showing the oldest occurrences of Cyprois *in North America.* Cyprois *occurs in the Paleocene and Eocene (P, E) of Nevada.*

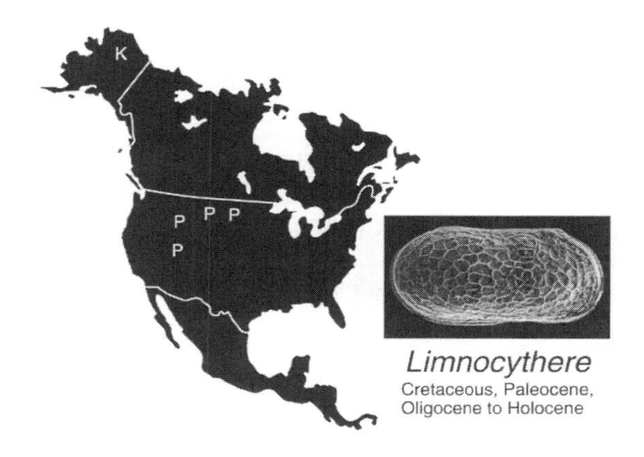

Figure 12. Map showing the oldest known occurrences of Limnocythere *in North America.* Limnocythere *occurs in the Paleocene (P) of Montana, North Dakota, and Idaho.*

Depositional Environment
The nonmarine rocks, sandwiched between the marginal-marine and shallow-marine sedimentary rocks (Fig. 9), were deposited in a marginal fluvial environment comparable to that inferred for the lower nonmarine rocks. Genera such as *Candona*, *Cyprois*, *Cypria*, *Timiriasevia*, and *Cytherissa* are indicative of ponds, streams, and other such habitats that exist in fluvial-deltaic complexes. The extensive root traces and organic-rich siltstones suggest that abundant vegetation occurred on the flood plain, and provided an abundant supply of organic detritus. Species of *Candonopsis* in this assemblage have been at least partly interstitial, as suggested by the mode of life of extant taxa belonging to the genus.

Marginal-Marine and Shallow-Marine Assemblages

Sediments
Marginal-marine deposits occur in the basal part of the Danian sequence, beginning at the base of measured section 18. These deposits grade upsection from a vegetated sand flat and channel sequence to a storm-dominated sequence (Phillips, 1990). Faunal diversity is moderate in both facies. Sediments contain calcareous benthic foraminifers, bivalves, brachiopods, marginal-marine ostracodes, and *Equisitites* root traces.

The uppermost sediments in the sequence are shallow-marine sandstones and siltstones with an abundant and diverse invertebrate fauna that includes bivalves, brachiopods, ostracodes, and calcareous benthic foraminifers. Sandstones and silty sandstones tend to occur with hummocky cross-stratification and include local shell lags. These horizons are probably storm-generated deposits. Finer sandy siltstones tend to form low-angle cross-beds and include bivalves in growth position. These sediments are presumably not related to storm deposition.

The contact between the shallow-marine and the underlying nonmarine rocks (Fig. 9) is distinctive. The uppermost nonmarine unit is an organic-rich siltstone that is immediately overlain by a pebble lag. The pebble lag in turn is overlain by marine sandstones and siltstones. The transgressive event recorded in the shallow-marine rocks is attributed by R. L. Phillips (1990, personal commun.) to fluctuations in sediment supply and sediment compaction as well as to sea-level changes.

Ostracodes
Ostracodes occur in most samples collected from the marginal-marine and shallow-marine facies. A plot of species richness through the sequence (Fig. 7) illustrates a number of peaks that are probably more related to aspects of sedimentation than to other environmental changes. Peaks in the marginal-marine facies are likely a consequence either of lag deposition or deposition from several adjacent environments. The numerous and sharp diversity peaks in the shallow-marine facies are believed to be a consequence of concentration and redeposition by repeated storm events.

The marginal-marine assemblage includes nine species, consisting predominantly of members of the genera *Paracyprideis* and *Cytheromorpha*, which are characteristic marginal-marine genera (Keen, 1978; Van Morkhoven, 1963). The occasional presence of *Bisulcocypridea*, inferred to be a nonmarine form, indicates that salinity must have been relatively low. Variability and water chemistry of an estuarine environment precludes most nonmarine and fully marine taxa, and this is seen in the composition of the marginal-marine assemblage. The restrictive physiochemical habitat of the Colville River marginal-marine environment is best illustrated by sample 127 (Fig. 9), which contains 705 valves of three species (*Cytheromorpha* sp. A, *Paracyprideis similis*, and *Paracyprideis* sp.). This very high abundance and low species richness is a distinctive phenomenon of a marginal-marine habitat, where stressed physiochemical conditions preclude most taxa, but those taxa capable of survival thrive in large numbers.

There are 44 species of ostracodes that occur in the shallow-marine assemblage (Marincovich and others, 1985, 1990), most of the species reflecting a strong endemic composition that is believed to be a result of Arctic Ocean isolation that occurred from near the end of the Cretaceous to the Eocene. Marincovich and others (1985, 1990) described the marine stratigraphic sequence, biostratigraphic age determinations based on ostracodes and mollusks, and paleogeographic reconstructions. The isolated Arctic Ocean probably induced the evolution of new taxa, evidenced by the first appearance of temperate-climate genera such as *Sarsicytheridea*, *Eucytheridea*, *Loxoconcha*, and *Paracyprideis*, which subsequently dispersed to the North Atlantic during the late Paleocene and Eocene.

The basal marine deposits (section 24; Fig. 9) contain a peak in species diversity and abundance that corresponds to a lag horizon. A number of nonmarine taxa (*Candona* and *Bisulcocypridea*) occur together with marine taxa (*Eucytheridea*, *Cytherura*, *Clithrocytheridea*(?), *Brachycythere*). This mixture is inferred to be a consequence of reworking the older nonmarine sediments into the marine environment with the erosive onset of the transgressive event. Repeated storm-generated deposition resulted in further reworking of previously deposited sediments and fossils.

The marine ostracode assemblage indicates normal marine salinity and shallow water depths (probably inner shelf). Differences in species composition through the marine sequence are believed to be more a function of storm deposition and spatial variation than of actual differences in habitat (see Marincovich and others, 1985, 1990, for a plot of the stratigraphic distribution of the marine ostracode species). Overall dominance and abundance relations remain the same throughout the marine sequence.

Depositional Environments

A simplified reconstruction of the environment of deposition for the Paleogene sedimentary rocks is shown in Figure 8 and is based on the fossil assemblages and sedimentary features. The deltaic-fluvial complex had shifted landward relative to its position during the Late Cretaceous, so that bay and shallow-marine facies rather than the deltaic fluvial complex were being deposited during the Paleocene. The deposition of marginal-marine and shallow-marine sediments may be related to an early Paleocene eustatic sea-level rise (Haq and others, 1987).

The marginal-marine environment, represented by the basal part of the Tertiary sequence, was very shallow and nearshore, and probably consisted of shallow bays that fronted the rivers and coastal regions. Salinities were low, indicated by the presence of salinity-tolerant nonmarine ostracode taxa and the absence of stenohaline marine ostracode and mollusk taxa. Low salinities are also indicated by the presence of abundant roots of *Equisetites* and other subaquatic plants (Parrish and Spicer, 1988; Phillips, 1988).

On the basis of the ostracode and mollusk assemblages, the marginal-marine and marine environment had a mild-temperate to cold-temperate marine climate. Marine faunas are taxonomically similar to coeval faunas from the Canadian Arctic Islands, northern Russia, and northern Europe (Marincovich and others, 1990). These northern high-latitude assemblages indicate the existence of an endemic temperate faunal province during the Paleocene, consisting of populations characterized by low richness, low abundance, and the conspicuous absence of warm water taxa.

Paleogeography

Reconstructions of paleogeography show that by Paleocene time the Arctic Ocean was almost completely isolated from the world's oceans (Marincovich and others, 1985). The only marine connections with the Arctic Ocean were intermittent seaways through Turgai Strait, the Norwegian-Greenland Sea, and the North Sea basin. The marginal marine and marine ostracodes have their strongest taxonomic affinities (both genus and species level) with Paleogene assemblages from the North Atlantic (Marincovich and others, 1990; M. C. Keen, 1991, personal commun.).

The marine ostracode assemblage consists of genera that are diachronous, including forms previously documented from Cretaceous to Danian strata, from upper Paleocene to Oligocene strata, and from Neogene and Quaternary strata (Marincovich and others, 1985). The presence of relict Cretaceous taxa implies that terminal Cretaceous extinction processes were only partly effective at high latitudes. Cretaceous holdovers probably made up a marginal population that may have been adapted to a harsh physiochemical environment and therefore was more capable of surviving an ecologic crisis. The precocious Paleogene genera presumably evolved in this high-latitude environment, because of geographic isolation and its temperate climate. These conditions in some ways were unique, at least for the Northern Hemisphere.

SUMMARY

Four ostracode assemblages occur in a 200 m sequence of nonmarine, marginal-marine, and shallow-marine rocks of Maastrichtian and Danian age. Spectacular bluff exposures along the northern Colville River contain a terrestrial and marine record of events near the Cretaceous-Tertiary boundary in a high-latitude environment. Near the end of the Cretaceous, organisms that dominated the Cretaceous, such as dinosaurs, ammonites, and inoceramids, became extinct in Alaska, along with typical Cretaceous ostracode genera such as *Mongolocypris*, *Cypridea*, and *Ziziphocypris*.

Genus-level extinctions typifying the Cretaceous-Tertiary boundary worldwide were less severe in northern Alaska. The nonmarine fauna shows some turnover at the species level across the boundary. Marginal-marine and shallow-marine faunas indicate that a number of "Cretaceous" genera persisted into the Tertiary and coexisted with so-called Tertiary genera. This suggests that northern high-latitude marine and nonmarine habitats may have been buffered somewhat from terminal Cretaceous events.

The nonmarine taxa survived the terminal Cretaceous ecologic crisis under inferred cold-temperature conditions that should have had a great effect on shallow aquatic habitats. Such cold conditions probably did not strongly affect the ostracode populations, because many of the nonmarine genera were capable of surviving freezing or desiccation by undergoing torpor or by egg-laying. In addition, the environmental conditions that later prevailed in the Tertiary in lower latitudes (e.g., seasonality, colder temperatures) already existed in Alaska. Many ostracode taxa are considered to be opportunistic species and capable of survival under substantial climatic and ecological contrasts. Those attributes surely made them good candidates for survivorship across the Cretaceous-Tertiary boundary.

Another factor that may have helped survival across the Cretaceous-Tertiary boundary in high latitudes is dissolved oxygen levels. When water temperatures are low, dissolved oxygen levels can be higher than in warm water bodies at lower latitudes. This phenomenon may explain in part the high abundance of ostracodes in organic-rich sediments such as the deltaic environment in Alaska, which are usually lower in dissolved oxygen. These high oxygen levels must have been important for ostracode survival in shallow-water environments in Alaska. Landis and others (this volume) describe a period of lowered atmospheric oxygen caused by mantle degassing and biologic controls just prior to the K-T boundary. The implied presence of greater dissolved oxygen at high latitudes may explain the lessened extinction effects in both the marine and nonmarine records.

The nonmarine ostracode genera *Cytherissa*, *Candonopsis*, *Candona*, *Cyprois*, and *Limnocythere* have their oldest occurrences in this sequence, suggesting that they evolved in northern Alaska during the Late Cretaceous or perhaps earlier in the Cretaceous, such as during the Cenomanian, when temperatures were warmer in the Arctic. The new taxa could have become established in high latitudes and adapted to progressively cooler climates.

ACKNOWLEDGMENTS

We thank Mike Keen and Janina Szczechura for their help in determining Paleogene distributions of various nonmarine taxa. Robin Whatley, Ken Bird, and Thomas Bown provided reviews of the manuscript. Larry Phillips has been helpful in providing sedimentological interpretations of the sequence. Rick Forester has been supportive and informative at various stages of this research. Financial support for the field work in Alaska was from the National Petroleum Reserve of Alaska (NPRA) program of the U.S. Geological Survey; support for preparing the material and subsequent papers is from the U.S. Geological Survey program on the Evolution of Sedimentary Basins.

REFERENCES CITED

Beznosov, N. V., Gorbatchik, T. N., Mikhailova, I. A., and Pergament, M. A., 1978, Soviet Union, *in* Moullade, M., and Nairn, A. E. M., eds., The Phanerozoic geology of the world II, The Mesozoic: Amsterdam, Elsevier, p. 5–53.

Bronstein, Z. S., 1947, Fresh-water Ostracoda, *in* Fauna of the USSR, Crustaceans, Volume 2: Academy of Sciences of the USSR, new ser. no. 31, p. 1–470.

Brosge, W. P., and Whittington, C. L., 1966, Geology of the Umiat-Maybe Creek region, Alaska: U.S. Geological Survey Professional Paper 303-H, p. 501–638.

Brouwers, E. M., and De Deckker, P., 1993, Late Maastrichtian and Danian ostracode faunas from northern Alaska: Reconstructions of environment and paleogeography: Palaios, v. 8, p. 140–154.

Brouwers, E. M., Clemens, W. A., Spicer, R. A., Ager, T. A., Carter, L. D., and Sliter, W. V., 1987, Dinosaurs on the North Slope, Alaska: High latitude, latest Cretaceous environments: Science, v. 237, p. 1608–1610.

Carter, L. D., and Galloway, J. P., 1985, Engineering-geologic map of northern Alaska, Harrison Bay quadrangle: U.S. Geological Survey Open-File Report 85-256, 49 p., 2 sheets.

Carter, L. D., Repenning, C. A., Marincovich, L. N., Jr., Hazel, J. E., Hopkins, D. M., McDougall, K., and Naeser, C. W., 1977, Gubik and pre-Gubik Cenozoic deposits along the Colville River near Ocean Point, North Slope, Alaska: U.S. Geological Survey Circular 751-B, p. B12–B14.

Clemens, W. A., and Nelms, L. G., 1994, Paleoecological implications of Alaskan terrestrial vertebrate fauna in latest Cretaceous time at high paleolatitudes: Geology, v. 21, p. 503–506.

Conrad, J. E., McKee, E. H., and Turin, B. D., 1990, K-Ar and ^{40}Ar/^{39}Ar ages of tuff beds at Ocean Point on the Colville River, Alaska: U.S. Geological Survey Bulletin 1946, p. 77–82.

Danielopol, D. L., Carbonel, P., and Colin, J. P., eds., 1990a, *Cytherissa* (Ostracoda)—The *Drosophila* of paleolimnology: Bulletin de Institute Geologie Bassin d'Aquitaine, no. 47, 297 p.

Danielopol, D. L., Olteanu, R., Loffler, H., and Carbonel, P., 1990b, Present and past geographical ecological distribution of *Cytherissa* (Ostracoda, Cytherideidae), *in* Danielopol, D. L., Carbonel, P., and Colin, J. P., eds., *Cytherissa* (Ostracoda)—The *Drosophila* of paleolimnology: Bulletin de Institute Geologie Bassin d'Aquitaine, no. 47, p. 97–118.

De Deckker, P., 1986, The use of ostracods in palaeolimnology in Australia: Palaeogeography, Palaeoclimatology, Palaeoecology, v. 62, p. 463–475.

De Deckker, P., and Forester, R. M., 1988, The use of ostracods to reconstruct continental palaeoenvironmental records, *in* De Deckker, P., Colin, J. P., and Peypouquet, J. P., eds., Ostracoda in the earth sciences: Amsterdam, Elsevier, p. 175–200.

Delorme, L. D., 1969, Ostracodes as Quaternary paleoecological indicators: Canadian Journal of Earth Sciences, v. 6, p. 1471–1476.

Delorme, L. D., and Zoltai, S. C., 1984, Distribution of an arctic ostracod fauna in space and time: Quaternary Research, v. 21, p. 67–73.

Feist, M., and Brouwers, E. M., 1991, A new *Tolypella* from the Ocean Point dinosaur locality, North Slope, Alaska, and the Late Cretaceous to Paleocene nitelloid charophytes, *in* Bird, K. J., ed., Evolution of sedimentary basins series, North Slope, Alaska: U.S. Geological Survey Bulletin 1990F, p. F1–F7.

Fouch, T. D., Hanley, J. H., Forester, R. M., Keighin, C. W., Pitman, J. K., and Nichols, D. J., 1987, Chart showing lithology, mineralogy, and paleontology of the nonmarine North Horn Formation and Flagstaff Member of the Green River Formation, Price Canyon, central Utah: A principal reference section: U.S. Geological Survey Miscellaneous Investigations Series Map I-1797-A.

Frederiksen, N. O., 1989, Changes in floral diversities, floral turnover rates, and climates in Campanian and Maastrichtian time, North Slope of Alaska: Cretaceous Research, v. 10, p. 249–266.

Frederiksen, N. O., 1991, Pollen zonation and correlation of Maastrichtian marine beds and associated strata, Ocean Point dinosaur locality, North Slope, Alaska: U.S. Geological Survey Bulletin 1990E, 24 p.

Frederiksen, N. O., Ager, T. A., and Edwards, L. E., 1988, Palynology of Maastrichtian and Paleocene rocks, lower Colville River region, North Slope of Alaska: Canadian Journal of Earth Sciences, v. 25, p. 512–527.

Fujita, K., and Newberry, J. T., 1983, Accretionary terranes and tectonic evolution of northeast Siberia, *in* Hashimoto, M., and Uyeda, S., eds., Accretion tectonics in the Circum-Pacific regions: Tokyo, Terra Scientific Publishing Co., p. 43–57.

Geiger, W., 1990, The role of oxygen in the disturbance and recovery of the *Cytherissa lacustris* population of Mondsee (Austria), *in* Danielopol, D. L., Carbonel, P., and Colin, J. P., eds., *Cytherissa* (Ostracoda)—The *Drosophila* of paleolimnology: Bulletin de Institute Geologie Bassin d'Aquitaine, no. 47, p. 167–189.

Hao, Yi-Chun, Su, De-Ying, and Li, You-Qui, 1982, Late Mesozoic nonmarine ostracodes in China, *in* Maddocks, R. F., ed., Applications of Ostracoda: Proceedings of the Eighth International Symposium on Ostracoda: Houston, Texas, Department of Geosciences, University of Houston, p. 372–380.

Haq, B. U., Hardenbol, J., and Vail, P. R., 1987, Chronology of fluctuating sea level since the Triassic: Science, v. 235, p. 1156–1166.

Hickey, L. J., West, R. M., Dawson, M. R., and Choi, D. K., 1983, Arctic terrestrial biota: Paleomagnetic evidence of age disparity with mid-northern latitudes during the Late Cretaceous and early Tertiary: Science, v. 221, p. 1153–1156.

Hillhouse, J. W., and Grommé, C. S., 1982, Limits to northward drift of the Paleocene Cantwell Formation, central Alaska: Geology, v. 10, p. 552–556.

Hou, You-Tang, 1983, Characteristics and significance of Late Cretaceous non-marine ostracods: Bulletin of Nanjing Institute of Geology and Paleontology, Academica Sinica, no. 6, p. 309–319.

Jones, P. B., 1983, The cordilleran connection—A link between Arctic and Pacific sea-floor spreading: Alaska Geological Society Journal, v. 2, p. 41–55.

Karczewska, J., and Ziembinska-Tworzydlo, M., 1970, Upper Cretaceous Charophyta from the Nemegt Basin, Gobi Desert, *in* Kielan-Jaworoska, Z., ed., Results of the Polish-Mongolian Palaeontological Expeditions—Part II: Palaeontologica Polonica, no. 21, p. 121–146.

Keen, M. C., 1972, The Sannoisian and some other upper Palaeogene Ostracoda from north west Europe: Palaeontology, v. 15, p. 267–325.

Keen, M. C., 1977, Ostracod assemblages and the depositional environments of the Headon, Osborne, and Bembridge Beds (upper Eocene) of the Hampshire Basin: Palaeontology, v. 20, p. 405–445.

Keen, M. C., 1978, The Tertiary-Palaeogene, *in* Bate, R., and Robinson, E., eds., A Stratigraphic index of British Ostracoda: Geological Journal Special Issue 8, p. 385–450.

Keller, G., Barrera, E., Schmitz, B., and Mattson, E., 1993, Gradual mass extinction, species survivorship, and long-term environmental changes across the Cretaceous-Tertiary boundary in high latitudes: Geological Society of America Bulletin, v. 105, p. 979–997.

Li, You-Gui, Su, De-Ying, and Zhang, Li-Jun, 1988, The Cretaceous ostracod fauna from Fuxin Basin, Liaoning Province, *in* Hanai, T., Ikeya, N., and Ishizaki, K., eds., Evolutionary biology of Ostracoda: Developments in Paleontology and Stratigraphy no. 11, p. 1173–1186.

MacBeth, J. I., and Schmidt, R. A. M., 1973, Upper Cretaceous foraminifers from Ocean Point, Alaska: Journal of Paleontology, v. 47, p. 1047–1061.

Marincovich, L., Jr., 1993, Danian mollusks from the Prince Creek Formation, northern Alaska, and implications for Arctic Ocean paleogeography: Paleontological Society Memoir 35, 35 p.

Marincovich, L., Jr., and Zinsmeister, W., 1991, The first Tertiary (Paleocene) marine mollusks from the Eureka Sound Group, Ellesmere Island, Canada: Journal of Paleontology, v. 65, p. 242–248.

Marincovich, L., Jr., Brouwers, E. M., and Hopkins, D. M., 1983, Paleogeographic affinities and endemism of Cretaceous and Paleogene marine faunas in the Arctic, *in* U.S. Geological Survey Polar Research Symposium, Abstracts with Program: U.S. Geological Survey Circular 911, p. 45–46.

Marincovich, L., Jr., Brouwers, E. M., Hopkins, D. M., and McKenna, M. C., 1990, Mesozoic and Cenozoic paleogeographic and paleoclimatic history of the Arctic Ocean Basin, based on shallow-water marine faunas and terrestrial vertebrates, *in* Grantz, A., Johnson, L., and Sweeney, J. F., eds., The Arctic Ocean region: Boulder, Colorado, Geological Society of America, The Geology of North America, v. L, p. 403–426.

Marincovich, L., Jr., Brouwers, E. M., and Carter L. D., 1985. Early Tertiary marine fossils from northern Alaska: Implications for Arctic Ocean paleogeography and faunal evolution: Geology, v. 13, p. 770–773.

Matthews, R. K., 1984, Oxygen isotope record of ice-volume history: 100 million years of glacio-eustatic sea-level fluctuation, *in* Schlee, J. S., ed., Interregional unconformities and hydrocarbon accumulation: American Association of Petroleum Geologists Memoir 36, p. 97–107.

McDougall, K., 1987, Maestrichtian benthic foraminifers from Ocean Point, North Slope, Alaska: Journal of Foraminiferal Research, v. 17, p. 344–366.

McKee, E. H., Conrad, J. E., and Turin, B. D., 1989, Better dates for Arctic dinosaurs: Eos (Transactions, American Geophysical Union), v. 70, p. 74.

Parrish, J. T., and Spicer, R. A., 1988, Late Cretaceous terrestrial vegetation: A near-polar temperature curve: Geology, v. 16, p. 22–25.

Parrish, J. M., Parrish, J. T., Hutchison, J. H., and Spicer, R. A., 1987, Late Cretaceous vertebrate fossils from the North Slope of Alaska and implications for dinosaur ecology: Palaios, v. 2, p. 377–389.

Peck, R. E., 1951a, Nonmarine ostracodes—The Subfamily Cyprideinae in the Rocky Mountain area: Journal of Paleontology, v. 25, p. 307–320.

Peck, R. E., 1951b, A new ostracode genus from the Cretaceous Bear River Formation: Journal of Paleontology, v. 25, p. 575–577.

Phillips, R. L., 1988, Measured sections, paleoenvironments, and sample locations near Ocean Point, Alaska: U.S. Geological Survey Open-File Report 88-40.

Phillips, R. L., 1990, Summary of Late Cretaceous environments near Ocean Point, North Slope, Alaska: U.S. Geological Survey Bulletin 1946, p. 101–106.

Smith, A. G., Hurley, A. M., and Briden, J. C., 1981, Phanerozoic paleocontinental world maps: London, Cambridge University Press, 102 p.

Spicer, R. A., 1987, The significance of the Cretaceous flora of Northern Alaska for the reconstruction of the climate of the Cretaceous: Geologische Jahrbuch, v. A96, p. 265–291.

Spicer, R. A., 1989, Plants at the Cretaceous-Tertiary boundary: Royal Society of London Philosophical Transactions, ser. B, v. 325, p. 291–305.

Spicer, R. A., and Parrish, J. T., 1986, Paleobotanical evidence for cool north polar climates in the mid-Cretaceous (Albian-Cenomanian): Geology, v. 14, p. 703–706.

Spicer, R. A., and Parrish, J. T., 1990, Latest Cretaceous woods of the central North Slope, Alaska: Palaeontology, v. 33, p. 225–242.

Swain, F. M., 1949, Early Tertiary Ostracoda from the Western Interior United States: Journal of Paleontology, v. 23, p. 172–181.

Sweeney, J. F., 1983, Evidence for the origin of the Canada Basin margin by rifting in the Early Cretaceous time: Alaska Geological Society Journal, v. 2, p. 17–23.

Szczechura, J., 1979, Fresh-water ostracodes from the Nemegt Formation (Upper Cretaceous) of Mongolia, *in* Kielan-Jaworoska, Z., ed., Results of the Polish-Mongolian Palaeontological Expeditions—Part VIII: Palaeontologia Polonica, no. 38, p. 65–121.

Szczechura, J., and Blaszyk, J., 1970, Fresh-water Ostracoda from the Upper Cretaceous of the Nemegt Basin, Gobi Desert, *in* Kielan-Jaworoska, Z., ed., Results of the Polish-Mongolian Palaeontological Expeditions—Part II: Palaeontologia Polonica, no. 21, p. 107–120.

Triebel, E., 1941, Fossile arten der ostracoden-gattung *Paracyprideis* Klie: Senckenbergiana, v. 23, p. 153–164.

Vail, P. R., Mitchum, R. M., Jr., and Thompson, S., 1977, Seismic stratigraphy and global changes of sea level, Part 4: Global cycles of relative changes in sea level, *in* Payton, C. E., ed., Seismic stratigraphy—Applications to hydrocarbon exploration: American Association of Petroleum Geologists Memoir 36, p. 83–97.

von Morkhoven, F. P. C. M., 1963, Post Paleozoic Ostracoda, Volume 2: Elsevier, Amsterdam, 478 p.

Vinogradov, A. P., Vereschagin, V., Nalivkin, V., Ronov, A., Khabakov, A., and Khain, V., 1968, Atlas of the lithological-palgeographical maps of the USSR, Triassic, Jurassic and Cretaceous: Moscow, USSR Ministry of Geology and Academy of Science, 20 maps.

Waller, T. R., and Marincovich, L., Jr., 1992, New species of *Camptochlamys* and *Chlamys* (Mollusca: Bivalvia: Pectinidae) from near the Cretaceous/Tertiary boundary at Ocean Point, North Slope, Alaska: Journal of Paleontology, v. 66, p. 215–227.

Williams, G. D., and Stelck, C. R., 1975, Speculations on the Cretaceous paleogeography of North America, *in* Caldwell, W. G. E., ed., The Cretaceous System in the western interior of North America, Geological Association of Canada Special Paper 13, p. 1–20.

Witte, W. K., Stone, D. B., and Mull, C. G., 1987, Paleomagnetism, paleobotany, and paleogeography of the Cretaceous, North Slope, Alaska, *in* Tailleur, I., and Weimer, P., eds., Alaskan North Slope geology, Volume 2: Society of Economic Paleontologists and Mineralogists and Alaska Geological Society, p. 571–579.

Wolfe, J. A., 1979, Temperature parameters of humid to mesic forests of eastern Asia and relation to forests of other regions of the Northern Hemisphere and Australasia: U.S. Geological Survey Professional Paper 1106, p. 1–37.

Wolfe, J. A., and Upchurch, G. R., Jr., 1986, Vegetation, climatic and floral changes at the Cretaceous-Tertiary boundary: Nature, v. 324, p. 148–152.

Wolfe, J. A., and Upchurch, G. R., Jr., 1987, North American nonmarine climates and vegetation during the Late Cretaceous: Palaeogeography, Palaeoclimatology, Palaeoecology, v. 61, p. 33–77.

Zinsmeister, W. J., and Feldmann, R. M., 1984, Cenozoic high latitude heterochroneity of Southern Hemisphere marine faunas: Science, v. 224, p. 281–283.

9

Maastrichtian Extinction Patterns of Caribbean Province Rudistids

Claudia C. Johnson, *Earth Systems Science Center, Pennsylvania State University, Pennsylvania* and
Erle G. Kauffman, *Department of Geological Sciences, University of Colorado, Colorado*

INTRODUCTION

Rudistid bivalves (Hippuritacea) initially inhabited inner carbonate platform environments of the Caribbean Tethys, but evolved reef-adapted morphologies and life habits, and became important components of paleotropical reef ecosystems during the Aptian and Albian (Kauffman and Johnson, 1988). By the middle to Late Cretaceous, rudistids numerically dominated the bioconstructor guild and most phases of ecological succession in reefs and other biogenic frameworks across Caribbean Cretaceous carbonate platform and upper slope environments (Kauffman and Sohl, 1979; Kauffman and Johnson, 1988, and references therein). Rudistids are thus the major proxy for these paleotropical, shallow-marine ecosystems. By the Cenomanian, certain eurytopic lineages of rudistids had spread into deeper water habitats, and into subtropical to warm temperate climate zones. Their geographic range was thus expanded, as was their resistance, as a group, to extinction.

Rudistids survived three major mass extinctions during their evolutionary history (end of the Jurassic, late Aptian, and across the Cenomanian-Turonian boundary), as well as regional extinctions in the Caribbean province during the late Turonian and middle Coniacian (Fig. 1; Johnson and Kauffman, 1990). Each of these mass-extinction intervals resulted in the loss of more primitive lineages among the families Diceratidae, Requieniidae, Caprotinidae, and, during the Cenomanian, the Caprinidae, and their subsequent replacement by radiation of morphologically more advanced, reef-adapted rudistids among the Caprinidae,

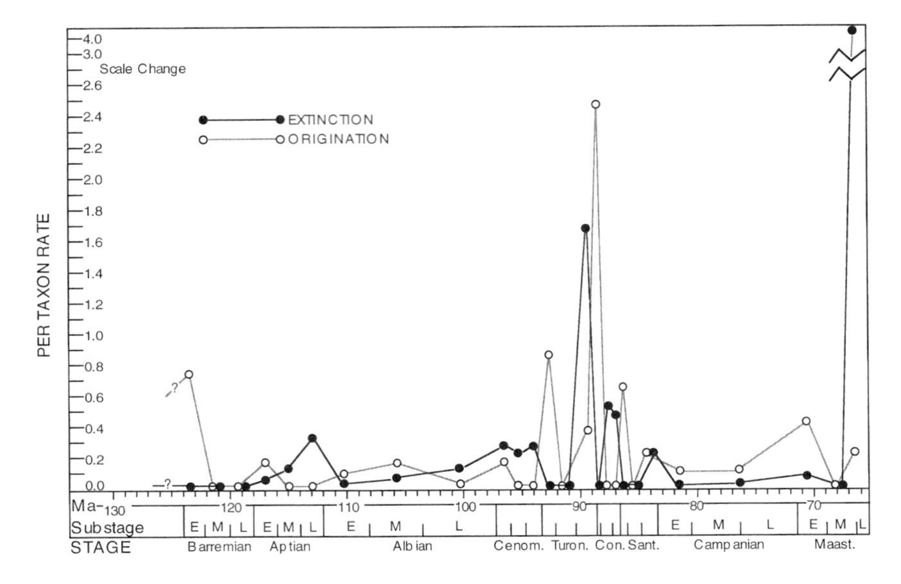

Figure 1. Per-taxon origination (open circles) and extinction rates (dark circles) of Cretaceous rudistid bivalves, based on all known species in the Caribbean province, plotted per substage (updated and modified from Johnson and Kauffman, 1990). Note expression of late Aptian, Cenomanian, and very strong Maastrichtian mass-extinction intervals (see Sepkoski, 1993), as well as strong late Turonian and middle to late Coniacian regional Caribbean extinctions. The graph utilizes macrofaunally defined substages, and thus shows the major rudistid extinction event near the middle-late Maastrichtian boundary (discussion in text).

Radiolitidae, and Hippuritidae. The net effect of these evolutionary turnovers was a marked increase in rudistid taxonomic diversity, adaptive range, and ecologic complexity between extinction intervals, especially during the Albian and Santonian through Maastrichtian radiations (see Johnson and Kauffman, 1990, Fig. 6).

In the Caribbean province, simple diversity among rudistids increased from 1 genus and one to two species in the Valanginian and Hauterivian to 30 genera and 116 species during the acme of rudistid evolution in the Maastrichtian (Johnson, 1993), just prior to their extinction in both tropical and temperate habitats. In well-documented Antillean examples, discussed subsequently, the Maastrichtian extinction of reef ecosystems and numerous lineages of successful rudistid bioconstructors was abrupt, and occurred within a meter or so of continuously deposited carbonate platform facies. Immediately following the demise of rudistid-dominated reef ecosystems, latest Cretaceous biostromal frameworks on these same carbonate platforms were composed of small coral-algal or algal-echinoderm associations, and subsequently of giant marine oysters and associated fully marine taxa, including rare isolated rudistids. The Maastrichtian extinction of the rudistid-dominated reef ecosystem was so complete that significant coral-algal reef communities did not recover to basic levels of ecological organization until the late Eocene, 8 m.y. later. The extinction of the subtropical to temperate-adapted Western Hemisphere rudistids occurred in the late Maastrichtian, after the extinction of reef ecosystems, but below the Cretaceous-Tertiary (K-T) boundary.

In this paper we discuss the major hypotheses proposed for mass-extinction intervals, and analyze the rates and patterns of rudistid extinction relative to each of these scenarios near the K-T boundary. We then review geologic events and environmental changes that characterized the Caribbean province during the Late Cretaceous, and utilize these data to evaluate causal mechanisms for the pre-K-T boundary extinction of the rudistid bivalve reef ecosystem. In summary, our high-resolution analyses confirm that the extinction of diverse, well-structured Maastrichtian rudistid reef and framework ecosystems, and many specialized rudistid lineages, was dramatic and occurred within continuously deposited carbonate platform facies prior to the K-T boundary, but at about the same stratigraphic level throughout the Caribbean province. Macrofaunal biostratigraphy (e.g., Sohl and Kollmann, 1985) suggests that the reef extinction took place near or at the middle-late Maastrichtian boundary. Some nannoplankton data suggest that it may have been earlier, near the early-middle Maastrichtian boundary (e.g. Jiang and Robinson, 1987; Verdenius, 1993); in either case, the Caribbean-wide demise of reef ecosystems predated the K-T boundary extinction by ~1.5, or 2.5–3.0 m.y. No rudistid bivalves are known to have survived to the K-T boundary in this region.

MASS-EXTINCTION HYPOTHESES AND THE ROLE OF RUDISTID BIVALVES

The extinction of the rudistid bivalves, as well as numerous other characteristic Mesozoic groups, was originally considered to be catastrophic at the K-T boundary (e.g., Schindewolf, 1962; Alvarez and others, 1980). However, as Late Cretaceous stratigraphic resolution increased and both biostratigraphic and isotopic dating methods were refined, the timing of the rudistid bivalve and overall Cretaceous reef extinctions relative to the K-T boundary were reevaluated. As a result, rudistids have recently been considered in context of both stepwise (Kauffman, 1988) and graded (Kerr, 1992) mass-extinction hypotheses. Arguments pertaining to each interpretation are as follows.

The Catastrophic Mass-Extinction Hypothesis

Many authors have envisioned the end of the Cretaceous extinction of rudistids and associated reef ecosystems as essentially catastrophic, occurring abruptly at a peak in their evolutionary and ecological development very near to or at the K-T boundary (e.g., Schindewolf, 1962; Dechaseaux, 1967; Kauffman and Sohl, 1979; Alvarez and others, 1980; Jones and Nicol, 1986). For example, Dechaseaux (1967, p. N765; Fig. E234) stated:

> Among the last surviving rudists were three radiolitids and a hippuritid which lived in Catalonia during the Maastrichtian Stage. They... formed large "rudist reefs." Their abrupt extinction is ... difficult to explain since in this region deposition of calcareous sediment had continued uninterruptedly from the Coniacian onward,

> some species nevertheless disappeared while others persisted and new ones appeared; then, at the same moment, apparently all became extinct....we are entirely ignorant of factors responsible for the extinction of the rudists.

Dechaseaux (1967, p. N765) also cited Tavani as having noted the abrupt extinction of rudistids in continuously deposited Maastrichtian–early Eocene carbonate platform facies of Somalia, without any evidence of a change in depositional environments or ecological conditions. Schindewolf (1962) listed the rudistids among many typical Cretaceous groups that became catastrophically extinct at the K-T boundary, an event which he attributed to rapid changes in atmospheric and ocean chemistry. Because these early views of rudistid extinction were in large part due to the lack of fine stratigraphic resolution of collections, and to the uncertainty in the position of the K-T boundary, it is not surprising that when Alvarez and others (1980) first proposed a catastrophic mass extinction at the end of the Cretaceous due to the impact of a large bolide on Earth, rudistids were listed as one of the primary victims.

In recent years, high-resolution stratigraphic studies of the Maastrichtian and K-T boundary interval worldwide have wholly contradicted the idea of a terminal Cretaceous catastrophic extinction of the rudistids. Data from both Caribbean and Mediterranean paleotropical sequences show that no rudistid reefs or frameworks are known from strata representing the last one million years or more of the Maastrichtian, and no rudistid individuals have been confirmed from highest Maastrichtian paleotropical strata. Individual rudistids from reef-forming lineages, and those adapted to subtropical and temperate paleoenvironments, persisted longer than the reef ecosystems. However, only a single specimen of the epibiont *Gyropleura* has been reported in the highest Maastrichtian condensed beds below the K-T boundary at Stevns Klint, Denmark (Heinberg, 1979). In the Western Hemisphere, the youngest rudistid specimen is a warm-temperate adapted species of *Titanosarcolites, T.* sp. aff. *T. oddsensis* Stephenson, which has been found by us in the Brazos River sections of Texas, in the upper part of the *Pseudotextularia deformis* (= *A. mayaorensis*) foraminiferal biozone (Keller, 1989b). It thus appears that the extinction of rudistid reefs, and of individual lineages, was essentially complete prior to the K-T boundary (65.5 Ma) in the Western Hemisphere.

The Stepwise Mass-Extinction Hypothesis

A number of recent studies (Alvarez and others, 1984; Kauffman, 1979, 1984a, 1984b, 1988, 1995; Johnson and Kauffman, 1990) have proposed that the rudistid reef ecosystem (including most actaeonellid and nerineid gastropods, many species of tropical bivalves, echinoids, corals, and larger foraminifers, and many specialized rudistid lineages) underwent a dramatic mass killing and short-term extinction somewhere between 1.5 and 2.5 m.y. below the K-T boundary as a first "step" or event in the K-T extinction interval. Within the same time frame there were significant extinctions among temperate molluscs (e.g. inoceramid bivalves and ammonites; Ward, 1988; Ward and others, 1991) and terrestrial plants (Johnson, 1992). Additional steps in molluscan extinction (Kauffman, 1988;

Hansen, and others, 1987, 1993), and specialized calcareous plankton (Keller, 1989a, 1989b), preceeded the K-T boundary event, and cool-temperate brachiopod and bryozoan extinction events postdated the K-T boundary to compose a 2.0–2.5-m.y.-long stepwise extinction interval (Kauffman, 1988, Fig. 3, and references therein).

The apparent stepwise nature of the K-T extinction, however, is manifested in observed data and has not yet been statistically tested for the Signor-Lipps effect (Signor and Lipps, 1982) utilizing equations for predicting stratigraphic ranges of Marshall (1991) or Koch and Morgan (1988). For the rudistid data, these calculations must await the difficult chronostratigraphic integration of observed data from isolated boundary sections in various parts of the Caribbean and Mediterranean provinces. Therefore, despite the observations of Dechaseaux (1967, and Tavani, cited therein), Kauffman (1979, 1988, 1995), Johnson and Kauffman (1990), Swinburne (1991), and those cited herein, that rudistid reefs and frameworks disappeared well below the K-T boundary in continuously deposited carbonate platform facies, the relevance of this to the stepwise extinction theory must remain a hypothesis until statistical tests confirm that this is not an artifact of preservation.

The Graded Mass-Extinction Hypothesis

The graded hypothesis was developed to reflect an increase in the rate of background extinction due to relatively rapid Earth-bound environmental changes such as sea level, oceanic oxygen levels, and large-scale volcanism (Kauffman, 1988). Kerr (1992) reported that Swinburne's (1991) rudistid data from the Mediterranean province suppported a gradualistic hypothesis for the extinction of rudistid reefs and lineages. Swinburne (1991) documented a dramatic decline in rudistids in the middle Maastrichtian of the European-African-Arabian Tethyan region—a decline that was at slightly different times in different places. Swinburne (1991) attributed the middle Maastrichtian rudistid extinction to a drop in sea level, which removed rudistid habitats. However, because few taxa survived the interval from the middle Maastrichtian to the K-T boundary in these regions, a stepwise extinction case could also be argued from Swinburne's (1991) interpretations.

The rudistid reef extinction has thus been cited as representative of three very different extinction scenarios. In this paper we present high-resolution Caribbean province data that seem to fit most closely the stepwise extinction hypothesis, initiating with the abrupt demise of reef ecosystems about 1.5 to 2.5–3.0 m.y. below the K-T boundary (depending on the biochronology used), and including final extinction of the remaining subtropical to temperate-adapted lineages just prior to the K-T boundary.

MAASTRICHTIAN EVENTS AND PALEOENVIRONMENTS OF THE CARIBBEAN PROVINCE

The Caribbean province was geologically dynamic during the Maastrichtian. Tropical paleoenvironments were broadly affected by tectonic and volcanic activity related to the immigration of the Caribbean plate between North and South America (Pindell and Barrett, 1990), fluctuating but generally falling global sea level (Haq and others, 1987), and lowering atmospheric CO_2 (Berner, 1994) coupled with increasing thermal gradients, which led to global cooling (Hay, 1988; Jenkyns and others, 1994). Although these factors affected the depositional environments and ecosystems of shallow-marine platforms, none of these factors acted rapidly enough, or at a sufficient magnitude over short time spans, to cause widespread ecological shock leading abruptly to mass killing, the collapse of reef ecosystems, and widespread extinction of rudistids within <100 ka—the apparent pattern recorded in the Maastrichtian rock record of Jamaica, Puerto Rico, and Mexico. Indeed, despite geologic and environmental trends thought to be stressful to reef ecosystems, the record of Caribbean rudistid evolution shows that the greatest peak in taxonomic and ecologic diversification was just prior to the sudden collapse of the tropical ecosystem more than 1.5 m.y. before the Chicxulub bolide impact event. Maastrichtian marine environments of the Caribbean province were characterized by the following changes (Fig. 2).

Tectonism, Volcanism, and Facies Development

Volcanic activity and outgassing from the Pacific superplume virtually ceased during the Late Cretaceous (Larson, 1991a, 1991b), as had sea-floor spreading between North and South America (Pindell and Barrett, 1990). The Caribbean

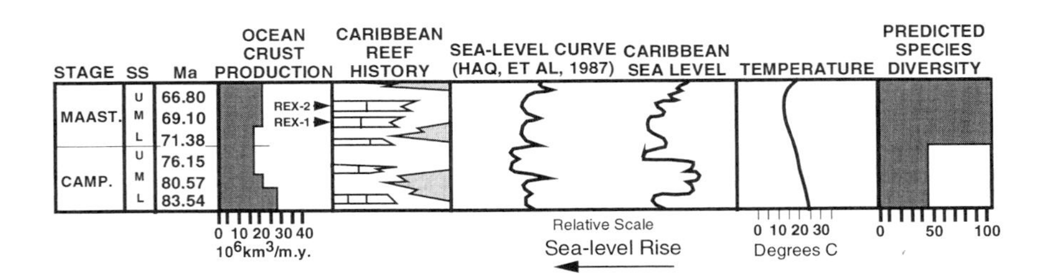

Figure 2. Summary of Campanian-Maastrichtian environmental changes in the Caribbean province (modified from Johnson and Kauffman, 1990) showing (left to right) latest radiometric ages for North America (Obradovich, 1993; Kauffman and others, 1993); ocean crust production (from Larson, 1991a); intervals of carbonate platform and reef development (limestone patterns) and levels of volcanism and volcaniclastic sedimentation (sand pattern) during Caribbean depositional history; the global eustatic sea-level curve (replotted from Haq and others, 1987) and the Caribbean relative sea-level curve (Kauffman and Johnson, in prep.; Fig. 3); a general sea-surface temperature curve (from Kauffman, 1988; Jenkyns and others, 1994); and the predicted rudistid species diversity from Johnson (1993). REX-1 marks the age of this abrupt extinction event as defined by planktonic microbiotas; REX-2 marks the macrofaunally defined level of the rudistid reef extinction (discussed in text). Note that in either case this extinction occurs midway through the latest Cretaceous sea-level fall and near the Maastrichtian thermal low, during a time of widespread carbonate platform development without much active volcanism. These conditions represent normal background fluctuations for the Cretaceous and neither their rate of change nor their magnitude are unusual.

plate migrated relatively northeastward from Cenomanian to Campanian time, but no major Caribbean plate movement is indicated between Pindell and Barrett's (1990) Campanian and Paleocene reconstructions. Arc volcanism remained at low levels until the latest Maastrichtian. This relatively quiet time in Caribbean plate tectonic history was reflected in diminished volcaniclastic sedimentation, relative stabilization and planation of island arc and tectonic platforms, and in renewed late Campanian–Maastrichtian carbonate platform growth in many parts of the Caribbean province (e.g. Puerto Rico, Jamaica, Cuba, and Mexico; Fig. 2). Rudistids and their biogenic frameworks flourished in Maastrichtian carbonate and mixed carbonate and siliciclastic settings. No evidence exists for the drowning or emergence of these last rudistid reefs, or for their pervasive inundation by volcaniclastic or siliciclastic sediments, as will be shown in detailed stratigraphic analyses of Jamaican and Puerto Rican sections. A major episode of renewed tectonic and volcanic activity affected the Antillean arc islands at the very end of the Maastrichtian, after the major reef extinction, as evidenced by the emplacement of stocks and the outpouring of basaltic lavas in Puerto Rico and Jamaica, and the spread of volcaniclastic sediments across some of the carbonate platforms (e.g., the Summerfield and Masemure volcanic rocks of Jamaica, Coates, 1977; the El Rayo Formation of Puerto Rico, Slodowski, 1956; Volckmann, 1984). However, well-developed carbonate platform facies with coral-algal communities containing rare rudistids, or lacking them altogether, separated the last rudistid frameworks from the initial late Maastrichtian volcanics in most areas; for example, the Jerusalem Mountain inlier of Jamaica, the El Rayo Formation of southern Puerto Rico, and the La Popa Formation of northeastern Mexico.

Sea Level

Haq and others (1987) noted an overall second-order eustatic fall in global sea level of 100–150 m that began in the early part of the late Campanian and proceeded through a series of 5 third-order fluctuations to lowstand at or just above the K-T boundary (Figs. 2 and 3), 10.7 m.y. later. Third-order sequence boundaries near peak sea-level fall were formed in the earliest, middle, and late Campanian, near the early-middle Maastrichtian boundary, and in the latest middle Maastrichtian. Third-order maximum flooding intervals occurred near the middle-late Campanian boundary, in the lower and middle part of the late Campanian, and in the lowest, middle, and early part of the late Maastrichtian. Many of the third-order late Campanian-Maastrichtian sea-level fluctuations proposed in the Haq and others (1987) global cycle chart are also recorded to varying degrees in sequence stratigraphic analyses based on facies changes throughout the Caribbean province (Fig. 3), although some are complicated in timing and magnitude by local or regional tectonic movements (e.g. the early Maastrichtian in parts of Jamaica).

All of the Antillean islands seem to have become emergent near to, or just above, the K-T boundary (Fig. 2), or else became inundated by very shallow water volcaniclastic and siliciclastic sediments. However, this occurred well above the widespread loss of rudistid reefs. In fact, the relatively slow rate of sea-level fall through the later Campanian and Maastrichtian was coeval with the greatest radiation and diversification of Caribbean rudistids for the entire Cretaceous, which peaked in the Maastrichtian. Third-order sea-level falls near

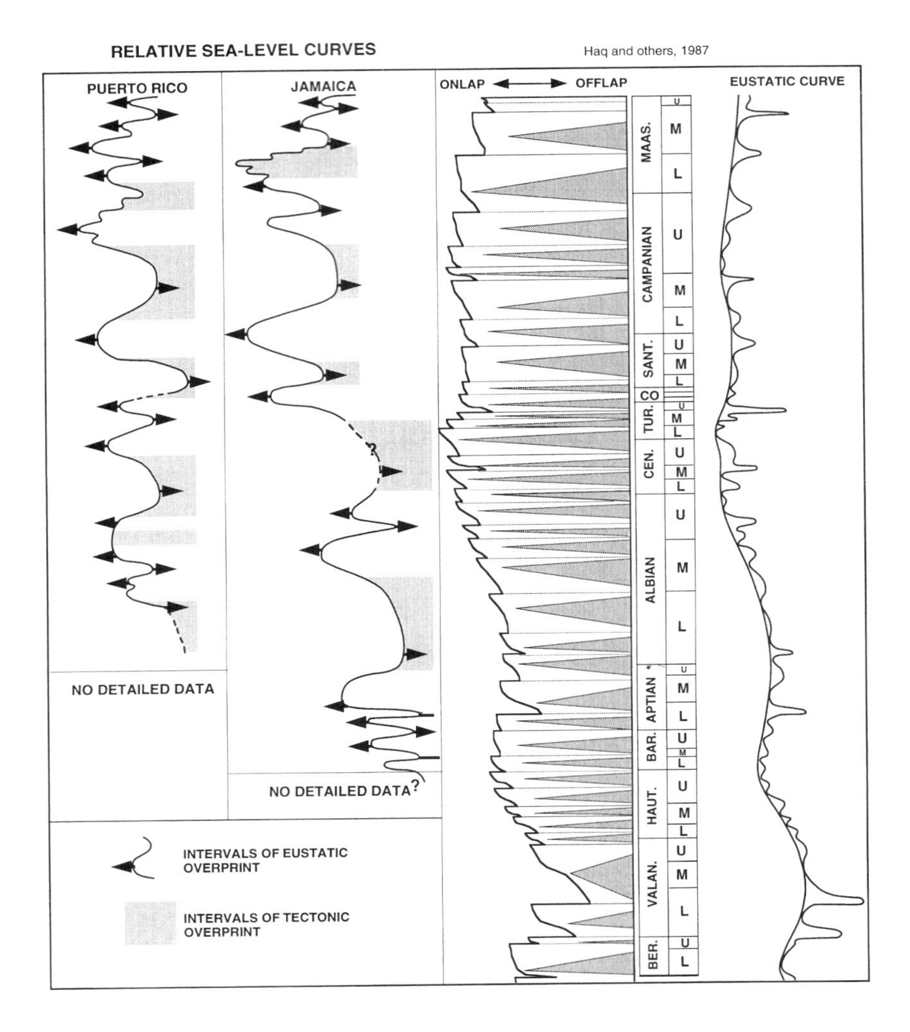

Figure 3. Relative Cretaceous sea-level curves for Jamaica and Puerto Rico (left), based on sequence stratigraphic analyses, and plotted against the Haq and others (1987) global onlap-offlap and eustatic curves. Arrows indicate sea-level fluctuations on those islands that correlate closely biostratigraphically to the Haq and others (1987) eustatic curve, and shaded areas indicate relative sea-level changes that do not correlate globally, and probably reflect local or regional tectonic control on relative sea level. Note that fluctuations around the middle Maastrichtian, the time of abrupt extinction of Caribbean reef ecosystems, reflect a third-order global pattern, and that the extinction level generally correlates with a minor episode of falling relative sea level similar to many other high-frequency Cretaceous fluctuations that had no prior effect on the evolutionary success of rudistid reef ecosystems.

the Maastrichtian substage boundaries (proposed levels of rudistid reef extinction, discussed below) were normal for the Cretaceous and were followed by eustatic rises to approximately the same levels worldwide, as were documented for the early and middle Albian (Fig. 3), a time of extensive rudistid diversification. Whereas lowering sea level would obviously have caused local extinction and loss of reefs as they became emergent or covered by coarse clastics derived from encroaching shorelines, carbonate platforms were still widespread after the collapse of the rudistid reef ecosystems in many areas. The claim by Hallam (1984, 1989) and Swinburne (1991) that sea-level fall was a primary earthbound cause for mass extinction among tropical marine ecosystems near the K-T boundary is not substantiated for the Caribbean province. Sea-level lowstands show no consistent statistical correlation with well-dated mass-extinction events throughout the Phanerozoic (Kauffman and Fagerstrom, 1993).

Paleoceanography

To date there is no continuous isotopic record for the Maastrichtian of the Caribbean province that would allow interpretation of water temperature and/or salinity, carbon cycling, or productivity. Isotopic work utilizing well-preserved Caribbean rudistids along north-south time slices is in progress at this time. Stratified, low-oxygen Maastrichtian seas were suggested by Hallam (1984, 1989) as a cause for global K-T extinction, but there is no evidence for this in the Caribbean province. No Maastrichtian rudistid reefs are associated with or drowned by organic carbon–rich strata, nor are any such facies known at this time from the Caribbean province. Even the Vidono Formation of northern Venezuela, a marine shale formed above an upwelling margin that shows continuous deposition across the K-T boundary, contains very low levels of total organic carbon (1%–2%) compared to shales in the same region that reflect Cretaceous ocean anoxic events (OAEs) 1c, II, and III (3.5%–5%) (Kauffman and others, 1995). We conclude that the Caribbean–Gulf of Mexico seas were oxygenated and circulating actively during the Maastrichtian. Johnson (1993) mapped surface circulation patterns for the Maastrichtian of the Caribbean province based on high indices of similarity calculated between localities containing rudistid genera. These inferred Maastrichtian patterns show larval dispersal routes extended between the southern U.S., Mexico, and the Greater Antillean islands as a series of active current gyres.

Caribbean Maastrichtian Paleoclimates

The Cretaceous climate model of Barron (1995) for 100 m.y. (Albian) simulated a zone of warmer than normal tropical waters in the southern Caribbean province, just north of the equator, and a zone of hypersalinity in the warm northern part of the province. This latter zone correlates to an area of exceptionally high rudistid diversity and extensive reef development that Kauffman and Johnson (1988) termed the Supertethyan marine climate zone. No comparable climatic simulations have been published for the Late Cretaceous, but rudistid diversity patterns in the Maastrichtian of the Caribbean province are comparable to those of the Albian, leading Johnson and Kauffman (1990) to speculate that the collapse of the Supertethyan zone may have caused the Late Cretaceous rudistid extinction.

In addition, little is known about the Maastrichtian temperature differential between oceanic surface and bottom water temperatures in the Caribbean region, although Hay (1988) showed it as broadening globally in Late Cretaceous oceans. A broad temperature decline marks the Late Cretaceous (Kauffman, 1978, and references therein; Jenkyns and others, 1994), and is illustrated in Figure 2. No temperature fluctuations correlate with the rudistid reef extinction at current levels of resolution.

Extraterrestrial Events

The Cretaceous-Tertiary boundary in the Caribbean province was marked by the impact of a major bolide off the northeast margin of the Yucatan Peninsula, Mexico (Hildebrand and others, 1991; Alvarez and others, 1992; Quezada-Muneton and others, 1992; Sharpton and others, 1992), 65.0–65.5 Ma (Swisher and others, 1992; Obradovich, 1993). Evidence for this impact includes the Chicxulub crater (Hildebrand and others, 1991; Quezada-Muneton and others, 1992); proximal and distal impact ejecta (Hildebrand and Boynton, 1990; Alvarez and others, 1992; Smit and others, 1992), including the global boundary clay (Alvarez and others, 1982); tektites and microtektites with broad global distribution (Izett, 1991; Alvarez and others, 1992; Smit and others, 1992); multilamellate shock metamorphism (Bohor, 1990; Alvarez and others, 1992; Smit and others, 1992); a widespread tsunami deposit (Bourgeois and others, 1988), the presence of which is debated (e.g., Savrda, 1993); geochemical signals, including abrupt enrichment of iridium (Alvarez and others, 1992) and an associated trace element suite suggesting meteoritic origin; and an abrupt negative excursion of the marine $\delta^{13}C$ signal, coupled with dramatic perturbations in the $\delta^{18}O$ signal (Boersma and Shackleton, 1979; Buckardt and Jorgensen, 1979). We accept the Chicxulub impact as a major factor in terminal Cretaceous mass extinction, but we have no evidence that it affected the destruction of the Caribbean paleotropical reef ecosystems.

To date, no rudistid bivalves have been found at or just below the K-T boundary in the Caribbean province. The closest specimen known from a complete boundary section is a large solitary *Titanosarcolites* sp. aff. *T. oddsensis* collected in situ by us within 2 to 3 m (~100 ka) below the K-T boundary in a new boundary section 1 km southwest of sections B2 and B3 on the Brazos River, Texas (Hansen and others, 1993, Fig. 1). In the Oyster Limestone member of the El Rayo Formation in Puerto Rico, which may be a complete or nearly complete late Maastrichtian section, on the basis of a thick brecciated horizon (tsunamite?) at the top of the member, two small *Radiolites* sp. were noted 0.5–1.0 m below the apparent K-T boundary. Thus, the observed Caribbean data suggest that the rudistids did not survive to the K-T boundary in this region. However, considering the wide stratigraphic spacing of late Maastrichtian rudistid occurrences, range expansions to the K-T boundary cannot be ruled out until the data base is correlated graphically and analyzed statistically, and the predicted ranges are calculated. The abrupt demise of the rudistid reef ecosystem below the K-T boundary does not rule out extraterrestrial causes, considering the clustering of known terrestrial impact craters (Grieve, 1982; McHone and Dietz, 1991) around the K-T boundary (possibly describing an impact shower in the sense of Hut and others, 1987). Furthermore, there is a high probability that many more bolides fell into the sea at this time (Kauffman, 1988, 1995), inasmuch as planetary impacting is a stochastic process (Grieve, 1982). However, there is no direct evidence, at present, for an extraterrestrial catalyst to Maastrichtian rudistid reef and framework extinction.

MAASTRICHTIAN BIOLOGICAL HISTORY

Extinction patterns within highly structured tropical reef ecosystems are influenced strongly by biotic factors such as evolutionary rates and patterns, biogeography, ecology, and adaptive range. Caribbean rudistid bivalves have been analyzed for their biogeographic distributions (Johnson, 1993), per-taxon rates of origination and extinction (Fig. 1; Johnson and Kauffman, 1990), and ecology in terms of their evolving adaptive morphologies and community associations in Cretaceous reefs and frameworks (Kauffman and Sohl, 1974, 1979; Kauffman and Johnson, 1988). The paleobiogeographic range of tropical to subtropical Cretaceous biotas extended from southern North America through Central and northern South America, and into the Caribbean islands. The paleobiogeographic dynamics of this, the Caribbean province, provide insights into the causes for the terminal extinction of the rudistids and their reef ecosystems.

Although Maastrichtian rudistid genera have been documented from the southern United States through the islands of the West Indies (Fig. 4), the highest diversity of genera and species is found in reefal associations distributed through southern Mexico, northern Guatemala, Belize, and the Greater Antillean islands (Johnson, 1993). The geographic range of Maastrichtian species occurrences mimics those of the genera, and are shown per locality (each a group of closely spaced localities) relative to Cretaceous paleolatitudes on the 80 Ma plate tectonic reconstruction of Pindell and Barrett (1990) (Fig. 5).

The shape of the paleotropics for the Maastrichtian has been defined biologically by the same procedure used for the modern tropics, that is, by mapping the northern and southern limits of reef or significant biogenic framework growth, called the "reef line" (Johnson, 1993). The southern Maastrichtian reef line occurs at 5°N and the northern line at 30°N of the Maastrichtian paleoequator, although the rudistids, including temperate-adapted species, were spread across more than 40° N of paleolatitude (Figs. 5 and 6). This describes a climatically unusual Cretaceous world in which the marine tropics lay entirely north of the paleoequator.

The high mid-tropical Maastrichtian diversity peak (Fig. 6), in conjunction with diversity highs in the Albian, Cenomanian, and Campanian, led to the recognition of a new paleobiogeographic subprovince, the Supertethyan subprovince of the Caribbean province (Johnson, 1993). This Maastrichtian diversity peak, which shows up both in the observed and statistically predicted biogeographic range data using the Koch and Morgan (1988) equation, emphasizes the ecological success of the group just prior to their abrupt extinction. Substage analyses of per-taxon origination rates (Fig. 1) show a buildup through the late Campanian, a peak in the early Maastrichtian followed by a drop in the middle

Maastrichtian, and a final radiation near the middle-late Maastrichtian boundary, just prior to the rudistid extinction. Substage analyses of the per-taxon extinction rate (Fig. 1) show low rates through the Campanian and into the early to middle Maastrichtian, followed by the highest known extinction rates near to and just above the middle-late Maastrichtian boundary (the age of the extinction of the reef ecosystem, as defined by macrofaunal biostratigraphy), but prior to the terminal extinction of surviving ecological generalists among the rudistids just below the K-T boundary.

Figures 4, 5, and 6 show that certain genera and species of rudistids occurred both to the north and south of the Maastrichtian paleotropical margins (reef lines), into subtropical and even warm-temperate climate zones (in the sense of Kauffman, 1973, 1984b). In these more temperate areas, rudistids did not form significant biogenic frameworks, but occurred as individuals or small clusters. These temperate, nonreefal occurrences also existed in Turonian, Coniacian, Santonian, and Campanian time (see maps of species occurrences compiled in Johnson, 1993). Ecologically, these geographically widespread rudistids extended the adaptive range of a group that was previously restricted to the paleotrop-

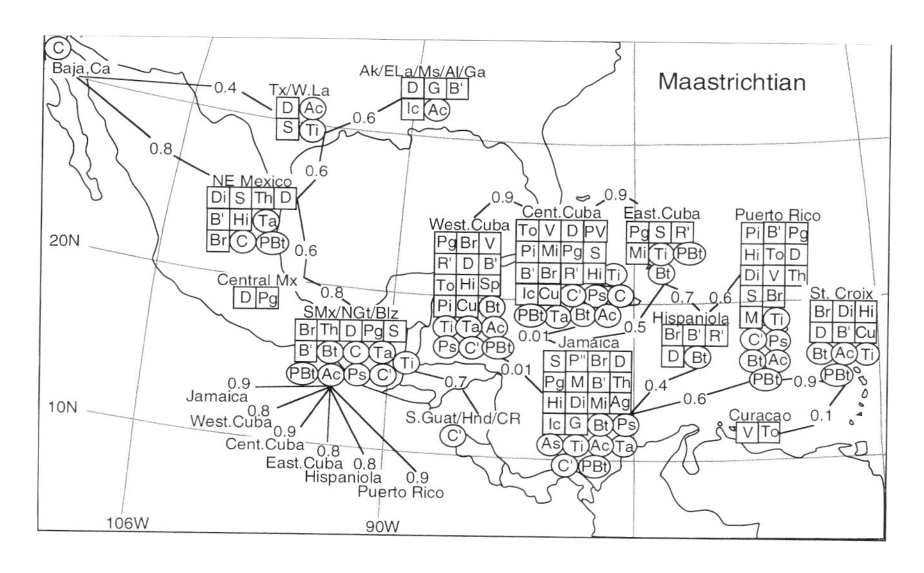

Figure 4. Mercator projection of the present world showing the distribution and diversity of Maastrichtian rudistid genera (letter codes in circles and squares) in the Caribbean province (from Johnson, 1993). Numbers and lines between localities are similarily coefficients of rudistid genera from different localities; these have been used to plot probable dispersal pathways of rudistid larvae (Johnson, 1993). Key to rudistid genera: Ac—Antillocaprina, As—Antillosarcolites, Ag—Agriopleura, B'—Biradiolites, Br—Bournonia, Bt—Barrettia, C—Coralliochama, C'—Chiapasella, Cu—Caprinula, D—Durania, Di—Distefanella, G—Gyropleura, Hi—Hippurites, Ic—Ichthyosarcolites, Mi—Mitrocaprina, M—Monopleura, PBt—Praebarrettia, Pg—Plagiotychus, Ps—Parastroma, P"—Praeradiolites, PV—Pseudovaccinites, Pi—Pironea, R'—Radiolites, S—Sauvagesia, Sp—Sphaerucaprina, Ta—Tampsia, Th—Thyrastylon, To—Torreites, Ti—Titanosarcolites, V—Vaccinites.

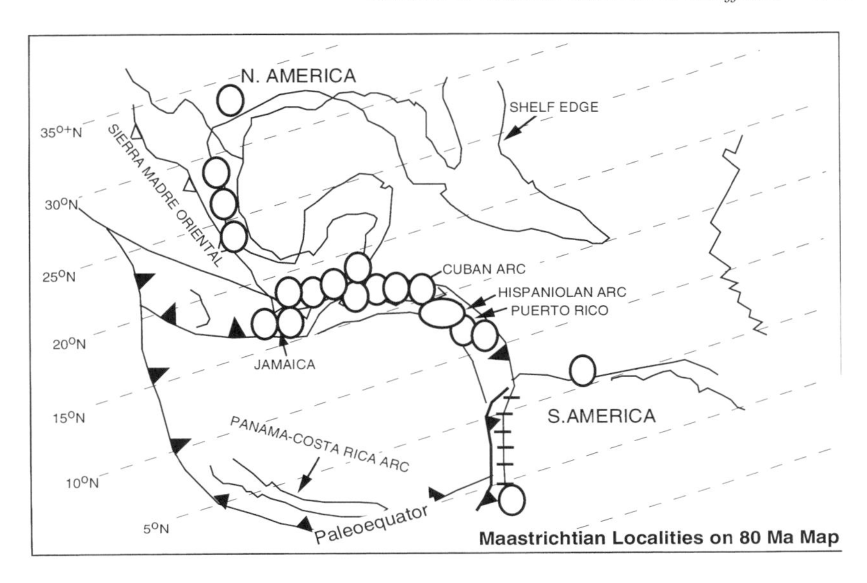

Figure 5. Known Maastrichtian rudistid localities plotted against the 80 Ma plate tectonic reconstruction of Pindell and Barrett (1990) for the Caribbean province. Each oval represents a cluster of many closely spaced localities. Paleolatitudes are averaged from numerous sources. Note that most rudistid localities lie within the paleotropics, as defined by the reef lines (lat 5° and 30°N; see Fig. 6). Localities outside the reef line yield temperate-adapted rudistids not occurring within significant biogenic frameworks (from Johnson, 1993).

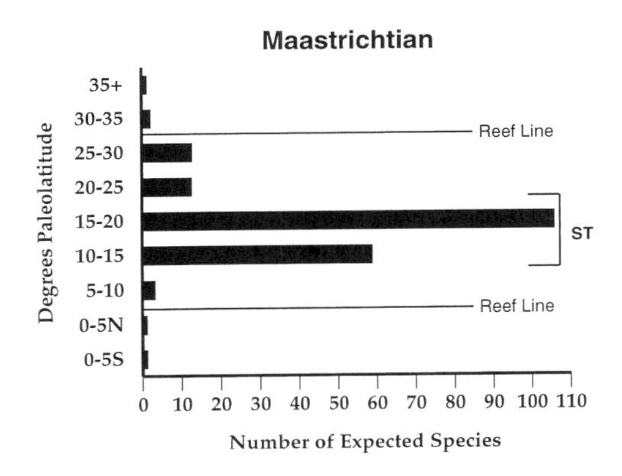

Figure 6. A plot of predicted Maastrichtian rudistid species diversity per 5° paleolatitude, relative to the reef lines (tropical climate boundaries) in the Caribbean province (Johnson, 1993). This is the highest diversity achieved by the rudistids during their evolution, and this peak occurred during falling sea level and cooling global temperatures, immediately before the abrupt demise of the reef ecosystem. Note that the biogeographically defined paleotropics lay wholly north of the paleoequator, and are characterized by an abrupt increase in tropical diversity within the reef line, as well as a second extraordinary increase in core tropical diversity. These peaks lie between lat 10° and 20°N—the projected position of Supertethys.

ics, and temporarily buffered them from extinction. As a group, the rudistids went extinct first in the tropics, 1.5–2.5 m.y. below the K-T boundary, but temperate-adapted forms survived selectively to within ~100 ka of the K-T boundary. The latitudinal extinction pattern shown by Caribbean rudistids contrasts with that presented by Raup and Jablonski (1993) for all Cretaceous bivalves, which showed no latitudinal extinction gradient at the K-T boundary.

ECOLOGY OF LATEST CRETACEOUS RUDISTID FRAMEWORKS

The ecologic complexity of Maastrichtian rudistid reefs was expressed in the Caribbean province through high levels of niche specialization across carbonate platforms, and through moderately complex levels of ecologic succession in rudistid-dominated reef and framework communities (Kauffman and Sohl, 1976; Kauffman and Johnson, 1988). The Maastrichtian rudistid frameworks of Jamaica provide the best examples.

The Guinea Corn Formation, continuously exposed in the bed of the Rio

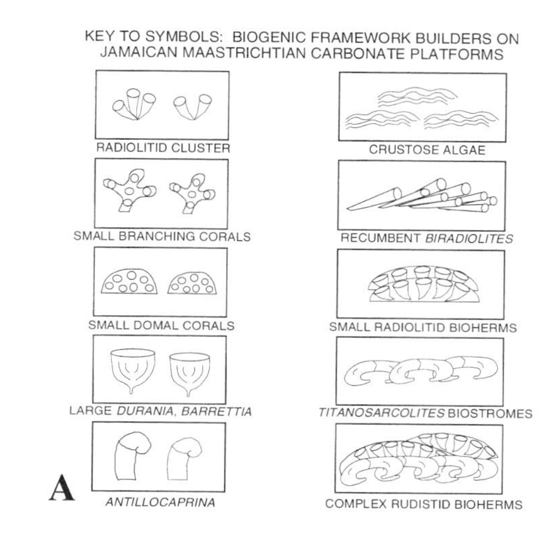

Figure 7. Stratigraphic section of the lower to middle Maastrichtian Guinea Corn Formation at the type section along the Rio Minho, Clarendon, Jamaica (modified after Coates, 1965; Kauffman and Sohl, 1974; and additional field work by us). A: Key to symbols for biogenic framework builders used in this and all subsequent figures; lithologic symbols in B are standard. B: High-resolution section along the Rio Minho showing predominance of rudistids in reef and framework building, and relative continuity of biogenic structures through the section, interrupted mainly by higher energy disturbance events. This is the best-developed sequence of rudistid frameworks in Jamaica. Termination of rudistid reefs at the top of the Guinea Corn Formation is due to increased regional volcaniclastic sedimentation (see Coates, 1977) and represents a local extinction that lies chronologically below the widespread rudistid reef extinction preserved in carbonate platforms of western Jamaica (Figs. 10–12).

Minho, in Clarendon, Jamaica, contains some of the most continuously developed and ecologically complex Caribbean Maastrichtian frameworks. The formation was described by Coates (1965, 1968) and Kauffman and Sohl (1974), and was subsequently restudied by us and N. F. Sohl. Figure 7 is a stratigraphic and paleoecologic composite of the Rio Minho rudistid reef complex based on these works. The Guinea Corn Formation at Rio Minho is 149.5 m thick and con-

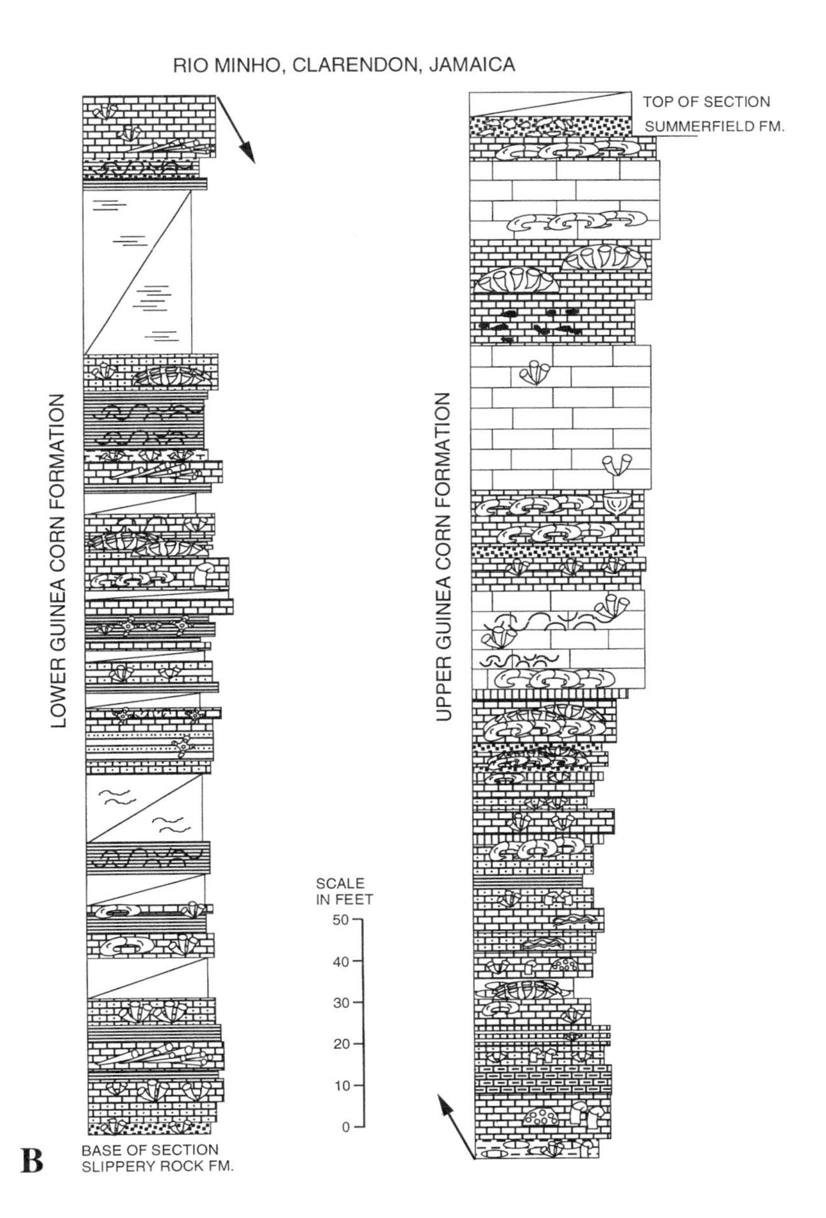

tains rudistid clusters, biostromes, and low biohermal and bafflestone mounds throughout, commonly stacked one on top of the other with evidence of disturbance events but little intervening sediment separating them. Of the Rio Minho section, 52%–55% is composed of rudistid frameworks; many of the intervening stratigraphic intervals have scattered rudistids and corals, abundant small marine molluscs, and echinoid debris. These frameworks can be classified into four community types, as defined by their taxonomic composition, biofabric (for example, packing, orientation) and geometry: (1) monospecific clusters of erect to semi-erect rudistids including giant *Durania*, large *Antillocaprina*, *Bournonia*, or *Chiapasella*, and small *Thyrastylon* or *Distefanella*, (2) tabular to lensoid biostromes of a single rudistid community of (erect-growing morphs) containing *Antillocaprina*, *Chiapasella*, and finger corals (± *Bournonia*), or monospecific biostromes of any one of these genera, with or without associated small corals; *Thyrastylon* biostromes; and *Distefanella* biostromes (in muddy substrates), (3) tabular to lensoid biostromes with simple two-community ecological succession, as follows, recumbent *Biradiolites jamaicaensis* mats overlain by small, erect *Thyrastylon*-dominated mounds; recumbent *Biradiolites jamaicaensis* mats overlain by an erect-growing, high-diversity community of *Antillocaprina*, *Chiapasella*, small finger corals, and in some cases *Bournonia*, in various combinations; and loose mats of large recumbent *Titanosarcolites* overlain by erect-growing rudistid communities dominated by *Thyrastylon*, *Orbignya*, or small *Bournonia*, with or without scattered larger *Chiapasella* and *Antillocaprina*. (4) The most complex rudistid frameworks of the Guinea Corn Formation are composed of three to four communities consistently ordered in ecological succession. (1) The first is a pioneer community of large, recumbent *Titanosarcolites*, most preferentially oriented with their convex margins facing into the current (Kauffman and Sohl, 1974). In some cases these are associated with small coral and hydrozoan or stromatoporoid heads and large benthic molluscs. (2) The next is a successor community of stacked, crowded, recumbent *Biradiolites jamaicaensis* mats, in some cases current oriented (apertures facing current), followed by (3) lensoid biostromes to low mounds of crowded, erect-growing small *Thyrastylon*, *Orbignya*, or less-commonly *Distefanella*, with individuals attached in protected crevices between recumbent colonizer and successor rudistids. (4) The last is a low, mound-shaped bioherm or bafflestone mound composed primarily of successive generations of erect, somewhat larger *Thyrastylon* or *Orbignya*, with lesser numbers of other rudistids, including large *Bournonia*, *Antillocaprina*, and/or *Chiapasella*. This represents the highest successional level of community development preserved in the Guinea Corn Formation. Figure 8 models this ecological succession, and is based on numerous outcrop and photograph maps. Large *Titanosarcolites* serve as anchors and stabilizers of mobile substrates in the pioneer community. Recumbent *Biradiolites* subsequently fill in empty space between pioneer *Titanosarcolites*, to form dense biostromes and create a sheetlike colonizing space in the successor community. Subsequent communities of erect-growing rudistids utilize this irregular but stable *Biradiolites* biostromal substrate for initial attachment and upward growth of the framework in successive stages. The morphologic adaptations of rudistids at each stage of succession are basically equivalent to vine, sheet, and treelike colonizing strategies of clonal organisms (Coates and Jackson, 1985). Acolonial rudistid framework communities mimicked the adaptive strategies of colonial corals and bryozoans in successional reef development.

MAASTRICHTIAN REEF EXTINCTIONS IN THE CARIBBEAN REGION

Maastrichtian carbonate platform sequences in Jamaica and Puerto Rico were studied extensively for stratigraphic and paleontologic information relating to the collapse of Cretaceous reef ecosystems and the extinction of the rudistids. More general investigations were carried out in Mexico. All studied localities lay with-

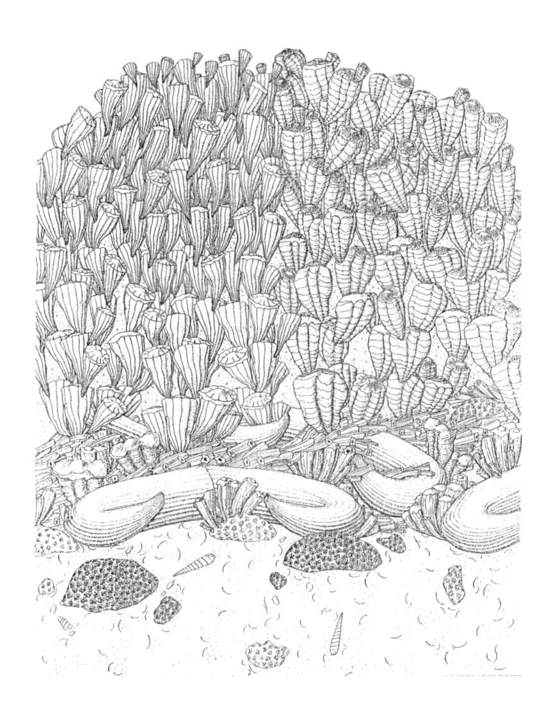

Figure 8. Reconstruction of the most fully developed ecological succession for Maastrichtian rudistid-dominated frameworks throughout Jamaica and Puerto Rico, based mainly on outcrop maps of frameworks from the Guinea Corn Formation along Rio Minho, Clarendon, and in the Jerusalem Mountain inlier of western Jamaica (see Fig. 7). The four stages of ecological succession, spread over 2–3 m of outcrop were (1) colonization of mobile mollusc-rich substrates (commonly the tops of small bars or sand waves; Kauffman and Sohl, 1974) by large recumbent Titanosarcolites. *(2) Infilling of the space between* Titanosarcolites *individuals by recumbent mats of* Biradiolites jamaicaensis. *(3) First upward growth of the framework with clusters of short, relatively broad radiolitids, mainly* Thyrastylon, Chiapasella, *and larger* Durania. *(4) Overgrowth of all earlier stages of succession by tightly packed clusters of erect-growing, conical shells of* Thyrastylon *and/or* Orbignya *(drawing by Carolyn Ensle).*

in the center of the paleoclimatic tropics, the Supertethyan climate zone described by Kauffman and Johnson (1988), Johnson (1993), and Kauffman and Fagerstrom (1993) as the primary habitat of diverse Cretaceous rudistids and biogenic frameworks.

The concept that many rudistid bivalve lineages, and all Caribbean rudistid-dominated reefs and frameworks, underwent widespread extinction 1.5–2.5 m.y. below the K-T boundary was first proposed by Kauffman (1984a) on the basis of field studies of Maastrichtian carbonate platform sequences in Puerto Rico and Jamaica (Kauffman and Sohl, 1974, and unpublished data). Johnson and Kauffman (1990) documented more fully this intra-Maastrichtian rudistid extinction by analyzing observed range data for all known Caribbean rudistid species. Physical stratigraphic evidence has suggested strongly that the demise of Caribbean rudistid reefs and frameworks within the Maastrichtian was a widespread, short-term event, possibly related to pervasive changes in the paleotropical ocean-climate system. Kauffman (1988) considered this event as initiating the K-T mass-extinction interval. In the Maastrichtian carbonate platform sequences of Jamaica, Puerto Rico, Cuba, and parts of Mexico, the stratigraphic sequence of events is essentially the same. Platform limestone units containing diverse rudistid frameworks and reefs (*Titanosarcolites* Limestone of Kauffman and Sohl, 1974) are overlain directly by similar limestones and/or marls and shales containing diverse normal marine molluscan faunas and, in some cases, echinoids, small corals, algae, and rare rudistid individuals. These, in turn, are overlain regionally by shallow-water limestones (or mixed carbonate-clastic facies) with biostromes and mounds of large normal marine oysters, in most cases associated with algae, echinoids, small molluscs, rare corals, and very rare, ecologically generalized rudistids (i.e., the Oyster Limestone of Kauffman and Sohl, 1974). The most complete of these sections have Maastrichtian mudstones, marls, or volcaniclastic facies above the Oyster Limestone units. In no case do rudistid bivalves reach the youngest preserved Maastrichtian strata in the Caribbean province.

The precise timing of this Maastrichtian extinction event remains an enigma, however. Kauffman (1984a) reported that the rudistid reef extinction occurred within the middle Maastrichtian *G. gansseri* planktonic biozone, on the basis of unpublished foraminiferal assemblage data provided by Ruth Todd of the U.S. Geological Survey. On the basis of molluscan biostratigraphy, Sohl and Kollmann (1985) showed that the range of *Titanosarcolites*, a major Maastrichtian framework builder in the *Titanosarcolites* Limestone, was restricted to the middle and late Maastrichtian. The rudistid reef extinction occurs near the middle of the biostratigraphic range of *Titanosarcolites*, and thus may be near the middle (*G. gansseri* biozone)-late Maastrichtian boundary (*A. mayaorensis* biozone of Sohl and Kollmann, 1985), a view accepted by Johnson and Kauffman (1990), and not incompatible with Kauffman's (1984) original observations, though occurring higher within the *gansseri* zone than he had proposed. Jiang and Robinson (1987) and Verdenius (1993), however, have placed the latest rudistid frameworks (upper *Titanosarcolites* Limestone and equivalents) near the early-middle Maastrichtian boundary on the basis of nannoplankton in overlying beds (*Q. trifidum* biozone). Hazel and Kamiya (1993) reviewed this biostratigraphic debate.

The important point is, no matter which biostratigraphic system of dating is applied, both indicate that the widespread, short-term extinction of rudistid-dominated reef and framework ecosystems, and the extinction of many rudistid lineages, occurred abruptly at some point between 1.5 and 3.0 m.y. below the K-T boundary in the Caribbean province.

Jamaica

Jamaica contains the most continuous record of the Late Cretaceous development and abrupt Maastrichtian demise of the rudistid reef ecosystem (Kauffman and Sohl, 1974; Coates, 1965, 1968; Schmidt, 1988) in the Central, Marchmont, Lucea, Green Island, and Jeruselum Mountain inliers (Fig. 9; Kauffman and Sohl, 1974). Maastrichtian strata of Jamaica containing the youngest well-developed rudistid reefs belong to what has historically been called the *Titanosarcolites* Limestone in the Antilles (Kauffman and Sohl, 1974). Rudistid frameworks disappear abruptly within or near the top of this limestone unit.

The *Titanosarcolites* Limestone is called the Guinea Corn Formation in the Central and Marchmont inliers of Jamaica (Coates, 1965, 1968; Robinson and Lewis, 1987; Jiang and Robinson, 1987), where it is 149 m thick at the type section exposed along the Rio Minho (Coates, 1965, 1968; Fig. 7). The Guinea Corn carbonate platform was inundated during the early or middle Maastrichtian by the Summerfield Volcanics, derived from eastern sources (Coates, 1977); no Oyster Limestone carbonate unit is preserved above it.

In the Maldon and Sunderland inliers of west-central Jamaica, the *Titanosarcolites* Limestone is called the Maldon and (overlying) Vaughansfield limestones (>244 m thick), consisting of two massive limestone units separated by thicker marine shales and mudstones; these are overlain by fine-grained marine volcaniclastic facies, the Garlands Formation, with an Oyster Limestone carbonate platform near the top (Coates, 1977, Fig. 5). In western Jamaica (Lucea, Green

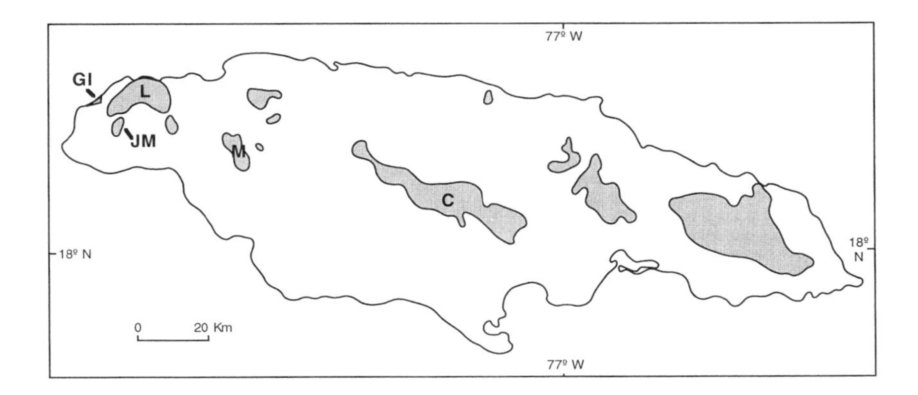

Figure 9. Generalized map of Jamaica showing Cretaceous inliers (patterned areas). Inliers with letters are those containing Maastrichtian rudistid reefs analyzed in this study. C—Central Inlier; GI—Green Island Inlier; JM—Jerusalem Mountain Inlier; L—Lucea Inlier; M—Marchmount Inlier. Base map modified from Coates (1977).

Island, and Jerusalum Mountain inliers) the *Titanosarcolites* Limestone sequence is considerably thinner (22.8–30.5 m) but temporally more complete than to the east, and is placed within the Jerusalem Mountain Limestone. In the Jerusalum Mountain inlier, Jiang and Robinson (1987) named the *Titanosarcolites* framework-bearing limestones the Thicket River Limestone Member, and named the younger oyster-bearing Maastrichtian limestones (*Ostrea arizpensis* beds, or Oyster Limestone of older literature; Kauffman and Sohl, 1974), the Jeruselum Limestone Member of the Jerusalem Mountain Formation.

The *Titanosarcolites* Limestone interval represents the initiation and peak development of a widespread, transgressing carbonate platform complex formed during the last major Maastrichtian sea-level rise (see Haq and others, 1987) (Fig. 3). The unit contains numerous, ecologically complex, rudistid reefs, bioherms, biostromes, and bafflestone frameworks. It is characterized by giant, recumbent pioneer species of *Titanosarcolites* (Caprinidae) at the base of elevated rudistid reefs, bafflestone frameworks, or dominating paucispecific biostromal mats in Jamaican carbonate platform settings. Kauffman and Sohl (1974, 1979) mapped the community ecology and epibiont and endobiont distribution on giant *Titanosarcolites*, and Kauffman and Johnson (1988) discussed its functional morphology. *Titanosarcolites* Limestone frameworks were ecologically the most complexly structured Cretaceous rudistid reefs of the Caribbean province, having three to four successional stages (Fig. 8), and *Titanosarcolites* itself was a temperate-adapted genus, so that the abrupt extinction of rudistid reef ecosystems at or near the top of the *Titanosarcolites* Limestone in western Jamaica was not predictable from an evolutionary or ecological perspective.

Dating the Reef Extinction in Jamaica

The biostratigraphic age of the rudistid reef extinction near the top of the *Titanosarcolites* Limestone and equivalent units in western Jamaica differs between macrofossil and microfossil schemes, even though it represents a single narrow stratigraphic interval. The genus *Titanosarcolites* is restricted to the Maastrichtian (Coates, 1965, 1968; Chubb, 1971; Kauffman and Sohl, 1974), and based on associated actaeonellid gastropods, probably to the middle and late Maastrichtian (Sohl and Kollmann, 1985; Kauffman, Johnson and Sohl, biostratigraphic compilation in prep). The *Titanosarcolites* Limestone and its Antillean equivalents span only the lower half of the biostratigraphic range of *Titanosarcolites*, and are thus probably middle Maastrichtian in age. This macrofaunal evidence suggests that the extinction of the rudistid reef ecosystem in Jamaica occurs close to or at the middle to late Maastrichtian boundary (Johnson and Kauffman, 1990). This would theoretically place it near the top of the *G. gansseri* planktonic foraminiferal biozone, between 1.0 and 1.5 m.y. below the K-T boundary. In the Jerusalem Mountain inlier, the rudistid reef extinction boundary occurs consistently at or very near the top of the *Titanosarcolites* Limestone, within carbonate facies, and small biostromes of coral, algae, probable stromatoporoids, and scattered radiolitid rudists characterize the immediate postextinction carbonate facies. Mollusc and microfossil-rich platform marls, massive to shaley limestone, and a higher massive marine-platform limestone dominated by large thick-shelled marine oyster biostromes and mounds

(*Arctostrea* or *Lopha*, *Crassostrea* spp.; the Oyster Limestone) overlie the last rudistid reef.

Micropaleontologic evidence suggests a somewhat older age for the rudistid reef extinction. Kauffman (1984a) reported a *G. gansseri* foraminiferal assemblage in marl beds overlying the *Titanosarcolites* Limestone in the Jerusalem Mountain inlier of western Jamaica (identifications by R. Todd, personal commun.), and therefore suggested that the reef extinction lay at or near the base of the *G. gansseri* zone (early and middle Maastrichtian), 2.5–3.0 m.y. below the K-T boundary. Hazel and Kamiya (1993) erected three ostracode zones for the *Titanosarcolites* Limestone and younger Cretaceous beds of Jamaica; they regard all of these as Maastrichtian in age. Ostracode zone 1 (interpreted as the middle part of the early Maastrichtian) and its subzones are restricted to the *Titanosarcolites* Limestone and younger beds of the Central, Maldon, and Marchmont inliers. In the Jerusalem Mountain inlier, the Thicket River Limestone Member (= *Titanosarcolites* Limestone) contains ostracode zone 2, to which Hazel and Kamiya (1993) assigned a late part of the early Maastrichtian age. The marls and mudstones above the Thicket River Limestone, and the Oyster Limestone and associated marls, fall into Hazel and Kamiya's (1993) highest ostracode zone, considered to be of the early part of the middle Maastrichtian in age, possibly in the lower part of the *G. gansseri* foraminifer biozone because of its cooccurrence with the nannofossil *Quadrum trifidum* in the same beds (Jiang and Robinson, 1987). Hazel and Kamiya (1993) noted about a 0.5 m.y. overlap of the upper *Q. trifidum* biozone with the lower *G. gansseri* biozone. These data suggest that the rudistid reef extinction lies near the early-middle Maastrichtian boundary, near or at the entry level of *G. gansseri* in the oceanic record. The main body of the Oyster Limestone (Jerusalem Member), which lies above the last occurrence of *Q. trifidum*, probably represents the upper *G. gansseri* zone (late part of the middle Maastrichtian), and the highest Cretaceous mudstones of this region (Masemure Formation of Jiang and Robinson, 1987, 1989) could possibly represent the late Maastrichtian (= *A. mayaroensis* zone), but these stratigraphic units have yielded no planktonic microfossils. Jiang and Robinson (1987, 1989) and Verdenius (1993) interpreted the age of the *Titanosarcolites* Limestone (Jerusalem Limestone Member) as ranging from late Campanian to early Maastrichtian at various sites in Jamaica on the basis of nannoplankton ages. Hazel and Kamiya (1993) argued effectively against a Campanian age (Hazel and Kamiya, 1993) but agreed that most of this limestone unit (their ostracode zone 2) lies within the early (but not earliest) Maastrichtian, on the basis of microfossils.

On the basis of these micropaleontologic data, the age of the top of the *Titanosarcolites* or Thicket River Limestone, just above the last rudistid reefs, lies within the early to middle Maastrichtian boundary interval, probably within the basal *G. gansseri* zone. Caron (1985) placed the entry of *G. gansseri* at ~68 Ma; Perch-Nielson (1985) placed the extinction of *Q. trifidum* at ~68.2 Ma. Using the new Ar^{40}-Ar^{39} time scale of Obradovich (1993) for the Western Interior Cretaceous basin of North America, as interpolated by Kauffman and others (1993), the base of the "middle" Maastrichtian is at ~68.7–69.0 Ma. These data suggest that the rudistid reef extinction event occurred ~2.5–3.0 m.y. below the K-T boundary, at 68–68.5 Ma; Kauffman (1988) originally dated this extinction globally at 67.5–67.75 Ma, using an older time scale. Macrofossil dating suggests a younger age (middle-late Maastrichtian boundary, 66.6–67.0 Ma). Regardless of the biochronology employed, the Jamaican data show that the Caribbean rudistid reef ecosystem became dramatically extinct at one narrow stratigraphic interval somewhere between 1.5 and 3.0 m.y. below the K-T boundary, within normal marine carbonate platform and shelfal facies. More precise dating of this unit and its equivalents throughout the Caribbean is difficult because of its carbonate platform setting, which environmentally seems to have excluded ammonites, inoceramids, and many oceanic plankton that might provide a more precise age.

THE JERUSALEM MOUNTAIN INLIER: A CASE HISTORY FOR RUDISTID REEF EXTINCTION

The Jerusalem Mountain inlier of western Jamaica contains the most complete, well-dated sections of the rudistid reef and framework extinction known from the Caribbean province. Detailed stratigraphic sections, recorded between 1966 and 1971 by N. F. Sohl, E. G. Kauffman, and J. E. Hazel, are summarized and correlated in Figure 10. Regionally (Fig. 10), the Thicket River Limestone Member (*Titanosarcolites* Limestone) overlies abruptly a series of volcaniclastic sandstones, siltstones, and argillaceous marls of the Summerfield Formation. From two to six thick-bedded, resistant, packstone-wackestone limestone units compose the 3.66–7.63 m Thicket River Limestone Member, with limestone beds up to 1.83 m thick intercalated with thinner marl, calcareous, and volcaniclastic shale-siltstone units (Figs. 10 and 11). The last rudistid frameworks consistently occur 2.1 m below the top of the member, and are overlain by coral-algal biostromes (Fig. 10). A marl and volcaniclastic siltstone unit 1–5 m thick separates the Thicket River from the Oyster Limestone or Jerusalem Limestone Member of Jiang and Robinson (1987). This younger member contains 2 to 3, 0.6–3.1-m-thick, massive-bedded limestone units separated by marl and volcaniclastic mudstone. Each contains large marine oyster beds, biostromes, and lenticular bioherms of *Lopha* and/or *Crassostrea*, with sparse small rudistids in one section. The Jerusalem Limestone Member is overlain by 4.52–9.33 m of almost totally unfossiliferous brown marl and volcaniclastic mudstone (the Masemure Member of Jiang and Robinson, 1987).

Schmidt (1988a, 1988b, and unpublished manuscript) and we conducted a high-resolution study of the extinction interval between the last rudistid framework in the Thicket River Limestone Member and the Jerusalem Member (Oyster Limestone). Collectively these studies demonstrate the following succession of events near or at the micropaleontologically determined early-middle Maastrichtian boundary.

1. A short episode of emergence or shallow-water cementation produced a carbonate hardground near the base of the uppermost massive *Titanosarcolites* Limestone unit of the Thicket River Formation. This was subsequently drowned and encrusted with a sparse *Plagioptychus* mat (Fig. 12, unit 1), and overlain by clay and marl (Fig. 12, units 2 and 3).

2. The cessation of significant clay input, the development of firm, micritic lime mud surfaces, and the initiation of rudistid growth led to stabilization of substrates by cemented *Plagioptychus* and recumbent mats of *Biradiolites jamaicaensis* to form a biostrome. The biostrome was overgrown in simple ecological succession by small *Orbignya* bioherms and mounds (Fig. 12, unit 4). These were again covered by an influx of clay and marl (Fig. 12, unit 5), a common type of disturbance event on Cretaceous carbonate platforms.

3. The dilution of clay and reworking of biogenic carbonates formed a condensed interval of thick-bedded calcarenite under conditions of moderately high wave and current energy. Large individuals and small mats of *Titanosarcolites*

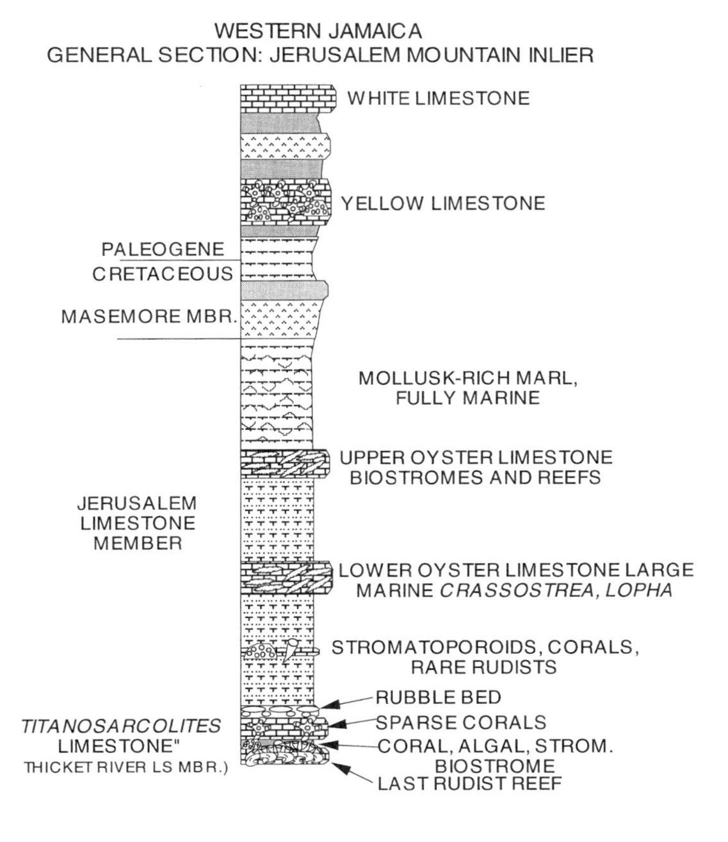

Figure 10. Correlation of major lithologic units within the late part of the early to middle Maastrichtian Jerusalem Limestone in the Jerusalem Mountain inlier, western Jamaica. Note that the last rudistid reef occurs consistently within shallow carbonate platform facies (limestones and marls) at approximately the same stratigraphic position, and that oyster-rich biogenic frameworks consistently dominate the upper two to three carbonate platform units in the Maastrichtian sequence.

stabilized this surface (Fig. 12, unit 6), initiating the development of the last major rudistid reef and/or framework.

4. A series of somewhat elevated, densely packed, rudistid bioherms and stacked biostromes in massive, micritic to calcarenitic carbonate facies then developed (Fig. 12, unit 7). At least four major, high-diversity, rudistid bioherms and many smaller structures were formed within this 18–20-m-thick reefal unit. Each bioherm was characterized by a pioneer community of *Titanosarcolites* biostromes, overgrown in some cases by a second community of recumbent *Biradiolites* forming biostromal units, and capped by several generations of erect, tightly packed *Thyrastylon* (mainly) and *Orbignya*. This was the typical ecological succession of Maastrichtian rudistid frameworks of the *Titanosarcolites* Limestone (Figs. 8 and 12).

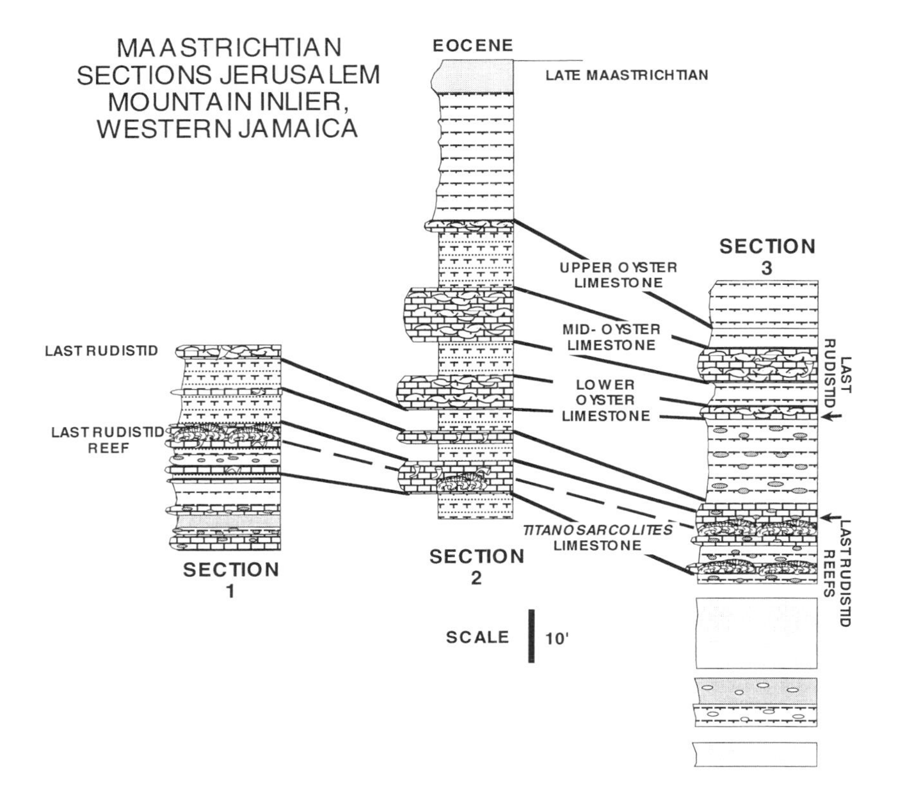

Figure 11. Generalized stratigraphic section of the late part of the early and middle Maastrichtian Jerusalem Limestone Member, Jerusalem Mountain inlier of western Jamaica, showing distribution of carbonate platforms and their main biogenic framework builders. Extinction of the rudistid reef ecosystem lies within the carbonate platform facies near the top of the Thicket River Limestone Member; the last rudistid reef is overlain directly by coral-algal biostromes. Large marine oyster mounds in platform limestones, and mollusc-echinoderm rich marls, overlie the last rudistid-dominated framework.

5. The rudistid reef extinction event was found at the top of this 18-20 m series of bioherms and biostromes. The reef disappeared within 1 to 2 m of carbonate strata, and no rudistid reefs formed again despite the continuation of carbonate deposition (Figs. 10 and 12). The upper part of unit 7 (Fig. 12) records this event

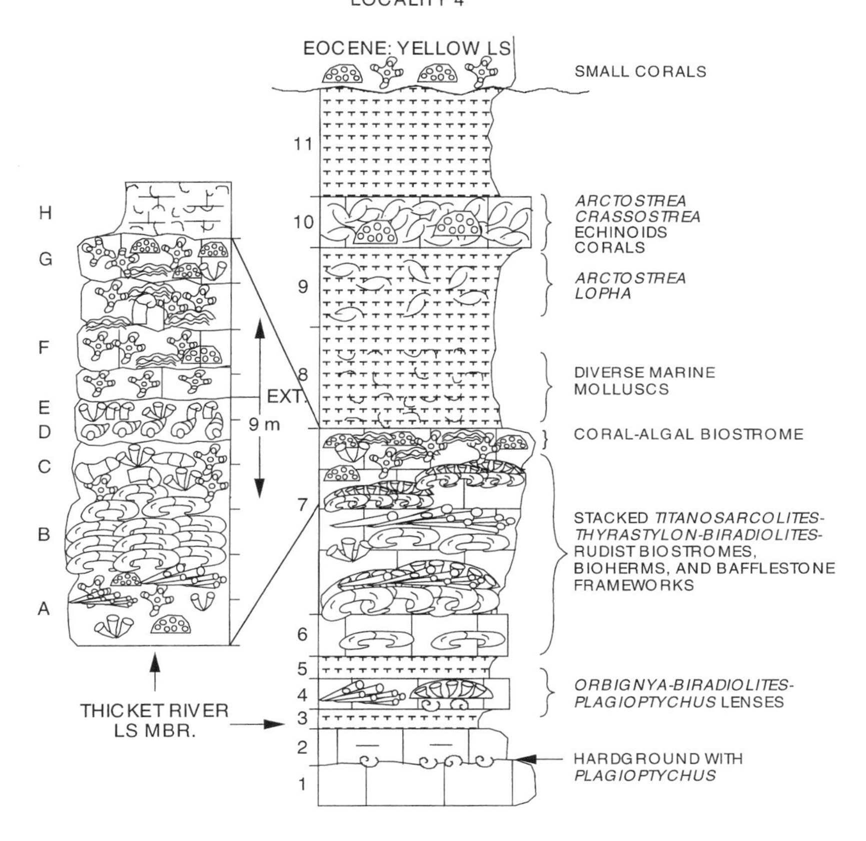

Figure 12. Stratigraphic detail of the middle Maastrichtian extinction of the rudistid reef ecosystem at locality 4, Jerusalem Mountain inlier. Expanded section of the upper 9 m of the Thicket River Limestone Member (left) shows (A) mixed coral and rudistid clusters and biostromes overlain by (B) a thick, dense Titanosarcolites *biostrome or low bioherm, above which is (C) a successor community of erect* Thyrastylon, Chiapasella, *and branching corals capped by a rudist-coral rubble bed (storm disturbance event). (D) A band of large actaeonellid gastropods (typical notophagous feeders on rudistid frameworks) is overlain by (E) biostromes and small mounds of erect* Thyrastylon *and* Antillocaprina *(the final rudistid framework). Successive biostromes and bafflestone mounds of small branching corals and crustose algae, with very rare rudistids (F, G), compose the biogenic frameworks of the upper 2.5 m of the Thicket River carbonate platform. Mollusk-rich marls and biostromes of large marine oysters with corals and echinoids characterize late development of the later middle Maastrichtian carbonate platform in the Jerusalem Mountain inlier.*

in the following manner. A thick *Titanosarcolites*-dominated biostrome, or lenticular bioherm, was disrupted by a high-energy wave and/or current event (but not unlike many others in older strata), which produced a *Titanosarcolites* rubble zone. This was overgrown and intergrown by a bafflestone mat or biostrome of small finger corals and sparse *Thyrastylon* clusters. A zone of densely packed actaeonellid gastropods (normal associates of rudist reefs) (Fig. 12, unit 7D) followed. This, in turn, was overgrown by a thin biostrome of erect *Thyrastylon* and *Antillocaprina*, composing the youngest rudistid framework (Fig. 12, unit 7E). Above this lay 3 to 4 m of massive limestone, bearing abundant small finger-coral thickets (unit F, Fig. 12), algal mats, and rare solitary or loosely clustered specimens of *Antillocaprina* and *Thyrastylon*. These thickets were reworked and toppled at the top of the platform limestone (Fig. 12).

6. Another influx of clay produced marls and claystones (Fig. 12, unit 8), bearing ostracods and molluscs representing diverse, normal marine, shallow-water assemblages. Rudistids in this unit are rare, and locally absent.

7. Calcarenitic mudstones marked decrease in clay deposition and allowed encrustation by abundant, large marine oysters (*Lopha* or *Arctostrea*) in small biostromes and clusters (Fig. 12, unit 9). These were overlain by massive carbonate platform limestones with oyster reefs and biostromes of *Arctostrea* or *Lopha*, *Crassostrea*, echinoids and smaller molluscs, but with only very rare rudistids (*Radiolites*). This marked the end of carbonate platform development as late Maastrichtian muds (Masemure Formation) covered the top of the last oyster reef.

Puerto Rico

The Maastrichtian carbonate platforms that contain the last rudistid reefs and overlying Oyster Limestone in southwestern Puerto Rico belong to the El Rayo Formation (Slodowski, 1956; redefined by Volckmann, 1984). The El Rayo comprises a series of dark purplish-gray to black porphyritic andesite and basalt lavas and tuffaceous breccias with interbedded shallow-water carbonate platform sequences; the formation is exposed intermittently in the southwestern portion of the island. Figure 13 shows a continuous, 68.64-m-thick section of the upper El Rayo Formation studied initially by us, N. F. Sohl, and H. Santos, and described subsequently by Santos (1993, personal commun.). The upper 48.4 m of this section are predominantly shallow-water limestones with thin volcaniclastic intercalations, representing the main body of the El Rayo carbonate platform. The rudistid reef extinction event occurred within a continuously deposited carbonate grainstone-packstone platform sequence situated near the outer edge of volcaniclastic sediment dispersal from the center of the arc island. The lower part of the carbonate sequence is composed predominantly of bioturbated volcaniclastic limestone (packstone) with scattered rudistids and other bivalves (including oysters), gastropods, algae, and echinoid debris. Between 8.6 and 11.1 m below the top of the section is a thick-bedded, resistant grainstone and packstone unit containing the last rudistid framework (Fig. 13). This rudistid framework is directly overlain first by a coral-rich layer, and subsequently by a biostrome of marine oysters. This sequence in Puerto Rico represents a condensed version of the one found in the Jerusalem Mountain inlier of Jamaica (Figs. 10 and 12).

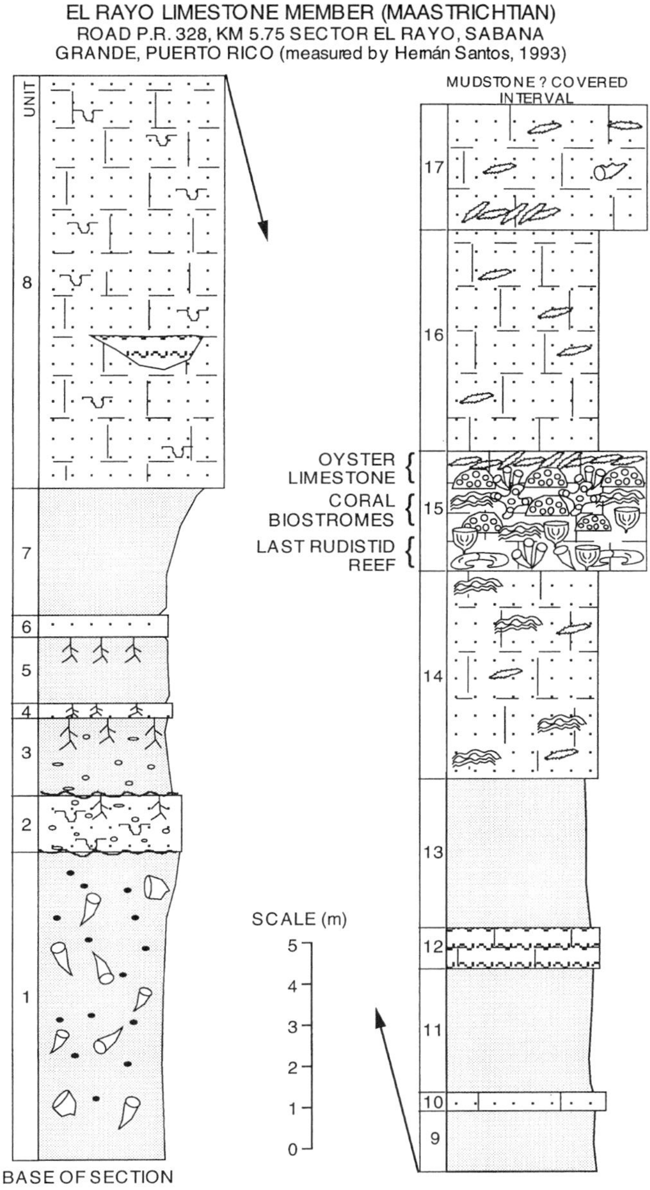

Figure 13. Section of the El Rayo Limestone (middle to late? Maastrichtian) in El Rayo sector, Sabana Grande, southwest Puerto Rico (measured by Hernan Santos; used with permission) showing a highly condensed but typical sequence of biogenic frameworks associated with the extinction of the rudistid reef ecosystem (right column). Within 3 m, a Titanosarcolites-Durania-Thyrastylon-Parastroma *framework is overlain directly by a coral bioherm with massive and branching corals, and this is overlain in turn by a large marine oyster biostrome within continuous carbonate platform facies.*

In detail, the lower 1.3 m of this thick-bedded, resistant, grainstone and packstone unit contains a bafflestone mound of large rudistids, with *Titanosarcolites* and coralline algae stabilizing the substrate, and anastomosing clusters of large, erect *Parastroma* and *Durania* growing up from this substrate. The upper 1.2 m of this limestone unit contains stacked biostromes and clusters of predominantly massive corals, some to 0.4 m in diameter, with extensive binding by coralline algae, and scattered large *Durania* and *Parastroma*, some of which are reworked and toppled. The contact between these two biofacies is abrupt. The framework is capped by a biostrome of densely to moderately packed, large marine oysters (the Oyster Limestone), probably large *Lopha* and *Crassostrea*. Carbonate platform facies continue for another 8.6 m above this framework complex, represented mainly by somewhat volcaniclastic-rich limestone (packstone) containing abundant laminae, small mounds, coated grains, fragments of coralline algae, and scattered large marine oysters, smaller mollusc fragments, and a single reworked radiolitid rudistid fragment, *Durania*, which belongs to a surviving, temperate-adapted Maastrichtian lineage.

A more expanded section of the Oyster Limestone unit at the top of the El Rayo Formation, several meters above the last rudistid reef, is exposed in the yard of the small farm Rayo Plata west of Muchachao, in southern Puerto Rico (Fig. 14). Here 9.5 m of volcanic mudstones and sandstones are overlain abruptly by a 5.75-m-thick medium-gray to medium-light-gray wackestone and packstone sequence consisting of five discrete units. In ascending order, these are: (1) a 0.75-m-thick, fine-grained limestone containing thin biostromes of large oysters, *Crassostrea*, *Lopha*, and/or *Arctostrea*, separated by sparsely fossiliferous intervals with echinoid and small mollusc debris; (2) a 0.8-m-thick unit of marl with large limestone rubble blocks containing oyster, small mollusc, and echinoid debris, probably a storm disturbance interval; and (3) a 2-m-thick, predominantly *Crassostrea-Lopha* oyster bioherm (oyster reef; actually a stacked series of lenticular biostromes), with individual thick-shelled oysters reaching 20–25 cm in length, and small mollusc, echinoid, and algal debris common in the matrix. Clionid sponge, worm, and lithophagid bivalve borings in the oysters suggest fully marine conditions. Two isolated, simple radiolitid rudistid specimens (probably *Radiolites*), 1–3 cm in diameter and 5–8 cm high, were found in this facies in growth position, 2.5–3.0 m below the top of the preserved Cretaceous section. These are the youngest known rudistids in Puerto Rico and they belong to temperate-adapted, generalized lineages that survived the reef extinction elsewhere. (4) A 0.8-m-thick bedded limestone unit (packstone) with moderate numbers of large scattered oysters and small mollusc-echinoid debris overlies (5) a 1.3-m-thick interval of sedimentary limestone breccia containing very angular blocks up to 0.3–0.4 m in diameter, many in point contact, floating in a clayey calcareous marl matrix with abundant shell fragments (mainly oysters, small molluscs), but lacking rudistids. This may represent the so-called K-T tsunami event in the Caribbean, reflecting the Chicxulub impact, but there is no independent chronologic information to confirm this. A similar deposit has been reported below the K-T boundary in Cuba (Iturralde-Vinent, 1992). Paleogene sponges encrust a firmground on the top of this debris bed, and are in turn overlain by reddish Paleogene mudstones and marlstones.

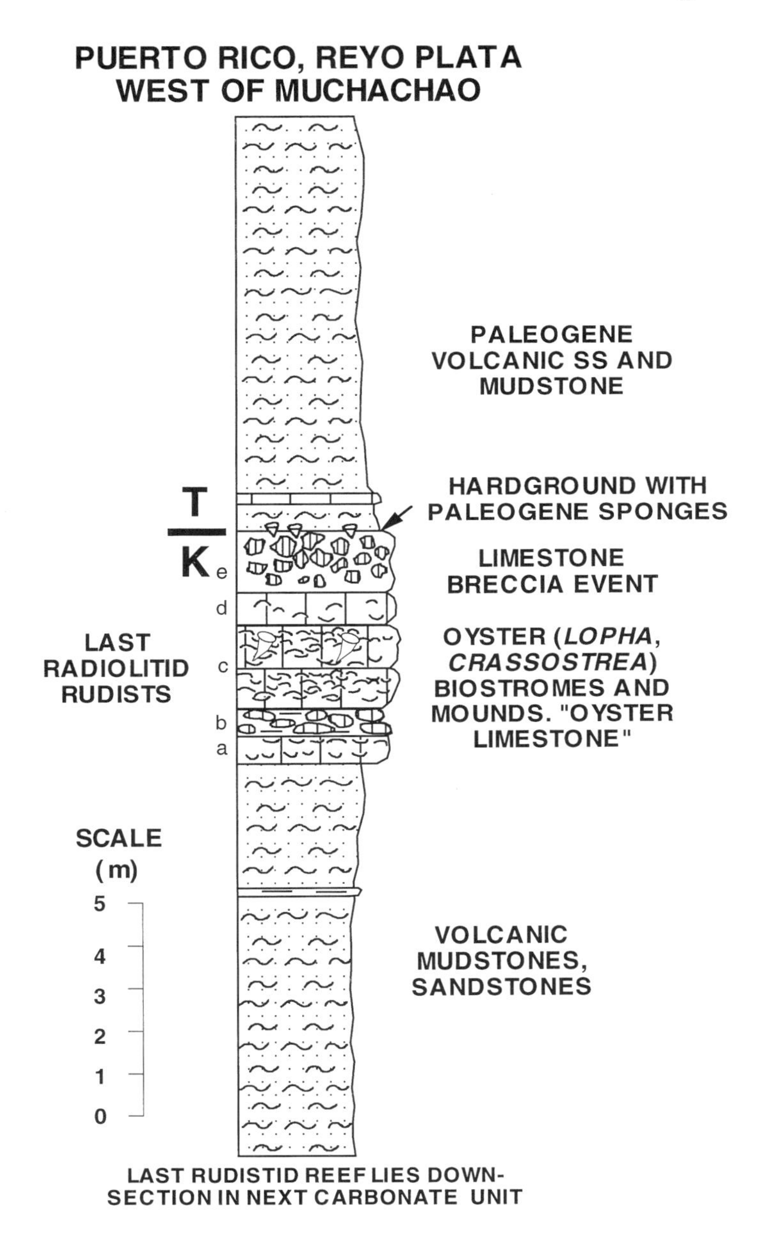

Figure 14. Section of the upper El Rayo Limestone (Oyster Limestone) at Reyo Plata, west of Muchachao, southwestern Puerto Rico. The last rudistid reef lies below this section. The final Maastrichtian carbonate platform is characterized by biostromes and low bioherms of large marine oysters with very rare small radiolitid rudistids in growth position 1–2 m below the top—the youngest rudistids known from Puerto Rico. The large limestone breccia event that overlies this platform may represent the Chicxulub impact tsunamite (in the sense of Bourgeois and others, 1988), but independent evidence of this is lacking. Section measured by E. Kauffman, C. Johnson, H. Santos, and N. Sohl, in 1989.

As in Jamaica, the extinction of the rudistid reef ecosystem in Puerto Rico takes place abruptly in continuous carbonate platform sequences, below the K-T boundary. The demise of the rudistid reefs is similarly marked by a sharp change from relatively diverse rudistid-dominated frameworks with *Titanosarcolites*, *Durania*, and *Parastroma*, to coral-algal dominated frameworks, algal-rich limestones, or massive wackestone-packstone sequences, all with rare rudistids; biostromes and bioherms of large marine oysters overlie directly or subsequently the post-rudistid reef carbonate platform facies on both islands. The age of the final demise of rudistid reefs in Puerto Rico is less certain than in Jamaica, but appears to be similar, on the basis of regional stratigraphic relations.

There are indications from the literature that relatively unstudied sections in central Puerto Rico may also contain evidence of an abrupt rudistid reef ecosystem extinction below the K-T boundary (see Sohl and Kollmann, 1985, for a stratigraphic summary). Berryhill and others (1960) reported that the Trujillo Alto Limestone Member near the top of the Coamo Formation in east-central Puerto Rico, and the equivalent Botijas Limestone of the Pozas Formation in the Barranquitas area of central Puerto Rico (Sohl and Kollmann, 1985), represent the highest reef and near-reef Cretaceous facies of this region. These limestones contain a middle and late Maastrichtian bivalve framework assemblage of *Durania nicholasi, Barrettia monilifera*, *Titanosarcolites* sp., and *Praebarrettia sparcilirata*. The last two taxa are restricted to the middle and late Maastrichtian according to Sohl and Kollmann (1985; Kauffman, Johnson, and Sohl, in prep.). *Anomia ornata* and *Pholadomya tippana* are found in the upper Pozas Formation, above the last rudistid reef, and are restricted to the upper part of the Maastrichtian sequence on the Atlantic and Gulf coastal plain (Sohl, *in* Berryhill and others, 1960). N. F. Sohl and we have collected the last isolated *Titanosarcolites* and *Barrettia* specimens (not in frameworks) in the late part of the middle to early part of the upper Maastrichtian El Reves Limestone Member, which lies just above the Botijas Limestone in the northwest Barranquitas quadrangle, and which in turn is overlain by mudstones with brackish-water Cretaceous molluscan faunas and early Tertiary carbonate platform facies. The rudistid reef extinction lies within the Botijas Limestone Member northwest of Botijas, where a cobbly limestone unit bearing patches of poritid-like corals is overlain by shaley beds with a diverse tropical bivalve fauna, and in turn by a rudistid framework of large *Durania nicholasi* (late Campanian through Maastrichtian range) overgrown by a bioherm of tightly clustered, "organ-pipe" morphs of Maastrichtian *Barrettia* spp. This framework is abruptly terminated within the same carbonate platform facies by a massive limestone without rudistids. Berryhill and others (1960) further noted that the highest (Trujillo Alta) limestone in this sequence lacks rudistids but bears a diverse marine molluscan, coral, and benthic foraminifer fauna that seems to span the K-T boundary.

In 1990, the late N. F. Sohl and we identified three other areas in southwestern and southern Puerto Rico that seem to have a relatively continuous record of the rudistid reef extinction within the middle to upper part of the Maastrichtian. In the San German quadrangle and in the Parguera area, in the Penuelas–Sabana Grande area, and in the Ponce area, the highest Cretaceous stratigraphic sequence includes a thick series of massive limestone units, interbedded with marls and volcaniclastic platform mudstones, which contain the high *Titanosarcolites*-dominated rudistid reef and framework community associated with *Praebarrettia sparcilirata*, taxa that characterize middle and late Maastrichtian strata throughout the Caribbean province (Sohl and Kollmann, 1985; Kauffman, Johnson, and Sohl, in prep.). These frameworks are capped by coral-algal-bearing argillaceous limestones overlain by marine mollusc-rich marls and mudstones, above which is a massive carbonate platform unit containing biostromes and lenses of thick-shelled marine oysters (*Crassostrea*, *Lopha*, *Arctostrea*?). This is the same sequence of facies and faunas as found in well-studied Puerto Rican and Jamaican sections; these unstudied sections are the focus of field work planned by us and H. Santos (University of Puerto Rico at Mayaguez).

Dating the Rudistid Reef Extinction in Puerto Rico

Collectively, these Puerto Rican data seem to indicate the widespread, abrupt disappearance of the middle and late Maastrichtian *Titanosarcolites* reef and framework fauna within continuous, normal marine, carbonate platform facies below the K-T boundary. The lack of independent biostratigraphic control from ammonites, inoceramid bivalves, and planktonic microbiotas in most of the Maastrichtian of Puerto Rico makes more precise dating of the reef ecosystem extinction difficult. The El Rayo Formation and its stratigraphic equivalents throughout Puerto Rico contain *Titanosarcolites* spp. throughout all but the uppermost portion (i.e., the Oyster Limestone Member). Sohl and Kollmann (1985), Sohl (1991, personal commun.), and recent biostratigraphic compilations by us and N. F. Sohl (in prep.) regard *Titanosarcolites* and a number of associated rudistids as restricted to the middle and late Maastrichtian; it has never been reported in the literature below the early Maastrichtian. The rudistid reef extinction event occurs near the middle of the biostratigraphic range of *Titanosarcolites* in Puerto Rico. Volckmann (1984, Fig. 2) reviewed the stratigraphy of the El Rayo Formation, which he regards as middle and late Maastrichtian in age on the basis of the rudistid assemblage of *Titanosarcolites* spp., *Parastroma* spp., *Chiapasella* spp., and *Distefanella* spp. (determinations by N. F. Sohl).

Existing molluscan biostratigraphic data (e.g. Sohl and Kollmann, 1985; Kauffman, Johnson, and Sohl, revised biostratigraphic range chart in prep.) thus suggest that the reef extinction in Puerto Rico occurred within the middle Maastrichtian, and possibly as late as the middle to late Maastrichtian transition. This is seemingly supported by the micropaleontologic analyses of Pessagno (1962), who placed the strata of the upper Rio Blanco and equivalent Rio Yauco formations, which disconformably underlie the uppermost *Titanosarcolites*-bearing carbonate platform facies of the San German Formation and equivalents in Puerto Rico, in the early Maastrichtian *Ganserina gansseri dicarinata* zonule. This implies a middle to early part of the late Maastrichtian age for the last rudistid frameworks in this region. Volckmann (1984) confirmed these stratigraphic relations in southwestern Puerto Rico. However, the lack of micropaleontologic data within and above the *Titanosarcolites* Limestone in Puerto Rico makes this age assignment speculative.

Mexico

Whereas high-resolution stratigraphic data are still being developed at other Maastrichtian localities in the Caribbean province and its margins, several areas are worthy of mention in regard to the rudistid extinction history, especially Cardenas and La Popa, Mexico.

Myers (1968) studied the Maastrichtian rudistids of the Cardenas Formation near Cardenas, Mexico, which occur with molluscs that are, in part, transitional between tropical and subtropical-warm temperate biotas. Siliciclastic facies containing scattered, relatively thin carbonate platform units, 1–13 m thick, dominate the Cardenas Formation (Fig. 15). Whereas there is no continuity of carbonate deposition across the rudistid reef extinction levels and overlying coral-rich carbonate facies, the rudistid extinction pattern is similar to that in Jamaica and Mexico. At section 1 (Fig. 15), the last rudistid framework—a hippuritid-radiolitid biostrome—lies 103 m below the approximate K-T boundary (Cardenas-Tabaco formation contact; questionably a complete section). A 12-m-thick marl unit 33 m higher in the section bears the last small coral-dominated frameworks containing rare hippuritid rudistids. The upper 60–65 m of the Cardenas Formation comprises siliciclastic facies without corals or rudistids, but yielding normal marine molluscan assemblages. At section 2, a more carbonate-rich sequence contains the last rudistid (hippuritid) biostrome at the top of a 13-m-thick massive carbonate platform unit. This is overlain by 40 m of siliciclastic facies with sparse thin limestone intercalations and a normal marine molluscan assemblage. The top of the preserved Maastrichtian section is composed of 22 m of marl with a 1–4-m-thick massive limestone beds. A 4-m-thick massive rubbly limestone, 14 m below the preserved K-T boundary, contains massive and branching corals, algae, and the last scattered rudistid individuals (hippuritids).

At La Popa, in northeastern Mexico, three major Campanian-Maastrichtian carbonate platforms, the northernmost of the Caribbean province, were formed around a rising salt dome in the rapidly subsiding La Popa basin. These are poorly studied, but a preliminary survey by us and Thor Hansen in 1989 revealed the following pattern of carbonate platform biofacies. Rudistid individuals and frameworks are scattered through the lower and middle limestones. The uppermost carbonate platform deposit, 30–50 m thick (Fig. 16), preserves the last known rudistid framework, apparently composed entirely of *Coralliochama* (Keith Young, 1987, personal commun.) near the middle of the unit. The upper 10 or more meters of the limestone contains no rudistids, insofar as is known, but instead is dominated by crustose algae, some small corals, and abundant echinoid and small mollusc debris. An unusual 1.0–1.5-m-thick limestone breccia, composed of highly angular blocks up to 1 m in diameter, lies 2–3 m below the top of the limestone unit. This breccia may be equivalent to a widespread chaotic sedimentary deposit found just below the K-T boundary in siliciclastic facies throughout the Parras basin, and as far north as Eagle Pass, Texas, and in Cuba (Iturralde-Vinent, 1992). It is possible that it represents the Chicxulub tsunamite (Bourgeois, and others, 1988). The upper surface of the highest La Popa limestone unit, very near to the K-T boundary, is densely encrusted by algae. A small mollusc fauna of Paleocene age was recovered in shales and mudstones several meters above the top of the last Maastrichtian limestone unit. The exact position

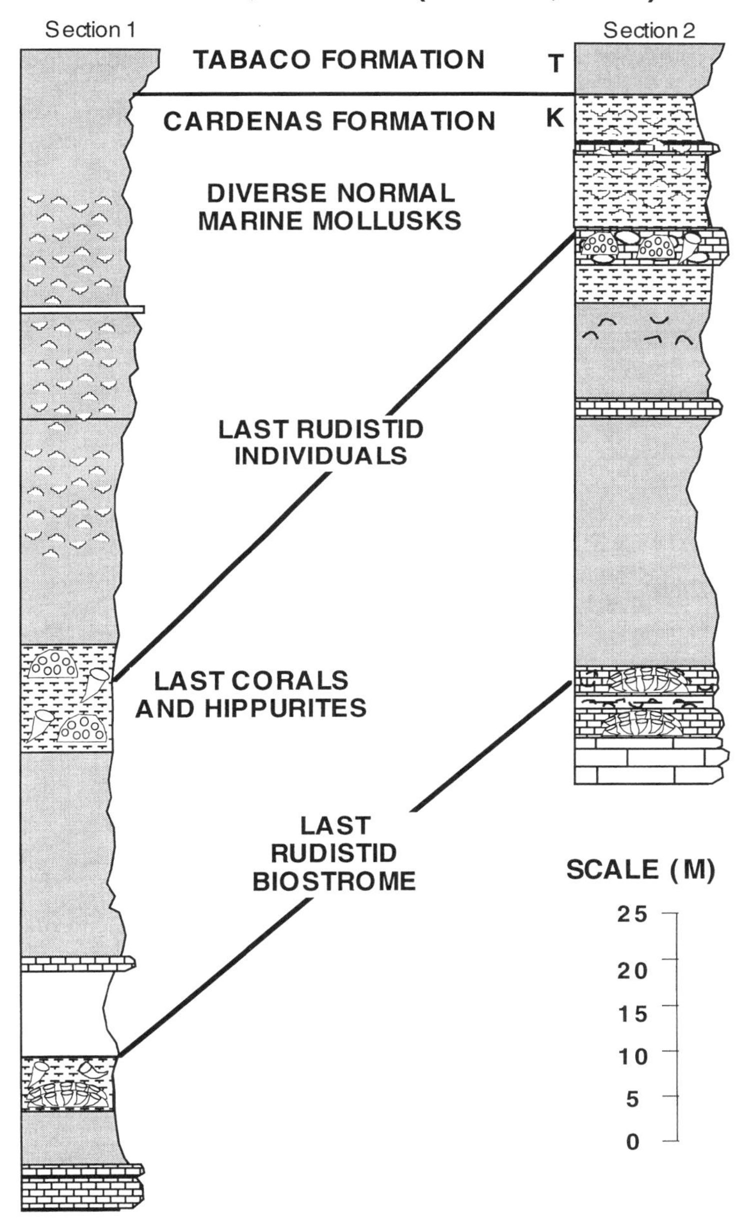

Figure 15. Generalized sections of the Maastrichtian sequence at Cardenas, Mexico (modified after Myers, 1968), showing the relative position of the last rudistid biostromes, and the last solitary rudistids, with corals, in isolated carbonate platform deposits within dominantly siliclastic and marly mudstones containing diverse marine molluscs of both tropical and temperate affinities.

of the K-T boundary is still unknown, and will be one focus of future field studies at La Popa.

SUMMARY AND CONCLUSIONS

Maastrichtian environments of the Caribbean region were characterized by slowly falling sea level, declining atmospheric CO_2 and global surface temperature, increased oceanic thermal gradients, a well-circulated upper water column without oxygen restriction, relatively low levels of volcanism, and Caribbean plate movement that allowed the spread of carbonate or mixed carbonate-siliciclastic and/or volcaniclastic platform facies and the final development of paleotropical Cretaceous reefs. Despite generally deteriorating Late Cretaceous environments,

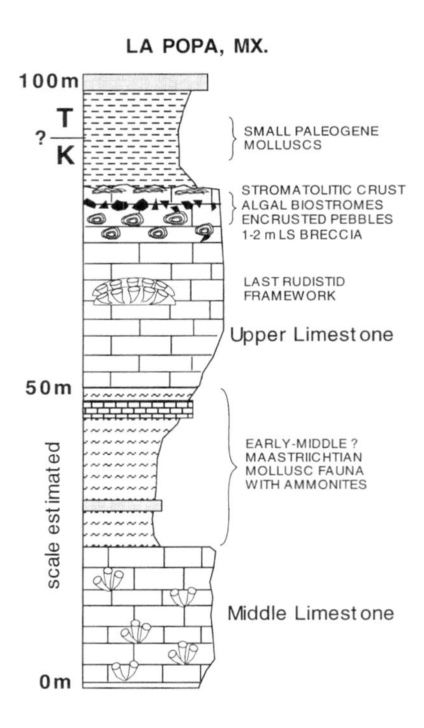

Figure 16. Generalized section of the upper two carbonate platforms (Maastrichtian) at La Popa, northeastern Mexico (thicknesses estimated), showing scattered rudistid clusters in the lower limestone, and the last rudistid framework (predominantly Coralliochama*; K. Young, 1987, personal commun.) in the middle of the upper platform limestone unit, with younger beds in the same carbonate platform sequence characterized by abundant crustose algae and echinoderm debris, scattered small molluscs and corals, and no rudistids. A 1–2-m-thick limestone breccia, with blocks >1 m in diameter (K-T tsunamite?; see Bourgeois and others, 1988), overlain by an algal and stromatoporoid-rich unit of unknown age, and subsequently by shales with small Paleogene molluscs, caps the sequence.*

rudistid-dominated reefs and biogenic frameworks reached their peak in diversity and ecological complexity during the early Maastrichtian, and rudistid lineages spread widely into subtropical and warm-temperate climate zones, as well as into deeper water. It would seem, therefore, that the rudistids were more resistant to extinction in the Maastrichtian than at any time in their evolutionary history. However, many rudistid lineages, and the reef ecosystem as a whole, underwent a dramatic, short-term regional extinction within carbonate platform facies, somewhere within the middle Maastrichtian. Subtropical and temperate-adapted lineages, and tropical generalists, survived until within 100–250 ka of the K-T boundary without forming biogenic frameworks. No rudistid specimens are known in the latest Maastrichtian of the Caribbean province. The youngest Maastrichtian rudistids are small *Radiolites* (Puerto Rico, Jamaica), hippuritids (southern Mexico), and a single large *Titanosarcolites* (Brazos River area, Texas).

The abrupt nature of the Maastrichtian extinction of reef ecosystems is suggested by Caribbean-wide patterns of physical stratigraphy and community development. Within carbonate platform sequences, rudistid reef and biogenic framework extinction takes place within 1 to 2 m of relatively uniform limestone facies; coral-algal or algal-echinoderm biostromes, with or without rare rudistids, overlie the last rudistid reefs, and these in turn are overlain by mollusc-rich and echinoderm-rich marls or limestones. The final biogenic frameworks in the carbonate platform sequence are composed almost entirely of large marine oysters (oyster banks or reefs) associated with a normal marine, shallow platform biota that may include very rare rudistids. Volcaniclastic and siliciclastic muds, silts, and sands associated with shoaling and renewal of tectonic and volcanic activity around the K-T boundary inundate these carbonate platforms in the latest Maastrichtian. The similarity of this sequencing of events from Mexico through Jamaica and Puerto Rico suggests broad oceanographic and climatic controls on Maastrichtian paleotropical environments, and thus relative synchroneity of depositional events between localities. However, this needs to be biostratigraphically tested.

The timing of the reef extinction event in the Caribbean province can be determined biostratigraphically; similar microbiotic and macrofaunal changes occur across the extinction event. Because there is poor integration at present of macrofaunal and microfossil concepts of Maastrichtian substage boundaries (see Hazel and Kamiya, 1993), these two biostratigraphic systems give somewhat different ages. Macrofaunal data throughout the Caribbean province suggest that the reef extinction occurs near the middle-late Maastrichtian boundary, 1.0–1.5 m.y. below the K-T boundary. This is based, however, on tropical bivalve and gastropod assemblages that are largely endemic to the Caribbean province; more cosmopolitan biostratigraphic indices among ammonites and inoceramid bivalves are rare in the Caribbean Maastrichtian. Microfossil data (planktonic foraminifers, nannoplankton, ostracodes) suggest an older age for the rudistid reef extinction, on the basis of data from Puerto Rico and Jamaica, at or near the base of the middle Maastrichtian, at the entry level of *G. gansseri* or in the lower part of the concurrent range zone of *Quadrum trifidum* and *G. gansseri*, and near the base of Hazel and Kamiya's (1993) ostracode zone 3, 2.5–3.0 m.y. below the K-T boundary.

Regardless of which biostratigraphic age is utilized, the similarity of lithologic and paleobiologic events between Jamaica, Puerto Rico, and generally with Cuba and Mexico, strongly suggest that the extinction of the reef ecosystem occurred during the same narrow time interval, and well below the K-T boundary. In no example considered herein does the reef extinction represent a stratigraphic artifact; younger Maastrichtian strata conformably overlie the last reefs in all studied sections. Nor can the extinction be related to facies changes in these sections. The elimination of the last rudistid reefs or frameworks occurs within single limestone units, and deposition of platform carbonates containing normal marine faunas, small coral-algal frameworks, and rare rudistids continued for more than 1 m.y. following the extinction. Refinement of the extinction pattern relative to the K-T boundary awaits better biostratigraphic correlation between sections, and statistical analyses of range data to test for sample biases.

Data presented from Jamaica, Puerto Rico, and Mexico support the stepwise extinction hypothesis for the Caribbean region, with the abrupt extinction of the reef ecosystem initiating the K-T interval, and the final loss of surviving rudistids in the tropical to temperate realms coinciding with the first loss of specialized planktonic assemblages and diverse molluscan faunas, just below the K-T boundary, on the Texas Gulf Coast. However, this conclusion must remain tentative pending graphic correlation and statistical analyses of range data in many more stratigraphic sections. At this time, however, we can reject the notion of both catastrophic and gradual extinction processes for the rudistid bivalves and their reef ecosystems in the Caribbean region. Local extinctions related to environmental changes have characterized the rudistids throughout their evolutionary history, especially in areas as dynamic as the Caribbean province (see Johnson and Kauffman, 1990), but the middle Maastrichtian extinction event is extraordinary in that it is Caribbean-wide (and also in the Mediterranean province; Swinburne, 1991), and occurs abruptly at the evolutionary peak in rudistid diversity and reef and/or framework development. The cause of this extinction event remains an enigma. No paleoenvironmental changes through the Maastrichtian correlate with the middle Maastrichtian reef extinction; changes in global temperature, sea level, and sedimentation regimes were slow, and did not exceed other Cretaceous fluctuations, which had little prior effect on the rudistid reef ecosystem. Whereas no evidence is known for extraterrestrial perturbations at this time, this does not rule out the possibility of extraterrestrial forcing related to a possible comet and/or meteorite shower that initiated 5 m.y. prior to the Chicxulub K-T impact at 65.5 Ma.

In regard to proposed causes for terminal Cretaceous extinctions, we conclude the following. The rudistid reef extinction and the final extinction of surviving rudistid lineages were not related to the large K-T boundary impact at Chicxulub; the best dating suggests that the reef extinction occurred consistently about 1.5 to 3.0 m.y. below the K-T boundary, at or near the base of the *G. gansseri* plankton zone boundary, without evidence, to date, of impact. No rudistids are known from K-T boundary strata in the Caribbean province.

Late Cretaceous sea-level fall may have affected shoreward reefs and brought about locally deleterious benthic environmental changes, but this is rejected as a primary cause. Whereas eustatic fall gradually restricted ecospace on continental margins and platforms during the Maastrichtian, carbonate platforms persisted after the reef extinction in many areas of the Caribbean province, and reef diversity was highest at the time of extinction.

Rudistid reef extinction was not clearly related to global cooling through the Late Cretaceous, although this may have contributed to biological stress. Even at the end of the Cretaceous, tropical Tethys was very well developed and reef-building rudistids reached a peak of diversification during this cooling trend. However, cooling might have restricted primary rudistid ecospace by constricting and/or eliminating the proposed Supertethyan climate zone. If so, the abrupt loss of the reef ecosystem suggests a very narrow lower thermal threshold for rudistids in reefal associations.

The extinction was not related to reef drowning by low-oxygen waters from the deep Caribbean or Atlantic. No Cretaceous oceanic anoxic events are recorded for the Maastrichtian; no organic-rich facies overlie or are laterally contiguous with the last Caribbean rudistid reefs, and total organic carbon values are low even in southern Caribbean upwelling areas.

The proposal that the Maastrichtian rudistid reef extinction might be related to upwelling and nutrification along carbonate platform margins in the Caribbean province is also rejected. Evidence for Cretaceous upwelling in this region is limited to the northern South American passive margin, where Villamil (1994, and personal commun.) noted a significant decrease in upwelling-related chert deposition between Santonian and Maastrichtian time in Colombia and Venezuela. Rudistids and reefs are rare or absent in this region during the latest Cretaceous and do not record the Maastrichtian extinction. Upwelling along the northern margin of South America is supported further by Cretaceous current circulation models for the Caribbean province (Barron and Peterson, 1990) that indicate westward-flowing longshore to north-flowing offshore currents along the northern South American passive margin. These currents, coupled with Ekman transport vectors, would produce extensive upwelling in the southern part of the province, but south of the major Cretaceous rudistid reef belts of Central America and the Antilles. Instead, Cretaceous reefs were best developed in areas that, at least in Albian (Barron, 1995) and Campian climate models (Yasuda, 1988), were characterized by unusually high Cretaceous temperatures and salinity (the Supertethyan subprovince), and possible downwelling of warm saline bottom waters. Furthermore, rudistids were bivalves and probably filter-feeders; living bivalves thrive in nutrient-rich waters.

Whereas aquatic impact as a causal mechanism for the widespread middle Maastrichtian extinction of rudistid reef ecosystems is feasible, the hypothesis that best fits available data is that a combination of accelerated changes in the ocean-climate system (e.g., temperature, sea level, circulation) reached a narrow ecological threshold for reef-building rudistids, possibly within hundreds of thousands of years, and caused the rapid disappearance of the proposed Supertethyan climate zone, the primary rudistid habitat. The abrupt change from rudistid reefs to coral-algal biostromes, algal biostromes, and marine oyster reefs within thousands of years supports this scenario, and demonstrates that normal tropical marine environments persisted well beyond the rudistid extinction event. The survival of normal tropical, subtropical, and temperate-adapted rudistids into the late Maastrichtian in non-reefal communities further supports this hypothesis.

ACKNOWLEDGMENTS

The late Norman F. Sohl shared much of his knowledge of rudistids and Caribbean stratigraphy with us, and was involved in Caribbean field studies with us. This research could not have been possible without his assistance; we dedicate this paper to his memory. In Mexico, Thor Hansen and Peter Harries assisted greatly in data collection; in Puerto Rico, Joseph E. Hazel and Hernan Santos worked closely with us in the field and provided valuable data in support of this research. The University of Puerto Rico (UPR) at Mayaguez provided transportation and other logistical support for this project; special thanks go to Alan Smith at UPR for his assistance. Work in Jamaica was greatly facilitated over the years by the scientific knowledge and logistical support given by Raymond Wright and his colleagues, Petroleum Corporation of Jamaica, and by Anthony G. Coates and Ted Robinson during their tenure at the University of the West Indies. Thanks also go to Winifred Schmidt for sharing information on the K-T boundary sections in the Jerusalem Montain Inlier of Jamaica. Keith Young at the University of Texas, Austin, Peter Skelton of the Open University, England, and Gloria Alencaster at the Universidad Nacional Autónoma de México all provided interesting discussions concerning the age, ecology, and taxonomy of Maastrichtian rudistids. Claudia Arango assisted in drafting the stratigraphic sections. This work was supported by National Science Foundation (NSF) grant EAR-8411202 to E. G. Kauffman and T. A. Hansen, NSF grant 94-18081 to C. C. Johnson, E. J. Barron, and M. A. Aurthur, NSF grant 93-04659 to E. G. Kauffman, and a grant (NSF EPSCOR) to the Department of Geology, University of Puerto Rico, Mayaguez, which provided funds to us for field studies between 1987 and 1992. To these individuals and organizations we offer our sincere thanks.

REFERENCES CITED

Alvarez, L. W., Alvarez, W., Asaro, F., and Michel, H. V., 1980, Extraterrestrial cause for the Cretaceous-Tertiary mass extinction: Science, v. 208, p. 1095–1108.

Alvarez, W., Alvarez, L., Asaro, F., and Michel, H. V., 1982, Current status of the impact theory for the terminal Cretaceous extinction, *in* Silver, L. T., and Schultz, P. H, eds., Geological implications of impacts of large asteroids and comets on Earth: Geological Society of America Special Paper 190, p. 305–315.

Alvarez, W., Kauffman, E. G., Surlyk, F., Alvarez, L., Asaro, F., and Michel, H. V., 1984, The impact theory of mass extinctions and the invertebrate fossil record across the Cretaceous-Tertiary boundary: Science, v. 223, p. 1135–1141.

Alvarez, W., Smit, J., Lowrie, W., Asaro, F., Margolis, S. V., Claeys, P., Kastner, M., and Hildebrand, A. R., 1992, Proximal impact deposits at the Cretaceous-Tertiary boundary in the Gulf of Mexico: A restudy of DSDP Leg 77, Sites 536 and 540: Geology, v. 20, p. 697–700.

Barron, E. J., 1995, Tropical climate stability and the implications for the distribution of life, *in* Stanley, S. M., and Usselman, T., eds., The effects of past global change on life: Washington, D.C., National Academy Press, National Research Council, p. 108–117.

Barron, E. J., and Peterson, W. H., 1990, Mid-Cretaceous ocean circulation: Results from model sensitivity studies: Paleoceanography, v. 5, p. 319–337.

Berner, R. A., 1994, GEOCARB II: A revised model for atmospheric CO_2 over Phanerozoic time: American Journal of Science, v. 294, p. 56–91.

Berryhill, H. L., Jr., Briggs, R. P., and Glover, L., 1960, Stratigraphy, sedimentation, and structure of Late Cretaceous rocks in eastern Puerto Rico—Preliminary report: American Association of Petroleum Geologists Bulletin, v. 44, p. 137–155.

Boersma, A., and Shackleton, N., 1979, Some oxygen and carbon isotope variations across the Cretaceous/Tertiary boundary in the Atlantic Ocean, *in* Christensen, W. K., and Birkelund, T., eds., Cretaceous-Tertiary boundary events (Symposium Proceedings) Volume 2: Copenhagen, University of Copenhagen, p. 50–53.

Bohor, B. F., 1990, Shocked quartz and more; impact signatures in Cretaceous/Tertiary boundary clays, *in* Sharpton, V. L., and Ward, P. D., eds., Global catastrophes in Earth history: An interdisciplinary conference on impacts, volcanism, and mass mortality: Geological Society of America Special Paper 247, p. 335–342.

Bourgeois, J., Wiberg, P. L., Hansen, T. A., and Kauffman, E. G., 1988, A tsunami deposit at the Cretaceous-Tertiary boundary in Texas: Science, v. 241, p. 567–570.

Buckardt, N., and Jorgensen, N. O., 1979, Stable isotope variations at the Cretaceous/Tertiary boundary in Denmark, *in* Christensen, W. K., and Birkelund, T., eds., Cretaceous-Tertiary boundary events (Symposium Proceedings), Volume 2: Copenhagen, University of Copenhagen, p. 54–61.

Caron, M., 1985, Cretaceous planktic foraminifera, *in* Bolli, H. M., Saunders, J. B., and Perch-Nielson, K., eds., Plankton stratigraphy: London, Cambridge University Press, p. 17–86.

Chubb, L. J., 1971, Rudists of Jamaica: Palaeontographica Americana, v. 7, no. 45, p. 162–257.

Coates, A. G., 1965, A new section in the Maestrichtian Guinea Corn Formation near Crawle River, Clarendon: Geological Society of Jamaica Quarterly Journal, v. VII, p. 28–33.

Coates, A. G., 1968, The geology of the Cretaceous central inlier around Arthur's seat, Clarendon, Jamaica, *in* Transactions, Caribbean Geological Conference, 4th, 1965: Port of Spain, Trinidad and Tobago, Geological Society of Trinidad, p. 309–315.

Coates, A. G., 1977, Jamaican coral-rudist frameworks and their geologic setting: American Association of Petroleum Geologists Studies in Geology, no. 4, p. 83–91.

Coates, A. G., and Jackson, J. B. C., 1985, Morphological themes in the evolution of clonal and aclonal marine invertebrates, *in* Jackson, J. B. C., Buss, L. W., and Cook, R. E., eds., Population biology and evolution of clonal organisms: New Haven, Connecticut, Yale University Press, p. 67–106.

Dechaseaux, C., 1969, Origin and extinction, *in* Moore, R. C., ed., Treatise on invertebrate paleontology, Part N, Mollusca 6, Bivalvia, Volume 2: Boulder, Colorado, Geological Society of America (and University of Kansas Press), p. N765.

Grieve, R. A. F., 1982, The record of impacts on Earth: Implications for a major Cretaceous/Tertiary impact event, *in* Silver, L. T., and Schultz, P. H., eds., Geological implications of impacts of large asteroids and comets on the Earth: Geological Society of America Special Paper 190, p. 25–38.

Hallam, A., 1984, The causes of mass extinction: Nature, v. 308, p. 686–687.

Hallam, A., 1989, The case for sea-level change as a dominant causal factor in mass extinction of marine invertebrates: Royal Society of London Philosophical Transactions, ser. B, v. 325, p. 437–455.

Hansen, T. A., Farrand, R. B., Montgomery, H. A., Billman,. H. G., and Blechschmidt, G., 1987, Sedimentology and extinction patterns across the Cretaceous-Tertiary boundary interval in east Texas: Cretaceous Research, v. 8, p. 229–252.

Hansen, T. A., Upshaw, Banks, III, Kauffman, E. G., and Gose, W., 1993, Patterns of molluscan extinction and recovery across the Cretaceous-Tertiary boundary in east Texas; report on new outcrops: Cretaceous Research, v. 14, p. 685–706.

Haq, B. V., Hardenbol, J., and Vail, P. R., 1987, Chronology of fluctuating sea levels since the Triassic: Science, v. 235, p. 1156–1167.

Hay, W. W., 1988, Paleoceanography: A review for the GSA Centennial: Geological Society of America Bulletin, v. 100, p. 1934–1956.

Hazel, J. E., and Kamiya, T., 1993, Ostracode biostratigraphy of the *Titanosarcolites*-bearing limestones and related sequences in Jamaica, *in* Wright, R. M., and Robinson, E., eds., Biostratigraphy of Jamaica: Geological Society of America Memoir 182, p. 65–76.

Heinberg, C., 1979, Bivalves from the latest Maastrichtian of Stevns Klint and their stratigraphic affinities, *in* Birkelund, T., and Bromley, R. G., eds., Cretaceous-Tertiary boundary events (Symposium Proceedings) Volume 1, The Maastrichtian and Danian of Denmark: Copenhagen, University of Copenhagen, p. 58–64.

Hildebrand, A. R., and Boynton, W. V., 1990, Proximal Cretaceous-Tertiary boundary impact deposits: Science, v. 248, p. 843–847.

Hildebrand, A. R., Penfield, G. T., Kring, D. A., Pilkington, M., Camargo, Z. A., Jacobsen, S. B., and Boynton, W. V., 1991, Chicxulub Crater: A possible Cretaceous/Tertiary boundary impact crater on the Yucatan Peninsula, Mexico: Geology, v. 19, p. 867–871.

Hut, P., Alvarez, W., Elder, W. P., Hansen, T. A., Kauffman, E. G., Keller, G., Shoemaker, E. M., and Weissman, P. R., 1987, Comet showers as a possible cause of stepwise extinctions: Nature, v. 329, p. 118–126.

Iturralde-Vinent, M., 1992, A short note on the Cuban late Maastrichtian megaturbidite (an impact-derived deposit?): Earth and Planetary Science Letters, v. 109, p. 225–228.

Izett, G. A., 1991, Tektites in Cretaceous-Tertiary boundary rocks on Haiti and their bearing on the Alvarez impact extinction hypothesis: Journal of Geophysical Research, v. 96, p. 20,879–20,905.

Jenkyns, H. C., Gale, A. S., and Corfield, R. M., 1994, Carbon- and oxygen-isotope stratigraphy of the English Chalk and Italian Scaglia and its palaeoclimatic significance: Geological Magazine, v. 131, p. 1–34.

Jiang, M.-J., and Robinson, E., 1987, Calcareous nannofossils and larger foraminifera in Jamaican rocks of Cretaceous to early Eocene age, *in* Aman, R., ed., Proceedings of a workshop on the status of Jamaican geology, March 1984: Geological Society of Jamaica Special Issue, p. 24–52.

Jiang, M.-J., and Robinson, E., 1989, Zonation of some Jamaican Cretaceous rocks using calcareous nanofossils, *in* Duque-Caro, H., ed., Transactions, Caribbean Geological Conference, 10th, 1983: Cartegena De Indices, Bogota, Columbia, Instituto Nacional de Investiaciones Geologico-Mineras (INGEOMINAS), p. 243–249.

Johnson, C. C., 1993, Cretaceous biogeography of the Caribbean region [Ph.D. thesis]: Boulder, University of Colorado, 651 p.

Johnson, C. C., and Kauffman, E. G., 1990, Originations, radiations and extinctions of Cretaceous rudistid bivalve species in the Caribbean Province, *in* Kauffman, E. G. and Walliser, O. H., eds., Extinction events in Earth history: Berlin, Heidelberg, New York, Springer-Verlag, p. 305–324.

Johnson, K. R., 1992, Foliar physiognomy of Maastrichtian leaf floras from the northern Great Plains: Implications for paleoclimate, *in* Abstracts SEPM 1992 Theme Meeting, Fort Collins: Tulsa, Oklahoma, Society for Sedimentary Geology, p. 36.

Jones, D. S., and Nicol, D., 1986, Origination, survivorship, and extinction of rudist taxa: Journal of Paleontology, v. 60, p. 107–115.

Kauffman, E. G., 1973, Cretaceous Bivalvia, *in* Hallam, A., ed., Atlas of paleobiogeography: Amsterdam, Elsevier, p. 353–384.

Kauffman, E. G., 1978, Cretaceous, *in* Robison, R. A., and Teichert, C., eds., Treatise on invertebrate paleontology, Part A, Fossilization (taphonomy), biogeography, and biostratigraphy: Boulder, Colorado, Geological Society of America (and University of Kansas Press) p. A418–A487.

Kauffman, E. G., 1979, The ecology and biogeography of the Cretaceous-Tertiary extinction event, *in* Christensen, W. K., and Birkelund, T., eds., Cretaceous-Tertiary boundary events (Symposium Proceedings), Volume 2: Copenhagen, University of Copenhagen, p. 29–37.

Kauffman, E. G., 1984a, The fabric of Cretaceous marine extinctions, *in* Berggren, W. A., and Van Couvering, J., eds., Catastrophes in Earth history: The new Uniformitarianism: Princeton, New Jersey, Princeton University Press, p. 151–246.

Kauffman, E. G., 1984b, Paleobiogeography and evolutionary response dynamic in the Cretaceous Western Interior Seaway of North America, *in* Westermann, G. E. G., Jurassic-Cretaceous biochronology and paleogeography of North America: Geological Association of Canada Special Paper 27, p. 273–306.

Kauffman, E. G., 1988, The dynamics of marine stepwise mass extinction, *in* Lamolda, M. A., Kauffman, E. G., and Walliser, O. H., eds., Paleontology and evolution: Extinction events. Revista Espanola de Paleontología, no. Extraordinario, p. 57–71.

Kauffman, 1995, Global change leading to biodiversity crisis in a greenhouse world: The Cenomanian-Turonian (Cretaceous) mass extinction, *in* Stanley, S. M., Knoll, A. H., and Kennett, J., eds., The effects of past global change on life: Washington, D.C., National Academy Press, National Research Council, p. 47–71.

Kauffman, E. G., and Fagerstrom, J. A., 1993, The Phanerozoic evolution of reef diversity, *in* Ricklefs, R. E., and Schluter, D., eds., Species diversity in ecological communities: Historical and geographical perspectives: Chicago, Illinois, University of Chicago Press, p. 315–329.

Kauffman, E. G., and Johnson, C. C., 1988, The morphological and ecological evolution of middle and Upper Cretaceous reef-building rudistids: Palaios, v. 3, p. 194–216.

Kauffman, E. G., and Sohl, N. F., 1974, Structure and evolution of Antillean Cretaceous rudist frameworks: Verhandlungen Naturforschende Gesellschaft, v. 84, p. 399–467.

Kauffman, E. G., and Sohl, N. F., 1979, Rudists, *in* Fairbridge, R. W., and Jablonski, D., eds., The encyclopedia of paleontology: Stroudsburg, Pennsylvania, Dowden, Hutchinson and Ross, p. 723–736.

Kauffman, E. G., Sageman, B. B., Kirkland, J. I., Elder, W. P., Harries, P. J., and Villamil, T., 1993, Molluscan biostratigraphy of the Western Interior Cretaceous Basin, North America, *in* Caldwell, W. G. E., and Kauffman, E. G., eds., Evolution of the Western Interior basin: Geological Association of Canada Special Paper 39, p. 397–434.

Kauffman, E. G., Johnson, C. J., and Villamil, T., 1995, High-resolution analysis of stratigraphy, chronology, and sea-level history in Cretaceous strata of northeastern Venezuela, *in* Pindell, J. L., and Drake, C. L., eds., Mesozoic-Cenozoic stratigraphy and tectonic evolution of the Caribbean region and northern South America: Geological Society of America Bulletin (in press).

Keller, G., 1989a, Extended period of extinctions across the Cretaceous/Tertiary boundary in planktonic foraminifera of continental shelf sections: Implications for impact and volcanism theories: Geological Society of America Bulletin, v. 101, p. 1408–1419.

Keller, G., 1989b, Extended Cretaceous-Tertiary boundary extinctions and delayed population change in planktonic foraminiferal faunas from Brazos River, Texas: Paleoceanography, v. 4, p. 287–332.

Kerr, R. A., 1992, Extinction with a whimper: Science, v. 258, p. 161.

Koch, C. F., and Morgan, J. P., 1988, On the expected distribution of species' ranges: Paleobiology, v. 142, p. 126–138.

Larson, R. L., 1991a, Latest pulse of the Earth: Evidence for a mid-Cretaceous superplume: Geology, v. 19, p. 547–550.

Larson, R. L., 1991b, Geological consequences of superplumes: Geology, v. 19, p. 963–966.

Marshall, C. R., 1991, Estimation of taxonomic ranges from the fossil record, *in* Gilinsky, N. L., and Signor, P. W., eds., Analytical paleobiology: Paleontological Society Short Courses in Paleontology no. 4, p. 19–38.

McHone, J. F., and Dietz, R. S., 1991, Multiple impact craters and astroblemes: Earth's record: Geological Society of America Abstracts with Programs, v. 23, no. 7, p. A183.

Myers, R. L., 1968, Biostratigraphy of the Cardenas Formation (Upper Cretaceous), San Luis Potosi, Mexico: Universidad Nacional Autónoma de Mexico Instituto de Geologia Paleontologia Mexicana, no. 24, p. 1–89.

Obradovich, J., 1993, A Cretaceous time scale, *in* Caldwell, W. G. E., and Kauffman, E. G., eds., Evolution of the Western Interior basin: Geological Association of Canada Special Paper 39, p. 379–396.

Perch-Nielson, K., 1985, Mesozoic calcareous nannofossils, *in* Bolli, H. M., Saunders, J. B., and Perch-Nielsen, K., eds., Plankton stratigraphy: London, Cambridge University Press, p. 329–426.

Pessagno, E. A., Jr., 1962, The Upper Cretaceous stratigraphy and micropaleontology of south-central Puerto Rico: Micropaleontology, v. 8, p. 349–368.

Pindell, J. L. and Barrett, S. F., 1990, Geological evolution of the Caribbean region; a plate-tectonic perspective, *in* Dengo, G. and Case, J. E., eds., The Caribbean region: Boulder, Colorado, Geological Society of America, The Geology of North America, v. H, p. 405–432.

Quezada-Muneton, J. M., Marin, L. E., Sharpton, V. L., Ryder, G., and Schuraytz, B. C., 1992, The Chicxulub impact structure: Shock deformation and target composition: Houston, Texas, Lunar and Planetary Science Conference Abstracts, v. 23, p. 1121–1122.

Raup, D. M., and Jablonski, D., 1993, Geography of end-Cretaceous marine bivalve extinctions: Science, v. 260, p. 971–973.

Robinson, E., and Lewis, J. F., 1987, Field guide to aspects of the geology of Jamaica: Mona, Jamaica, University of the West Indies, Department of Geology, 52 p.

Schindewolf, O., 1962, Neokatastrophismus?: Zeitschrift der Deutschen Geologischen Gesellschaft, v. 114, n. 2, p. 430–445.

Schmidt, W., 1988a, Stratigraphy and depositional environment of the Lucea Inlier, western Jamaica: Geological Society of Jamaica Journal, v. 24, p. 15–35.

Schmidt, W., 1988b, The middle Maastrichtian faunal break in Jamaica (Abstract, L. C. Chubb Centenial Conference, 1987): Geological Society of Jamaica Journal, v. 25, p. 43.

Sepkoski, J. J., Jr., 1993, Ten years in the libraray: New data confirm paleontological patterns: Paleobiology, v. 19, p. 43–51.

Sharpton, V. L, Dalrymple, G. B., Marin, L. E., Ryder, G., Schuraytz, B. C., and Urrutia-Fucugauchi, J., 1992, New links between the Chicxulub impact structure and the Cretaceous-Tertiary boundary: Nature, v. 359, p. 819–821.

Signor, P. W., and Lipps, J. H., 1982, Sampling bias, gradual extinction patterns, and catastrophes in the fossil record, *in* Silver, L. T., and Schultz, P. H., eds., Geological implications of impacts of large asteroids and comets on the Earth: Geological Society of America Special Paper 190, p. 291–296.

Slodowski, T. R., 1956, Geology of the Yauco area, Puerto Rico [Ph. D. thesis]: Princeton, New Jersey, Princeton University, 177 p.

Smit, J., Montanari, A., Swinburne, N. H. M., Alvarez, W., Hildebrand, A. R., Margolis, S. V., Claeys, P., Lowrie, W., and Asaro, F., 1992, Tektite-bearing, deep-water clastic unit at the Cretaceous-Tertiary boundary in northeastern Mexico: Geology, v. 20, p. 99–133.

Sohl, N. F., and Kollmann, H. A., 1985, Cretaceous actaeonellid gastropods from the Western Hemisphere: U.S. Geological Survey Professional Paper 1304, 104 p.

Svarda, C. E., 1993, Ichnosedimentologic evidence for a non-catastrophic origin of Cretaceous-Tertiary boundary sands in Alabama: Geology, v. 21, p. 1075–1078.

Swinburne, N .H. M., 1991, Tethyan extinctions, sea-level changes and the Sr-isotope curve in the 10 m.a. preceding the K/T boundary: Eos (Transactions, American Geophysical Union), v. 72, p. 267.

Swisher, C. C., Grajales-Nishimure, J. M., Montanari, A., Margolis, S. V., Claeys, P., Alvarez, W., Renne, P., Cedillo-Pardo, E., Maurrasse, F. J. M. R., Curtis, G. H., Smit,

J., and McWilliams, M. O., 1992, Coeval $^{40}Ar/^{39}Ar$ ages of 65.0 million years ago from Chicxulub crater melt rock and Cretaceous-Tertiary boundary tektites: Science, v. 257, p. 954–958.

Verdenius, J. G., 1993, Late Cretaceous calcareous nannoplankton zonation of Jamaica, *in* Wright, R. M., and Robinson, E., eds., Biostratigraphy of Jamaica: Geological Society of America Memoir 182, p. 1–18.

Villamil, T., 1994, High-resolution stratigraphy, chronology and relative sea level of the Albian-Santonian (Cretaceous) of Columbia [Ph.D. thesis]: Boulder, Colorado, University of Colorado, 462 p.

Volckmann, R. P., 1984, Upper Cretaceous stratigraphy of southwest Puerto Rico: A revision: U.S. Geological Survey Bulletin 1537-A, p. A73–A83.

Ward, P. D., 1988, Maastrichtian ammonite and inoceramid ranges from Bay of Biscay Cretaceous-Tertiary boundary sections, *in* Lamolda, M. A., Kauffman, E. G., and Walliser, O. H., eds., Paleontology and evolution: Extinction events: Revista Espanola de Paleontología, no. extraordinario, p. 116–126.

Ward, P., Kennedy, W. J., MacLeod, K. G., and Mount, J., 1991, End Cretaceous molluscan extinction patterns in Bay of Biscay K/T boundary sections: Two different patterns: Geology, v. 19, p. 1181–1184.

Yasuda, M. K., 1988, Geographic control of ocean circulation during the Late Cretaceous: Comparison of results of an ocean general circulation model with oxygen isotope paleotemperatures [M.S. thesis]: Los Angeles, California, University of Southern California, 176 p.

10

Survivorship of Mesozoic Mollusks in the Paleocene Arctic Ocean

Louie Marincovich, Jr., *U.S. Geological Survey, Menlo Park, California*

INTRODUCTION

The Arctic Ocean is the last of the world's oceans whose geologic and biotic history is largely unknown. The combination of geographic remoteness, inclement climate, expensive logistics, and the convergence of political boundaries in this vast region have challenged research efforts. However, significant insights into Arctic Ocean history have been gained in recent years across a range of disciplines (see Grantz and others, 1990a, 1990b).

The composition of earliest Tertiary marine biotas that immediately followed the terminal Cretaceous extinctions has been studied in a number of regions worldwide. One of the regions where such studies have only recently begun is the Arctic Ocean basin (Fig. 1), where the Paleocene interval previously has been known from only a few fossils. A number of bivalve and gastropod genera in Paleocene faunas of the Arctic Ocean are now known to have dwelled only in Mesozoic faunas of other ocean basins (Marincovich, 1993a). The survival of well-known and characteristically Mesozoic mollusks into the Paleocene in the Arctic Ocean basin implies that the terminal Cretaceous impact event had less effect on some mollusks in high northern latitudes than it did elsewhere (Marincovich, 1993b).

Paleogeography strongly influenced the composition of Arctic Ocean faunas. During the Late Cretaceous and the earliest Tertiary, the Arctic Ocean was about half of its present size (Harbert and others, 1990; Lawver and Scotese, 1990). The tectonostratigraphic terrane that includes northern Alaskan Paleocene deposits was much farther to the north than now, possibly as high as lat 80°N, according to paleomagnetic measurements and inferred tectonic movements of the present Alaskan North Slope (Harbert and others, 1990). A distinctive early Tertiary Arctic Ocean biogeographic province has been documented by molluscan faunas with shared species and genera in Alaska, North and South Dakota, Ellesmere Island, and Spitsbergen (Marincovich, 1992, 1993a, 1993b). The presence of endemic taxa in the Paleocene Arctic Ocean not only emphasizes the distinctive character of this northernmost faunal province, but was a consequence of its restricted biotic interchange with the world ocean. The existence of Paleocene molluscan faunas composed of mixed Mesozoic and Cenozoic genera supports the concept, which was derived earlier from plate tectonic models (Lawver and Scotese, 1990), of a geographically isolated or nearly isolated Arctic Ocean during the early Paleogene (Marincovich and others, 1983, 1985; Marincovich, 1993a, 1993b). The genus-level and species-level endemism of Arctic Ocean Paleocene mollusks reinforces this inferred isolation. Several mollusks of the Arctic Ocean province also occur in independently dated faunas of northwestern Europe and allow additional correlations of Arctic Ocean faunas with Paleocene faunas of the world ocean (Marincovich, 1993a). A reconstruction of Arctic Ocean paleogeography during late Paleocene time, based on genus-level and species-level affinities of northern molluscan faunas, is shown in Figure 2.

Cosmopolitan Mesozoic mollusks such as ammonites and inoceramid bivalves are conspicuously absent from latest Cretaceous faunas of the Arctic Ocean. The isolated northern ocean evidently was not a Paleocene refuge for these and some other major groups of mollusks. It is important to keep in mind that early Tertiary mollusks of the Arctic Ocean are known from a few widely separated exposures. Sites that contain well-preserved Arctic Ocean Paleocene marine faunas (Fig. 1) are the upper part of the Prince Creek Formation near Ocean Point, northern

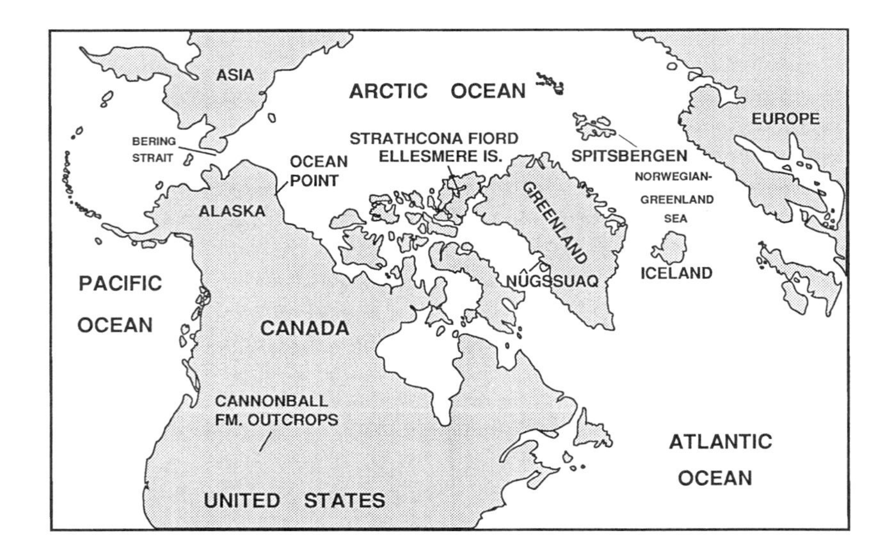

Figure 1. Locations of major place names used in text.

Alaska (70°05'N, 151°22'W), the Mount Moore Formation of the Eureka Sound Group at Strathcona Fiord, Ellesmere Island (78°45'N, 82°30'W), the Cannonball Formation (in the sense of Cvancara, 1966) of North and South Dakota (45°N, 102°W), and Spitsbergen (78°N, 15°E) (Ravn, 1922; Cvancara, 1966; Miall, 1981; Marincovich and Zinsmeister, 1991; Marincovich, 1993a, 1993b).

FAUNAL AND ISOTOPIC AGES OF ARCTIC OCEAN BIOTAS

The most thoroughly documented Arctic Ocean Paleocene molluscan fauna is in the upper part of the Prince Creek Formation near Ocean Point, northern Alaska (Fig. 3) (Marincovich, 1993a). These strata are ~45 m thick and have been referred to informally as the Ocean Point beds by Marincovich and others (1983) and later workers. These marine beds are the stratigraphically highest and youngest exposed strata of the Colville Group, which principally occurs subsurface. Phillips (1988, 1995) described the stratigraphy and interpreted the sedimentology of the uppermost 178 m of Colville Group outcrops, including the Ocean Point beds. Dinosaur bones and dated tephras (Fig. 3) are present in the lower portion of this 178-m-thick sequence (Clemens and Allison, 1985; Phillips, 1988, 1995; Nelms, 1989; Conrad and others, 1990). The presence of relict Jurassic, Early Cretaceous, and Late Cretaceous genera, together with some Cenozoic genera not previously known in faunas as old as Paleocene (Fig. 3), characterizes the marine molluscan fauna of the Ocean Point beds. The marine environment of the geographically isolated Arctic Ocean evidently allowed some well-known Mesozoic genera to live well past their occurrences elsewhere in the world ocean, and allowed other genera to evolve there and to disperse into the world ocean later in the Tertiary (Marincovich, 1993a). If a terminal Cretaceous bolide impact occurred during a Northern Hemisphere winter (or even during summer), then organisms already adapted to months-long winter darkness and relative cold may not have been affected as strongly as those living at lower lat-

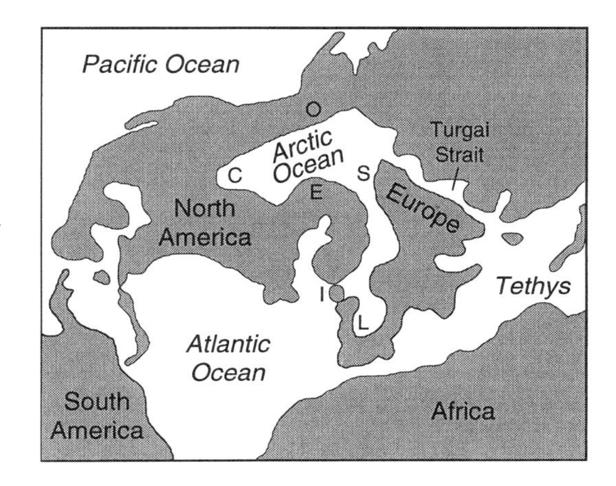

Figure 2. Late Paleocene (Thanetian) paleogeography of the Arctic Ocean, based on molluscan biogeographic affinities. O=Ocean Point, Alaska; C=Cannonball Formation; E=Ellesmere Island; S=Spitsbergen; I=Iceland; L=London basin (type Thanetian). From Marincovich (1993a).

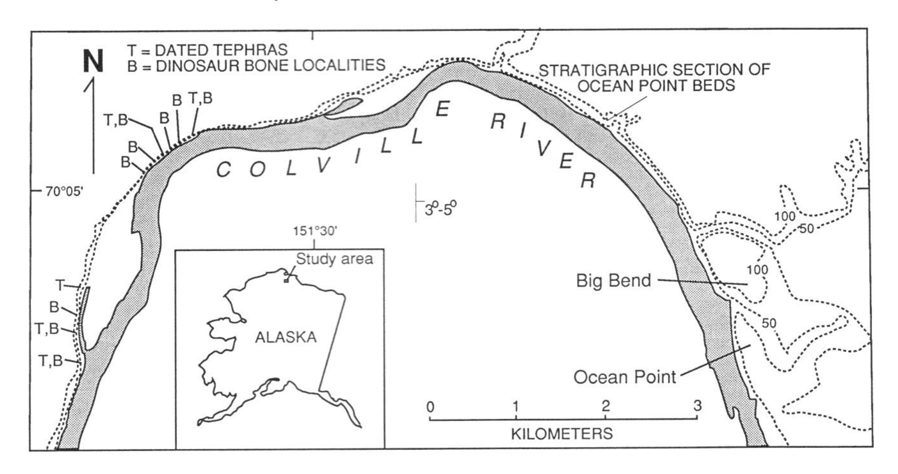

Figure 3. Location of Paleocene marine beds in the upper part of the Prince Creek Formation, northern Alaska. Dated tephras (T) and dinosaur bones (B) are downsection. Modified from Waller and Marincovich (1992). Note regional dip of 3°–5°C. Contour interval in feet.

itudes. The darkness that prevails for much of the year at 80°N would have masked the effects of an impact that reduced insolation, especially if such an event were centered in a low latitude during a Northern Hemisphere winter.

The age of the Prince Creek fauna near Ocean Point is critical to the concept that relict Mesozoic mollusks lived in the Paleocene Arctic Ocean. Paleontological age determinations for this fauna have ranged from Campanian to late Paleocene or early(?) Eocene, variously based on benthic foraminifers, palynomorphs, marine mollusks, marine and nonmarine ostracodes, and brachiopods. An assessment of the age interpretations based on these fossil groups is given below.

MacBeth and Schmidt (1973) inferred a Campanian age for benthic foraminifers from the Ocean Point beds. They emphasized that, except for a single species, the arenaceous foraminifers characterizing all other North Slope Cretaceous faunas are absent from the Ocean Point beds. These authors also noted numerous differences in test size and morphology from previously described northern Alaskan Late Cretaceous faunas. In contrast to their absence in the Paleocene Ocean Point beds (Marincovich, 1993a), arenaceous foraminifers of inferred late Campanian or early Maastrichtian age (W. V. Sliter, *in* Brouwers and others, 1987) are present in one thin marginal-marine bed of the Prince Creek Formation, interbedded with dinosaur-bearing nonmarine deposits at least 100 m stratigraphically below the Ocean Point beds. McDougall (1987) reexamined the calcareous benthic foraminifers of the Ocean Point beds and inferred an early Maastrichtian age. Moreover, this study confirmed the morphologic and test size differences with other northern Alaskan Late Cretaceous faunas originally noted by Macbeth and Schmidt (1973). McDougall (1987) also cited the presence of benthic foraminiferal species with ranges extending into the

Cenozoic. Among the 20 species used to infer an early Maastrichtian age (McDougall, 1987, Fig. 3), 8 are cited as also occurring in Cenozoic faunas. However, none of the species used for age inferences are cited in independently dated early Maastrichtian faunas elsewhere, either associated with planktonic microfossils, megafossils, isotopic ages, or other age-diagnostic criteria (McDougall, 1987). Consequently, the inferred early Maastrichtian age of benthic foraminifers from the Ocean Point beds is questionable.

Ages based on terrestrial pollen and spores have been assigned to the Ocean Point beds, but they are difficult to assess because (1) the palynomorphs used for dating do not have well-constrained ages in their type areas in the conterminous United States, and (2) the apparent but poorly understood diachroneity of pollen occurrences with latitude. Palynomorphs from the Ocean Point beds have been assigned ages within the Maastrichtian by Frederiksen and others (1988) and Frederiksen (1991). These authors differentiated presumed Maastrichtian and Paleocene palynofloras near Ocean Point on the basis of angiosperm pollen diversity, and of inferred climatic cooling deduced from an assumed shift from insect- to wind-pollinated vegetation. These changes were thought to coincide with the Cretaceous-Tertiary (K-T) boundary. The age significance of such changes can be challenged by the work of Nichols and others (1986), who noted that such floristic changes are likely attributable to local events but are not aspects of the K-T transition on a global scale. Frederiksen (1991) assigned palynomorphs from the Ocean Point beds to the *Wodehouseia spinata* assemblage zone of Nichols and others (1982), which was proposed for the central and northern Rocky Mountain region of the United States. A late Maastrichtian age of "approximately 66 to 70 Ma" was assigned to this terrestrial assemblage zone by Nichols and others (1982, p. 729), on the basis of an inferred but undocumented age equivalence with the biozone "of the ammonite *Haploscaphites nicolleti* and stratigraphically overlying biozones." The biozones referred to, including that of the dinosaur *Triceratops*, have a combined age range of 67 to 69 Ma according to Obradovich (1993), which agrees with the estimated age of Nichols and others (1982). Frederiksen (1991, p. E18) noted the presence of the nominate species and other species of this assemblage zone in the Ocean Point beds, but inferred a "middle Maastrichtian" age for the beds. His intent was to infer an early-late Maastrichtian age for the Ocean Point beds (N. O. Frederiksen, personal commun., *in* Marincovich, 1993a). Nichols and others (1982) cautioned against applying their regional zonal scheme for the central and northern Rocky Mountain area of the United States to other regions, and noted that Campanian and Maastrichtian biozones might be difficult to identify even in adjacent Alberta, owing to the influence of paleoclimate and floral provincialism.

The discussion above suggests that palynofloral ages for the Ocean Point beds are not well founded, because of the number of interrelated assumptions on which they depend. Nichols and others (1990) documented declining similarities northward of both Maastrichtian and Paleocene floral communities. Similarities among northern (i.e., Arctic) Maastrichtian communities were found to be generally low, "suggesting that northern palynofloras are not closely similar to one another" (Nichols and others, 1990, p. 357). Changes in palynofloral composition were observed to take place gradually from south to north (New Mexico to northern Alaska and Canada), and no palynoflora was identical with any other. As a consequence, Frederiksen (1991, p. E15–E17) noted the difficulty in making detailed correlations between northern Alaska and the nearest region with comparable palynofloras (northwestern Canada), as well as the lack of independent age control for Maastrichtian palynofloras in northwestern Canada, and the apparent "diachronous range tops and (or) range bases" of northern Alaskan and northwestern Canadian taxa when compared to their ages in middle-latitude palynofloras of the U.S. Western Interior. These observations coincide with those of Lerbekmo and others (1987), Sweet (1988), and Sweet and others (1990), that plant communities in northwestern Canada at paleolatitudes of 60° to 75°N underwent few species extinctions at the K-T boundary as defined by iridium enrichment, compared to lower latitude floras, and that major changes in relative abundance of angiosperm palynomorphs began below and continued above the boundary. Sweet (1988) concluded that latitudinal and habitat changes were stronger influences on palynofloral communities than was a single boundary event, and that there was no single, continent-wide floral response to the K-T boundary event. A review of palynofloral studies of North America by Keller and Barrera (1990) concluded that the effects of the boundary event were localized and most severe in middle latitudes of the western United States, decreasing into northern latitudes.

Wiggins (1976, p. 57) emphasized the diachroneity of northern Alaskan oculata pollen (which includes *Wodehouseia*). He noted that Late Cretaceous taxa in that region "are known to persist in relict floral communities, and subsequently exhibit minor periods of proliferation during upper Paleocene–lower Eocene and Oligocene–Lower Miocene? time. ... It is presumed that a few plants could have survived in these relict communities, thus range extension, at least as young as Paleocene, is not unexpected." Spicer and others (1987) reinforced this observation by documenting the first occurrences of Cretaceous and Tertiary floras at progressively younger ages northward from the middle latitudes of North America to northern Alaska and Canada. They concluded that the time lag for the first appearances of floristic elements in high latitudes versus middle latitudes may be several million years. Their study implies that floras inferred to be of late Maastrichtian age in low and middle latitudes may be of Paleocene age in the Arctic. The northward-younging first appearances of plant taxa was partly substantiated by Nichols and Sweet (1993), who documented the presence of *Wodehouseia spinata* in lowermost Paleocene beds of Alberta and Saskatchewan. This species, after which the floral assemblage zone used to assign an age to the Ocean Point beds was named, had previously been found only in Maastrichtian floras (Nichols and others, 1982).

The Ocean Point beds are a favorable setting for evaluating the age(s) of their palynoflora. Ages of palynofloras, whether in marine or terrestrial deposits, typically are determined by associations or correlations with age-diagnostic marine megafossils and microfossils, and with isotopically dated rocks. All of these elements are present in the sequence represented by the Ocean Point beds and underlying parts of the Prince Creek Formation. The 69.1 Ma age of tephras interbedded with strata containing late Maastrichtian dinosaurs (Conrad and others, 1990) implies that the overlying Prince Creek strata, including the Ocean

Point beds, are no older than late Maastrichtian. The cooccurrence in the Ocean Point beds of Paleocene marine mollusks and ostracodes suggests a Paleocene age for the contained palynoflora. Accordingly, palynomorphs from the Ocean Point beds are inferred to be of Paleocene age, on the basis of cooccurring Paleocene mollusks and ostracodes, stratigraphically concordant tephra ages, and $\delta^{13}C$ analyses of mollusks (discussed below).

Marine and nonmarine ostracodes from the Ocean Point beds are of inferred early Paleocene (Danian) age, on the basis of their occurrences in independently dated faunas elsewhere, including northern Europe, Mongolia, and the Western Interior of North America (Brouwers, 1988; Brouwers and DeDeckker, 1991, 1993). The inferred Danian age for ostracodes agrees with that for the marine mollusks. The inferred Paleocene age is supported by $\delta^{13}C$ analyses of Ocean Point marine mollusks and is stratigraphically consistent with the late-early Maastrichtian K/Ar and $^{40}Ar/^{39}Ar$ isotopic ages of underlying tephra beds. Two new brachiopod genera have also been recognized in the Ocean Point beds (Cooper, 1995). Because one genus is related to Late Cretaceous taxa and the other to early Tertiary taxa, these data are consistent with either age interpretation.

The marine molluscan fauna in the upper part of the Prince Creek Formation near Ocean Point has been dated as Danian, on the basis of comparisons with faunas of the North American Western Interior, the Arctic Ocean, and the North Atlantic (Marincovich, 1993a) (Fig. 3). Among the best age indicators are two bivalve species that previously have been found only in Paleocene faunas. The Prince Creek species *Arctica ovata*, formerly known only in the Cannonball Formation of North and South Dakota (Fig. 1), was dated as Danian on the basis of planktonic foraminifers (Fox and Olsson, 1969) or Thanetian (late Paleocene) on the basis of bivalve mollusks (Cvancara, 1966). Another once-unique Cannonball mollusk, the gastropod *Drepanochilus pervetus*, has been found in Danian strata of the Mount Moore Formation on Ellesmere Island (Marincovich and Zinsmeister, 1991). One of the most common Prince Creek bivalves, *Cyrtodaria rutupiensis*, occurs with *D. pervetus* in the Mount Moore Formation (Marincovich and Zinsmeister, 1991). This species also occurs in the Danian fauna of the Barentsburg Formation on Spitsbergen and in late Paleocene (Thanetian) faunas of the Thanet and Oldhaven formations in the London basin, England (Figs. 1 and 2) (Strauch, 1972). The occurrence of Ocean Point so-called Mesozoic bivalves in articulated and closed life positions, the excellent preservation of microscopic sculpture, and the presence of color patterns on some specimens argue strongly against the reworking of older specimens into a Paleocene deposit to explain their occurrence (Marincovich, 1993a).

The most dramatic example of a relict species in the Prince Creek fauna is the scallop *Camptochlamys alaskensis*. *Camptochlamys* was previously known only in Jurassic (Aalenian to Tithonian) faunas of northern Europe, and possibly Asia and Africa (Johnson, 1984), and was thought to have become extinct more than 70 m.y. before the Prince Creek fauna (Waller and Marincovich, 1992). The occurrence of *C. alaskensis* in a Paleocene fauna indicates that *Camptochlamys* dwelled in the Arctic Ocean long after its disappearance at lower latitudes. This is a conspicuous example of a Lazarus species (Jablonski, 1986), which persisted as an endemic taxon in an isolated environment, long after its formerly broad geographic distribution. The presence of this Lazarus species in northern Alaska reinforces the concept of a significantly isolated Arctic Ocean during much of the Late Cretaceous, which allowed the survival into the Paleocene of genera that had died out earlier in the world ocean.

Chronological ranges of other so-called Mesozoic mollusks that continued to live into the Tertiary in the Arctic Ocean are shown in Figure 4. The majority of Arctic Ocean species belonging to these genera are related to Late Cretaceous ancestral species in the Western Interior seaway of North America and in northwestern Europe (Marincovich, 1993a).

There are 12 tephra samples associated with dinosaur bones that underlie the Ocean Point beds (Fig. 3) that have a late-early Maastrichtian age, on the basis of K/Ar and $^{40}Ar/^{39}Ar$ analyses (McKee and others, 1989; Conrad and others, 1990). A best-age estimate of 69.1 ± 0.3 Ma was calculated using the weighted mean of analyses. The stratigraphically highest dated tephra is at least 100 m below the Ocean Point marine molluscan fauna (Conrad and others, 1990; Phillips, 1988, 1990). The Maastrichtian epoch lasted ~5.9 m.y., from 71.3 to 65.4 Ma (Obradovich, 1993), so these tephras were deposited during the late-early Maastrichtian, ~3.7 m.y. before the end of the Cretaceous. These dated tephras are within the lower, nonmarine, part of the exposed depositional sequence (Phillips, 1990). The overlying marine Ocean Point beds are younger than 69.1 ± 0.3 Ma (late-early Maastrichtian). This isotopic age is compatible with the Paleocene age of the Ocean Point beds inferred from marine mollusks, and marine and nonma-

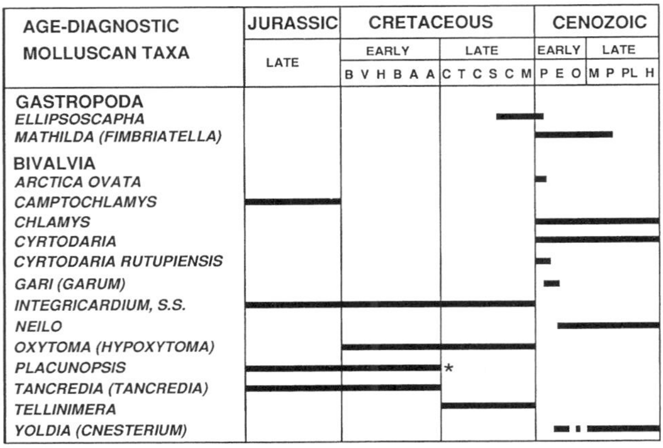

Figure 4. Chronostratigraphic ranges of age-diagnostic mollusks from the upper part of the Prince Creek Formation near Ocean Point, northern Alaska. Asterisk indicates range only of northern lineage of Placunopsis, *not entire range of genus. Abbreviations of geologic ages, from oldest to youngest, represent the following: Early Cretaceous: Berriasian, Valanginian, Hauterivian, Barremian, Aptian, Albian; Late Cretaceous: Cenomanian, Turonian, Coniacian, Santonian, Campanian, Maastrichtian; early Cenozoic: Paleocene, Eocene, Oligocene; late Cenozoic: Miocene, Pliocene, Pleistocene, Holocene. Modified from Marincovich (1993a).*

rine ostracodes, and the early-late Maastrichtian age of palynomorphs, but not with the early Maastrichtian age inferred for benthic foraminifers.

Recent studies of stable isotopes suggest a late Paleocene (Thanetian) age for some Arctic molluscan faunas that Marincovich (1993a, 1993b) thought to be Danian. Values of $\delta^{13}C$ for the Prince Creek pectinid bivalve *Camptochlamys* are in a range that was reached in the early Tertiary only during the Thanetian, and in the Late Cretaceous only at the Cenomanian-Turonian boundary (M. A. Arthur, 1994, personal commun.). It is theoretically possible that vital effects of the mollusk species influenced concentration of the carbon isotope in the mollusk shell layers. However, the extent of this effect, if any, is unknown for this group of bivalves, and modern pectinids do not seem to impart a large overprint on carbon isotope values (M. A. Arthur, 1994, 1995, personal commun.). No fossil groups studies from the Ocean Point beds imply an age nearly as great as Cenomanian or Turonian (90 Ma), which is represented by the underlying Seabee Formation. Therefore, these northern Alaskan mollusks are regarded as being of Thanetian age.

A $\delta^{13}C$ Thanetian age for the Ocean Point mollusks is compatible with the Danian faunal age inferred by Marincovich (1993a) and Brouwers and DeDeckker (1991, 1992, 1993). The difference between these age interpretations is attributable to the "monographic effect" that may influence paleontologic age estimates. In this case, many studies have been done of Danian mollusks, in part owing to the debate over terminal Cretaceous extinction effects, and curiosity about the composition of earliest Tertiary recovery faunas. In comparison, little work has been done on Thanetian faunas: there is, for example, no published monograph on the type Thanetian molluscan fauna of the Thanet Sands in England. As a result, comparing unidentified mollusk specimens of suspected Paleocene age with Danian faunas is possible because of numerous published monographs, whereas making comparisons with largely undescribed Thanetian faunas is not possible. As a result, the age interpretation is unavoidably biased toward the age of the better documented faunas, which in this case are Danian faunas. The Ocean Point beds are considered herein to be of Thanetian age, based on the combination of molluscan faunal data and stable isotope data.

A Thanetian age for the Ocean Point beds implies that there is an unconformity separating them from the underlying, dinosaur-bearing nonmarine sediments of the Prince Creek Formation. At least the lower Paleocene (Danian) strata, and perhaps a portion of the Maastrichtian strata, are not present in the Ocean Point stratigraphic section. The Ocean Point beds are estuarine at their base and progress upsection to a fully marine environment no deeper than storm wave base (Phillips, 1988, 1990; Marincovich, 1993a). This marine transgressive event probably was what produced the unconformity at the base of the Ocean Point beds.

Little is known about the marine environment in which Arctic Ocean Paleocene mollusks dwelled. The Ocean Point fauna is thought to have lived in a shallow, inner shelf setting no deeper than storm wave base, on the basis of sedimentological (Phillips, 1988, 1990, 1995) and molluscan evidence (Marincovich, 1993a). Marine temperatures are thought to have been in the temperate realm (Marincovich, 1993a). Recent $\delta^{18}O$ analyses of two Ocean Point bivalves indicate a mean annual water temperature of 14 °C ± 5 °C, assuming a possible range in ambient water salinity of 30‰ to 34‰ (K. Bice, 1994, personal commun.). Ongoing stable isotope studies are expected to refine these temperature estimates.

CONCLUSIONS

The molluscan and ostracode faunal evidence of a Paleocene age for the Ocean Point beds is reinforced and refined to late Paleocene (Thanetian) by $\delta^{13}C$ stable isotope analyses of bivalve shells. This age is consistent with the late-early Maastrichtian K/Ar and $^{40}Ar/^{39}Ar$ analyses of tephras that are at least 100 m stratigraphically below the Ocean Point beds. These faunal and isotopic data imply that relict molluscan genera once known only in Mesozoic faunas survived in the Arctic Ocean until the late Paleocene. The presence of these relict taxa suggests that a terminal Cretaceous impact event was less effective as an extinction mechanism in the Arctic Ocean than in lower latitudes.

REFERENCES CITED

Brouwers, E. M., 1988, Late Maestrichtian and Danian faunas from northern Alaska: Reconstructions of environment and biogeography: Geological Society of America Abstracts with Programs, v. 20, no. 7, p. A371.

Brouwers, E. M., and DeDeckker, P., 1991, Late Maastrichtian and Danian (?) ostracode faunas from northern Alaska: Reconstructions of environment and biogeography: Eleventh International Symposium on Ostracoda (Ostracoda in the Earth and life sciences): Warrnambool, Australia, Programme and Abstracts, p. 24.

Brouwers, E. M., and DeDeckker, P., 1992, Late Maastrichtian and Danian ostracode faunas from northern Alaska: Reconstructions of environment and biogeography: International Conference on Arctic Margins, Anchorage, Alaska, Abstracts, p. 8.

Brouwers, E. M., and DeDeckker, P., 1993, Late Maastrichtian and Danian ostracode faunas from northern Alaska: Reconstructions of environment and paleogeography: Palaios, v. 8, p. 140–154.

Brouwers, E. M., Clemens, W. A., Spicer, R. A., Ager, T. A., Carter, L. D., and Sliter, W.V., 1987, Dinosaurs on the North Slope, Alaska: Science, v. 237, p. 1608–1610.

Clemens, W. A., and Allison, C. W., 1985, Late Cretaceous terrestrial vertebrate fauna, North Slope, Alaska: Geological Society of America Abstracts with Programs, v. 17, p. 548.

Conrad, J. E., McKee, E. H., and Turrin, B. D., 1990, K-Ar and $^{40}Ar/^{39}Ar$ ages of tuff beds at Ocean Point on the Colville River, Alaska: U.S. Geological Survey Bulletin 1946, p. 77–82.

Cooper, G. A., 1995, Two new brachiopod genera from Ocean Point beds, North Slope, Alaska: U.S. Geological Survey Bulletin 1990 (in press).

Cvancara, A. V., 1966, Revision of the fauna of the Cannonball Formation (Paleocene) of North and South Dakota: University of Michigan, Museum of Paleontology Contributions, v. 20, 97 p.

Fox, S. K., and Olsson, R. K., 1969, Danian planktonic foraminifera from the Cannonball Formation in North Dakota: Journal of Paleontology, v. 43, p. 1397–1404.

Frederiksen, N. O., 1991, Pollen zonation and correlation of Maastrichtian marine beds and associated strata, Ocean Point dinosaur locality, North Slope, Alaska: U.S. Geological Survey Bulletin 1990E, 24 p.

Frederiksen, N. O., Ager, T. A., and Edwards, L. E., 1988, Palynology of Maastrichtian and Paleocene rocks, lower Colville River region, North Slope of Alaska: Canadian Journal of Earth Sciences, v. 25, p. 512–527.

Grantz, A., Johnson, L., and Sweeney, J. F., eds., 1990a, The Arctic Ocean region: Boulder, Colorado, Geological Society of America, The Geology of North America, v. L, 644 p.

Grantz, A., May, S. D., and Hart, P. E., 1990b, Geology of the Arctic continental margin of Alaska, *in* Grantz, A., Johnson, L., and Sweeney, J. F., eds., The Arctic Ocean region: Boulder, Colorado, Geological Society of America, The Geology of North America, v. L, p. 257–288.

Harbert, W., Frei, L., Jarrard, R., Halgedahl, S., and Engebretson, D., 1990, Paleomagnetic and plate-tectonic constraints on the evolution of the Alaskan–eastern Siberian Arctic, *in* Grantz, A., Johnson, L., and Sweeney, J. F., eds., The Arctic Ocean region: Boulder, Colorado, Geological Society of America, The Geology of North America, v. L, p. 567–592.

Jablonski, D., 1986, Causes and consequences of mass extinctions: A comparative approach, *in* Elliott, D. K., ed., Dynamics of extinction: New York, Wiley and Sons, p. 183–229.

Johnson, A. L. A., 1984, The paleobiology of the bivalve families Pectinidae and Propeamussiidae in the Jurassic of Europe: Zitteliana, v. 11, 235 p.

Keller, G., and Barrera, E., 1990, The Cretaceous/Tertiary boundary impact hypothesis and the paleontological record, *in* Sharpton, V. L. and Ward, P. D., eds., Global catastrophes in Earth history: An interdisciplinary conference on impacts, volcanism, and mass mortality: Geological Society of America Special Paper 247, p. 563–575.

Lawver, L. A., and Scotese, C. R., 1990, A review of tectonic models for the evolution of the Canadian basin, *in* Grantz, A., Johnson, L., and Sweeney, J. F., eds., The Arctic Ocean region: Boulder, Colorado, Geological Society of America, The Geology of North America, v. L, p. 593–618.

Lerbekmo, J. F., Sweet, A. R., and St. Louis, R. M., 1987, The relationship between the iridium anomaly and palynological floral events at three Cretaceous-Tertiary boundary localities in western Canada: Geological Society of America Bulletin, v. 99, p. 325–330.

Macbeth, J. I., and Schmidt, R. A. M., 1973, Upper Cretaceous foraminifera from Ocean Point, northern Alaska: Journal of Paleontology, v. 47, p. 1047–1061.

Marincovich, L., Jr., 1992, Earliest Cenozoic (Danian) paleogeography of the Arctic Ocean: International Conference on Arctic Margins, Anchorage, Alaska, September 2–4, 1992, Abstracts, p. 34.

Marincovich, L., Jr., 1993a, Danian mollusks from the Prince Creek Formation, northern Alaska, and implications for Arctic Ocean paleogeography: Paleontological Society Memoir 35, 35 p.

Marincovich, Louie, Jr., 1993b, Delayed extinction of Mesozoic marine mollusks in the Paleocene Arctic Ocean basin: Geological Society of America Abstracts with Programs, v. 25, no. 7, p. A–295.

Marincovich, L., Jr., and Zinsmeister, W. J., 1991, The first Tertiary (Paleocene) marine mollusks from the Eureka Sound Group, Ellesmere Island, Canada: Journal of Paleontology, v. 65, p. 242–248.

Marincovich, L., Jr., Brouwers, E. M., and Hopkins, D. M., 1983, Paleogeographic affinities and endemism of Cretaceous and Paleocene marine faunas in the Arctic: U.S. Geological Survey Circular 911, p. 45–46.

Marincovich, L., Jr., Brouwers, E. M., and Carter, L. D., 1985, Early Tertiary marine fossils from northern Alaska: Implications for Arctic Ocean paleogeography and faunal evolution: Geology, v. 13, p. 770–773.

Marincovich, L., Jr., Brouwers, E. M., Hopkins, D. M., and McKenna, M. C., 1990, Late Mesozoic and Cenozoic paleogeographic and paleoclimatic history of the Arctic Ocean

Basin, based on shallow-water marine faunas and terrestrial vertebrates, *in* Grantz, A., Johnson, L., and Sweeney, J. F., eds., The Arctic Ocean region: Boulder, Colorado, Geological Society of America, The Geology of North America, v. L, p. 403–426.

McDougall, K., 1987, Maestrichtian benthic foraminifers from Ocean Point, North Slope, Alaska: Journal of Foraminiferal Research, v. 17, p. 344–366.

McKee, E. H., Conrad, J. E. and Turin, B. D., 1989, Better dates for Arctic dinosaurs: Eos (Transactions, American Geophysical Union), v. 70, p. 74.

Miall, A. D., 1981, Late Cretaceous and Paleogene sedimentation and tectonics in the Canadian Arctic Islands, *in* Miall, A. D., ed., Sedimentation and tectonics in alluvial basins: Geological Association of Canada Special Paper 23, p. 221–272.

Nelms, L. G., 1989, Late Cretaceous dinosaurs from the North Slope of Alaska: Journal of Vertebrate Paleontology, v. 9, p. 34A.

Nichols, D. J., and Sweet, A. R., 1993, Biostratigraphy of Upper Cretaceous non-marine palynofloras in a north-south transect of the Western Interior Basin, *in* Caldwell, W. G. E., and Kauffman, E. G., eds., Evolution of the Western Interior Basin: Geological Association of Canada Special Paper 39, p. 539–584.

Nichols, D. J., Jacobson, S. R., and Tschudy, R. H., 1982, Cretaceous palynomorph biozones for the central and northern Rocky Mountain region of the United States, *in* Powers, R. B., ed., Geologic studies of the Cordilleran thrust belt: Denver, Colorado, Rocky Mountain Association of Geologists, p. 721–733.

Nichols, D. J., Jarzen, D. M., Orth, C., and Oliver, P. Q., 1986, Palynological and iridium anomalies at Cretaceous-Tertiary boundary, south-central Saskatchewan: Science, v. 231, p. 714–717.

Nichols, D. J., Fleming, R. F., and Frederiksen, N. O., 1990, Palynological evidence of the terminal Cretaceous event on terrestrial floras in western North America, *in* Kauffman, E. G., and Walliser, O. H., eds., Extinction events in Earth history: Lecture Notes in Earth Sciences, v. 30, p. 351–364.

Obradovich, J. D., 1993, A Cretaceous time scale, *in* Caldwell, W. G. E., and Kauffman, E. G., eds., Evolution of the Western Interior Basin: Geological Association of Canada Special Paper 39, p. 379–396.

Phillips, R. L., 1988, Measured sections, paleoenvironments, and sample locations near Ocean Point, Alaska: U.S. Geological Survey Open-File Report 88-40, 1 sheet.

Phillips, R. L., 1990, Summary of Late Cretaceous environments near Ocean Point, North Slope, Alaska: U.S. Geological Survey Bulletin 1946, p. 101–106.

Phillips, R. L., 1995, Measured sections and paleoenvironments near Ocean Point, Alaska, U.S. Geological Survey Bulletin 1990 (in press).

Ravn, J. P. J., 1922, On the Mollusca of the Tertiary of Spitsbergen collected by Norwegian and Swedish expeditions: Videnskapsselskapet i Kristiania, Resultater av de Norske Statsunderstøttede Spitsbergenekspeditioner (Skrifter om Svalbard og Ishavet), v. 1, 28 p.

Spicer, R. A., Wolfe, J. A., and Nichols, D. J., 1987, Alaskan Cretaceous-Tertiary floras and Arctic origins: Paleobiology, v. 13, p. 73–83.

Strauch, F., 1972, Phylogenese, Adaptation und Migration einiger nordischer mariner Molluskengenera (*Neptunea*, *Panomya*, *Cyrtodaria*, und *Mya*): Abhandlungen der Senckenbergischen Naturforschenden Gesellschaft, no. 531, 211 p.

Sweet, A. R., 1988, A regional perspective on the palynofloral responses to K-T boundary event(s) with emphasis on variations imposed by the effects of sedimentary facies and latitude, *in* Global catastrophies in Earth history: An interdisciplinary conference on impacts, volcanism, and mass mortality, Snowbird, Utah, October 20–23, 1988, Abstracts: Houston, Texas, Lunar and Planetary Institute Contribution 673, p. 190–191.

Sweet, A. R., Braman, D. R., and Lerbekmo, J. F., 1990, Palynofloral response to K/T boundary events; a transitory interruption within a dynamic system, *in* Sharpton, V. L. and Ward, P. D., eds., Global catastrophes in Earth history: An interdisciplinary conference on impacts, volcanism, and mass mortality: Geological Society of America Special Paper 247, p. 457–469.

Waller, T. R., and Marincovich, L., Jr., 1992, New species of *Camptochlamys* and *Chlamys* (Mollusca: Bivalvia: Pectinidae) from near the Cretaceous/Tertiary boundary at Ocean Point, North Slope, Alaska: Journal of Paleontology, v. 66, p. 215–227.

Wiggins, V. D., 1976, Fossil oculata pollen from Alaska: Geoscience and Man, v. 15, p. 51–76.

11

Ammonite Extinctions and Environmental Changes across the Cretaceous-Tertiary Boundary in Central Chile

Wolfgang Stinnesbeck, *Facultad de Ciencias de la Tierra, Universidad Autónoma de Nuevo León, Mexico*

INTRODUCTION

The Upper Cretaceous Quiriquina Formation in central Chile is known worldwide for its classic Indo-Pacific invertebrate fauna. The biostratigraphic distribution of this important fauna is less well understood, and Senonian, Campanian, Campanian-Maastrichtian, early Maastrichtian, early to late and late Maastrichtian ages have been proposed by the different authors. These stratigraphical uncertainties are in part due to the lack of a widely applicable ammonite zonation for the Campanian and Maastrichtian (Kennedy, 1984, 1989; Ward, 1990; Kennedy and Henderson, 1992a; Ward and Kennedy, 1993). Belemnites provide the best macrofossil stratigraphic control during this period but their use is limited to northern Europe and parts of the former U.S.S.R. (e.g., Christensen, 1988; Ward and Kennedy, 1993). Direct correlations of the ammonite or belemnite record with zonations of planktonic foraminifera (e.g., Caron, 1985) or nannofossils (e.g., Perch-Nielsen, 1985) have not been fully accomplished.

In Southern Hemisphere localities of the Upper Cretaceous, belemnites are usually very scarce, and they have never been encountered in any of the sedimentary basins along the coast of central Chile. (Note that *Naefia neogaeia* Wetzel, 1930, is not a belemnite, but belongs to the Groenlandibelidae, a family of the sepiids; Stinnesbeck, 1986; Doyle, 1986.) In the Quiriquina Formation, direct micropaleontological correlation to Tethyan pelagic facies localities or the boreal classic areas of Western Europe is restrained further by the absence of planktonic foraminifera or nannofossils.

Ammonite ranges based on comparative stratigraphic data from micropaleontologically dated faunas provide the most promising opportunity for precise biostratigraphic correlation of the Chilean coastal sediments. Many of the ammonite species present in the Quiriquina Formation are also common in other Indo-Pacific localities, notably in the upper Valudayur Formation of the Pondicherry district in southern India, the Miria Formation in Western Australia, and the upper López de Bertodano Formation of Seymour Island on the Antarctic Peninsula. Others show notable affinities to European species of the Biscay region in southern France and northern Spain, as well as other localities in France and the type area of the Maastrichtian stage in the Netherlands. These faunal affinities of the Quiriquina ammonites were last examined by Stinnesbeck (1986), who assigned the fossiliferous interval of the formation to the Maastrichtian, leaving only the transgressive basal conglomerate biostratigraphically undated. The biostratigraphy of the Quiriquina Cretaceous-Tertiary (K-T) transition is refined further in this study.

Although ammonite biostratigraphy is relatively well constrained, information regarding their mass extinction at the end of the Maastrichtian is still scarce. Did ammonites range up to the K-T boundary, where they suddenly went extinct, or did they decline gradually in abundance and diversity prior to the K-T boundary? Are the patterns of extinctions in high and low latitudes the same, or do they differ across latitudes as observed in some other fossil groups (e.g., planktonic foraminifera)? To date, only isolated data are available from South America. Thus, the current study of the Quiriquina Formation with its abundant invertebrates provides a glimpse into the history of the mass extinction of ammonites in austral high latitudes.

PREVIOUS INVESTIGATIONS

Macrofossils from the Upper Cretaceous of central Chile were first discovered in the middle of the nineteenth century. Famous early descriptions include those of d'Orbigny (1842, 1847), Forbes (*in* Darwin, 1846), Hupé (1854), Gabb (1860), and Philippi (1887). The first stratigraphic work was that of Steinmann (1895). He described a section on the west coast of Quiriquina Island (called "Saurierbucht"), introduced it as "locus typicus" of the "Quiriquina stage" and classified it as upper Senonian. In a companion paper on the paleontology, Steinmann and others (1895) described more molluscs and compared the ammonites with the faunas of India and New Zealand, hence establishing the Indo-Pacific relation of the Quiriquina fauna.

Further collections and a revision of the bivalves and gastropods were carried out by Wilckens (1904) and Wetzel (1930). Wetzel (1930) also described the ecological conditions and added sedimentological data. Supplementary faunal descriptions were written by Lambrecht (1929), Broili (1930), Wetzel (1960), Hünicken and Covacevich (1975), Perez and Reyes (1978, 1980), Maeda and others (1981), Biro-Bagoczky (1982b, 1982c), Gasparini and Biro-Bagoczky (1986), Stinnesbeck (1986), and Förster and Stinnesbeck (1987). Hünicken and Covacevich (1975), Biro-Bagoczky (1982a), and Stinnesbeck (1986) also referred to the biostratigraphic sequence and paleoecological conditions.

GEOLOGIC SETTING

In central Chile marine sediments from the Upper Cretaceous occur only locally along the Pacific coast, forming small fore-arc basins along the western margin of the coastal cordillera. Outcrops are known from Algarrobo near Valparaiso, Topocalma, the area of Chanco south of Constitucion, the district of Concepcion, and the eastern part of the peninsula of Arauco (Fig. 1). These sediments are detrital and usually transgress on the folded Hercynian basement of the Chilean coastal cordillera. These basins open to the west. Several wells have penetrated Upper Cretaceous sediments of enormous thickness in the western part of Arauco and the Chanco area, as well as offshore at the latitude of Concepcion (Mordojovich, 1975, cited *in* Biro-Bagoczky, 1982a). The basins are thought to extend to the south as far as Mocha Island (S. Cespedes, 1994, personal commun.). North of Valparaiso and south of Mocha Island, all known sedimentary basins are younger, ranging from the Oligocene-Miocene to recent.

Two of the best known and stratigraphically most complete lithological sequences are found in the vicinity of Concepcion. One outcrop is the type section of the Quiriquina Formation, as formally defined by Biro-Bagoczky (1982a) in Las Tablas Bay on Quiriquina Island (Figs. 1 and 2). This section was studied in detail by Hünicken and Covacevich (1975), Biro-Bagoczky (1982a), and Stinnesbeck (1986). The second section is situated at the coastal cliffs north of

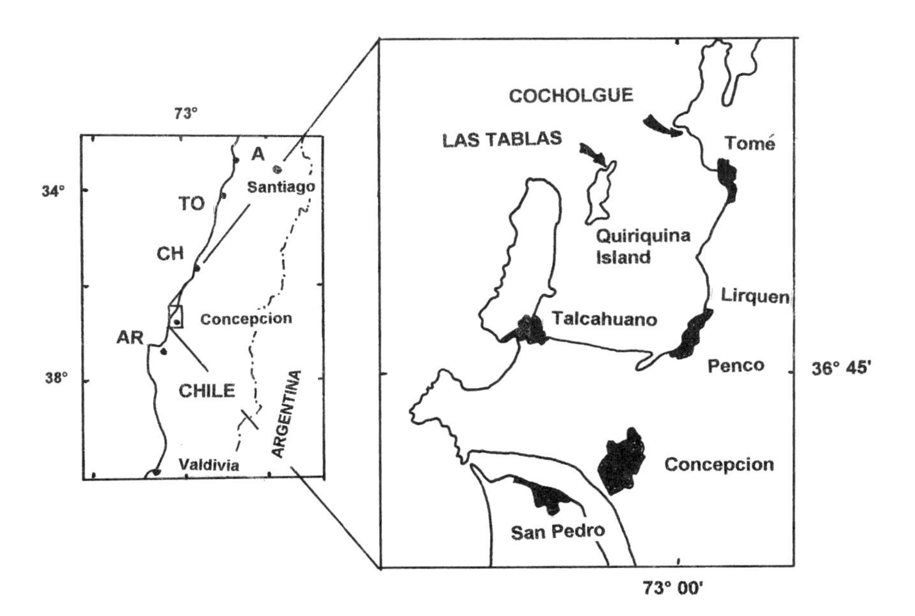

Figure 1. Map showing location of major localities mentioned in the text. Left: A = Algarrobo; TO = Topocalma; CH = Chanco area; AR = Arauco Peninsula. Right: Position of the Cocholgue and Las Tablas sections in the area of Concepcion.

Cocholgue, a small fishing harbor northwest of Tomé (Figs. 1 and 2): this section was studied by Biro-Bagoczky (1982a) and Stinnesbeck (1986). Biro-Bagoczky (1982a) proposed this sequence as a paratype locality of the Quiriquina Formation.

Both sequences show nearly all the sedimentological characteristics and faunal communities that are important in the Upper Cretaceous of the region, and thus favor geologic and paleontological studies. Macrofossils are abundant throughout and are normally well preserved, whereas microfaunal and macrofloral elements are scarce. In addition to rich bivalve and gastropod associations, ammonite occurrences allow biostratigraphic correlation within the Indo-Pacific region as well as with the European type sections.

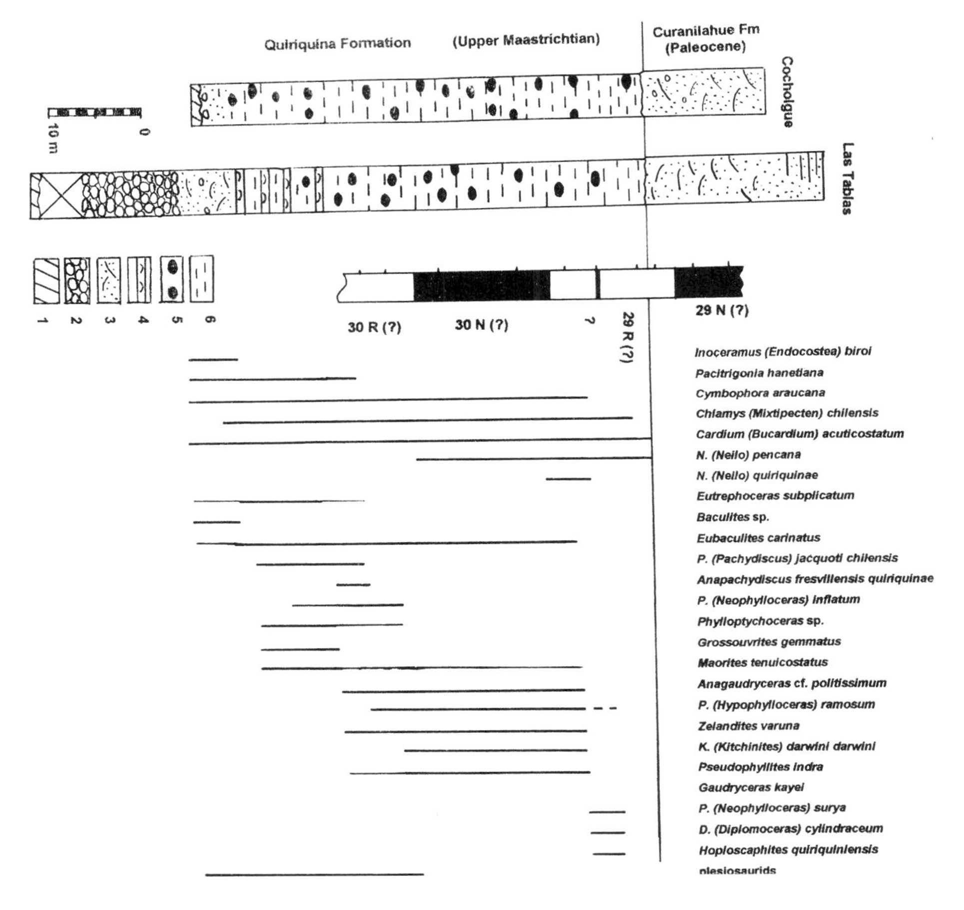

Figure 2. Las Tablas and Cocholgue sections showing paleomagnetic and biostratigraphic results. Lithology: (1) folded Hercynian basement (schists); (2) breccias and conglomerates; (3) cross-bedded sandstones; (4) coquinite layers; (5) calcareous sandstone concretions; (6) strongly bioturbated sandstone and siltstone. Magnetostratigraphy: sequence of polarity zones in Las Tablas, with tentative assignment of polarity chronozones (black = normal; white = reversed). Biostratigraphy: ranges of important ammonites, selected bivalves, and marine reptiles. Results from Las Tablas and Cocholgue are combined. Thick line = range certain; thin line = range inferred.

Magnetostratigraphic work has been carried out in Las Tablas on the harder concretionary horizons and coquinite layers (Stinnesbeck, 1986, p. 158–161; Fig. 2), whereas geochronologic studies are still lacking despite the presence of abundant glauconite. Metamorphism or structural complications have not been observed. The generally low inclination of the strata allows relatively easy sampling.

BIOSTRATIGRAPHY

Considering their traditional association with K-T boundary extinction event scenarios, it is ironic that no generally accepted international standard for Maastrichtian ammonite biochronology exists. Moreover, there is considerable disagreement regarding the placement of the base and top of the Maastrichtian and the relative positions of macrofossil and microfossil zonal boundaries within. Stratigraphic data on Maastrichtian ammonite ranges prior to their extinction remain incomplete, and correlations with belemnite, planktonic foraminifera, or nannofossil zonations are rare and restricted to only a few sections (Kennedy, 1984, 1989; Ward, 1990; Kennedy and Henderson, 1992a; Ward and Kennedy, 1993).

Unfortunately, no microfossils appear to be preserved in the Quiriquina Formation This further complicates age assignments for ammonite faunas. Relative age assignments can be obtained, however, on the basis of comparison of biostratigraphic ranges of ammonite faunas worldwide. For this reason, the ranges of ammonite taxa present in the Quiriquina Formation are compared with ranges of these taxa globally. Nomenclature of the Quiriquina ammonite species listed below follows the revision by Stinnesbeck (1986) and includes later revisions from other areas, notably by Macellari (1986, 1988a), Kennedy (1986a, 1986b, 1987), Kennedy and others (1986), Kennedy and Henderson (1992a, 1992b), Henderson and others (1992), Kennedy and Hancock (1993), and Ward and Kennedy (1993).

OCCURRENCE AND DISTRIBUTION OF MAASTRICHTIAN AMMONITES

Phylloceras (Neophylloceras) ramosum (Meek) is known to occur in the Biscay region to the top of the *A. mayaroensis* zone (if sedimentation had been continuous; Ward and Kennedy, 1993). At Zumaya, one individual has been recovered at a level 15 cm below the K-T boundary from within the *Micula prinsii* nannofossil zone (Ward, 1990; Ward and Kennedy, 1993). *Phylloceras (Hypophylloceras) surya* (Forbes) is restricted to the Maastrichtian, extending to the top of the *junior* belemnite zone in Denmark (Birkelund, 1993). In the Pondicherry district of southern India, this species probably belongs to the lower part of the *Abathomphalus mayaroensis* planktonic foraminiferal biozone. In the Biscay region, this taxon ranges to the upper part of the late Maastrichtian, to the base of the highest ammonite zone (*Anapachydiscus terminus* zone, correlative to the upper part of the *A. mayaroensis* zone; Ward and Kennedy, 1993).

Pseudophyllites indra (Forbes) ranges from the upper Santonian to Campanian to uppermost Maastrichtian *A. mayaroensis* zone in the Biscay region of France and Spain. In the Bidart section near Biarritz on the French Atlantic coast, this is the highest reported ammonite occurrence, collected 14 cm below the K-T boundary from the *M. prinsii* nannofossil zone (Ward, 1990; Ward and Kennedy, 1993). *Gaudryceras kayei* (Forbes) ranges from the Santonian-Campanian to high in the Maastrichtian (Kennedy and Henderson, 1992a).

Zelandites varuna probably belongs to the lower part of the *A. mayaroensis* zone in southern India (Kennedy and Henderson, 1992a). The species seems to be restricted to the upper parts of the Maastrichtian on Seymour Island, with a last occurrence below the glauconite interval considered to contain the K-T boundary (Macellari, 1986; Zinsmeister and others, 1989).

Diplomoceras notabile Whiteaves appears to be a junior synonym of *D. cylindraceum* (Defrance) (Kennedy and Henderson, 1992b; and Ward and Kennedy, 1993), and ranges to the very top of the Maastrichtian.

Phylloptychoceras sp., in Stinnesbeck (1986), is a possible synonym of *Phylloptychoceras sipho* (Forbes) (see Kennedy and Henderson, 1992b, p. 709): this latter species spans the upper *Gansserina gansseri* and lower *A. mayaroensis* zones in the Biscay region and in India (Ward, 1990; Kennedy and Henderson, 1992a; Ward and Kennedy, 1993).

Eubaculites carinatus (Morton 1834) is a senior synonym of *Eubaculites lyelli* (d'Orbigny 1847), as established by Kennedy (1987). The species is widely distributed within the Quiriquina Formation; it appears in the yellow sandstones directly overlying the transgressive basal conglomerate and extends to the upper part, but not the top of the formation (Fig. 2). *Eubaculites carinatus* is widespread in the low and middle latitudes, and is a useful indicator of middle to late Maastrichtian age (Henderson and others, 1992). In the Biscay sections, this taxon is restricted to the upper part of the *Anapachydiscus fresvillensis* and lower *A. terminus* ammonite zones (*A. mayaroensis* planktonic foraminiferal zone; Ward and Kennedy, 1993).

Baculites huenickeni Stinnesbeck (1986) seems to be closely related to *Eubaculites simplex* (Kossmat) (see Henderson and others, 1992, p. 158). *Eubaculites simplex* seems to be restricted to the Maastrichtian in Western Australia and India, but does not range into the upper part of this stage (Henderson and others, 1992). In the Quiriquina Formation, *B. huenickeni* seems to be restricted to the lower part of the sections (Fig. 2).

Baculites vicentei Stinnesbeck (1986) is related to a distinctive group of small baculitids, including *Fresvillia constricta* Kennedy; *F. teres* (Forbes), *Baculites columna* (Morton), and probably *B. paradoxus* (Pervinquiere), that are homeomorphic with the middle Cretaceous genera *Lechites* Nowak and *Sciponoceras* Hyatt (see Henderson and others, 1992). Stratigraphic ranges of these species indicate a Maastrichtian age; *Fresvillia constricta* and *F. teres* are restricted to the upper Maastrichtian (Kennedy and Henderson, 1992b). *Baculites vicentei* was detected only near the base of the Quiriquina Formation (Fig. 2).

Hoploscaphites quiriquiniensis (Wilckens) seems to be closely allied to *Hoploscaphites constrictus* (J. Sowerby), the index fossil of the Maastrichtian stage (see Kennedy, 1984). The *niedzwiedzkii* forms of *H. constrictus*, which are characteristic of the *casimirovensis* belemnite zone of the latest Maastrichtian (see van der Tuuk, 1987), show close morphological resemblance with the Chilean species.

Maorites tenuicostatus Marshall closely resembles *M. densicostatus* (Kilian and Reboul), to which it may be a synonym (Macellari, 1986; Henderson and others, 1992, p. 153). The species was considered by Macellari (1986) to be restricted to the upper Maastrichtian of Seymour Island, and was also recognized in the upper Maastrichtian Miria Formation of Western Australia (Henderson and McNamara, 1985). In both cases, the age assignments are substantiated by microfossil evidence. In Patagonia, *M. densicostatus* is considered an index species of the highest kossmaticeratid assemblage (Macellari, 1988a). In Antarctica, this species is one of the last in situ ammonites that was detected only 3.5 m below the K-T boundary in the basal part of the glauconitic K-T boundary interval (Elliot and others, 1994).

Type specimens of *Grossouvrites gemmatus* (Hupé) from Quiriquina Island closely resemble the Antarctic (Macellari, 1986) and Australian material (Henderson and McNamara, 1985) in ornamentation, but are slightly more involute (umbilicus/diameter [U/D] ratio = ~0.2 compared to 0.235 in the material from Seymour Island and 0.25 to 0.35 for the material from the Miria Formation) and show a less-inflated whorl section (W/H = 0.65 to 0.69 compared to 0.75 to 0.83 in the Seymour Island and 0.72 to 1.0 in the Miria specimens). These differences are, however, minor and favor a close relation. In Western Australia, the species is of late Maastrichtian age (Henderson and others, 1992).

The Antarctic *Grossouvrites gemmatus* ranges through most of the Maastrichtian to the base of the glauconitic interval that contains the K-T boundary (Macellari, 1986, 1988b).

Kitchinites darwini darwini (Steinmann) is known from Seymour Island, where it occurs in Maastrichtian beds of the López de Bertodano Formation (Macellari, 1986, 1988b).

Pachydiscus (Pachydiscus) jacquoti chilensis Stinnesbeck closely resembles *P. (P.) jacquoti dissitus* Henderson and McNamara (1985), and was regarded as a synonym by Kennedy and Henderson (1992a, p. 426): the latter subspecies is of late Maastrichtian age in the Miria Formation of Western Australia. The nominal species, the European *P. (P.) jacquoti* Seunes 1890, is known to range from the upper Maastrichtian *junior* belemnite zone to the top of the uppermost Maastrichtian *A. terminus* ammonite zone (*A. mayaroensis* planktonic foraminiferal zone) in the Biscay region (Ward and Kennedy, 1993).

Anapachydiscus fresvillensis quiriquinae (Steinmann) is closely allied with or even synonymous to *A. fresvillensis* (Seunes 1890), which is of late Maastrichtian age on Cotentin Peninsula, the Pyrénées-Atlantiques in France, and the Maastricht area in the Netherlands (e.g., Kennedy, 1986b, 1987; Kennedy and others, 1986; Kennedy and Henderson, 1992a), and ranges from the upper *G. gansseri* zone into the *A. mayaroensis* zone in the Biscay region (Ward, 1990; Ward and Kennedy, 1993).

AGE OF THE QUIRIQUINA FORMATION

Hünicken and Covacevich (1975) and Stinnesbeck (1986) considered the Quiriquina Formation to be of Maastrichtian age on the basis of ammonite ranges. This biostratigraphic assignment can now be refined further based on comparison of ammonite ranges globally, as described above and as suggested by Ward (1990). Many of the Quiriquina ammonites are common to the upper López de Bertodano Formation of Seymour Island (Macellari, 1986, 1988b; Zinsmeister and others, 1989), the Miria Formation of Western Australia (Henderson and McNamara, 1985; Henderson and others, 1992), the upper Valudayur Formation of the Pondicherri district in southern India (Kennedy and Henderson, 1992a, 1992b), and sections from the Biscay region in Europe (Ward and Kennedy, 1993). These units are of late Maastrichtian age, as also indicated by the presence of planktonic foraminifera of the *G. gansseri* and *A. mayaroensis* zones. A similar age is now attributed to the Quiriquina Formation, as evidenced by the composition of the fauna and in agreement with the magnetostratigraphic record (Fig. 2).

PLACEMENT OF THE CRETACEOUS-TERTIARY BOUNDARY

Steinmann (1895) was the first to place the K-T boundary on Quiriquina Island at the sedimentary contact and unconformity between the glauconitic marine siltstones of the top of the Quiriquina Formation and the yellow fluviatile sandstones of the basal Curanilahue Formation. This placement was shared by all later authors (including myself), although no ammonite or other unequivocally Maastrichtian index fossils have been detected above 5 m below the lithological contact. At about this level, the last ammonites, *P. (Neophylloceras) surya*, *Diplomoceras cylindraceum,* and *Hoploscaphites quiriquiniensis*, were identified by me in the Las Tablas and Cocholgue sections. Upsection, only bivalves have been described, including sporadic *Cardium (Bucardium) acuticostatum* and *Neilo (Neilo) pencana*, which extend to just below the contact in the Tomé section (see Fig. 1 for location). Isolated specimens of *Tellina largillierti*, *Chlamys (Mixtipecten) chilensis*, *Nucula (Leionucula) ceciliana*, *N. (Neilo) quiriquinae*, *Yoldia levitestata,* and *Cymbophora araucana* are present in the top 5 m of the Quiriquina Formation at Cocholgue. *Echitriporites triangulariformis*, *Gleicheniidites feronensis*, *Hymenophyllumsporites deltoidus*, and *Triletes tuberculiformis* characterize the microflora of this interval. All these species are typical components of the Quiriquina Formation, but their biostratigraphic ranges are not known precisely.

Above the lithological contact, at the base of the Curanilahue Formation, charcoaled wood and an isolated shark tooth (*Isurus* sp.) are the only megafossils detected thus far. These, along with the microflora (e.g., *Triletes tuberculiformis*, *Pterocarya stellatus*, *Pityosporites microalatus*, *Stereisporites psilatus*, *Hymenophyllumsporites deltoidus*, *Brachyphyllum mamilare*, *Laevigatosporites discordatus*, and *Lygistipollenites florinii*), include only long-ranging taxa. The first microflora of unequivocally lower Tertiary age (probably Paleocene) were found 18 m upsection by Frutos (1984, personal commun.).

The lithological contact between the Quiriquina and the Curanilahue Formation is characterized by an irregular surface and flute casts. Vertical burrows of boring bivalves and U-shaped tubes of *Rhizocorallium* initiate in this bedding plane and are filled with the yellow sandstone of the basal Curanilahue Formation. These trace fossils characterize the *Glossifungites* ichnofacies (Seilacher, 1984; Ekdale and others, 1984) and indicate consolidation or partial lithification of the Quiriquina glauconitic siltstone prior to the deposition of the Curanilahue sediments. Erosion of the Quiriquina sediments has been limited, because no reworked fossils, glauconitic siltstone, or calcareous sandstone pebbles have been observed.

Both the lithological features and the observed trace fossil association strongly suggest that a sedimentary hiatus is present at the contact between the two formations. This omission plane probably spans the K-T boundary. A hiatus commonly characterizes the K-T boundary in shallow-water sections as a result of a sea-level lowstand near the end of the Maastrichtian and sea-level rise across the K-T boundary (Haq and others, 1987, 1988; MacLeod and Keller, 1991a, 1991b; Keller and Stinnesbeck, this volume).

In the uppermost meters of the Quiriquina Formation, detritus feeding bivalves (*Nucula, Neilo, Yoldia, Tellina*) are the prevailing elements of the invertebrate fauna, whereas filter feeders are scarce (*Chlamys, Cymbophora*). This faunal assemblage suggests quiet-water conditions in an offshore shelf setting and high supply of organic matter. Conditions near the sediment-water interface were probably reducing, as also suggested by the abundance of glauconite and small pyrite concretions. Additional environmental deterioration is also indicated for the topmost 2 to 3 m of the Quiriquina Formation, where shelly macrofaunas are very rare and bioturbation by *Teichichnus* and *Zoophycos* is particularly intensive.

AMMONITE EXTINCTION PATTERN AND DEPOSITIONAL ENVIRONMENT

In Las Tablas the Cretaceous series starts with a 15-m-thick basal conglomerate composed of debris from the surrounding Hercynian paleocliffs (Fig. 2). Imbrication of the phyllite breccia and intensive cross-bedding of the sandstone lenses suggest an ancient littoral environment. Although these transgressive beds have not yielded fossils, all other horizons of the Quiriquina Formation in Las Tablas contain abundant macrofaunas, principally mollusks. Their faunal succession indicates a transgressive storm-influenced shelf sequence, from intertidal to deeper subtidal, and is considered to be characteristic of the development of fauna and sedimentation in the area of Concepcion.

Overlying the transgressive breccia beds are 6.5 m of yellow cross-bedded sandstones and intercalated conglomeratic lenses. Marine faunas including *Mytilus primigenius*, *Inoceramus (Endocostea) biroi*, *Ostrea* sp., *Dentalium (Antalis) chilensis,* and *Baculites* sp. occur here for the first time. Upsection, the character of the sediment changes toward intensively bioturbated glauconitic sand and siltstones. Beds of coquinite characterize facies 1 of the lower 10 m, and calcareous sandstone concretions characterize facies 2 of the overlying 35 m. The faunas of the two facies differ considerably.

For example, bivalves of facies 1, especially *Pacitrigonia hanetiana* and *Cardium (Bucardium) acuticostatum* and less-abundant *Cymbophora araucana* and *Aphrodina quiriquinae*, indicate shallow water and high-energy conditions with respect to the coquinite facies. The shell layers themselves are considered tempestites. Additional faunas include diverse gastropods, the nautilid *Eutrephoceras subplicatum*, crabs (*Callianassa saetosa*), vertebrate bones (pliosaurids), and shark teeth. *Ophiomorpha* is the prevailing trace fossil.

Ammonites of this lower horizon include *Eubaculites carinatus*, *Anapachydiscus fresvillensis quiriquinae*, *P. (Pachydiscus) jacquoti chilensis*, *Maorites tenuicostatus*, *Grossouvrites gemmatus*, *P. (Neophylloceras) inflatum*, and *Phylloptychoceras* sp.

Facies 2 is dominated by *Eubaculites carinatus*, *Chlamys (Mixtipecten) chilensis*, and *Solariella unio*; all of these are particularly abundant in concretions. *Gyrodes euryomphala*, and *Cymbophora araucana* are also abundant locally. The sediments are completely bioturbated, *Teichichnus* and *Zoophycos* predominating. Paleoecologic analyses suggest a quiet offshore shelf environment. *Eubaculites carinatus* probably lived in shoals near the sea floor, hidden in loosely distributed patches of seaweed. Seaweed served *Chlamys (Mixtipecten) chilensis* as a substrate and *Solariella unio* as a grazing surface.

Ammonites other than *Eubaculites carinatus* are *P. (Neophylloceras) ramosum*, *K. (Kitchinites) darwini*, *Maorites tenuicostatus*, *Pseudophyllites indra*, *Zelandites varuna*, and *Gaudryceras kayei*. Hünicken and Covacevich (1975, p. 163) also mentioned *Grossouvrites gemmatus* from their fossiliferous level 12. I examined their two specimens in 1984, and regard them as conspecific with *Kossmaticeras (Natalites?) erbeni* Stinnesbeck.

Fossiliferous concretions of calcareous sandstone are still abundant in the uppermost 10 m of the Quiriquina Formation in Las Tablas and other sections in the area, but no specimen of *Eubaculites carinatus* has been detected during my collecting, whereas other ammonites, *D. (Diplomoceras) cylindraceum*, *P. (Hypophylloceras) surya*, and *Hoploscaphites quiriquiniensis* (in Cocholgue), are still present, but rare. This pattern of ammonite abundance decline and extinction in the upper 5 to 10 m below the K-T boundary indicates a major environmental change. This is also indicated by the presence of detritus feeding bivalves such as *Nucula (Leionucula) ceciliana*, *N. (Neilo) pencana*, *N. (Neilo) quiriquinae*, *Yoldia levitestata*, *Nuculana cocholguei*, and *Tellina largillierti*, which mark still-water conditions and sediments rich in organic matter. Stagnant water and a reducing environment are suggested for the upper 2 to 3 m of the Quiriquina Formation (uppermost Maastrichtian?), where small pyrite concretions and ichnofaunas are particularly abundant, and macrofaunas are scarce. The K-T boundary is marked by the sudden lithological change and unconformity between the marine sequence of the Quiriquina Formation and the gravelly yellow sandstones of the Curanilahue Formation, which reflect brackish to fluviatile conditions.

In Cocholgue the transgressive horizon of the basal Quiriquina Formation is only 1.5 m thick (Fig. 2) and contains *Ostrea* sp., *Trochus ovallei*, and *Crepidula* sp. A gradual transition occurs to the overlying 45-m-thick bioturbated sand and siltstones with calcareous sandstone concretions of the Quiriquina Formation. Bivalves such as *Pacitrigonia* and *Cardium* are scarce throughout this interval, whereas the succession of faunas and lithologies in the upper part of the sequence and the contact to the Curanilahe Formation are very similar to those of Las Tablas (see Biro-Bagoczky, 1982a; Stinnesbeck, 1986).

CONCLUSIONS

Ammonite species present in the Quiriquina Formation show close affinities to faunas from southern Europe and the Indo-Pacific area, notably India, Australia, and Antarctica, where the planktonic foraminifera indicate a late Maastrichtian (*G. gansseri* to the *A. mayaroensis* zone) age. Ammonite ranges in the Quiriquina Formation suggest an equivalent age. Deposition was probably a consequence of the early-late Maastrichtian sea-level rise, which continued into the uppermost Maastrichtian (sea-level cycle 4.5 of supercycle UZA-4 of the upper Maastrichtian; Haq and others, 1987, 1988). Near the end of the Cretaceous a sea-level drop ended marine sedimentation. The K-T boundary coincides with this hiatus and is marked by an unconformity between the bioturbated marine siltstones of the Quiriquina Formation and the fluviatile conglomeratic sandstones of the Curanilahue Formation.

Ammonite abundance declined and their extinction in central Chile appears to have been gradual, the last ammonite species occurring 5 m below the K-T boundary. *Eubaculites carinatus*, the most common Quiriquina ammonite, is absent in the top 10 m of the formation, whereas *P. (Neophylloceras) ramosum*, *P. (N.) surya*, *D. (Diplomoceras) cylindraceum*, and *Hoploscaphites quiriquiniensis* range several meters upsection. Thus, patterns of ammonite extinction in the Quiriquina Formation indicate the presence of major late Maastrichtian environmental changes that appear to be unrelated to the K-T boundary event in the strict sense.

ACKNOWLEDGMENTS

I thank E. Olivero and V. Covacevich for valuable criticism on an earlier version of this paper and reviewers P. D. Ward, W. J. Zinsmeister, and G. Keller for many comments and suggestions. A. R. Ashraf kindly identified the microflora. Field work in central Chile during 1981, 1982, and 1984, was supported by the Deutsche Forschungsgemeinschaft (Er 4/46-1) and the Deutscher Akademischer Austauschdienst (DAAD- 322/504/504/4).

REFERENCES CITED

Birkelund, T., 1993, Ammonites from the Maastrichtian White Chalk in Denmark: Geological Society of Denmark Bulletin, v. 40, p. 40–81.

Biro-Bagoczky, L., 1982a, Revisión y Redefinición de los "Estratos de Quiriquina", Campaniano-Maastrichtiano, en su localidad típo en la Isla Quiriquina, 36° 35' Lat Sur, Chile, Sudamerica, con un perfil complementário en Cocholgue: Actas III Congreso Geológico Chileno, v. 1, p. A29–A64.

Biro-Bagoczky, L., 1982b, *Hoploscaphites constrictus* (J. Sowerby) en la Formación Quiriquina, Campaniano-Maastrichtiano, Región del Bio Bio, Chile, Sudamerica: Actas III Congreso Geológico Chileno, v. 1, p. A1–A16.

Biro-Bagoczky, L., 1982c, Contribución al conocimiento de Naefia neogaeia Wetzel, Coleoidea, en la Formación Quiriquina, Campaniano-Maastrichtiano, Región del Bio Bio, Chile, Sudamerica: Actas III Congreso Geológico Chileno, v. 1, p. A1–A28.

Broili, F., 1930, Plesiosaurierreste von der Insel Quiriquina: Neues Jahrbuch für Mineralogie, Geologie und Paläontologie, Beilage Band v. 63, p. 497–514.

Caron, M., 1985, Cretaceous planktic foraminifera, *in* Bolli, H., Saunders, J., and Perch-Nielsen, K., eds., Plankton stratigraphy: London, Cambridge University Press, p. 17–87.

Christensen, W. K., 1988, Upper Cretaceous belemnites of Europe: State of the art, *in* Streel, M., and Breel, M. J. M., eds., The Chalk District of the Euregio Meuse-Rhine. Selected papers on the Upper Cretaceous: Maastricht, Nederlands, Liège University, p. 1–16.

Darwin, C., 1846, Geological observations on South America: London, Smith, Elder and Co., 279 p.

d'Orbigny, A., 1842, Voyage dans l'Amérique méridionale 1826–1833: Paris, t. 3, 3e partie, Géologie, 4e partie Paléontologie, 188 p.

d'Orbigny, A., 1847, Voyage au Pol Sud et dans l'Océanie sur les Corvettes l'Astrolabe et la Zélée, 1838–1840, *in* Dumont D'Urville, M. J., ed., Géologie (Paléontologie) Atlas de Géologie: Paris, Imprimavie de J. Claye et Cie., pl. 1–9.

Doyle, P., 1986, Naefia (Coleoidea) from the Late Cretaceous of southern India: British Museum (Natural History) Bulletin, Geology, v. 40, p. 133–139.

Ekdale, A. A., Bromley, R. G., and Pemberton, S. G., 1984, Ichnology: The use of trace fossils in sedimentology and stratigraphy: Society of Economic Paleontologists and Mineralogists Short Course Notes 15, p. A1–A28.

Elliot, D. H., Askin, R. A., Kyte, F. T., and Zinsmeister, W. J., 1994, Iridium and dinocysts at the K-T boundary on Seymour Island, Antarctica: Implications for the K-T event: Geology, v. 22, p. 675–678.

Förster, R., and Stinnesbeck, W., 1987, Zwei neue Krebse, *Callianassa saetosa* n. sp. und *Homolopsis chilensis* n. sp. (Crustacea, Decapoda) aus der Oberkreide Zentral-Chiles: Mitteilungen der Bayerischen Staatssammlung für Paläontologie und historische Geologie, v. 27, p. 51–65.

Gabb, J. C., 1860, Description of some new species of Cretaceous fossils from South America, in the collection of the Academy: Academy of Natural Sciences of Philadelphia Proceedings, 1860, p. 197–198.

Gasparini, Z., and Biro-Bagoczky, L., 1986, *Osteopygis* sp. (Reptilia, Testudines, Toxochelyidae) tortuga fósil de la Formación Quiriquina, Cretácico Superior, Sur de Chile: Revista Geológica de Chile, v. 27, p. 85–90.

Haq, B. U., Hardenbol, J., and Vail, P., 1987, Chronology of fluctuating sea levels since the Triassic: Science, v. 235, p. 1156–1166.

Haq, B. U., Hardenbol, J., and Vail, P. R., 1988, Mesozoic and Cenozoic chronostratigraphy and eustatic cycles of sea-level change, *in* Wilgus, C. K., Hastings, B. S., Kendall, C. G., Posamentier, H. W., Ross, C., and Van Wagoner, J. C., eds., Sea-level changes: An integrated approach: Society of Economic Paleontologists and Mineralogists Special Publication 42, p. 71–108.

Henderson, R. A., and McNamara, K. J., 1985, Maastrichtian nonheteromorph ammonites from the Miria Formation, Western Australia: Palaeontology, v. 28, p. 35–88.

Henderson, R. A., Kennedy, J. W., and McNamara, K. J., 1992, Maastrichtian heteromorph ammonites from the Carnarvon Basin, Western Australia: Alcheringa, v. 16, p. 133–170.

Hünicken, M. A., and Covacevich, V., 1975, Baculitidae en el Cretácico sup. de la Isla Quiriquina, Chile, y consideraciones paleontológicas y estratigráficas: Actas del Primer Congreso Argentino de Paleontología y Bioestratigrafía, v. 2, p. 141–172.

Hupé, L. H., 1854, Zoologia, *in* Gay, C., ed., Historia física y política de Chile: Paris, Imprimavia Maulde y Renau, 499 p.

Kennedy, W. J., 1984, Ammonite faunas and the "standard zones" of the Cenomanian to Maastrichtian Stages in their type areas, with some proposals for the definition of the stage boundaries by ammonites: Geological Society Denmark Bulletin, v. 33, p. 147–161.

Kennedy, W. J., 1986a, The ammonite fauna of the Calcaire à Baculites (upper Maastrichtian) of the Cotentin Peninsula (Manche, France): Palaeontology, v. 29, p. 25–83.

Kennedy, W. J., 1986b, Campanian and Maastrichtian ammonites from northern Aquitaine, France: Palaentological Association of London Special Papers in Palaeontology 36, 145 p.

Kennedy, W. J., 1987, The ammonite fauna of the type Maastrichtian with the revision of Ammonites colligatus Binkhorst, 1861: Institut Royal des Sciences Naturelles de Belgique Bulletin, v. 56, p. 151–267.

Kennedy, W. J., 1989, Thoughts on the evolution and extinction of Cretaceous ammonites: Geologists Association Proceedings, v. 100, p. 251–279.

Kennedy, W. J., and Hancock, J. M., 1993, Upper Maastrichtian ammonites from the Marnes de Nay between Gan and Rébénacq (Pyrénées Atlantiques), France: Geobios, v. 26, p. 575–594.

Kennedy, W. J., and Henderson, R. A., 1992a, Non-heteromorph ammonites from the Upper Maastrichtian of Pondicherry, Southern India: Palaeontology, v. 35, p. 381–442.

Kennedy, W. J., and Henderson, R. A., 1992b, Heteromorph ammonites from the Upper Maastrichtian of Pondicherry, Southern India: Palaeontology, v. 35, p. 693–731.

Kennedy, W. J., Bilotte, M., Lepicard, B., and Segura, F., 1986, Upper Campanian and Maastrichtian ammonites from the Petites-Pyrenees, southern France: Eclogae Geologicae Helvetiae, v. 79, p. 1001–1037.

Lambrecht, K., 1929, *Neogeornis wetzeli* n. gen. n. sp., der erste Kreidevogel der südlichen Hemisphäre: Paläontologische Zeitschrift, v. 11, p. 121–129.

Macellari, C. E., 1986, Late Campanian–Maastrichtian ammonite fauna from Seymour Island (Antarctic Peninsula): Paleontological Society Memoir 18, 55 p.

Macellari, C. E., 1988a, Late Cretaceous Kossmaticeratidae (Ammonoidea) from the Magellanes Basin, Chile: Journal of Paleontology, v. 62, p. 889–905.

Macellari, C. E., 1988b, Stratigraphy, sedimentology and paleoecology of Upper Cretaceous/Paleocene shelf-deltaic sediments of Seymour Island, *in* Feldmann, R. M., and Woodburne, M. O., eds., Geology and paleontology of Seymour Island, Antarctic Peninsula: Geological Society of America Memoir 169, p. 25–53.

MacLeod, N., and Keller, G., 1991a, Hiatus distribution and mass extinctions at the Cretaceous/Tertiary boundary: Geology, v. 19, p. 497–501.

MacLeod, N. and Keller, G., 1991b, How complete are Cretaceous/Tertiary boundary sections: A chronostratigraphic estimate based on graphic correlation: Geological Society of America Bulletin, v. 103, p. 1439–1457.

Maeda, S., Fuller, R. C., Corvalan Diaz, J., Tazke, H., and Kawabe, T., 1981, On Pacitrigonia hanetiana from the Quiriquina Formation, Quiriquina Island, Chile: Chiba, Japan, Chiba University, Paleontological Studies on the Andes 2, 72 p.

Perch-Nielsen, K., 1985, Mesozoic calcareous nannofossils, *in* Bolli, H., Saunders, J., and Perch-Nielsen, K., eds., Plankton stratigraphy: London, Cambridge University Press, p. 327–426.

Perez, E. d'A., and Reyes, B. R., 1978, Las Trigonias del Cretácico superior de Chile y su valor crono-estratigrafico: Intituto de Investigaciones Geológicas de Chile, Boletin, v. 34, 67 p.
Perez, E. d'A., and Reyes, B. R., 1980, *Buchotrigonia (B.) topocalmensis* sp. nov. (Trigoniidae) del Cretácico superior de Chile: Revista Geológica de Chile, v. 9, p. 37–55.
Philippi, R. A., 1887, Die tertiären und quartären Versteinerungen Chiles: Leipzig, Brockhaus, 266 p.
Seilacher, A., 1984, Bathymetrie von Spurenfossilien, *in* Luterbacher, H. P., ed., Paläobathymetrie: Paläontologische Kursbücher, v. 2, p. 104–123.
Steinmann, G., 1895, Das Auftreten und Alter der Quiriquina Schichten: Neues Jahrbuch für Mineralogie, Geologie und Paläontologie, Abhandlungen 1895, p. 1–31.
Steinmann, G., Deeke, W., and Möricke, W., 1895, Das Alter und die Fauna der Quiriquina Schichten in Chile: Neues Jahrbuch für Mineralogie, Geologie und Paläontologie, Abhandlungen 1895, p. 32–118.
Stinnesbeck, W., 1986, Zu den faunistischen und palökologischen Verhältnissen in der Quiriquina Formation (Maastrichtium) Zentral-Chiles: Palaeontographica, Abt. A, v. 194, p. 99–237.
van der Tuuk, L. A., 1987, *Scaphitidae* (Ammonoidea) from the Upper Cretaceous of Limburg, the Netherlands: Paläontologische Zeitschrift, v. 61, p. 57–79.
Ward, P. D., 1990, A review of Maastrichtian ammonite ranges, *in* Sharpton, V. L., and Ward, P. D., eds., Global catastrophes in Earth history; an interdisciplinary conference on impacts, volcanism, and mass mortality: Geological Society of America Special Paper 247, p. 519–530.
Ward, P. D., and Kennedy, W. J., 1993, Maastrichtian ammonites from the Biscay region (France, Spain): Paleontological Society Memoir 34, p. 1–58.
Wetzel, W., 1930, Die Quiriquina Schichten als Sediment und paläontologisches Archiv: Palaeontographica, v. 73, p. 49–101.
Wetzel, W., 1960, Nachtrag zum Fossilarchiv der Quiriquina Schichten: Neues Jahrbuch für Geologie und Paläontologie Monatshefte, 1960, p. 439–446.
Wilckens, O., 1904, Revision der Fauna der Quiriquina Schichten: Neues Jahrbuch für Mineralogie, Geologie und Paläontologie, Beilage Band v. 18, p. 181–284.
Zinsmeister, W. J., Feldmann, R. M., Woodburne, M. O., and Elliot, D. H., 1989, Latest Cretaceous/earliest Tertiary transition on Seymour Island, Antarctica: Journal of Paleontology, v. 63, p. 731–738.

12

Late Cretaceous Faunal Changes in the High Southern Latitudes: A Harbinger of Global Biotic Catastrophe?

William J. Zinsmeister, *Department of Earth and Atmospheric Sciences, Purdue University, West Lafayette, Indiana,* and Rodney M. Feldmann, *Department of Geology, Kent State University, Kent, Ohio*

INTRODUCTION

Considerable hyperbole has been generated during the past decade about the terminal Cretaceous extinction event. With the publication of the impact hypothesis (Alvarez and others, 1980), a vocal segment of the scientific community and the media has accepted as factual the impact scenario as the cause for the global extinction at the end of the Cretaceous, and for the extinction of the dinosaurs in particular. Because of the high-profile nature of the Cretaceous-Tertiary (K-T) extinction, a number of investigators have joined the debate and have purportedly proved spectacular scenarios based on tenuous assumptions with little and, sometimes, no supporting data (Wolbach and others, 1988; Hsü, 1986; D'Hondt, 1994). Unfortunately, the scientific discussion has become polarized and has led to the rise of what might be termed "scientific McCarthyism" with many within the impact community questioning the scientific abilities of those who have raised questions concerning the impact hypothesis. Within the media reporter bias in favor of the impact hypothesis has exacerbated this polarization (see Kerr, 1994; Keller, 1994; N. MacLeod, 1994).

It is evident that a major biotic crisis occurred at the end of the Cretaceous, but for many in the geologic community, the cause or causes of the crisis are far from certain. If an impact event is shown to have occurred, it proves only that there was an impact at the end of the Cretaceous. Contrary to what many have said, the presence of an impact crater does not provide the final piece of evidence to explain the demise of the dinosaurs or the wholesale loss of life at the end of the Cretaceous. Understanding the nature and causes of the extinction will come only from the thorough examination of the biotic record during the latest Cretaceous.

If the K-T extinction event resulted from a cataclysmic impact, the fossil record provides little direct positive evidence to support this mode of extinction. Virtually all of the paleontologic arguments used by proponents of the impact hypothesis are based on negative evidence—species disappearance. The scientific method demands that a hypothesis be both testable and tested with positive evidence. Thus far, impact proponents have provided little direct evidence to support an impact-driven K-T extinction.

Accordingly, and in a purely rhetorical sense, one must ask the question, Where is the layer of burnt and twisted dinosaurs bones? Not one "burnt stump" has been offered to support the global forest fires that are alleged to have swept (Wolbach and others, 1988) across the Earth, consuming the last of the dinosaurs. For that matter, if it is possible to preserve a faint trace of iridium enrichment in marine rocks, why is there no associated extirpation layer of ammonite shells and marine reptile bones? Although some sedimentary structures in the boundary sequence in northeastern Mexico (Bohor and Betterton, 1994) are alleged to be evidence for impact-generated tsunamis following the impact, it is surprising that such a widespread impact-induced catastrophe produced no geomorphic signature, such as increased erosion of the denuded world or corresponding deposition signal in the marine record. If the impact event was as catastrophic and pervasive as portrayed by the impact community, where are the bodies?

Another paleontologic observation that demands to be addressed, but has been largely ignored, is the well-documented decline in biotic diversity during the late Campanian and early Maastrichtian (Clemens, 1982; Archibald and Bryant, 1990; Kauffman, 1984). The extinction event has been portrayed by some as though the world simply had a very bad day, ignoring the long decline in diversity leading up to the extinction event. If biotic diversity had remained unchanged until the end of the Cretaceous, the impact scenario would be far more appealing. However, the Earth's biotic diversity did decline throughout the Late Cretaceous.

Another important phenomenon that has been largely ignored is the discrepancy in the timing and intensity of extinctions between the low and high latitudes (Zinsmeister, 1994; Sweet and others, 1993; Johnson, 1993; Johnson and Greenwood, 1993). To suggest that the decline in biodiversity was distinct and unrelated to the K-T extinction verges on special pleading. It is more likely that the Late Cretaceous diversity decline is a harbinger of impending doom and approaching global catastrophe.

If we are to resolve the cause of one of nature's great catastrophes, we must integrate all available physical and biological data. The purpose of this paper is to examine the Late Cretaceous—earliest Tertiary shallow-marine record from the Antarctic Peninsula and compare it with the temporal record from the temperate and tropical latitudes.

MASS EXTINCTIONS

Mass extinctions have played an integral role in the evolution of life. The fossil record during the Phanerozoic shows five major extinction events (the so-called Big Five—Late Ordovician, Late Devonian, Permian-Triassic, Late Triassic, and the Cretaceous-Tertiary boundary) in which the biosphere was traumatized severely (Raup and Sepkoski, 1984). Each of these mass-extinction events had a profound effect on the biosphere. Initially, global diversity declined, but each extinction episode was followed quickly by a burst of evolutionary diversification. These postextinction radiations resulted in fundamental changes in the Earth's biosphere. The rise of mammals following the terminal Cretaceous event is a classic example of how mass-extinction events have altered the biosphere. Examination of the major Phanerozoic extinctions fails to reveal a single common causative factor for all five events, thereby suggesting that each is unique. These events may have been caused by one dominant mechanism (Late Ordovician—glaciation; Stanley, 1984, 1988; Sheehan, 1973) or by the conjunction of several physical factors (Permian-Triassic; Erwin, 1993, 1994). Whereas most of the big five seem to be driven by terrestrial geologic processes, extraterrestrial processes must also be considered as potential extinction mechanism for the last of the "Big Five" Phanerozoic mass extinctions at the end of the Cretaceous (Alvarez and others, 1980).

The Phenomenon of Extinction

With any discussion of the role of catastrophic extinctions in the history of life, it is necessary to examine first the requisite phenomena of extinction; that is, the various types of extinction events, plausible underlying causes, and the rates at which these events take place. Simpson (1983) pointed out that extinction of species is neither simple nor straightforward. Species evolve and persist for a period of time and eventually become extinct. The abundance and distribution of each species depends on the environmental history of the region it occupies. As a consequence, each species has its own unique evolutionary history.

Types of Extinction

Simpson (1944, 1983) recognized two types of extinction, pseudoextinction and terminal extinction (Fig. 1). Pseudoextinction is the consequence of evolutionary change in which morphologic changes are so great that the descendent no longer resembles the ancestral form. The ancestral species has become extinct, but the evolutionary lineage continues. Terminal extinction occurs when highly specialized forms cannot adapt to new conditions, and thus become extinct, terminating the evolutionary lineage (Simpson, 1983; Stanley, 1987). Few authors in the K-T extinction debate have attempted to distinguish between these two very different types of extinction events.

Rates of Extinction

The rate of extinction has been divided into two categories, background extinction and mass extinction (Fig. 2). During the Phanerozoic there have been

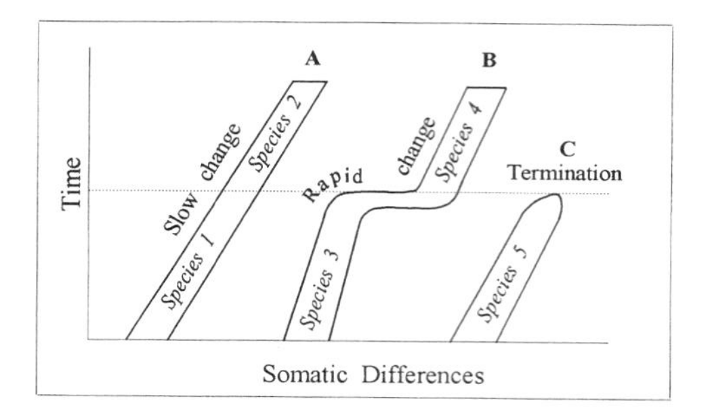

Figure 1. Types of species extinctions. A and B = pseudoextinction; C = terminal extinction.

extended periods when some taxa became extinct while others appeared in approximately equal numbers. This has been termed background extinction, where the number of species becoming extinct and appearance of new species are approximately equal. Mass extinction, however, encompasses periods when the numbers of taxa becoming extinct far exceed the numbers of new species appearing, resulting in a reduction of diversity in the biosphere. The phenomenon of mass extinction also may be subdivided into two categories (Fig, 2), depending upon the rate of extinction. Accelerated mass extinction is herein defined as occurring over a geologically short period of time such as tens of thousands of years, hundreds of thousands of years, or as long as several million years. Instantaneous catastrophic mass extinction is herein defined as virtually instantaneous, occurring over a period of only days, weeks, months, or, at most, a few years. Because mass extinction can be either geologically short and gradual or catastrophic and instantaneous, the problem lies in our ability to discriminate between these geologically short term events. The farther we go back in time, the greater the difficulty in distinguishing small increments of time. Because of this inability, it becomes extremely difficult to distinguish between accelerated and

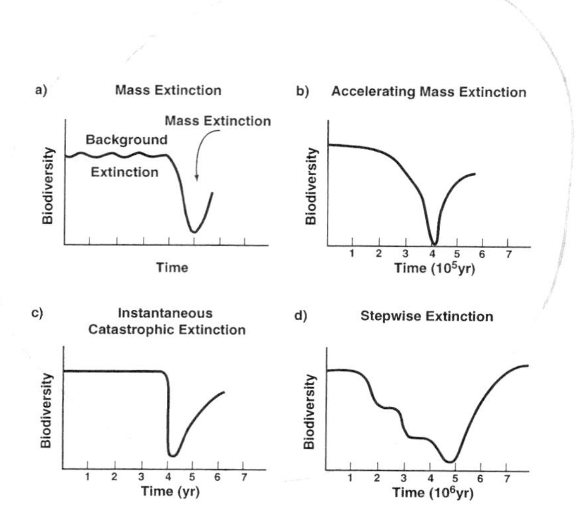

Figure 2. Types of major biospheric extinction events.

catastrophic extinction events and has led some to view all mass-extinction events in the fossil record as instantaneous catastrophic phenomena.

A third type of extinction phenomenon, stepwise extinction (Fig. 2), was proposed by Kauffman (1988) for the episodic extinction events resulting in a stepwise decline in biodiversity during the Late Cretaceous. Kauffman suggested that the stepwise extinction pattern was produced by a series of asteroid or cometary impacts.

LATE CRETACEOUS-EARLY TERTIARY SHALLOW-MARINE RECORD OF SEYMOUR ISLAND AND THE JAMES ROSS BASIN

The James Ross basin region of the northeastern tip of the Antarctic Peninsula contains a remarkable middle to late Mesozoic and early Cenozoic record extending from the Late Jurassic to the Eocene (Fig. 3). Upper Jurassic and Lower Cretaceous rocks crop out along Prince Gustav Channel and on the west side of James Ross Island, and Upper Cretaceous rocks occur to the east at a number of localities along the east and northeast coasts of James Ross Island, Cape Lamb, Vega islands, and on Snow Hill and Seymour islands. The lower Tertiary is restricted to Seymour and Cockburn islands. Although no single continuous section exists from the Late Jurassic through the Cretaceous into the Tertiary in the James Ross region, field work during the past two decades has established a fairly detailed composite section through the late Mesozoic and ear-

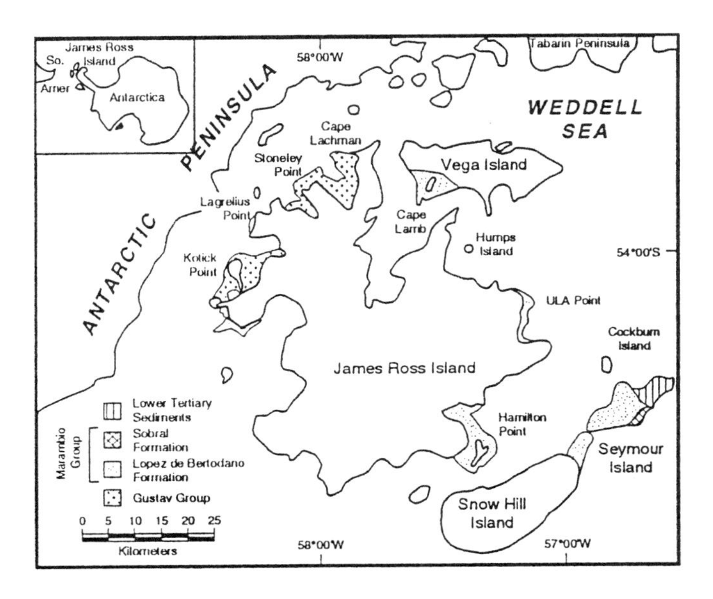

Figure 3. Location map of James Ross Island region, Antarctica.

liest Tertiary sequence for the basin. Most of the work done on James Ross and Vega islands has been accomplished by the British Antarctic Survey (Bibby, 1966; Crame and others, 1991; Pirrie and others, 1991) and the Instituto Antartico Argentino (Rinaldi and others, 1978; Rinaldi, 1982; Olivero, 1984; Olivero and others, 1986), while the U.S.-supported field programs have focused on Seymour Island (see Feldmann and Woodburne, 1988, for an extended list of stratigraphic studies on Seymour Island).

The field activities of these workers has resulted in the establishment of a detailed biostratigraphic record of the shallow-water marine sequence from the Late Jurassic into the Tertiary. Although questions of stratigraphic nomenclature still remain, examination of the stratigraphic data reveals a close correspondence between the three organizations. Uncertainties remain concerning the exact stratigraphic relations between the Santa Marta Formation and the lowest part of the López de Bertodano Formation. Because of the large number of field programs in the James Ross basin, our knowledge of the Late Cretaceous–early Tertiary biota has reached a level where we can draw a number of conclusions about the biotic history of Antarctica during this important interval of Earth's history.

The geologic history of the James Ross Island region can be traced back to the Late Jurassic, when the Antarctic Peninsula was a volcanic arc of low relief and deposition was dominated by quiet deep-water anoxic mudstones and tuffs of the Nordenskjld Formation (Farquharson, 1983). During the Early Cretaceous, tectonic and volcanic activity increased during the uplift of the Antarctic Peninsula. Sediments deposited in the James Ross Island basin varied from coarse conglomerates and sandstones of Gustav group (Crame, 1983; Pankhurst, 1982; Elliot, 1988). Rates of uplift diminished during the Late Cretaceous and erosion began to reduce the high relief that characterized the Antarctic Peninsula during the Early Cretaceous. Although tectonism decreased along the northern end of the peninsula, volcanic activity continued and even increased in the Maastrichtian (Elliot, 1988). Late Cretaceous deposition in the James Ross basin was dominated by silty sandstones and siltstones with local conglomerates.

The present-day geography of the James Ross basin provides insight into location of James Ross, Vega, Snow Hill, and Seymour islands during the early part of the Late Cretaceous. During the Campanian, sediments on James Ross and Vega islands were deposited ~20 km from the nearest land while the sequence on Seymour Island was deposited as much as 80 to 100 km offshore. The discovery of rafted dinosaur skeletons in the Santa Marta Formation (Olivero and others, 1991) and the lower part of the López de Bertodano Formation on Vega Island (Hooker and others, 1991) supports their proximal location near land during the Campanian and Maastrichtian. During the latter part of the Cretaceous the western part of the James Ross basin shallowed, and by the late Maastrichtian the strand line shifted to the east, to the vicinity of Vega Island. Eastward progradation of the strand line reduced the distance of Seymour Island from the nearest land to less than 40 km. Although the James Ross basin was shallowing during the Late Cretaceous, the absence of shallow-water sedimentary structures suggests that deposition of latest Maastrichtian and earliest Tertiary strata on Seymour Island took place below wave base (Macellari, 1988; Harwood, 1988; Zinsmeister and others, 1989).

Campanian-Maastrichtian Faunal Changes

Climatic conditions in the James Ross region during the Campanian and Maastrichtian were characterized by temperate to cool-temperate conditions (Askin, 1988, 1989, 1990; Dettmann, 1989) with pronounced seasonal fluctuations (Francis, 1991). Recent stable isotopic data (Pirrie and Marshall, 1990) have suggested that nearshore oceanic temperatures declined from 13.6 °C in the Santonian-Campanian to 11.7 °C in the late Campanian–Maastrichtian. The cooling and intensification of seasonal extremes reflect the global cooling trends discussed by Stanley (1984, 1988).

Several cosmopolitan groups of molluscs in the James Ross Island region exhibit changes in either decline in diversity or disappearance that preceded similar trends in the middle and low latitudes. It is our belief that these faunal changes were responses to a global cooling first felt in the high latitudes.

Inoceramidae

The bivalve family Inoceramidae was a widespread element of the marine fauna from the polar to tropical latitudes during the late Mesozoic. In a review of the Inoceramidae, Dhondt (1992) summarized the diversity and global distribution of the family through the Cretaceous. Dhondt noted that inoceramid diversity declined throughout the Late Cretaceous. By the middle Maastrichtian this family was represented by only a few species surviving in the middle and low latitudes. Kauffman (1988) also discussed this pattern of middle Maastrichtian extinction.

Potential causes for the decline during the latest Cretaceous have been attributed to a combination of habitat loss, reorganization of ocean circulation, and the oxygenation and cooling of bottom waters (K. G. MacLeod, 1994; MacLeod and Ward, 1990; MacLeod and Orr, 1993). Crame (1983), in a review of the biostratigraphic distribution of Antarctic Inoceramidae, noted that they composed an important component of the marine fauna from the Berriasian to the Campanian (Table 1). The youngest reported inoceramids are two large undescribed species

TABLE 1
Stratigraphic distribution of Principal Antarctic Inoceramids
(Came, 1983; Medina and Buatois, 1992)

SPECIES	OCCURRENCE	PROBABLE AGE RANGE
Giant *Inoceramus* sp. 1	James Ross Island	Campanian
Giant *Inoceramus* sp. 2	James Ross Island	Campanian
I. neocaledonicus grp.	James Ross Island	Coniacian/Campanian
I. madagascariensis	James Ross Island	Late Turonian–Early Coniacian
I. lamarcki grp.	Cape Longing	Early–Middle Turonian
Anopaea cf. *A. mandibula*	Alexander Island	Albian
I. angilicus grp.	Alexander Island	Albian
I. concentricus grp.	James Ross and Alexander Is.	Middle–Late Albian
Anopaea sp.	James Ross Island	Aptian–Albian
I. neocomiensis grp.	James Ross and Alexander Is.	Barremian/Aptian
I. heretropterus grp.	Annenkov Island	Haut.–Albian?
I. cf. *I. anomiaeformis*	Annenkov Island	Haut.–Berremian
Anopaea trapezoidalis	Alexander Island	Neocomian
I. ovatus grp.	Alexander Island	Berriasian
Retroceramus everesti	Alexander Island	Berriasian

from the late Campanian of Rabot Point (Crame, 1985; Marenssi and others, 1992). At present, no inoceramids have been found from either the late Campanian or the Maastrichtian on Seymour Island.

Although it is possible that inoceramids might still be found in the Campanian part of the section on Seymour Island, it is unlikely that any will be found in the Maastrichtian part of the section. Their absence in the Maastrichtian is based on the fact that this part of the Upper Cretaceous on the island, which is richly fossiliferous, has been studied intensively for nearly a decade and no inoceramids have been encountered. The latest Campanian–early Maastrichtian sequence on Seymour was located well offshore and in a water depth similar to that of the Late Campanian inoceramid-bearing beds on James Ross Island. Their absence in the López de Bertodano Formation on Seymour Island is not due to local environmental factors, but to the disappearance of the Inoceramidae in the high southern latitudes sometime during the early Campanian. A similar pattern is seen in the Antarctic with the last belemnites disappearing during the latest Campanian or earliest Maastrichtian (Doyle and Zinsmeister, 1988; Crame, 1992).

Ammonites

The Cretaceous sediments of the James Ross region are richly endowed with a diverse and well-preserved ammonite assemblage. Tables 2 and 3 list the stratigraphic distribution of ammonite species and families from the Santonian through the Campanian to the end of the Maastrichtian (Spath, 1953; Howarth, 1966; Macellari, 1986, Olivero, 1992). The richest ammonite assemblages occur in the Santa Marta Formation on the east side of James Ross Island. Olivero (1992) divided the Santa Marta Formation into three lithologic members (Alpha—Santonian, Beta—early to late Campanian, and Gamma—latest Campanian to earliest Maastrichtian). Stratigraphically overlying the Santa Marta Formation on Vega, Snow Hill, and Seymour islands is the López de Bertodano Formation, which ranges in age from late Campanian through Danian (Rinaldi, 1982; Macellari, 1984b). The López de Bertodano Formation also crops out at a number of localities along the eastern coast of James Ross Island. The diversity of ammonite species increased from 14 species representing 8 families in the Santonian Alpha member of the Santa Marta Formation to a maximum of 45 species in 11 families in the early to late Campanian Beta member. In the latest Campanian and earliest Maastrichtian, the diversity drops to 26 species in 7 families. In the Maastrichtian part of the López de Bertodano Formation, ammonite diversity drops to 14 species in 6 families (Table 4). Although the diversity of the ammonite assemblage dropped dramatically between the lower and upper Campanian of the Santa Marta Formation and the Maastrichtian part of the López de Bertodano Formation, the total number of species in the Maastrichtian is slightly misleading. Whereas the total number of ammonites taxa reported from the Maastrichtian is 12, the maximum number of taxa at any locality is 7. This number decreases as the local K-T boundary is approached.

Although the stratigraphic relations of the upper Campanian part of the López de Bertodano Formation on Vega and Snow Hill islands with that on Seymour Island are still not understood, two distinct ammonite assemblages are recognizable (Olivero, 1992; Pirrie and others, 1991). The lower assemblage is dominat-

TABLE 2
Stratigraphic distribution of Campanian/Maastrichtian ammonites in the James Ross Basin, Antarctica, based on Olivero, 1992, Olivero, personal communication 1994; Howarth, 1966; Macellari 1986; Spath, 1953 (cont'd. page 312)

James Ross Island	
SANTONIAN	E. CAMPANIAN TO L. CAMPANIAN
Lower Santa Marta Formation,	Middle Santa Marta Formation,
Alpha member	Beta member
Family Phylloceratidae	Family Phylloceratidae
Neophylloceras sp.	*Neophylloceras* sp.
Phylloceras sp.	*Phylloceras* sp.
Family Tetragonitidae	Family Tetragonitidae
Tetragonites sp.	*Tetragonites* sp.
Family Guadrycertidae	*Pseudophylites* sp.
Guadryceras sp.	Family Guadrycertidae
Anaguadryceras sp.	*Guadryceras* sp.
Family Scaphitidae	*Anaguadryceras* sp.
Scaphites sp.	*Vertebrites* sp.
Hoploschities sp. 1	*Zelandites* sp.
Yeozoites sp. 1	Family Scaphitidae
Yeozoites sp. 2	*Hoploschities* sp. 2
Family Baculitidae	*Yeozoites* sp. 3
B. aff. *B. kirki*	*Yeozoites* sp. 4
Family Nostoceratidae	Family Baculitidae
Eubostrychoceras aff. *E. elongatum*	*Baculites bailyi*
Family Diplomoceratidae	*B. rectus*
Polyptychoceras aff. *P. obstrictum*	*B.* aff. *B. kirki*
"Scalarites" sp.	*B.* aff. *B. subanceps pacificus*
Family Pachydiscidae	*B.* sp. 1
Anapachydiscus sp. 2	*B.* sp. 2
	Family Nostoceratidae
	Eubostrychoceras medinai
	Ainoceras zinsmeisteri
	Nostoceratidae spp.
	Parasolenoceras sp.
	Family Diplomoceratidae
	Polyptychoceras sp.
	Ryugasella antarctica
	R. sp.
	Family Desmoceratidae
	Desmites cf. *D. hetonaiensis*
	Hauericeras (Gardeniceras) sp.
	Kitchenites cf. *K, angolaensis*
	Oiophyllites decipiens
	Oiophyllites sp.
	Family Kossmaticeratidae
	Natalites rossensis
	N. sp. 1
	N. sp. 2
	Caledonites vilidus
	Grossouvrites occultus
	Neograhamites aff. *N. morenoi*
	Neograhamites taylori
	Neograhamites sp. 1
	Maorites sp.
	"Karapadites" aff. *"K. centinelaensis"*
	"Karapadites" sp.
	Family Pachydiscidae
	Anapachydiscus constrictus
	Anapachydiscus sp. 2
	Anapachydiscus sp. 3
	Eupachydiscus paucituberculatus
	Family Placenticeratidae
	Placenticeras sp.
	Metaplacenticeras aff. *M. subtillistriatum*
	Hoplitoplacenticeras? sp.

TABLE 2 (CONTINUED)

	Snow Hill, Vega, and Seymour Islands
LATEST CAMPANIAN/EARLIEST MAASTRICHTIAN Upper Santa Marta Formation, Gamma member Family Phylloceratidae *Phylloceraidae* sp. Family Tetragonitidae *Tetragonites* sp. Family Guadrycertidae *Guadryceras* sp. *Anaguadryceras* sp. Family Diplomoceratidae *Polyptychoceras* sp. 2 "*Scalarites*" sp. *Diplomoceras lambi* *Astrepticeras?* sp. Family Desmoceratidae *Hauericeras* sp. *Mesopuzosia* sp. Family Kossmaticeratidae *Neograhamites* aff. *N. kiliani* *N.* sp. 2 *Gunnarites* spp. Family Pachydiscidae *Anapachydiscus* spp. *constictus* *Eupachydiscus* sp.	LATEST CAMPANIAN Lower Lopez de Bertodano Formation Family Kossmaticeratidae *Gunnarites antarcticus* *G. gunnari* *G. bhavaniformis* *G. kalika* *Maorites tuberculatus* *M.* sp. *Jacobites anderssoni* *J. crofti* *Grossuvrites gemmatus* *Neograhamites kiliani* Family Pachydiscidae *?Eupachydiscus* sp. *Puzosia* sp. Family Desmoceratidae *Desmites loryi.* *Hauericeras* sp. Family Diplomoceratidae *Diplomoceras lambi* Family Tetragonitidae *Tetragonites* cf. *epigonus* *T. (Saghalinites) cala* *Pseudophyllites* sp. Family Phylloceratidae *Neophylloceras ramosum* *Epiphylloceras surya* Family Guadrycertidae *Gaudryceras* sp. MAASTRICHTIAN Upper Lopez de Bertodano Formation, Family Kossmaticeratidae *Maorites densicostatus* *M. seymourianus* *M. weddellensis* *Grossourvites gemmatus* Family Pachydiscidae *Pachydiscus ootacodensis* *P. ultimus* *P. riccardi* Family Diplomoceratidae *Diplomoceras lambi* *D. maximus* Family Tetragonitidae *Pseudophylites loryi* Family Guadrycertidae *Zelandites varuna* *Anagaudryceras seymourlensis* Family Desmoceratidae *Kitchenites darwini* *K. laurea*

TABLE 3
Stratigraphic distribution of Santonian through Maastrichtian ammonite families on James Ross, Vega, Snow Hill, and Seymour islands (Olivero, 1992, Olivero, personal communication 4/03/94; Howarth, 1966; Macellari, 1986; Spath, 1953)

LATE SANTONIAN Lower Santa Marta Formation	LATEST CAMPANIAN/EARLIEST MAASTRICHTIAN Upper Santa Marta and Lower López de Bertodano Formations
Phylloceratidae Tetragonitidae Guadrycertidae Scaphitidae Baculitideae Nostoceratidae Diplomoceratidae Pachydiscidae	Phylloceratidae Tetragonitidae Guadrycertidae Diplomoceratidae Desmoceratidae Kossmaticeratidae Pachydiscidae
E. CAMPANIAN TO L. CAMPANIAN Middle Santa Marta Formation	MAASTRICHTIAN Upper López de Bertodano Formation
Phylloceratidae Tetragonitidae Guadrycertidae Scaphitidae Baculitidae Nostoceratidae Diplomoceratidae Desmoceratidae Kossmaticeratidae Pachydiscidae Placenticeratidae .	Kossmaticeratidae Pachydiscidae Diplomoceratidae Tetragonitidae Guadrycertidae Desmoceratidae

TABLE 4
Number of ammonite species and families from the Late Cretaceous of the James Ross Island region (Olivero, 1992; Olivero, personal communication 4/03/94; Howarth, 1966; Macellari, 1986; Spath, 1953)

AGE	NO. SPECIES	NO. FAMILIES
Maastrichtian	14	6
Latest Campanian	21	7
Early to Late Campanian	45	11
Santonian	14	8

ed by the kossmaticeratids *Gunnarites* and *Jacobites*, and the upper assemblage is characterized by *Maorites*, *Grossouvrites*, and pachydiscids. The *Gunnarites-Jacobites* assemblage is present in the lower part of the section on Vega and Snow Hill islands, but is apparently absent in the Campanian on Seymour Island. The presence of *Gunnarites bhavaniformis* near Cape Bodman suggests that the assemblage may be present in the lowest levels of the López de Bertodano Formation of Seymour Island. The upper assemblage referred to as the *Maorites-Grossouvrites* assemblage is restricted to the Maastrichtian part of the López de Bertodano Formation on Vega and Seymour islands. A total of 13 ammonite species together with two morphotypes of *Maorites densicostatus* on Seymour Island were described from the López de Bertodano Formation by Macellari (1986, 1988). An additional species of *Diplomoceras* was described by Olivero and Zinsmeister (1989), bringing the total ammonite fauna to 14 described species. Although Macellari (1986, 1988) provided the first detailed biostratigraphic range data, subsequent stratigraphic study and collecting have added new data to his original data set.

The most perplexing aspect of these new data is the apparent occurrence of ammonites above the K-T boundary. Five species of ammonites (Fig. 4) have been found above the iridium horizon. Unfortunately, none have been found in place, and Askin (1988) reported only Maastrichtian palynomorphs in these ammonite-bearing concretions. The presence of Maastrichtian palynomorphs would seem to indicate that these ammonites were reworked into the Tertiary sediments. However, there is no field evidence for early Tertiary reworking of Cretaceous concretions into the Danian part of the López de Bertodano Formation. There is also no convincing evidence for surficial reworking of the

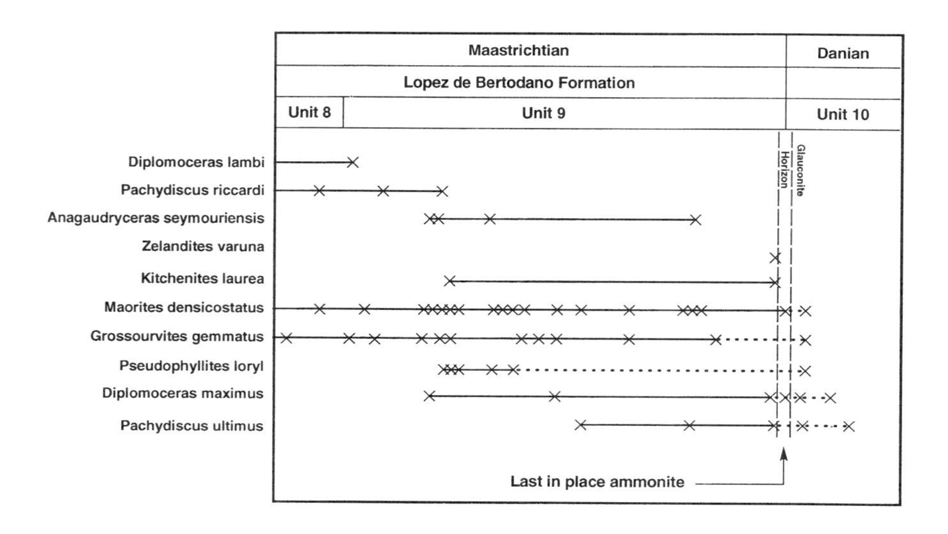

Figure 4. Late Maastrichtian biostratigraphic distribution of ammonite species in the López de Bertodano Formation. Xs indicate individual occurrences of each ammonite. All ammonite occurrences above the glauconitic sand were found on the surface and may be reworked (see text). Range bars for each taxon that has been found above the glauconitic sand have been dashed from their last unambiguously in place occurrence in unit 9. Arrow indicates the last location of demonstrably in place ammonites in the López de Bertadano Formation (modified from Macellari, 1986; Zinsmeister and others, 1989, with addition of previously unpublished data).

concretions from the Cretaceous outcrops to the site of their discovery in the Tertiary. Zinsmeister and others (1989) plotted the stratigraphic distribution of all large-bodied invertebrates and noted the questionable occurrence of "Tertiary" ammonites, and distinguished them from the last "in place ammonite." These "Tertiary" ammonites distort the changes in ammonite diversity at the end of the Cretaceous. If we assume that they are in place, the maximum diversity of the ammonite assemblage (7 species) was at the very end of the Maastrichtian, and their complete disappearance came shortly after the K-T boundary. If, however, we accept that they are reworked, then the number of ammonites had dropped to 4 species at the end of the Cretaceous, the maximum ammonite diversity (7 species) occurring ~100 m below the K-T boundary.

Ward (1990a, 1990b) presented Macellari's data along with ammonite data sets from the European Chalk regions and the Bay of Biscay. Macellari's original 1986 range data were included in Ward's discussion: unfortunately, they were plotted incorrectly. Although Ward (1990a) quoted Macellari (1986) as saying that the ammonites decline to four species at the boundary, he confused the presentation of Macellari's ammonite diversity trend on Seymour Island by incorrectly plotting the data in graphical comparison (Ward, 1990a, Fig. 8) with the European Chalk regions and Bay of Biscay. Ward's (1990a) Figure 8 shows an increase in diversity to 10 species with an abrupt truncation at the K-T boundary rather than a decline to 4 species.

Examination of diversity of ammonite families also reveals a similar decline from the latest Campanian through the Maastrichtian (Table 3). The cosmopolitan families—Baculitidae, Scaphitidae, Nostoceratidae, and Placenticeratidae, which composed 30% of the ammonite assemblage from the Alpha and Beta members—disappear from the Antarctic faunas by the latest Campanian, followed by the disappearance of the Phylloceratidae by the earliest Maastrichtian. Although these five families survive to the end of the Maastrichtian in the middle and low latitudes (Ward and Signor, 1983), their disappearances from the Antarctic faunas strongly suggest the development of climatic conditions that precluded their presence in the high southern latitudes.

Late Cretaceous–Early Tertiary Biotic Record on Seymour Island

During the austral summers of 1981, 1983–1984, and 1985, field parties supported by the United States began a comprehensive survey of Seymour Island (Zinsmeister, 1982, 1984, 1985; Feldmann and Woodburne, 1988). The López de Bertodano Formation, as exposed on Seymour Island, consists of about 1190 m of poorly consolidated concretionary sandy siltstones and sandstones. Macellari (1984a, 1988) divided the formation into 10 lithologic units. On the basis of these initial studies, it was recognized that the López de Bertodano Formation ranged in age from late Campanian through the Maastrichtian and into the Paleocene, with continuous deposition across the K-T boundary. The K-T boundary was placed initially at a thin glauconite horizon between Macellari's lithologic units 9 and 10 (1984a, 1984b), which, during the early studies, also appeared to coincide with a marked change in composition of the faunas. During the course of these studies in the early 1980s, it was recognized that the Upper Cretaceous and lower Paleocene interval of the López de Bertodano Formation contained an abundant and well-preserved macrofauna, in contrast to many other described K-T boundary sections where the preservation of megafossils is generally poor, such as the Brazos River section (Hansen and others, 1987) and Braggs, Alabama (Jones and others, 1987).

The abundance of marine life, the uniform bioturbated nature of the rocks, and the absence of wave-generated current or tidal bedding structures in units 8, 9, and 10 indicate that this part of the López de Bertodano Formation was deposited below wave base, and the sediment source was located less than 40 km to the west in the present location of the Antarctic Peninsula. The presence of volcaniclastic components in the upper part of unit 9 and in unit 10 (Macellari, 1984a, 1988) indicates the onset of volcanic activity during the latest Maastrichtian and earliest Tertiary along the Antarctic Peninsula.

K-T BOUNDARY ON SEYMOUR ISLAND

The presence of a well-preserved macrofauna from the late Maastrichtian into the Paleocene (Zinsmeister and Macellari, 1983) presented an unusual opportunity to study the history of larger invertebrates across the K-T boundary. Macellari (1986) originally placed the K-T boundary on Seymour Island at the point that he believed represented the last stratigraphic occurrence of ammonites and a number of other molluscs. His placement also coincided with a glauconitic sandy horizon. During the 1982 field season, series of 50-cm-interval samples were collected for palynological study. Samples across the glauconitic interval were sent to Frank Kyte (UCLA) for iridium analysis. He reported that one of the samples contained significant iridium enrichment (Zinsmeister and others, 1989). The occurrence of iridium provided additional support for the K-T boundary occurring in the glauconitic sand. Unfortunately, the 50 cm sampling interval prevented precise location of the iridium horizon. During the austral summer of 1989–1990, series of samples with a spacing of 10 cm were collected from a well-exposed section across the glauconitic sand about 2 km to the southeast from the 1982 section. Analysis of these samples revealed a strong iridium spike (Elliot and others, 1994). It is clear that the K-T boundary on Seymour Island as defined by the presence of iridium and faunal changes is located within the glauconitic sand.

Chronostratigraphic boundaries represent unique instants in time and are recognized by geochronologic events. The placement of the K-T boundary has been subject to a great deal of discussion and its placement depended on what criterion (fossil group) was chosen as defining the boundary. Furthermore, virtually all boundary sections reported in the literature are characterized either by a hiatus or greatly reduced rate of sedimentation (MacLeod and Keller, 1991). Jablonski (1985) coined the term "Truncation Effect" for an apparent abrupt extinction as a consequence of the truncation of species ranges by a hiatus. A range truncation effect can also be produced by greatly reduced rate of sedimentation resulting in a condensed section in which a significant interval of time is recorded in a few centimeters. In a condensed section the truncation effect on the ranges of megafossils will appear identical to those produced by the presence of a hiatus (Fig. 5).

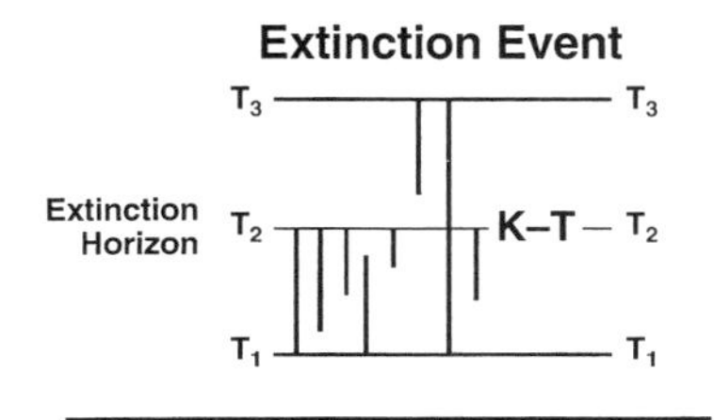

Pseudo—Extinction Events

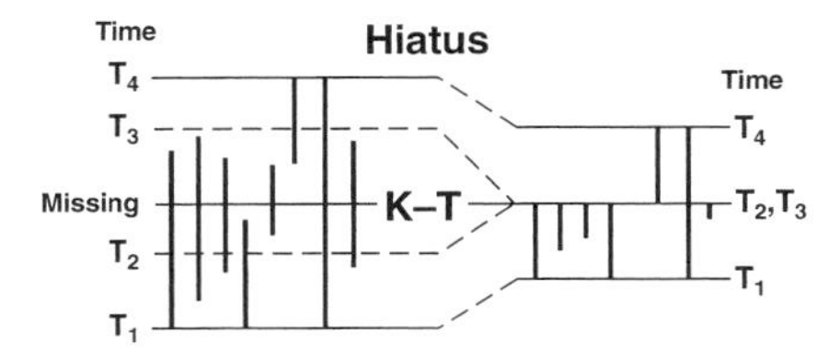

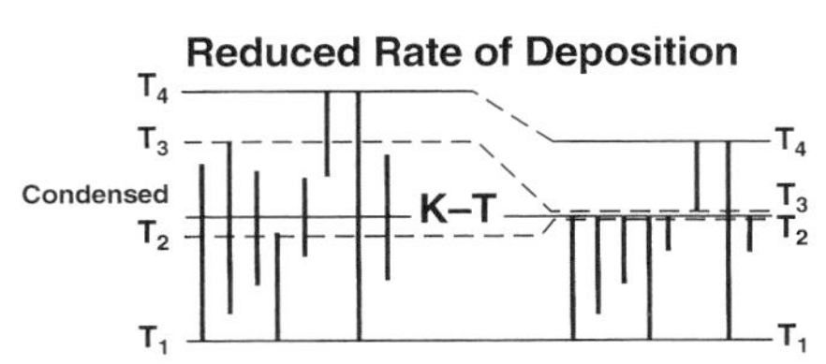

Figure 5. Pseudoextinction horizon resulting from the presence of a hiatus or interval of condensed sediment accumulation.

Seymour Island offers the first opportunity to examine the faunal transition from the latest Cretaceous to the earliest Tertiary in a relatively thick sequence of high-latitude, shallow, continuous siliciclastic shelf rocks. Although several other shallow, siliciclastic shelf facies containing the K-T boundary have been discussed in the literature (Hansen and others, 1987; Jones and others, 1987), each is characterized by sedimentological disruption at the boundary. Because of the relatively rapid rate of sediment accumulation on Seymour Island, the interval spanning the latest Cretaceous and the earliest Paleocene is represented by ~50 m of section. Zinsmeister and others (1989) referred to this sequence on Seymour Island as an expanded K-T boundary interval, in contrast to the condensed interval characteristic of other described boundary sections, such as in the deep sea (MacLeod and Keller, 1991). If the sudden disappearance of many forms of life at the end of the Cretaceous was the result of an instantaneous catastrophic event, an abrupt extirpation horizon should be present, even in an expanded sequence such as the one on Seymour Island.

Although Signor and Lipps (1982) suggested that the fossil record will tend to preserve gradual stepwise histories of disappearances as a result of the nature of the fossil occurrences, and Sadler (1981) observed that the resolution of the geologic record is not adequate to identify unequivocally an instantaneous event, it is equally true that biostratigraphic determinations continue to be useful in defining boundaries. The relevant observation from the standpoint of defining the K-T boundary is the mutual exclusion of groups that are key indicator species for the time intervals on either side of the boundary. Appearances of new taxa are as important as the loss of older lineages. It would be naive to expect that all geologic boundaries necessarily represent moments of time during which catastrophic changes occurred; however, it would be equally naive to argue that such a catastrophe would be reflected, in an expanded section, by a kind of long-term, gradual turnover of biotic elements. Such a change is illustrated within the Seymour Island K-T boundary section.

THE K-T FAUNAL TRANSITION

Biotic changes from the latest Cretaceous into the earliest Tertiary can be viewed from the disappearance of old taxa or the appearance of new taxa. The pervasive approach in invertebrate paleontology has been to focus on extinction; little regard is given to survivors or to the appearance of new taxa. Even when taxa have been recognized as surviving the terminal Cretaceous extinction event, figures have been drawn to emphasize local disappearances at the boundary (Surlyk and Johansen, 1984). Alternatively, when taxa disappear from the geologic record within the Maastrichtian, but not precisely at the boundary, their ranges are normally reported to span the entire stage (Kauffman, 1984). Recent studies by Kauffman (1984, 1988) and Ward (1990a) have illustrated how this approach has distorted the pattern of K-T extinction data (Fig. 6).

Data from Seymour Island (Zinsmeister and Macellari, 1988; Zinsmeister and others, 1989) show that the upper 200 m of units 9 and 10 are characterized by

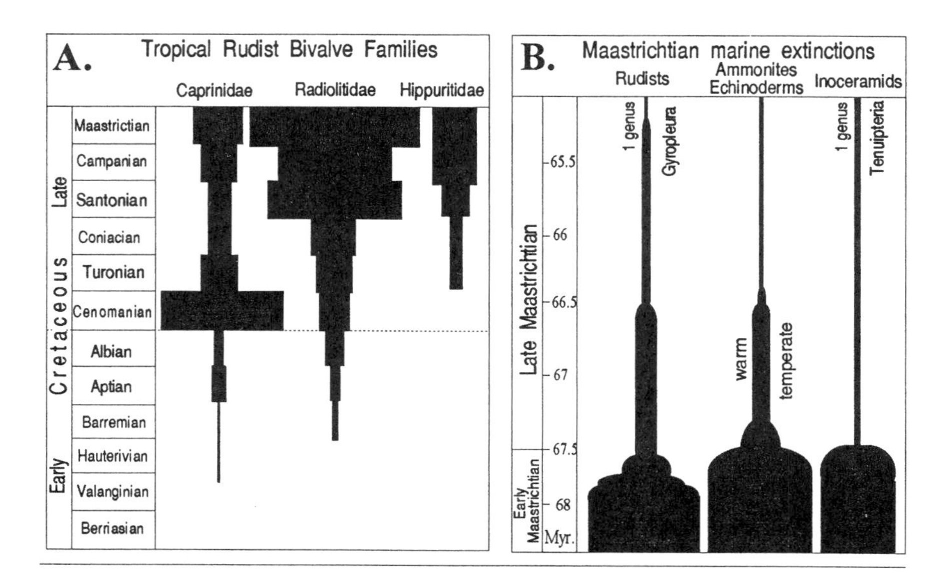

Figure 6. A: Classical mass (pseudo)extinction pattern produced by artificially extending stratigraphic ranges to stage boundaries (redrawn from Kauffman, 1984, 1989). B: Actual diversity changes during the Maastrichtian of the major molluscan groups (redrawn from Ward, 1990a).

an abundant and diverse fauna of larger invertebrates and vertebrates with carbonized wood and log debris present at several horizons. Palynomorphs are also well represented throughout this interval (Askin, 1988, this volume). The richness of the fauna for the first 100 m below the transition averages between 28 to 35 taxa of larger invertebrates and vertebrates. At 199 m, the faunal richness begins to decline and a richness minimum of 11 taxa is reached at 249 m. The low faunal richness continues for the next 40 m, at which point diversity begins to increase. From 269 m to the top of the section, total overall richness increases gradually to about 16 taxa. Eight long-ranging taxa pass across the K-T boundary completely unaffected by events at the end of the Cretaceous (Fig. 7).

Stratigraphic distributions of the larger invertebrates show progressive disappearances and appearances throughout the late Maastrichtian and early Tertiary, and an increase during the 40 m below the K-T transition interval. At no place within the interval from 199 to 269 m is there a single horizon characterized by an unusually large number of last occurrences (extinctions), as would be expected from a catastrophic event.

CONCLUSIONS

The Late Cretaceous–early Tertiary biostratigraphic record of shelf faunas from the high southern latitudes shows two important diversity patterns that must be addressed in any causal discussion of the mass-extinction event at the end of the Cretaceous. The first is the diachronous latitudinal disappearance of inoceramid bivalves and belemnites, and the dramatic decline in diversity of ammonite faunas of the high southern latitudes. Why did the inoceramid and belemnites disappear three to four million years earlier in the high southern latitudes than their counterparts in the low latitudes? This latitudinal disappearance of these ammonite families from the Antarctica faunas strongly supports the hypothesis that the development of new environmental conditions associated with global cooling precluded their survival in the high southern latitudes, well before their disappearance at the end of the Cretaceous. Raup and Jablonski (1993) reported in an analysis of bivalve extinctions that if the tropical reef-building rudists are omitted from their data set, there was no statistical latitudinal extinction pattern on the generic level during the Maastrichtian. Unfortunately, Raup and Jablonski's restriction of their analysis to the Maastrichtian does not address demonstrable biotic changes occurring during the Campanian that are inseparable from climatic events that began in the Campanian.

The expanded nature of the K-T section on Seymour Island presents a unique opportunity to observe an uninterrupted biotic record from the latest Cretaceous across the K-T boundary. The transitional nature of the faunal changes at the species level across the K-T boundary on Seymour Island is in marked contrast to sections from the middle and lower latitudes. This raises fundamental questions regarding the terminal Cretaceous extinction event. Why does the record in the high southern latitudes differ from those of the Northern Hemisphere and from the low latitudes? Are we looking at a high-latitude phenomenon or a Southern Hemisphere phenomenon? Are the latitudinally controlled diachronous extinctions and diversity declines during the Campanian and Maastrichtian related to the mass extinction at the end of the Cretaceous? These are some of the questions that have not been addressed thus far in the debate about the extinction at the end of the Cretaceous, which saw the demise of the dinosaurs.

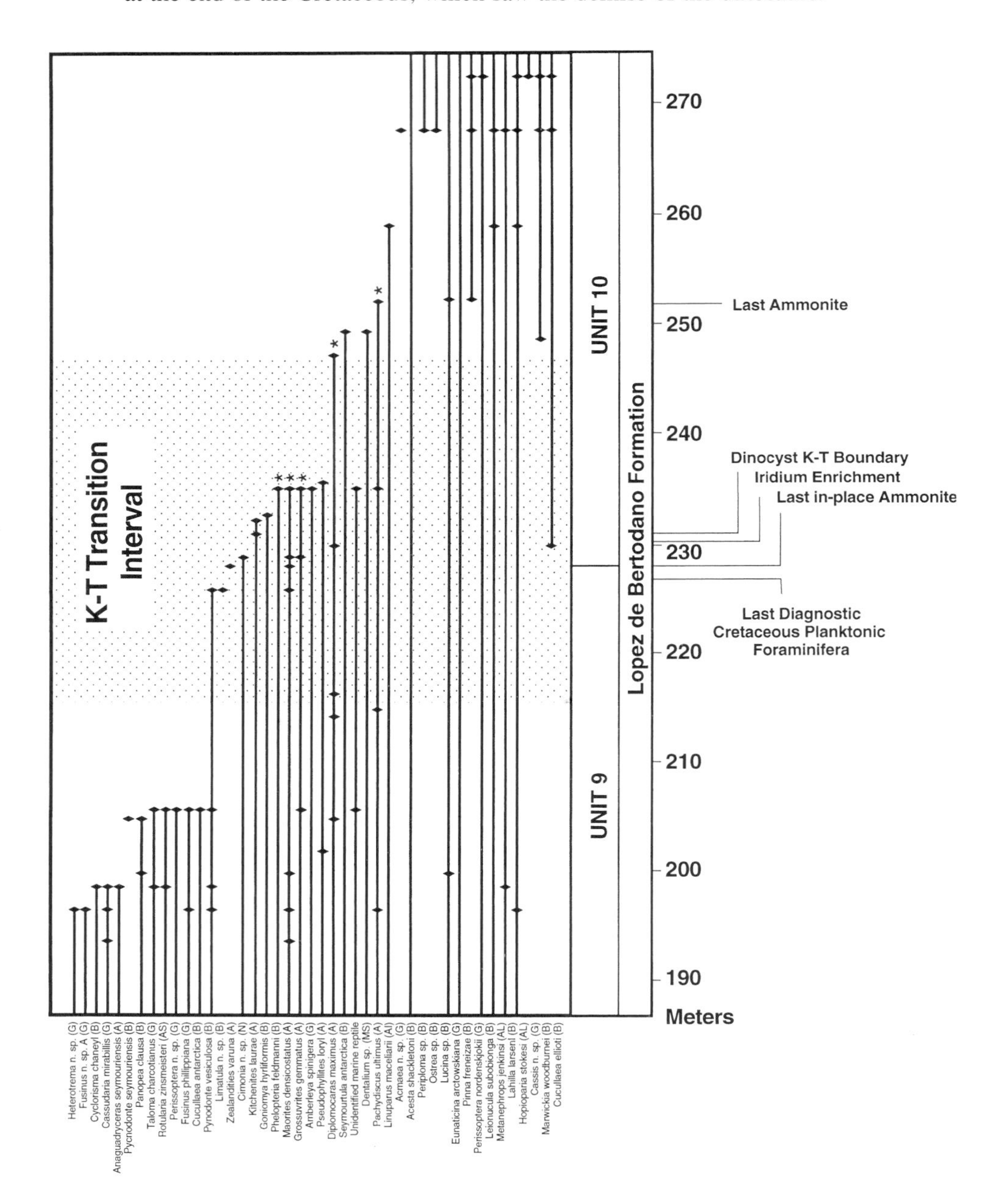

Figure 7. Last occurrence of megafossil species across the K-T boundary on Seymour Island (after Zinsmeister and others, 1989).

ACKNOWLEDGMENTS

We would like to express our appreciation to the Office of Polar Programs for the continuous support of our field work on Seymour Island and to both of us over the years. Zinsmeister would also like thank Carlos Rinaldi and Rudy del Valle for providing logistical support for work on Seymour and Vega islands during the austral summers of 1989–1990 and 1992–1993. Special thanks are given Edwardo Olivero for graciously providing his updated biostratigraphic data for Campanian ammonite faunas from James Ross and Vega islands.

REFERENCES CITED

Alvarez, L. W., Alvarez, W., Asaro, F., and Michel, H. V., 1980, Extraterrestrial cause for the Cretaceous/Tertiary extinction: Science, v. 208, p. 1095–1108.

Archibald, J. D., and Bryant, L. J., 1990, Differential Cretaceous/Tertiary extinctions of nonmarine vertebrates: Evidence from northeastern Montana, *in* Sharpton, V. L., and Ward, P. D., eds., Global catastrophes in Earth history: An interdisciplinary conference on impacts, volcanism, and mass mortality: Geological Society of America Special Paper 247, p. 549–562.

Askin, R. A., 1988, Campanian to Paleocene palynological succession of Seymour and adjacent islands, northeastern Antarctic Peninsula, *in* Feldmann, R. M., and Woodburne, M. O., eds., Geology and paleontology of Seymour Island, Antarctic Peninsula: Geological Society of America Memoir 169, p. 131–153.

Askin, R. A., 1989, Endemism and heterochroneity in the Late Cretaceous (Campanian) to Paleocene palynofloras of Seymour Island, Antarctica: Implication for origins, dispersal and palaeoclimates, *in* Crame, J. A., ed., Origins and evolution of the Antarctic biota: Geological Society of London Special Publication 47, p. 107–119.

Askin, R. A., 1990, Campanian to Paleocene spore and pollen assemblages of Seymour Island, Antarctica: Review of Palaeobotany and Palynology, v. 65, p. 105–113.

Bibby, J. S., 1966, The stratigraphy of part of the north-east Graham Land and the James Ross Island group: British Antarctic Survey Scientific Report 53, 37 p.

Bohor, B. F., and Betterton, W. J., 1994, Debris flow/turbities clastic units at the K-T boundary, northeastern Mexico [abs.], *in* New developments regarding the K-T event and other catastrophes in Earth history, February 9–12, 1994 [abs]: Houston, Texas, Lunar and Planetary Institute Contribution 15, p. 13–14.

Clemens, W. A., 1982, Patterns of extinction and survival of the terrestrial biota during the Cretaceous/Tertiary, *in* Silver, L. T., and Schultz, P. H., eds., Geological implications of impacts of large asteroids and comets on the Earth: Geological Society of America Special Paper 190, 407–422.

Crame, J. A., 1983, Cretaceous inoceramid bivalves from Antarctica, *in* Oliver, R. L., James, P. R., and Jago, J. B., eds., Antarctic Earth science: Proceeding of the fourth International Symposium on Antarctic Earth Sciences: Canberra, Australian Academy of Science, p. 298–302.

Crame, J. A., 1992, Late Cretaceous palaeoenvironments and biotas: An Antarctic perspective: Antarctic Science, v. 4, p. 371–382.

Crame, J. A., Pirrie, D., Riding, J. B., and Thomson, M. R. A., 1991, Campanian-Maastrichtian (Cretaceous) stratigraphy of the James Ross Island area, Antarctica: Geological Society of London Journal, v. 148, p. 1125–1140.

Dettmann, M. E., 1989, Antarctica: Cradle of austral temperate rainforests, *in* Crame, J. A., ed., Origins and evolution of the Antarctic biota: Geological Society of London Special Publication 47, p. 89–105.

Dhondt, A. V., 1992, Cretaceous inoceramid biogeography: A review: Palaeogeography, Palaeoclimatology, Palaeoecology, v. 92, p. 1217–1232.

D'Hondt, S., 1994, The impact of the Cretaceous-Tertiary boundary: Palaios, v. 9, p. 221–223.

Doyle, P., and Zinsmeister, W. J., 1988, The new dimitobelid belemnite from the Upper Cretaceous of Seymour Island, Antarctic Peninsula, *in* Feldmann, R. M., and Woodburne, M. O., eds., Geology and paleontology of Seymour Island, Antarctic Peninsula: Geological Society of America Memoir 169, p. 285–290.

Elliot, D. H., 1988, Tectonic setting and evolution of the James Ross Basin, northern Antarctic Peninsula, *in* Feldmann, R. M., and Woodburne, M. O., eds., Geology and paleontology of Seymour Island, Antarctic Peninsula: Geological Society of America Memoir 169, p. 541–555.

Elliot, D. H., Askin, R. A., Kyte, F. T., and Zinsmeister, W. J., 1994, Iridium and dinocysts at the Cretaceous-Tertiary boundary on Seymour Island, Antarctica: Implications for the K-T event: Geology, v. 22, p. 675–678.

Erwin, S. H., 1993, The great Paleozoic crisis, life and death in the Permian: New York, Columbia University Press, 327 p.

Erwin, D. H., 1994, The end-Permian mass extinction: A complex, multicausal extinction [abs.], *in* New developments regarding the K-T event and other catastrophes in Earth history, February 9–12, 1994 [abs]: Houston, Texas, Lunar and Planetary Institute Contribution 15, p. 33–34.

Farquharson, G. W., 1983, Evolution of late Mesozoic sedimentary basins in the northern antarctic Peninsula, *in* Oliver, R. L., James, P. R., and Jago, J. B., eds., Antarctic Earth science: Proceeding of the Fourth International Symposium on Antarctic Earth Sciences: Canberra, Australian Academy of Science, p. 323–327.

Feldmann, R. M., and Woodburne, M. O., eds., 1988, Geology and paleontology of Seymour Island, Antarctica Peninsula: Geological Society of America Memoir 169, 557 p.

Francis, J. E., 1991, Palaeoclimatic significance of Cretaceous–early Tertiary fossil forests of the Antarctic Peninsula, *in* Thomson, M. R. A., Crame, J. A., and Thomson, J. W., eds., Geological evolution of Antarctica: London, Cambridge University Press, p. 623–627.

Hansen, T. A., Farrand, R. B., Montgomery, H. A., Billman, H., and Blechschmidt, G., 1987, Sedimentology and extinction patterns across the Cretaceous-Tertiary boundary interval in east Texas: Cretaceous Research, v. 19, p. 251–265.

Harwood, D. M., 1988, Upper Cretaceous and lower Paleocene diatom and silicoflagellate biostratigraphy of Seymour Island, eastern Antarctic Peninsula, *in* Feldmann, R. M., and Woodburne, M. O., eds., Geology and paleontology of Seymour Island, Antarctic Peninsula: Geological Society of America Memoir 169, p. 55–130.

Hooker, J. J., Milner, A., and Sequeira, S. E. K., 1991, An ornithopod dinosaur from the Late Cretaceous of West Antarctica: Antarctic Sciences, v. 3, p. 331–332.

Howarth, L. K., 1966, Ammonites from the Upper Cretaceous of the James Ross Island Group: British Antarctic Survey Bulletin, v. 10, p. 55–69.

Hsü, K. J., 1986, The great dying: New York, Harcourt Brace Jovanovich, 292 p.

Jablonski, D., 1985, Causes and consequences of mass extinctions: A comparative approach, *in* Elliott, D. K., ed., Dynamics of extinction: New York, Wiley, p. 183–229.

Johnson, K., 1993, Megafloral biostratigraphy of the Cretaceous-Tertiary boundary: North Dakota, USA and South Island, New Zealand: Geological Association of Canada Proceeding and Abstracts, v. 25, p. A50.

Johnson, K. R., and Greenwood, D., 1993, High-latitude decideous forest and Cretaceous-Tertiary boundary, New Zealand: Geological Society of America Abstracts with Programs, v. 25, no. 7, p. 295.

Jones, D. S., Mueller, P. A., Byran, J. R., Dobson, J. P., Channell, J. E. T., Zachos, J. C., and Arthur, M. A., 1987, Biotic, geochemical, and paleomagnetic changes across the Cretaceous/Tertiary boundary at Braggs, Alabama: Geology, v. 15, p. 311–315.

Kauffman, E. G., 1984, The fabric of Cretaceous extinctions, *in* Berggren, W. A., and van Couvering, J. A., eds., Catastrophes and Earth history: Princeton, New Jersey, Princeton University Press, p. 151–245.

Kauffman, E., 1988, The dynamics of marine stepwise extinction, *in* Lamolda, M., Kauffman, E., and Walliser, O., eds., Paleontology and evolution; extinction events: Revista Española de Paleontología, extraordinario, p. 57–71.

Keller, G., 1994, K-T boundary issues: Science, v. 264, p. 641.

Kerr, R. A., 1994, K-T boundary issues: Science, v. 264, p. 642.

Macellari, C. E., 1984a, Late Cretaceous stratigraphy, sedimentology, and macropaleontology of Seymour Island, Antarctic Peninsula [Ph.D. thesis]: Columbus, Ohio State University, 599 p.

Macellari, C. E., 1984b, Revision of serpulids of the genus Rotularia (Annelida) and their value in stratigraphy: Journal of Paleontology, v. 58, p. 1098–1116.

Macellari, C. E., 1986, Late Campanian–Maastrichtian ammonites from Seymour Island, Antarctic Peninsula: Journal of Paleontology, v. 60, p. 1–55.

Macellari, C. A., 1988, Stratigraphy, sedimentology, and paleoecology of Upper Cretaceous/Paleocene shelf-deltaic sediments of Seymour Island, *in* Feldmann, R. M., and Woodburne, M. O., eds., Geology and paleontology of Seymour Island, Antarctic Peninsula: Geological Society of America Memoir 169, p. 25–54.

MacLeod, K. G., 1994, Bioturbation, inoceramid extinction and mid-Maastrichtian ecological change: Geology, v. 22, p. 139–142.

MacLeod, K. G., and Orr, W. N., 1993, The taphonomy of Maastrichtian inoceramids in the Basque region of France and Spain and the pattern of their decline and disappearance: Paleobiology, v. 19, p. 235–250.

MacLeod, K. G., and Ward, P. D., 1990, Extinction patterns of Inoceramus (Bivalvia) based on shell fragment biostratigraphy, *in* Sharpton, V. L., and Ward, P. D., eds., Global catastrophes in Earth history: An interdisciplinary conference on impacts, volcanism, and mass mortality: Geological Society of America Special Paper 247, p. 509–518.

MacLeod, N., 1994, K-T boundary issues: Science, v. 264, p. 641–642.

MacLeod, N., and Keller, G., 1991, How complete are Cretaceous?Tertiary boundary sections? A chronostratigraphic estimate based on graphic correlation: Geological Society of America Bulletin, v. 103, p. 1439–1457.

Marenssi, S. A., Liro, J. M., Santilluna, S. N., Martinioni, D. R., and Palamarczuk, S., 1992, The Upper Cretaceous of southern James Ross Island, Antarctica, *in* Isla James Ross: Instituto Antarctico Argentino, p. 89–100.

Olivero, E. B., 1984, Nuevos amonites campanianos de la isla Ross, Antartida: Ameghiniana, v. 21, p. 53–84.

Olivero, E. B., 1992, Asociaciones de amonites de la formacion Santa Marta (Cretacio Tardio), Isla James Ross, Antarctica, *in* Isla James Ross: Instituto Antarctico Argentino, p. 47–76.

Olivero, E. B., and Zinsmeister, W. J., 1989, Large heteromorph ammonite from the Upper Cretaceous of Seymour Island, Antarctica: Journal of Paleontology, v. 63, p. 626–636.

Olivero, E. B., Scasso, R. A., and Rinaldi, C. A., 1986, Revision of the Marambio Group, James Ross Island, Antarctica: Contribucion del Instituto Antartico Argentino, v. 331, 28 p.

Olivero, E. B., Gasparini, Z., Rinaldi, C. A., and Scasso, R., 1991, First record of dinosaurs in Antarctica (Upper Cretaceous, James Ross Island): Palaeogeographical implications, *in* Thomson, M. R. A., Crame, J. A., and Thomson, J. W., eds., Geological evolution of Antarctica: London, Cambridge University Press, p. 617–622.

Pankhurst, R. J., 1982, Rb-Sr geochronology of Graham Land, Antarctica: Geological Society of London Journal, v. 139, p. 701–712.

Pirrie, D., and Marshall, J. D., 1990, High-paleolatitude Late Cretaceous paleotemperatures: New data from James Ross Island, Antarctica: Geology, v. 18, p. 31–34.

Pirrie, D., Crame, J. A., and Riding, J. B., 1991, Late Cretaceous stratigraphy and sedimentology of Cape Lamb, Vega Island, Antarctica: Cretaceous Research, v. 12, p 227–258.
Raup, D. M., and Jablonski, D., 1993, Geography of end-Cretaceous marine bivalve extinctions: Science, v. 260, p. 971–973.
Raup, D. M., and Sepkoski, J. J., 1984, Periodicity of extinctions in the geologic past: National Academy of Sciences Proceedings, v. 81, p. 109–125.
Rinaldi, C. A., 1982, The Upper Cretaceous in the James Ross Island Group, *in* Craddock, C., ed., Antarctic geosciences: Madison, Wisconsin University Press, p. 281–286.
Rinaldi, C. A., Massabie, A., Morelli, J. R., and Rosenman, H. L., 1978, Geologia de la Isla Vicecomodoro Marambio: Contribuciones Instituto Antartico Argentino, v. 217, 37 p.
Sadler, P. M., 1981, Sediment accumulation rates and the completeness of stratigraphic sections: Journal of Geology, v. 89, p. 569–584.
Sheehan, P. M., 1973, The relation of Late Ordovician glaciation to the Ordovician-Silurian changeover in North America brachiopod faunas: Lethaia, v. 6, p. 147–154.
Signor, P. W., and Lipps, J. H., 1982, Sampling bias, gradual extinction patterns, and catastrophes in the fossil record, *in* Silver, L. T., and Schultz, P. H., eds., Geological implications of large asteroids and comets on the Earth: Geological Society of America Special Paper 190, p. 291–298.
Simpson, G. G., 1944, Tempo and mode in evolution: New York, Columbia University Press, 237 p.
Simpson, G. G., 1983, Fossils and the history of life: New York, Scientific American Library, 239 p.
Spath, L. F., 1953, The Upper Cretaceous cephalopod fauna of Graham Land: Falkland Islands Dependencies Survey Scientific Reports, v. 3, 60 p.
Stanley, S. M., 1984, Temperature and biotic crisis in the marine realm: Geology, v. 12, p. 205–208.
Stanley, S. M., 1987, Extinction: New York, Scientific American Library, 242 p.
Stanley, S. M., 1988, Paleozoic mass extinctions: Shared patterns suggest global cooling as a common cause: America Journal of Science, v. 288, p. 334–352.
Surlyk, F., and Johansen, M. B., 1984, End-Cretaceous brachiopod extinction in the chalk of Denmark: Science, v. 223, p. 1174–1177.
Sweet, A., Braman, D. R., and Lerbekmo, J. F., 1993, Northern mid-continental Maastrichtian and Paleocene palynofloristic extinction events: Geological Association of Canada Proceedings and Abstracts, p. A103.
Ward, P. D., 1990a, The Cretaceous/Tertiary extinctions in the marine realm; a 1990 perspective, *in* Sharpton, V. L., and Ward, P. D., eds., Global catastrophes in Earth history: An interdisciplinary conference on impacts, volcanism, and mass mortality: Geological Society of America Special Paper 247, p. 425–432.
Ward, P. D., 1990b, A review of Maastrichtian ammonite range, *in* Sharpton, V. L., and Ward, P. D., eds., Global catastrophes in Earth history: An interdisciplinary conference on impacts, volcanism, and mass mortality: Geological Society of America Special Paper, 247 p. 519–530.
Ward, P. D., and Signor, P. W., 1983, Evolutionary tempo in Jurassic and Cretaceous ammonites: Paleobiology, v. 9, p. 183–198.
Wolbach, W. S., Gilmour, I., Anders, E., Orth, C. J., and Brooks, R. R., 1988, A global fire at the Cretaceous-Tertiary boundary: Nature, v. 334, p. 665–669.
Zinsmeister, W. J., 1982, Review of the Upper Cretaceous–lower Tertiary stratigraphy of Seymour Island: Geological Society of London Journal, v. 139, p. 776–786.
Zinsmeister, W. J., 1984, Geology and paleontology of Seymour Island, Antarctica: Antarctic Journal of the United States, v. 19, p. 1–5.
Zinsmeister, W. J., 1985, Seymour Island expedition: Antarctic Journal of the United States, v. 20, p. 41–42.

Zinsmeister, W. J., 1994, What can the fossil record tell us about the terminal Cretaceous extinction event and the disappearance of the dinosaurs?, *in* Rosenberg, G. D., and Wolberg, D. L., eds., Dino fest: Proceedings of a conference for the general public, March 24–26, 1994: Paleontological Society Special Paper 7, p. 487–500.

Zinsmeister, W. J., and Macellari, C. E., 1983, Changes in the macro-fossil faunas across the Cretaceous-Tertiary boundary on Seymour Island, Antarctic Peninsula: Antarctic Journal of the United States, v. 17, p. 68–69.

Zinsmeister, W. J., and Macellari, C. A., 1988, Bivalvia (Mollusca) from Seymour Island, Antarctica Peninsula, *in* Feldmann, R. M., and Woodburne, M. O., eds., Geology and paleontology of Seymour Island, Antarctic Peninsula: Geological Society of America Memoir 169, p. 253–284.

Zinsmeister, W. J., Feldmann, R. M., Woodburne, M. O., and Elliot, D. H., 1989, Latest Cretaceous/Tertiary transition on Seymour Island, Antarctica: Journal of Paleontology, v. 63, p. 731–738.

13

Biological Consequences of Mesozoic Atmospheres: Respiratory Adaptations and Functional Range of *Apatosaurus*

Richard A. Hengst, *Department of Biological Sciences, Purdue University North Central, Westville, Indiana,*
J. Keith Rigby, Jr., *Department of Civil Engineering and Geological Sciences, University of Notre Dame, Notre Dame, Indiana,*
Gary P. Landis, *U.S. Geological Survey, Denver, Colorado,*
and Robert L. Sloan, *Department of Geology and Geophysics, University of Minnesota, Minneapolis, Minnesota*

INTRODUCTION

Atmospheric gases are related intimately to life processes and oxygen is especially important to life function. Sustained activity demands oxygen to power respiratory processes. Carbon dioxide is basic to photosynthesis in plants and also has profound effects on animal metabolism. Limits of physical performance are largely determined through complex relations within the oxygen delivery pathway that supplies this nutrient to tissue mitochondria. Physical performance is also linked to atmospheric oxygen concentration in exercising vertebrates (Jones and others, 1993). The artificially enriched atmospheres provided by oxygen masks are used routinely by pilots and mountaineers to extend the environmental limits of physical activity. Performance may be increased if more oxygen is available or decreased if it is not. If oxygen limits are applied to the struggle for life, then O_2 availability may well affect survival, distribution, and evolution, because these activities are limited, at least in part, by the maximal sustained activity of an organism.

Although we recognize that CO_2 almost certainly affects evolutionary processes, except for illustrating a few points, we discuss animal respiration and oxygen delivery because the combined effects of O_2 and CO_2 on physiological performance await experimental verification in the vertebrate classes pertaining to this analysis. We discuss factors limiting the oxygen delivery pathway and their possible consequences in physical performance limits in the sauropod dinosaur, *Apatosaurus*.

Several authors have presented evidence that CO_2 and O_2 concentrations were elevated throughout the Mesozoic. Cerling's (1991) analysis of root paleosols found CO_2 concentrations from two to ten times that of present atmospheric levels (PAL) in the Mesozoic. Carbon dioxide concentrations in the low end of this range are known to differentially promote plant growth in modern plants, but concentration effects in the upper range in modern plants or effects of elevated CO_2 on plants of Mesozoic origin are not complete at this time.

Ambient oxygen may have been elevated above PAL during the Cretaceous and possibly the Jurassic. Landis and others (1993) analyzed gases trapped within amber bubbles. Their data indicate that oxygen composed ~35% of the atmosphere throughout most of the Cretaceous and Late Jurassic. This value dropped rapidly to 28% at the close of the Cretaceous. If this inference is correct, such a change in atmospheric composition would exert strong selective pressure on animals with the smallest safety margins in respiratory delivery pathways, especially animals long adapted to an oxygen-rich environment. More important, adaptations to breathing high-oxygen atmospheres should persist in the dinosaur fossil record in that breathing mechanisms should reflect the smaller air volumes used to support activity. Because a low-performance breathing system would be sufficient to support a large animal if each breath were supercharged with oxygen, analysis of performance limits possible with dinosaur thoracic structure may be used to test the proposition that atmospheric oxygen concentrations were elevated during the Mesozoic.

Indirect evidence for varied oxygen concentrations has been found in two places. Erben and others (1979) analyzed dinosaur eggshell thickness for possible pathologies and changes in Late Cretaceous specimens. The changes they found coincide with the chronology of O_2 decreases noted by Landis and coworkers. Although Erben's analysis approached this subject for other reasons entirely, their results are in agreement with those expected if oxygen declined. McAlester (1970) discussed oxygen usage as a potential evolutionary selection factor. Using evidence from a wide variety of animal families, he demonstrated a high positive correlation between oxygen consumption rates in living animals and the extinction rates of their ancestral forms. Schopf and others (1971) disputed his interpretation of these data. However, even taking their arguments into consideration, it is evident that the oxygen delivery path has acted in a selective manner in the past.

Biological effects resulting from changes in atmospheric oxygen would most likely be rapid but not cataclysmic. Rigby and others (1987) and Sloan and others (1986) presented evidence that dinosaurs continued to exist into early Paleocene time in the western United States. If true, this argues for an environmental selection factor to have separated dinosaurs from vertebrate survivors of the Upper Cretaceous faunal transition, including the Cretaceous-Tertiary (K-T) boundary event. The degree to which oxygen demands are supported by oxygen transport pathways must be examined.

Evolution hones physical adaptations to fit environmental conditions. This assumes, of course, that such adaptations fall within realistic biological limits. The cardio-respiratory system in particular is developed to meet the maximum demands an animal is likely to place upon the system, and little more (Taylor and Weibel, 1981). Jones and Lindstedt (1993) and Lindstedt and Jones (1987) presented considerable evidence that respiratory system design tends toward optimality. Adaptations to breathing oxygen-rich atmospheres should be most evident in animals fitted to that environment. Because our most complete data were derived from Late Jurassic through Late Cretaceous amber sources, we chose to analyze a dominant and anatomically well characterized vertebrate animal from this era for its respiratory characteristics and limits. Because it is the first study of its type, it was important to consider a dinosaur species that may have been well adapted to the oxygen-rich environment. If our evidence that environmental stresses existed at the end of the Cretaceous is correct, then pressures for adaptive change would have been operating, and analysis would be made far more difficult.

A review of the literature showed a single reference (Daniels and Pratt, 1992) analyzing the physiology of dinosaur respiration. These authors analyzed CO_2 retention in *Mamenchisaurus* through engineering models of air flow and concluded that sauropods had bird lungs and an ectothermic metabolism. No mention was made of ventilation mechanics such as depth or frequency. This is an important point, because an animal cannot use more oxygen than is supplied by ventilation. Spotila and others (1991) and Dunham and others (1989) analyzed energy consumption in dinosaurs of various mass, including that needed for locomotor costs and growth rates. Their formulas, derived from Calder (1984), permitted us to calculate oxygen needs according to mass and walking speed. The large body of empirical and theoretical literature that exists for respiratory physiology for a wide variety of contemporary vertebrates allows for comparisons with modern animals. A complete review of physiological limitations and performance is beyond the scope of this paper: see Jones and Lindstedt's (1993) review of the current literature on this subject.

We chose to analyze the sauropod dinosaur *Apatosaurus* because a reasonably complete specimen was available for study and because it represents a well-derived type that may have lived in oxygen-rich conditions. We also felt that some of the principles derived in this analysis might be applied to our ongoing studies of ornithischian and non-diplodocene sauropod dinosaurs.

MATERIALS

The specimen used in this study was the articulated *Apatosaurus* skeleton of the Field Museum, Chicago, Illinois, and was made available for study through the cooperation of the staff and curators of the museum. It is an adult specimen that was measured and photographed with particular attention to the morphology of ribs, sternum, dorsal vertebrae, and pelvis. Osteological descriptions from the literature were also used in deriving measurements used in this study (Riggs, 1903; Gilmore, 1936; Hatcher, 1901; Ostrom and McIntosh, 1966). Details of a cast of a *Diplodocus* skull were used in this study because of its similarity to *Apatosaurus* and because it was not possible to make direct measurements of nostril size and shape from the skull casting on the articulated *Apatosaurus* specimen. Nostril size and shape were determined from the *Diplodocus* casting and later modified from photographs, personal communications with McIntosh, and examinations of Gilmore's (1936) and Hatcher's (1901) descriptions of diplodocene dinosaurs. All other measurements and observations were taken from the specimen.

ASSUMPTIONS

A number of assumptions are necessary to establish the respiratory capabilities of *Apatosaurus*, covering such physiological features as metabolic rate, ventilation constraints, thoracic volume, lung capacity, and blood transport capability. These assumptions are listed, discussed, and justified below.

Metabolic Rate

Oxygen usage at any moment is a mix of that needed to support tissues for life and that needed to support activity. Various authors have argued that the metabolic rates of sauropods were of an endothermic ("warm-blooded"; see Bakker, 1986) or ectothermic nature ("cold-blooded"; see Dodson, 1990). We assumed intermediate oxygen needs for the following reasons: first, trackway evidence indicates that most dinosaurs were reasonably active over extended periods of time; second, modern organisms with endurance exhibit either endothermy or gigantothermy (Bennett and Ruben, 1979; Bennett, 1980; Spotila and others, 1991; Paladino and others, 1990) accompanied by increased O_2 demands; and third, it has been suggested that metabolic rates of ectotherms and endotherms converge as size increases (Farlow, 1976; Alexander, 1989). That is, metabolic differences are less in large animals.

Spotila and others (1991) advocated gigantothermy as an advantageous thermoregulatory mechanism in discussing energy consumption in migrating dinosaurs. Bennett (1980) argued that endothermy is a consequence of expanded aerobic support for physical activity, wheras McNab (1978) argued that size increases resulted in homoiothermy, which led to endothermy as size decreased. Irrespective of the endothermy-ectothermy debate, it is likely that dinosaurian oxygen needs were greater than expected from predictions based on ectothermic metabolism in small reptiles. Trackway data indicate that dinosaurs had some levels of extended activity, but the issue of endothermy remains unresolved. It is therefore conservative to assume that a somewhat aerobic state intermediate between true ectothermy and true endothermy existed in dinosaurs. All calculations of oxygen need discussed in this article assume low intermediate metabolic rates.

Locomotor work in terrestrial animals is more a function of an animal's mass than of its metabolic type (Alexander, 1989; Bakker, 1972; Pennycuick, 1992; Schmidt-Nielsen, 1984; Calder, 1984; Spotila and others, 1991). The mass of an adult *Apatosaurus* has been estimated between 33,000 kg and 17,000 kg by various authors. Anderson and others (1985) calculated a mass of 25,000 kg for *Apatosaurus* based on humerus and femur mid-shaft circumference. We employed this estimate because we believe it to represent a reasonable consensus of published values.

Airway Size

The volume of air moved to the lung per breath is limited to the amount thoracic volume can change. The actual amount available for exchange is less than the total breath because of air remaining in the trachea and bronchial passages (dead space). Dead space is a constant in respiratory calculations. This constitutes a large percentage of shallow breaths but only a small part of deeper ones. Nostril length was used as the smallest possible tracheal diameter, on the basis of anatomical dissections and general experience. Nostrils may limit air flow during vigorous ventilation, so it is common to open the mouth when at high breathing rates. Both mouth open and closed breathing situations had to be accounted for in our analysis of *Apatosaurus* performance.

Diplodocene dinosaurs are unique in possessing small, well-defined nostrils (Fig. 1). Their nostrils are wedge-shaped, ~8.0 cm in length by 3 cm in width at the upper end. It is clear from an inspection of the aperture that the nostrils were separated by a septum in life. If a reasonable thickness for a septum and membranes is deducted from nostril dimensions, then the combined area of both nostrils approximates a rectangle 2.5 x 7.5 cm. It is further presumed that this aperture represents the narrowest point in the airway. This opening, approximately equal to that of modern horses, was unlikely to have had any unusual nasal organs due to the exceptionally small aperture size. High airflow rates would have been limited by the excessive turbulence generated when air rapidly flowed through this small port.

Tracheal Diameter and Respiratory Dead Space

Tracheal diameter is unknown for *Apatosaurus* (or indeed any dinosaur) because these tissues do not fossilize. However, in all modern terrestrial vertebrates, nostril size is less than tracheal diameter. This relation provides us a minimal value for probable tracheal diameter. Probable tracheal length was measured at 6.25 m, the length from thorax to nostrils. It was recognized that bronchial contributions to dead space could not be estimated because lung morphology is purely specu-

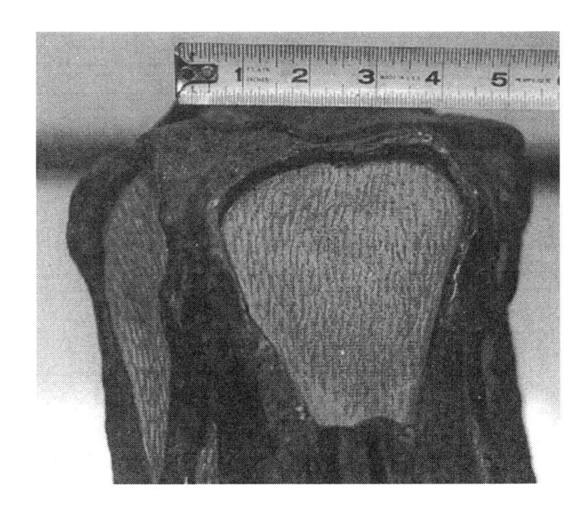

Figure 1. Nostril size and shape of Diplodocus *sp. Scale is in inches.*

lative. Tracheal volume was considered to be equivalent to the total dead space because the bulk of dead space in modern animals is found here.

Maximal rates of air delivery for *Apatosaurus* can be estimated if the volume of air moved per single breath is known. The method used to determine this value is detailed in the next paragraph. Maximal breathing depth was determined to be 330–400 l. Animals normally use only slightly more than half that when working hard (Leath, 1994, personal commun.). This is ~225–250 l for a working breath.

In order to estimate apatosaurid airflow rates, a physical model of the breathing system was constructed. This consisted of a 6.25 m tube of various simulated tracheal diameters, connected to a series of barrels to approximate the thoracic volume of *Apatosaurus*, and a 250 l bellows to create inspiratory forces. The minimum time needed to move 250 l of air allowed us a benchmark against which to compare air delivery for "tracheae" of various diameters. The open end of the tube was fitted with a small chamber that could be covered with a plate containing simulated nostrils. This allowed us to test air delivery rates of open mouth and nostril breathing. A diagram of the apparatus is shown in Figure 2.

The results of these tests (Fig. 3) were compared with dead space of tube diameter: optimal diameter was near 12 cm for *Apatosaurus*. This diameter allows for reasonable breathing depths (tidal volumes) without excessive dead space at rest and during activity. It also provides us with a means of estimating the maximum air delivery possible each minute.

Expiration lasts as long as or longer than inspiration in modern animals. Assuming that breathing cycles are continuous, it is possible to estimate the maximum possible air delivery per minute. A volume of ~1600 l/ min for a 12 cm trachea appears to represent the optimum compromise between excessive dead space and maximum air flow.

Thoracic Volume and Lung Volume

A transverse section of an *Apatosaurus* thorax is irregular in shape but similar enough to an ellipse to justify using this shape in calculations of thoracic volume.

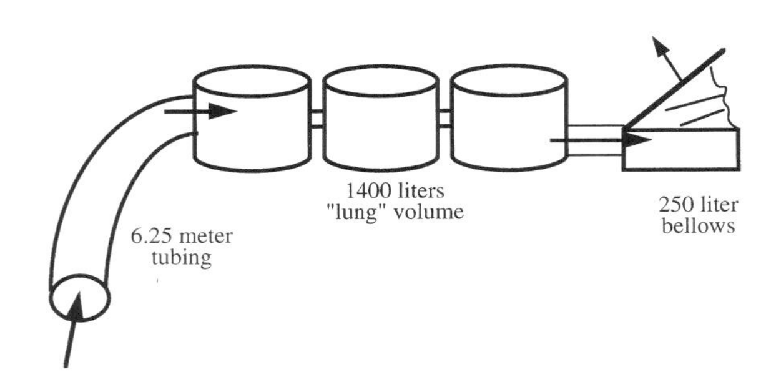

Figure 2. Diagram of apparatus for determining maximum airflow rates in simulated tracheae of various test diameters.

The horizontal axis of each cross section was assumed to be the widest point between internal surfaces of opposite ribs. The vertical axis was somewhat more difficult to measure at each level because, in life, boney ribs were connected to the sternal bones through cartilaginous ribs. This arrangement was confirmed by inspecting the sternal margin and from the literature. Marsh's 1879 reconstruction (as shown in Ostrom and MacIntosh, 1966) showed reports of thoracic cartilage in *Apatosaurus*. Vertical thoracic dimensions were estimated because cartilaginous structures were missing. Marsh's sketches and fluoroscopic and skeletal specimen studies of the thorax of cattle, dogs, cats, rats, iguanas, alligators, and monitor lizards allowed for reasonable estimates of chest profile. Vertical measurement increased progressively from 155 cm at the level of the first rib to 185 cm at the fifth rib. The length of the thorax was assumed to be the distance from the thoracic inlet where the neck joins the trunk to the posterior surface of the fifth rib. This pattern was arrived at by an examination of rib and vertebra structure. Bone responds to mechanical stress by increasing bone mass. Ribs three, four, and five are distinctly heavier than the sixth and ensuing ribs. This is most likely a response to stresses on the rib from breathing motions. It is likely that such stresses are increased further by the demands of moving rib cartilage in addition to respiratory movements.

Vertebrae associated with ribs 3, 4, and 5 have expanded processes which are consistent with support of rib articulation sites. Processes posterior to vertebra 5 are much less developed than those anterior to 5. This indicates that greater forces operate on the first five (emphasis on 3–5) vertebrae than on the following dorsal vertebrae (6–11). The average length of the thorax was 125 cm, as measured at the midpoint of ribs 1 and 5.

Thoracic volume was modeled as a truncated elliptical cone. Although the thorax is, in fact, a series of five truncated cones, it was modeled as a single cone using the area at the level of the first ribs and the area at the level of the fifth rib in approximating the shape of the chest. It was felt that this shape was reasonably close to a five-cone model for our purposes. Thoracic volume was estimated at 1580 l.

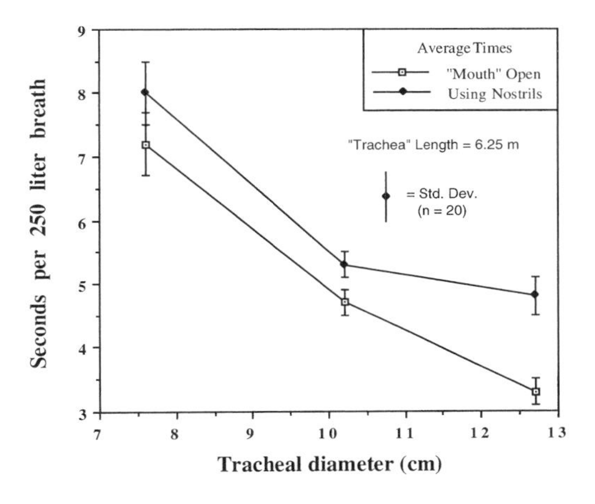

Figure 3. Average time to "inspire" 250 l of air with simulated tracheae 6.25 m in length of various diameters. Vertical bars represent standard deviations of 20 trials. Closed circles represent average times with nostril plate in place. Open squares represent average times of open tubing without simulated nostrils.

Air movement is a function of changes in lung volume during breathing. These changes are limited by changes in thoracic cavity volume. In dinosaurs, rib and vertebra meet at two joints. Thus, ribs are constrained to move in a hingelike fashion in a single plane of motion. The positions of the vertebral articulations indicate that the fifth rib moved mostly laterally, while more anterior thoracic ribs moved progressively more anteriorly. The lateral displacement of the ribs during breathing movements was estimated by considering each rib to move through an average angle of 45° to the long axis of the body. The lateral and ventral displacement was calculated as the cosine of this angle. Rib movement during ventilation may be estimated as a proportion of muscle length. The space between adjacent ribs 1–5 was measured to determine the probable limits of rib movement during ventilation. The measured distances are listed in Table 1.

Lung volume is always less than thoracic volume due the presence of thoracic viscera, including lung tissue itself, heart and blood vessels, esophagus, and trachea. The mass of the heart scales as a constant percentage of body mass in all but the smallest endothermic vertebrates (Schmidt-Nielsen, 1984). However, this percentage varies by vertebrate group. Mammalian hearts are ~0.6% of body mass; birds are slightly greater than this percentage; reptiles are ~0.4% of the total. It is reasonable to assume a value of 0.5% in estimating cardiac size, in keeping with our assumption of intermediate metabolism and endurance, and assuming that dinosaurs fell somewhere between the extremes of bird and reptile. For a 25,000 kg sauropod, this corresponds to a cardiac mass of 124 kg with a volume of 125 l.

Lung Characteristics

Dinosaur lungs were very likely to have been variations of the modern crocodilian model. Crocodilian lungs are more divided and include more surface area than other reptiles. Perry (1987) analyzed the crocodilian lung and noted that the distribution and structure of air passageways was similar to avian lung structure. Bird and crocodilian lungs are dissimilar in that crocodilian lungs lack air sacs and are elastic rather than rigid. The lung structure exhibited by crocodilians is probably very ancient and common to both crocodilian and dinosaur groups. Because

TABLE 1
Thoracic Measurements in *Apatosaurus* sp.

Rib Number	Intercostal Length (cm)	Thoracic Length
1-2	35	125
2-3	15	
3-4	15	
4-5	16	

Thoracic Volume = 1580 liters
Estimated Lung Volume = 1400 liters
Max. Inspiration = 330–400 liters

crocodilians coexisted with dinosaurs and birds are phylogenetic descendants of dinosaurs, it is likely that a form similar to the crocodilian lung was present in the earliest dinosaurs and that this lung structure was perpetuated. It is unlikely that dinosaurs exhibited the compact and highly efficient lung of modern birds, because they lacked a means of mechanically ventilating it (see below).

It is assumed that a highly septate form of reptilian-crocodilian lung was found in dinosaurs. Using data from Perry (1987), the volume of solid lung tissue predicted for a thoracic space of 1580 l is ~53.5 l. Because cardiac tissue occupies about 125 l, a combined lung and heart tissue displacement reduces thoracic volume by about 179 l. *Apatosaurus* lung capacity is estimated at 1400 l.

Oxygen Transport by Blood

Blood transports oxygen to body tissues. Oxygen temporarily binds to hemoglobin, the transport pigment of all terrestrial vertebrate blood cells. A lack of hemoglobin or of blood cells results in various degrees of anemia with reduced capacity for oxygen delivery. It is important then to consider our assumptions concerning dinosaur blood. Birds and reptiles both have nucleated red blood cells (erythrocytes), whereas most mammalian erythrocytes lack nuclei (Burton, 1972). This allows mammalian cells to assume a smaller size and pack more hemoglobin-containing cells into a given blood volume. If the maximum oxygen transport capabilities of reptile, bird, and mammal blood are compared, mammals transport the greatest quantity of oxygen per blood volume, birds slightly less, and reptiles the least. This division is greatest between ectotherms and endotherms (Fig. 4). Reptiles and birds show an overlap in oxygen transport

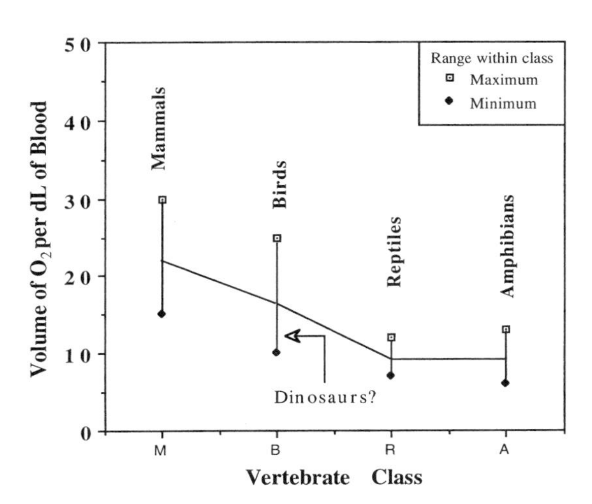

Figure 4. Oxygen-transport capacity of vertebrate blood. Vertical bars represent the range of oxygen-transport capacity of each class of living vertebrate. Lines are drawn through the median point of each range to aid comparison. Ectothermic animals (amphibians and reptiles) exhibit lower median oxygen-transport capability than do the endothermic animals (birds and mammals) in keeping with their lower oxygen demands (dl = deciliters).

characteristics. That both metabolic groups share a common functional region suggests that dinosaurs very likely had blood characteristics appropriate to this overlap region as well. Although intermediate-level rather than high-level endothermy was an assumption, high metabolic rates place more demands on the ventilation system, not less. Likely blood characteristics include nucleate cells similar in size to large bird or small reptilian erythrocytes and a somewhat low oxygen-carrying capacity, at least by avian standards.

INTERPRETATION AND DISCUSSION

Apatosaurus shares much of the general thoracic structure of ornithiscians and its fellow sauropods of Jurassic and Cretaceous origin. Therapod dinosaurs are unique and are not included in ensuing discussions. However, we speculate that some ideas presented here apply to all dinosaurs, because a similar physiology is likely shared in the larger sense while differing through specific adaptation. Sustained physical activities require continuous cardiac and respiratory support of muscle metabolic needs, including uninterrupted supplies of energy sources and oxygen. More than anything else, oxygen delivery rates establish the limits of performance (Jones and Lindstedt, 1993). Taylor and Weibel (1981) and Weibel and others (1981) proposed the principle of symmorphosis to explain structural relations between physical demands and system design. Essentially, no part of a system is structured to deliver more than the maximal demands normally required of it. For example, animals do not have a heart capable of pumping 100 l of blood per minute if their blood vessels are capable of receiving only 20 l per minute. From a design viewpoint, successful evolutionary change is most possible in systems wherein reasonably increasing that which is already present is the solution to resisting stress. For example, it is more likely that a heart wall would be increased in mass rather than developing additional chambers if both solved a physiological problem equally well. This is because new structures require simultaneously adaptation or evolution of more factors if survival is to occur.

What factors should we examine in determining performance characteristics of dinosaurs? In the broadest terms, Figure 5 diagrams the four major steps of oxygen delivery. First, air must be moved efficiently to and from lung surfaces (ventilation). Second, O_2 must cross lung surfaces to reach the blood (diffusive capacity). Third, O_2 must be transported by blood, especially by hemoglobin (oxygen transport capacity). Fourth, time is needed for blood to shuttle between tissue and lung (circulation time). Circulation time need not be considered here because the sheer size of the animal precludes this as an adaptational strategy.

Discussion of ventilation effectiveness must consider three questions. What are the minimal and maximal oxygen needs of an organism? What volume of air fills the system? What proportion of this volume is refreshed by each breath? Minimal demands for oxygen are defined by the nature of the most basic life-supporting activities. Endothermy creates much greater demands for oxygen than does ectothermy, at least for resting animals. However, when an animal is foraging, fighting for territory or survival, climbing hills, or engaged in other vigorous

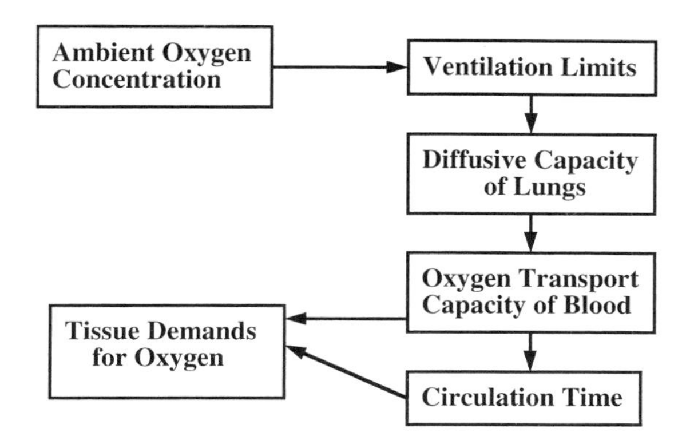

Figure 5. Diagrammatic representation of factors in the oxygen delivery pathway. Tissue activity determines the demand for oxygen, but the actual supply of oxygen for tissue use has several steps, any or all of which determine performance limits of the animal. Each step is dependent on the prior one if oxygen flow is matched to tissue need.

activities, oxygen needs are more dominated by muscle activity than by basic life processes. Therefore, it is useful to address maximal performance rather than minimal when considering demands of ectothermy or endothermy because basal functions contribute less to the overall energy budget of the organism. Calder (1984) used the formula $NCT = 10.7\ (M)^{0.68}$ for estimating locomotor energy requirements in endotherms while ectothermic energy requirements are scaled to $NCT = 10.3\ (M)^{0.64}$, where the net cost of transit is in kilojoules per kilometer and mass (M) is in kilograms. Assuming intermediate values of $NCT = 10.5(M)^{0.66}$ as discussed earlier, a typical minimal activity for *Apatosaurus* requires an oxygen consumption rate of 69 l of O_2 per minute. For purposes of comparison, maximal oxygen uptake was extrapolated from mammalian data cited in Jones and Lindstedt (1993). Their figures predict values of 1200 l of O_2 per minute. This value is slightly high due to the source animals and their size, but metabolic rates of reptiles and mammals may converge as size increases. Oxygen needs of *Apatosaurus* at various speeds have been calculated from Calder's formula and are shown in Table 2, assuming 4.825 l O_2 per calorie.

TABLE 2
Air Deliveries Needed to Support *Apatosaurus* at Various Walking Speeds

SPEED (km/h)	CALORIES	O_2 CONSUMED (l/m)	VOL. NEEDED (21% O_2; l/m)	VOL. NEEDED (35% O_2; l/m)
1	334	69	329	197
2	668	138	657	394
3	1002	208	990	594
5	1670	346	1648	989
7	2338	485	2310	1386

The ability to deliver these volumes varies with the ventilation mechanism utilized. The possible methods for changing thoracic volume are (1) lateral and ventral rib movement (reptiles, especially lizards); (2) internal expansion by posterior movement of diaphragm (or liver in the case of crocodilians), which may be enhanced by outward rib movement (as in mammals); and (3) use of large, widely distributed, sacs to hold air for exchange in rigid tubelike exchange surfaces (birds).

The latter two systems are very efficient ventilation mechanisms and are clearly adequate for supporting high levels of activity in endothermic animals and probably animals as large as *Apatosaurus*. We must examine the fossil evidence for the most likely breathing mechanism.

It is unlikely that *Apatosaurus* possessed a diaphragm. Diaphragmatic movements require abdominal contents to be displaced posteriorly, which forces them outward due to the physical barrier formed by the pelvis. The well-developed abdominal ribs in *Apatosaurus* (MacIntosh, 1990, and personal communs.) would resist lateral movements and argue against the presence of a diaphragm. Diaphragm contraction occurs in a transverse plane, whereas abdominal contents are compressed and displaced in a posterior and lateral movement during inspiration. Diaphragmatic contraction can direct only a small portion of the total inspiratory force in a posterior direction. Rib motion permits a more efficient breathing because muscular forces are free of problems of visceral displacement and the abdominal rib resistance that would have to be overcome by a sauropod's diaphragm. The thoracic profile also contributes to the efficiency of the mammalian diaphragm (Decramer, 1993). This shape is well known and consistent in mammals. *Apatosaurus* lacks this distinctive thoracic shape. In summary, it is unlikely that *Apatosaurus* had a diaphragm.

Avian pulmonary systems have been suggested by a number of authors, including Bakker (1986), Daniels and Pratt (1992), and Britt (1993). This hypothesis must be considered because birds are closely related to dinosaurs and because this is an extremely efficient form of respiratory system. In particular, the avian lung is the most efficient system for extracting oxygen from air passed to it. The assertion of dinosaurs possessing avian lung respiratory systems is inferred because many dinosaurs have pleurocoels (extensive depressions) or pneumocoels (hollow cavities) associated with anterior vertebrae and ribs. This structure is well developed in living birds. Hollow bone structure may be an adaptation for reducing non-load-bearing bone mass (Pennycuick, 1992). This does not mean that the pneumocoelous surfaces are used for respiratory exchange. In birds, pneumocoels are linked to air passages of the respiratory system but do not participate in respiratory gas exchange (Fedde, 1976).

Ventilation of avian pulmonary systems depends upon nearly simultaneous expansion or compression of extensive air sacs distributed over much of the body's length. Pumping continually flushes respiratory surfaces with fresh air via pressure generated through a bellows-levering action of the elongated sternum characteristic of birds. The sternum acts as a third-class lever as it pivots about its anterior end against a fulcrum of corocoid bones. This action compresses and expands the entire ventral surface, thus extending the regions in which air sacs can function. Rib structures guide and stabilize the anterior-posterior–directed movements by uncinate processes that limit lateral movement of ribs. An avian skeleton (pigeon) is illustrated in Figure 6 to compare avian thoracic structure with other vertebrate groups.

Sauropods lack comparable sternal or rib structures capable of compressing widely scattered air sacs, as birds do. It is difficult to see how avian lungs could explain dinosaur respiration without a supportive system for ventilation of air sacs. This structural feature also applies to ornithischian dinosaurs and other sauropods. For these reasons, avian lungs, even if present in sauropods, would be limited to the air volumes and flow rates possible with reptilelike rib movements.

Lateral movement of anterior ribs, a mechanism similar to that seen in many lizards, is the most likely breathing system in sauropods. Other options require positing exotic systems for which no experiential basis exists. Rib movement can be understood if consideration is given to the axis formed by the pair of rib-vertebra articulations. Dinosaur ribs have two functional heads, the tuburculum and the capitulum. These features articulate with the vertebrae at two points deter-

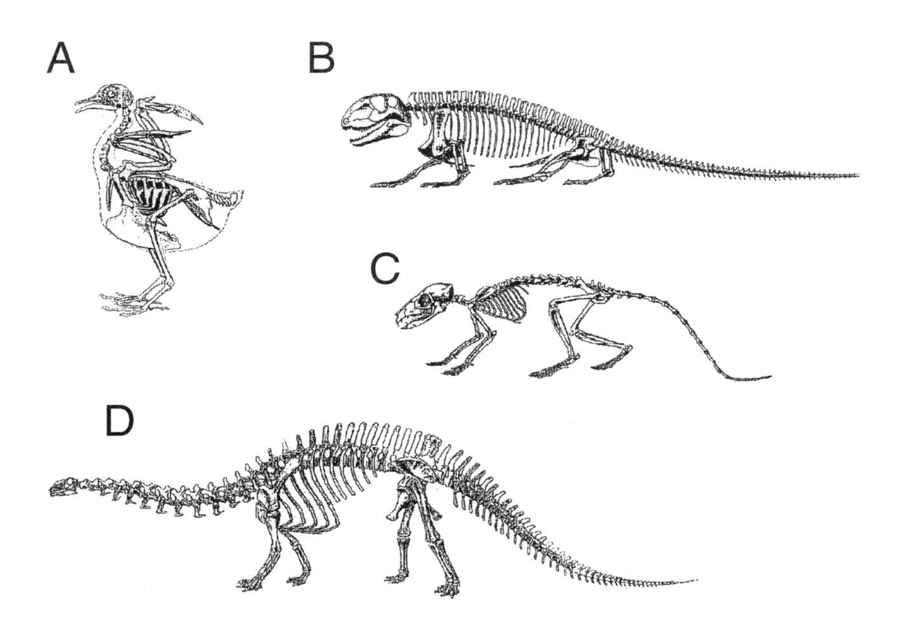

Figure 6. Thoracic structure in generalized vertebrates. Skeletons shown are of (A) bird, (B) primitive reptile, (C) mammal, and (D) Apatosaurus. *A, B, and C are after Romer (1956); D is after Ostrom and McIntosh (1966).*

mined by the vertebrae themselves (Fig. 7). Thus, the plane of rib movement is determined by the geometry of the articulation sites. Ribs 1, 2, and 3 have an axis of movement directed forward (see Fig. 7), rib 4 moves somewhat forward and somewhat laterally, and rib 5 has an axis of motion directed almost laterally. This sequence of angles is in contrast to that seen in primitive reptiles in which all thoracic ribs move in the same direction regardless of position (i.e., forward and slightly laterally). For sauropods, the advantage of this sequence is that each rib passes through a separate plane of space, thus increasing the efficiency with which thoracic dimensions are changed with breathing motions.

Modern mammals have a somewhat similar sequence of rib-vertebra angles, and this similarity was used to test the efficiency of this ventilation mechanism. Ruben and others (1987) examined the effects of rib-only breathing in rats, an animal having a high metabolic demand. The phrenic nerve was blocked from stimulating the diaphragm. Rib movements were the sole means of ventilation. Respiratory and metabolic measurements were made under resting and exercise conditions. In resting animals, Ruben and others (1987) found tidal air volumes and metabolic rates unchanged from those having diaphragms still intact. That is, at low air volumes, rib breathing is as efficient as diaphragmatic breathing. However, when exercised, rats with paralyzed diaphragms had small reserves and quickly became exhausted. Rib breathing is adequate to supply the needs of an endotherm with a high metabolic rate when at rest or during mild exercise, but is limited when demands for air are large.

Respiratory rate varies predictably with vertebrate class and with size. Mammals breathe most frequently among similarly sized vertebrates. Reptiles are expected to breathe slowly due to their ectothermic metabolism and low oxygen needs, but birds also breathe more slowly than comparable mammals even though their metabolic rate is high. This avian slowness is due to their deep ventilation of air sac volume. A standing African elephant breathes 10 times per minute (Adams, 1981), but its breathing rate during exercise is unknown. From

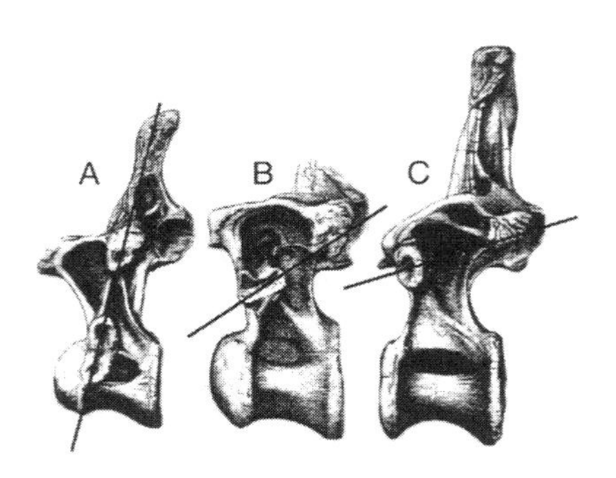

Figure 7. Dorsal vertebrae (thoracic region) 3–5 of Apatosaurus *showing axes formed at rib-vertebra joints. Rib movements resulting from differences in axis of rotation enhance thoracic expansion in a manner similar to that occurring in mammals. Drawings modified from Ostrom and McIntosh (1966).*

the foregoing, it is expected that a dinosaur the size of an elephant would take fewer than 10 breaths per minute due to its metabolic state (probably lower than modern endotherms) and its long neck (rapid breathing leaves a large proportion of the total volume in the trachea).

Large animals breathe less often than small animals of the same general type. Thus, an animal the size of *Apatosaurus* would very likely breathe approximately 5 times per minute, which is similar to the rate exhibited by elephants when lying down (Adams, 1981). This is in agreement with values we found from our own studies of air flow in elongated tracheae. Maximal effective forced breathing rates are roughly double the normal awake, but quiet, breathing rates. This is in agreement with our maximum for 12 cm tracheal diameter. Ruben and others (1987) indicate that rib breathing at high air volumes is both inefficient and exhausting. It is likely that the upper limit of ventilation rates is somewhere between 8 and 12 breaths per minute, because increasing the depth of respiration is a better strategy than increasing rate.

Apatosaurus's long neck would have diminished the effectiveness of rapid breathing as a strategy for enhanced oxygen delivery. As respirations increase, a point is reached where the constant volume of dead space air composes a large portion of total air moved but the time available for inspiration is decreasing. Thus, less air reaches the lungs. Increasing tracheal diameter to reduce resistance and increase breathing rates is also limited as a solution because dead space increases geometrically with diameter. *Apatosaurus* was probably confined to slow deep ventilation rates of ~5–10 per minute.

How much air could a dinosaur move with a single breath? Earlier, we estimated the lung volume of *Apatosaurus* at 1400 l. By calculating the increase in thoracic volume we know the maximum air moved per breath. Perry (1987) determined the compliance (or give) of the crocodile lung used in our model to be much greater than that of the thoracic wall. This means that under normal breathing forces the lung closely follows chest wall movement. It is reasonable to assume that thoracic expansion equals inspiratory volume of air. Because each rib moves in a different direction from its neighbor, this direction was averaged as 45° to the longitudinal axis of the chest, and thoracic expansion was approximated as the cosine of 45° of rib excursion.

Limits of movement were determined from average intercostal distances. Physiologists have long known that the length over which a skeletal muscle can contract is a constant fraction of its relaxed (starting) length. Muscle reaches maximum force and efficiency when shortened by about 20% of its relaxed length (McMahon, 1984). However, repetitive 20% contractions lead quickly to muscle fatigue, as occurred with the rats in Ruben and others' (1987) rib-breathing experiment. Shortening of 20% represents only temporary extreme exertions, but not normal tidal breathing. Repeated contractions generally involve 4% to 10% of resting length. By averaging the distance between ribs, the average distance moved by a contracting intercostal muscle could be derived. Assuming 10% contraction, this was about 1.25 cm.

Scalene muscles are a second set of muscles that assist intercostal muscles in inspiration. The contribution of the scalene muscles depends upon the geometry of the thorax in addition to the muscle's length. These muscles insert on the first two ribs in most animals and we would expect this to be the situation in sauropods. However, the small size of the first two ribs (the most likely point of insertion) argues against these powerful muscles as participating in respiratory movements. In mammals, scalene muscles become more effective when used in combination with postural movements. The flexible vertebral columns of mammals allow the scalene origin to move while contracting, thus extending the distance of respiratory movements. Dinosaur vertebral columns, in contrast, were inflexible due to the great overlap of prezygopophyses and postzygopophyses. Scalene muscles originate on neck vertebrae and would have pulled against the leverlike neck of *Apatosaurus* as the chest was drawn forward. Excessive effort must have been needed to keep the neck up if scalene muscles and gravity were pulling down at the same time. Further studies are needed to assess the contributions of this muscle set to overall chest movement, but it is not likely that scalene muscles were important inspiratory muscles in sauropods. It is likely that the most important respiratory movements were made with intercostal muscles.

Breathing sequences were also considered. Mammals breathe by active inspiration followed by passive expiration. In contrast, reptiles and birds breathe with active inspiratory and active expiratory phases. The most likely respiratory cycle for a dinosaur is one in which expiration is followed by a passive return to starting position: then an active inspiration phase is begun, followed by passive return to the neutral starting position. This biphasic breathing allows separate muscle groups to contract by 20%, which has the effect of doubling the distance of excursion distance. For *Apatosaurus*, this expansion is about 2.5–3.0 cm of movement, resulting in radial expansive movements of 1.8–2.1 cm. This corresponds to 330 l per breath, of which 78 l must be deducted for dead space of the trachea. Normal breathing was more likely to deliver about 80 l per breath, although less is needed if atmospheric oxygen were higher than PAL. Normal quiet breathing would require ~15–18 cycles to refresh lung air; humans require about 7 to 8 breaths to do this.

It appears that *Apatosaurus* would retain respiratory CO_2 unless metabolic rates were low, a conclusion reached by Daniels and Pratt (1992). Lower metabolic rates not only demand less of the respiratory system, but also contribute less CO_2 load to the system. This supports the argument for lower metabolic rates in sauropods presented in Dodson (1990) or in dinosaurs in general by Spotila and others (1991) based on thermal exchange rates and ecological grounds. Retained CO_2 might also have affected blood pH in sauropods, although it might have been less of a problem for dinosaurs with shorter necks.

The frequency of tidal breathing was probably reduced to slower patterns by the small nostrils. Slow and mildly turbulent air movement allows both for humidifying inspired air and recovery of moisture from expired air (Hillenius, 1992). The combination of biphasic breathing and long neck almost surely dictates a strategy of deep slow breathing. Maximal airflow through nostrils was distinctly lower than with mouth open as determined by results from our airflow experiments. Increases in ventilation frequency rather than depth reduces air volume moved per breath to exchange surfaces, even if the mouth were opened to reduce airway resistance and turbulence. Future analyses of breathing rhythm should account for the effects of locomotor activity. Stride action can enhance breathing efficiency and may alter frequency (Bramble and Carrier, 1983; Bramble and Jenkins, 1993; Alexander, 1993).

Mammalian breathing varies from relatively small tidal volumes in quiet breathing to the large vital capacities associated with deep forceful breaths. In humans, this ratio is about 1 to 10; in *Apatosaurus* this appears to be about 1 to 5 or less based on energy needs when breathing 21% oxygen. In 35% oxygen this ratio becomes ~1 to 7–8, a value compatible with life. When these limits are exceeded, anaerobic energy sources must be relied upon. The duration and intensity of possible activities become limited. When energy needs for two basic activities, walking and climbing a gentle slope, are compared with our data for air delivery, it becomes obvious that *Apatosaurus* cannot deliver oxygen sufficient for these activities, even at slow speeds. Lockely (1991) calculated walking speeds in *Apatosaurus* of 4 km/hr based on trackway evidence. If our ventilation calculations are correct, *Apatosaurus* could not reach these speeds in 21% O_2, but was capable of 5 to 6 km/hr in 32% O_2. When exotic ventilation mechanisms are excluded from consideration, the simplest explanation is that atmospheric oxygen levels were near or above 30%.

The rationale for assuming a refined crocodilian type of lung was discussed earlier. Oxygen can only diffuse through the thin exchange surfaces for eventual blood transport to body tissues. Although a crocodilian-type lung is proposed for dinosaurs, the partitioning of tissue for gas exchange as opposed to airways cannot be known; however, qualitative speculation is certainly warranted. In all likelihood, a far greater proportion of tissue was devoted to gas exchange than is known in modern crocodilians. Alligators and crocodiles have brief periods of activity followed by long periods of inactivity. Their low metabolic rate and sporadic activity patterns make it more efficient to tolerate anaerobic metabolic episodes than to carry the lung apparatus necessary for supporting sustained activity, which is seldom used. In mammals, the proportion of lung surface in large mammals is greater than would be expected from allometric calculations (Schmidt-Nielsen, 1984; Taylor and Weibel, 1981; Weibel and others, 1981). Such proportions may have been true for dinosaurs as well. Jones and Lindstedt (1993) found that only rarely is lung surface area less than adequate to maximal activity. At this point, it is probably best to analyze overall oxygen delivery capabilities and assume that, with some reservations, lung surface was generally adequate to the task.

One significant exception to this surface adequacy is observed in thoroughbred racehorses, which are capable of vigorous exercise but have relatively small lung surface area (Jones and Lindstedt, 1993). The high pulmonary pressures needed to circulate blood at rates needed to support maximum performance also causes pulmonary bleeding. It appears that pulmonary blood flow in thoroughbreds is at design limits. It also emphasizes adaptational limits. It is interesting to note that thoroughbreds increase performance by an average of 15% when breathing 25% O_2. If dinosaurs breathed 30% O_2 or above, it may have considerably extended their range of activities, as demonstrated in horses with their small lung surface area or in humans when the lung surface is reduced through emphysema.

Blood transport of oxygen is the final major factor in supporting active tissues. Figure 4 was used in comparing the oxygen-carrying capacity of vertebrate blood. Oxygen transport is a function of the number of erythrocytes per volume of total blood (the hematocrit of the blood) and the amount of hemoglobin per cell. This is presuming that red cells are uniform and not pathological in nature. At high cell densities, cells deform, blood viscosity increases dramatically, and oxygen transport decreases (Burton, 1972). There is an optimum cell concentration at which oxygen can be transferred for a given cell size. In vertebrates, the optimal hematocrit for efficient oxygen transport is very predictable and is affected by the size of the cells. Blood cell size and shape limit possible increases in hematocrit to compensate for low ambient oxygen, and the amount of hemoglobin that may be packed into a single cell is likewise limited.

Dinosaurs probably had nucleated blood cells similar to bird or smaller reptilian erythrocytes. This is because both reptiles and bird have nucleate erythrocytes. In general, reptiles have larger erythrocytes than birds. Dinosaur cells were probably closer to bird dimensions than reptilian because of reportedly high activity levels as evidenced by trackway data (Lockley, 1991), and possibly evidence for migration (Spotila and others, 1991). This demands greater oxygen-delivery capabilities and argues for a smaller, more avian, cell size.

The efficiency of these nucleated erythrocytes in oxygen delivery must also be considered. Nucleate erythrocytes are whole functioning cells that use oxygen to live. Oxygen usage during transport decreases oxygen available to the tissues, particularly when the long distances and times from dinosaur heart to tissues are considered. How substantial this loss would be has yet to be determined, but could be important when tissue demands were great and even small amounts of additional oxygen would be useful.

We have argued that *Apatosaurus* could exist only at low activity levels if it had lived in modern atmospheres. Our results suggest that under modern atmospheric conditions *Apatosaurus* would have had limited metabolic range or scope with which to meet the challenges of life. However, if the Jurassic atmosphere contained elevated oxygen concentrations, our results indicate that *Apatosaurus* would have been able to sustain the active metabolism inferred for it from independent studies of skeletal structure. These results are in agreement with the atmospheric composition put forward by Landis and others (1993). However, the postulation of elevated oxygen is the most parsimonious explanation of sauropod ventilation and support of normal activity levels, irrespective of uncertainties in the physical measurement of fossilized Mesozoic gas samples.

It is important to address the applicability of this study to an understanding of other dinosaur groups. Ornithischians and other sauropods have rib-vertebra angles similar to that shown for *Apatosaurus*. We do not know whether other dinosaurs used rib-breathing exclusively. However, given the diversity of dinosaur morphology we regard the existence of alternative breathing system designs as likely. What seems apparent is that oxygen-supply pathways and mechanisms are not as well developed as in modern endotherms. Although arguments may be advanced about lung type or hemoglobin and blood characteristics, these must necessarily be confined to the realm of speculation. The ventilation method remains the most addressable and significant aspect of the respiratory system. Mass-dependent oxygen needs are independent of the ectothermy-endothermy debate and enable estimated performance limits to be compared over a broad range of possible atmospheric compositions.

If future analyses of performance yield similar results for other dinosaur groups, we expect that the usual adaptive responses to decreased oxygen would be limited. This has great appeal in accounting for survival patterns across the K-T boundary. Low metabolic demands require little of respiratory systems. Amphibians that lack lungs but breathe through their skin are but one example. Thus, small ectotherms (along with diving animals) would be expected to survive. Endothermic birds and mammals have efficient ventilation systems linked to systems with an excellent ability to transport and absorb oxygen. Endotherms too would largely be expected to survive. Dinosaurs were obviously well adapted to their environment, possibly one with an oxygen-rich atmosphere. Any reduction of oxygen would have reduced their viability through reduction in respiratory system reserves available for stressful situations.

CONCLUSIONS

From the Late Jurassic or Early Cretaceous, *Apatosaurus* appears to have had a respiratory system inadequate for active lifestyle in a modern atmosphere. *Apatosaurus* probably had a metabolic type intermediate between ectothermy and endothermy. Ventilation was by rib movements that were limited by the mechanical constraints of muscle length. The primary respiratory muscles were intercostal and activated in biphasic respiration patterns. Rapid respiration was probably hindered by the small size of the nostrils and the elongated trachea. Limits of performance were compatible with life if elevated atmospheric oxygen is assumed. These results are consistent with the biotic predictions of McAlester (1970) and Erben and others (1979), along with the atmospheric composition data of Landis and coworkers. The lung type was probably an advanced form of crocodilian lung, and was probably adequate for a moderately active terrestrial life. Blood cells were probably nucleated with oxygen-carrying capacities similar to those of the lower ranges of modern birds. Cell size was large and had the potential to strongly affect blood viscosity. Changes in atmospheric oxygen may have provided the sieve that reduced fitness in dinosaurs but had lesser effects on other vertebrate groups. Surviving vertebrates include two groups: ectothermic animals with low oxygen demands and endothermic animals with efficient respiratory systems.

ACKNOWLEDGMENTS

We thank the staff and curators of the Field Museum, Chicago, Illinois, for permitting the use of the museum collections for this study; Olivier Reipel, Ray Leo, Bill Simpson, Alan Resetar, and Tamara Biggs of the Field Museum for their technical expertise and for the loan of equipment and skeletal specimens; Nancy Machin of Purdue University North Central for her assistance in and critiques of animal experiments; and John S. McIntosh and Peter Dodson of the University of Pennsylvania, and James Hobson of the University of Chicago for the many useful suggestions and their detailed knowledge of vertebrates that helped to crystallize many of the key ideas expressed in this paper.

REFERENCES CITED

Adams, J., 1981, Wild elephants in captivity: Carson, California, Center for the Study of Elephants, 201 p.
Alexander, R. M., 1989, Dynamics of dinosaurs and other extinct giants: New York, Columbia University Press, 167 p.
Alexander, R. M., 1993, Breathing while trotting: Science, v. 262, p. 196–197.
Anderson, J. F., Hall-Martin, A., and Russell, D. A., 1985, Long-bone circumference and weight in mammals, birds and dinosaurs: Zoological Society of London Journal, v. 207, p. 53–61.
Bakker, R. T., 1972, Locomotor energetics of lizards and mammals compared: Physiologist, v. 15, p. 278–283.
Bakker, R. T., 1986, The dinosaur heresies: New York, William Morrow, 481 p.
Bennett, A. F., 1980, The metabolic foundations of vertebrate behavior: BioScience, v. 30, p. 452–456.
Bennett, A. F., and Ruben, J. A., 1979, Endothermy and activity in vertebrates: Science, v. 206, p. 649–654.
Bramble, D. M., and Carrier, D. R., 1983, Running and breathing in mammals: Science, v. 219, p. 251–256.
Bramble, D. M., and Jenkins, F. A., Jr., 1993, Mammalian locomotor-respiratory integration: Implications for diaphragmatic and pulmonary design: Science, v. 262, p. 235–240.
Britt, B., 1993, Pneumatic postcranial bones in dinosaurs and other archosaurs [Ph.D. thesis]: Calgary, Alberta, University of Calgary, 383 p.
Burton, A. C., 1972, Physiology and biophysics of the circulation: Chicago, Illinois, Year Book Medical Publishers, 226 p.
Calder, W. A., 1984, Size, function and life history: Cambridge, Massachusetts, Harvard University Press, 431 p.
Cerling, T., 1991, Carbon dioxide in the atmosphere: Evidence from Cenozoic and Mesozoic paleosols: American Journal of Science, v. 291, p. 77–90.
Daniels, C. B., and Pratt, J., 1992, Breathing in long necked dinosaurs: Did the sauropods have bird lungs?: Comparative Biochemistry and Physiology, v. 101A, p. 43–46.
Decramer, M., 1993, Respiratory muscle interaction: News in Physiological Science, v. 8, p. 121–124.
Dodson, P., 1990, Sauropod paleoecology, *in* Weishampel, D. B., Dodson, P., and Osmolska, H., eds., The dinosauria: Berkeley, University of California Press, p. 402–407.
Dunham, A. E., Overall, K. L., Porter, W. P., and Forster, C. A., 1989, Implications of ecological energetics and biophysical and developmental constraints for life-history variation in dinosaurs, *in* Farlow, J. O., ed., Paleobiology of the dinosaurs: Geological Society of America Special Paper 238, p. 1–19.
Erben, H. K., Hoefs, J., and Wedepohl, K. H., 1979, Paleobiological and isotopic studies of eggshells from a declining dinosaur species: Paleobiology, v. 5, p. 380–414.
Farlow, J. O., 1976, A consideration of the trophic-dynamics of a Late Cretaceous large-dinosaur community, Oldman Formation: Ecology, v. 57, p. 841–857.
Fedde, M. R., 1976, Respiration, *in* Sturkie, P. D., ed., Avian physiology (third edition): New York, Springer Verlag, p. 122–145.
Gilmore, C. W., 1936, Osteology of Apatosaurus with special reference to specimens in the Carnegie Museum: Carnegie Museum Memoirs, v. 11, p. 175–300.
Hatcher, J. B., 1901, Diplodocus Marsh, its osteology, taxonomy and probable habits, with a restoration of the skeleton: Carnegie Museum Memoirs, v. 1, p. 1–64.
Hillenius, W. J., 1992, The evolution of nasal turbinates and mammalian endothermy: Paleobiology, v. 18, p. 17–29.

Jones, J. H., and Lindstedt, S. L., 1993, Limits to maximal performance: Annual Review of Physiology, v. 54, p. 547–69.

Landis, G. P., Rigby, J. K., Sloan, R. E., and Hengst, R., 1993, Pele hypothesis: A unified model of ancient atmospheres and biotic crisis: Geological Society of America Abstracts with Programs, v. 25, no. 7, p. A-362.

Lindstedt, S. L., and Jones, J. H., 1987, Symmorphosis: The concept of optimal design, *in* Feder, M., and others, eds., New directions in ecological physiology: London, Cambridge University Press, p. 290–310.

Lockley, M., 1991, Tracking dinosaurs: A new look at an ancient world: London, Cambridge University Press, 238 p.

McAlester, A. L., 1970, Animal extinctions, oxygen consumption, and atmospheric history: Journal of Paleontology, v. 44, p. 405–409.

McIntosh, J. S., 1990, Sauropoda, *in* Weishampel, D. B., Dodson, P., and Osmolska, H., eds., The dinosauria: Berkeley, California, University of California Press, p. 345–401.

McMahon, T. A., 1984, Muscles, reflexes and locomotion: Princeton, New Jersey, Princeton University Press, 331 p.

McNab, B. K., 1978, The evolution of endothermy in the phylogeny of mammals: American Naturalist, v. 112, p. 1–21.

Ostrom, J. H., and McIntosh, J. S., 1966, Marsh's dinosaurs; the collections from Como Bluff: New Haven, Connecticut, Yale University Press, 388 p.

Paladino, F. V., O'Connor, M. P., and Spotila, J. R., 1990, Metabolism of Leatherback turtles, gigantothermy, and thermoregulation of dinosaurs: Nature, v. 344, p. 858–860.

Pennycuick, C. J., 1992, Newton rules biology: Oxford, Oxford University Press, 111 p.

Perry, S. F., 1987, Functional morphology of the lungs of the Nile crocodile, *Crocodylus niloticus*: Non-respiratory parameters: Journal of Experimental Biology, v. 134, p. 99–117.

Rigby, J. K., Jr., Newman, K. R., Smit, J., van der Kaars, S., Sloan, R. E., and Rigby, J. K., 1987, Dinosaurs from the Paleocene Part of the Hell Creek Formation, McCone County, Montana: Palaios, v. 2, p. 296–302.

Riggs, E. S., 1903, Structure and relationships of opisthocoelian dinosaurs. Part 1. *Apatosaurus* Marsh: Field Columbian Museum, Geological Series II, p. 165–196.

Romer, A. S., 1956, Osteology of the reptiles: Chicago, Illinois, University of Chicago Press, 772 p.

Ruben, J. A., Bennett, A. F., and Hisaw, F. L., 1987, Selective factors in the origin of the mammalian diaphragm: Paleobiology, v. 13, p. 54–59.

Schmidt-Nielsen, K., 1984, Scaling: Why is animal size so important?: London, Cambridge University Press, 241 p.

Schopf, T. J., Farmanfarmaian, M. A., and Gooch, J. L., 1971, Oxygen consumption rates and their paleontologic significance: Journal of Paleontology, v. 45, p. 247–252.

Sloan, R. E., Rigby, J. K., Jr., Van Valen, L. M., and Gabriel D., 1986, Gradual dinosaur extinction and simultaneous ungulate radiation in the Hell Creek Formation: Science, v. 232, p. 629–633.

Spotila, J. R., O'Connor, M. P., Dodson, P., and Paladino, F., 1991, Hot and cold running dinosaurs: Body size, metabolism and migration: Modern Geology, v. 16, p. 203–227.

Taylor, C. R., and Weibel, E. R., 1981, Design of the mammalian respiratory system. I. Problem and strategy: Respiratory Physiology, v. 44, p. 1–10.

Weibel, E. R., Taylor, C. R., Gehr, P., Hoppeler, H., Mathieu, O., and Maloiy, G. M. O., 1981, Design of the mammalian respiratory system. IX. Functional and structural limits for oxygen flow: Respiratory Physiology, v. 44, p. 151–164.

14

The Cretaceous-Tertiary Boundary in the Nanxiong Basin (Continental Facies, Southeast China)

Johannes Stets, *Geological-Paleontological Institute, University of Bonn, Bonn, Germany*, Abdul-Rahman Ashraf, *Geological-Paleontological Institute, University of Bonn, Bonn, Germany*, Heinrich Karl Erben, *Geological Paleontological Institute, University of Bonn, Bonn, Germany*, Gabriele Hahn, *Geological-Paleontological Institute, University of Cologne, Cologne, Germany*, Ulrich Hambach, *Geological-Paleontological Institute, University of Cologne, Cologne, Germany*, Klaus Krumsiek, *Geological-Paleontological Institute, University of Cologne, Cologne, Germany*, Jean Thein, *Geological-Paleontological Institute, University of Bonn, Bonn, Germany*, and Paul Wurster, *Geological-Paleontological Institute, University of Bonn, Bonn, Germany*

INTRODUCTION

Biological causalities of the dinosaur extinction near the Cretaceous-Tertiary (K-T) boundary were the background of our research up to the late 1970s (Erben, 1969, 1970, 1972; Erben and others, 1979). This work gained new impetus when in 1980 the bolide impact hypothesis was presented with the discovery of the iridium anomaly at the K-T boundary in the Gubbio section (Alvarez and others, 1980). Interdisciplinary research in southern France and northern Spain by our group started at that time (Erben, 1985; Erben and others, 1983). Meanwhile, the discussion concerning the nature and causes of K-T boundary faunal and floral changes continued. This debate was invigorated further by redescription of the buried Chicxulub structure on the Yucatán Peninsula of Mexico (Hildebrand and others, 1991; Hildebrand and Stansberry, 1992; Sharpton and others, 1992). On the basis of radiometric dating, this structure is thought to be contemporaneous with the K-T boundary events.

The controversy on the pros and cons of the bolide impact theory has resolved into several alternative models. In contrast to the catastrophic extraterrestrial impact theory, the K-T boundary extinction has also been attributed to long-term ecologic changes resulting from terrestrial processes such as volcanism or mantle superplume degassing (Landis and others, 1993).

In the 1980s, rapid floral changes were discovered in continental red beds at the K-T boundary (Krassilow, 1981; Hickey, 1984; Tschudy, 1984; Tschudy and Tschudy, 1986), in the western Mediterranean (Ashraf and Erben, 1986), in Chile (Stinnesbeck, 1986), and in Brazil (Ashraf and Stinnesbeck, 1988). If this event is truly worldwide in scope, contemporaneous floral changes should have occured in eastern Asia as well.

Since 1982, research work in China was made possible by a geoscientific cooperation agreement between Max Planck Gesellschaft, Munich (Germany), and Academia Sinica, Beijing (Peoples Republic of China). The Nanxiong basin (Guangdong Province, southeastern China), ~200 km north of Guangzhou (Canton) was chosen for research (Fig. 1). Here, rather continuous outcrops in different sections provide representative observations for the K-T boundary interval.

This paper deals with the results gained in southeastern China during 1983 to 1986 in the field as well as in the laboratories at Bonn and Cologne universities. It takes into account a palynological change caused by a change in paleoclimate, the local extinction of a dinosaur population, geochemical anomalies caused by the enrichment of trace elements, and magnetostratigraphic investigations concerning magnetochrons 29 to 30. In addition, the geological situation had to be explored carefully, because such a serious event should have left traces in the sedimentary

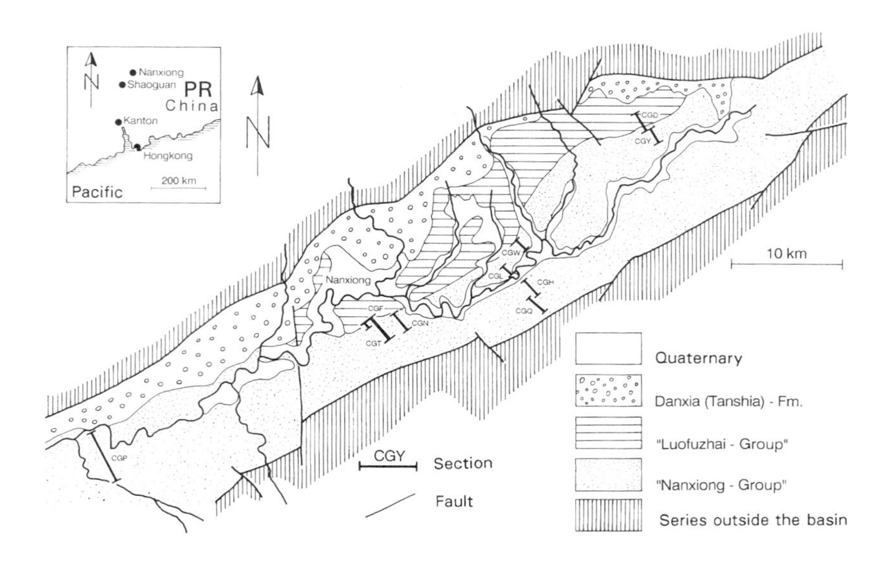

Figure 1. Geologic map of the Nanxiong basin, southeastern China. Section abbreviations are as follows: CGY = Yan Mei Ken, CGD = Datang, CGQ = Quiang Tao, CGH = Han Ping, CGL = Long Feng Tang, CGW = Wu Tai Gang, CGN = Nanxiong, CGT = Tan Tien, CGF = Fong Men Au, CGP = Pubei.

record. Conclusions presented here were developed by an interdisciplinary group of earth scientists who published their results in detail in Erben and others (1994).

REGIONAL SETTING AND STRATIGRAPHY

The Nanxiong basin is a grabenlike intramontane structure 9 to 18 km wide, its long axis extending northeast-southwest (Figs. 1 and 2). The basin is bounded by normal faults on both sides. All strata dip gently toward the northwest (0°–25°). The graben is filled with a >3000-m-thick sequence of Upper Cretaceous to lower Tertiary continental red beds. Marine sediments are entirely lacking. Subsidence that took place syngenetically beginning in the Late Cretaceous (Campanian to Maastrichtian) along the bordering normal faults explains the enormous sediment thickness generated in this narrow basin.

The blocks on both sides of the graben structure consist mainly of fine- to coarse-grained Jurassic granites and related rocks. Two generations can be distinguished, the 175–185 Ma Yanshanian granitoids, and younger intrusions of 135–160 Ma (Yang Zunyi and others, 1986). Both blocks are built up of Paleozoic (Cambrian-Ordovician) siliciclastic strata (Fig. 2) up to 8000 m in thickness that contain slates and quartzites of flyshoid character. In places, these strata are overlain unconformably by Upper Carboniferous limestones. Remnants of all these rock types are found as detrital components of the Cretaceous to Tertiary sediments that filled the Nanxiong basin, and are now conglomerates and sandstones. In addition, these sed-

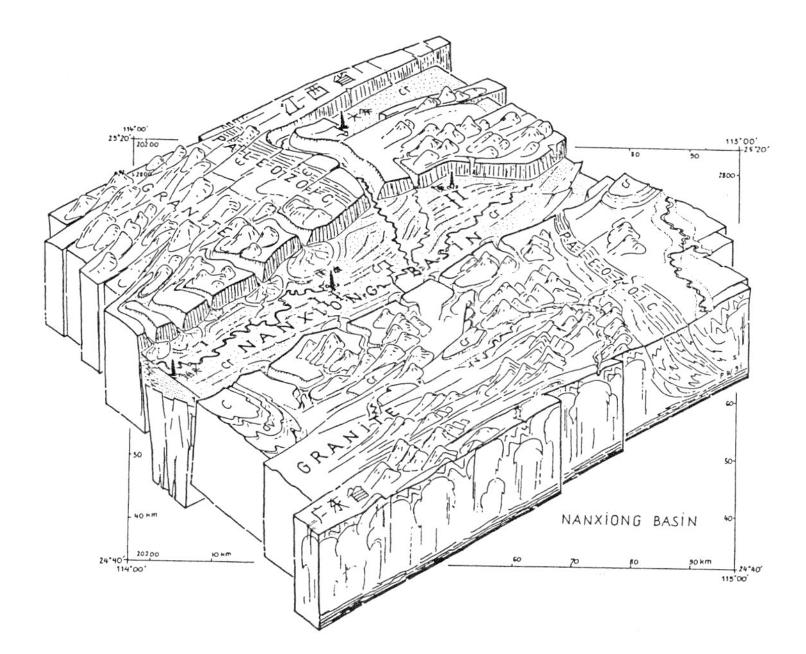

Figure 2. Block diagram showing the intramontane Nanxiong basin and the surrounding blocks. Cr = Cretaceous sedimentary rocks.

iments contain mainly feldspars and quartz, and thus form many types of arcosic rocks. There are no volcanic rocks in the Nanxiong basin except in the southwestern part at Pubei, where several basalt flows have been reported in the uppermost Cretaceous. Whole-rock $^{40}Ar/^{39}Ar$ radiometric dating yielded a maximum age of 66.7 ± 0.3 Ma for these basalts (Rigby and others, 1993).

Stratigraphic subdivision of the red beds started in 1928, following lithostratigraphic criteria. In addition to ostracodes (Hao Yi Chun and others, 1979), indications for age determination are remnants of dinosaurs in the lower part, and mammalian fossils in the upper part of the thick sedimentary pile. Up to 1962, two series were mainly recognized by Chinese authors as well as by the Nanshiung and Kwangtung geological parties. In 1991, Zhao Zikui and others created a new subdivision following the 1983 to 1986 field work of our joint Sino-German project. This subdivision is shown in Figure 3. According to the Chinese opinion (Zhao Zikui and others, 1991), the K-T boundary is identified according to field data and lithological criteria. This boundary separates the Nanxiong and the Luofuzhai groups, specifically the Pingling and Shanghu formations. The boundary coincides with the more sandy series in the lower part and more silty to muddy series in the upper part of the section. This boundary is characterized by a slight color change within the red beds, the size of carbonate nodules, and the extinction of dinosaurs within a reversed magnetochron. Starting with the Shanghu Formation, mammals of *Bemalambda* type occur that indicate a Paleocene age for the overlying series.

From the field geologist's point of view, the K-T boundary (in the sense of Zhao Zikui and others, 1991) is a very reliable boundary that can be followed throughout the Nanxiong basin. However, we do not believe that the criteria men-

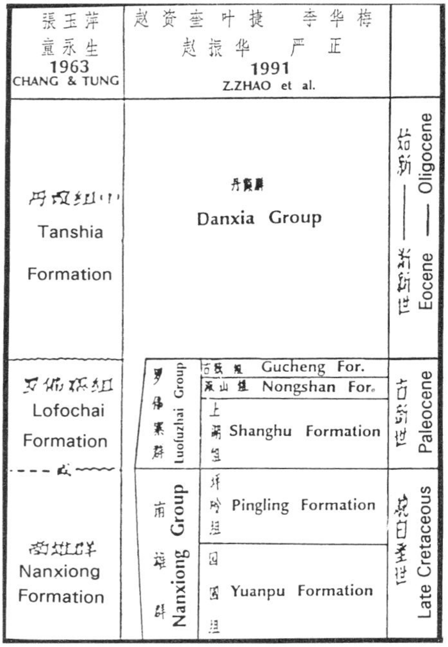

張玉萍 童永生 1963 CHANG & TUNG	赵资奎 叶捷 李华梅 赵振华 严正 1991 Z.ZHAO et al.		
Tanshia Formation	Danxia Group		Eocene — Oligocene
Lofochai Formation	Luofuzhai Group	Gucheng For.	Paleocene
		Nongshan For.	
		Shanghu Formation	
Nanxiong Formation	Nanxiong Group	Pingling Formation	Late Cretaceous
		Yuanpu Formation	

Figure 3. Stratigraphy of the continental red beds of the Nanxiong basin, southeastern China (Chang and Tung, 1963).

tioned above are significant at a continental scale, or that they bear a critical relation to the internationally accepted K-T boundary (Erben and others, 1994).

GEOLOGICAL AND SEDIMENTOLOGICAL SETTING

Geological and sedimentological field work pursued two main objectives. (1) Registration and characterization of basin development in space and time from its beginning up to the K-T boundary. Taking all the data available along one complete transect through the basin, an event of the magnitude proposed for the K-T boundary should be recognizable in the sedimentary record. In a continental paleoenvironment, large catastrophes should have left behind traces even if the section contains gaps or the rocks are reworked or altered by diagenesis. (2) Registration of the K-T boundary interval itself within a representative columnar section based on geological, lithological, and sedimentological criteria. In this way, a basis was to be established for paleontological, geochemical, and paleomagnetic research as well as for the interpretation of their results.

Several sections in the Nanxiong basin (Fig. 1) were chosen (together with our Chinese colleagues of Academia Sinica) to accomplish these objectives. All formations up to the Paleocene were measured at a scale of 1:200 using a Jacobs staff (Kummel, 1943; Wurster and Stets, 1979). Field data were presented in Erben and others (1995). Along these cross sections, neither serious faults nor greater gaps or even unconformities were encountered. Thus, a reliable correlation of the local geology was set up for basin analysis and environmental studies.

Using lithological and sedimentological criteria, the following seven formations can be distinguished in the red sediments of the Nanxiong basin (Fig. 4).

Formation I consists mainly of eroded granitic material forming a coarse-grained basal arcose of ~150 m. It looks like a deeply weathered granite. Pebbles of quartz and quartzite, however, show that the material was reworked several times. Bedding is of minor importance. This formation represents the proximal part of an alluvial fan near the southern margin of the basin.

Formation II contains well-bedded conglomerates, sandstones, and siltstones up to 250 m in thickness in the Quiang Tao section (CGQ) (Figs. 1 and 4). Several fining-up cycles document fluvial influence in the proximal to mid-fan environment at the southern margin of the basin.

Formation III is as thick as 935 m in the Quiang Tao (CGQ) and Han Ping (CGH) sections. It consists mainly of red siltstones and mudstones representing the distal portion of the alluvial fan mentioned above. In that large flood plain, dessication cracks occur as well as calcareous nodules, and trace fossils indicate low-energy conditions that allowed organisms to colonize the sediment.

Farther toward the interior of the basin (toward the northwest) Formation IV represents the fluvial facies of a low-sinuosity meandering river system overlying the mud flat facies. Coarse-grained red to white sandstones and arcoses as well as conglomerates represent channel fill facies alternating with crevasse splay sediments and flood-plain mud and siltstones. Formation IV is as thick as 450 m. This fluvial system forms one of the drainage systems, possibly the central drainage system, of the Nanxiong depression at that time. Fluvial transport

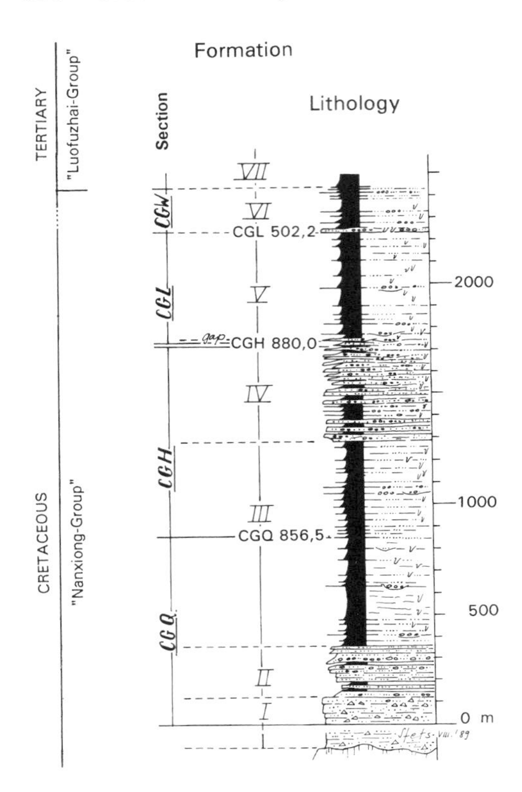

Figure 4. Generalized stratigraphic succession of the Upper Cretaceous to lower Tertiary sediments in the Nanxiong basin; the Cretaceous-Tertiary interval is indicated, abbreviations as in Figure 1.

was directed toward the west, according to the sparse cross-bedding. Within this fluvial environment, the first indications of the presence of dinosaurs consist of footprints and eggshell fragments.

Formation V is up to 470 m thick in the Long Feng Tang section (CGL, Fig. 4). It consists mainly of mudstones. It is a younger facies equivalent of Formation III. Calcareous nodules and dessication cracks indicate periods of dry conditions, with calcrete formation in the flood-plain environment. In contrast, ripple marks, calcareous layers containing ostracodes and characean *oogonia*, as well as bioturbation represent intervals of humid conditions and the presence of lakes on the mud flat. Dinosaur remains are rare, except for some eggshell fragments and footprints.

In the Wu Tai Gang section (CGW, Fig. 4), Formation VI coincides with Formation IV in respect to lithology and paleoenvironment. Here, fragments of dinosaur eggshells are more frequent and widespread in the sediments of the meandering fluvial system.

Formation VII indicates a fluvial flood-plain environment, as do Formations III and V. Up section, several channel fills with coarse sands show that the mud flat was nourished by a widespread channel system.

There is only one horizon of larger calcareous nodules that forms the top of Formation VI throughout the basin. Dinosaur eggshell fragments are only found up to this horizon. Within Formation VII remnants of dinosaurs are totally lacking, whereas mammal fossils, such as a lower jaw bone of *Bemalambda* sp. found in 1984, are present.

Estimates of sedimentation rates for the Nanxiong basin fill are diffcult because a clear chronostratigraphic classification of Formations I to VII is not possible. Combining paleontological data with results of the paleomagnetic research, an average sediment accumulation rate of ~40 cm/ka seems realistic. This value lies in the range of values typically calculated for intraplate basins (Einsele, 1992).

Other than the marginal facies of Formations I and II, successive stratigraphic sections described from the southern margin toward the central part of the basin document the change from a high-energy channel environment to the low-energy flood-plain facies of a meandering river–alluvial fan system (Fig. 5). This seems applicable to Formations III to VII, as well as to Formation VI at a smaller scale. According to our results, the K-T boundary lies within Formation VI, which is characterized for the most part by high-energy sedimentation. To compensate for reworking as well as for rapid lateral facies changes, several sections were taken across the K-T boundary interval at different locations (Fig. 1). Without exception, the same conditions were found everywhere. As predicted by the bolide impact theory, a grave hiatus and/or sudden change in the sedimentary record should be expected, caused by the consequences of such a worldwide catastrophic event (Hsü, 1980; Tollmann and Tollmann, 1993). However, no such indications have been found in the Nanxiong basin. Nevertheless, based on all stratigraphic data, the K-T boundary must be located somewhere in the interval of Formations VI and VII (e.g., in the Pingling and Shanghu formations interval; Fig. 6).

RESULTS

Results of our interdisciplinary study on the K-T boundary are discussed in reference to the Yan Meiken (CGY) and Datang (CGD) sections (Fig. 6), taken about 30 km east of the city of Nanxiong (Fig. 1). These sections proved to be the most complete, and span Formations V to VII. The Chinese scientists locate

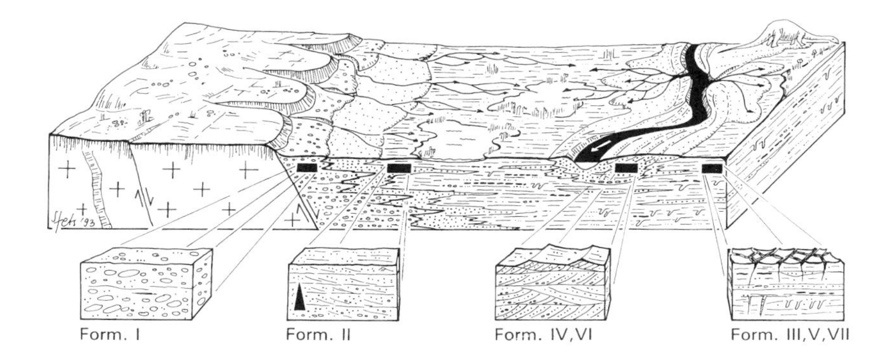

Figure 5. Block diagram showing sedimentary environment and processes within the Nanxiong basin during the Late Cretaceous to early Tertiary; I–VII correspond to formations and facies listed in Figure 4.

the K-T boundary at CGD meter 161, at the Pingling/Shanghu formations boundary (Zhao Zikui and others, 1991).

Palynology

We collected 425 samples for palynological research in 1983 and 1984. Preparation followed the principles and method of Kaiser and Ashraf (1974). All the specimens prepared are deposited in the paleobotanical department of the Paleontological Institute at Bonn University. Duplicates of the samples are stored at the Institute of Geology and Paleontology of Academia Sinica at Nanjing. Due to the sedimentary environment the recovery of palynomorphs was lower than expected. This is due to the lack of floral remnants and coal seams, as well as to reworking and diagenesis in the red beds.

We found 32 genera and 41 species, including some dinoflagellates and acritarchs (Fig. 7). These taxa suggested placement of the K-T boundary in the

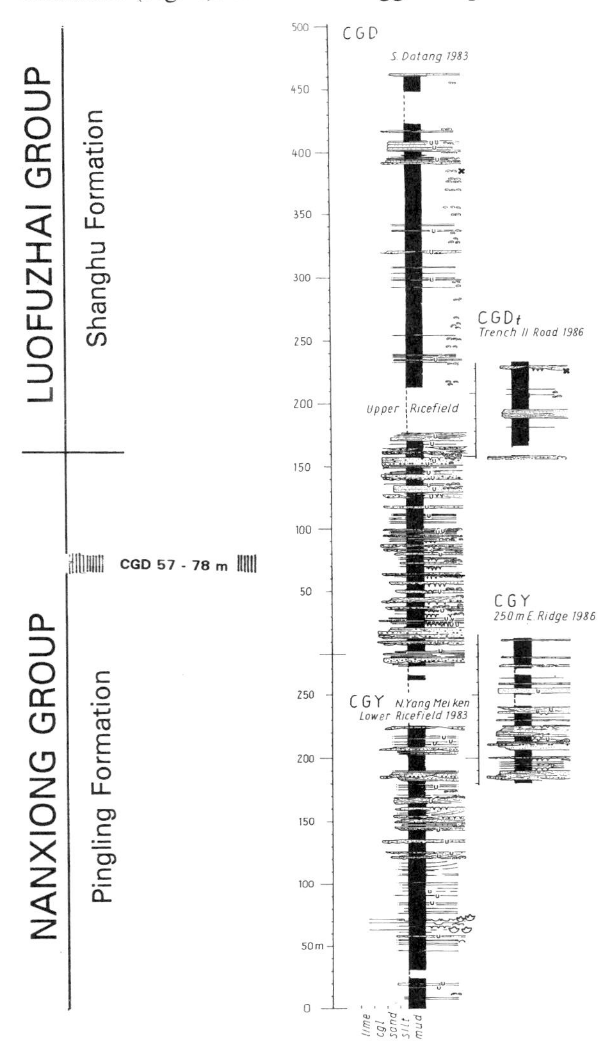

Figure 6. Generalized standard sections Yan Mei Ken (CGY) and Datang (CGD) including the Cretaceous-Tertiary boundary interval at CGD, 57–78 m.

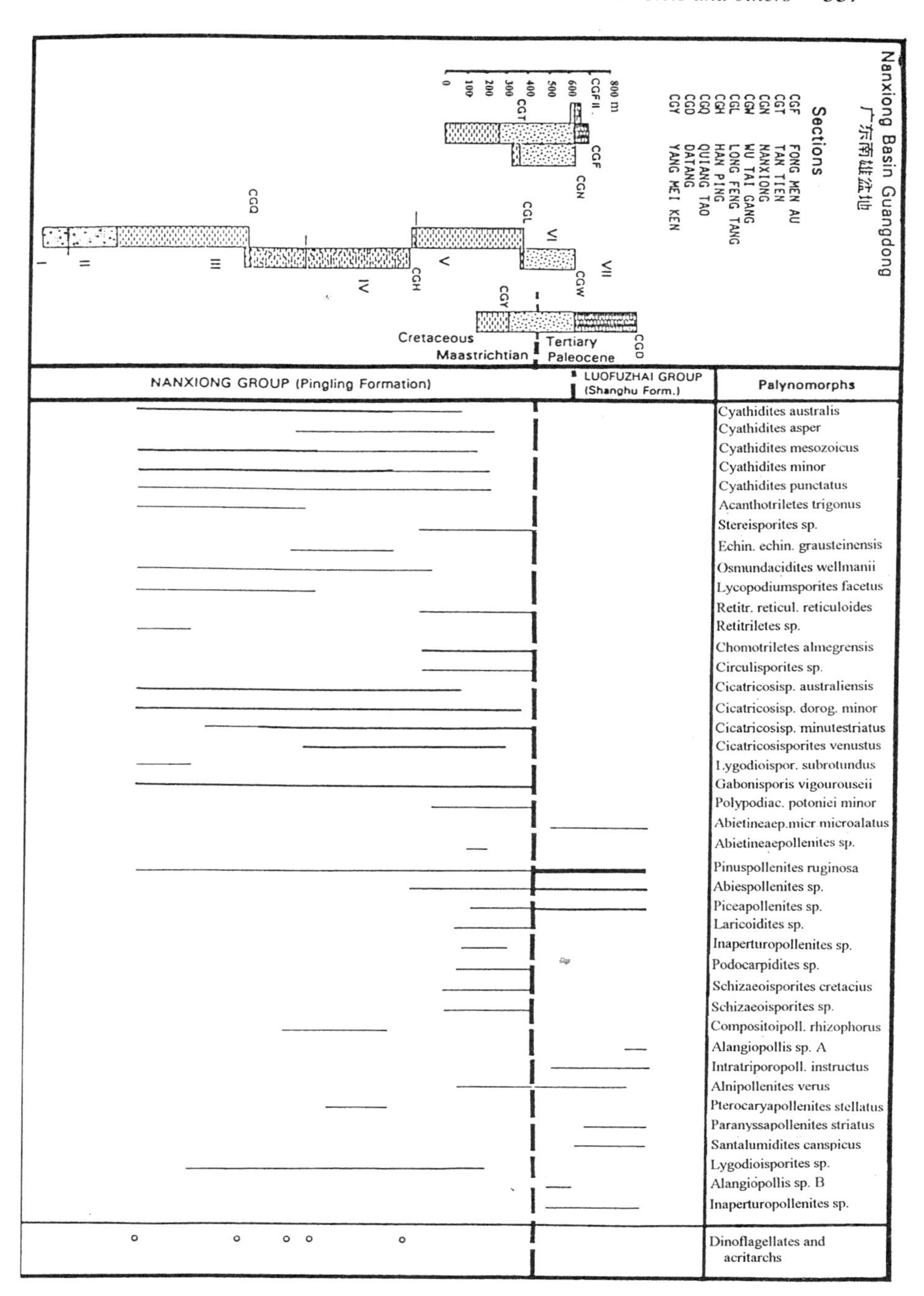

Figure 7. Results of palynological investigations in the red sediments of the Nanxiong basin. The Cretaceous-Tertiary boundary is indicated by the dashed horizontal line.

Datang section, somewhere within the interval from CGD meters 57 to 78 (Fig. 6). Taxa indicating a Late Cretaceous age are found up to CGD meter 57, and a mixed floral assemblage is as high as CGD meter 78. Palynomorphs of solely Tertiary age occur above meter 78. Therefore, the K-T boundary must lie within this 21 m interval. However, no distinct boundary layer or any sediment of anomalous character was found here or elsewhere in the Nanxiong basin within equivalent strata. The paleoenvironment did not change, not even for a short time.

A floral change took place in the interval mentioned above, making possible a statement with respect to climatic change. The red beds of Formations II to VI up to the K-T boundary interval (Datang section: CGD meters 57–78) yielded 24 genera and 32 species. They belong to the Cyathacea, Selaginellacea, Osmun-

TABLE 1

Pollen and spores of the Late Cretaceous and early Tertiary of the Nanxiong basin sediments, southeastern China

Pollen and Spores
Acanthotriletes trigonus NILSSON 1958
Chomotriletes almegrensis POCOCK 1962
Cicatricosisporites australiensis (COOKSON) POTONIÉ 1956
Cicatricosisporites dorogensis minor KEDVES 1961
Cicatricosisporites minutestriatus (BOLKHOVITINA 1961) POCOCK 1964
Cicatricosisporites venustus DEAK 1963
Circulisporites sp.
Cyathidites australis COUPER 1953
Cyathidites asper (BOLKHOVITINA 1953) DETTMANN 1963
Cyathidites mesozoicus (THIERGART 1949) POTONIÉ
Cyathidites minor COUPER 1953
Cyathidites punctatus (DEL. & SPRUMONT 1955) DELCOURT, DETTMANN & HUGHES 1963
Echinatisporis echinoides grausteinensis KRUTZSCH 1963
Gabonisporis vigourouseii BOLTENHAGEN 1967
Lycopodiumsporites facetus DETTMANN 1963
Lygodioisporites sp.
Lygodioisporites subrotundus
Osmundacidites wellmanii COUPER 1953
Polypodiaceaesporites potoniei minor
Retitriletes reticuloides reticuloides KRUTZSCH 1963
Retitriletes sp.
Schizaeoisporites cretacius (KRUTZSCH 1954) POTONIÉ 1956
Schizaeoisporites sp.
Stereisporites sp.
Abiespollenites sp.
Abietineaepollenites microalatus POTONIÉ 1951 microalatus
Abietineaepollenites sp.
Alangiopollis sp. A and B
Alnipollenites verus ex. POTONIÉ 1934
Compositoipollenites (R. POTONIÉ 1934) rhizophorus POTONIÉ 1960
Inaperturopollenites sp.
Intratriporopollenites instructus (R. POTONIÉ & VENITZ 1934) THOMSON & PFLUG 1952, 1953
Laricoidites sp.
Paranyssapollenites striatus (SUNG & LEE 1976) SONG & LI 1986
Piceapollenites sp.
Pinuspollenites ruginosa (STANLEY 1965) OLTZ 1969
Podocarpidites sp.
Pterocaryapollenites stellatus (R. POTONIÉ 1931) THIERGART 1937
Rhoipites megadolium SONG & LI 1986
Santalumidites canspicus SONG & LI 1986
Triatrioppollenites bituitus (POTONIÉ 1931) THOMSON & PFLUG 1953

dacea, Lycopodiacea, Marsiliacea, Asteracea, and Juglandacea, suggesting a warm and humid, perhaps tropical, climate (Table 1).

In contrast, the flora above the boundary interval (Datang section: above CGD meter 78), in the uppermost part of the Pingling Formation (Nanxiong Group) and the lowermost Shanghu Formation (Luofuzhai Group), contains only 14 genera and 14 species. Thus, an extinction of the majority of the Cretaceous taxa occurred (Fig. 7). The Paleocene taxa belong to the Betulacea, Juglandacea, Myrioacea, and Tilia. *Pinacea* occurs more frequently, and at a larger quantity than below. With respect to climate, all tropical forms had disappeared. According to Li Manying (in prep.), the flora increased again in diversity in the upper part of the Luofuzhai Group, and changed toward subtropical to tropical.

TABLE 1 (CONTINUED)

Botanical affinity	Occurrence of recent representatives	Range
?	?	Jurassic-Cretaceous
Schizaeazeae	tropical and temperate	Jurassic-Cretaceous
Schizaeazeae	tropical and temperate	Cretaceous
Schizaeazeae	-	Cretaceous
Schizaeazeae	-	Cretaceous
Schizaeazeae	-	Upper Cretaceous
?	?	
? Cyathaceae	tropical and warm temperate	Permian-Tertiary
Cyathaceae	-	Cretaceous
? Cyathaceae	-	Rhaetian-Tertiary
Cyathaceae	-	Jurassic-Tertiary
Cyathaceae	-	Cretaceous
Selaginellaceae	tropical and warm temperate	Upper Cretaceous-Tertiary
Marsileaceae	temperate and subtropical	Upper Cretaceous-Tertiary
Lycopodiaceae	tropical	Jurassic-Cretaceous
Lycopodiaceae	?	Cretaceous
Lycopodiaceae	tropical	Cretaceous
Osmundaceae	tropical and temperate	Jurassic-Cretaceous
? Polypodiaceae	cosmopolitan	Maastrichtian
Lycopodiaceae, Lycopodium	?	Upper Cretaceous-Tertiary
Lycopodiaceae	?	
Schizaeaceae	tropical and temperate	Cretaceous
Schizaeaceae	-	
Sphagnaceae, Sphagnum	?	
Pinaceae	temperate and boreal	Cretaceous-Tertiary
Pinaceae	temperate and boreal	U. Cret. (Maastr.)-Paleocen.
?	?	
Betulaceae	subtropical	Tertiary
Betulaceae	subtropical	Tertiary (Danian)
Asteraceae	cosmopolitan	Upper Cretaceous-Paleogene
Taxodiaceae	warm temperate	U. Cret. (rare), Tert. (freq.)
Tiliaceae	temperate	Tertiary (Paleoc., "Danian")
?	?	
?	?	Tertiary
Pinaceae	temperate and boreal	Cretaceous-Tertiary
Pinaceae	temperate and boreal	Upper Cretaceous-Tertiary
Podocarpaceae	tropical-temperate	
Juglandaceae, Pterocarya	subtropical and warm temperate	Cretaceous-Tertiary
?	?	Tertiary
?	?	Tertiary
Myricaceae	subtropical	Tertiary

The floral assemblage can be interpreted as typifying an evergreen mixed forest assemblage.

Vertebrate Paleontology

Within Formation VI (in the Pingling Formation, Fig. 6), numerous fragments of dinosaur eggshells were collected. Most of the material is deposited in the Institute of Vertebrate Paleontology and Paleoanthropology (IVPP) of Academia Sinica at Beijing. Beyond that, single complete eggs were found as well as entire nests containing several eggs. Frequently, eggs and eggshell fragments were collected above sandstones and mudstones penetrated by crustaceans, the burrows of which resemble *Ophiomorpha*.

On the basis of raster electron microscope (REM) investigations of the shell structure, these shells belong to type B (in the sense of Erben, 1970), and are believed to be ornithischian dinosaurs. Following Mikhailov (1991), the shells belong to Theropoda and/or Ornithopoda, according to Zhao's terminology to Elongatoolithidea and Spherolithidea (Zhao Zikui and others, 1991). The most frequent types found in the Nanxiong basin belong to *Macroolithus yautunensis*, *M. rugustus*, and *Elongatoolithus elongatus*. Less frequent are *Elongatoolithus andrewsi*, *Nanshiungoolithus chuetienensis*, *Stromatoolithus pinglingensis*, *Ovaloolithus laminadermus*, *O. chinkangkouensis*, and *Shixingoolithus erbeni* (Zhao Zikui and others, 1991). The frequency of eggshell fragments differs within the K-T transition interval, but *M. yautunensis*, *M. rugustus*, and *E. elongatus* extend across the boundary at Datang CGD meters 57–78, up to the top of the Pingling Formation (Formation VII, Nanxiong Group; Fig. 8). Thus, the dinosaur population does not show any decline or complete extinction in the interval characterized by the floral change.

In southern France (Provence), a significant reduction in the thickness of dinosaur eggshells was found by Erben (1972, 1975) and Erben and others (1979), as well as in southeastern China by Zhao Zikui and others (1991). In the Nanxiong basin, the eggshells generally are thinner than in southern France, but a similar thinning tendency can be noticed in the K-T boundary interval. Following the results of Zhao Zikui and others (1991; Fig. 3), the eggshell thickness of *M. yautunensis* shows an aberrant reduction at meters 50 to 90 in the Datang (CGD) section. In addition, pathologic malformation of eggshells of *M. yautunensis* occurs in the CGD meters 40–110 interval in a frequency up to 75%: these observations can be interpreted as being caused by environmental stress produced by the disappearance of food plants, as indicated by the palynologic analysis. However, eggshell thinning and pathology coincides with an anomaly of the $\delta^{18}O$ values (Zhao Zikui and others, 1991; also see below).

Close to the base of the Shanghu Formation (Luofuzhai Group; Fig. 3), nearly coinciding with the beginning of the red siltstones and mudstones of Formation VII, the remains of tortoises and bones of advanced mammals were abundant. These indicate a Torrejonian (middle-early Paleocene) age for Formation VII, from the beginning up section. According to the sedimentologic results, no severe hiatus or unconformity is observed between Formations VI and VII within any of the sections (Fig. 6). Only the horizon containing numerous calcareous nodules (Datang section: CGD meters 159–161) suggests a short break in the

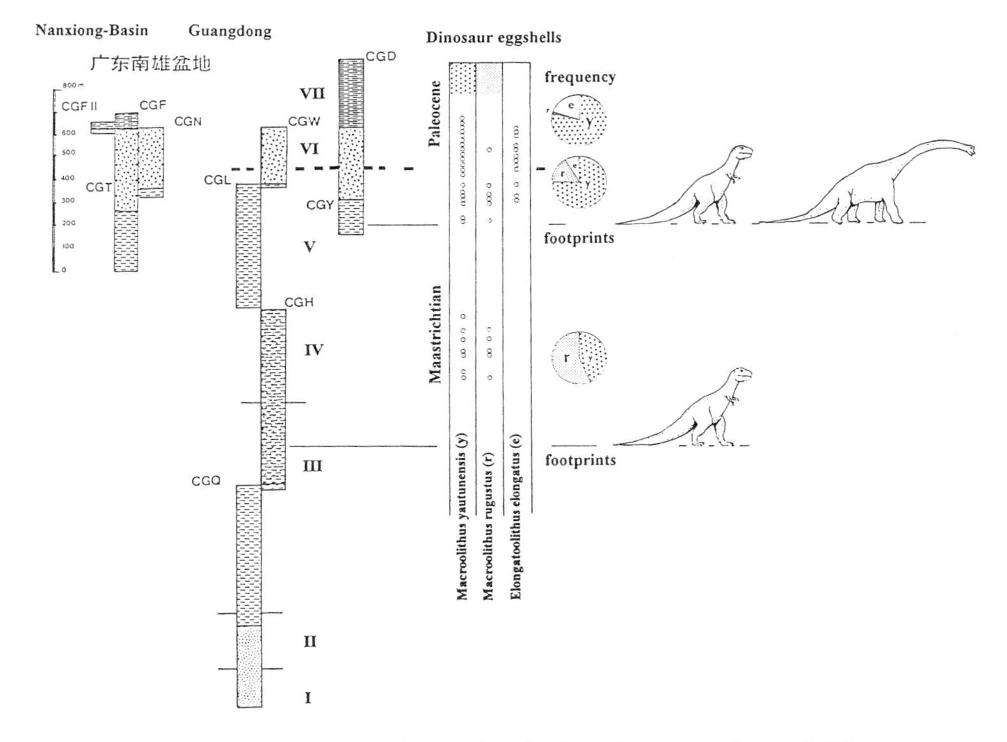

Figure 8. Occurrence of dinosaurs in the Nanxiong basin as documented by eggshell fragments and footprints; abbreviations of sections and formations as in Figure 1.

sedimentation process during which calcrete formation predominated. After this short interruption, sedimentation continued as before. This is mentioned because a greater gap or even an unconformity has been proposed because of the close proximity of dinosaurs and advanced mammalian fossils. Our data thus support the conclusion that dinosaurs survived the K-T boundary in southeastern China. Rigby and others (1993) corroborated this interpretation via radiometric dating of basalt flows and tuffs in the Nanxiong basin. The interval between the K-T boundary (as defined above) and the last occurrence of dinosaur eggshell fragments together with the first mammalian fossils amounts to at least 83 m of section (Datang section: CGD meters 78–161). Assuming an average sedimentation rate of 40 cm/ka, this interval corresponds to at least 208 ka.

Geochemistry

In addition to paleontological data, geochemical anomalies often characterize the K-T boundary. These have been found in marine as well as in continental environments, thus providing evidence for a possible K-T boundary event. In most cases, one thin layer rich in trace metals occurs that also shows a major anomaly of REE (rare earth elements). A drastic change in the ratio of stable isotopes (mainly ^{18}O and ^{13}C) may also be present. These anomalies have been observed in sediments as well as in carbonaceous shells and other calcareous fossil remnants.

In the Nanxiong basin, the most difficult problem to be solved was the identification of a well-defined geochemical boundary layer within the 760-m-thick Yan Mei Ken (CGY) and Datang (CGD) sections. Therefore, numerous samples were taken systematically at specific distances. All critical intervals (e.g., the Chinese K-T boundary at CGD meter 161 or the Datang interval at CGD meters 57–78) as well as lithologically aberrant layers were sampled at shorter distances and at different locations.

Sediments of the Yan Mei Ken and Datang sections (CGY/CGD) show only little deviation from the standard geochemical content of continental red beds. Relatively high Ca contents are caused by cabonate cements, calcareous nodules, or bioarenitic lake sediments. The lack of Ca may be due to leaching by pedogenic processes. Concentrations in heavy metals and nonferrous metals normally are low, reaching maximum values of 63 ppm for chromium (Cr), 178 ppm for vanadium (V), 92 ppm for copper (Cu), 60 ppm for nickel (Ni), 44 ppm for cobalt (Co), 178 ppm for zinc (Zn), and 110 ppm for lead (Pb). Cr, Ni, and Co are enriched locally, probably by pedogenic processes, but no specific anomaly could be found within the sediments of the CGY and CGD standard sections.

The chemical composition of dinosaur eggshell fragments was investigated at Bonn and Göttingen together with Zhao Zenhua. Although the concentration of trace elements in the eggshells is less than in the sediments, a striking enrichment of nearly all trace metals was observed half-way up the CGD section (Fig. 9). Zhao Zikui and others (1991) reported nearly the same results using their abundant material of *Macroolithus yautunensis* from the Datang (CGD) section. An anomaly interval for all trace elements lies between meters CGD 45 and 80 (Zhao Zikui and others, 1991, Fig. 4).

In addition, Zhao Zikui and others (1991) also described a marked anomaly of the amino acid composition in the eggshell samples of Datang section at meters

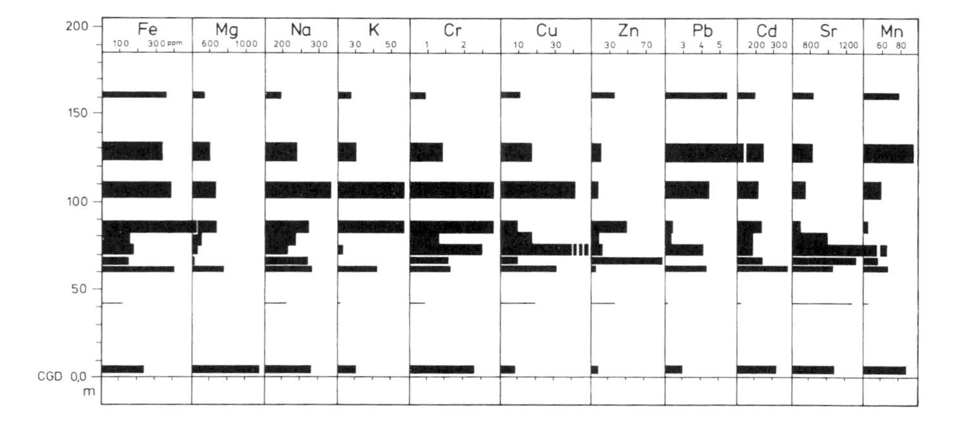

Figure 9. Trace element distribution (ppm) in dinosaur eggshells in the Datang (CGD) section.

CGD 45 and at a smaller rate at CGD meters 106–110. This coincides with the pathologically deformed eggshells and, according to these authors, suggests a severe environmental change.

Erben and others (1979) showed that toward the K-T boundary, $\delta^{18}O$ values of dinosaur eggshells shift toward extreme ratios in southern France. Assuming that diagenesis was nearly equal throughout the sedimentary pile, $\delta^{18}O$ values also indicate a potential climatic change. Samples of eggshell fragments were investigated in China and Göttingen by Zhao Zenhua and Hoefs. Figure 10 shows a drastic shift of $\delta^{18}O$ of the Nanxiong basin samples toward positive values beginning in the interval of floral change mentioned above. Results for $\delta^{13}C$, however, did not reveal any significant anomaly.

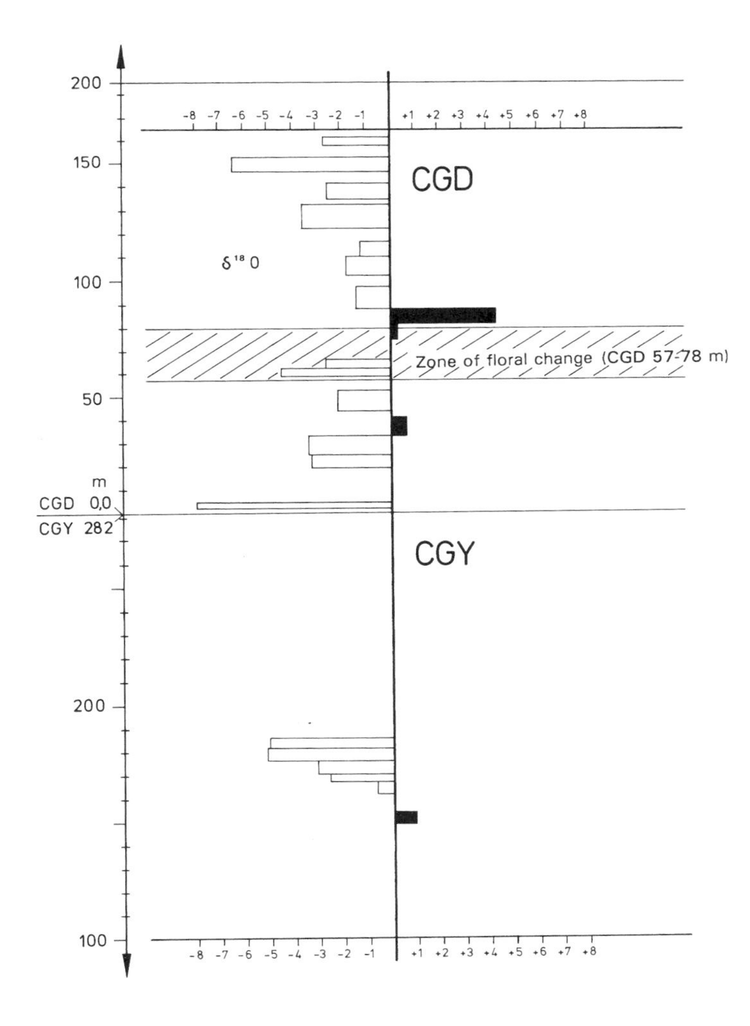

Figure 10. Variation of $\delta^{18}O$ ratios of dinosaur eggshell fragments at Yan Mei Ken (CGY) and Datang (CGD). Zone of floral change corresponds to the K-T transition interval.

Paleomagnetics

According to the geomagnetic polarity time scale (GPTS, Harland and others, 1990), the K-T boundary is located within magnetochron 29R (Fig. 11). From the paleomagnetic point of view, all the boundary intervals in question in the Nanxiong basin had to be analyzed to determine their magnetic polarity.

In the Yan Mei Ken (CGY) and Datang (CGD) sections, 712 samples from 180 lithological units were drilled in 1984. Results were obtained after a complicated procedure of demagnetization by heating the cores to 700 °C in well-defined steps, as well as measurement of magnetic polarity and a vectorial component analysis according to Kirschvink (1980). Details of the procedure along with a complete description of all results were given in Erben and others (1994).

Apart from gaps in the sample pattern, a significant subdivision of the CGY

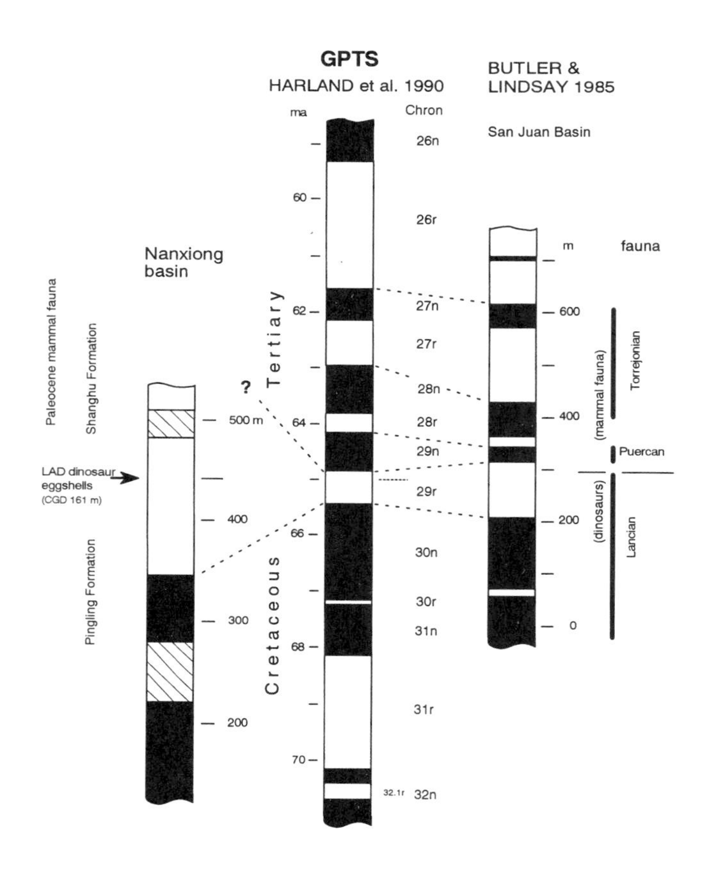

Figure 11. Pattern of magnetic polarity in the Late Cretaceous to early Tertiary; correlation of Nanxiong basin results (CGY and CGD sections) with the geomagnetic polarity time scale (GPTS) and San Juan basin. Black = normal polarity, white = reverse polarity, hachures = no samples available. LAD is last appearance datum.

and CGD sections is evident (Fig. 11). Normal polarity prevails in the lower part, whereas reverse polarity was found in the upper part. A change of polarity occurs between CGD meters 49–57 in the Datang section. Based on the sedimentological model and the Paleocene mammalian fossils from the Shanghu Formation (Formation VII), correlation of the polarity pattern with chrons 30N and 29R of the GPTS is obvious. In addition, the transition from Cretaceous to Tertiary floral elements occurs within chron 29R. Figure 11 shows the correlation of the polarity pattern in the K-T boundary interval of the Nanxiong basin with results from the San Juan basin (Butler and Lindsay, 1985), and the GPTS (Harland and others, 1990). Proportions are similar in both cases, and the correlation of the data seems evident.

Apart from their stratigraphic evidence, the paleomagnetic data allow estimates to be made with respect to maximal average sediment accumulation rate. The section under paleomagnetic consideration may represent a time span of approximately one million years. Thus, the average sediment accumulation rate was 0.40 m/k.y. This value corresponds well with rates estimated in southern France (Westphal and Durand, 1990) and in the Siwaliks (Appel and others, 1991), where values of 0.42 m/ka are typical.

DISCUSSION

The Chinese group first tried to solve the problem of positioning the K-T boundary by means of magnetostratigraphy in the Datang (CGD) section (Zhao Zikui and others, 1991). However, interpreting the rather long interval of reverse polarity as magnetochron 29R, they had to place the position of the K-T boundary "0.20 m above the base of the Shanghu Formation" (Zhao Zikui and others, 1995, p. 14). However, this was impossible because the mammalian fossils found in that formation indicate a Paleocene age. Therefore, they selected the Pingling/Shanghu formations boundary (the boundary between Formations VI and VII), in the Datang section at CGD meter 161, as the K-T boundary. This procedure seemed justified because of lithological and color change, the size of concretions, the total extinction of the dinosaur population, and the Paleocene age of the mammalian fossils in the overlying series.

In our opinion, these arguments are not fully convincing because the boundary is selected only on the basis of lithology and does not correspond to international chronostratigraphic standards. For example, the extinction of the dinosaur population obviously does not coincide with the K-T boundary, while the age of the Shanghu mammals corresponds to the Torrejonian 1 (To_1: R. E. Sloan, 1986, personal commun.), and not to the Puercan (lowermost Paleocene; Fig. 11).

Reviewing all the data collected by our group, we are convinced that in contrast to the Chinese scientists' opinion, the K-T boundary is situated farther down in the Datang section at CGD meters 57–78, as mentioned above. This boundary is seen as conformable to the international K-T boundary standards, because (1) floral change occurs at Datang (CGD meters 57–78), (2) a geochemical anomaly is documented by our own results for this interval as well as by those of the Chinese group, (3) a marked change of $\delta^{18}O$ values is observed, and (4) these

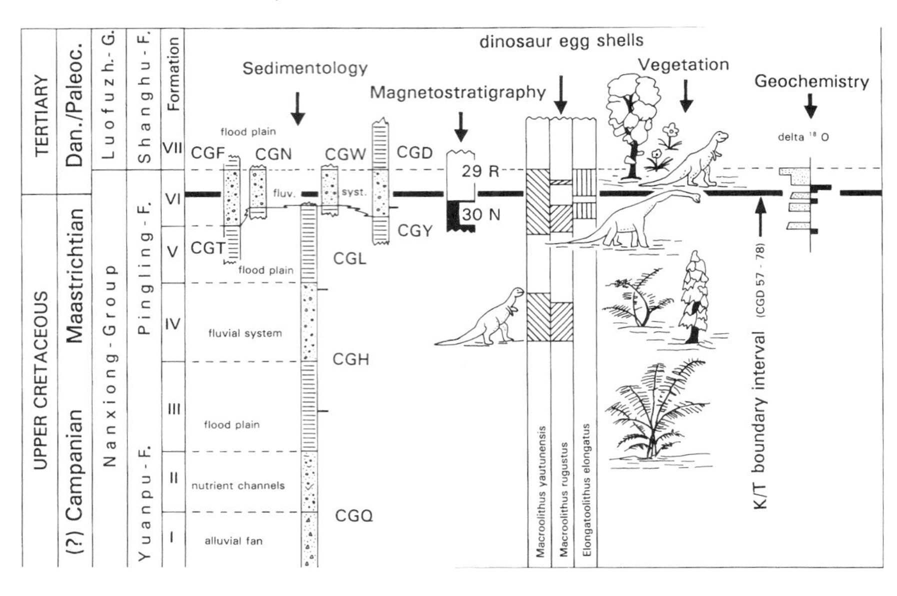

Figure 12. Correlation chart of all Cretaceous-Tertiary boundary stratigraphic results from the Nanxiong basin, southeastern China; abbreviations as in Figure 1.

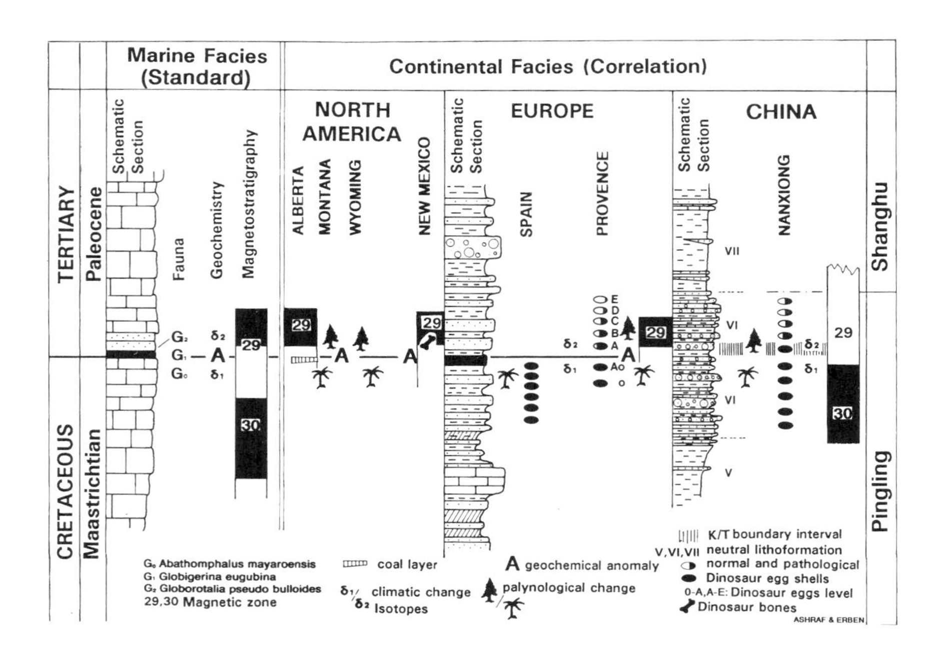

Figure 13. Correlation chart of Cretaceous-Tertiary boundary results in marine and continental environment.

three changes occur within a zone of reverse magnetic polarity inferred to be chron 29R (Fig. 12).

In Alberta, Montana, southern France, and northern Spain (Fig. 13), all these changes and anomalies characterizing the boundary under discussion also occur within one thin and well-defined boundary layer. We believe that this may be the result of a low sediment accumulation rate at that particular time. In the Nanxiong basin, however, a high sediment accumulation rate apparently compensated for the effects of condensation. Therefore, the K-T boundary lies within a poorly defined transition zone of ~20 m corresponding to a time span of ~50 ka.

It is possible that the boundary layer may yet be found, and its apparent absence recognized to result from local removal due to erosion. However, the total absence of such a distinct layer throughout the whole basin strongly suggests that it is not present.

The three main arguments for positioning the K-T boundary in the Datang (CGD) section at meters 57–78—floral change, geochemical anomaly, and change in $\delta^{18}O$ values—do not manifest themselves within one clear boundary layer or anomaly (Fig. 12). However, the thick red sediments of the Nanxiong basin surely underwent changes in chemical composition during sedimentation as well as by diagenesis. Subsequent alteration is indicated by the products of pedogenic processes. Palynomorphs are preserved rarely, other floral remnants are destroyed, and no distinct geochemical anomaly was preserved in the sediment.

Extinction of dinosaurs can no longer be a major criterion for defining the K-T boundary. Some paleontologists have reported single dinosaur populations vanishing long before the K-T boundary, whereas others have reported that dinosaurs survived into the early Tertiary (Sloan, 1976; Fasset, 1982; Erben and others, 1983): thus, no sudden and general extinction occurred. This is in agreement with the results of Rigby and others (1993), who also found dinosaur remnants in lower Tertiary sediment of the Nanxiong basin.

Considering the broad and animated international discussion about a catastrophic K-T boundary event, special attention has been on investigations that specifically addressed this question. No direct evidence for a bolide impact has been found thus far within the Nanxiong basin sediments. Neither shocked quartz grains nor microtektites or even particles of soot or soot flakes have been reported. We do not deny that a bolide impact may have occurred at the K-T boundary. However, if it did occur, it had no decisive and direct sudden influence on the paleoenvironment, or on flora and fauna in the Nanxiong basin in southeastern China.

CONCLUSIONS

According to our data, the internationally accepted K-T chronostratigraphic boundary is not represented by a sharp horizon or by a thin layer within the thick continental red beds of the Nanxiong basin (Guangdong province, southeastern China). Instead, transitional beds ~20 m thick occur that correspond to a time span of ~50 k.y., based on the assumption of an average sedimentation rate of 40 cm/k.y.

In contrast, a rather sharp K-T boundary represented by a single well-defined layer has been established in marine sections (Fig. 13). According to MacLeod and Keller (1991, p. 1439), however, this may represent "artifacts of a temporally incomplete (or extremely condensed) deep-sea stratigraphic record." A similar situation may exist in those continental paleoenvironments where a coal seam or a similar formation forms the boundary layer. According to MacLeod and Keller (1991), the record seems more complete in sections with a higher sediment accumulation rate (e.g., El Kef), and the boundary tends toward a transition zone. The K-T boundary interval in the Datang (CGD) section of the Nanxiong basin obviously belongs to this type.

In our opinion, the most convincing evidence for positioning the K-T boundary in the Nanxiong basin is a significant floral change recorded by palynomorphs (Fig. 13). In addition, this change occurs within an interval of reverse magnetic polarity that can be correlated to chron 29R. Geochemical analysis of dinosaur eggshell fragments yielded distinct anomalies in the concentration of trace elements in the boundary interval. Trace elements and their anomalous distribution are preserved in the eggshell fragments only because the calcitic biocrystallites were protected by tiny membranes of organic matter. Equivalent geochemical anomalies have not been observed in the sedimentary record despite many efforts and a large number of samples analyzed repeatedly. Within the fluvial red bed environment, including lacustrine sediments as well as pedogenic concretions, geochemical anomalies may have been destroyed by diagenetic processes.

According to our observations, dinosaur remnants, especially eggshells, are present even above the K-T boundary (Figs. 12 and 13). In addition, it seems significant that a mammalian fauna of early Paleocene age (*Bemalambda* fauna) follows almost immediately above the last dinosaur finds. According to geological field data, no severe gap or even unconformity has been observed in the boundary interval itself or nearby. All these facts suggest that the last dinosaurs of the Nanxiong basin survived into the early Paleocene.

Pathologic eggshells close to the boundary interval suggest that the final extinction of the dinosaur population occurred stepwise, and was triggered by environmental stress. Change from a warm, perhaps tropical, climate to temperate humid conditions caused the disappearance of the food plants, as indicated by palynological data. In addition, a slow environmental poisoning has to be taken into account, as indicated by the trace element analysis of eggshells.

All these data and observations do not support the hypothesis of a sudden catastrophic event at the K-T boundary in southeastern China. Instead, other reasons and complex mechanisms must be invoked for a slow and fluctuating climatic change that ultimately led to the fundamental change in climatic conditions of the late Tertiary and Pleistocene.

ACKNOWLEDGMENTS

The authors gratefully acknowledge the financial support granted by Deutsche Forschungsgemeinschaft (DFG, grant Er-4/49) as well as by Max Planck Gesellschaft, Munich, and Academia Sinica, Beijing. The Sino-German team included the following scientists: Zhao Zikui, Zhang Yuping, and Ye Jie (IVPP of Academia Sinica, Beijing); Li Huamei and Zhao Zenhua (Geochemical Institute of Academia Sinica, Guiyang); and Li Manying (Geological and Paleontological Institute of Academia Sinica, Nanjing). Thanks are due to James Nebelsick, Tübingen, and Martin Sander, Bonn, who vetted a first draft of this manuscript.

REFERENCES CITED

Alvarez, L., Alvarez, W., Asaro, F., and Michel, H., 1980, Extraterrestrial cause for the Cretaceous extinctions: Science, v. 208, p. 1095–1108.

Appel, E., Rösler, W., and Corvinus, G., 1991, Magnetostratigraphy of the Miocene-Pliocene Surai Khola Siwaliks in West Nepal: Geophysical Journal International, v. 105, p. 191–198.

Ashraf, A. R., and Erben, H. K., 1986, Palynologische Untersuchungen an der Kreide/Tertiär-Grenze West-Mediterraner Regionen: Paläontographica, v. 200, ser. B, p. 111–163.

Ashraf, A. R., and Stinnesbeck, W., 1988, Pollen und Sporen an der Kreide/Tertiär-Grenze im Staat Pernambuco, NE-Brasilien: Paläontographica, v. 208, ser. B, p. 39–51.

Butler, R. F., and Lindsay, E. H., 1985, Mineralogy of magnetic minerals and revised magnetic polarity stratigraphy of continental sediments, San Juan Basin, New Mexico: Journal of Geology, v. 93, p. 535–554.

Chang, Y.-P., and Tung, Y.-Sh., 1963, Subdivision of "redbeds" of Nanxiong basin: Vertebrata Palasiatica, v. 7, p. 249–260.

Einsele, G., 1992, Sedimentary basins; evolution, facies and sediment budget: Heidelberg, Springer-Verlag, 628 p.

Erben, H. K., 1969, Dinosaurier: Pathologische Strukturen ihrer Eischale als Lethalfaktor: Umschau in Wissenschaft und Technik, v. 17, p. 552.

Erben, H. K., 1970, Ultrastrukturen und Mineralisation rezenter und fossiler Eier bei Vögeln und Reptilien: Biomineralisation (Research Report): Forschungsberichte, v. 1, p. 2–66.

Erben, H. K., 1972, Ultrastrukturen und Dicke pathologischer Eischalen: Akademie der Wissenschaften und der Literatur Mainz, Abhandlungen der Mathematisch-Wissenschaftlichen Klasse, v. 1972, p. 191–216.

Erben, H. K., 1975, Die Entwicklung der Lebewesen: Spielregeln der Evolution: München, Piper, 62 p.

Erben, H. K., 1985, Faunal mass extinctions by ecocatastrophes: Paleontological Society of Korea Journal, v. 1, p. 1–18.

Erben, H. K., Hoefs, J., and Wedepohl, K. H., 1979, Palaeobiological and isotopic studies of eggshells from a declining dinosaur species: Paleobiology, v. 5, p. 380–414.

Erben, H. K., Ashraf, A. R., Krumsiek, K., and Thein, J., 1983, Some dinosaurs survived the Cretaceous "final event": Terra Cognita, (European Union of Geoscientists Journal), v. 3, p. 211–212.

Erben, H. K., Ashraf, A. R., Böhm, H., Hahn, G., Hambach, U., Krumsiek, K., Stets, J., Thein, J., and Wurster, P., 1995, Die Kreide/Tertiär-Grenze im Nanxiong-Becken (Kontinentalfazies; Südostchina) -eine interdisziplinäre geowissenschaftliche Studie: Akademie der Wissenschaften und der Literatur, Mainz, Franz Steiner Verlag, Erdwissenschaftliche Forschung, v. 32, 245 p.

Fasset, J. E., 1982, Dinosaurs in the San Juan Basin, New Mexico, may have survived the event that resulted in creation of an iridium-enriched zone near the Cretaceous/Tertiary boundary, *in* Silver, L. T., and Schultz, P. H., eds., Geological implications of impacts of large asteroids and comets on the Earth: Geological Society of America Special Paper 190, p. 435–447.

Hao Yi-Chun, Yu Jing-Xian, Guan Shao-Zheng, and Sun Meng-Rong, 1979, Some Late Cretaceous and early Tertiary assemblages of ostracoda, spores and pollen in China, *in*

Christensen, W. K., and Birkelund, T., eds., Proceedings, Cretaceous-Tertiary Boundary Events Symposium, Volume 2: Copenhagen, University of Copenhagen, p. 251–255.

Harland, W. B., Armstrong, R. L., Cox, A. V., Craig, L. E., Smith, A. G., and Smith, D. G., 1990, A geologic time scale 1989: London, Cambridge University Press, 263 p.

Hickey, L. J., 1984, Changes in the angiosperm flora across the Cretaceous-Tertiary boundary, *in* Berggren, W. A., and Couvering, J. A., eds., Catastrophes and earth history, the new uniformitarism: Princeton, New Jersey, Princeton University Press, p. 279–311.

Hildebrand, A. R., and Stansberry, J. A., 1992, K/T boundary ejecta distribution predicts size and location of Chicxulub crater: Lunar and Planetary Science, v. 23, p. 537–538.

Hildebrand, A. R., Penfield, G. T., Kring, D. A., Pilkingdon, M., Carmago, Z. A., Jacobsen, S. B., and Boynton, W. V., 1991, Chicxulub crater; A possible Cretaceous/Tertiary boundary impact crater on the Yucatán Peninsula, Mexico: Geology, v. 19, p. 867–871.

Hsü, K. J., 1980, Terrestrial catastrophe caused by cometary impact at the end of Cretaceous: Nature, v. 285, p. 201–203.

Kaiser, H., and Ashraf, A. R., 1974, Gewinnung und Präparation fossiler Sporen und Pollen sowie anderer Palynomorphae unter besonderer Betonung der Siebmethode: Geologisches Jahrbuch, v. A, no. 25, p. 85–114.

Kirschvink, J. L., 1980, The least-squares line and plane and the analysis of paleomagnetic data: Royal Astronomical Society Geophysical Journal, v. 62, p. 699–718.

Krassilow, V. A., 1981, Changes of Mesozoic vegetation and the extinction of dinosaurs: Palaeogeography, Palaeoclimatology, Palaeoecology, v. 34, p. 207–224.

Kummel, B., Jr., 1943, New technique for measurement of stratigraphic units: American Association of Petroleum Geologists Bulletin, v. 27, p. 220–222.

Landis, G. P., Rigby, J. K., Jr., Sloan, R. E., and Hengst, R., 1993, Pele hypothesis: A unified model for ancient atmosphere and biotic crisis: Geological Society of America Abstracts with Programs, v. 25, no. 7, p. A362.

MacLeod, N., and Keller, G., 1991, How complete are Cretaceous/ Tertiary boundary sections? A chronostratigraphic estimate based on graphic correlation: Geological Society of America Bulletin, v. 103, p. 1439–1457.

Mikhailov, K. E., 1991, Classification of fossil eggshells of amniotic vertebrates: Acta Palaeontogica Polonica, v. 36, p. 193–238.

Rigby, J. K., Jr., Sneel, L. W., Unruh, D. M., Harlan, S. S., Guan, J., Li, F., Rigby, J. K., and Kowalis, B. J., 1993, $^{40}Ar/^{39}Ar$ and U-Pb dates for dinosaur extinction, Nanxiong basin, Guang-dong Province, Peoples Republic of China: Geological Society of America Abstracts with Programs, v. 25, no. 7, p. A296.

Sharpton, V. L., Dalrymple, L. E., Ryder, G., Schuraytz, C., and Urrutia-Fucugauchi, J., 1992, New links between the Chixculub impact structure and the Cretaceous/Tertiary boundary: Nature, v. 359, p. 819–820.

Sloan, R. E., 1976, The ecology of dinosaur extinction: Toronto, Royal Ontario Museum, Athlon, p. 134–153.

Stinnesbeck, W., 1986, Zu den faunistischen und palökologischen Verhältnissen in der Quiriquina Formation (Maastrichtium) Zentral-Chiles: Palaeontographica, ser. B, v. 194, p. 99–237.

Tollmann, A., and Tollmann, E., 1993, Und die Sintflut gab es doch. Vom Mythos zur historischen Wahrheit: München, Droemer/Knaur, 560 p.

Tschudy, R. H., 1984, Palynological evidence for change in continental floras at the Cretaceous-Tertiary boundary, *in* Berggren, W. A., and van Couvering, J. A., eds., Catastrophes and Earth history; the new uniformatarianism: Princeton, New Jersey, Princeton University Press, p. 315–337.

Tschudy, R. H., and Tschudy, B. D., 1986, Extinction and survival of plants following the Cretaceous/Tertiary boundary event, Western Interior, North America: Geology, v. 14, p. 667–670.

Westphal, M., and Durand, J.-P., 1990, Magnétostratigraphie des séries continentales fluvio-lacustres du Crétacé supérieur dans le synclinal de l'Arc (région Aix-en-Provence, France): Bulletin de la Société Géologique de France, v. 8, p. 609–620.

Wurster, P., and Stets, J., 1979, Der Bonner Profilstab (BPS) - ein geländegeologisches Grundgerät: Neues Jahrbuch für Geologie und Paläontologie Monatselfte, v. 1979, p. 560–576.

Yang Zunyi, Cheng Yuqi, and Wang Hongzhen, 1986, The geology of China: Oxford, Clarendon, p. 153–167.

Zhao Zikui, Ye Jie, Li Huamei, Zhao Zhenhua, and Yan Zheng, 1991, Extinction of the dinosaurs across the Cretaceous-Tertiary boundary in Nanxiong basin, Guangdong province: Vertebrata Palasiatica, v. 29, p. 1–20.

15

Testing Extinction Theories at the Cretaceous-Tertiary Boundary Using the Vertebrate Fossil Record

J. D. Archibald, *Department of Biology, San Diego State University, San Diego, California*

INTRODUCTION

The synergy between hypothesis and data provides the basis for testing and accepting or rejecting theories of how the world operates today. Testing in science is fortunately not limited to whether we performed an experiment. Thus the interplay between hypothesis and data is applicable to natural events that occurred in the very distant past, although our database may be a bit frayed by the wear and tear of time. This means that in the historical sciences our ability to test and reject various theories is limited by the quality of our database.

One of the continuing debates concerning events near the Cretaceous-Tertiary (K-T) boundary involves the quality of the geologic and paleontologic database for vertebrates, including nonavian dinosaurs. The debates are unfortunately not always limited to varying interpretations of the scientific data, but also are clouded by misinformation. The most pernicious of these is the myth of a worldwide record of K-T vertebrate extinction, usually made as a pronouncement of a global record of nonavian dinosaur extinction. Thus statements in the scientific and popular press such as "the impact theory has been buttressed by researchers who have age-dated dinosaur fossils around the world..." (Perlman, 1993, p. A5) assert explicitly that a global record of nonavian dinosaurs placed within a well-constrained stratigraphic framework exists. This is patently false, although we hope for such a record in the near future.

We also read conflicting views on how fast the extinctions of nonavian dinosaurs were. Some authors (e.g., Sloan and others, 1986) have argued that the extinction of nonavian dinosaurs and other vertebrates was gradual or stepwise, while others (Sheehan and others, 1991) have argued that the extinction patterns are commensurate with catastrophic events. Unfortunately, at present we are unable to address issues surrounding rates of extinction for vertebrates at the K-T boundary. Present evidence indicates that most of the gradual or stepwise patterns are the result of the reworking of fossils (e.g., Lofgren and others, 1990), while the patterns that seem to agree with catastrophic events are the result of statistical errors (Hurlbert and Archibald, 1995).

In spite of the fact that, for now, the K-T vertebrate record is not global in extent and cannot tell us anything about the rate of extinctions, we can use more regionally limited fossil data to frame limitations for various extinction theories. In this paper I discuss the single best-documented regional vertebrate fossil record across the K-T boundary and examine how various extinction theories fit these data. This fossil database comes from the Western Interior of North America and is especially well known from eastern Montana.

Although this fossil database is limited in its geographic coverage, any theory of extinction at the K-T boundary must explain patterns of vertebrate extinction and survival in the Western Interior, especially if it claims to have been global in its effects. If such a theory is not compatible with this vertebrate record, it fails, and must either be rejected or modified. If we can integrate the vertebrate database with what we know of other terrestrial species, most notably plants, we can begin to frame reasonable scenarios of what occurred 65 million years ago at the K-T boundary.

LATEST CRETACEOUS VERTEBRATE DATABASE, WESTERN INTERIOR

Previous Studies

Our knowledge of latest Cretaceous vertebrates in the Western Interior began with U.S. government surveys in the late nineteenth century. It was not until the late 1960s and early 1970s with the publications of Clemens (1964, 1966, 1973) and Estes (1964) that a much more complete picture of the latest Cretaceous vertebrate fauna began to emerge. The fossils in these studies were recovered from the uppermost Cretaceous Lance Formation in eastern Wyoming. Subsequently, papers on similarly aged faunas from the Hell Creek Formation of eastern Montana appeared (Sloan and Van Valen, 1965; Estes and others, 1969; Archibald, 1982; Bryant, 1989). Portions of approximately contemporaneous faunas were also described from the Western Interior—the mammals were described from the Scollard Formation of Alberta by Lillegraven (1969) and from the Frenchman Formation of Saskatchewan by Fox (1989)—but neither described the complete vertebrate fauna (although Fox, 1989, listed all vertebrates).

In 1990 a comprehensive tally of almost all species of latest Cretaceous vertebrates from part of the Western Interior was published (Archibald and Bryant, 1990) for the purpose of examining vertebrate turnover across the K-T boundary. The only two major groups of vertebrates not included were birds and pterosaurs.

Both birds and pterosaurs have extremely fragile bones because of requirements for flight. This renders them very susceptible to destruction before they can be fossilized. Pterosaurs were also on their evolutionary "last legs." At most, one family of pterosaurs is known from the latest Cretaceous of North America. Birds were quite the opposite. They were expanding evolutionarily, although their record is poor in the latest Cretaceous.

As most now know, birds are dinosaurs, specifically coelurosaurs, the smaller cousins of carnosaurs such as *Tyrannosaurus*. This is why one sometimes sees the larger, ground-dwelling species more accurately called nonavian dinosaurs. Throughout this paper I refer to nonavian dinosaurs as dinosaurs for the sake of simplicity. When I use the formal taxonomic term Dinosauria, however, it is correctly understood to include birds.

Laurie Bryant's and my compilation of 111 species of vertebrates came from the uppermost Cretaceous Hell Creek Formation of eastern Montana (Garfield and McCone counties). A major reason for the compilation was for use in examining various extinction theories. Our results showed species survival across the K-T boundary of between 53% and 64%, depending upon how three important artifacts of the fossil record were considered.

The first artifact Archibald and Bryant (1990) considered was survival outside the study area. If a species survives the K-T boundary elsewhere in North America but disappears locally in eastern Montana, the species obviously survived even if its geographic range was diminished. Thus such local disappearances must not be counted as true extinctions.

Second, if a species disappears from the fossil record through the process of speciation, either through cladogenesis (splitting) or anagenesis (change within a single lineage), its disappearance was not true extinction: rather, it is one form of pseudoextinction. If such pseudoextinctions are not recognized, extinction rates can be greatly exaggerated. In a study of early Tertiary mammals Archibald (1993b) found that in ~25% of the cases, species disappearances were the result of speciation events, not true extinction events. We do not have detailed phylogenetic analyses for most latest Cretaceous mammals, but common sense tells us that their apparent 97% disappearance is not all true extinction. There is no question that mammals were beginning a tremendous radiation that occurred after the disappearance of nonavian dinosaurs and thus many of these disappearances are pseudoextinctions resulting from speciation. Such pseudoextinction is undoubtedly masquerading as true extinction in other vertebrate groups. However, until we have better phylogenies for these other animals, the extent of pseudoextinction remains hidden.

The third and final artifact was rarity of taxa. This is the least tractable of the three artifacts. Archibald and Bryant (1990) found a much higher level of disappearance of rare species (91% for rare versus 54% for common), suggesting a strong preservational bias. The great disparity of survival for common versus rare species suggested that rare species were missed in the analysis or that the disparity was real because rare species truly were rare and more easily suffered extinction. How to treat rare species in the analysis remains a problem—exclusion ignores part of the fauna, while inclusion may overemphasize the importance of truly rare taxa.

One approach to dealing with these artifacts was to include only common latest Cretaceous vertebrates that survived both within and outside the study area (Archibald and Bryant, 1990, Table 2, column five). This resulted in a survival level of 53% (36 of 68). When the artifact of pseudoextinction for mammals (both common and rare) was included, survival increased to almost 64% (47 of 74 species). These two estimates provided the range of between 53% (now 50%) and 64% of vertebrate survival used in subsequent papers.

Archibald and Bryant (1990) argued that although some patterns of extinction and survival could be recognized (e.g., aquatic versus terrestrial, large versus small), no single, dominant pattern emerged. In reanalyzing a culled portion of the original data set, Sheehan and Fastovsky (1992) found a 90% survival for freshwater versus only 12% for terrestrial species. Using a complete data set, Archibald (1993a) calculated that the difference in freshwater versus terrestrial species was 78% and 28%, respectively. Thus, Sheehan and Fastovsky's culling resulted in a 30% exaggeration of freshwater species' survival.

Present Study

The original database (Archibald and Bryant, 1990) is updated in Table 1, and will be further refined and expanded in Archibald (1996). There are three general differences between Table 1 here, and Table 1 in Archibald and Bryant (1990). First, there are some taxonomic changes and additions. Second, the table is here condensed to show only whether a particular species or species lineage survived the K-T boundary and whether it was rare. For further detail see Archibald and Bryant (1990, Table 1). Third, eight problematic taxa are excluded. Changes from Archibald and Bryant are footnoted in the explanation of Table 1.

The only major difference in approach between this paper and Archibald and Bryant (1990) is in the treatment of rare species. In Archibald and Bryant (1990) one approach was, as noted above, to exclude rare species. This is the most difficult of the three artifacts with which to deal, but it was decided to include them in estimations of faunal turnover here because this seems the most conservative approach. The other two artifacts included by Archibald and Bryant (1990) in calculations of turnover are also included for the reasons discussed above.

Before automatically including all rare species extinctions in this study, the 28 of 43 rare species in Archibald and Bryant (1990) that completely disappear from the record at the K-T boundary were more fully examined relative to possible taxonomic and preservational problems. I found that 8 of these 28 species (see explanation in Table 1 for a listing) were either so poorly known or were based on such questionable taxonomy that they still should be excluded, leaving 20 of 28 rare species disappearances. An example of questionable taxonomy is the ceratopsian *Ugrosaurus olsoni* (CoBabe and Fastovsky, 1987), which is based on a nasal fragment that is not distinguishable from the same structure in species of *Triceratops* (Foster, 1993). An example of preservational problems is the baenid turtle *Thescelus insiliens*. Of the 75 specimens of baenid turtles from the uppermost Hell Creek Formation and the lowermost Paleocene Tullock Formation in the study area that were referable to 1 of 6 species, only one Hell Creek speci-

TABLE 1
Vertebrate species from the uppermost Cretaceous Hell Creek Formation, eastern Montana (cont'd. p. 378)

Taxon	Survival
Class ELASMOBRANCHII	
Suborder BATOIDEA	
Myledaphus bipartinus	O
Family HYBODONTIDAE	
Lissodus selachos	O
Family ORECTOLOBIDAE	
Brachaelurus estesi	O
Squatirhina americana	X
Family PRISTIDAE	
Ischyrhiza avonicola	O
Number & % Survival	**1/5 (20%)**
Class OSTEICHTHYES	
Order ACIPENSERIFORMES	
Family ACIPENSERIDAE	
"Acipenser" albertensis	X
"Acipenser" eruciferus	X
Protoscaphirhynchus squamosus	r O
Family POLYODONTIDAE	
undescribed Polyodontidae	X
Order AMIIFORMES	
Family AMIIDAE	
Kindleia fragosa	X
Melvius thomasi	O
Order LEPISOSTEIFORMES	
Family LEPISOSTEIDAE	
Lepisosteus occidentalis	X
Number & % Survival	**5/7 (71%)**
Infraclass TELEOSTEI	
"Family" ASPIDORHYNCHIDAE	
Belonostomus longirostris	r O
Belonostomus sp.	X
Order ELOPIFORMES	
Teleostei incertae sedis	
Suborder PACHYRHIZODONTOIDEI	
Pachyrhizodontoidei, indet.	r O
Family PALAEOLABRIDAE	
Palaeolabrus montanensis	X
Family PHYLLODONTIDAE	
Phyllodus paulkatoi	X
Family indeterminate	
Platacodon nanus	X
Number & % Survival	**4/6 (67%)**
Class AMPHIBIA	
Order ANURA	
Family DISCOGLOSSIDAE	
Scotiophryne pustulosa	X
Order CAUDATA	
Family BATRACHOSAUROIDIDAE	
Opisthotriton kayi	X
Prodesmodon copei	X
Family PROSIRENIDAE	
Albanerpeton nexuosus	X
Family SCAPHERPETONTIDAE	
Lisserpeton bairdi	X
cf. Piceoerpeton sp.	X
Scapherpeton tectum	X
Family SIRENIDAE	
Habrosaurus dilatus	X
Number & % Survival	**8/8 (100%)**
Class REPTILIA	
Order CHORISTODERA	
Family Champsosauridae	
Champsosaurus sp. indet.	X
Number & % Survival	**1/1 (100%)**
Order CROCODILIA	
Family CROCODYLIDAE	
Subfamily ALLIGATORINAE	
Brachyclampsa montana	O
undescribed alligatorine A	X
undescribed alligatorine B	X
Subfamily CROCODYLINAE	
Leidyosuchus sternbergi	X
Subfamily THORACOSAURINAE	
Thoracosaurus neocesariensis	X
Number & % Survival	**4/5 (80%)**
Order ORNITHISCHIA	
Family ANKYLOSAURIDAE	
Ankylosaurus magniventris	r O
Family CERATOPSIDAE	
3 *Torosaurus ?latus*	r O
Triceratops horridus	O
Family HADROSAURIDAE	
Edmontosaurus annectens	O
1 *Anatotitan copei*	O

Table shows which species or species lineages that do not survive (O), or do survive (X), the K-T boundary in the region or elsewhere. An "r" indicates apparent extinction of rare species. Some taxa in Archibald and Bryant (1990) were excluded because of poor taxonomy or because they were too fragmentary to confidently recognize (Dermatemydinae indet., Boidae indet., Neoplagiaulacidae gen. et sp. indet., Paleopsephurus wilsoni, Avisaurus archibaldi, Thescelus insiliens, Ugrosaurus olsoni, Cimolestes stirtoni*). The following species or equivalents in this table are modified from Archibald and Bryant (1990) as follows: (1)* Anatosaurus copei = Anatotitan copei*: Horner, 1992; (2)* Aublysodon sp. = A. cf. A. mirandus*: Molnar and Carpenter, 1989; (3) Modifications after Weishampel and others, 1990; (4)* Alphadon lulli = Protalphadon lulli*: Cifelli, 1990; (5)* Alphadon rhaister = Turgidodon rhaiste*: Cifelli, 1990.*

Table 1 (continued)

Taxon	
Family "HYPSILOPHODONTIDAE"	
3 *Thescelosaurus neglectus*	O
Family NODOSAURIDAE	
3 *?Edmontonia sp.*	r O
Family PACHYCEPHALOSAURIDAE	
Pachycephalosaurus wyomingensis	r O
Stegoceras validus	r O
Stygimoloch spinifer	r O
Number & % Survival	**0/10 (0%)**
Order SAURISCHIA	
Family AUBLYSODONTIDAE	
2 *Aublysodon cf. A. mirandus*	r O
Family DROMAEOSAURIDAE	
3 *Dromaeosaurus sp.*	O
3 *?Veliceraptor sp.*	O
Family ELMISAURIDAE	
3 *?Chirostenotes sp.*	O
Family ORNITHOMIMIDAE	
3 *Ornithomimus sp.*	O
Family TROODONTIDAE	
3 *Troodon formosus*	O
3 *Paronychodon lacustris*	O
Family TYRANNOSAURIDAE	
Tyrannosaurus rex	O
3 *Albertosaurus lancensis*	O
Number & % Survival	**0/9 (0%)**
Superorder SQUAMATA	
Order SAURIA	
Family ANGUIDAE	
Odaxosaurus piger	X
Family ?HELODERMATIDAE	
Paraderma bogerti	r O
Family NECROSAURIDAE	
Parasaniwa wyomingensis	O
Family SCINCIDAE	
Contogenys sloani	X
Family TEIIDAE	
Chamops segnis	O
Haptosphenus placodon	O
Leptochamops denticulatus	O
Peneteius aquilonius	r O
Family ?VARANIDAE	
Palaeosaniwa canadensis	r O
Family XENOSAURIDAE	
Exostinus lancensis	X
Number & % Survival	**3/10 (30%)**
Order TESTUDINES	
Family BAENIDAE	
Eubania cephalica	X
Neurankylus cf. N. eximius	X
Palatobaena bairdi	X
Plesiobaena antiqua	X
Stygiochelys estesi	X
Family CHELYDRIDAE	
Chelydriae indet.	X
Emarginochelys cretacea	X
Family DERMATEMYDIDAE	
Adocus sp.	X
Basilemys sinuosa	O
Family KINOSTERNIDAE	
Kinosternidae indet.	X
Family MACROBAENIDAE	
"Clemmys" backmani	X
Family TRIONYCHIDAE	
Heloplanoplia distincta	O
"Plastomenus" sp. A	X
"Plastomenus" sp. C	X
Trionyx (Aspideretes) sp.	X
Trionyx (Trionyx) sp.	X
Cryptodira incertae sedis	
Compsemys victa	X
Number & % Survival	**15/17 (88%)**
Class MAMMALIA	
Subclass ALLOTHERIA	
Order MULTITUBERCULATA	
Family CIMILODONTIDAE	
Cimolodon nitidus	X
Family CIMOLOMYIDAE	
Cimolomys gracilis	r O
Meniscoessus robustus	O
Family indet.	
Essonodon browni	O
Paracimexomys priscus	O
Cimexomys minor	X
Family NEOPLAGIAULACIDAE	
Mesodma formosa	X
Mesodma hensleighi	O
Mesodma thompsoni	X
?Neoplagiaulax burgessi	X
Number & % Survival	**5/10 (50%)**
Subclass THERIA	
Infraclass EUTHERIA	
"PROEUTHERIANS"	
Family GYPSONICTOPIDAE	
Gypsonictops illuminatus	X
Family PALAEORYCTIDAE	
Batodon tenuis	X
Cimolestes cerberoides	X
Cimolestes incisus	X
Cimolestes magnus	X
Cimolestes propalaeoryctes	X
Number & % Survival	**6/6 (100%)**
Infraclass METATHERIA	
Order MARSUPICARNIVORA	
Family DIDELPHODONTIDAE	
Didelphodon vorax	O
Family PEDIOMYIDAE	
Pediomys cooki	r O
Pediomys elegans	r O
Pediomys florencae	O
Pediomys hatcheri	r O
Pediomys krejcii	r O
Family PERADECTIDAE	
Alphadon marshi	X
Alphadon wilsoni	O
4 *Protalphadon lulli*	r O
5 *Turgidodon rhaister*	r O
Family indet.	
Glasbius twitchelli	O
Number & % Survival	**1/11 (9%)**
Total Number & % Survival	**53/105 (50%)**

men was assignable to *Thescelus insiliens* (Hutchison and Archibald, 1986). Identification of this species, which is based only on shells, requires the associated front end of both the upper shell (carapace) and lower shell (plastron), an unlikely find even where turtles are common.

As shown in Table 1, of the of the remaining 105 vertebrate species or species lineages, 50% (53 of 104) survived across the K-T boundary in the Western Interior. If all the extinctions of rare species are actually artifactual, survival would rise to 70% (73 of 105). This gives a range of 51%-70% species survival. I doubt that many of these rare species did survive the K-T boundary in the Western Interior (especially nonavian dinosaurs), but some of the extinctions of very rare species are certainly artifactual. The real question is which apparent extinctions are artifacts and which are not? At this time our data are such that we cannot answer this question with confidence.

Furthermore, at the present time there are no other well-studied vertebrate faunal turnovers with which to compare this K-T boundary record in eastern Montana. Archibald and Bryant (1990) noted, however, that for the intervals preceding and just following the K-T boundary, generic level survivorship was, at most, 10% higher than at the K-T boundary (Fig. 1). Thus, levels of extinctions for vertebrates do not appear extremely high at the K-T boundary, but two other important survival patterns are especially notable.

The first of these is the general reorganization of the terrestrial component of the fauna. With the apparent extinction of all nonavian dinosaurs, mammals begin to dominate the large body vertebrate niches within less than one million years, probably sooner.

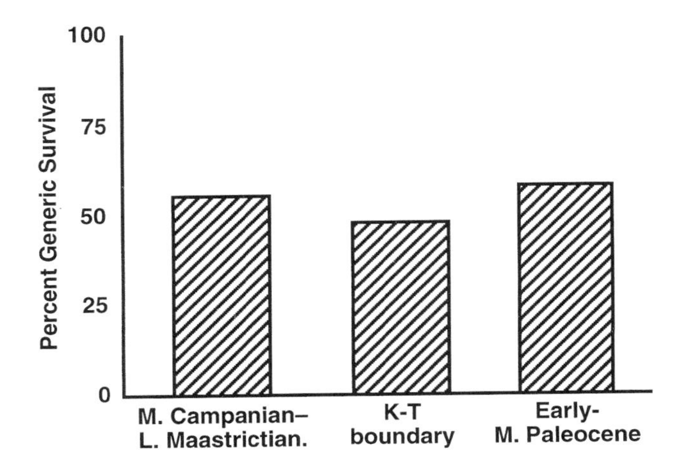

Figure 1. Comparison of generic-level survivorship for vertebrates in the Western Interior of North America from Late Cretaceous through middle Paleocene. Data taken from Archibald and Bryant (1990).

The second pattern is that extinctions are not evenly distributed across the various vertebrate groups. This pattern becomes clear when the level of extinction is compared across higher taxa in Table 1. It is this disparity of extinctions, emphasized by Archibald and Bryant (1990), that is most informative in assessing alternative extinction scenarios.

COMPARISON OF EXTINCTION AND SURVIVAL PATTERNS AND VARIOUS K-T EXTINCTION HYPOTHESES

Table 1, which shows species-level percentage survival, is meant only as a database from which other analyses can be made. A very prominent pattern that emerges from Table 1 is that levels of extinction are not uniform across the twelve major taxa. Rather, as Figure 2 shows, extinction was concentrated in just five of the major taxa; sharks and relatives, lizards, bird- and reptile-hipped

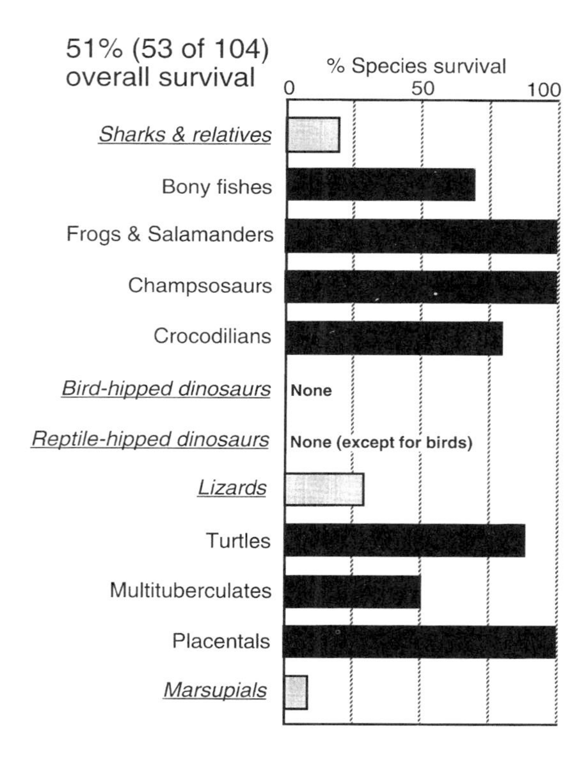

Figure 2. Species-level survivorship for major groups of vertebrates at the K-T boundary in the Western Interior. Taxa with blackened bars have 50% or greater survival for included species. Taxa with stippled bars have 30% or less survival. Those indicated as "none" have no survivals, except for the undetermined number of survivals among avian saurischians (reptile-hipped dinosaurs).

dinosaurs, and marsupials, all of which show 30% or less survival. There is no biologically compelling reason to emphasize a 30% cut-off for survival. These five groups, however, do stand out as having comparatively lower levels of survival. In fact, these 5 groups account for 78% (40 of 51) of the total K-T vertebrate extinctions. The next question is, What event(s) could cause this sort of differential pattern? The most appropriate way in which to approach this question is to use this pattern as a template against which predictions of the various K-T extinction theories and their various extinction scenarios can be compared.

Bolide Impact and its Corollaries

A basic premise of the bolide impact theory of extinction (Alvarez and others, 1980; Alvarez, 1986), is that on a global scale, a dust cloud shrouded the sun, cutting off photosynthesis, killing herbivores, and finally killing carnivores. Although many variants of this scenario have been proposed, a common but incorrect perception is that all major taxa show very high levels of extinction across the ecological spectrum. Extinction estimates of 75% (Glen, 1990) or more are suggested, but these figures are probably apocryphal, because I know of no species-level studies that substantiate such high estimates. The scenario of high levels of extinction across most environments is so broad a spectrum and tries to explain so much that it is difficult to test (see Williams, 1994, for an extended discussion of this issue). Such "Chicken Little" models argue that if a major environmental catastrophe occurs, all extinctions must be somehow related to the event. If applied to contemporary events, such a perspective would lead to the absurd conclusion that all deaths occurring during an earthquake were the direct or indirect result of the earthquake. In the fossil record, we cannot reject the possibility that events such as the impact of an asteroid or comet at the K-T boundary caused all extinctions, but in order to test this assertion, more specific, testable hypotheses of causes of extinction must be provided. "A whole Dante's Inferno of appalling environmental disturbances" (Alvarez, 1986, p. 653) have been suggested as corollary effects of the impact. Some of the major corollaries are a short, sharp decrease in temperature, highly acidic rains, global wildfire, and increased detritus feeding. Some of these can be tested more specifically against the fossil record.

On the basis of what we know of extant vertebrates, if a short, sharp decrease in temperature occurred, the vertebrates most likely to be affected are ectothermic tetrapods, that is, reptiles and amphibians. In these taxa, body temperature is regulated by environmental conditions. Endotherms such as birds and mammals that control body temperature through metabolic means, and ectotherms such as fish, are less-severely affected by a sudden temperature drop. Some amphibians and reptiles inhabiting areas with low winter temperatures or of severe drought have evolved methods of torpor to survive. There is not, however, any basis to believe that most Late Cretaceous ectothermic tetrapods in subtropical to tropical climates had such abilities. Moreover, torpor is most often preceded by decreases in ambient temperature, changes in light regimes, and decreases in food supply. Ectotherms could not have anticipated a short, sharp decrease in temperature. Except for a decline in lizards, the idea of temperature decrease does not accord well with the vertebrate data at the K-T boundary because only

4 of 12 major taxa show extinction or survival patterns in agreement with this scenario (Table 2).

Clemens and Nelms (1993) discussed the lack of ectothermic tetrapods at Late Cretaceous sites in northern Alaska. Endothermic tetrapods, dinosaurs and mammals, and fishes could deal with the lower temperatures and thus are present. In more southern faunas such as in eastern Montana all of these faunal components are present as well as the ectothermic tetrapods (Fig. 3). If lower temperatures

TABLE 2

	Sharks and relatives	Bony fish	Amphibians	Champsosaurs	Crocodilians	Bird-hipped dinosaurs
Number L K vertebrate species	5	13	8	1	5	10
Number of K-T survivals	1	9	8	1	4	0
Extinction observed ≥70%	**YES**	**NO**	**NO**	**NO**	**NO**	**YES**
Short, sharp temperature decrease	no	**NO**	yes	yes	yes	no
Acid rain	**YES**	yes	yes	yes	yes	no
Global wildfire	**YES**	yes	yes	yes	yes	**YES**
Local wildfire	no	**NO**	**NO**	**NO**	**NO**	**YES**
Detritus feeding	no	**NO**	**NO**	**NO**	**NO**	**YES**
BOLIDE IMPACT	**YES**	yes	yes	yes	yes	**YES**
Habitat fragmentation	no	**NO**	**NO**	**NO**	**NO**	**YES**
Loss of freshwater connection	**YES**	**NO**	**NO**	**NO**	**NO**	no
Competition	no	**NO**	**NO**	**NO**	**NO**	no
REGRESSION/ HABITAT FRAGMEN- TATION	**YES**	**NO**	**NO**	**NO**	**NO**	**YES**

Number of Late Cretaceous vertebrate species from eastern Montana (first row) compared to how many survived across the K-T boundary (second row). Third row (bold caps) show levels of observed exteinction (≥70% = YES; ≤50% = NO). Next ten rows compare specific predictions of the bolide impact and regression/habitat fragmentation extinction models as well as the combined predictions for each of these more general excluded ectothermic tetrapods in Alaska during the Late Cretaceous, then a severe drop in global temperature should have devastated ectothermic tetrapods at middle latitudes at the K-T boundary: this clearly did not occur (Table 2).

The next corollary of a bolide impact is acid rain. It is now possible to tabulate some of the effects of acid rain caused by human activity (Cox, 1993). Pure water has a pH of 5.6 (neutral is 7.0), and rain below 5.0 is considered unnaturally acidic. Each unit drop in pH represents a 10-fold drop in the acidity. Rain as low

TABLE 2 (CONTINUED)

Reptile-hipped dinosaurs	Lizards	Turtles	Multituberculates	Placentals	Marsupials	Total
9	10	17	10	6	11	105
0	3	15	5	6	1	53
YES	**YES**	**NO**	**NO**	**NO**	**YES**	n/a
no	**YES**	yes	**NO**	**NO**	no	4 of 12
no	no	yes	**NO**	**NO**	no	3 of 12
YES	**YES**	yes	yes	yes	**YES**	5 of 12
YES	**YES**	**NO**	yes	yes	**YES**	9 of 12
YES	**YES**	**NO**	yes	yes	**YES**	9 of 12
YES	**YES**	yes	yes	yes	**YES**	5 of 12
YES	no	**NO**	**NO**	**NO**	no	9 of 12
no	no	**NO**	**NO**	**NO**	no	8 of 12
no	no	**NO**	**NO**	**NO**	**YES**	8 of 12
YES	no	**NO**	**NO**	**NO**	**YES**	11 of 12

models. A "yes" for any of the specific predictions is counted as a "yes" for the general model. Capitalized, bold "YES" and "NO" signify agreement between predictions of theory and observed patterns, while lower case signifies disagreement. Numbers on right indicate for how many of 12 major taxa predicted agrees with observed. See text for discussion.

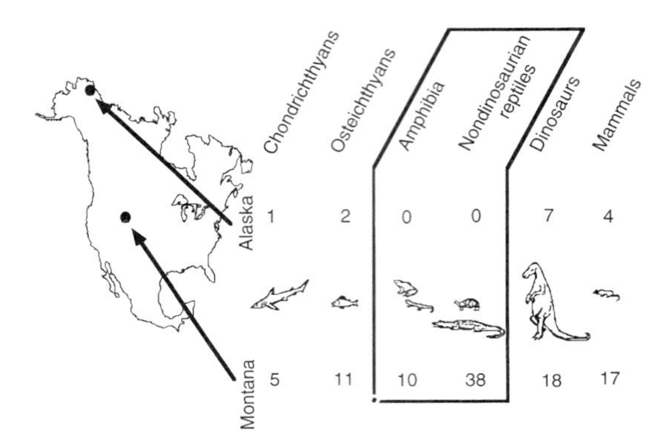

Figure 3. Species-level comparison of vertebrates from northern Alaska and eastern Montana that shows the lack of ectothermic amphibians and nondinosaurian reptiles in Alaska and their abundance in eastern Montana. Modified after Clemens and Nelms (1993).

as pH 2.4 has been recorded, but annual averages in areas affected by acid rain range from pH 3.8 to 4.4. Acid fogs and clouds from pH 2.1 to 2.2 have been recorded in southern California and have been known to bathe spruce-fir forest in North Carolina. Although different vertebrates react differently, aquatic species (fish, amphibians, and some reptiles) are the first and most drastically affected. Those reproducing in water are the first to suffer, but if pH becomes low enough (e.g., lower than about 3.0) deaths of adults may occur.

Estimates of the pH of acid rain following a bolide impact vary. Prinn and Fegley (1987) suggested that, depending upon the impacting object, the global rains could have a pH as low as 0–1.5. D'Hondt and others (1994, p. 30) suggested that global effects could have "driven the pH of near-surface marine and fresh water below 3." D'Hondt and others seem to be aware that the patterns of vertebrate extinction do not fit their scenario, yet they seem to hedge by saying that the rains would have been "rapid" and "nonuniform."

If such low pHs were reached, the consequences for aquatic vertebrates across the K-T boundary would have been horrendous. In some areas the effects of acid rain could have been buffered, to some degree, by bedrock high in carbonates such as limestone. But there was no such bedrock underlying the aquatic systems in the latest Cretaceous of eastern Montana. More important, if the rainwater reached the pH of battery acid, as suggested above, not only could this factor kill large nonaquatic vertebrates such as nonavian dinosaurs outright, any buffering in aquatic systems would also have been rendered useless.

When we compare the effects of acid rain on modern vertebrate species to what is observed at the K-T boundary, we find almost no correlation. Only 3 of 12 major taxa have extinction-survival patterns matching the acid rain corollary of bolide impact (Table 2), and thus the likelihood of such low-pH rain is highly implausible.

Another corollary that receives various levels of support from proponents of bolide impact is global wildfire (Wolbach and others, 1990, and references therein). During a global wildfire, most of the above-ground biomass all over the world, both plants and animals, would have been reduced to ashes. In freshwater, those plants and animals not boiled outright, would have faced a rain of organic and inorganic matter unparalleled in human experience. These organisms would have literally choked on the debris or suffocated as oxygen was suddenly depleted with the tremendous influx of dead organic matter. The global wildfire scenario is so broad in its killing effects that it could not have been selective. Yet as shown above, patterns of vertebrate of extinction and survival are highly selective. It is therefore no surprise that global wildfire only agrees with 5 of the 12 cases of extinction-survival (Table 2).

The problem with this corollary is not only its complete lack of biological selectivity, but the physical basis for the event having occurred. It is argued that there is a global charcoal and soot layer that coincides with the K-T boundary, the time of emplacement of which is measured in months (Wolbach and others, 1990). A necessary corollary of this scenario is that the layer containing the charcoal and soot was also deposited in only months. This is demonstrably not the case for at least one K-T section that continues to be cited in these studies—the Fish Clay of the Stevns Klint section on the coast of Denmark. As Officer and Ekdale (1986, p. 263) showed, the Fish Clay is a laterally discontinuous, complexly layered, and burrowed clay reflecting "the environmental conditions at the time of its deposition." It is not the result of less than a year of deposition caused by an impact-induced global wildfire. Thus, carbon near the K-T boundary at Stevns Klint as well as in other sections is likely the result of much longer term accumulation during normal sedimentation.

Although global wildfires are dismissed on both biological and geological grounds, the importance of more localized wildfires should not be dismissed. Such fires would be more detrimental to terrestrial species and might produce a pattern that agrees better with known extinction patterns for both vertebrates and lower latitude plants. Thus, assuming that terrestrial species would be more adversely affected by local wildfires, there is agreement in 9 of 12 extinction-survival patterns (Table 2).

The final corollary of an impact is increased detritus feeding in the freshwater aquatic realm. Sheehan and Fastovsky (1992) argued this possibility as a result of their reexamination of the study of Archibald and Bryant (1990). As noted earlier, using a selectively culled database Sheehan and Fastovsky (1992) found a 90% survival for freshwater species and only 12% for terrestrial species. When the complete rather than a culled dataset was used, Archibald (1993a) found that the freshwater and terrestrial species survivals were 78% and 28%, respectively. The argument of greater survival among freshwater species was certainly not a new observation, because Archibald and Bryant (1990) also recognized this survival-extinction pattern among other patterns at the K-T boundary. As Archibald (1993a, p. 92) noted, the "pattern of higher aquatic survivorship is not even unique to the K-T boundary. In *all* purported mass extinctions including data for both terrestrial and aquatic vertebrates, aquatic vertebrates *always* fare better (Bakker, 1977; Padian and Clemens, 1985), yet at only the K-T boundary is there convincing evidence of an asteroid impact."

The basic thesis of Sheehan and Fastovsky (1992) is that land-based communities depend largely upon primary production, whereas riverine communities obtain their organic carbon from detritus. According to Sheehan and Fastovsky, with the catastrophic cessation of primary production on land as the result of an impact, land-based species would suffer greatly while freshwater species would do well because of an influx of detritus. This hypothesis is at first quite appealing in its simplicity and it agrees with 9 of 12 extinction-survival patterns for major taxa (Table 2). The increased detritus feeding hypothesis, however, has two major assumptions that need to be addressed. First is the assumption that freshwater communities, especially streams, are primarily detritus based. Second, if the catastrophic scenario were correct, there should be a noticeable increase in detrital material in terrestrially derived sediments at the K-T boundary.

As Sheehan and Fastovsky (1992) admit, riverine systems can vary considerably in their energy sources derived from primary production versus detritus. Sheehan (1994, oral comm.) argued more strongly that detrital feeding is of primary importance in streams, citing the work of Closs and Lake (1993). The stream studied by Closs and Lake (1993) is an upland, intermittently flowing stream, whereas those flowing through eastern Montana during the latest Cretaceous were lowland, permanently flowing streams. Thus, the degree to which one can draw reasonable parallels between these different kinds of streams remains in question.

As Vannote and others (1980) noted in their classic paper on river systems, primary production in streams is generally of less importance near the headwaters, increases near the mid-size of the streams, and then decreases in larger streams where depth and turbidity are important. Streams flowing through eastern Montana in the latest Cretaceous were middle to larger size streams; thus Sheehan and Fastovsky (1992) may be correct that detrital feeding is important in this portion of Late Cretaceous streams. This is of far less import than the claim of these authors that detrital feeding would have differentially favored aquatic species over terrestrial species on a global scale. Just as important is the fact that any buffering of aquatic systems could as easily have been the result of marine regression as an impact. As discussed later, marine regression results in the proliferation and lengthening of riverine systems, and thus could also have favored freshwater systems.

A worldwide increase in detrital feeding at the K-T boundary would have been accompanied by a tremendous influx in detrital material as the catastrophic deaths of terrestrial plants and animals occurred. This does not mean that we should expect large accumulations of dead nonavian dinosaurs or other animals or plants, because where they lived and died was a fairly restricted area and thus far less likely to be preserved (e.g., see Cutler and Behrensmeyer, 1994). If the idea of mass killing, especially by global wildfires, were true, however, we would certainly expect to see the detrital residue in most terrestrially derived sediments that accumulated at the K-T boundary. No such concentrations have ever been recognized. Most K-T boundary sections in terrestrial sections are thin, often carbonaceous units showing no great influx of organic or inorganic material. Thus, the detritus-feeding hypothesis has no physical evidence in its support. In fact, this scenario appears to be based entirely on the pattern of survival and extinction that is itself the subject of analysis.

When all of these corollaries of a bolide impact are summed as shown in Table 2, the agreement of observed versus expected extinction-survival patterns is only 5 of 12 for vertebrates. Thus, the actual pattern of extinction and survival for vertebrates at the K-T boundary in eastern Montana is in relatively poor agreement with the corollaries of bolide impact. Without special pleading, these corollaries as currently proposed (except possibly more local wildfires and localized effects of detritus feeding in streams) are unlikely causes of vertebrate extinction. This does not mean all should be rejected even if some are incorrect. It is imperative, however, that those proposing the different corollaries separate those that are supported by the vertebrate fossil from those that are not.

Global Marine Regression and Habitat Fragmentation

It has been known for more than 100 years that many areas of the modern terrestrial realm were inundated by shallow epicontinental seas during the Late Cretaceous. Only recently, however, has it become clear just how dramatic the loss of these seas was from the latest Cretaceous into the early Tertiary. During this interval, the nonmarine area increased from 109 million km^2 to 138 million km^2—almost a 27% increase (Smith and others, 1994). As Smith and others showed (1994, Fig. 2), this is the single greatest increase in nonmarine area during the past 250 m.y. of Earth's history.

Nowhere was this change more dramatic than in North America. Near the end of the Cretaceous, maximum transgression divided North America into two continents (Fig. 4). As regression began and continued until at or near the K-T boundary, coastal plains decreased in size and became fragmented; stream systems multiplied and lengthened; and as sea level fell, land connections were established or reestablished (Fig. 5). The placement of the shoreline during maximum transgression, and the subsequent movement of the shoreline indicating marine regression, can be tracked during the latest Cretaceous in some areas of the Western Interior such as in the Dakotas, Montana, and Wyoming by strandline positions of various marine invertebrates (e.g., Waage, 1968; Gill and Cobban, 1973). Thus, general patterns of interior sea versus land shown in Figure 4 are becoming more accurate. We know less about the very latest Cretaceous or

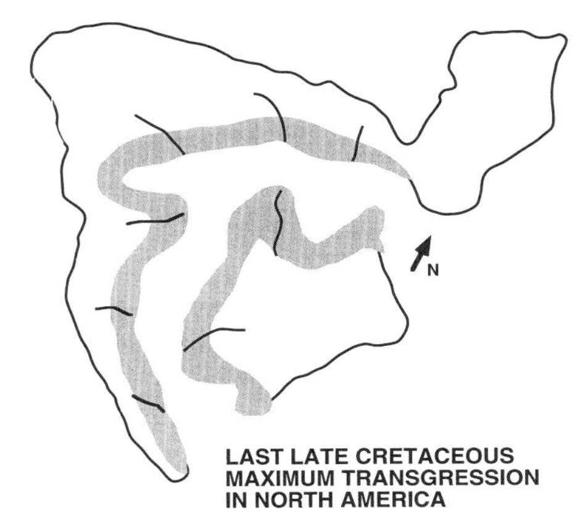

Figure 4. North America during the last Late Cretaceous maximum transgression, emphasizing the presence of low coastal areas (gray shading) and rivers (lines). The extents of the coastal areas are approximations and the actual positions of rivers are hypothetical.

earliest Paleocene maximum regression as suggested in Figure 5, because in the center of the Western Interior, the seas completely left as the southern shoreline pushed into Texas. Invertebrate and vertebrate marine species did not reappear until latest early Paleocene or middle Paleocene in the Dakotas (Cvancara and Hoganson, 1993).

With the loss of the epicontinental seaway there is no question that the total land area increased dramatically. It does not follow, however, that all terrestrial environments and their biotic communities increased similarly. Areas open to freshwater communities increased with the increase and lengthening of watercourses, unless there was a dramatic shift to drier conditions (see 2 in Fig. 5). Fastovsky and McSweeney (1987) argued that, at least in eastern Montana and western North Dakota, the opposite occurred, and the amount of ponded water increased across the K-T boundary. Thus in freshwater habitats, as stream systems increased in size with marine regression, freshwater vertebrates did well except for those with close marine ties—sharks and some bony fishes. These are fishes that spend at least a portion of their life in a marine environment, in many instances to reproduce. Especially among the sharks and relatives, it is not clear whether the four of five species that disappear (Table 2) from the Western Interior are actually extinctions at the K-T boundary or whether they survived elsewhere in marine environments. Cvancara and Hoganson (1993) argued that these were extinctions among sharks and relatives. They also showed in their range chart that new Paleocene species appeared in the Western Interior at the K-T boundary. Their own data and discussion showed, however, that this is not the case, because the definitively oldest marine sediments that postdate the K-T boundary in the Western Interior (the Cannonball Formation) are at most late-early Paleocene in age. This means a gap in marine sedimentation in the Western Interior of a million years or more. Such a pattern suggests strongly that, as marine regression continued, sharks and relatives departed because connections to the sea became attenuated. They did not reappear until a smaller transgression reached the Western Interior at or just before middle Paleocene time (Fig. 6).

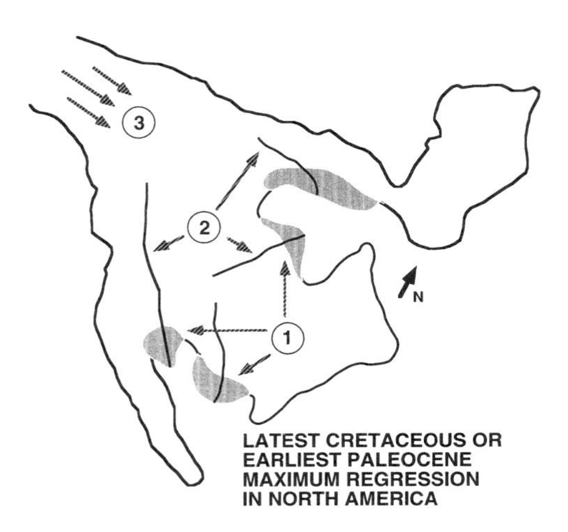

Figure 5. North America during the latest Cretaceous or earliest Paleocene maximum regression emphasizing the decrease of low coastal areas (gray shading) and increase in length of rivers (lines). The extents of the coastal line and coastal areas are approximations and the actual positions of rivers are hypothetical. See text for discussion of numbers.

Other environments such as coastal plains and the land-based portion of the vertebrate communities decreased in size as the shoreline receded (see 1 in Fig. 5). This is the vertebrate community that we are sampling up to the K-T boundary. It could be argued that organisms hit the hardest by events at the K-T boundary, such as nonavian dinosaurs, would have survived in other environments. We know with certainty that dinosaurs (both avian and nonavian) did live in other environments, such as the higher, drier Gobi Desert in Mongolia. At present, however, all large, well-studied vertebrate communities at the K-T boundary, including nonavian dinosaurs, are coastal. Thus arguments about what nonavian dinosaurs and other vertebrates may or may not have done in other environments are moot.

The actual placement of the coastline during the latest Cretaceous and earliest Paleocene remains speculative (Fig. 5), but the overall physical consequences are not. The geological record shows a decrease in marine rocks and an increase in terrestrially derived rocks in the waning time before the K-T boundary. Therefore, the argument that freshwater environments increased while coastal habitats decreased is not speculative. Based on what we know of modern biotas, it is a reasonable extrapolation that as the coastal environments decreased in size (compare Figs. 4 and 5), the largest land vertebrates, the nonavian dinosaurs, would be the most affected by loss and fragmentation of habitats.

Some earth scientists have suggested that habitat fragmentation is vague (Buffetaut, 1994) or even untestable in the geologic record (Fastovsky and Sheehan, 1994) in the context of marine regression. Although habitat fragmentation is not a well understood phenomenon among earth scientists, contrary to the above assertions, it is all too real among biologists studying the effects of human activity in modern rainforests and in urban settings. In San Diego, for example,

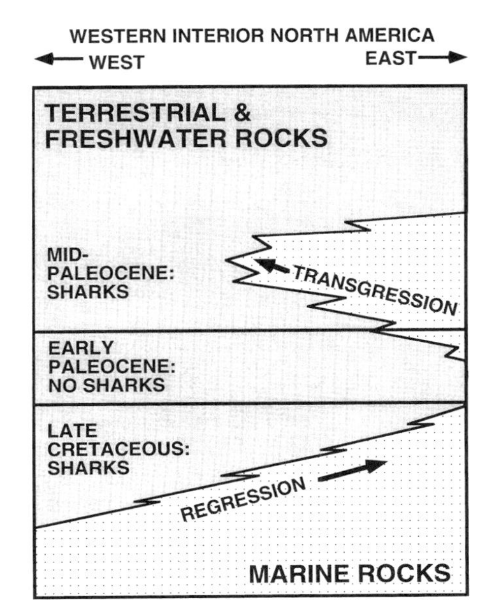

Figure 6. Late Cretaceous and Paleocene regression and transgression cycle in the northern part of the Western Interior emphasizing why partially marine species such as sharks and their relatives disappear from terrestrially derived sediments in eastern Montana. See text for discussion.

Soulé and colleagues (*in* Bolger and others, 1991) documented the decline of birds and mammals as urban development divides and isolates habits in canyon areas. One would not expect that the natural equivalent of habitat fragmentation would be easily, if at all, preserved in the rock. The forcing factor for habitat fragmentation in the latest Cretaceous—marine regression—is a thoroughly documented fact during the waning years of the Late Cretaceous in North America. Globally, marine regression occurred within this same general time frame, although how close in time it occurred in various regions is a matter of debate.

As sea level lowered, new land areas were exposed. This included the establishment or reestablishment of intercontinental connections. One such connection was the Bering Land Bridge (see 3 in Fig. 5). At various times during the Late Cretaceous this bridge probably appeared and then disappeared. This is suggested by similarities in parts of the Late Cretaceous vertebrate faunas in Asia and North America, especially the better studied nonavian dinosaurs and mammals. Unlike the case for loss of coastal habitats and the increase in freshwater systems, the establishment of land bridges is less predictive as to which species would be affected. Although one cannot predict in general which species will experience increased competition through the establishment of land bridges, the appearance of archaic ungulates in North America as marsupials declined fits this pattern well (Fig. 7).

The above scenario agrees with 11 of 12 extinction-survival patterns seen for major vertebrate taxa (Table 2), although if the specific effect on marsupials is not included because it is not a general prediction, then 10 of 12 extinction-survival patterns are in agreement with the above scenario. This suggests that regression is hardest on terrestrial species, especially of larger size, while aquatic forms fare better by comparison. The next step is to investigate whether the opposite would be true during a transgressive phase. Do freshwater species undergo greater levels of extinction than do terrestrial species during transgression? We know of four major transgressive-regressive cycles in North America during the Late Cretaceous through the earliest Tertiary (Fig. 8). Unfortunately, we currently have detailed vertebrate records only near the K-T boundary. There are, however, two other examples that bear further scrutiny.

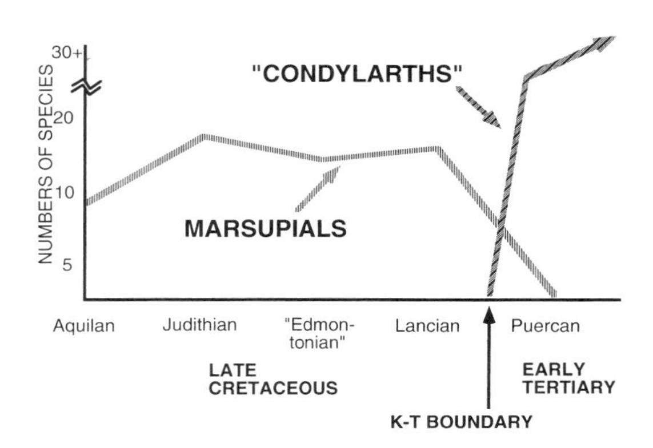

Figure 7. Species diversity of marsupials during the Late Cretaceous of the Western Interior compared to that for "condylarths." See text for further discussion.

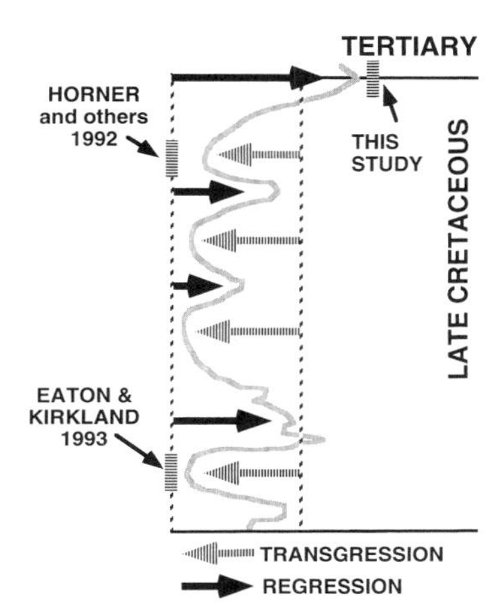

Figure 8. The four transgression-regression cycles of the Late Cretaceous in North America showing the positions of three vertebrate studies discussed in the text. Modified after Hallam (1992).

Transgression and Regression as Forcing Factors in Evolution and Extinction

Horner and others (1992) argued that during the five million year deposition of the Judith River Formation, much of this occurring while there was a lowered sea-level stillstand, little change occurred in nonavian dinosaurian and mammalian lineages (Fig. 9). As marine transgression began and preceded to the west, Horner and others (1992) documented anagenetic appearances within four separate lineages of nonavian dinosaurs during the last transgressive phase of the Late Cretaceous (Fig. 9). Thus, it appears that transgression was favorable for evolutionary change in the four different nonavian dinosaur lineages.

Eaton and Kirkland (1993; Kirkland, 1987) reported that across the Cenomanian-Turonian boundary in southern Utah, which encompasses the first

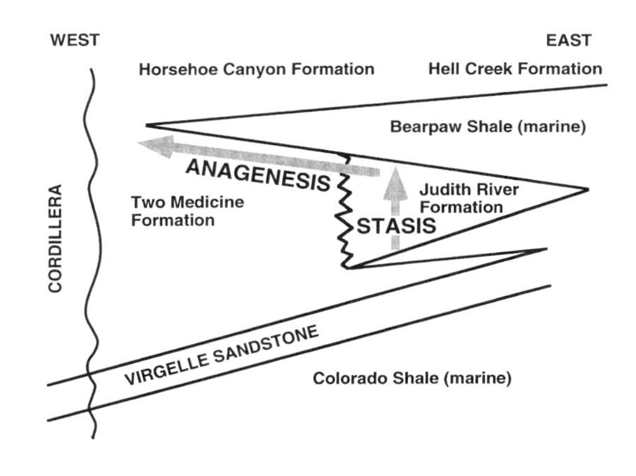

Figure 9. Schematic stratigraphic section of the Upper Cretaceous in the northern Western Interior modified from Horner and others (1992), indicating the evolutionary stasis of mammals and dinosaurs that they observed in the Judith River Formation compared to the anagenetic changes they observed in four different lineages of dinosaurs during the transgression that deposited the Bearpaw Shale.

maximum transgression in the Late Cretaceous of North America (Fig. 8), the riverine-riparian vertebrate fauna shows notable extinction, while groups of terrestrial vertebrates continued to radiate. Eaton (1994, personal commun.) noted that "there is no extinction 'event' recorded in the brackish water faunas. Freshwater and riparian faunas were strongly affected as indicated by the extinction of some turtles, crocodilians, and fish. Fully terrestrial organisms such as mammals and nonavian dinosaurs appear to diversify across the boundary."

These two earlier transgressions, combined with the K-T marine regression, argue that regressions are hardest on terrestrial species, but favor freshwater forms, whereas transgressions are the hardest on freshwater species, but favor terrestrial forms. These three different studies, one of regression and two of transgression, are at this point only suggestive of the role that these cycles may play in vertebrate turnover. The evidence is sufficient to suggest that transgression and regression may be forcing factors in the evolution of at least vertebrates and, furthermore, that during intervening highstands and lowstands of the Western Interior seaways, much less evolution or extinction occurred.

CONCLUSIONS AND AN ALTERNATIVE EXTINCTION SCENARIO

There is no longer any doubt that leading up to and crossing the K-T boundary a number of physical events, especially marine regression, bolide impact(s), and increased volcanic activity occurred. I have not discussed volcanic activity in this paper, not because I don't feel that it is important, but rather because there have been far fewer assessments of the direct role(s) it may have played in vertebrate turnover at the K-T boundary as compared to bolide impact or marine regression. If volcanic activity were as catastrophic in effect as usually proposed for an impact, then many of the same scenarios discussed earlier for an impact would pertain equally to cataclysmic volcanism.

Even with increased documentation of specific marine regressions, impacts, and volcanism, not a single major extinction event can be correlated to one of these geologic events alone. This argues strongly that, if considered as isolated factors, such events are not sufficient to explain mass extinction in the fossil record in general and the K-T boundary in particular. The sum of their effects may explain the extinctions of many plant and animal species in the fossil record. This is especially germane to the K-T boundary, where the documentation is as good as it gets for marine regression, bolide impact(s), and increased volcanic activity.

Unifying the fossil and geologic data with the presumed effects of marine regression and bolide impact (but setting aside volcanic activity because of a lack of specified affects on vertebrates), it is possible to estimate how events unfolded in the terrestrial realm in the waning tens of thousands of years of the Cretaceous in the Western Interior.

I begin with the marine regression that had been occurring for tens of thousands of years. At this point none of the vertebrate species show obvious populational stress. As marine regression continues, and the coastal habitats decrease ever more rapidly, large vertebrates—the nonavian dinosaurs—are the first to experience population declines. They certainly are capable of migrating from one shrinking coastal habitat to another, but at some point even this cannot stop further decline. They are not alone. The Komodo dragon–sized lizards and the single exclusively terrestrial turtle, *Boremys*, are also undergoing population declines. Among the smaller terrestrial vertebrates some populations are also declining, but those that are adapting more quickly to the environmental changes because of their shorter life spans and quicker turnover rates are faring well.

Invaders appear as land bridges are established. In this case the luck of the draw positions the newly invading diminutive archaic ungulates ("condylarths") against the equally diminutive marsupials. In the Western Interior, at least, the "condylarths" are successful over the marsupials who had flourished for 20 million years or longer. In South America we know events are different. Both groups of mammals appear in South America near the K-T boundary, but here they divide the guilds, marsupials becoming the carnivores and "condylarths" becoming the herbivores. This coevolutionary arrangement lasts for almost 50 million years in South America, with only an infusion of rodents and primates from the outside world.

In the freshwater habitats, most species are not stressed, especially because the size of their habitat is increasing. Notable exceptions are the sharks and relatives, whose forays into rivers are less and less frequent as the distance to the sea expands from tens to hundreds and then thousands of miles.

As in Dickens' Europe, it is the best of times and worst of times—depending upon which species you are and in which environment you live. A sudden, earth-shattering event magnifies the differences between those doing well and those not doing well. An asteroid or comet has struck what will be present-day Yucatan. Material injected into the upper atmosphere forms a blanket of darkness blanking the sun to the point that photosynthesis ceases or diminishes, depending upon where you are, for many weeks. The effects are especially acute for lower latitude plants unaccustomed to lower light regimes caused by seasonal changes in the sun's position, and also because the bolide hit in the lower latitudes. Higher latitude plants survive much better as presumably do some animals. The extinction rates for coastal plants are also exaggerated because of continued habitat losses.

Evidence is beginning to appear for megafloral (leaves) studies that suggest substantially less extinction in the high-latitude plants such as in New Zealand compared to the classic sites in the Western Interior, including eastern Montana (Johnson, 1993). Although Johnson (1994, personal commun.) noted that it is premature to specify percentage differences of extinction for these megafloras, the percentage of extinction based upon pollen indicates the magnitude of the differences. For New Zealand, he estimates extinction using pollen at less than 5%, while in the Western Interior it is on the order of 35%. The percentage of megafloral extinctions is undoubtedly higher for both New Zealand and the Western Interior because megafloral studies sample at a finer taxonomic scale compared to pollen.

Through the loss of sunlight, regional wildfires at lower latitudes with substantial dead plant biomass, and the very rapid loss of plant habitats through marine regression, plant extinctions soar. With the dramatic and rapid loss of plant species and biomass in an already highly stressed ecosystem, some vertebrate species, most notably large herbivorous nonavian dinosaurs, finally succumb. The predaceous nonavian dinosaurs follow quickly, the larger species going first. For the first time in more than 150 million years, no large land vertebrates grace the Earth. The landscape is open and waiting for evolution's next gambit—mammals.

ACKNOWLEDGMENTS

I thank Gerta Keller and Norm MacLeod for inviting me to participate in their theme session at the 1993 Geological Society of America meeting in Boston and to contribute a paper to this volume. Many thanks to P. Abbott, W. A. Clemens, L. McClenaghan, and N. MacLeod for reading and commenting on the manuscript, and to colleagues at San Diego State University for sundry ecological information.

REFERENCES CITED

Alvarez, L. W., Alvarez, W., Asaro, F., and Michel, H., 1980, Extraterrestrial cause for the Cretaceous-Tertiary extinction: Science, v. 208, p. 1095–1108.

Alvarez, W., 1986, Toward a theory of impact crises: Eos (Transactions, American Geophysical Union), v. 67, p. 649, 653–655, 658.

Archibald, J. D., 1982, A study of Mammalia and geology across the Cretaceous-Tertiary boundary in Garfield County, Montana: University of California Publications in the Geological Sciences, v. 122, p. 1–286.

Archibald, J. D., 1993a, Major extinctions of land-dwelling vertebrates at the Cretaceous-Tertiary boundary, eastern Montana: Comment: Geology, v. 21, p. 90–92.

Archibald, J. D., 1993b, The importance of phylogenetic analysis for the assessment of species turnover: A case history of Paleocene mammals in North America: Paleobiology, v. 19, p. 1–27.

Archibald, J. D., 1996, Dinosaur extinctions: What the fossils say: New York, New York, Columbia University Press (in press).

Archibald, J. D., and Bryant, L., 1990, Differential Cretaceous-Tertiary extinctions of nonmarine vertebrates: Evidence from northeastern Montana, *in* Sharpton, V. L., and Ward, P., eds., Global catastrophes in Earth history: An interdisciplinary conference on impacts, volcanism, and mass mortality: Geological Society of America Special Paper 247, p. 549–562.

Bakker, R. T., 1977, Tetrapod mass extinctions—A model of the regulation of speciation rates and immigration by cycles of topographic diversity, *in* Hallam, A., ed., Patterns of evolution as illustrated by the fossil record: Amsterdam, Elsevier, p. 439–468.

Bolger, D. T., Alberts, A. C., and Soulé, M. E., 1991, Occurrence patterns of bird species in habitat fragments: Sampling, extinction, and nested species subsets: American Naturalist, v. 137, p. 155–166.

Bryant, L. J., 1989, Non-dinosaurian lower vertebrates across the Cretaceous-Tertiary boundary in northeastern Montana: University of California Publications in Geological Sciences, v. 134, p. 1–107.

Buffetaut, E., 1994, Paleoecological implications of Alaskan terrestrial vertebrate fauna in latest Cretaceous time at high paleolatitudes: Comment: Geology, v. 22, p. 191.

Cifelli, R. L., 1990, Cretaceous mammals of southern Utah. I. Marsupials from the Kaiparowits Formation (Judithian): Journal of Vertebrate Paleontology, v. 10, p. 295–319.

Clemens, W. A., 1964, Fossil mammals of the type Lance Formation, Wyoming: Part I. Introduction and Multituberculata: University of California Publications in Geological Sciences, v. 48, p. 1–105.

Clemens, W. A., 1966, Fossil mammals of the type Lance Formation, Wyoming: Part II. Marsupialia: University of California Publications in Geological Sciences, v. 62, p. 1–122.

Clemens, W. A., 1973, Fossil mammals of the type Lance Formation, Wyoming: Part III. Eutheria and summary: University of California Publications in Geological Sciences, v. 94, p. 1–102.

Clemens, W. A., and Nelms, L. G., 1993, Paleoecological implications of Alaskan terrestrial vertebrate fauna in latest Cretaceous time at high paleolatitudes: Geology, v. 21, p. 503–506.

Closs, G. P., and Lake, P. S., 1993, Spatial and temporal variation in the structure of an intermittent-stream food web: Ecological Monographs, v. 64, p. 1–21.

CoBabe, E. A., and Fastovsky, D. E., 1987, *Ugrosaurus olsoni*, a new ceratopsian (Reptilia: Ornithischia) from the Hell Creek Formation of eastern Montana: Journal of Paleontology, v. 61, p. 148–154.

Cox, G. W., 1993, Conservation ecology biosphere and biosurvival: Dubuque, Iowa, Wm. C. Brown, 352 p.

Cutler, A. H., and Behrensmeyer, A. K., 1994, Bone beds at the boundary: Are they a real expectation?, *in* New developments regarding the K/T event and other catastrophes in Earth history: Houston, Texas, Lunar and Planetary Institute Contribution 825, p. 28.

Cvancara, A. M., and Hoganson, J. W., 1993, Vertebrates of the Cannonball Formation (Paleocene) in North and South Dakota: Journal of Vertebrate Paleontology, v. 13, p. 1–23.

D'Hondt, S., Sigurdsson, H., Hanson, A., Carey, S., and Pilson, M., 1994, Sulfate volatilization, surface-water acidification, and extinction at the KT boundary, *in* New developments regarding the K/T event and other catastrophes in Earth history: Houston, Texas, Lunar and Planetary Institute Contribution no. 825, p. 29–30.

Eaton, J. G., and Kirkland, J. I., 1993, Faunal changes across the Cenomanian-Turonian (Late Cretaceous) boundary, southwestern Utah: Geological Society of America Abstracts with Programs, v. 25, no. 5, p. 33–34.

Estes, R., 1964, Fossil vertebrates from the Late Cretaceous Lance Formation eastern Wyoming: University of California Publications in Geological Sciences, v. 49, p. 1–187.

Estes, R., Berberian, P., and Meszoely, C., 1969, Lower vertebrates from the Late Cretaceous Hell Creek Formation, McCone County, Montana: Breviora, no. 337, p. 1–33.

Fastovsky, D. E., and McSweeny, K., 1987, Paleosols spanning the Cretaceous-Paleogene transition, eastern Montana and western North Dakota: Geological Society of America Bulletin, v. 99, p. 66–77.

Fastovsky, D. E., and Sheehan, P. M., 1994, Habitat vs. asteroid fragmentation in vertebrate extinctions at the KT boundary: The good, the bad, and the untested, *in* New developments regarding the K/T event and other catastrophes in Earth history: Houston, Texas, Lunar and Planetary Institute Contribution no. 825, p. 36–37.

Foster, C. A., 1993, Taxonomic validity of the ceratopsid dinosaur *Ugrosaurus olsoni* (CoBabe and Fastovsky): Journal of Paleontology, v. 67, p. 316–318.

Fox, R. C., 1989, The Wounded Knee local fauna and mammalian evolution near the Cretaceous-Tertiary boundary, Saskatchewan, Canada: Palaeontographica, v. 208, p. 11–59.

Gill, J. R., and Cobban, W. A., 1973, Stratigraphic and geologic history of the Montana Group and equivalent rocks, Montana, Wyoming, and North and South Dakota: United States Geological Survey Professional Paper 776, p. 1–37.

Glen, W., 1990, What killed the dinosaurs?: American Scientist, v. 78, p. 354–370.

Hallam, A., 1992, Phanerozoic sea-level changes: New York, Columbia University Press, 266 p.

Horner, J. R., 1992, Cranial morphology of *Prosaurolophus* (Ornithischia: Hadrosauridae) with descriptions of two new hadrosaurid species and an evaluation of hadrosaurid phylogenetic relationships: Museum of the Rockies Occasional Paper no. 2, p. 1–119.

Horner, J. R., Varricchio, D. J., and Goodwill, M., 1992, Marine transgressions and the evolution of Cretaceous dinosaurs: Nature, v. 358, p. 59–61.

Hurlbert, S. H., and Archibald, J. D., 1995, No statistical support for sudden (or gradual) extinction of dinosaurs: Geology (in press).

Hutchison, J. H., and Archibald, J. D., 1986, Diversity of turtles across the Cretaceous/Tertiary boundary in northeastern Montana: Palaeogeography, Palaeoclimatology, Palaeoecology, v. 55, p. 1–22.

Johnson, K. R., 1993, High-latitude deciduous forests and the Cretaceous-Tertiary boundary in New Zealand: Geological Society of America Abstracts with Programs, v. 25, no. 6, p. 114.

Kirkland, J. I., 1987, Upper Jurassic and Cretaceous lungfish tooth plates from the Western Interior, the last dipnoan faunas of North America: Hunteria, v. 2, p. 1–16.

Lillegraven, J. A., 1969, Latest Cretaceous mammals of upper part of Edmonton Formation of Alberta, Canada, and review of marsupial-placental dichotomy in mammalian evolution: University of Kansas Paleontological Contributions, Art. 50, Vert. 12, p. 1–122.

Lofgren, D. L., Hotton, C. L., and Runkel, A. C., 1990, Reworking of Cretaceous dinosaurs into Paleocene channel deposits, upper Hell Creek, Montana: Geology, v. 18, p. 874–877.

Molnar, R. E., and Carpenter, K., 1989, The Jordan theropod (Maastrichtian, Montana, U.S.A.) referred to the genus *Aublysodon*: Geobios, no. 22, p. 445–454.

Officer, C. B., and Ekdale, A. A., 1986, Cretaceous extinctions: Evidence for wildfires and search for meteoritic material: Comment: Science, v. 234, p. 262–263.

Padian, K., and Clemens, W. A., 1985, Terrestrial vertebrate diversity: Episodes and insights, *in* Valentine, J. W., ed., Phanerozoic diversity patterns: Profiles in macroevolution: Princeton, New Jersey, Princeton University Press, p. 41–96.

Perlman, D., 1993, Volcanoes may have stolen dinosaur's oxygen supply: San Francisco Chronicle, Final Edition, Thursday, October 28, 1993, p. A5.

Prinn, R. G., and Fegley, B., Jr., 1987, Bolide impacts, acid rain, and biospheric traumas at the Cretaceous-Tertiary boundary: Earth and Planetary Science Letters, v. 83, p. 1–15.

Sheehan, P. M., and Fastovsky, D. E., 1992, Major extinctions of land-dwelling vertebrates at the Cretaceous-Tertiary boundary, eastern Montana: Geology, v. 20, p. 556–560.

Sheehan, P. M., Fastovsky, D. E., Hoffman, R. G., Berghaus, C. B., and Gabriel , D. L., 1991, Sudden extinction of the dinosaurs: Latest Cretaceous, upper Great Plains, U.S.A.: Science, v. 254, p. 835–839.

Sloan, R. E., and Van Valen, L., 1965, Late Cretaceous mammals from Montana: Science, v. 148, p. 220–227.

Sloan, R. E., Rigby, J. K., Jr., Van Valen, L. M., and Gabriel, D. L., 1986, Gradual dinosaur extinction and simultaneous ungulate radiation in the Hell Creek Formation: Science, v. 234, p. 1173–1175.

Smith, A. G., Smith, D. G., and Funnell, B. M., 1994, Atlas of Mesozoic and Cenozoic coastlines: London, Cambridge University Press, 99 p.

Vannote, R. L., Minshall, G. W., Cummins, K. W., Sedell, J. R., and Cushing, C. E., 1980, The river continuum concept: Canadian Journal of Fisheries and Aquatic Science, v. 37, p. 130–137.

Waage, K. M., 1968, The type Fox Hills Formation, Cretaceous (Maestrichtian), South Dakota. Part 1. Stratigraphy and paleoenvironments: Yale University, Peabody Museum of Natural History, Bulletin, v. 27, p. 1–175.

Weishampel, D. B., Dodson, P., and Osmólska, H., eds., 1990, The Dinosauria: Berkeley, University of California Press, 733 p.

Williams, M. E., 1994, Catastrophic versus noncatastrophic extinction of the dinosaurs: Testing, falsifiability, and the burden of proof: Journal of Paleontology, v. 68, p. 183–190.

Wolbach, W. S., Gilmour, I., and Anders, E., 1990, Major wildfires at the Cretaceous/Tertiary boundary, *in* Sharpton, V. L., and Ward, P., eds., Global catastrophes in Earth history: An interdisciplinary conference on impacts, volcanism, and mass mortality: Geological Society of America Special Paper 247, p. 391–400.

16

Stratigraphic and Sedimentologic Characteristics of the Cretaceous-Tertiary Boundary in the Ager Basin, Lleida Province, Northeastern Spain

Ferran Colombo, *Departamento de Geológia Dinámica, Geofísica y Paleontologia, Universidad de Barcelona, Barcelona, Spain*

INTRODUCTION

The sedimentary strata that form the transition between the Mesozoic and the Cenozoic in the Ager basin accumulated in different depositional environments in continental areas, as evidenced by abundant nonmarine fossils. For this reason the Cretaceous and the Tertiary sediments in the Ager-Tremp region have been the subject of a number of reports in recent years. The first discoveries of Mesozoic and Cenozoic fossils in this area more than a century ago prompted many geological and paleontological studies (e.g., Leymerie, 1863, 1877; Vidal, 1873; Lapparent and Aguirre, 1956; Bataller, 1958). However, because these sediments contain Mesozoic fossils in the lower part and Cenozoic fossils in the upper part, separated by an intermediate zone considered to be unfossiliferous, their chronostratigraphic significance remained unclear.

At the turn of the century, there was some confusion concerning the chronostratigraphic value of some fossil groups. As a result, some authors considered the nonmarine sediments to be Cretaceous whereas others considered them Tertiary. With designation of the "Garumniense" (Leymerie, 1863, 1877; Bataller, 1958), a compromise was reached with proposition of a "geologic stage" corresponding to the Cretaceous-Tertiary transition. This compromise, however, is no longer accepted. Nevertheless, the Garumniense (Garumnian) has achieved widespread acceptance in regional studies and has also been used in some geological studies of the Lleida Province. Some authors have used this stage name to denote a stratigraphic stage (El Garumniense), whereas others used it to denote the sedimentary rocks in continental facies present at the Cretaceous-Tertiary (K-T) boundary (facies garumnienses). Subsequent studies by Bataller (1959), Rosell (1967), and Liebau (1973) have identified those continental sediments deposited without major discontinuities between them during the latest Mesozoic and those sediments deposited during the earliest Cenozoic. Unfortunately, this has added to the vagueness of identifying these stratigraphic units, because some stratigraphic intervals characterized by lithology only had been given a chronostratigraphic age. Thus, a stratigraphic subdivision with an earliest Garumnian (Cretaceous) and a latest Garumnian (Tertiary) separated by a carbonate unit that extends all over the Ager basin (Rosell, 1967; Rosell and Llompart, 1988; Liebau, 1973), was proposed. Accordingly, for many years it was assumed that the K-T boundary was related to a particular carbonate layer, thereby perpetuating the erroneous belief that a lithologic horizon can be equated with a chronostratigraphic horizon.

The aim of this paper is to describe the limits between the Mesozoic and the Tertiary in the Ager basin, northeastern Spain (Fig. 1). For such a study it is necessary to determine the biostratigraphic and chronostratigraphic limits currently accepted and then apply these to the studied area. Moreover, it is well known that in continental sedimentary environments, where no iridium, shocked quartz, or other K-T boundary-defining signatures have been found, there is no consensus as to where, how, and when the transition between the Mesozoic and the Tertiary (K-T boundary) occurred. Earlier studies have shown that students of different fossil groups place the K-T boundary interval at different stratigraphic levels

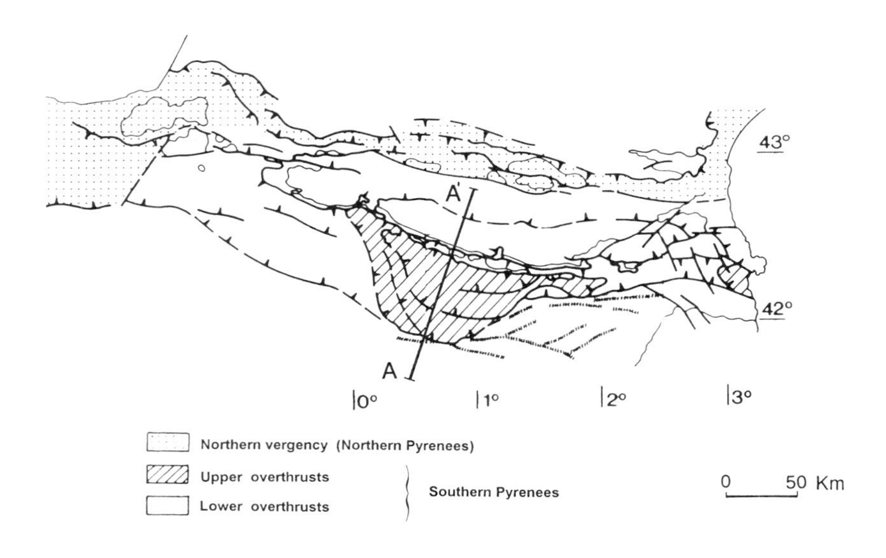

Figure 1. General structural setting of the Eastern Pyrenees (modified after Puigdefábregas and others, 1989). Structural cross section in Figure 2 is A-A', and the star indicates the study area.

(Feist and Colombo, 1983; Ashraf and Erben, 1986), although it has been proposed that the boundary should be defined by the disappearance of dinosaurs.

On the basis of the suggestion by Alvarez and others (1977, 1982), to search for the K-T boundary iridium anomaly as the Cretaceous-Tertiary boundary marker, numerous studies were carried out (Medus and others, 1988, 1992; Medus and Colombo, 1991; Galbrun and others, 1993). Although none of these studies located the iridium anomaly, they resulted in a higher degree of precision in locating the K-T boundary in the Ager basin by combining stratigraphic and paleomagnetic studies.

GEOLOGIC SETTING

The Ager basin in the central South Pyrenees is characterized by an imbricated system of southward-verging overthrust sheets (Seguret, 1972; Garrido-Megías and Rios, 1972; Cámara and Klimowitz, 1985; Martínez-Peña and Pocoví, 1988; Roure and others, 1988; Choukroune and others, 1989). It is located in the Serres Marginals area, bordered to the north by the Montsec overthrust front. The Ager basin can be considered a typical example of a piggyback basin (Ori and Friend, 1984) that formed in the latest part of the Serres Marginals overthrust during the evolution of the south Pyrenean foreland basin, which developed between the latest Cretaceous and the Eocene (Puigdefábregas and others, 1989). Major structures principally exhibit west-east orientation, but in some cases the structural trend has a predominately north-south component (e.g., Sant Mamet anticline) probably associated with the lateral ramps of the main thrust sheets (Fig. 2).

Initial studies covered a large area (Feist and Colombo, 1983; van Eeckhout and others, 1991). Detailed work was performed in an area surrounding Fontllonga village, from the Noguera Pallaresa River on the west, to the Coll d'Orenga and the Serra de Sant Mamet area in the east. The Ilerdian (Ypresian) marine sediments of the Ager syncline are to the north and the latest Cretaceous marine sediments of the Serra de Montroig are to the south.

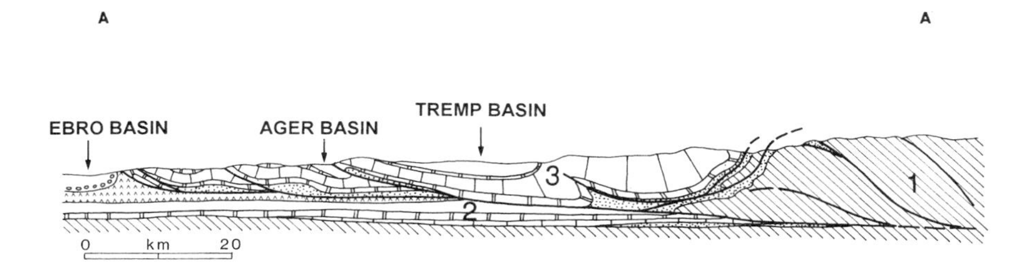

Figure 2. Ager basin is located in the upper part of one of the imbricated overthrust sheets (Montsec overthrust sheet) that characterize the Southern Pyrenees structure (modified after Cámara and Klimowitz, 1985). 1: Paleozoic; 2: Triassic and Jurassic; 3: Cretaceous. Cross section is located in Figure 1.

STRATIGRAPHIC AND SEDIMENTOLOGIC SETTING

The sedimentary continental infilling of the southern sector of the Ager basin is composed of different lithostratigraphic units. The Fontllonga Group consists of the nonmarine sediments between the sandy carbonate-rich marine Cretaceous rocks of the Bona Formation, and the carbonate-rich marine Ilerdian (Ypresian) strata. The following four main units (Figs. 3, 4, and 5) are distinguished in this group.

The La Massana Formation

The earliest limestones are known as La Massana Formation (Puigdefábregas and others, 1989). These limestones reach a thickness of ~80 m and interfinger with

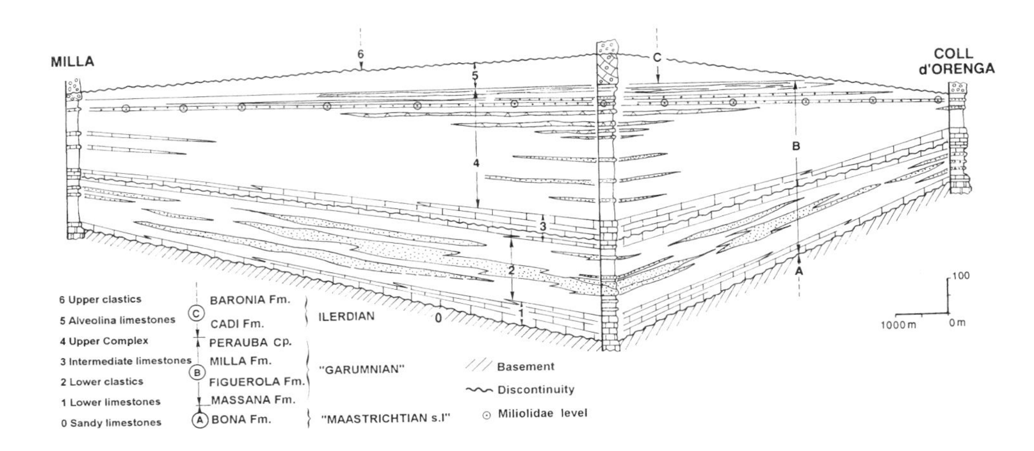

Figure 3. General stratigraphic relations in the Ager basin and distribution of main lithostratigraphic units in this study (modified after Colombo and others, 1986). See text for discussion.

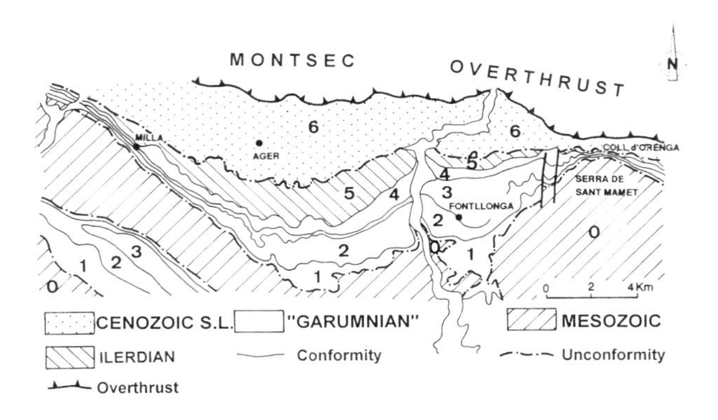

Figure 4. General cartographic scheme of the Ager basin. The numerical designation is the same as in Figure 3.

lutitic layers that are more abundant to the top, which change laterally in thickness. They are well represented in the La Massana area located near the Camarasa Dam. The base of the limestone unit is sharp and conformably grades into the terrigenous deposits of the overlying Figuerola Formation, which contains freshwater fossils (e.g., ostracodes, charophytes, gastropods). The lower part of the La Massana Formation is characterized by laminated dark gray limestones representing variations in the organic content. These limestones may be internally brec-

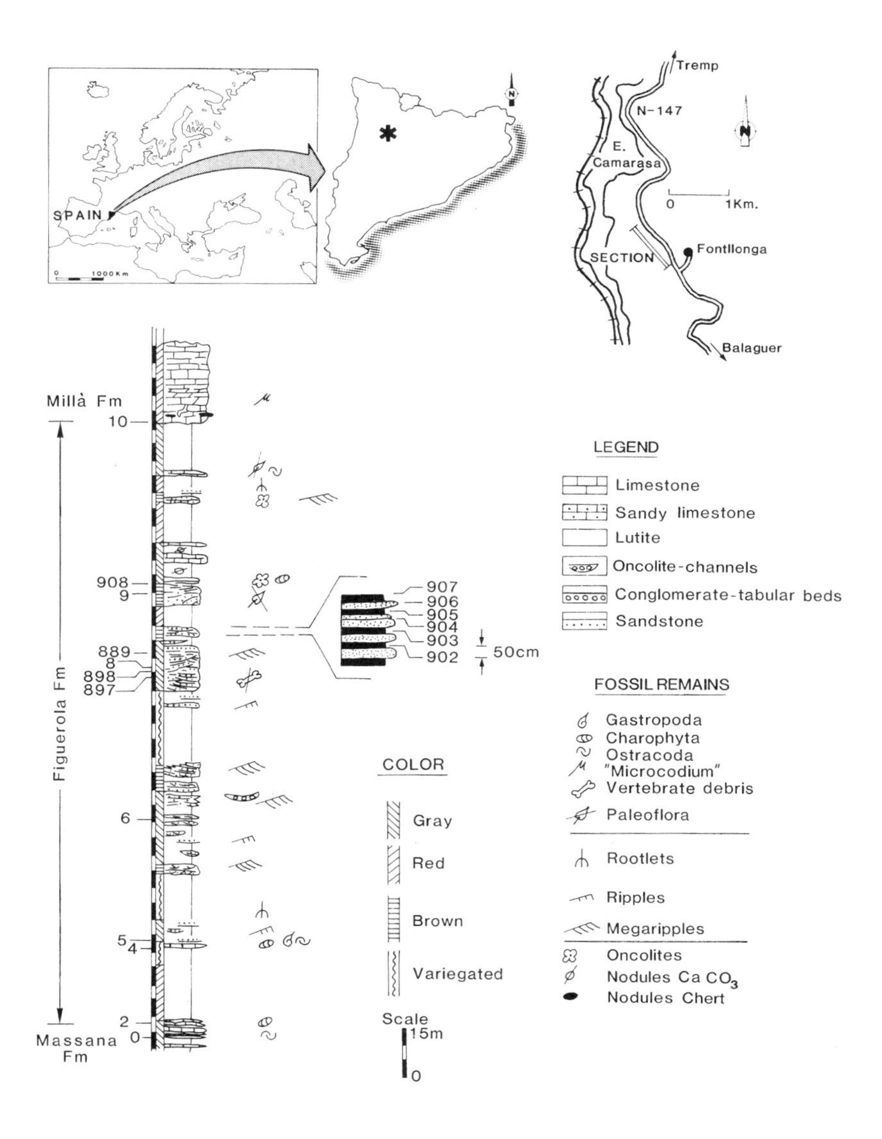

Figure 5. Location and log of measured section in the Fontllonga area showing the position of paleontologic samples.

ciated. Different levels of light colored limestones exhibiting local bioclastic laminations overlie the dark gray limestones. Gray lutite facies overlie these light colored limestones. These lithologies suggest that the sediments were deposited in a shallow and relatively isolated lacustrine environment (Fig. 6A) with low-oxygen conditions that favored the preservation of organic material. Intermittently, probably due to fluctuations in the water level or to lacustrine retraction periods, abundant vegetation produced pedogenic brecciation. This appears to have been a shallow lacustrine-palustrine environment because there is no evidence of mixing waters. The vertical distribution of sedimentary facies suggests various episodes of shallowing with some episodes of lacustrine regression.

The Figuerola Formation

The Figuerola Formation corresponds to the earliest terrigenous unit and is ~190 m thick near Figuerola de Meià village (Puigdefábregas and others, 1989). Red lutitic sediments are predominant in the earliest part of the formation, where different paleosols are locally well developed. The red lutite contains sandy lenses that increase in abundance upsection. The latest terrigenous levels reach a maximum thickness of 80 m and are laterally continuous. They contain abundant fos-

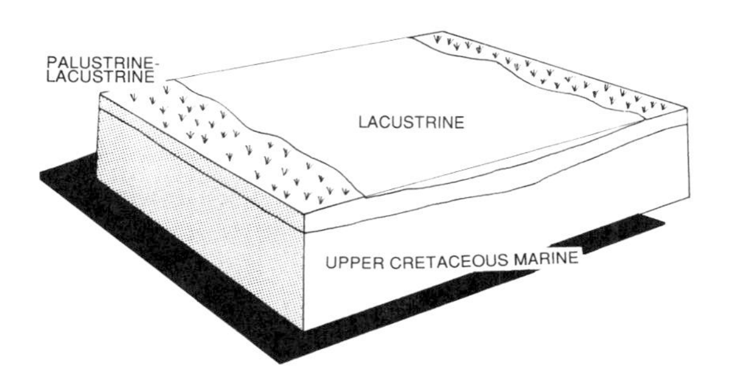

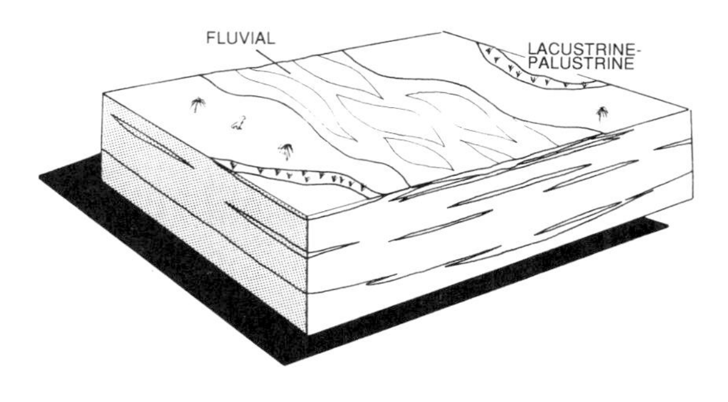

Figure 6. Depositional model of the main sedimentary environments related with the distinguished lithostratigraphic units. A: La Massana Formation; B: Figuerola Formation; C: Millà Formation; and D: Perauba Complex.

sil remains and are also known as "Reptilian Sandstones" (Masriera and Ullastre, 1983, 1990). The coarse terrigenous sediments consist of arkosic sandstones with rare mud clasts associated with some basal scars. There are also sporadic thin carbonate intercalations. In vertical distribution, the sandy facies show predominantly tabular beds with planar cross-bedding and abundant reactivation surfaces near the base of the unit. These facies are overlain by thicker beds of sandstones with metric trough cross-bedding and are topped by thin, fine-grained sandstone levels with subhorizontal laminations. Locally, there are large curved discontinuities that define bodies of different grain size and lithology. In some places the mud clasts are very abundant, whereas in others there are some dispersed quartzitic granules along with fragments of dinosaur bones that appear to have been transported as clasts. All sandy levels contain laterally extensive, carbonate-rich layers that contain freshwater fossils. These carbonate-rich layers reach a thickness of 3 m and are interfingered with lutitic levels which increase in abundance toward the latest part of the formation.

These lithologies suggest the development of a wide flood plain (Fig. 6B) that underwent periods of extreme aridity, as indicated by the presence of different types of paleosols along with poorly developed, sporadic gypsum (alabastrine)

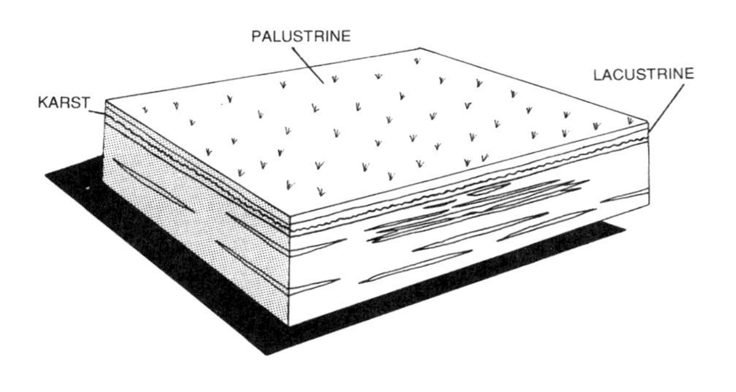

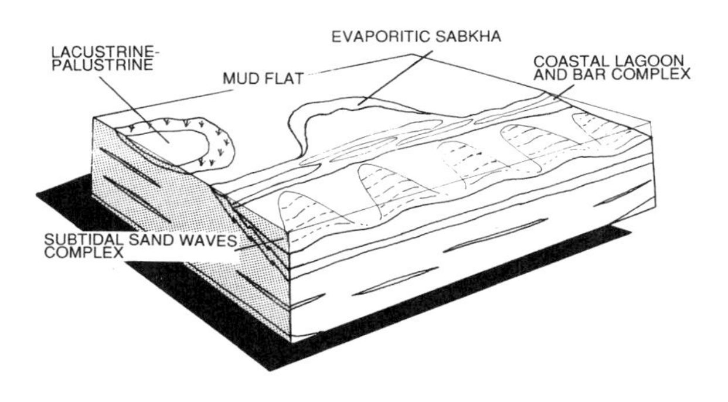

D **PERAUBA Complex.**

nodules. The major terrigenous deposits seem to have been deposited in a complex fluvial, braided-type environment with paleocurrents toward the north and northwest (Cuevas and Mercadé, 1986). In the main channels, the sandy sediments were transported as diverse types of bedforms (dunes), and in the intermediate zones sandy sediments were deposited as small bedforms. Gray lutites with faunal remains and some oncoids are locally very abundant, suggesting episodes during which the channels were inactive. This type of intermittent active-inactive channel system ends with the formation of extensive, relatively shallow lakes and deposition of carbonate-rich and lutitic sediments.

The Millà Formation

Intermediate limestones ~70 m thick form the Millà Formation and are well represented in the area surrounding Millà village. Basal levels within the formation consist of massive dolmicrites containing nodules of chert having maximum diameters of 80 cm, beds of bioclastic limestones with abundant ostracodes and charophyte remains, and beds of laminated carbonates. Massive brecciated limestones containing colonies of *Microcodium* are very abundant: these were deposited in a lacustrine environment with a narrow palustrine belt, which suggests brecciation due to well-developed pedogenesis.

Vertical facies distribution suggests that deposition occurred in a predominantly lacustrine environment in the earliest part of the Millà Formation, and in a palustrine environment in the latest part of the formation (Fig. 6C). These two sedimentary environments are separated by red clays. At the contact between the lower carbonates and the red clay are irregular depressions filled with brecciated irregular thin-laminated carbonates. These are interpreted as representing a karstification period. The presence of laminated limestones (caliche) suggests the existence of a relatively prolonged immersion period characterized by nondeposition.

The Perauba Complex

The Perauba Complex consists of a thick lithostratigraphic unit containing different lithotypes that have complex geometric relations. The unit is 550 m thick and well exposed in the Collada de Perauba area. The main lithology consists of a reddish variegated lutitic unit with intercalated sandstone beds. This unit also contains sandy bodies with lenticular geometries and trough cross-bedding. These are generally amalgamated and contain some oncolitic concretions, as well as rare remains of thick-shelled bivalves. Some carbonate layers are intercalated and have abundant colonies of *Microcodium*. Some centimetric and vertically lengthened gypsum nodules appear within the lutites. These sedimentary characteristics (Fig. 6D) suggest that deposition occurred in a poorly drained area where carbonate sediments were deposited in small ponds that show mainly palustrine facies. This unit corresponds to sedimentation during an arid climatic event in a muddy continental plain (mud flat) that may have been connected with the more distal parts of a fluvial system.

Toward the upper part of the Perauba Complex evaporites are locally very well developed. The sediments are characterized by lutite and evaporite beds with few interlayers of grayish carbonates. The gypsum is massive, nodular, laminated, and rarely enterolithic, suggesting deposition below a sheet of water (laminated facies) and vadose (above of the phreatic nappe) deposition (nodules) in an evaporitic environment. The upper part of the Perauba Complex represents a transition to carbonate-rich sediments containing remains of marine fauna (Alveolina Limestone). The general sedimentary context represents (Fig. 6D) a facies belt of mud flats with some carbonate-dominated lakes related with coastal evaporitic lagoons (sabkhas), where the repetitive oscillations of the water level suggest several lacustrine contractions and expansions. These series are overlain by the Ilerdian (Ypresian) carbonate platform of the Alveolina Limestone (Cadí Formation), characterized by bioclastic sediments and oyster banks deposited under tidal influence.

The Fontllonga Group reaches a total thickness of ~900 m measured along the Noguera Pallaresa stratigraphic section, which lies in the central part of a small sedimentary basin. Evaporitic sediments are restricted to this central basin, and the overall thickness of the deposits (Fig. 3) decreases toward the east (Coll d'Orenga area) and west (Millà area).

MAASTRICHTIAN TO PALEOCENE DEPOSITIONAL SETTING

Sediment deposition in the Ager basin suggests a regressive depositional sequence between the upper part of the marine Maastrichtian (Bona Formation) and the upper part of the intermediate carbonate sediments (Millà Formation). Subsequently, a transgressive sequence (Perauba Complex) was deposited in an environment span from dry mud flats to coastal sabkhas (Fig. 3). This was followed by renewed transgression and deposition of the upper marine sediments (Cadí Formation) of the Ilerdian (Ypresian). The maximum regressive phase corresponds to the sandy bodies of the Figuerola Formation which attain a thickness of 40 m. The medial portion of the Figuerola Formation is informally called the Reptilian Sandstones unit. The maximum transgressive interval is in the upper limestones with Alveolina fossils (Cadí Formation).

The geometry of the Ager basin is characterized by the subsidence of the depocenter located (Fig. 3) at the Fontllonga stratigraphic section. Active tectonic movements appear to have occurred from the Maastrichtian until the Paleocene.

Biostratigraphy indicates that the transition between the Mesozoic and the Tertiary is located in the lower clastic unit (Figuerola Formation). However, it is not possible to locate the K-T boundary accurately because the facies are similar and the chronostratigraphic resolution of the different fossil groups is coarse. However, paleomagnetic data help constrain the location of the K-T boundary and aid in correlation of different fossil biostratigraphies.

THE K-T BOUNDARY

The K-T boundary has been variously placed at the upper limestone beds of the Millà Formation (Rosell, 1967; Rosell and Llompart, 1988) and in the upper part of the "Reptilian Sandstones" (Fig. 3) of the Figuerola Formation, placed clearly below the Millà Formation (Feist and Colombo, 1983). In order to clarify placement of the K-T boundary, a team of sedimentologists and paleontologists has restudied these sections based on the same set of samples (Fig. 5). The results, published by Medus and others (1988, 1992), Medus and Colombo (1991), and Galbrun and others (1993), have not produced agreement on a single stratigraphic horizon for the K-T boundary. The placement of the K-T boundary differs among palynofloras, charophytes, and gastropods (Medus and others, 1988, 1992; Medus and Colombo, 1992). Among these fossil groups, the palynoflora may be most suitable in correlating the Ager basin to K-T boundary sequences in North America, where similar floras have been described from the Western Interior (Nichols and others, 1986; Tschudy and Tschudy, 1986). The two floras differ largely in that some forms considered typical of the Tertiary are present in the Maastrichtian, whereas characteristic forms from the Maastrichtian are present in Tertiary sediments of the Ager basin. This may be largely due to the very condensed interval that characterizes the Cretaceous-Tertiary transition (Medus and Colombo, 1991; Medus and others, 1992) as well as to the climatic changes at this time.

Medus and others (1992) observed that a major environmental change occurred in three phases in the Ager basin: (1) the decrease in and eventual extinction of Senonian fossils (Upper Cretaceous), (2) the expansion of the Danian flora, many of which originated in the earliest Maastrichtian levels (Medus, 1977), and (3) the introduction of more modern floras in the Tertiary and their expansion in the earliest Thanetian.

Ostracodes and gastropods are rare and provide imprecise biostratigraphic information. Different assemblages of ostracodes and gastropods that coexist in southeastern France with dinosaur remains have been found in Pyrenean strata that are clearly Cretaceous. Moreover, some gastropods and ostracodes seemingly extend across the K-T boundary into the Paleocene. Charophyte assemblages do not show any major faunal changes between the latest Maastrichtian and the latest Danian (Danian-Montian, Riveline, 1986), an interval that spans approximately 5 m.y., in which five subzones were identified (Fig. 7). Within this transitional interval Maastrichtian species coexist with Danian species and no abrupt floral changes mark the K-T boundary (Galbrun and others, 1993).

In order to locate the K-T boundary in the Fontllonga section, samples were collected systematically for iridium analysis across the K-T transitional interval defined by paleontological analysis: 35 samples were analyzed for iridium concentrations between 50 and 100 picograms/gram, but no anomalous concentrations were detected (Galbrun and others, 1993). Thus, on the basis of paleontological and geochemical analysis, the K-T boundary in the Fontllonga section cannot be identified as one specific horizon similar to the North American K-T boundary sections in the Raton basin and Wyoming (Orth and others, 1981; Bohor and others, 1987).

Paleomagnetic analysis of the K-T transition in the Fontllonga section helped to constrain the location of the K-T boundary. A total of 72 samples were analyzed resulting in a magnetostratigraphic zonation (Fig. 7) comparable with the paleomagnetic polarity time scale of Haq and others (1988; see discussion in Galbrun and others, 1993). By integrating the paleomagnetic stratigraphy with the paleontological data, the Maastrichtian to earliest Paleocene history of the Ager basin can be deciphered. The last dinosaur footprints and dinosaur eggshells were discovered in the upper part of the middle *Microcara cristata* zone, which corresponds to chron 31N at an estimated age of 69 Ma (Fig. 7).

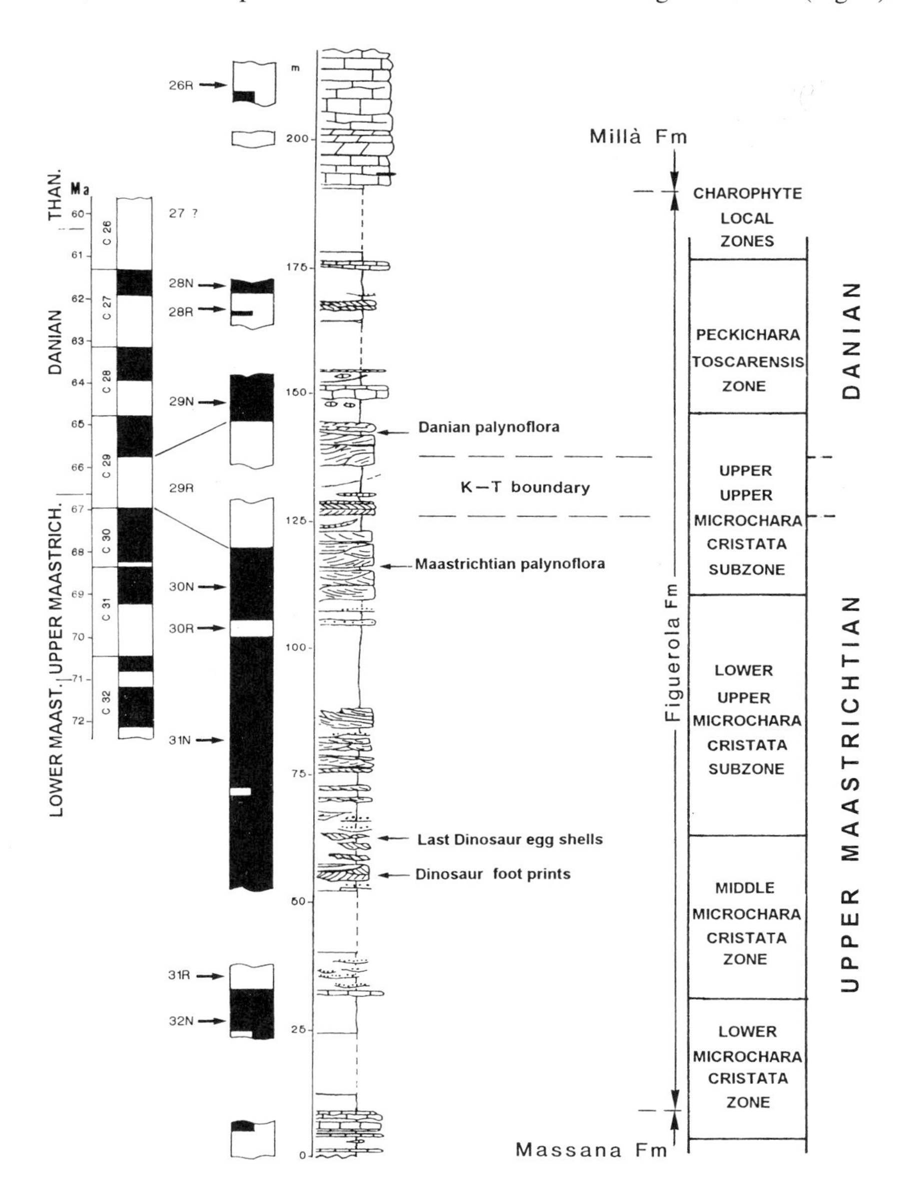

Figure 7. Stratigraphic log from Fontllonga showing the different paleomagnetic units. The magnetic polarity standard is from Haq and others (1988). The charophyte local biozones and the uncertain interval of the Maastrichtian-Danian boundary are modified from Galbrun and others (1993).

Maastrichtian palynofloras are present up through chron 30N and into the lower part of chron 29R. Near the top of chron 29R and in 29N a Danian palynoflora is present. The K-T boundary occurs within chron 29R, an interval which in the Ager basin is marked by transitional faunas and floras, rather than a sudden mass extinction or climatic change, as observed in marine sequences (Keller, 1988, 1993; Smit, 1990; Canudo and others, 1991). Thus, paleontological and paleomagnetic data bracket the K-T boundary in the Ager basin as a transitional zone that provides no evidence of a severe sudden catastrophic event. Instead, these data indicate that a gradual transition occurred in this terrestrial environment between the Maastrichtian and the Tertiary.

On the basis of chronostratigraphy, the disappearance of Cretaceous charophytes and dinosaurs (last appearance of eggshells and footprints) seems to have occurred ~2 m.y. prior to the K-T boundary in this region. Charophyte assemblages also changed throughout this interval. These changes are also observed in southeastern France (Westphal and Durand, 1990; Galbrun and others, 1991). Therefore these biotic changes are not related to K-T boundary crisis. Rather, the faunal turnover and extinctions appears to be related to the climatic deterioration and global cooling that characterizes the Campanian and Maastrichtian (Barrera and Huber, 1991; Barrera and Keller, 1995).

ACKNOWLEDGMENTS

I wish to thank Robert Scott, Gerta Keller, and one anonymous reviewer for their helpful comments on the manuscript. This study was supported by the Spanish Government Project DIGICYT PB91-0805 and DIGICYT PB84-0871, and EEC Project JOU CT92-0110, and a Catalan Government grant ("Pla de recera de Catalunya, ajuts CIRIT"), no. 10481/1994.

REFERENCES CITED

Alvarez, W., Arthur, M. A., Fisher, A. G., Lowrie, W., Napoleone, G., Premoli-Silva, I., and Roggenthen, W. M., 1977, Upper Cretaceous–Paleocene magnetic stratigraphy at Gubbio, Italy. V. Type section for the Late Cretaceous–Paleocene geomagnetic reversal time scale: Geological Society of America Bulletin, v. 88, p. 367–389.

Alvarez, W., Alvarez, L. W., Asaro, F., and Michel, H. V., 1982, Current status of the impact theory for the terminal Cretaceous extinctions, *in* Silver, L. T. and Schultz, P. H., eds., Geological implications of impacts of large asteroids and comets on the Earth: Geological Society of America Special Paper 190, p. 305–315.

Ashraf, A. R., and Erben, H. K., 1986, Palynologische untersuchungen and der Kreide/Tertiär-Grenze West-Mediterraner Regionen: Palaeontographica, v. 200, p. 111–163.

Barrera, E., and Huber, B. T., 1991, Paleogene and early Neogene oceanography of the Southern Indian Ocean, *in* Proceedings of the Ocean Drilling Program, Scientific results, Volume 119: College Station, Texas, Ocean Drilling Program, p. 693–717.

Barrera, E., and Keller, G., 1995, Foraminiferal stable isotope evidence for gradual decrease of marine productivity and Cretaceous species survivorship in the lower Danian: Paleoceanography (in press).

Bataller, J. R., 1958, El Garumniense español y su fauna: Notas y Comunicaciones del Instituto Géolologico y Minero de España, v. 50, p. 1–41.

Bataller, J. R., 1959, Paleontologia del Garumnense: Estudios Geológicos, v. 15, p. 39–53.

Bohor, B. F., Triplehorn, D. M., Nichols, D. J., and Millard, H. T., Jr., 1987, Dinosaurs, spherules, and the "magic" layer; a new K-T boundary clay site in Wyoming: Geology, v. 15, p. 896–899.

Cámara, P., and Klimowitz, J., 1985, Interpretación geodinámica de la vertiente centro-occidental surpirenaica (Cuenca de Jaca-Tremp): Estudios Geológicos, v. 41, v. 391–404.

Canudo, J. I., Keller, G., and Molina, E., 1991, Cretaceous-Tertiary boundary extinction pattern and faunal turnover at Agost and Caravaca, S.E. Spain: Marine Micropaleontology, v. 17, p. 319–341.

Choukroune, F., and Ecors team, 1989, The ECORS Pyrenean deep seismic profile reflection data and the general structure of an orogenic belt: Tectonics, v. 8, p. 23–29.

Colombo, F., Cuevas, J. L., and Mercadé, L., 1986, La facies garumnense del flanco sur del sinclinal de Ager: análisis sedimentológico: XI Congreso Español de Sedimentologia, Resúmenes de Comunicaciones, p. 50.

Cuevas, J. L., and Mercadé, L., 1986, Estudio sedimentológico y estratigráfico de la facies garumnense del flanco sur del sinclinal de Ager: Universidad de Barcelona, Departamento Geológia Dinámica, Geofísica y Paleontologia, 46 p.

Feist, M., and Colombo, F., 1983, La limite Crétacé-Tertiaire dans le nord-est de l'Espagne, du point de vue des charophytes: Géologie Méditerranéenne, v. 20, p. 303–326.

Galbrun, B., Rasplus, L., and Durand, J. P., 1991, La limite Crétacé-Tertiaire dans le domaine provençal: étude magnétostratigraphique du passage Rognacien-Vitrollien à l'Ouest du synclinal de l'Arc: Paris, Académie des Sciences Comptes Rendus, v. 312, p. 1467–1473.

Galbrun, B., Feist, M., Colombo, F., Rocchia, R., and Tambareau, Y., 1993, Magnetostratigraphy and biostratigraphy of Cretaceous-Tertiary continental deposits, Ager Basin, Province of Lerida, Spain: Palaeogeography, Palaeoclimatology, Palaeoecology, v. 102, p. 41–52.

Garrido-Megias, A., and Rios, L. M., 1972, Síntesis geológica del Secundario y Terciario entre los ríos Cinca y Segre: Bolotín Geológico y Minero, v. 83, p. 147.

Haq, B. U., Hardenbol, J., and Vail, P. R., 1988, Mesozoic and Cenozoic chronostratigraphy and cycles of sea-level change, *in* Wilgus, C. K., Hastings, B. S., Kendall, C. G., Posamentier, H. W., Ros, C., and Van Wagoner, J. C., eds., Sea-level changes: An integrated approach: Society of Economic Paleontologists and Mineralogists Special Publication 42, p. 71–108.

Keller, G., 1988, Extinction, survivorship and evolution of planktic foraminifera across the Cretaceous-Tertiary boundary at El Kef, Tunisia: Marine Micropaleontology, v. 13, p. 239–263.

Keller, G., 1993, K/T boundary mass extinctions restricted to low latitudes?: Geological Society of America Abstracts with Programs, v. 25, no. 7, p. A296.

Lapparent, A, F. De, and Aguirre, E., 1956, Algunos yacimientos de Dinosaurios en el Cretácico de la Cuenca de Tremp: Estudios Geológicos, v. 12, p. 377–382.

Leymerie, A., 1863, Note sur la système Garumnien: Bulletin de la Société Géologique de France, v. 20, p. 483–488.

Leymerie, A., 1877, Mémoire sur le type Garumnien comprenant une description de la Montaigne d'Ausseing, un aperçu des principaux gîtes du Département de la Haute-Garonne et une notice sur la faune d'Auzas: Annales des Sciences Géologiques, v. 9, p. 1–63.

Liebau, A., 1973, El Maastrichtiense lagunar ("Garumniense") de Isona: XIII Coloquio Europeo de Micropaleontología, p. 87–100.

Martínez-Peña, M. B., and Pocoví, A., 1988, El amortiguamiento frontal de la estructura de la cobertera surpirenaica y su relación con el anticlinal de Barbastro-Balaguer: Acta Geológica Hispanica, v. 23, p. 81–94.
Masriera, A., and Ullastre, J., 1983, Essai de synthèse stratigraphique des couches continentales de la fin du Crétacé des Pyrénées Catalanes (NE de l'Espagne): Géologie Méditerranéenne, v. 20, p. 283–290.
Masriera, A., and Ullastre, J., 1990, Yacimientos inéditos de carófitas que contribuyen a fijar el límite Cretácico-Terciario en el Pirineo Catalán: Rev. Soc. Geol. España, v. 3, p. 33–42.
Medus, J., 1977, Quelques palynoflores du Tertiaire inférieur du Sud de la France et du Nord de l'Espagne: Revista Española Micropaleontología, v. 7, p. 113–126.
Medus, J., and Colombo, F., 1991, Succession climatique et limite stratigraphique Crétacé-Tertiaire dans le N.E. de l'Espagne: Acta Geológica Hispanica, v. 26, p. 173–180.
Medus, J., Feist, M., Rocchia, R., Boclet, D., Colombo, F., Tambareau, Y., and Villatte, J., 1988, Prospects for recognition of the palynological Cretaceous/Tertiary boundary and an iridium anomaly in nonmarine facies of the eastern Spanish Pyrenees: A preliminary report: Newsletters in Stratigraphy, v. 18, p. 123–138.
Medus, J., Colombo, F., and Durand, J. P., 1992, Pollen and spore assemblages of the uppermost Cretaceous continental formations of south-eastern France and north-eastern Spain: Cretaceous Research, v. 13, p. 119–132.
Nichols, D. J., Jarzen, D. M., Orth, D. J., and Oliver, P. Q., 1986, Palynological and iridium anomalies at Cretaceous-Tertiary boundary. South-central Saskatchewan: Science, v. 231, p. 714–717.
Ori, G. G., and Friend, P. F., 1984, Sedimentary basins formed and carried piggy-back on active thrust sheets: Geology, v. 12, p. 475–478.
Orth, C. J., Gilmore, J. S., Knight, J. D., Pillmore, C. L., Tschudy, R. H., and Fasset, J. E., 1981, An iridium abundance anomaly at the palynological Cretaceous-Tertiary boundary in northern New Mexico: Science, v. 214, p. 1341–1343.
Puigdefábregas, C., Collinson, J., Cuevas, J. L., Dreyer, T., Marzo, M., Mercade, L., Nijman, W., Verges, J., Mellere, D., and Muñoz, J. A., 1989, Alluvial deposits of the successive foreland basin stages and their relation to the pyrenean thrust sequences, *in* Marzo, M., and Puigdefábregas, C., eds., 4th International Conference on Fluvial Sedimentology, (Guidebook Series, Excursion no. 10): Barcelona, Educacion Servei Geològica de Catalunya, 175 p.
Riveline, J., 1986, Les Charophytes du Paléogène et du Miocène infèrieur d'Europe Occidentale: Cahiers Paléontologie, p. 1–227.
Rosell, J., 1967, Estudio geológico del sector del Prepirineo comprendido entre los ríos Segre y Noguera Ribagorzana (Prov. de Lérida): Pirineos, v. 21, p. 9–225.
Rosell, J., and Llompart, C., 1988, Guia Geològica del Montsec i de la Vall d'Ager: Barcelona, Montblanc-Martin-CEC, 168 p.
Roure, F., Choukroune, P., Berastegui, X., Muñoz, J. A., Villien, A., Matheron, P., Bayret, M., Seguret, M., Camara, P., and Deramond, J., 1988, ECORS deep seismic data and balanced cross-sections: Geometric constraints to trace the evolution of the Pyrenees: Tectonics, v. 8, p. 41–50.
Seguret, M., 1972, Etude tectonique des nappes et séries decollées de la partie centrale du versant Sud des Pyrénées. Publ. USTELA, Montpellier: Serie Géologie Structurale, v. 2, p. 1–155.
Smit, J., 1990, Meteorite impact, extinctions and the Cretaceous-Tertiary boundary: Geologie en Mijnbouw, v. 69, p. 187–204.
Tschudy, R. H., and Tschudy, B. D., 1986, Extinction and survival of plant life following

the Cretaceous/Tertiary boundary event. Western interior, North America: Geology, v. 14, p. 667–670.

Van Eeckhout, J. A., Gimenez, J., Martinez, A., Mato, E., Ramos, E., Saula, E., Busquets, P., Colombo, F., and Permanyer, A., 1991, Variaciones geométricas de la cuenca de antepaís surpirenaica relacionadas con los episodios de progradación-retrogradación de los sistemas deposicionales aluviales, transicionales y marinos en la zona Ripollès-Berguedà, *in* Colombo, F., ed., I Congreso del Grupo Español del Terciario. Guia de Campo, Excursión, n 3: Victoria, El Grupo Español del Terciario, p. 1–88.

Vidal, L. M., 1873, Datos para el conocimiento del terreno Garumniense de Cataluña: Boletín Comision del Mapa Geológica de España, v. 1, p. 209–247.

Westphal, M., and Durand, J. P., 1990, Magnétostratigraphie des séries continentales fluvio-lacustres du Crétacé supérieur dans le synclinal de l'Arc (région d'Aix-en-Provence, France): Bulletin de la Société Géologique de France, v. 6, p. 609–620.

17

Sea-Level Changes, Clastic Deposits, and Megatsunamis across the Cretaceous-Tertiary Boundary

Gerta Keller, *Department of Geological and Geophysical Sciences, Princeton University, Princeton, New Jersey,* and Wolfgang Stinnesbeck, *Facultad de Ciencias de la Tierra, Universidad Autónoma de Nuevo León, Linares, Mexico*

INTRODUCTION

Along basin margins and continental shelves, marine sedimentation is frequently packaged as unconformity-bounded clastic depositional strata. These sequences are generally related to eustatic sea-level changes and in particular to the sea-level fall and rise inflection points when unconformities and condensed sedimentation, respectively, form (Posamentier and others, 1988; Posamentier and Vail, 1988). Sequence stratigraphic models provide a conceptual framework for understanding eustatic control on sedimentation and for applying these models to field observations (Posamentier and others, 1988; Posamentier and Vail, 1988; Donovan and others, 1988; Baum and Vail, 1988; Van Wagoner and others, 1990). On the basis of these models, sequences frequently encountered in continental shelf to basin margin settings of Cretaceous-Tertiary (K-T) transitions or near the K-T boundary in the gulf states (Texas, Alabama, Georgia), Mexico, and the Gulf of Mexico are traditionally interpreted as sea-level lowstand followed by sea-level rise deposits (Donovan and others, 1988; Baum and Vail, 1988; Mancini and others, 1989; Keller, 1989a; Savrda, 1991, 1993; Habib and others, 1992; Moshkovitz and Habib, 1993; Stinnesbeck and others, 1993; Keller and others, 1993a, 1994a).

With the suggestion of the circular gravity anomaly structure at Chicxulub in northern Yucatan as a probable K-T boundary bolide impact crater (Pope and others, 1991; Hildebrand and others, 1991; Swisher and others, 1992; Sharpton and others, 1992, 1993), these same clastic deposits have been reinterpreted by some workers as impact-generated megatsunami deposits (Hildebrand and Boynton, 1990; Hildebrand and others, 1994; Smit and others, 1992, 1994a, 1994b; Alvarez and others, 1992; Bohor and Betterton, 1993). This reinterpretation is based mainly on the theory that the Chicxulub structure is the K-T boundary impact crater, and the extrapolation that such an impact created megatsunami waves which left their marks in the sedimentary strata, as hypothesized earlier by Bourgeois and others (1988) for clastic deposits along the Brazos River, Texas. Proponents of impact-generated tsunami deposition generally agree that it is not known what such deposits look like, that is, whether bedding would be chaotic or graded, locally channelized, or sheet-like over large geographic areas; or whether they would look substantially different from sea-level lowstand deposits. They agree that such tsunami deposits would be emplaced during a single catastrophic event lasting no more than a few days and ideally that it would contain impact products such as iridium, Ni-rich spinels, shocked quartz grains, and glassy microspherules or microtektites.

To date, no K-T clastic deposit has been identified unequivocally as of impact origin, although Hildebrand and Boynton (1990), Alvarez and others (1992), Hildebrand and others (1994), and Smit and others (1992, 1994a, 1994b) have claimed such an origin for virtually all known K-T or near-K-T clastic deposits. For example, coarse-grained sands in sections from Braggs, Mussel Creek, Millers Ferry, and Moscow Landing in Alabama (known as Clayton sands), long interpreted as sea-level lowstand deposits (Donovan and others, 1988; Baum and Vail, 1988; Mancini and others, 1989; Savrda, 1991, 1993; Habib and others, 1992; Moshkovitz and Habib, 1993), are now postulated to be impact-generated tsunami deposits by Smit and others (1994a, 1994b) and Hildebrand and others (1994), largely on the basis of their stratigraphic proximity to the K-T boundary and geographic proximity to Chicxulub. Similarly, coarse-grained deposits of near-K-T boundary age at the Brazos River section, otherwise devoid of impact indicators, were interpreted as tsunami deposits by Bourgeois and others (1988).

In some sections from northeastern and east-central Mexico (Mimbral, Peñon, Mulato, Lajilla, and Tlaxcalantongo), near K-T boundary clastic deposits contain thin layers or lenses of unusual spherule-rich sediments with very rare glass fragments on a basal unconformity. These spherule-rich sediments are overlain by a thick laminated sandstone, which is overlain by alternating sand and siltstone beds (Stinnesbeck and others, 1993, this volume; Keller and others, 1994a). These sediments have also been postulated as impact-generated megatsunami deposits by Smit and others (1992, 1994a, 1994b). This interpretation was challenged by Bohor and Betterton (1993), who prefer turbiditic deposition, and by Stinnesbeck and others (1993, 1994b, this volume) and Keller and others (1994a), who argue for deposition during a sea-level lowstand followed by a transgression preceding the K-T boundary. Deep-water clastic deposits in Deep Sea Drilling Project (DSDP) Site 540 in the Gulf of Mexico, originally interpreted as gravity-flow deposits or submarine slumps (Worzel and others, 1973), were postulated by Alvarez and others (1992) as Chicxulub impact-generated tsunami deposits. However, restudy of these deposits reveals no independently age controlled K-T boundary sediments at Site 540, or elsewhere in the Gulf of Mexico and Caribbean (Keller and others, 1993a).

Are there any K-T boundary impact-generated clastic deposits? Such deposits, of course, are technically possible. We believe, however, that the K-T and near-K-T boundary clastic deposits must be interpreted within the context of their stratigraphic position, geographic location (proximity to sediment source), and paleodepth of emplacement, and within the framework of eustatic sea-level changes and their effects on deposition of sedimentary strata. In this study we address these issues not only for selected K-T boundary sequences that contain siliciclastic strata, but for K-T boundary sequences worldwide (Fig. 1), and emphasize the integration of hitherto neglected global aspects. We begin by reviewing evidence for eustatic sea-level changes across the K-T transition based on K-T sequences worldwide. Clastic deposits in all known marine K-T sequences are then reviewed for biostratigraphy, paleodepth of deposition, and sedimentary characteristics, and correlated to the global eustatic sea-level curve.

BIOSTRATIGRAPHY

Over the past decade the K-T boundary debate has indirectly resulted in the accumulation of the most extensive high-resolution paleontological database of any geologic interval, including more than 45 biostratigraphic zone complete (all biozones present) or nearly complete K-T boundary sections in addition to numerous K-T successions that contain major hiatuses (MacLeod and Keller, 1991a, 1991b; Keller and others, 1993a). Our results are based upon the boundary sec-

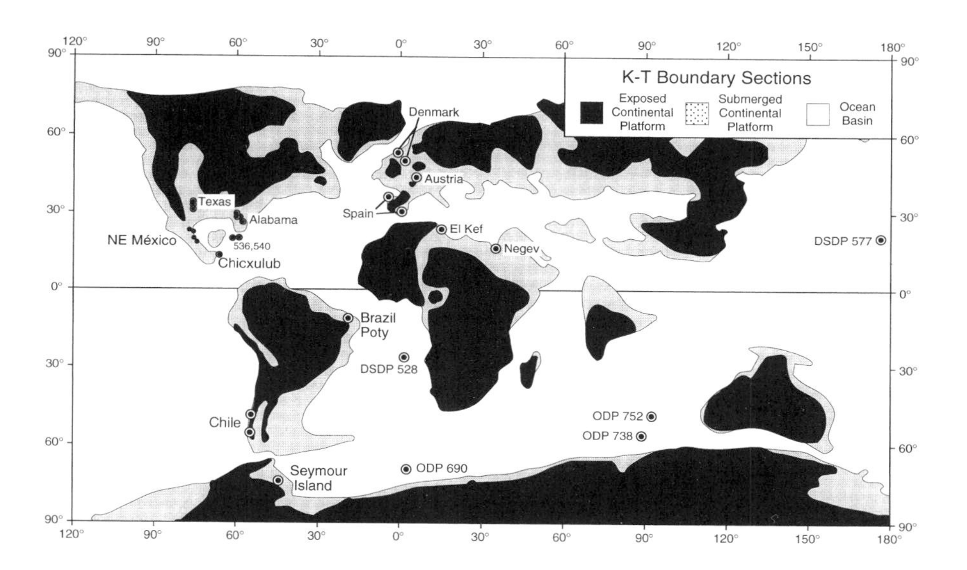

Figure 1. Locations of complete and near-complete K-T boundary sections plotted on a paleogeographic reconstruction of continental positions at the time of the K-T boundary. White = ocean basins; light stippling = continental platforms; black = inferred extend of terrestrial exposure (from MacLeod and Keller, 1991a). DSDP = Deep Sea Drilling Project; ODP = Ocean Drilling Program.

tions shown in Figure 1, where many location points represent multiple sections (e.g., 10 sections for northeastern Mexico, 3 sections each for Brazos and Alabama, 5 sections for the Negev). Each of these sections contains an abundant, well-preserved and well-documented planktonic, foraminiferal, nannoplankton, or dinoflagellate record. In addition to geochemical markers (e.g., Ir anomaly, Ni-rich spinels, $\delta^{13}C$ shift), these biostratigraphies provide the necessary high-resolution time control for accurate placement of the K-T boundary, along with dating of sea-level changes and coarse-grained clastic units.

Many recent high-resolution biostratigraphic studies have lead to significant refinements of the K-T boundary biozonations, especially for the Danian interval. Consequently, the original planktonic foraminiferal zonation of Keller (1988a) was modified and updated in Canudo and others (1991) and Keller (1993) to reflect the improved database and increased confidence in additional and alternative zonal markers. Figure 2 shows the updated planktonic foraminiferal biozonation with first and last appearance datums and comparison with biozonations of Smit (1982) and Berggren and Miller (1988). Based on the Agost and El Kef sections, Pardo and others (this volume) defined the

PLANKTONIC FORAMINIFERAL ZONATIONS

	Datum events	Keller, 1993 Pardo et al, 1996		Keller & Benjamini, 1991	Canudo et al., 1991	Keller, 1989a	Keller, 1988a	Berggren & Miller, 1988
Early Paleocene (Danian)		P1d		P1d	P1d	no data	P1d	P1c
	⊥ M. trinidadensis	P1c	Plc(2)	P1c	P1c		P1c	P1a & P1b
	⊥ M. inconstans		P1c(1)					
	⊤ G. conusa; ⊥ S. varianta						G.taurica	
		P1b		P1b	P1b	P1b	P1b(2)	
	⊤ P. eugubina, P. longiapertura	P1a	P1a(2)	P1a	P1a	P1a	P1b(1)	
	⊥ P. compressus; ⊥ E. trivialis; ⊥ G. pentagona; ⊥ S. pseudobulloides; ⊥ S. triloculinoides; ⊥ G. daubjergensis; ⊥ S. moskvini; ⊥ P. planocompressus; ⊥ G. taurica; ⊥ C. midwayensis		P1a(1)				Eoglobigerina	Pα
							P1a	
	⊥ P. eugubina, P. longiapertura; ⊥ E. eobulloides; ⊥ E. edita, W. hornerst.; ⊥ E. fringa, E. simplicis.; ✝ G. conusa, P. hantkeninoides	P0		P0	P0b / P0a	P0	P0b / P0a	unzoned
L. Maast.		P. hantkeninoides		K/T boundary — A. mayaroensis	A. mayaroensis	P. deformis	P. deformis	A. mayaroensis
	⊥ P. hantkeninoides	A. mayaroensis					A. mayaroensis	

*Figure 2. The most commonly used planktonic foraminiferal biozonations for the K-T boundary transition showing successive refinements as a larger global database was developed. Zonation follows Keller (1993); the uppermost Maastrichtian biozone (*Plummerita hantkeninoides*) is added to mark the topmost 170–200 k.y. of the Maastrichtian.*

Plummerita hantkeninoides biozone (see also Masters, 1984, 1993) to characterize the topmost (~170–200 k.y.) of the Maastrichtian and to be within chron 29R below the K-T boundary.

The age and duration of early Danian biozones can be estimated relatively accurately. The K-T boundary is currently placed within the upper half of paleomagnetic anomaly C29R at 65 Ma based on $^{40}Ar/^{39}Ar$ dating of melt rock from K-T boundary sections at Beloc, Haiti, and Chicxulub (Swisher and others, 1992; Sharpton and others, 1993). The portion of C29R below the K-T boundary is estimated as 350 k.y., and the portion above the boundary is estimated as 230 k.y. (Herbert and D'Hondt, 1990; Berggren and others, 1985). The C29R-C29N boundary corresponds closely to the top of zone P1a (extinction of *P. eugubina*), as observed in sections from Brazos River and DSDP Sites 577 and 528 (Keller, 1989a; Chave, 1984; Bleil, 1985). Biozones P0 and P1a thus span a total of 230 k.y. and of this time interval, zone P0 is estimated to span ~40 to 50 k.y. (MacLeod and Keller, 1991b; D'Hondt and Herbert, 1991).

RELATIVE SEA-LEVEL CHANGES

Relative changes in sea level frequently determine the nature and characteristics of sediment deposition. It is therefore important to determine paleodepths of deposition of each K-T boundary sequence and to trace the relative sea-level changes across the boundary horizon. Figure 3 illustrates the average paleodepth

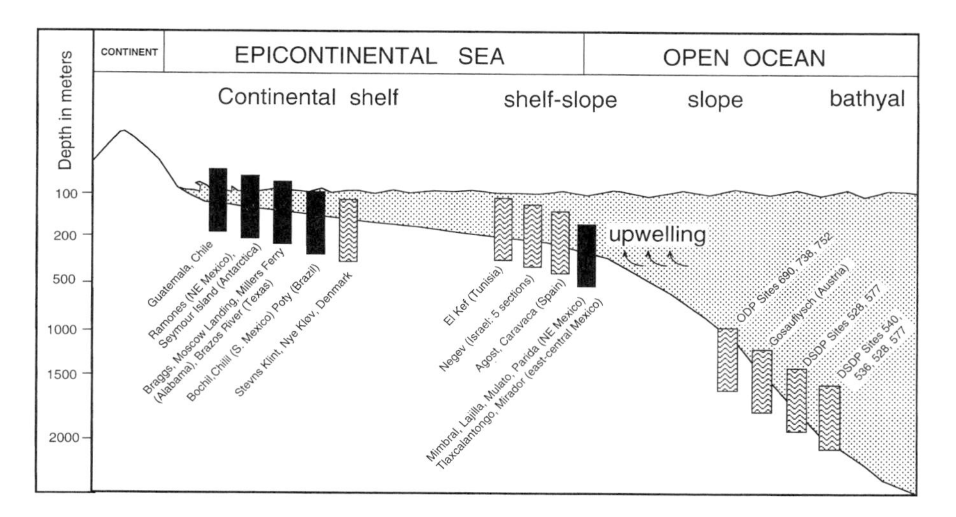

Figure 3. Depositional environments and average paleodepths of primary (complete or near-complete) K-T boundary sections from nearshore (inner neritic), middle shelf, outer shelf to upper slope, and the deep sea. Note that clastic deposits (black) are generally common in nearshore environments and rare in deeper waters unless locations are near continental margins such as the northeastern Mexico sections. Black columns mark sections with near K-T clastic deposits. Patterned columns mark sections without near K-T clastic deposits.

of deposition of all primary (complete or near-complete) K-T boundary sections. These sections span depositional environments from nearshore (inner neritic), middle shelf, outer shelf to upper slope, to the deep sea. Note that clastic sediments and breccias are deposited primarily in inner to middle shelf regions and on continental margins, but are absent in the deep sea. The estimated depth of deposition is based upon benthic foraminifera, ostracodes, dinoflagellates, invertebrates, and paleodepth backtracking of the deep-sea sections.

Determining relative sea-level change across the K-T boundary is more difficult and involves detailed quantitative studies of benthic foraminifera, shallow-water planktonic foraminifera, spores and pollen, dinoflagellates, and macrofossils. Such studies have been done for many K-T boundary sections, and they reveal a consistent pattern of global sea-level changes (Fig. 4) based on sections from northern to southern high latitudes (MacLeod and Keller, 1991a, 1991b; Schmitz and others, 1992; Keller, 1988b, 1992, 1993; Brinkhuis and Zachariasse, 1988; Keller and others, 1993b; Askin, 1992; Askin and Jacobson, this volume; Stinnesbeck and Keller, this volume). This database indicates generally rising seas after a sea-level lowstand, which Baum and others (1982) and Haq and others (1987) estimated occurred 0.5 m.y. below the K-T boundary, and Pardo and others (this volume) estimate at 0.2 to 0.3 m.y. below the K-T boundary based on

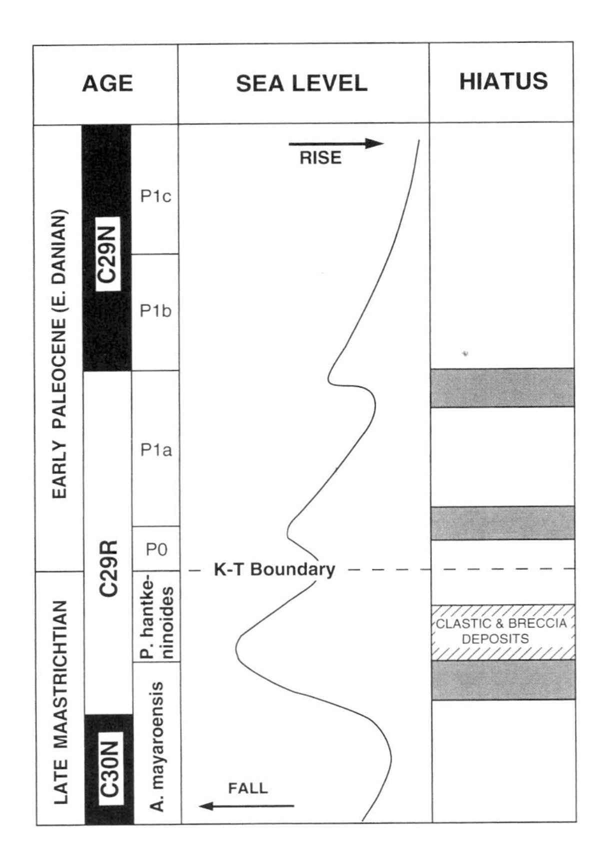

Figure 4. Sea-level changes, hiatuses, and clastic or breccia deposits across the K-T transition based on examination of K-T sections globally. Paleodepth interpretations are based on planktonic and benthic foraminifera, spores and pollen, dinoflagellates, and macrofossils. Biostratigraphy and paleomagnetic stratigraphy are from the Agost section in Spain (Pardo and others, this volume).

paleomagnetic stratigraphy. This rising sea level, beginning during the last 50 to 100 k.y. of the Maastrichtian, was interrupted by at least two short-term sea-level lowstands in the early Danian. In the stratigraphic record, these short-term sea-level lowstands are generally marked by hiatuses that removed part or entire biozones, and/or by condensed sedimentation (as expressed by a type-Z sequence boundary). Thus, they most likely represent sea-level fall and rise inflection points (Posamentier and others, 1988).

We recognize that local tectonics, especially along plate margins and on continental shelves, may amplify, obscure, or obliterate global sea-level signals when subsidence or uplift rates exceed the eustatic (global) rate. To avoid major tectonic overprints in our database, we have searched for consistent sea-level-change signals across latitudes that would indicate global controls. The sea-level changes identified are consistent in both magnitude and timing and thus suggest mainly global controls, any local tectonic overprint being relatively minor.

Most continental shelf and slope sections indicate a major sea-level lowstand, often accompanied by a hiatus, during the latest Maastrichtian preceding the K-T boundary. This sea-level lowstand occurred about 200–300 k.y. below the K-T boundary (Keller, 1989a; Keller and others, 1993b; Pardo and others, this volume), as indicated by its position in the lower part of C29R just below the *P. hantkeninoides* biozone. This latest Maastrichtian sea-level lowstand and hiatus is global in extent. At Stevns Klint and Nye Kløv in Denmark, this hiatus and sea-level lowstand is marked by a disconformity between white chalk beds that contain only rare bryozoans and the overlying undulating gray-white chalks that are rich in bryozoans (Surlyk, 1979; Hultberg and Malmgren, 1986; Schmitz and others, 1992; Keller and others, 1993b). At Brazos River, the sea-level lowstand and unconformity occurs at the base of a clastic deposit and within the portion of C29R below the K-T boundary (Keller, 1989a, 1989b). In a K-T boundary section at the Poty Quarry in Brazil (near Recife), a sea-level lowstand and unconformity occurs 70 cm below the K-T boundary within the *P. hantkeninoides* zone and is marked by a micritic limestone that is overlain by a 10–20-cm-thick limestone breccia (Stinnesbeck, 1989; Stinnesbeck and Keller, this volume). In K-T boundary sections from east-central Mexico (Mirador near La Ceiba) and southern Mexico (Bochil in Chiapas), evidence for a latest Maastrichtian sea-level lowstand is found in deposition of the clastic and breccia deposits 100 cm below the K-T boundary (Macias Pérez, 1988; Montanari and others, 1994; Stinnesbeck and others, 1994a). In most other K-T sections in northeastern Mexico, coarse-grained clastic deposits have been reported at or below the K-T boundary (Smit and others, 1992; Stinnesbeck and others, 1993, this volume; Keller and others, 1994a), as will be discussed below.

After the latest Maastrichtian sea-level lowstand, a marine transgression marks the interval up to and across the K-T boundary and through the earliest Paleocene zone P0. At Stevns Klint and Nye Kløv, the pre-K-T boundary transgression is marked by an influx of rugoglobigerinids from lower latitudes, as well as by benthic foraminiferal changes indicating deeper waters and climatic warming (Schmitz and others, 1992; Keller and others, 1993b). Deeper water benthic and palynofloral assemblages are also present in earliest Danian sediments of Spain (Agost and Caravaca), Tunisia (El Kef), Israel (Negev), and Texas (Brazos River)

(Peypoquet and others, 1986; Brinkhuis and Zachariasse, 1988; Keller, 1988b, 1992; Beeson, 1992). Biozone P0, which nearly always consists of a gray to black organic-rich clay with low-oxygen-tolerant faunas, appears to mark a sea-level highstand. Zone P0 is generally absent in settings below 1000 m depth because of nonaccumulation during the transgression (MacLeod and Keller, 1991a, 1991b).

The transition from the black clay of zone P0 to the shales of zone P1a is generally marked by a short hiatus and marks the first short-term Paleocene sea-level lowstand (Fig. 4). Erosion or condensed sedimentation due to decreased planktonic productivity, reduced upwelling, or decreased terrestrial input, frequently results in all or part of zone P0 being missing, as observed in sections from Brazos River (Keller, 1989a), Spain (Canudo and others, 1991), Israel (Keller and Benjamini, 1991), Denmark (Schmitz and others, 1992; Keller and others, 1993b), Mexico (Keller and others, 1994a, 1994b), the Gulf of Mexico (Keller and others, 1993a), and the southern Indian Ocean (Keller, 1993). Transgressing seas marked zone P1a (*P. eugubina*) with the deposition of increasingly shaly sediments that are frequently rich in glauconite (e.g., Brazos, Negev, Spain, Tunisia, Seymour Island).

The second early Paleocene hiatus and sea-level lowstand nearly coincides with the zone P1a-P1b boundary just prior to the extinction of *Parvularugoglobigerina eugubina* (Fig. 4). In many sections, erosion has removed part or all of zone P1a, including sections in Denmark, Spain, Israel, Mexico, Texas, and the southern Indian Ocean (Keller, 1989a, 1989b; MacLeod and Keller, 1991a, 1991b; Canudo and others, 1991; Keller and Benjamini, 1991; Keller and others, 1993b, 1994a, 1994b), and frequently removed sediments into the upper Maastrichtian (Keller and others, 1993a)

Increasingly carbonate rich marls or limestones are deposited in zones P1b and P1c across latitudes, and benthic foraminifera and palynofloras indicate deeper water environments approaching depths similar to the latest Maastrichtian prior to the sea-level fall (Brinkhuis and Zachariasse, 1988; Keller, 1988b, 1992).

Although these hiatuses mark the physical expression of sea-level fall and rise inflection points, the same relative changes in sea level can be observed in sections where no hiatuses are apparent in the biostratigraphic record. For example, the El Kef stratotype represents the most complete K-T boundary sequence known to date with no apparent hiatuses (Donce and others, 1985, 1994; Brinkhuis and Zachariasse, 1988; Keller, 1988a; MacLeod and Keller, 1991a, 1991b; Keller and others, in prep.), although the same relative changes in sea level are indicated in benthic foraminiferal and dinoflagellate assemblages (Brinkhuis and Zachariasse, 1988; Brinkhuis and Leereveld, 1988; Keller, 1988b, 1992). Similar sea-level fluctuations based on dinoflagellate assemblages are also reported from Alabama boundary sections (Habib and others, 1992), Brazos, Texas (Beeson, 1992), and Seymour Island, Antarctica (Askin, 1992; Askin and Jacobson, this volume). Thus, the sea-level lowstands of the latest Maastrichtian and early Danian (Fig. 4) record fluctuations in eustatic sea levels, and their effects on sedimentation must be considered in interpreting K-T clastic deposits.

Relative changes in sea level, timing of sea-level lowstands, and the relative magnitude of sea-level falls across the K-T transition, based on the global database discussed above, are shown in Figure 4. The latest Maastrichtian sea-level lowstand occurred within the lower part of paleomagnetic chron 29R below the K-T boundary and just below the *P. hantkeninoides* zone, which spans the last 170–200 k.y. of the Maastrichtian (Pardo and others, this volume). Thus, this sea-level drop must have occurred within ~100 k.y. or less, and, on the basis of benthic foraminiferal data, reached a magnitude of ~70 to 100 m (Keller, 1992; Keller and others, 1993b; Schmitz and others, 1992; Stinnesbeck and Keller, this volume). At this magnitude, the sea-level fall would have occurred at a rate of 0.7 to 1.0 m/k.y. (or less than 1 mm/yr).

Generally rising seas characterize the top of the Maastrichtian across the K-T boundary and through the early Danian, reaching the pre-K-T boundary sea-level high by zone P1c (C29N), ~500 k.y. after the K-T boundary. This trend is interrupted by two short-term sea-level lowstands at the zone P0–P1a boundary, ~40 to 50 k.y. above the K-T boundary, and near the top of zone P1a, ~230 k.y. above the K-T boundary. The magnitude of these short-term sea-level drops is significantly less than during the latest Maastrichtian.

What caused the relatively rapid sea-level drop (less than 1 mm/yr) near the end of the Maastrichtian? A sea-level change due to mid-ocean ridge activity, which averages 1 cm/k.y., is too slow to be a causal agent. Changes in continental ice volume must be considered. Sea-level change due to buildup in continental ice volume during the Pleistocene averages 1 cm/yr. In comparison, the late Maastrichtian sea-level change averaged less than 1 mm/yr, or less than one-tenth of Pleistocene rates.

There is evidence for significant continental glaciation during the middle Maastrichtian in Weddell Sea Ocean Drilling Program (ODP) Site 690, in a marked cooling and increase in $\delta^{18}O$ values associated with a major sea-level drop (Barrera and Huber, 1990; Barrera, 1994). The magnitude of the $\delta^{18}O$ increase is similar to that of the middle Eocene cooling and major Antarctic glaciation, which is associated with ice-rafted debris (Barrera and Huber, 1991). This suggests that significant continental glaciation may also have occurred during the middle Maastrichtian (Barrera, 1994). Stable isotope values indicate that relatively cool temperatures prevailed through the late Maastrichtian, suggesting that the sea-level drop just preceding the *P. hantkeninoides* zone is likely the result of increased continental glaciation. At ODP Site 690, stable isotope values record a rapid warming (greenhouse warming?) just preceding the K-T boundary and following the maximum late Maastrichtian cooling (Stott and Kennett, 1990; Barrera, 1994), and at Stevns Klint and Nye Kløv a rising sea level and incursion of tropical species also mark this pre-K-T boundary climatic warming (Schmitz and others, 1992; Keller and others, 1993b).

The cause for this climatic warming is still unclear. It is most likely related to the extensive global volcanism (Deccan Traps, India) and the associated increase in atmospheric carbon dioxide levels at that time, although a similar atmospheric carbon dioxide increase and climatic warming could have been triggered by a pre-K-T boundary bolide impact.

CLASTIC DEPOSITS

Clastic coarse-grained or breccia deposits have been described from many K-T boundary sequences: the best known are from Texas, Alabama, Georgia, Cuba, Haiti, and Mexico (Fig. 5), where they have been variously related to sea-level changes or impact-generated megatsunami deposits. The geographic concentration of such K-T sections is not accidental, but rather the result of many years of intensive search for evidence of a bolide impact in the Caribbean in the form of tsunami deposits. A similarly intensive search elsewhere may yield similar K-T or near K-T clastic deposits in other marginal seas beyond the reach of a proposed Caribbean megatsunami. In this section we review the age, biostratigraphy, lithology, and depositional environment of these clastic deposits to evaluate whether they are coeval, as assumed by proponents of the tsunami model and the late Maastrichtian sea-level lowstand model, or whether they represent different sea-level lowstand ages below, at, or above the K-T boundary.

Alabama: Braggs, Mussell Creek, Moscow Landing

Clastic deposits in Alabama are restricted to the Clayton Sands, which are widespread though intermittent throughout the region. They occur near the K-T boundary and have been interpreted as transgressive infilling of incised valleys cut during a preceding sea-level lowstand (Donovan and others, 1988; Baum and Vail, 1988; Mancini and others, 1989; Savrda, 1991, 1993; Moshkovitz and Habib, 1993; Habib and others, 1992) and as high-energy impact-generated megatsunami deposits (Hildebrand and Boynton, 1990; Olsson and Liu, 1993; Habib, 1994; Smit and others, 1994a, 1994b).

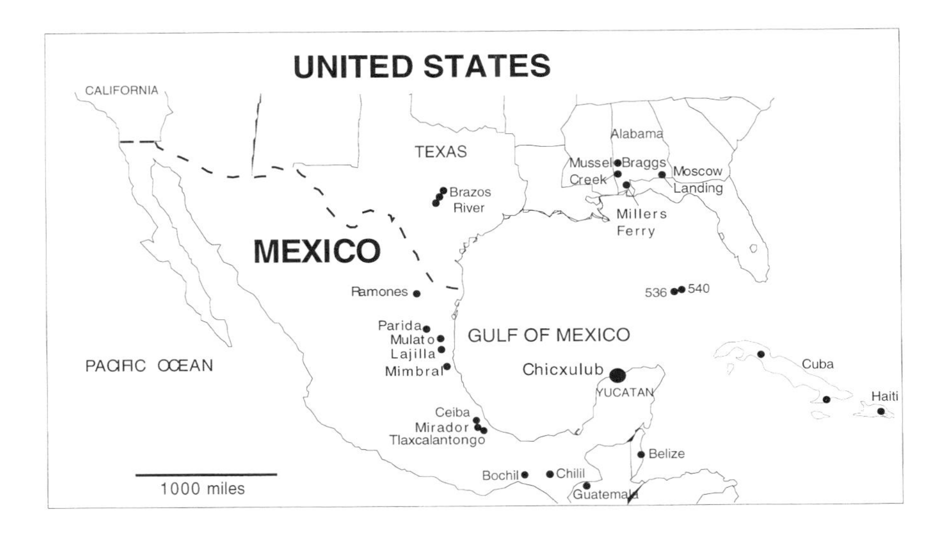

Figure 5. Location of K-T boundary sections in the Caribbean and Central America from which coarse-grained clastic or breccia deposits have been described and interpreted as impact-generated megatsunami deposits.

The lithology and stratigraphic sequences are similar in the Alabama sections, including Braggs, Mussell Creek, Millers Ferry, and Moscow Landing (Mancini and others, 1989; Savrda, 1993; Olsson and Liu, 1993). Here, the gray to olive-gray silty to sandy marls of the late Maastrichtian Prairie Bluff Chalk are intensively burrowed and contain abundant Cretaceous body fossils (principally bivalves). On the basis of the absence of *Abathomphalus mayaroensis*, Mancini and others (1989) concluded that the upper Maastrichtian is missing at the unconformity at the top of the Prairie Bluff Chalk. However, the presence of the latest Maastrichtian *Micula prinsii* nannofossil zone in the uppermost part of the Prairie Bluff Chalk indicates that this "hiatus" is much shorter (Moshkovitz and Habib, 1993; Olsson and Liu, 1993). On the basis of the irregular geometry of consolidated marlstone and truncation of thick-shelled fossils, Savrda (1993) estimated at least 2–3 m of erosion. Furthermore, he estimated an extended period of subaerial exposure on the basis of microkarstification (solution-enhanced erosion).

At Mussell Creek and Moscow Landing, and many other localities in central and western Alabama, the Prairie Bluff is unconformably overlain by the Clayton Sands, which consist of irregular bodies of uncemented yellowish-brown silty sands. At Moscow Landing, this member is characterized by inclined parallel and ripple cross-laminations, rip-up clasts of the Prairie Bluff Chalk, and reworked Cretaceous fossils. The resident ichnofauna includes rare *Ophiomorpha* burrows in the lower beds and *Thalassinoides* and *Planolites* burrows in the middle and upper beds of the Clayton Sands (Savrda, 1993). Sediments and fossils both indicate that deposition occurred in a marginal marine tidally influenced environment (Savrda, 1993; Moshkovitz and Habib, 1993).

The Clayton Sands are now generally considered to be of early Danian age. The Clayton Sands exposed at Mussell Creek and Moscow Landing contain planktonic foraminifera indicative of zones P1b and P1c (*Subbotina pseudobulloides*, *S. triloculinoides*, *S. trivialis*, and *Globoconusa daubjergensis*; Mancini and others, 1989), whereas at Millers Ferry the earliest Danian zones P0 and P1a (*P. eugubina*) span the entire Clayton Sand, as reported by Olsson and Liu (1993) and Liu and Olsson (1992). This suggests that the older (zones P0 and P1a) Clayton Sand is present at Millers Ferry, but absent at Mussell Creek, Braggs, and Moscow Landing, whereas the younger (zone P1b or P1c) Clayton Sand occurs in the latter outcrops but not at Millers Ferry. This, in turn, implies that the Clayton Sands are not coeval in all sections and may represent a time-transgressive facies spanning the interval from zone P0 to at least zone P1b in the early Danian.

At all four Alabama sections (Mussell Creek, Braggs, Moscow Landing, Millers Ferry), the Clayton Sands and/or Prairie Bluff Chalk are overlain by the basal limestone bed of the Clayton Formation, which truncates the Clayton Sands and Prairie Bluff Chalk. This contact is sharp and clearly erosional at Mussell Creek, Braggs, and Moscow Landing (Mancini and others, 1989; Savrda, 1993), and the abrupt lithological change at Millers Ferry may also indicate discontinuous sedimentation.

Is deposition of the Clayton Sands the result of a K-T boundary impact-generated tsunami? The presence of Danian microfossils in the Clayton Sands alone argues against deposition caused by a K-T boundary impact. Moreover, no shocked quartz, tektite glass, or other impact indicators have been found. The presence of microkarstification and a firmground at the Prairie Bluff–Clayton Sands contact argues for an extended period of subaerial exposure rather than a catastrophic event (Savrda, 1993). As indicated by zones P0–P1a and P1b–P1c planktonic foraminifera in the Clayton Sand, deposition occurred over an extended time period spanning at least 200–300 k.y. An extended period of deposition is also indicated by the resident ichnofauna, characterized by the presence of multiple phases of habitation and permanent domiciles (Savrda, 1993).

Sedimentological and biological indicators strongly indicate that deposition of the Clayton Sands occurred over an extended time period bounded by disconformities formed during two sea-level lowstands. The erosional surface at the top of the Prairie Bluff most likely represents the latest Maastrichtian sea-level lowstand (Baum and Vail, 1988; Donovan and others, 1988) characterized by microkarstification and prolonged subaerial exposure. By early Danian time sediment deposition resumed with the Clayton Sands; deposition occurred during the sea-level transgression and was probably intermittent. The disconformity between the top of the Clayton Sands and the basal limestone of the Clayton Formation marks a sea-level lowstand at or near the top of zone P1a as indicated by the presence of *P. eugubina* at Millers Ferry (Olsson and Liu, 1993). At Mussell Creek and Moscow Landing, deposition of Clayton Sands may have continued into zone P1b or P1c as suggested by the absence of *P. eugubina* (Mancini and others, 1989).

Donovan and others (1988) and Baum and Vail (1988) interpreted the upper Prairie Bluff Chalk as a prograding highstand systems tract (HST) that was followed by a major sea-level fall that truncated the sequence, forming an irregular locally incised topography that was temporarily exposed (type 1 sequence boundary). During the subsequent transgression, these incised valleys were filled with marginal marine sands and gravels (lowstand systems tract, LST). Further rise of sea level formed the transgressive erosional surfaces and firmgrounds at the base of the Clayton limestone (transgressive systems tract, TST) and was followed by the return to normal marine conditions. This interpretation is consistent with the biostratigraphic and paleontological evidence.

Texas: Brazos River

Outcrops along the Brazos River in central-east Texas contain a nearly continuous record across the K-T boundary (Jiang and Gartner, 1986; Keller, 1989a, 1989b; Hansen and Upshaw, 1990; MacLeod and Keller, 1991a, 1991b; Hansen and others, 1993; Beeson, 1992). The K-T transition consists of calcareous claystones of the Kinkaid Formation, with a thin (<20 cm) clastic unit at its base. The clastic unit rests unconformably upon a scoured surface of the Maastrichtian Corsicana Formation and consists of coarse sand containing shell hash and shale intraclasts capped by a thin limestone layer. This clastic event deposit has been variously interpreted as an impact-generated tsunami deposit (Bourgeois and others, 1988; Hildebrand and Boynton, 1990) linked to the proposed Chicxulub impact structure (Smit and others, 1992, 1994a, 1994b), or to noncatastrophic sedimentation related to the latest Maastrichtian sea-level lowstand and subsequent transgression (Keller, 1989a, 1992). No shocked quartz, Ni-rich spinels, or tektite glass have been found within the clastic unit. A recent study indicates an Ir anomaly 15–17 cm above the limestone layer that marks the top of the clastic unit (Rocchia, 1994, written commun.).

There has been disagreement in the past as to whether the K-T boundary should be placed at the unconformity at the base of the clastic deposit, at the top of the clastic deposit, or between 15 and 22 cm above the deposit. The general consensus seems to be to place the K-T boundary 17 to 20 cm above the top of the clastic deposit at the first appearance of Tertiary foraminifera (Keller, 1989a, 1989b), nannofossils (Jiang and Gartner, 1986), and palynomorphs (Beeson, 1992; Figure 6), coincident with the Ir anomaly as observed by Rocchia (1994, written communication).

All planktonic foraminifera observed by Keller (1989a) in the clastic deposits of three Brazos River sections are Late Cretaceous forms. Also present is the latest Maastrichtian index species *Plummerita hantkeninoides*, which indicates that deposition occurred during the last 170–200 k.y. of the Maastrichtian. Montgomery and others (1992), however, reported late Danian planktonic foraminifera, suggesting that deposition occurred during the late Danian. Examination of Montgomery and others' (1992) data indicates, however, that the species illustrated as Danian

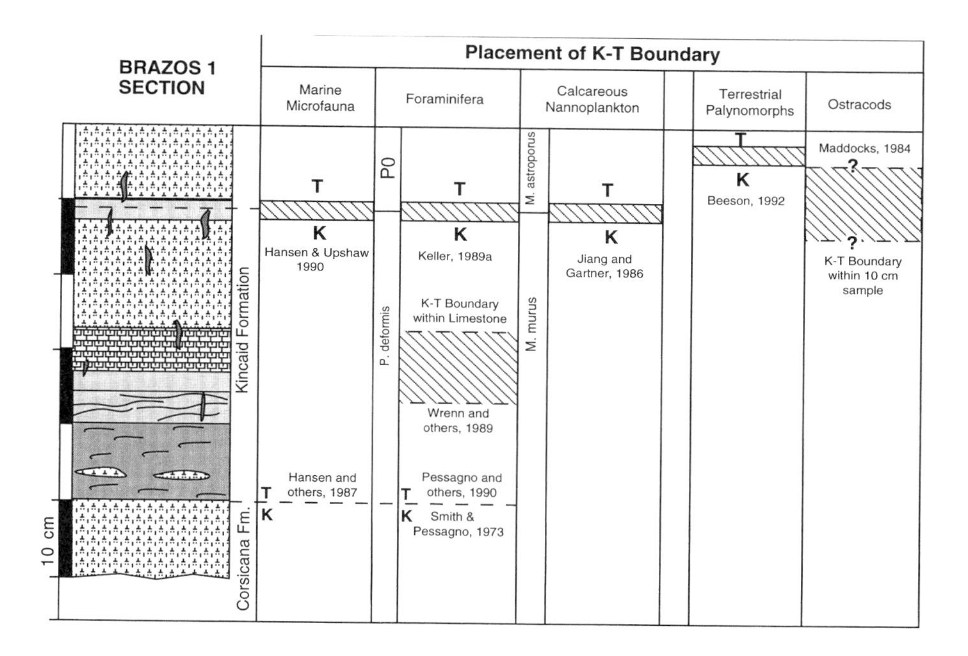

Figure 6. Placement of the K-T boundary in the Brazos-1 section, Falls County, Texas. Note the general biostratigraphic consensus between macrofossils, planktonic foraminifera, nannofossils, palynomorphs, and ostracodes in placing the K-T boundary between 15–20 cm above the clastic deposit. The paleontological boundary placement coincides with the iridium anomaly 15–17 cm above the clastic deposit as reported by Rocchia (1994, written commun.).

taxa are either too poorly preserved for positive species identification or lack precise location data relative to the clastic deposit, and in some cases are from Danian strata above the clastic deposit (MacLeod and Keller, 1994). Thus, no confirmed Danian microfossils are known in this clastic deposit.

Between the clastic deposit and the K-T boundary, the marly sediments contain well-preserved late Maastrichtian shallow-water benthic foraminifera and surface-dwelling planktonic foraminifera (heterohelicids, guembelitrids, rugoglobigerinids, hedbergellids, globigerinellids, and *P. hantkeninoides*). There is no evidence of size sorting by currents or upward fining of sediments. Above the K-T boundary, the same well-preserved late Maastrichtian assemblages continue (except *P. hantkeninoides*) with the addition of the evolving new Tertiary species. All early Danian zones are present, including zones P0, P1a, P1b, and P1c. A short hiatus is present at the zone P1a–P1b boundary, indicated by a lithological change, abundant glauconite, and the sudden truncation of species ranges (Keller, 1989a). A short hiatus or condensed sedimentation is also present at the P0–P1a boundary (Fig. 7; MacLeod and Keller, 1991a, 1991b).

At the Brazos River section, catastrophic tsunami deposition is difficult, if not impossible, to reconcile with the biostratigraphy and the absence of tektite glass and shocked quartz. Proponents of this theory explain the sediments between the top of the clastic bed and the K-T boundary as the result of settling through the water column after the tsunami (Bourgeois and others, 1988; see also Keller, 1991; Smit and others, 1994b), and a similar explanation is given for the 5–100-cm-thick Maastrichtian sediments above the clastic deposits in Mexico (Smit and others, 1992, 1994a, 1994b; Montanari and others, 1994). Because there is no evidence of impact-derived material or size sorting, and the dwarfing of some

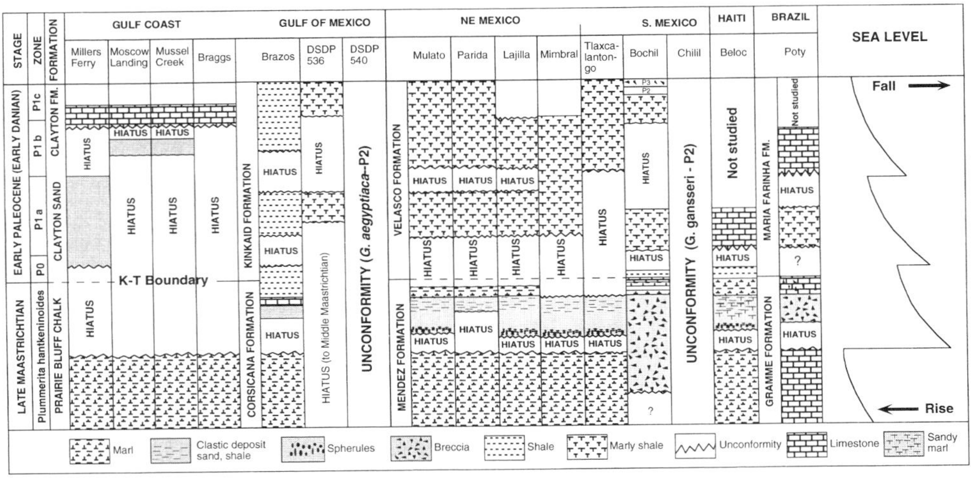

Figure 7. Biostratigraphy, lithostratigraphy, and hiatuses (white intervals) of K-T sections with clastic or breccia deposits correlated with sea-level fluctuations. Note that the clastic or breccia deposits, which have been reinterpreted by some workers as impact-generated tsunami deposits, occur stratigraphically below the K-T boundary and correlate with the latest Maastrichtian sea-level lowstand.

species (e.g., heterohelicids) begins below the clastic deposit at the Brazos sections, catastrophic impact-generated tsunami deposition is unlikely. Prevailing evidence suggests that the unconformity and clastic deposit are related to the latest Maastrichtian sea-level regression and subsequent transgression across the K-T boundary. On the basis of magnetostratigraphy at the Brazos core section, the unconformity formed in C29R just below the K-T boundary, and an estimated 300 k.y. of the latest Maastrichtian may be missing (Keller, 1989a). Sediment deposition at Brazos thus differs from Alabama sections where Montgomery and others (1992) suggested that a major Upper Cretaceous to Danian unconformity is present based on the absence of *A. mayaroensis*. Stable isotopic data indicate, however, that this species is a deeper water dweller probably living at depths of 200 m or deeper (Boersma and Shackleton, 1981; Barrera and Huber, 1990) and that its absence in the shallow waters of the Gulf of Mexico coastal plain is likely due to ecologic exclusion, rather than a hiatus. This is also indicated by the presence of the latest Maastrichtian nannofossil index species *Micula prinsii* (Savrda, 1993; Olsson and Liu, 1993) and planktonic foraminifer *P. hantkeninoides*.

A latest Maastrichtian sea-level fall at Brazos is indicated by middle to inner neritic benthic foraminifera (Keller, 1992), a decline in invertebrate diversity, and an increase in oysters (Hansen and others, 1993). These shallowing waters may have been accompanied by reduced salinity as suggested by the decrease in stenohaline carnivorous snails (Hansen and others, 1993). Paleobathymetric trends among benthic foraminifera and palynomorph assemblages indicate that the sea-level lowstand was followed by rapidly rising seas across the K-T boundary (Keller, 1992). Short-term sea-level lowstands appear to have occurred at the P0–P1a and P1a–P1b boundary (Fig. 7), as indicated by abrupt changes in planktonic foraminifera at both horizons, and a lithological shift to sand and glauconite at the P1a–P1b boundary (Keller, 1989a; MacLeod and Keller, 1991a, 1991b).

Cuba

A big boulder bed present in the uppermost Cretaceous sediments of Cuba has been interpreted as a proximal ejecta blanket of a nearby K-T boundary bolide impact (Bohor and Seitz, 1990). This boulder bed has been shown, however, to be the result of exfoliation weathering in the coarse calcarenaceous part of a megaturbidite (Brönnimann and Rigassi, 1963; Dietz and McHone, 1990; Iturralde-Vincent, 1992). This megaturbidite forms a sheet-like unit that is widely known in Cuba as the Cacarajicara, Amaro, and Peñaver formations (Pszolkowski, 1986; Iturralde-Vincent, 1992).

Whatever the cause of the sudden collapse of sediment prisms on the platform and slope of Cuba, whether earthquake, bolide impact, or sea-level lowstand, it did not coincide with the K-T boundary. Planktonic foraminifera from the top of the Cacarajicara Formation indicate an early to middle Maastrichtian age (Pszolkowski, 1986), whereas the top of the Peñaver and Amaro formations, which contain the megaboulder bed, are of late Maastrichtian *A. mayaroensis* zone age, and contain the index taxon (Brönnimann and Rigassi, 1963; Iturralde-Vincent, 1992). Moreover, the Amaro Formation and megaboulder bed is overlain by a 150–250-m-thick limestone unit that also contains late Maastrichtian planktonic foraminifera (Iturralde-Vincent, 1992). Thus, the megaboulder bed of

Cuba is of late Maastrichtian age and unrelated to the K-T boundary event. Continuous K-T boundary sequences are present in marly sediments in eastern and southern Cuba that contain no evidence of the megaboulder bed, or any other clastic deposit (Iturralde-Vincent, 1992; Fernandez and others, 1991).

Haiti: Beloc

The K-T boundary sections of Beloc in Haiti contain a 40–60-cm-thick clastic sandstone deposit rich in calcareous and glassy spherules that have been variously reported as of tektite origin (Izett, 1990; Sigurdsson and others, 1991; Blum and Chamberlain, 1992; Koeberl and Sigurdsson, 1992) or of volcanic origin (Jéhanno and others, 1992; Lyons and Officer, 1992; Koeberl, 1994; and Robin and others, 1994). The clastic bed rests upon marls containing latest Cretaceous planktonic foraminiferal (*A. mayaroensis* zone) and nannofossil (*Micula murus* zone) assemblages (Maurasse and Sen, 1991; Sigurdsson and others, 1991). The spherule-rich sandstone is graded in size and abundance and has cross-laminations toward the top of the unit.

In most Beloc sections, the spherule-rich sandstone and overlying sediments are incomplete due to redeposition. However, two sections have been reported with a less disturbed and more expanded sedimentary record (Jéhanno and others, 1992). In these two sections, a 25–30-cm-thick bed of Maastrichtian sandy marl conformably overlies the clastic deposit, similar to sections in northeastern Mexico. This sandy marl layer is succeeded by a thin clay layer that contains the cosmic markers Ni-rich spinels and the Ir anomaly that characterize the K-T boundary elsewhere. Thus, the sandy marls below the K-T boundary indicate that a period of normal hemipelagic sedimentation followed deposition of the spherule-rich clastic unit which must predate the K-T boundary event by at least several thousand years. Moreover, faunal assemblages indicate that water depth decreased in this sandy marl as compared to the water depth below the spherule-rich clastic bed (Jéhanno and others, 1992). On the basis of these data, we suggest that the Haiti spherule-rich clastic sediments were deposited during the latest Maastrichtian sea-level lowstand (Fig. 7) that may have triggered gravity flow or turbidite deposition containing spherules from an earlier or coeval widespread Caribbean volcanic event or ejecta from a pre-K-T boundary impact.

Northeastern and East-Central Mexico

More than a dozen K-T boundary sections containing siliciclastic deposits at or near the boundary are known from northeastern Mexico and two are known from east-central Mexico (Mirador and Tlaxcalantongo near La Ceiba, Fig. 5). Most of these sections show comparable depositional sequences, although clastic units are variable in lithology and thickness, ranging from only 2 cm to a maximum of 11 m thick. Additional sections have no clastic deposit. The geographic distribution of these sections spans over 500 km in a north-northwest–south-south-east–trending area that is parallel to and 40–80 km east of the front range of today's Sierra Madre Oriental (Keller and others, 1994a).

The regionally developed clastic member has been described by a number of investigators (Smit and others, 1992, 1994a, 1994b; Longoria and Gamper, 1992; Stinnesbeck and others, 1993, 1994b, this volume; Bohor and Betterton, 1993;

Longoria and Grajales Nishimura, 1993; Keller and others, 1994a, 1994b; Adatte and others, 1994) and has been interpreted as a K-T boundary deposit produced by an impact-generated megatsunami (Smit and others, 1992, 1994a), a series of normal high-energy turbidity currents triggered by an impact (Bohor and Betterton, 1993), and a series of gravity flows or turbidity currents related to the latest Maastrichtian sea-level lowstand and tectonic activity (uplift of Sierra Madre) (Stinnesbeck and others, 1993, 1994a, this volume; Keller and others, 1994a; Adatte and others, 1994).

In sections where no clastic deposit is present at the K-T boundary (e.g., Parida where clastic deposit disappears), the K-T boundary occurs between the fine-grained marls of the Maastrichtian Mendez Formation and the overlying marly shales of the Tertiary Velasco Formation. Although the contact between these formations is transitional and no erosional surface is recognizable, the earliest Danian zone P0 is generally not present and the lower part of zone P1a also appears to be missing, either due to a hiatus or condensed sedimentation (Fig. 7, type Z sequence boundary, Lopez-Oliva and Keller, 1994). The clastic deposit is stratigraphically near the K-T boundary. In the most complete sections, a 5–10-cm-thick layer of Maastrichtian marls overlies the clastic deposit (e.g., Lajilla, Mulato, Parida, in northeastern Mexico), and a 1-m-thick Maastrichtian age sequence of sand, shale and clay overlies a breccia deposit at Bochil in Chiapas, Mexico. In some sections (e.g., Mimbral), Tertiary shales of the Velasco Formation directly overlie the clastic deposit and in still other sections, Tertiary sediments are missing (e.g., Peñon, Ramones, Sierrita; Keller and others, 1994a). The latest Maastrichtian index species *P. hantkeninoides* has been oberved in all sections (except Mulato) within, below, or above the clastic member, but always below the K-T boundary (Lopez-Oliva and Keller, 1994). The presence of this species indicates that deposition of the clastic member occurred within the last 170–200 k.y. of the Maastrichtian.

Although the thickness and character of the clastic member vary between individual outcrops, three distinct lithologic units can be recognized in the thickest deposits (e.g., Mimbral, Peñon, Mulato, Sierrita; Stinnesbeck and others, 1993, this volume; Keller and others, 1994a). The basal unit 1, which consists of a carbonate-spherule-rich weathered sediment, is the key evidence for linking the clastic deposit to a bolide impact. This layer is of variable thickness, but generally no more than 10–30 cm thick, and contains abundant spherules. The spherules are 1–5 mm in diameter, commonly infilled with blocky calcite, and surrounded or partly filled by mixed layers of illite/smectite, chlorite, palagonite, or mica (Stinnesbeck and others, 1993; Keller and others, 1994a), or altered to kaolinite, replaced by smectite, pyrite, and glauconite. Many large spherules contain several smaller spherules. Some spherules contain apatite concretions, rutile crystals, clasts of limestone, or foraminiferal tests filled with glauconite. Many spherules have a tan organic coating, indicating that they are calcite infilled algal resting cysts (Stinnesbeck and others, 1993, 1994b; Keller and others, 1994a). Thus, the spherules of unit 1 have multiple origins, including organic, authigenic, volcanic, and accretionary oolites that formed at neritic depths and were probably subsequently transported. Proponents of tsunami deposition (Smit and others, 1992, 1994a) believe that all of these spherules are altered, glassy microtektites.

Rare glass particles in the spherule-rich layer are similar in chemical composition to glass spherules from Beloc, Haiti (Smit and others, 1992; Bohor and Betterton, 1993; Stinnesbeck and others, 1993). Many workers consider these glass particles and spherules as indicative of impact origin (Izett, 1990; Blum and Chamberlain, 1992; Sigurdsson and others, 1991; Koeberl and Sigurdsson, 1992), whereas others consider them to be indicative of volcanic origin (Jéhanno and others, 1992; Lyons and Officer, 1992; Leroux and others, 1995). For a discussion of this controversy see Robin and others (1994) and Keoberl (1994). If proven to originate from impact, these rare glass shards represent the only potential evidence to date of an impact in the clastic deposits of northeastern and east-central Mexico. A small iridium anomaly (0.8 ppb; Smit and others, 1992; Keller and others, 1994a; Stinnesbeck and others, 1993) is present above the clastic deposit in the basal shales of the Tertiary Velasco Formation.

A 10–20-cm-thick sandy limestone layer containing microlayering of fine and coarse-grained sediments and few spherules (W. Ward, 1994, personal commun.) is present within this spherule-rich layer at the Mimbral, Peñon, and Lajilla sections. We suggest that this sandy limestone layer represents a distinct and separate event during deposition of the spherule-rich sediments of unit 1. It seems unlikely that these distinct depositional events are the products of air-borne fallout, settling through the water column and reworking by wave action, all prior to the arrival of the first tsunami wave within two hours of the bolide impact event, as suggested by Smit and others (1994a). Deposition of these sediments probably occurred over a longer time period and by multiple events, as also suggested by the presence of abundant material from shallow-water areas that was transported and redeposited in the deeper water areas of Mimbral, Peñon, Lajilla, Mulato, Sierrita, and Tlaxcalantogo.

In northeastern Mexico sections, an erosional surface marks the contact between unit 1 and the overlying unit 2, which consists of a weakly laminated sandstone devoid of spherules, glass, or iridium. Plant debris is present in distinct layers at the base of unit 2 at the Mimbral outcrop and mud clasts are frequently present at the unconformity in all sections. Moreover, burrows, some infilled with the underlying spherule-rich sediment, are present and indicate habitation by invertebrates during deposition of unit 2 (Ekdale and Stinnesbeck, 1994). An erosional unconformity also marks the contact between unit 2 and the overlying unit 3, which consists of alternating sand, silt, and shale layers topped by a rippled sandy limestone. Ripple marks, flaser bedding, convolute bedding, climbing ripples, and small-scale cross-bedding are commonly found. Unit 3 is lithologically, mineralogically, and petrographically most variable, and contains thin layers of hemipelagic sediments with rich late Maastrichtian foraminiferal assemblages and two distinct layers rich in zeolites that are correlatable over the region and represent an influx of volcanogenic sediments (Adatte and others, 1994, 1995). The rippled sandy limestone layer that caps unit 3 is bioturbated in all outcrops, and burrows of *Chondrites*, *Zoophycos*, *Thalassinoides*, and *Ophiomorpha* commonly found in two to three discrete layers indicate different levels and different times of habitation. There is no downward mixing of Tertiary sediments (Stinnesbeck and others, 1993; Keller and others, 1994a, 1994b). Benthic foraminifera indicate that deposition of the clastic member occurred at a shallower depth than either the underlying Maastrichtian Mendez marls or the overlying Tertiary Velasco shales (Keller and others, 1994a, 1994b).

Our studies of more than a dozen K-T boundary outcrops in northeastern Mexico thus indicate that deposition of the clastic member occurred during the latest Maastrichtian sea-level lowstand within the last 170–200 k.y. of the Maastrichtian, but preceding the K-T boundary by at least several thousand years (Stinnesbeck and others, 1993, 1994b; Keller and others, 1994a, 1994b; Lopez-Oliva and Keller, 1994). Moreover, deposition occurred over an extended time period that allowed repeated recolonization of the substrate after erosion by burrowing organisms. Sedimentary processes also indicate multiple event deposition by debris flows, gravity flows, and periods of normal hemipelagic sedimentation. If the rare glass shards in unit 1 prove to be of impact origin, this impact would have preceded the K-T boundary. However, there is no evidence of a major pre-K-T boundary bolide impact in the biotic, stratigraphic, or geochemical records.

Southern Mexico: Chiapas

To date, there are no detailed studies of K-T boundary sequences of southern Mexico, despite visits by numerous geologists (including us) in search of outcrops with megatsunami deposits or continuous sedimentation. The problem is in the complex regional tectonics, which have resulted in repeated uplift and erosion and, hence, a major hiatus, or repeated flysch deposition (Quezada Muñeton, 1990; Michaud and Fourcade, 1989; Stinnesbeck and others, 1994a). We report here on the Chilil and Bochil sections of Chiapas (Fig. 5), which are representative of these two conditions.

The Chilil section, near Cristobal de las Casas, shows the transition between the Late Cretaceous Angostura Formation and the Paleocene Soyalo Formation. Our investigations indicate that the Angostura Formation consists of bioclastic shallow-water limestones with corals, rudists, echinoderms, and large benthic foraminifera. This unit is capped by a thin layer of bioturbated marly limestone with abundant early-late Maastrichtian (*G. gansseri* zone) planktonic foraminifera, indicating drowning of the carbonate platform. The top of this marly limestone contains crusts of large iron- and manganese-oxide nodules that formed during a long period of nondeposition. This surface is overlain by rhythmically bedded pelagic marls and shales of the Paleocene Soyalo Formation, which contains abundant planktonic foraminifera of the late Paleocene zone P2. The Chilil section thus contains a hiatus from the lower-upper Maastrichtian *G. gansseri* zone to the upper Paleocene zone P2 (Fig. 7). We found no evidence of the continuous K-T boundary deposition reported by Longoria and Gamper (1992).

The Bochil section north of Tuxtla Guiterrez contains a more complete K-T boundary record (Montanari and others, 1994; Stinnesbeck and others, 1994a). However, as at Chilil this K-T transition must be viewed within the tectonic and depositional history of the region. Flysch deposition began in the late Campanian and continued into the lower Tertiary, depositing many hundreds of meters of rhythmically bedded shales, silts, and sandstones (Quezada Muñeton, 1990). This flysch sequence is interrupted by numerous debris flows of well-rounded conglomerates in the lower part, and thick breccia beds of platform carbonates and isolated olistolithic blocks in the upper part. These conglomerate and breccia deposits mark the gradual uplift, tilting, and final collapse of the nearby platform carbonates (Michaud and Fourcade, 1989) and sea-level changes. Within this sequence of repeated breccia deposits, the thick clastic breccia containing angular platform limestone clasts (Montanari and others, 1994) near the K-T boundary does not mark an unusual sedimentological or lithological change, but rather a recurring phenomena, and hence provides no evidence for a single marine disturbance triggered by the proposed Chicxulub impact. These near K-T breccia deposits are likely the results of local tectonic activity during the latest Maastrichtian sea-level lowstand. This is supported by the fact that the breccia is overlain by a succession consisting of a 1-m-thick sandstone, silt, reddish clay, silty chalk, 5–10-cm-thick micritic limestone, thin clayey arenite, and white laminated chalk, the latter containing abundant early Danian zone P1a (*P. eugubina*) species (Montanari and others, 1994). Enriched iridium values possibly marking the K-T boundary were reported by Montanari and others (1994) from a reddish clay layer 6–10 cm below the micritic limestone of early Danian zone P1a.

These data indicate that K-T boundary age sediments are present at Bochil, although it is unclear whether sedimentation was continuous. At present, there is no evidence (no shocked quartz or glass) to link the breccia deposit 1 m below the K-T boundary to a K-T impact event (Fig. 7). Because the Bochil section contains many similar breccia deposits throughout the Campanian, Maastrichtian, and early Paleocene, the near K-T breccia layer, similar to the earlier and later deposits, most likely reflects sea-level changes and regional tectonic activity.

Guatemala and Belize

Cretaceous-Tertiary boundary deposits of Guatemala and Belize are still poorly understood, and no detailed studies exist. Preliminary investigations suggest that the depositional setting of the Petén region of Guatemala is similar to that of Chilil in Chiapas, Mexico. Michaud and Fourcade (1989) described Maastrichtian platform limestones and platform margin breccias overlain by rhythmically bedded Tertiary marls and shales in both regions. Neither they nor Hildebrand and others (1994) provided detailed biostratigraphic descriptions of the Petén sections. Hildebrand and others (1994, p. 49) stated, however, that "a dramatic deepening" occurred at K-T boundary time, giving evidence "of the dramatic erosional effects near the point of impact" of the Yucatan bolide. No explanation is provided how a bolide impact on Yucatan would cause dramatic deepening in Guatemala sections, nor did Hildebrand and others (1994) provide biostratigraphic evidence that the deepening is of K-T boundary age.

Our preliminary investigation of Guatemalan K-T boundary sections shows an 8–50-m-thick breccia deposit that is disconformably overlain by lower Tertiary sediments. The breccia consists of thick beds of large rounded clasts, angular clasts and interlayers of smaller-sized clasts, and pebbles in a muddy matrix. The top 2 m of the breccia unit consists of beds with smaller-sized breccia clasts and pebbles. Deposition of limestone clasts within the breccia occurred in a shallow-marine environment at depth of less than 20 m, as indicated by the predominance of miliolid benthic foraminifera and absence of planktonic foraminifera. Subsequent breccia deposition occurred at neritic depth via transport. The age of the breccia is unknown. Limestone or marl layers disconformably overlie the breccia unit, the contact being marked by an undulating erosional surface and breccia clasts in the overlying sediments. These sediments are of early Danian zone P1a age and benthic foraminiferal fauna indicates deposition in an outer neritic to uppper bathyal environment. Further studies are necessary before the age and nature of breccia deposition in Guatemala can be determined.

A K-T boundary section was reported by Ocampo and Pope (1994) from Albion Island in northern Belize. These authors described a K-T boundary marked by a major erosional unconformity separating crystalline Cretaceous dolomites and a poorly sorted dolomitic and carbonaceous breccia which they interpreted as the "product of ballistic sedimentation from the Chicxulub impact" (Ocampo and Pope, 1994, p. 86). It is interesting that, by the authors' own admission, their K-T boundary age interpretation is not based on any biostratigraphic evidence, either supporting a Maastrichtian age of the dolomite or the breccia deposits. Moreover, they did not observe Tertiary strata overlying the breccia deposit, nor did they observe them in other parts of the island. Ocampo and Pope's (1994, p. 86) interpretation of a breccia deposit as the "product of ballistic sedimentation from the Chicxulub impact" is a speculation based solely on a breccia of unknown age and the Chicxulub impact scenario.

Northeastern Brazil: Poty Quarry

The Poty Quarry section north of Recife in northeastern Brazil is outside the region of the disputed megatsunami deposits, but contains a limestone breccia bed 70 cm below the K-T boundary, coincident with the latest Maastrichtian sea-level lowstand. This section contains the most complete marine K-T boundary section known to date in South America (Stinnesbeck, 1989; Stinnesbeck and Keller, this volume). The upper Maastrichtian sediments (Gramame Formation) consist of micritic marly limestones (wackestones) that unconformably underlie a 20-cm-thick limestone breccia bed 70 cm below the K-T boundary (Fig. 7). The breccia bed contains reworked parts of the underlying marly limestone, calcispheres, planktonic foraminifera, bone fragments, phosphatic particles, glauconite, pyrite concretions, and abundant serpulids. In the 50 cm above the breccia bed, the grain size of limestone clasts diminishes from rudite to arenite size fraction, and in the top 10 cm changes into marly limestone similar to the marly limestones below the breccia bed and unconformity. The K-T boundary is placed at a thin clay layer, which contains high concentrations of iridium, between the top of the marly limestone and the overlying 5-cm-thick marly layer. This marl contains abundant planktonic foraminifera indicative of the earliest Tertiary zone P1a (*Parvularugoglobigerina eugubina*). A latest Maastrichtian age of the breccia as well as limestones below and above is indicated by the presence of *P. hantkeninoides*, which marks the last 170–200 k.y. of the Maastrichtian.

The presence of ammonites and neritic benthic foraminifera, the abundance of shallow surface-dwelling planktonic foraminifera (rugoglobigerinids, hedbergellids, guembelitrids, heterohelicids), and near absence of the deeper dwelling globotruncanids in the marly limestones indicate shallow-marine shelf deposition (<150 m depth) during the last 170–200 k.y. of the Maastrichtian. Different paleoenvironmental data appear 70 cm below the K-T boundary coincident with deposition of the limestone breccia which marks a sea-level lowstand followed by transgression. Benthic foraminifera suggest shallowing from middle neritic to inner neritic depths. At the same time, palynomorph taxa indicate a change from tropical to temperate climatic conditions (Ashraf and Stinnesbeck, 1988; Stinnesbeck and Keller, this volume). The sea-level transgression, which continued across the K-T boundary, reached middle neritic depths by K-T boundary time. The earliest Paleocene zone P0 has not been detected and may be missing at the Poty section, although an iridium anomaly is present in the thin clay layer (Albertao and others, 1994). The last 200 k.y. in the Poty Quarry section thus record a major global sea-level lowstand during the latest Maastrichtian, which was accompanied by deposition of a breccia layer, followed by a sea-level transgression and deposition of marly limestone and clay by K-T boundary time.

Albertao and others (1994) interpreted the Poty Quarry K-T boundary sequence as indicating two closely spaced bolide impacts: one at the K-T boundary which they placed at the base of the breccia, and the second in the Danian marked by the iridium anomaly. They based this interpretation on the putative identification of Danian species and microspherules, which they considered to be microtektites in the breccia and overlying marls and limestones. Examination of these sediments does not confirm their observations. Several of the species they considered to be of Danian age actually range into the Maastrichtian and others seem misidentified (for a discussion see Stinnesbeck and Keller, this volume). Moreover, the microtektite-like spherules that are abundantly present throughout the section and appear to be amber colored and glassy, are phosphatic and dissolve in strong acid. They are likely infillings of algal resting cysts which are commonly present in K-T transitions (Stinnesbeck and Keller, this volume).

Chile

Along the Pacific coast of central Chile, late Maastrichtian marine detrital sediments of the Quiriquina Formation are preserved in isolated basins. Late Maastrichtian fossils including ammonites [*Eubaculites carinatus*, *Hoploscaphites quiriquiniensis*, *D. (Diplomoceras) cylindraceum*, and *P. (Neophylloceras) surya*] are abundant (Stinnesbeck, 1986; this volume) and disappear between 5 to 10 m below the K-T boundary.

In the type area near Concepción, glauconitic sand and siltstones with calcareous sandstone concretions characterize the upper part of the late Maastrichtian sediment sequence and indicate deposition in a quiet offshore shelf setting. Bioturbation is intensive, and *Teichichnus* and *Zoophycos* are the prevailing trace fossils. Upsection, the abundance of detritus feeding bivalves suggests stillwater conditions and sediments rich in organic matter for the uppermost 4 to 5 m of the Quiriquina Formation. This suggests the latest Maastrichtian sea-level transgression and maximum flooding surface. Above this interval, the marine sequence is disconformably truncated by (prograding?) gravelly yellow sandstones of the Curanilahue Formation, reflecting brackish to fluviatile conditions.

The K-T boundary is placed at the unconformity between the bioturbated marine siltstones of the Quiriquina Formation and the fluviatile conglomeratic sandstones of the Curanilahue Formation The last unequivocally Maastrichtian index taxon is present 5 m below this lithological contact. Above this contact, only long-ranging microfloral taxa, charcoaled wood, and an isolated shark tooth have been found at the base of the Curanilahue Formation. The first palynoflora of unequivocal early Tertiary age (probably Paleocene) was found 18 m above the lithological contact that marks the K-T boundary (Frutos, 1984, personal commun.).

These lithological and faunal data indicate that, during the latest Maastrichtian sea-level regression, marine conditions gradually changed to shallower neritic and eventually brackish to fluviatile conditions in the earliest Tertiary. This is evident in the deposition of glauconite-rich sands near the top of the late Maastrichtian that are overlain by the fluviatile conglomeratic bed near the K-T boundary. There is no evidence of catastrophic deposition at or near the K-T boundary; only normal sedimentary processes associated with a sea-level regression that changes a marine environment to a nearshore depositional environment.

Antarctica: Seymour Island

The K-T boundary on Seymour Island is located in an expanded, shallow-marine clastic sequence consisting of loosely consolidated glauconitic silty sandstones of the Lopez de Bertodano Formation. Except for palynomorphs, the glauconitic interval is sparsely fossiliferous: planktonic foraminifera and calcareous nannoplankton are absent. Invertebrate faunas indicate a gradual faunal turnover throughout this interval, the stratigraphically highest ammonite occurrence (*Maorites densicostatus*) being near the base of the glauconite-rich interval (Zinsmeister and others, 1989; Zinsmeister and Feldmann, this volume). Based on a marked decrease of most invertebrate groups, Macellari (1986, 1988) originally placed the K-T boundary at the base of the glauconite-rich interval.

Dinoflagellate cysts provide the most precise biostratigraphic control for placing the K-T boundary in a 20–30 cm transitional interval just above a sharp iridium peak and ~3.5 m above the last ammonite occurrence (Askin and Jacobson, this volume; Elliot and others, 1994).

Both macrofossil and microfossil assemblages indicate that biotic and environmental changes on the Seymour shelf began well before the iridium layer was deposited at the end of the Cretaceous. Invertebrate species show a rapid turnover and declining ammonite diversity in the upper 40 m of Maastrichtian strata (*Pachydiscus ultima* and *Zelandites varuna* ammonite zones, Macellari, 1986, 1988; Zinsmeister and others, 1989; Zinsmeister and Feldmann, this volume). Dinoflagellates also show an accelerated turnover shortly before the end of the Maastrichtian, marked by an influx of new forms (Askin, 1988; Askin and Jacobson, this volume).

These biotic and lithological changes can be explained, at least in part, by the latest Maastrichtian sea-level regression, followed by transgression across the K-T boundary. In the Seymour Island sections, the glauconite-rich interval marks the sea-level regression during the latest Maastrichtian that temporarily reduced water depths from an offshore marine shelf to a marginal marine environment. At or before K-T boundary time, sea level began to rise, reaching a maximum flooding surface shortly after (30 cm above) the K-T boundary and iridium peak (Askin and Jacobson, this volume; Elliot and others, 1994). Thus, Seymour Island sections provide no evidence for unusual chaotic sediment deposition across the K-T boundary event, but rather show normal marine sedimentary processes associated with sea-level changes in shallow-marine shelf settings.

DISCUSSION AND CONCLUSIONS

We have examined K-T boundary sections from low to high latitudes, spanning depositional environments from continental shelf to shelf-slope and the deep sea (Figs. 1 and 3). Across these latitudes the overwhelming majority of deep-sea sections are marked by condensed sedimentation, nondeposition, or erosion across the K-T boundary transition (MacLeod and Keller, 1991a, 1991b; Keller, 1993; Keller and others, 1993a; Peryt and others, 1993) and, except for flysch deposition near continental margins, no clastic deposits are present. Outer shelf to upper slope environments, currently known from the Tethys region (Tunisia, Israel, Spain, and Mexico, Fig. 3), are characterized by generally higher rates of sedimentation, more continuous deposition and very short intervals of nondeposition or hiatuses (MacLeod and Keller, 1991a, 1991b; Keller and others, 1994a, 1994b). Clastic deposits are present only in northeastern and east-central Mexico sections and the southern United States, where they vary in nature and thickness, ranging from several meters to 10 cm and locally disappear (Keller and others, 1994a, 1994b). Middle shelf to inner neritic environments are known from many sections spanning low to high latitudes (Fig. 3). Sediment accumulation rates are generally high and sea-level fluctuations are marked by erosion, deposition of clastic deposits, and glauconite formation. Of the 12 shallow-water sections examined, only the Danish sections (Stevns Klint and Nye Kløv) lack clastic deposits, although they contain hiatuses (Schmitz and others, 1992; Keller and others, 1993b). Clastic deposits, including coarse sands, glauconite, and breccias, are thus common occurrences in continental shelf or platform settings worldwide at times of sea-level regression to transgression inflection points.

Are near K-T boundary clastic deposits the result of an impact-generated megatsunami wave? This sensational interpretation has received such popular support in some circles that a K-T boundary impact origin has been invoked for breccia deposits of unknown age in Belize (Ocampo and Pope, 1994), Guatemala (Hildebrand and others, 1994), and Chiapas (Montanari and others, 1994), and even for exfoliated sandstone boulders in Cuba (Bohor and Seitz, 1990), without any supporting evidence of impact origin or evidence of K-T boundary age. Clastic coarse sandstone deposits in Texas, Alabama, and Georgia have been reinterpreted as K-T impact-generated tsunami deposits regardless of their stratigraphic positions below or even above the K-T boundary, microkarstification or bioturbation indicating deposition over a longer time interval than a few days (Savrda, 1993), and the absence of any impact material (Bourgeois and others, 1988; Hildebrand and Boynton, 1990; Habib, 1994; Smit and others, 1994a, 1994b). Moreover, breccia, sandstone, and channel-fill deposits of near K-T boundary age in Mexico have been reinterpreted as Chicxulub impact-generated megatsunami deposits (Hildebrand and others, 1994; Smit and others, 1992, 1994a, 1994b; Montanari and others, 1994), although typical late Maastrichtian hemipelagic sediments top the clastic units and discrete layers of bioturbation are present within (e.g., Lajilla, Mulato, Parida and Bochil in Mexico, Poty Quarry in Brazil), indicating pre-K-T boundary deposition. Independent impact evidence is generally lacking (a small iridium anomaly found at Mimbral occurs within the basal Danian, not in the underlying clastic deposit). Some sections (e.g., La

Parida, see Stinnesbeck and others, this volume) lack any clastic deposit showing normal sedimentation across the K-T boundary.

Of all the sections for which K-T impact-generated tsunami deposits have been claimed, sections in northeast and east-central Mexico and Haiti deserve special attention because of their unique spherule-rich deposits. In the Beloc section of Haiti, this spherule-rich deposit contains many glassy spherules that have been variously interpreted as of impact origin (Izett, 1990; Sigurdsson and others, 1991; Blum and Chamberlain, 1992; Koeberl and Sigurdsson, 1992; Keoberl, 1994) or of volcanic origin (Jéhanno and others, 1992; Lyons and Officer, 1992; Robin and others, 1994). The spherule-rich layer at the base of the clastic deposits at Mimbral and Lajilla contains very rare glass fragments similar in chemical composition to those from Haiti, suggesting a common origin. If these glass spherules and fragments are of impact origin, then impact-triggered earthquake or tsunami deposition of the overlying clastic sediments in the northeastern and east-central Mexico sections becomes probable. However, this bolide impact would not have been of K-T boundary age, but preceded the boundary event by some thousands of years. This is suggested by the 25–30 cm of late Maastrichtian sandy marls followed by the K-T boundary iridium anomaly, overlying the clastic deposit in the undisturbed Beloc sections (Jéhanno and others, 1992), and similarly by the 5–10-cm-thick late Maastrichtian marls above the clastic deposits in sections at Lajilla, Mulato, and Parida, and 100 cm at Bochil (Keller and others, 1994a; Montanari and others, 1994; Lopez-Oliva and Keller, 1995; Macias Pérez, 1988; Stinnesbeck and others, this volume). In these sections as well as others, the Ir anomaly, Ni-rich spinels, first appearance of Tertiary planktonic foraminifera, and the $\delta^{18}O$ shift that are used to mark the K-T boundary worldwide occur well above the clastic deposits.

We do not believe sufficient evidence is present to identify any of the known K-T or near K-T clastic deposits as of a single impact origin because of the following. (1) They are of variable ages and frequently predate or postdate the K-T boundary. (2) They contain no unequivocal impact ejecta (with the possible exception of glass spherules from Haiti and rare glass shards from some northeastern Mexico sections, the origin of which, whether volcanic or impact, is still in dispute; see Keoberl, 1994, and Robin and others, 1994). (3) They do not represent a single event deposit over a few days, but multiple events over a longer time period as indicated by disconformities, different mineralogical contents between strata, layers of hemipelagic sedimentation, microkarstification, and microlayering of fine- and coarse-grained sediments. (4) They are frequently burrowed and contain multiple horizons of resident ichnofauna truncated by erosion. This repeated recolonization indicates deposition over an extended time period, rather than rapid catastrophic accumulation. (5) A tsunami interpretation fails to account for the effects of sea-level changes upon sediment deposition across the K-T transition. (6) A tsunami interpretation ignores the fact that microfossils and/or macrofossils generally indicate a sea-level lowstand at the time of breccia or clastic deposition, which was followed by rising seas. (7) Normal hemipelagic sedimentation of Maastrichtian age above the clastic deposits invalidates a K-T boundary age for these deposits. Ignoring these factors and prematurely placing an impact-generated tsunami interpretation upon clastic deposits that are selected primarily on the basis of their geographic proximity to the proposed Chicxulub impact crater inevitably leads to spurious scenario-driven interpretations.

Our biostratigraphic, lithological, and depositional analyses of K-T boundary transitions for which impact-generated clastic deposits have been claimed are illustrated in Figure 7 along with the sea-level curve derived from high-resolution studies of numerous K-T boundary transitions worldwide (see earlier discussion). In all sections where near K-T boundary clastic deposits were identified, the underlying unconformity is determined to be of latest Maastrichtian age (*P. hantkeninoides* zone, chron 29R) and coeval with the sea-level regression, as indicated by benthic foraminifera, ostracodes, invertebrates, and dinoflagellates (Keller, 1988b, 1992; Moshkovitz and Habib, 1993; Olsson and Liu, 1993; Savrda, 1993; Stinnesbeck and Keller, this volume; Stinnesbeck and others, 1993; Keller and others, 1993b; Schmitz and others, 1992).

This sea-level regression resulted in erosion and nondeposition that varied depending on the depositional environment (increased erosion landward). In the shallowest Gulf Coast sections (Braggs, Mussell Creek, Moscow Landing), sediment deposition (Clayton Sand) did not resume until the later Danian sea-level transgression (Mancini and others, 1989; Savrda, 1993; Moshkovitz and Habib, 1993), whereas in the Millers Ferry section it resumed earlier (Liu and Olsson, 1992), possibly because of its somewhat deeper water environment. Brazos River sections differ in that the clastic deposit is of Maastrichtian age (*P. hantkeninoides* zone), followed by shale deposition, interrupted only by a short interval of nondeposition at the P0–P1a boundary and by a short hiatus and glauconite deposition at the P1a–P1b boundary (Keller, 1989a, 1989b) coincident with short-term sea-level lowstands. This suggests that Brazos sections may have been located in somewhat deeper waters than the Alabama sections.

In the deeper water (400 m) northeastern and east-central Mexico sections, sediment deposition is similar to that of the Brazos section. The basal spherule-rich unit of the clastic deposits rests unconformably upon marls of the latest Maastrichtian Mendez Formation (*P. hantkeninoides* zone), coincident with the sea-level lowstand (Fig. 7). The top of the clastic deposit is also marked by a disconformity followed by a thin layer of Maastrichtian marls ranging from 5–10 cm thick in northeastern Mexico sections (Lajilla, Mulato, Parida) to 100 cm thick in southern Mexico (Bochil, Chiapas) (Keller and others, 1994a, 1994b; Montanari and others, l994; Macias Peréz, 1988). In most sections, the earliest Danian zone P0 and the lower part of zone P1a are missing, and a short hiatus is present at the P1a–P1b boundary, as also observed at the Brazos section. On the basis of available stratigraphic information, sediment deposition in the Haiti sections appears to be similar to that in northeastern Mexico (Fig. 7).

Stratigraphically, the platform breccia deposits of the Bochil section in Chiapas and the Poty section in Brazil are of late Maastrichtian (*P. hantkeninoides* zone) age. In these locations, the breccia beds are overlain by latest Maastrichtian and early Paleocene sediments. The Chilil section, which has also been claimed to have impact-related clastic deposits, in fact, contains no sediments of K-T boundary age because an erosion surface cuts from the upper Paleocene (zone P2) to the lower upper Maastrichtian *G. gansseri* zone.

Gulf of Mexico DSDP Sites 536 and 540 have also been claimed to contain K-T impact-generated megatsunami deposits (Alvarez and others, 1992). However, DSDP Site 540 has no sediment deposition between the middle Maastrichtian (*G. aegyptiaca* zone) and the early Paleocene zone P2. Likewise, DSDP Site 540 has a hiatus that spans from the middle or lower upper Maastrichtian to the early Danian zone P1a, followed by a second short hiatus at the zone P1a–P1b interval (Keller and others, 1993a).

Thus, Figure 7 shows that, based on current biostratigraphic information, not all near K-T clastic deposits are coeval. Their stratigraphic positions vary systematically with paleodepth and proximity to shorelines. In upper slope to outer shelf environments, clastic deposition began with the onset of the latest Maastrichtian sea-level regression following the global sea-level lowstand and unconformity. In the stratigraphically more complete sections, normal hemipelagic sedimentation resumed prior to the K-T boundary, such as at Lajilla, Mulato, and Parida in northeastern Mexico, the Mirador section near La Ceiba in east-central Mexico, the Bochil section in Chiapas, southern Mexico, the Beloc section in Haiti, the Poty section in Brazil, and the Brazos section in Texas. In these localities, sediment deposition during the latest Maastrichtian (after deposition of the clastic members) and early Danian occurs in generally deepening environments. Deposition is generally interrupted by short hiatuses or condensed intervals in the early Danian, indicating short-term sea-level lowstands at the P0–P1a and P1a–P1b boundaries. Deposition of clastic deposits in shallow neritic environments such as Braggs, Mussel Creek, Moscow Landing, and Millers Ferry in Alabama seems directly related to water depth. In these shallow-water sections, a major erosional unconformity is associated with the latest Maastrichtian sea-level lowstand, and sediment deposition (in this case of clastic deposits) resumes only with the early Danian sea-level rise (Fig. 7).

Our investigation thus concludes, in agreement with many previous studies, that variations in clastic deposition of near K-T boundary age are associated directly with a sea-level lowstand (Donovan and others, 1988; Baum and Vail, 1988; Mancini and others, 1989; Keller, 1989; Savrda, 1991, 1993; Habib and others, 1992; Moshkovitz and Habib, 1993). At this time, we see no convincing evidence that any of these clastic deposits are generated by a megatsunami wave as a result of a K-T boundary bolide impact on Yucatan. We remain open to this interpretation for at least some deposits (e.g., glass at Mimbral and Haiti). An impact origin for this glass, however, would indicate a pre-K-T boundary bolide impact for which there is no biotic, stratigraphic, or geochemical evidence at this time. Likewise, a volcanic origin for this glass suggests that major volcanic activity preceded the K-T boundary. In either case, these catastrophic events would have occurred during the last 200 k.y. of the Maastrichtian, and possibly coincided with the latest Maastrichtian sea-level lowstand, which influenced the nature of sediment deposition.

ACKNOWLEDGMENTS

We gratefully acknowledge support from National Science Foundation grants INT-9314080 and OCE-9021338, the Petroleum Research Fund ACS-PRF grant

26780-AC8, National Geographic Society grant 4620-91, and Conacyt Grant L120-36-36. We are very grateful to Steve Jacobson, Rosemary Askin, and John Cooper for critical reviews, advice, and numerous suggestions for improvement of this paper. Our investigations have also benefited from many discussions with colleagues, including Thierry Adatte, Bill Ward, Don Lowe, Ferran Colombo, and Norman MacLeod. Our gratitude to all of them for advice, critique, and a patient ear.

REFERENCES CITED

Adatte, T., Stinnesbeck, W., and Keller, G., 1994, Mineralogical correlations of near-K/T boundary deposits in northeastern Mexico: Evidence for long-term deposition and volcanoclastic influence, *in* New developments regarding the K/T event and other catastrophes in Earth history: Houston, Texas, Lunar and Planetary Institute Contribution 825, p. 1.

Adatte, T., Stinnesbeck, W., and Keller, G., 1995, Lithostratigraphic and mineralogical correlations of near K/T boundary clastic sediments in NE Mexico: Implications for origin and nature of deposition, *in* Ryder, G., Fastovsky, D., and Gartner, S., eds., New developments regarding the K/T event and other catastrophes in Earth history: Geological Society of America Special Publication (in press).

Albertao, G. A., Koutsoukos, E. A. M., Regali, M. P. S., Attrep, M., Jr., and Martins, P. P., Jr., 1994, The Cretaceous-Tertiary boundary in southern low-latitude regions: Preliminary study in Pernambuco, northeastern Brazil: Terra Nova, v. 6, p. 366–375.

Alvarez, W., Smit, J., Lowrie, W., Asaro, F., Margolis, S. V., Claeys, P., Kastner, M., and Hildebrand, A. R., 1992, Proximal impact deposits at the Cretaceous-Tertiary boundary in the Gulf of Mexico: A restudy of DSDP Leg 77 Sites 536 and 540: Geology, v. 20, p. 697–700.

Ashraf, A. R., and Stinnesbeck, W., 1988, Pollen und Sporen an der Kreide-Tertiärgrenze im Staate Pernambuco, NE Brasilien: Palaeontographica, v. 208, p. 39–51.

Askin, R. A., 1988, The palynological record across the Cretaceous/Tertiary transition on Seymour Island, Antarctica, *in* Feldmann, R. M., and Woodburne, M. O., eds., Geology and paleontology of Seymour Island, Antarctic Peninsula: Geological Society of America Memoir 169, p. 155–162.

Askin, R. A., 1992, Preliminary palynology and stratigraphic interpretation from a new Cretaceous/Tertiary boundary section from Seymour Island: Antarctic Journal of the United States, v. 25, p. 42–44.

Baum, G. R., and Vail, P. R., 1988, Sequence stratigraphic concepts applied to Paleogene outcrops, Gulf and Atlantic basins, *in* Wilgus, C. K., Hastings, B. S., Kendall, C. G., Posamentier, H. W., Ross, C., and Van Wagoner, J. C., eds., Sea-level changes: An integrated approach: Society of Economic Paleontologists and Mineralogists Special Publication 42, p. 309–328.

Baum, G. R., Vail, P. R., and Hardenbol, J., 1982, Unconformities and depositional sequences in relationship to eustatic sea-level changes, Gulf and Atlantic Coastal Plains: Interregional Geological Correlation Program, Project 174, International Field Conference: Baton Rouge, Louisiana State University, 3 p.

Barrera, E., 1994, Global environmental changes preceding the Cretaceous-Tertiary boundary: Early-late Maastrichtian transition: Geology, v. 22, p. 877–880.

Barrera, E., and Huber, B. T., 1990, Evolution of Antarctic waters during the Maastrichtian: Foraminifer oxygen and carbon isotope ratios, ODP Leg 113, *in* Proceedings of the Ocean Drilling Program, Scientific results, Volume 113: College Station, Texas, Ocean Drilling Program, p. 813–827.

Barrera, E., and Huber, B. T., 1991, Paleogene and early Neogene oceanography of the southern Indian Ocean: Leg 119 foraminifer stable isotope results, *in* Proceedings of the Ocean Drilling Program, Scientific results, Volume 119: College Station, Texas, Ocean Drilling Program, p. 693–717.

Beeson, D. C., 1992, High resolution palynostratigraphy across a marine Cretaceous-Tertiary boundary interval, Falls County, Texas [Ph.D. thesis]: University Park, Pennsylvania State University, 341 p.

Berggren, W. A., and Miller, K. G., 1988, Paleogene tropical planktic foraminiferal biostratigraphy and magnetobiochronology: Micropaleontology, v. 34, p. 362–380.

Berggren, W. A., Kent, D. V., and Flynn, J. J., 1985, Paleogene geochronology and chronostratigraphy, *in* Snelling, N. J., ed., The chronology of the geological record: Geological Society of London Memoir 10, p. 141–195.

Bleil, U., 1985, The magnetostratigraphy of Northwest Pacific sediments, Deep Sea Drilling Project Leg 86, *in* Initial reports of the Deep Sea Drilling Project, Volume 86: Washington, D.C., U.S. Government Printing Office, p. 444–458.

Blum, J. D., and Chamberlain, C. P., 1992, Oxygen isotope constraints on the origin of impact glasses from the Cretaceous-Tertiary boundary: Science, v. 257, p. 1104–1107.

Boersma, A., and Shackleton, N. J., 1981, Oxygen and carbon isotopic variations in planktonic foraminiferal depth habitats: Late Cretaceous to Paleocene, Central Pacific DSDP Sites 463 and 465, Leg 65, *in* Initial reports of the Deep Sea Drilling Project, Volume 65: Washington, D.C., U.S. Government Printing Office, p. 513–526.

Bohor, B. F., and Betterton, W. J., 1993, Arroyo el Mimbral, Mexico, K/T unit: Origin as debris flow/turbidite, not a tsunami deposit [abs.]: Lunar and Planetary Science, v. 24, p. 143–144.

Bohor, B. F., and Seitz, R., 1990, Cuban K/T catastrophe: Nature, v. 344, p. 593.

Bourgeois, J., Hansen, T. A., Wiberg, P. L., and Kauffman, E. G., 1988, A tsunami deposit at the Cretaceous-Tertiary boundary in Texas: Science, v. 241, p. 567–570.

Brinkhuis, H., and Leereveld, H., 1988, Dinoflagellate cysts from the Cretaceous/Tertiary boundary sequence of El Kef, N.W. Tunisia: Review of Paleobotany and Palynology, v. 56, p. 5–19.

Brinkhuis, H., and Zachariasse, W. J., 1988, Dinoflagellate cysts, sea level changes and planktonic foraminifera across the Cretaceous-Tertiary boundary at El Haria, northwest Tunisia: Marine Micropaleontology, v. 13, p. 153–190.

Brönnimann, P., and Rigassi, D., 1963, Contribution to the geology and paleontology of the area of the city of La Habana, Cuba, and its surroundings: Ecologae Geologicae Helvetiae, v. 56, p. 194.

Canudo, I., Keller, G., and Molina, E., 1991, K/T boundary extinction pattern and faunal turnover at Agost and Caravaca, SE Spain: Marine Micropaleontology, v. 17, p. 319–341.

Chave, A. D., 1984, Lower Paleocene–Upper Cretaceous magnetostratigraphy, Sites 525, 527, 528 and 529, Deep Sea Drilling Project Leg 74, *in* Initial reports of the Deep Sea Drilling Project, Volume 74: Washington, D.C., U.S. Government Printing Office, p. 525–532.

D'Hondt, S., and Keller, G., 1991, Some patterns of planktic foraminiferal assemblage turnover at the Cretaceous-Tertiary boundary: Marine Micropaleontology, v. 17, p. 77–118.

Dietz, R. S., and McHone, J., 1990, Isle of Pines (Cuba), apparently not K/T boundary impact site: Geological Society of America Abstracts with Programs, v. 22, no. 7, p. A79.

Donce, P., Jardine, S., Legoux, O., Masure, E., and Méon, H., 1985, Les évènements à la limite Crétacé-Tertiaire: au Kef (Tunisie septentrionale), l'analyse palynoplanctologique montre qu'un changement climatique est décelable à la base du Danian: Tunis, Tunisia, Actes du Premier Congrès National des Sciences de la Terre, p. 161–169.

Donce, P., Méon, H., Rocchia, R., Robin, E., and Groget, L., 1994, Biological changes at the K/T stratotype of El Kef (Tunisia), *in* New developments regarding the K/T event and other catastrophes in Earth history: Houston, Texas, Lunar and Planetary Institute Contribution 825, p. 30–31.

Donovan, A. D., Baum, G. R., Blechschmidt, G. L., Loutit, T. S., Pflum, C. E., and Vail, P. R., 1988, Sequence stratigraphic setting of the Cretaceous/Tertiary boundary in central Alabama, *in* Wilgus, C., K., Hastings, B. S., Kendall, C. G., Posamentier, H. W., Ross, C., and Van Wagoner, J. C., eds., Sea-level changes: An integrated approach: Society of Economic Paleontologists and Mineralogists Special Publication 42, p. 300–307.

Ekdale, A. A., and Stinnesbeck, W., 1994, Sedimentologic significance of trace fossils in KT "Mimbral Beds" of northeastern Mexico: Geological Society of America Abstracts with Programs, v. 26, no. 7, p. A395.

Elliot, D. H., Askin, R. A., Kyte, F. T., and Zinsmeister, W. J., 1994, Iridium and dinocysts at the Cretaceous-Tertiary boundary on Seymour Island, Antarctica: Implications for the K-T event: Geology, v. 22, p. 675–678.

Fernandez, G., Quintas, C., Sánchez, J. R. and Cobiella, R. J. L., 1991, El limite Cretacico-Terciario en Cuba: Revista Minera y Geología de Cuba, v. 8, p. 69–85.

Habib, D., 1994, Biostratigraphic evidence of the K/T boundary in the eastern Gulf Coastal Plain, north of the Chicxulub crater, *in* New developments regarding the K/T event and other catastrophes in Earth history: Houston, Texas, Lunar and Planetary Institute Contribution 825, p. 45.

Habib, D., Moshkowitz, S., and Kramer, C., 1992, Dinoflagellate and calcareous nannofossil response to sea-level change in Cretaceous-Tertiary boundary sections: Geology, v. 20, p. 165–168.

Hansen, T. A., Farrand, R., Montgomery, H., Billman, H., and Blechschmidt, G., 1987, Sedimentology and extinction patterns across the Cretaceous-Tertiary boundary interval in east Texas: Cretaceous Research, v. 8, p. 229–252.

Hansen, T. P., and Upshaw, B., 1990, Aftermath of the Cretaceous/Tertiary extinction: Rate and nature of early Paleocene molluscan rebound, *in* Kauffman, E. G., and Walliser, O. H., eds., Extinction events in Earth history: New York, New York, Springer Verlag, p. 402–409.

Hansen, T. P., Farrell, B.R., and Upshaw, B., 1993, The first 2 million years after the Cretaceous-Tertiary boundary in east Texas: rate and paleoecology of the molluscan recovery: Paleobiology, v. 19, p. 251–265.

Haq, B. U., Hardenbol, J., and Vail, P. R., 1987, Chronology of fluctuating sea levels since the Triassic: Science, v. 235, p. 1156–1166.

Herbert, T. D., and D'Hondt, S., 1990, Environmental dynamics across the Cretaceous-Tertiary extinction horizon measured 21 thousand year climate cycles in sediments: Earth and Planetary Science Letters, v. 99, p. 263–275.

Hildebrand, A. R., and Boynton, W. V., 1990, Proximal Cretaceous/Tertiary boundary impact deposits in the Caribbean: Science, v. 248, p. 843–847.

Hildebrand, A. R., Penfield, G. T., Kring, D. A., Pilkington, M., Camargo, A. Z., Jacobson, S. B., and Boynton, W. V., 1991, Chicxulub crater: A possible Cretaceous/Tertiary boundary impact crater on the Yucatan Peninsula, Mexico: Geology, v. 19, p. 867–869.

Hildebrand, A. R., and 14 others, 1994, The Chicxulub crater and its relation to the KT boundary ejecta and impact-wave deposits, *in* New developments regarding the K/T event and other catastrophes in Earth history: Houston, Texas, Lunar Planetary Institute Contribution 825, p. 49.

Hildebrand, A. R., Bonis, S., Smit, J., and Attrep, M., Jr., 1993, Cretaceous/Tertiary boundary deposits in Guatemala: Evidence for impact waves and slumping on a platform scale? A.C. IV Congreso Nacional de Paleontologia (Oct. 1993), Mexico City, Mexico: Sociedad Mexicana de Paleontologia, p. 133–137.

Hultberg, S. U., and Malmgren, B. A., 1986, Dinoflagellate and planktonic foraminiferal paleobathymetrical indices in the boreal uppermost Cretaceous: Micropaleontology, v. 32, p. 316–323.

Iturralde-Vincent, M. A., 1992, A short note on the Cuban late Maastrichtian megaturbidite (an impact derived deposit?): Earth and Planetary Science Letters, v. 109, p. 225–229.

Izett, G. A., 1990, Tektites in Cretaceous-Tertiary boundary rocks on Haiti and their bearing on the Alvarez extinction hypothesis: Journal of Geophysical Research, v. 96, p. 20,879–20,905.

Jéhanno, C., Boclet, D., Froget, L., Lambert, B., Robin, E., Rocchia, R., and Turpin, L., 1992, The Cretaceous-Tertiary boundary at Beloc, Haiti: No evidence for an impact in the Caribbean area: Earth and Planetary Science Letters, v. 109, p. 229–241.

Jiang, M. J., and Gartner, S., 1986, Calcareous nannofossil succession across the Cretaceous/Tertiary boundary in east-central Texas: Micropaleontology, v. 32, p. 232–255.

Keller, G., 1988a, Extinction, survivorship and evolution of planktic foraminifera across the Cretaceous/Tertiary boundary at El Kef, Tunisia: Marine Micropaleontology, v. 13, p. 239–263.

Keller, G., 1988b, Biotic turnover in benthic foraminifera across the Cretaceous/Tertiary boundary at El Kef, Tunisia: Palaeogeography, Palaeoclimatology, Palaeoecology, v. 66, p. 153–171.

Keller, G., 1989a, Extended Cretaceous/Tertiary boundary extinctions and delayed population changes in planktonic foraminifera from Brazos River, Texas: Paleoceanography, v. 4, p. 287–332.

Keller, G., 1989b, Extended period of extinctions across the Cretaceous/Tertiary boundary in planktonic foraminifera of continental shelf sections: Implications for impact and volcanism theories: Geological Society of America Bulletin, v. 101, p. 1408–1419.

Keller, G., 1991, Extended period of extinctions across the Cretaceous-Tertiary boundary in planktic foraminifera of continental shelf sections: Implications for impact and volcanism theories: Discussion and reply: Geological Society of America Bulletin, v. 103, p. 434–436.

Keller, G., 1992, Paleoecologic response of Tethyan benthic foraminifera to the Cretaceous-Tertiary boundary transition, *in* Takayanagi, Y., and Saito, T., eds., Studies in benthic foraminifera: Sendai, Tokyo University Press, p. 77–91.

Keller, G., 1993, The Cretaceous-Tertiary boundary transition in the Antarctic Ocean and its global implications: Marine Micropaleontology, v. 21, p. 1–45.

Keller, G., and Benjamini, C., 1991, Paleoenvironment of the eastern Tethys in the early Paleocene: Palaios, v. 6, p. 439–464.

Keller, G., MacLeod, N., Lyons, J. B., and Officer, C. B., 1993a, Is there evidence for Cretaceous/Tertiary boundary-age deep-water deposits in the Caribbean and Gulf of Mexico?: Geology, v. 21, p. 776–780.

Keller, G., Barrera, E., Schmitz, B., and Mattson, E., 1993b, Gradual mass extinction, species survivorship, and long-term environmental changes across the Cretaceous/Tertiary boundary in high latitudes: Geological Society of America Bulletin, v. 105, p. 979–997.

Keller, G., Stinnesbeck, W., Adatte, T., MacLeod, N., and Lowe, D. R., 1994a, Field guide to Cretaceous-Tertiary boundary sections in northeastern Mexico: Houston, Texas, Lunar and Planetary Institute Contribution 827, 110 p.

Keller, G., Stinnesbeck, W., and Lopez-Oliva, J. G., 1994b, Age, deposition and biotic effects of the Cretaceous/Tertiary boundary event at Mimbral NE Mexico: Palaios, v. 9, p. 144–157.

Koeberl, C., 1994, Deposition of channel deposits near the Cretaceous-Teriary boundary in northeastern Mexico: Catastrophic or "normal" sedimentary deposits?: Comment: Geology, v. 22, p. 957.

Koeberl, C., and Sigurdsson, H., 1992, Geochemistry of impact glasses from the K/T boundary in Haiti: Relation to smectites and new types of glass: Geochimica et Cosmochimica Acta, v. 56, p. 2113–2129.

Liu, G., and Olsson, R. K., 1992, Evolutionary radiation of microperforate planktonic foraminifera following the K/T mass extinction event: Journal of Foraminiferal Research, v. 22, p. 328–346.

Longoria, J. F., and Gamper, M. A., 1992, Planktonic foraminiferal biochronology across the KT boundary from the Gulf coastal plain of Mexico: Implications for timing the extraterrestrial bolide impact in the Yucatan: Boletín de la Asociación Mexicana de Geólogos Petroleros, v. 42, p. 19–40.

Longoria, J. F., and Grajales Nishimura, J. M., 1993, The Cretaceous/Tertiary event in Mexico, field trip guide to selected KT boundary localities in Tamaulipas and Nuevo León, northeast Mexico: Mexico City, Sociedad Mexicana de Paleontología, 93 p.

Lopez-Oliva, J. G., and Keller, G., 1994, Biotic effects of the K/T boundary event in northeastern Mexico, *in* New developments regarding the K/T event and other catastrophes in Earth history: Houston, Texas, Lunar and Planetary Institute Contribution 825, p. 72–73.

Lopez-Oliva, J. G. and Keller, G., 1995, Age and stratigraphy of near-K-T boundary siliciclastic deposits in northeastern Mexico, *in* Ryder, G., Fastovsky, D., and Gartner, S., eds., New developments regarding the K/T event and other catastrophes in Earth history: Geological Society of America Special Publication (in press).

Lyons, J. B., and Officer, C. B., 1992, Mineralogy and petrology of the Haiti Cretaceous/Tertiary section: Earth and Planetary Science Letters, v. 109, p. 205–224.

Macellari, C. E., 1986, Late Campanian-Maastrichtian ammonite fauna from Seymour Island (Antarctic Peninsula): Journal of Paleontology Memoir 18, 55 p.

Macellari, C. E., 1988, Stratigraphy, sedimentology and paleoecology of Upper Cretaceous/Paleocene shelf-deltaic sediments of Seymour Island (Antarctic Peninsula), *in* Feldmann, R. M., and Woodburne, M. O., eds., Geology and paleontology of Seymour Island, Antarctic Peninsula: Geological Society of America Memoir 169, p. 25–53.

Macias Pérez, F. J., 1988, Estratigrafia detallada del límite Cretácico Terciario en Xicotepec, Puebla: una alternative al catastrofismo [Ph.D. thesis]: Mexico City, Instituto Politécnico Nacional, 112 p.

MacLeod, N., and Keller, G., 1991a, Hiatus distribution and mass extinction at the Cretaceous/Tertiary boundary: Geology, v. 19, p. 497–501.

MacLeod, N., and Keller, G., 1991b, How complete are Cretaceous/Tertiary boundary sections? A chronostratigraphic estimate based on graphic correlation: Geological Society of America Bulletin, v. 103, p. 1439–1457.

MacLeod, N., and Keller, G., 1994, Comparitive biogeographic analysis of planktic foraminiferal survivorship across the Cretaceous/Tertiary (K/T) boundary: Paleobiology, v. 20, p. 143–177.

Maddocks, R. F., 1984, Ostracoda of continuous and discontinuous Cretaceous-Tertiary contact sections in central Texas: Geological Society of America Abstracts with Programs, v. 16, p. 106.

Mancini, E. A., Tew, B., and Smith, C. C., 1989, Cretaceous-Tertiary contact, Mississippi and Alabama: Journal of Foraminiferal Research, v. 19, p. 93–104.

Masters, B., 1984, Comparison of planktonic foraminifers at the Cretaceous-Tertiary boundary from the El Haria shale (Tunisia) and the Esna shale (Egypt). Proceedings of the Seventh Exploration Seminar: Cairo, Egyptian General Petroleum Corporation, p. 310–324.

Masters, B., 1993, Re-evaluation of the species and subspecies *Plummerita* Brönimann and a new species of *Rugoglobigerina* Brönimann (Foraminiferida): Journal of Foraminiferal Research, v. 23, p. 267–274.

Maurasse, F. J-M., and Sen, G., 1991, Impacts, tsunamis, and the Haitian Cretaceous-Tertiary boundary layer: Science, v. 252, p. 1690–1693.

Michaud, F., and Fourcade, E., 1989, Stratigraphie et paléogéographie du Jurassique et du Crétacé du Chiapas (Sud-Est du Mexique): Bulletin de la Société Géologique de France, v. 8, p. 639–650.

Montanari, A., Claeys, P., Asaro, F., Bermudez, J., and Smit, J., 1994, Preliminary stratigraphy and iridium and other geochemical anomalies across the K/T boundary in the Bochil section (Chiapas, southeastern Mexico), *in* New developments regarding the K/T event and other catastrophes in Earth history: Houston, Texas, Lunar and Planetary Institute Contribution 825, p. 84.

Montgomery, H., Pessagno, E., Soegaard, K., Smith, C., Muñoz, I., and Pessagno, J., 1992, Misconceptions concerning the Cretaceous/Tertiary boundary at the Brazos River, Falls County, Texas: Earth and Planetary Science Letters, v. 109, p. 593–600.

Moshkovitz, S., and Habib, D., 1993, Calcareous nannofossil and dinoflagellate stratigraphy of the Cretaceous-Tertiary boundary, Alabama and Georgia: Micropaleontology, v. 39, p. 167–191.

Ocampo, A. C., and Pope, K. O., 1994, A K/T boundary section from northern Belize, *in* New developments regarding the K/T event and other catastrophes in Earth history: Houston, Texas, Lunar and Planetary Institute Contribution 825, p. 86.

Olsson, R. K., and Liu, G., 1993, Controversies on the placement of the Cretaceous-Paleogene boundary and the K/T mass extinction of planktonic foraminifera: Palaios, v. 8, p. 127–139.

Pessagno, E. A., Jr., Longoria, J. F., Montgomery, H., and Smith, C., 1990, Misconceptions concerning the Cretaceous-Tertiary boundary, Brazos River, Falls County, Texas: Geological Society of America Abstracts with Programs, v. 22, no. 7, p. A277.

Peryt, D., Lahodynsky, R., Rocchia, R., and Boclet, D., 1993, The Cretaceous/Paleogene boundary and planktonic foraminifera in the Flyschgosau (Eastern Alps, Austria): Palaeogeography, Palaeoclimatology, Palaeoecology, v. 104, p. 239–252.

Peypouquet, J. P., Grousset, F., and Mourguiart, P., 1986, Paleoceanography of the Mesogean Sea based on ostracods of the northern Tunisian continental shelf between the Late Cretaceous and early Paleogene: Geologische Rundschau, v. 75, p. 159–174.

Pope, K. O., Ocampo, A. C., and Duller, C. E., 1991, Mexican site for K/T impact crater: Nature, v. 351, p. 105–108.

Posamentier, H. W., and Vail, P. R., 1988, Eustatic controls on clastic deposition II, *in* Wilgus, C. K., Hastings, B. S., Kendall, C. G., Posamentier, H. W., Ross, C., and Van Wagoner, J. C., eds., Sea-level changes: An integrated approach: Society of Economic Paleontologists and Mineralogists Special Publication 42, p. 125–153.

Posamentier, H. W., Jervey, M. T., and Vail, P. R., 1988, Eustatic controls on clastic deposition I, *in* Wilgus, C. K., Hastings, B. S., Kendall, C. G., Posamentier, H. W., Ross, C., and Van Wagoner, J. C., eds., Sea-level changes: An integrated approach: Society of Economic Paleontologists and Mineralogists Special Publication 42, p. 109–123.

Pszolkowski, A., 1986, Megacapas del Maestrichtiano en Cuba occidental y central: Bulletin of the Polish Academy of Sciences, Earth Sciences, v. 34, p. 81–94.

Quezada Muñeton, J. M., 1990, El Cretácio Medio-Superior, y el Limite Cretácio Superior-Terciario Inferior en la Sierra de Chiapas: Boletin de la Asociación Mexicana de Geólogos Petroleros, v. 39, p. 3–98.

Robin, E., Rocchia, R., Lyons, J. B., and Officer, C. B., 1994, Deposition of channel deposits near the Cretaceous-Teriary boundary in northeastern Mexico: Catastrophic or "normal" sedimentary deposits?: Reply: Geology, v. 22, p. 958.

Savrda, C. E., 1991, Ichnology in sequence stratigraphic studies: An example from the Lower Paleocene of Alabama: Palaios, v. 6, p. 39–53.

Savrda, C. E., 1993, Ichnosedimentologic evidence for a noncatastrophic origin of Cretaceous-Tertiary boundary sands in Alabama: Geology, v. 21, p. 1075–1078.

Schmitz, B., Keller, G., and Stenvall, O., 1992, Stable isotope and foraminiferal changes across the Cretaceous/Tertiary boundary at Stevns Klint, Denmark: Arguments for long-term oceanic instability before and after bolide impact: Palaeogeography, Palaeoclimatology, Palaeoecology, v. 96, p. 233–260.

Sharpton, V. L., Dalrymple, G. B., Marin, L. E., Ryder, G., Schuraytz, B. C., and Urrutia-Fucugauchi, J., 1992, New links between the Chicxulub impact structure and the Cretaceous/Tertiary boundary: Science, v. 359, p. 819–821.

Sharpton, V. L., Burke, K., Camargo-Zanoguera, A., Hall, S. A., Lee, S., Marin, L., Suárez-Reynoso, G., Quezada-Muñeton, J. M., Spudis, P. D., and Urrutia-Fucugauchi, J., 1993, Chicxulub multiring impact basin: Size and other characteristics derived from gravity analysis: Science, v. 261, p. 1564–1567.

Sigurdsson, H., D'Hondt, S., Arthur, M. A., Bralower, T. J., Zachos, J. C., Fossen, M., and Channell, J. E. T., 1991, Glass from the Cretaceous/Tertiary boundary in Haiti: Nature, v. 349, p. 482–486.

Smit, J., 1982, Extinction and evolution of planktonic foraminifera after a major impact at the Cretaceous/Tertiary boundary, *in* Silver, L.T., and Schultz, P. H., eds., Geological implications of impacts of large asteroids and comets on the Earth: Geological Society of America Special Paper 190, p. 329–352.

Smit, J., Montanari, A., Swinburne, N. H. M., Alvarez, W., Hildebrand, A., Margolis, S. V., Claeys, P., Lowrie, W., and Asaro, F., 1992, Tektite bearing deep-water clastic unit at the Cretaceous-Tertiary boundary in northeastern Mexico: Geology, v. 20, p. 99–103.

Smit, J., Roep, T. B., Alvarez, W., Claeys, P., and Montanari, A., 1994a, Cretaceous-Tertiary boundary sediments in northeastern Mexico and the Gulf of Mexico: Comment: Geology, v. 22, p. 953–954.

Smit, J., Roep, T. B., Alvarez, W., Montanari, A., and Claeys, P., 1994b, Stratigraphy and sedimentology of K/T clastic beds in the Moscow Landing (Alabama) outcrop: Evidence for impact related earthquakes and tsunamis, *in* New developments regarding the K/T event and other catastrophes in Earth history: Houston, Texas, Lunar and Planetary Institute Contribution 825, p. 119.

Smith, C. A., and Pessagno, E. A., Jr., 1973, Planktonic foraminifera and stratigraphy of the Corsicana Formation (Maastrichtian) north-central Texas: Cushman Foundation for Foraminiferal Research Special Publication 12, 68 p.

Stinnesbeck, W., 1989, Fauna y microflora en el limite Cretácio-Terciario en el estado de Pernambuco, Noreste de Brasil: Contribucionas a los Simposios sobre el Cretacio de America Latina, Parte A: Eventos y Registro Sedimentario, Buenos Aires, Argentina, p. 215–230.

Stinnesbeck, W., Barbarin, J. M., Keller, G., Lopez-Oliva, J. G., Pivnik, D. A., Lyons, J. B., Officer, C. B., Adatte, T., Graup, G., Rocchia, R., and Robin, E., 1993, Deposition of channel deposits near the Cretaceous-Tertiary boundary in northeastern Mexico: Catastrophic or "normal" sedimentary deposits?: Geology, v. 21, p. 797–800.

Stinnesbeck, W., Keller, G., and Adatte, T., 1994a, K/T boundary sections in southern Mexico (Chiapas): Implications for the proposed Chicxulub impact site, *in* New developments regarding the K/T event and other catastrophes in Earth history: Houston, Texas, Lunar and Planetary Institute Contribution 825, p. 120–121.

Stinnesbeck, W., Keller, G., Adatte, T., and MacLeod, N., 1994b, Cretaceous-Tertiary sediments in NE Mexico and the Gulf of Mexico: Reply: Geology, v. 22, p. 955–956.

Stott, L. D., and Kennett, J. P., 1990, The paleoceanographic and climatic signature of the Cretaceous/Paleogene boundary in the Antarctic: Stable isotopic results from ODP Leg 113, *in* Proceedings of the Ocean Drilling Program, Scientific results, Volume 113: College Station, Texas, Ocean Drilling Program, p. 829–848.

Surlyk, F., 1979, Guide to Stevns Klint, *in* Birkelund, T., and Bromley, R. G., eds., Cretaceous-Tertiary boundary events, I. The Maastrichtian and Danian of Denmark: University of Copenhagen, p. 164–170.

Swisher, C. C., Grajales, Nishimura, J. M., Montanari, A., Margolis, S. V., Claeys, P., Alvarez, W., Renne, P., Cedillo-Pardo, E., Maurasse, F. J-M., Curtis, G. H., Smit, J., and McWilliams, M. O., 1992, Coeval ^{40}Ar/^{39}Ar ages of 65.0 million years ago from Chicxulub crater melt rock and Cretaceous-Tertiary boundary tektites: Science, v. 257, p. 954–958.

Van Wagoner, J. C., Mitchum, R. M., Campion, K. M., and Rahmanian, V. D., 1990, Siliclastic sequence stratigraphy in well logs, cores, and outcrops: Concepts for high-resolution correlation of time and facies: American Association of Petroleum Geologists Methods in Exploration Series, no. 7, 53 p.

Worzel, J. L., Bryant, W., and others, 1973, Initial reports of the Deep Sea Drilling Project, Volume 10: Washington, D.C., U.S. Government Printing Office, 747 p.

Wrenn, J. H., Stein, J. A., Breard, S. Q., and White, R. J., 1989, Biostratigraphy of the Cretaceous-Tertiary boundary, Brazos River section, Texas: Palynology, v. 13, p. 288–289.

Zinsmeister, W. J., Feldmann, R. M., Woodburne, M. O., and Elliot, D. H., 1989, Latest Cretaceous/earliest Tertiary transition on Seymour Island, Antarctica: Journal of Paleontology, v. 63, p. 731–738.

18

Environmental Changes across the Cretaceous-Tertiary Boundary in Northeastern Brazil

Wolfgang Stinnesbeck, *Facultad de Ciencias de la Tierra, Universidad Autónoma de Nuevo León, Linares, México,*
and Gerta Keller, *Department of Geological and Geophysical Sciences, Princeton University, Princeton, New Jersey*

INTRODUCTION

One of the most fascinating debates in the geosciences centers around causal mechanisms for the global mass extinctions at the end of the Cretaceous; whether extraterrestrial and catastrophic, or Earth-derived and gradual. The catastrophic hypothesis suggests that a bolide collided with the Earth 65 Ma (Alvarez and others, 1980). In this scenario, the dust carried aloft by the impact blocked out sunlight, suppressed photosynthesis for weeks or months, and thereby interrupted the food chain, leading to the outright extinction of most microplankton and affecting all higher organisms, including invertebrates and dinosaurs. Under this hypothesis the fossil record should reveal a stratigraphically instantaneous global mass extinction. The alternative hypothesis requires no extraterrestrial intervention. Instead, long-term environmental changes, including volcanism, sea-level fluctuations, and associated climate change are regarded as triggers for major environmental instabilities leading to a mass extinction (Courtillot and Cisowski, 1987; Courtillot, 1994). In this scenario, the fossil record should reveal an extended and selective species extinction pattern that is not necessarily global at any one instant. Unfortunately, the existing paleontological database that must be used to test these hypotheses is still poor for most groups other than planktonic foraminifera. Nevertheless, most microfossil groups show no major extinctions across the Cretaceous-Tertiary (K-T) boundary (e.g., diatoms, radiolarians, dinoflagellates, ostracodes, and benthic foraminifera; Hollis, 1993; Méon, 1990; Donce and others, 1985; Keller, 1992; Kaiho, 1992). For the vast majority of invertebrate and vertebrate taxa, including dinosaurs, the current database is insufficient to document the short-term extinction-related phenomena across the K-T boundary.

Only few and geographically isolated data points exist from South America for the K-T boundary mass extinction (Stinnesbeck, 1986): among these, the Poty Quarry in northeastern Brazil contains the most complete K-T boundary transition known to date. Both microfossils and invertebrates are abundant, providing the opportunity for high-resolution geochemical, petrographic, biostratigraphic, and extinction studies.

We report here on the lithology, sedimentology, biostratigraphy, and the depositional environment across the K-T boundary at the Poty section in northeastern Brazil. Planktonic foraminiferal biostratigraphy suggests that sediment accumulation across the K-T boundary was relatively continuous at this location. Lithologic and faunal studies indicate two major environmental changes, one below and the other across the boundary horizon. These appear to be related to sea-level fluctuations that had profound effects on the local depositional environment.

LITHOSTRATIGRAPHY

Outcrops of K-T boundary sediments occur locally within the Pernambuco-Paraiba coastal basin. One of the best exposures is located in the Poty cement quarry near the village of Paulista, northeast of Olinda (Fig. 1).

The lithological column of the Poty Quarry outcrop is shown in Figure 2. Here the Gramame Formation reaches a maximum of 9.5 m, underlying ~5 m of the Maria Farinha Formation. Both the Gramame and the Maria Farinha formations are characterized by alternating thinly bedded marls, calcareous marls, and marly limestones that are ash colored and show variable amounts of bioturbation, principally by callianassid crabs. The carbonate content of the sediment varies gradually and lithological units are separated by erosion surfaces. Macrofossils are

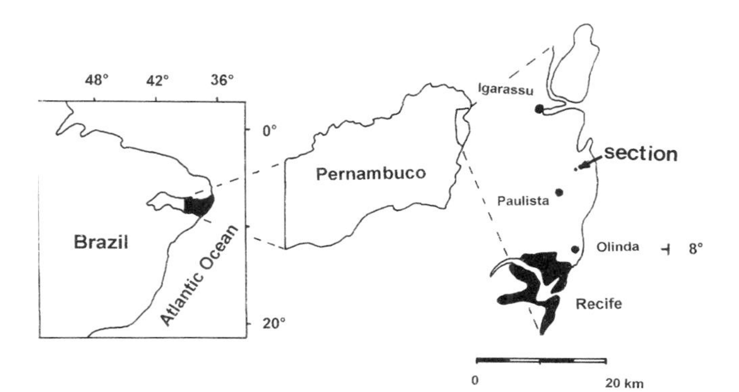

Figure 1. Location of the Poty K-T boundary section in northeastern Brazil.

abundant, and gastropods, cephalopods, bivalves, crustaceans, echinoids, annelids, and fish debris predominate. The ichnofauna is dominated by *Thalassinoides*, which forms large burrowing networks on various bedding surfaces. The microfaunas of the Gramame Formation are principally calcispheres and planktonic foraminifera, whereas benthic foraminifera and ostracodes are more abundant in the overlying Maria Farinha Formation.

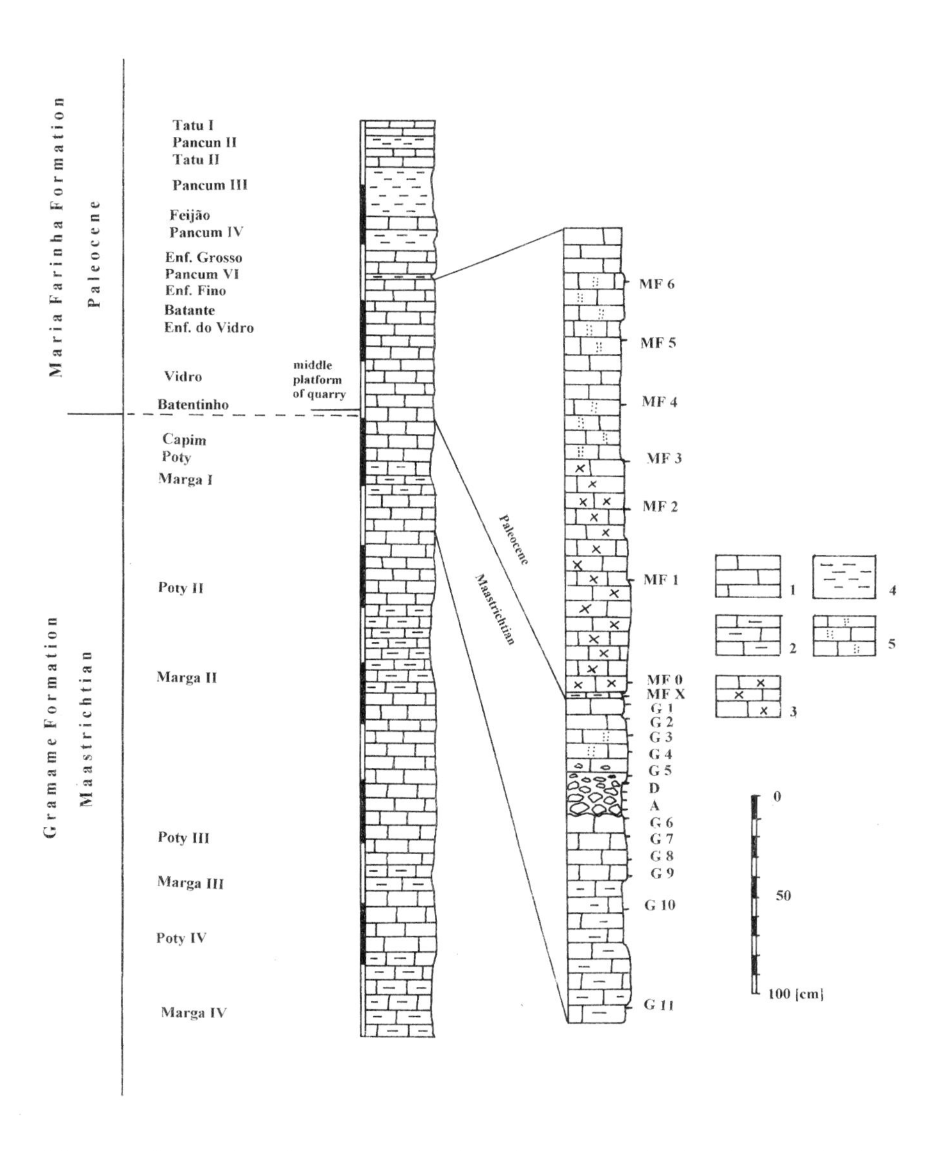

Figure 2. Lithostratigraphic column of the Poty Quarry including local names of individual sediment units and sampled horizons. 1: limestones; 2: marly limestone; 3: recrystallized limestone; 4: shales; 5: sandy limestone.

Figure 3 shows the K-T transition beginning with the bioturbated limestone (unit A), followed by an 80-cm-thick micritic limestone (unit B). These limestones contain variable abundances of planktonic foraminifera and calcispheres (samples G12 to G6, Table 1). Globotruncanids are rare, whereas rugoglobigerinids and heterohelicids are common along with echinoids (*Hemiaster*) and annelids (*Hamulus*). This micritic limestone (unit B) is unconformably overlain by a 20-cm-thick breccia bed (unit C) that contains reworked components of the biomicritic limestones with calcispheres and planktonic foraminifera. This breccia unit also contains bones, phosphatic lumps, phosphatized foraminifera, glauconite, small pyrite concretions, and abundant serpulid worm tubes and inner neritic benthic foraminifera.

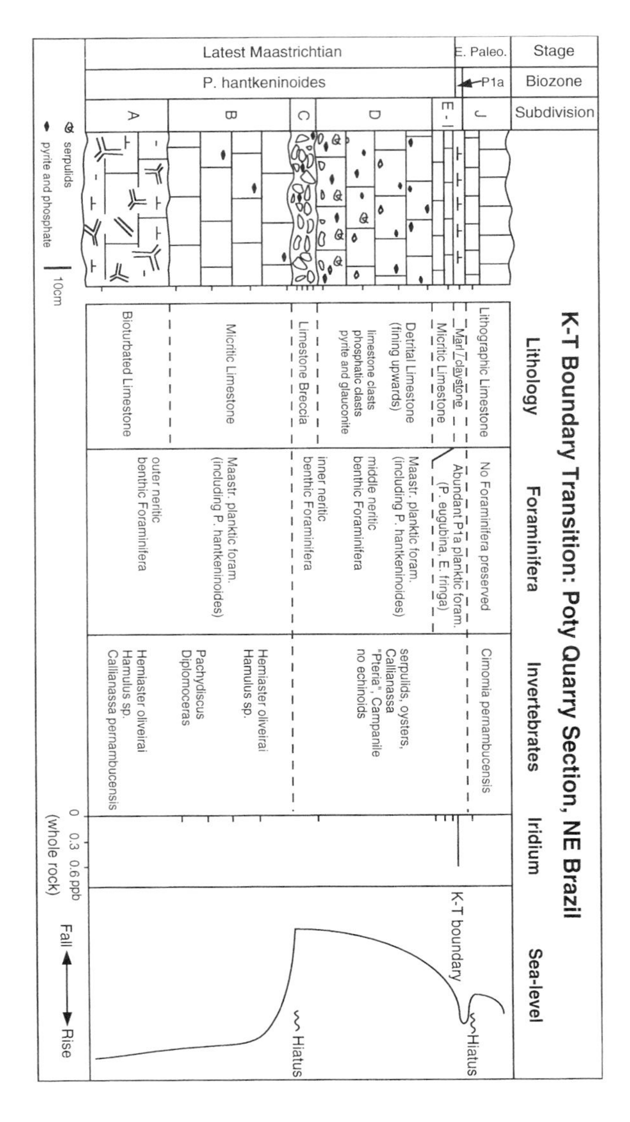

Figure 3. K-T boundary transition at Poty showing lithological column, faunal events, and sea-level changes. Iridium data from Albertao and other (1994).

The breccia bed (samples G5 to G1, Table l) is overlain by a 50-cm-thick detrital limestone (unit D) with limestone clasts that diminish upward in size from rudite to arenite. This unit is overlain by a l0-cm-thick micritic limestone (unit E), which is similar to the micritic limestone below the breccia layer (Fig. 3). Pyrite and phosphatic lumps are still present within this interval. Megafaunal constituents consist of serpulids, occasional oysters, pteriid bivalves, and gastropods, whereas echinoids are generally absent. Foraminiferal faunas consist of abundant phosphatized middle neritic benthic foraminifera and common poorly preserved Maastrichtian heterohelicids, pseudoguembelinids, hedbergellids, globigerinellids and rugoglobigerinellids (samples G5-G2, Table 1). In the topmost 10 cm, the sediments are more marly, benthic foraminifers show a deepening environment, and Maastrichtian planktonic foraminifera are more abundant and well preserved, marking an influx of species including the deeper dwelling globotruncanids. Also, in sample G0, calcite spherules (probably algal resting cysts) are common. This marl layer, with its influx of warm-water and deeper-dwelling species and algal remains, marks the sea-level transgression and probable warming that was observed at the end of the Maastrichtian in the Danish sections (Keller and others, l993; Schmitz and others, l992). The K-T boundary appears to be between the top of the micritic limestone (unit E) and the overlying 5-cm-thick marl-claystone layer (unit I) that contains abundant Tertiary (zone P1a) planktonic foraminifera, including the index species *Parvularugoglobigerina eugubina*. An iridium enrichment was reported from this marl-claystone layer by Albertao and others (l994).

Dense layers of recrystallized limestones (unit J, banco vidro, samples MF1 to MF3, Table l) form the base of the Paleocene Maria Farinha Formation. These limestone layers are recognized easily in the outcrop by their resistance to weathering. Few fossils were identified in these limestone layers, and these are micritized, detectable only as shadowlike discolorations. Isolated vertical burrows of unknown origin contain abundant benthic foraminifera with shells that are partly phosphatized. Bioturbation of the entire sedimentary column is present above sample MF3 (Table l). Upsection (samples M4 to MF7, Table l) detrital limestones are rich in echinoderms, bivalves, gastropods, and chlorophycean algae.

BIOSTRATIGRAPHY AND PALEONTOLOGY

Samples for micropaleontological analyses were collected beginning 3 m below to 3 m above the K-T boundary (see above). Samples from 1 m below the boundary to the "banco vidro" (~5 cm above the boundary) were taken at 5 to 10 cm intervals, as compared to the 20 to 50 cm intervals for the remainder of the section. In addition, megafossils were collected stratigraphically layer by layer.

Foraminifera

The planktonic foraminiferal zonation used in this paper is that of Keller (1993), with the addition of the new *Plummerita hantkeninoides* zone that marks the last l70–200 k.y. of the Maastrichtian (Pardo and others, this volume). The presence of *P. hantkeninoides* along with characteristic late Maastrichtian planktonic

Table 1
Species ranges and relative abundances of planktonic foraminifera during the latest Maastrichtian at Poty Quarry, Brazil

Samples	G10	G9	G8	G7	G6	GA
Globigerinelloides aspera						
Guembelitria cretacea						
Globotruncanella monmouthensis	R		R			R
Globotruncanella petaloidea	C	E	R	R	R	R
Hedbergella monmouthensis						R
Rosita contusa	C					
Archaeoglobigerina cretacea			R		R	
Globotruncana aegyptiaca	C	R	F	R	F	R
Globotruncana duwi		R				
Globotruncana insignis	R	R		R	R	
Globotruncana rosetta	C	R	F	R	F	R
Globotruncanita stuarti	R	R				R
Globotruncana arca	R					
Heterohelix globulosa	C	C	C	F	F	R
Heterohelix striata						
Pseudoguembelina costulata	F		F	R	F	R
Pseudoguembelina palpebra	C	F	R	R	F	R
Pseudoguembelina punctulata	F	F		R	F	R
Pseudoguembelina kempensis			R			R
Planoglobulina carseyae	F				F	R
Pseudotextularia elegans	C	F	R		R	
Pseudotextularia deformis	F					
Planoglobulina brazoensis				R		
Rugoglobigerina hexacamerata	F	R	R	F		R
Rugoglobigerina macrocephala	C	C	C	C	C	R
Rugoglobigerina rugosa	C	C	C	C	C	R
Rugoglobigerina pennyi	F	F	F		F	
Rugoglobigerina scotti	C	R	F	F	F	R
Plummerita hantkenoides	C	R	F	R		C
Racemiguembelina powelli	R					
Racemiguembelina intermedia						
Rosita patelliformis	R	R				

Note: C = Common, F = Few, R = Rare.

foraminiferal assemblages in units B to E marks deposition during the last 200 k.y. years of the Maastrichtian. Albertao and others (1994) reported the presence of rare Tertiary species within the breccia and overlying sediments and therefore placed the K-T boundary at the base of the breccia layer. We could not confirm the presence of Tertiary species in these sediments (units B to E) despite repeated sample washings and analyses. There are several possible reasons for this discrepancy, including a disconformity where erosion is variable over the distance sampled by us and Albertao and others (1994), as well as misidentification of species and sample contamination. The latter was, in fact, the case for an earlier interpretation by Stinnesbeck (1989), who placed the K-T boundary at the breccia layer.

Koutsoukos (1995) stated that Danian species are so rare in samples of units C to E that he searched a single sample for two to three days in order to find a single Danian specimen. Admittedly, we did not search beyond a couple of hours

TABLE 1 (CONTINUED)

GB	GC	GD	G5	G4	G3	G2	G1	G0	MX
	R			R					R
							A		R
							R	R	
	R			R				R	R
			R						
								R	R
R			R						
R	R	R	R	R		R	R	R	C
							R		R
R	R	R		R			R	R	C
	R					R	R	R	C
R	R	R		R		R	R		
R	R	R					R	R	F
R	R	R		R		R	R	R	F
					R	R	R	R	R
	R		R				R		F
	R				R		R	R	R
R			R	R		R	R	R	R
R	R		R	R	R	R	R	R	R
								R	
R			R			R	R	R	R
R		R					R		R
			R						

per sample, which is a higher time investment than ordinarily spent for biostratigraphic analysis. When Danian species are as rare as Koutsoukos indicates, one can never know whether the single specimen found was not the result of contamination. As a working principal it is best not to base biostratigraphic interpretations on single isolated specimens.

Albertao and others (1994) reported the presence of such rare and isolated specimens of *Woodringina hornerstownensis*, *W. claytonensis*, *Eoglobigerina fringa*, and *Guembelitria irregularis* within an abundant late Maastrichtian fauna in breccia unit C. However, *G. irregularis* originated in the late Maastrichtian, and illustrations of the other morphotypes (shown at a meeting in Angers, France, in July 1994) suggested that they may be *Heterohelix globulosa*, *Chiloguembelina waiparaensis*, and *Hedbergella monmouthensis*, three small Maastrichtian species. In the overlying bed D, Albertao and others (1994) reported very rare additional Danian species, *Parasubbotina pseudobulloides*, *Eoglobigerina eobulloides*, *Praemurica taurica*, and *Parvularugoglobigerina eugubina*. Our investigation of the same stratigraphic interval failed to find any Danian species, but we did find the presence of abundant heterohelicids, hedbergellids, and globigerinellids, which were not recorded as being present by Albertao and others (1994). It appears that these authors may have considered the small hedbergellid species, including the four-chambered morphotype of *H. monmouthensis*, as the somewhat similar Danian species *P. taurica*, *P. eugubina*, *P. subbotina*, and *E. eobulloides*. Alternatively, it is possible that the breccia unit and overlying marly limestone layers are disconformable over the region with Tertiary sediments overlying breccia in the area sampled by Albertao and others (1994). Our investigation of this stratigraphic interval reveals a diverse and typical Maastrichtian assemblage with 32 species, including the index species *P. hantkeninoides*. This indicates that the breccia unit C was deposited during the latest Maastrichtian, but prior to the K-T boundary, as indicated by the overlying 70-cm-thick micritic and detrital limestone layers (Fig. 3; Stinnesbeck and Keller, 1996).

We found the first Tertiary planktonic foraminifers, including *P. eugubina* and *E. fringa*, in the 5-cm-thick marl-claystone (unit I). On the basis of the biostratigraphy, the K-T boundary is thus placed between the micritic limestone (unit E) and the overlying claystone-marl (unit I, Fig. 3). Placement of the K-T boundary at this interval is supported by the presence of an iridium anomaly in unit I, as reported by Albertao and others (1994). Thus, the K-T boundary at the Poty section is marked by biomarkers and an iridium anomaly similar to those in complete K-T sections worldwide. We find no evidence in support of Albertao and others (1994) suggestion that this iridium anomaly represents a second Danian bolide impact. The restricted presence of *P. eugubina* within the claystone-marl layer (zone Pla, unit I) and the absence of the lowermost Danian zone P0, however, strongly indicate that the K-T boundary is incomplete, with zone P0 and part of zone P1a missing.

Above the claystone-marl layer of unit I, which represents zone Pla, *Subbotina pseudobulloides* is commonly present up to 120 cm above the K-T boundary. In addition, rare specimens of *Guembelitria* spp., *Globanomalina taurica*, *Subbotina pseudobulloides*, *S. varianta*, and *Morozovella inconstans* are present and indicate a zone Plb-Plc interval.

The depositional environment of the Poty section can be inferred from benthic and planktonic foraminifera. Planktonic foraminifera below the breccia layer suggest middle to outer neritic depths, as indicated by the presence of few globotruncanids that thrive in deeper waters and the abundance of shallow-water surface dwellers (e.g., hedbergellids, heterohelicids, rugoglobigerinids). *Abathomphalus mayaroensis*, the upper Maastrichtian index species, is absent from the late Maastrichtian of the Poty section. Because *A. mayaroensis* is a deep-water dweller (Barrera and Huber, 1990), this absence is probably due to the relatively shallow depth of the Poty section, similar to the absence of this species in the relatively shallow water sediments of El Kef (Tunisia), Brazos River (Texas), and Nye Kløv and Stevns Klint (Denmark) (Keller, 1988, 1989; Keller and others, 1993; Schmitz and others, 1992).

Benthic foraminifera also indicate deposition in middle to outer neritic environments below the breccia unit, as suggested by abundant *Discorbinella*, *Lenticulina*, and few *Nodosaria*, *Globulina*, and *Vaginulinopsis*. The breccia unit is marked by a sudden influx of benthic foraminifera typical of inner neritic environments, including abundant *Discorbinella burlingtonensis*, *Lenticulina midwayensis* along with a few *Quinqueloculina*, *Alabamina*, *Nonionella*, and *Elphidium* spp. Planktonic foraminifera are few to rare and include the surface-dwelling taxa *Rugoglobigerina* and *Heterohelix*.

Above the breccia and across the K-T boundary (units C to I), benthic foraminifera indicate deepening to middle and outer neritic depths, as suggested by the presence of a more diverse assemblage dominated by *Anomalinoides acuta*, common *Alabamina midwayensis*, *Discorbinella burlingtonensis*, and *Lenticulina midwayensis,* and few *Cibicidoides*, *Nonionella*, *Gaudryina*, *Spiroplectammina*, and *Dentalina*. Planktonic foraminiferal assemblages are also more diverse in the top 10 cm of the marly limestone below the marl-claystone layer, and include abundant rugoglobigerinids, heterohelicids, hedbergellids, globigerinellids, and pseudotextularids, and the return of globotruncanids (Table 1). The reappearance of the latter two groups also suggests a deepening shelf, reflecting a sea-level transgression associated with climatic warming at the end of the Maastrichtian.

Ostracodes

Ostracodes are rare in the Poty Quarry sediments and include *Bythocypris*, *Cytherella*, and *Bairdia* in the Gramame Formation. The ostracode fauna of the overlying Maria Farinha Formation consists of *Brachycythere* and *Dahomeya* (in sample MF6). In addition, the West African species *Soudanella laciniosa* was identified in sample levels GA, G0, and MF4. This species is known from Mali, Nigeria, and Libya, but has also been described from the Paleocene of northeastern Brazil (Neufville, 1979). Bertels (1969) reported the genus *Soudanella* from Argentina, within sediments spanning the K-T transition. Our findings confirm and strengthen the evidence for interfaunal exchange between West Africa and Brazil in the latest Maastrichtian and early Paleocene (Stinnesbeck and Reyment, 1988).

Calcareous Nannofossils

Coccoliths are present in samples G13 to G9, rare above this interval in samples G8 to G5, and only questionably present in the interval G4, G2, G1, MF1, and MF5. All assemblages are dominated by *Watznaueria barnesae* and/or *Micula decussata*. In addition, ~15 species are present in G13, including *Lithraphidites quadratus*, a characteristic "middle" to late Maastrichtian marker. In G11, 20 species were identified, and several late Maastrichtian index species, including *Micula murus*, were among them. *Micula prinsii*, the youngest representative of its genus, was not detected during our study, but this might be due to the recrystallization of this delicate species. No species diagnostic of the early Paleocene have been identified and the local K-T boundary cannot be recognized on the basis of nannofossils (Stinnesbeck and others, 1993a).

Palynomorphs

A well-preserved microflora consisting of 19 species of 14 genera was determined in the Poty K-T boundary sequence (Ashraf and Stinnesbeck, 1988). Samples G10 to G6 of the upper Gramame Formation yielded 12 species that are assigned to the *Schizeazeae*, *Gleicheniaceae*, *Cyathaceae*, and *Marsileaceae*. Today, these groups are typically found in tropical and subtropical regions. Above G6 (breccia bed and higher), representatives of the genera *Cicatricosisporites (Schizeazeae)*, *Gabonisporis (Marsileaceae)*, and other tropical and/or subtropical forms are conspicuously absent. Species of the basal Maria Farinha Formation (MF2 to MF7) belong to the *Myriaceae*, *Asteraceae* as well as pollen of "*Tilia*" and *Pinaceae*, and indicate a temperate climate for the early Paleocene.

Macrofossils

The Gramame Formation is well known for its abundant megafauna (Ruthbun, 1875; Maury, 1930; Oliveira, 1957; Beurlen, 1967b). More than 38 species of mollusks have been described in the literature, in addition to numerous crustaceans, annelids, echinoids, fish, reptiles, and plant remains. At least three different facies are generally differentiated by faunal associations and lithology (e.g., Beurlen, 1967a, 1967b, 1970; Mabesoone and Oliveira, 1993). These include the following.

(1) A basal member of yellow detrital limestones is characterized by thick-shelled bivalves and gastropods (*Cucullea*, *Pseudocucullea*, *Veniella*, *Turritella*, *Pugnellus*, *Tibia*, *Volutomorpha*, and others) that indicate littoral environments.

(2) A special facies of phosphate-rich detrital limestones that contains a typical molluscan fauna consisting of bivalves and gastropods, with abundant *Plicatula*, *Venericardia*, *Lucina*, *Xenomorpha*, *Helicaulax*, and *Cypraea*. Both members characterize the marine transgressive interval of the Gramame Formation overlying the terrestrial clastic sediments of the Beberibe Formation.

(3) Above these units biomicritic limestones reach a thickness of 35 to 40 m (Dantas, 1980) and contain abundant ammonites (various species of *Pachydiscus*, *Sphenodiscus*, *Pseudophyllites*), echinoids (*Hemiaster*), and only few bivalves and gastropods, except for *Atrina regiomaris*, *Volutomorpha brasiliensis*, and *Pyrazus brasiliensis* (Beurlen, 1967b, 1970; Mabesoone and Oliveira, 1993).

Only the upper 9.5 m of this last member of the Gramame Formation are present in the Poty Quarry section. Despite an intensive search and excellent outcrops, only a few invertebrate species have been encountered in the biomicritic marls and limestones well below the breccia unit: *D. (Diplomoceras)* sp., *P. (Pachydiscus) neubergicus*, *P. (Pachydiscus)* cf. *euzebioi*, *Pachydiscus* sp., ?*Pecten gramamensis*, *Hamulus* sp., *Hemiaster oliveirai*, and *Callianassa pernambucana*. This fauna is indicative of normal marine shelf conditions. Only *Hamulus* sp. and *Hemiaster oliveirai*, however, are abundant. The faunal association changes below and within the breccia unit where abundant *Serpula* sp., *Pteria* cf. *invalida*, unidentified snails and oysters, and the absence of echinoids and ammonites suggest a shallow-water environment perhaps with reduced salinity conditions (Fig. 3). Invertebrate faunas are also impoverished in the basal Maria Farinha Formation (units I and above, Paleocene). The abundant and diversified fauna described by White (1987), Oliveira (1953), Cassab (1983) and other authors is from the upper part of this sediment sequence. Near the K-T boundary we found only *Callianassa pernambucana*, *Cimomia pernambucensis*, and *Serratocerithium buarquianum*.

The scarcity of megafossils in the K-T boundary transition at Poty cannot be explained by diagenetic processes alone, although selective dissolution of aragonite may have eliminated some specimens. Ammonites, gastropods, and bivalves with aragonitic shells are usually preserved as molds, whereas oysters, annelids, and sea urchins often exhibit original calcite shells. Impressions and internal molds of aragonitic shells are, however, present in various sediment beds of the Poty section and can usually be identified to species or genus level. Their scarcity below and above the K-T boundary most likely reflects adverse environmental conditions. The shelf sea of northeastern Brazil appears to have been characterized by low species richness and low numbers of individuals well before the end of the Maastrichtian.

Among the terminal Maastrichtian invertebrate fauna of Brazil are two ammonite genera, *Pachydiscus* and *Diplomoceras*. Both genera are also known to occur in the latest Maastrichtian sediments of Denmark (Birkelund, 1979), the Biscay region of Spain (Ward and Kennedy, 1993), and the Antarctic Peninsula (Macellari, 1986). In central Chile, *Diplomoceras* is one of the last surviving ammonites (Stinnesbeck, 1986). Some authors regard *Pachydiscus neubergicus* as an index fossil of the lower Maastrichtian (e.g., Wright, 1957; Wiedmann, 1979). However, the presence of this species in late Maastrichtian strata of northeastern Brazil, 4.5 m below the K-T boundary, and in sediments containing well-defined late Maastrichtian planktonic foraminiferal assemblages indicates that this ammonite taxon survived into the late Maastrichtian. This observation is confirmed by Kennedy and Henderson (1992) and Ward and Kennedy (1993).

DISCUSSION

The Gramame Formation is of Maastrichtian age as determined by *Sphenodiscus*, *Eubaculites*, and *Neophylloceras surya*. These ammonites were collected from the basal transgressive part of the Gramame Formation and are generally considered Maastrichtian index fossils including *P. (Pachydiscus) neubergicus*. The first megafossils of undoubtedly early Tertiary age *(Cimomia pernambucensis)* were detected in unit J (sample MF0), the base of the banco vidro at the Poty section. *Cimomia pernambucensis* is related to *C. vaughani* from the Paleocene Midway Formation (Oliveira, 1953).

Planktonic foraminifera, palynoflora, and nannofossils are more abundant and provide better age diagnosis across the Poty K-T boundary than do the rare invertebrates, whose stratigraphic ranges are not yet well established. Both coccoliths and planktonic foraminifera indicate the presence of upper Maastrichtian sediments, and the presence of *Plummerita hantkeninoides* suggests that the topmost meters of the Gramame biomicritic marls and limestones are of latest Maastrichtian age. The breccia unit and detrital limestones directly overlying the micritic limestone member (unit B) also contain only Maastrichtian microfossils; earlier determinations of isolated early Tertiary planktonic foraminifera in these levels (Stinnesbeck, 1989; Stinnesbeck and others, 1993a; Albertao and others, 1994; Koutsoukos, 1995; Stinnesbeck and Keller, 1996) were not confirmed. The first undoubtedly early Tertiary foraminiferal species occur in a marl-claystone layer (unit I) 5 cm below the "banco vidro" (unit J), which consists of a dense recrystallized limestone that forms the base of the Maria Farinha Formation (Fig. 3). *Parvularugoglobigerina eugubina* and *Eoglobigerina fringa* are present in these sediments and are indicative of zone P1a of the early Paleocene. The basal Tertiary zone P0 has not been identified and zone Pla is restricted to a 5-cm-thick layer. This indicates that at least part of zone Pla and zone P0 are missing at the Poty section. The breccia bed, located ~60 to 70 cm below the first Tertiary planktonic foraminifera and iridium anomaly that characterize the K-T boundary, is of latest Maastrichtian age and apparently precedes the K-T boundary by many thousands of years.

Deposition of the micritic limestones and limestones of the upper Gramame Formation occurred under normal marine shelf conditions. This is indicated by faunas as well as the uniform sediment microfacies of exclusively micritic wackestones. Pelagic elements such as ammonites, planktonic foraminifera, radiolaria, and coccolithophorids prevail, but the presence of benthic foraminifera, ostracodes, annelids, endobenthonic echinoids (*Hemiaster*), callianassid crabs, and rare bivalves (*Atrina*) demonstrate that the sea floor was well oxygenated and that water depths were relatively shallow (inner neritic at the time of breccia deposition and deepening to middle neritic thereafter). Stable isotope ranking of planktonic foraminifera indicates that during the Late Cretaceous, large taxa (globotruncanids, planoglobulinids, pseudotextularids) generally lived in deeper waters at or below the thermocline, whereas small taxa (rugoglobigerinids, heterohelicids, guembelitrids, globigerinellids, hedbergellids) were surface dwellers that lived within the photic zone in the upper 100 to 150 m of the water column (e.g., Douglas and Savin, 1978; Boersma and Shackleton, 1981; Keller and others, 1993). The scarcity of deeper dwelling globotruncanids in the upper Gramame Formation combined with the abundance of biserials planktonic taxa and rugoglobigerinids suggests that shallow water depths were responsible for the absence of deeper water planktonic foraminifera from the Poty region. The climate of the Poty section was tropical to subtropical, as indicated by palynomorphs with schizacean, gleicheniacean, cyathacean, and marsilacean spores as the prevailing taxa.

Environmental conditions changed about 70 cm below the K-T boundary, nearly coincident with the deposition of the limestone breccia which marks a sea-level lowstand (Fig. 3). The brecciated marls and limestones contain bones and teeth, phosphatic lumps, phosphatized foraminifera, glauconite, pyrite concretions, and serpulid tubes, suggesting erosion and reworking from nearshore areas. The benthic foraminiferal fauna of these sediments also suggests shallow water depths. The breccia layer marks a sea-level lowstand during the latest Maastrichtian that may have reduced the water depths from outer-middle neritic (>150 m) to shallow middle-inner neritic (<100 m). All tropical or subtropical palynomorph taxa present below the breccia layer disappeared and do not reappear above this layer, which may indicate climatic cooling.

Upsection from the breccia layer, the relatively coarse grained detrital limestones show upward-fining grain sizes along with evidence for terrigenous influx. The megafauna of this interval consists of serpulid tubes, occasional oysters, pteriid bivalves, and gastropods, whereas ammonites and echinoids are generally absent. This also suggests shallow water with perhaps reduced salinity—a diagnosis that is consistent with the abundant middle neritic benthic foraminiferal fauna. The uppermost 10 cm of this lithological sequence grades into micritic limestones, although phosphate and glauconite particles are still abundant here. Benthic foraminifera are indicative of a deepening middle neritic environment: planktonic foraminifera are represented by biserials, rugoglobigerinids, and a few globotruncanids, similar to the marly limestone below the breccia layer. This suggests deeper waters, a rising sea level, and possibly climatic warming at the end of the Maastrichtian (Fig. 3) as also observed in Danish sections (Schmitz and others, 1992; Keller and others, 1993).

Above unit E a 5-cm-thick marl-claystone layer is encountered that contains the first Paleocene (zone P1a) foraminiferal species in an outer neritic environment. The environmental conditions change again in the overlying "banco vidro," where the only fossils recognizable in thin sections of this recrystallized limestone are gastropods, bivalves, and phosphatized benthic foraminifera. This fauna suggests that sea level dropped again following the K-T boundary and the early Danian transgression. Palynomorph assemblages in this unit are dominated by *Intratriporopollenites* ("*Tilia*") and conifers (*Pinuspollinites*) that suggest a temperate early Paleocene climate.

The reduced thickness of the zone P1a marl-claystone layer immediately below the "banco vidro" suggests condensed sedimentation or a hiatus during this biozone or shortly thereafter. Unfortunately, the extent of this possibly condensed interval is not known due to poor preservation of fossils in the limestone. In general, the microfacies and abundance of benthic foraminiferal assemblages indicate relatively low sea levels for the sediments of the basal Maria Farinha Formation overlying the "banco vidro." The age of these sediments is probably zone P1b or P1c of the early Paleocene, but the exact biostratigraphic position is not known, once again as a result of the rare occurrence of planktonic foraminifera.

Lithologic and faunal analyses across the K-T boundary transition at Poty indicate oceanic instability associated with sea-level fluctuations (Fig. 3). The maximum sea-level lowstand is marked by the breccia bed ~70 cm below the K-T boundary. A gradual sea-level rise occurred below and continued across the boundary, where there is evidence for another drop within or directly above zone P1a. Similar paleobathymetric changes were reported from the K-T boundary sections in Nye Kløv and Stevns Klint in Denmark (Schmitz and others, 1992; Keller and others, 1993), Braggs in Alabama (Donovan and others, 1988), El Kef, Tunisia (Keller, 1988), the Negev of Israel (Keller and Benjamini, 1991), Agost, Spain (Canudo and others, 1991), Brazos, Texas (Keller, 1989), and Mimbral, along with other sections in northeastern Mexico (Stinnesbeck and others, 1993b; Keller and others, 1994; Keller and Stinnesbeck, this volume). These studies strongly support a global sea-level lowstand just below the K-T boundary followed by a sea-level rise across the K-T boundary. Sea level seems to have been generally rising throughout the early Danian (Brinkhuis and Zachariasse, 1988), although short sea-level regressions are indicated at the zone P0–P1a and

P1a–P1b boundaries (see MacLeod and Keller, 1991a, 1991b; Keller and Stinnesbeck, this volume).

The late Maastrichtian sea-level drop at Poty was accompanied by climatic cooling from tropical toward more temperate conditions, as indicated by the last appearance of tropical palynomorphs directly below the breccia unit. This tendency continued into the early Paleocene, with conifers and "*Tilia*" predominating. A comparable cooling tendency was reported by Stott and Kennett (1990) and Barrera and Keller (1994) on the basis of $\delta^{18}O$ isotope data from planktonic foraminifera. Similar observations based on the microflora have been reported from southern France and northern Spain (Ashraf and Erben, 1986), different regions of North America (Tschudy and others, 1984; Tschudy and Tschudy, 1986; Hickey, 1984), and eastern Siberia (Krassilov, 1981).

Whereas we interpret the presence of the breccia layer in the Poty section as the result of a sea-level regression, Albertao and others (1994; see also Koutsoukos, 1994; 1995; Albertao and Martin, 1994) have recently come to a different interpretation. These authors suggest that the K-T boundary is located at the top of the micritic limestone underlying the breccia layer. They further suggest that the detrital limestone with its upward fining bioclastic limestone grains is a tsunami deposit triggered by a bolide impact on Yucatan. By their interpretation, a second bolide impact occurred in the early Danian between 150 and 300 k.y. after the K-T boundary, and is marked by the iridium anomaly (0.69 ppb) within the marl-claystone layer of unit I where we observed the first appearance of Danian species including *P. eugubina*.

Their interpretation is based on the presumed presence of rare Danian species (*Guembelitria irregularis*, *Woodringina claytonensis*, *W. hornerstownensis*, *Eoglobigerina simplicissima*, and *E. fringa*) within the breccia unit C and the overlying detrital limestone. We were unable to confirm these observations and suggest that these putative Danian species were either misidentified small Maastrichtian species or contaminants, or that in our sampled section K-T erosion was less severe, resulting in Maastrichtian sediments preserved above the breccia. As supporting evidence for two bolide impacts, Albertao and others (1994) cited the presence of shocked quartz and microspherules they believe to be microtektites. However, the microtektites they report from units C to I appear to be phosphatic spherules, which we observed to be abundant throughout the sequence. They dissolved when treated with a strong acid. We believe them to be algal resting cysts or infilled calcispheres. Quartz grains that they illustrate as shocked quartz show curved, bifurcated, and irregularly spaced lamellae that are more characteristic of tectonic or volcanic origins than a bolide impact (for a discussion see Stinnesbeck and Keller, 1995).

CONCLUSIONS

The Poty Quarry north of Recife contains a well-exposed K-T boundary section with a nearly continuous record of fossiliferous marls and limestones. Maastrichtian sediments contain the late Maastrichtian coccolithophorid index species *Micula murus* as well as typical latest Maastrichtian planktonic foraminiferal assemblages, including *Plummerita hantkeninoides*. The latter characteristically appears in the topmost 3 to 6 m of the Cretaceous in the K-T boundary type section of El Kef in Tunisia and other continuous K-T boundary sections (e.g., Agost and Caravaca in Spain and Mimbral and Lajilla in Mexico) where it ranges through the top 170–200 k.y. of the Maastrichtian (Pardo and others, this volume).

Invertebrate megafossils other than annelids (*Hamulus*) and echinoids (*Hemiaster*) are rare in these sediments, especially when compared to the rich faunas reported from the basal and middle Gramame Formation. This impoverished fauna suggests that adverse environmental conditions started well within the Maastrichtian. *Diplomoceras* and *Pachydiscus* are the last surviving ammonite genera in the Poty section, but their last observed occurrences take place 100 and 80 cm, respectively, below the K-T boundary.

Sediments and faunal assemblages provide evidence of a sea-level drop during the latest Maastrichtian about 70 cm below the K-T boundary. This regression reduced water depth from a middle to inner neritic environment in this region. The sea-level drop resulted in truncation of the marly limestone sequence and deposition of a layer of brecciated marls and limestones with abundant phosphate, glauconite, and pyrite concretions. Terrigenous vertebrate debris and the lithology indicate a high-energy environment characterized by erosion and reworking. Upsection, grain size gradually diminishes, terrigenous influx decreases, and sediment deposition with lithologies similar to the marly (micritic) limestone below the breccia resumes. In the top 10 cm of the Maastrichtian benthic foraminifera and megafaunal constituents indicate a gradually rising sea level that reestablishes middle neritic water depths. This transgression appears to continue into the earliest Tertiary, with deposition of a 5-cm-thick marl layer containing *E. fringa* and *P. eugubina*, which are index fossils of early Paleocene zone P1a.

The Poty sequence gives thus evidence of a sharp drop in sea level during the latest Maastrichtian followed by transgressive seas across the K-T boundary. Similar bathymetric changes have been reported from K-T boundary sections in Denmark, Tunisia, Alabama, Texas, and Mexico, and agree with the global sequence stratigraphic curves for this time period (Haq and others, 1987, 1988; Donovan and others, 1988). At Poty, the sea-level lowstand was accompanied by a significant change in temperature. Most tropical and subtropical palynomorphs show last appearances or a decrease in abundance at or just below the breccia bed. Apparently, climatic conditions deteriorated further into the early Paleocene, as suggested by the abundant presence of conifer pollen (*Pinuspollenites*) and "*Tilia*." Our results suggest that the environmental crisis at the end of the Maastrichtian was not a sudden event at the K-T boundary, but started much earlier in the Maastrichtian. Oceanic instability during the latest Maastrichtian and early Paleocene accompanied by temperature decrease seem to be the primary causes for the K-T boundary faunal transition. The iridium anomaly that marks the K-T boundary in the Poty section is in a thin marl-claystone layer that also contains the first Tertiary planktonic foraminifera of zone P1a age. This suggests the presence of a hiatus or very condensed K-T boundary transition that does not allow evaluation of the immediate biotic effects of a K-T boundary impact event.

ACKNOWLEDGMENTS

This study was supported in part by National Science Foundation grant EAR-9115044 to Keller, and DFG grant Er4/51-1 and CONACYT grant L.120-36-36 to Stinnesbeck. We thank E. Koutsoukos for reviewing this manuscript and N. Ortiz, T. Adatte, and N. MacLeod for discussions.

REFERENCES CITED

Albertao, G. A., and Martin, P. P., Jr., 1994, Stratigraphic record and geochemistry of the Cretaceous-Tertiary (K-T) boundary in the Pernambuco/Paraiba (PE/PB) Basin, northeastern Brazil [abs.]: 12 Colloque Africain de Micropaleontologie and 2 Colloque de Stratigraphie et Paleogeographie de l'Atlantique Sud, Angers, 16–20 Juli, 1994, p. 4–5.

Albertao, G. A., Koutsoukos, E. A. M., Regali, M. P. S., Attrep, M., Jr., and Martins, P. P., Jr., 1994, The Cretaceous-Tertiary boundary in southern low latitude regions: Preliminary study in Pernambuco, northeastern Brazil: Terra Nova, v. 6, p. 366–375.

Alvarez, L. W., Alvarez, W., Asaro, F., and Michel, H., 1980, Extraterrestrial cause for the Cretaceous-Tertiary extinction: Science, v. 208, p. 1095–1108.

Ashraf, A. R., and Erben, H. K., 1986, Palynologische Untersuchungen an der Kreide-Tertiär-Grenze West-Mediterraner Regionen: Palaeontographica, v. 194, p. 111–163.

Ashraf, A. R., and Stinnesbeck, W., 1988, Pollen und Sporen an der Kreide-Tertiärgrenze im Staate Pernambuco, NE Brasilien: Palaeontographica, v. 208, p. 39–51.

Barrera, E., and Huber, B. T., 1990, Evolution of Antarctic waters during the Maastrichtian: Foraminifer oxygen and carbon isotope ratios, Leg 113, *in* Proceedings of the Ocean Drilling Program, Scientific results, Volume 113: College Station, Texas, Ocean Drilling Program, p. 813–827.

Barrera, E., and Keller, G., 1990, Stable isotope evidence for gradual environmental changes and species survivorship across the Cretaceous-Tertiary boundary: Paleooceanography, v. 5, p. 867–890.

Barrera, E., and Keller, G., 1994, Productivity across the Cretaceous/Tertiary boundary in high latitudes: Geological Society of America Bulletin, v. 106, p. 1254–1266.

Bertels, A., 1969, Estratigrafía del límite Cretácico-Terciario en Patagonia septentrional: Revista de la Asociacion Geológica Argentina, v. 24, p. 41–54.

Beurlen, K., 1967a, Estratigrafía da feixa sedimentar còsteira Recife-João Pessoa: Boletím da Sociedade Geologia de Brasil, v. 16, p. 43–53.

Beurlen, K., 1967b, Paleontología da feixa sedimentar costeira Recife-João Pessoa: Boletím da Sociedade Geologia de Brasil, v. 16, p. 73–79.

Beurlen, K., 1970, Geologie von Brasilien: Berlin, Gebrueder Bornträger, p. 444.

Birkelund, T., 1979, The last Maastrichtian ammonites, *in* Christensen, W. K., and Birkelund, T., eds., Cretaceous-Tertiary boundary events, Volume 1, The Maastrichtian and Danian of Denmark: Copenhagen, University of Copenhagen, p. 51–57.

Boersma, A., and Shackleton, N. J., 1981, Oxygen and carbon isotope variations and planktonic foraminiferal depth habits: Late Cretaceous to Paleocene, central Pacific, DSDP sites 463 and 465, Leg 65, *in* Initial reports of the Deep Sea Drilling Project, Volume 65: Washington, D.C., U.S. Government Printing Office, p. 513–526.

Brinkhuis, W., and Zachariasse, W. J., 1988, Dinoflagellate cysts, sea level changes and planktonic foraminifera across the Cretaceous-Tertiary boundary at El Haria, northwest Tunisia: Marine Micropaleontology, v. 13, p. 153–190.

Canudo, I., Keller, G., and Molina, E., 1991, K/T boundary extinction pattern and faunal turnover at Agost and Caravaca, SE Spain: Marine Micropaleontology, v. 17, p. 319–341.

Cassab, R., 1983, Moluscos fosseis da Formação Maria Farinha, Paleoceno de Pernambuco-Gastropoda: Anais de Academia Brasileira de Ciencias, v. 55, p. 385–393.

Courtillot, V. E., 1994, Mass extinctions in the last 300 million years: One impact and seven flood basalts?: Israel Journal of Earth Sciences, v. 43, p. 255–266.

Courtillot, V. E., and Cisowski, S., 1987, The Cretaceous-Tertiary boundary events: External or internal causes?: Eos (Transactions, American Geophysical Union), v. 68, p. 193–200.

Dantas, A. J. R., 1980, Mapa geológico do Estado de Pernambuco: Departamento Nacional de Producao Mineral, Mapas e Cartas de Síntese, 1, Seção Geologica 1, Recife.

Donce, P., Jardine, S., Legoux, O., Masure, E., and Méon, H., 1985, Les évènements à la limite Crétacé-Tertiaire: au Kef (Tunisie septentrionale), l'analyse palynoplanctologique montre qu'un changement climatique est decélable à la base du Danian: Actes du Premier Congres National des Sciences de la Terre, p. 161–169.

Donovan, A. D., Baum, G. R., Blechschmidt, G. L., Loutit, L. S., Pflum, C. E., and Vail, P. R., 1988, Sequence stratigraphic setting of the Cretaceous-Tertiary boundary in central Alabama, *in* Wilgus, C. K., Hastings, B. S., Kendall, C. G., Posamentier, H. E., Ros, C. A., and Van Wagoner, J. C., eds., Sea-level changes: An integrated approach: Society of Economic Paleontologists and Mineralogists Special Publication 42, p. 299–307.

Douglas, R. G., and Savin, S. M., 1978, Oxygen isotope evidence for depth stratification of Tertiary and Cretaceous planktic foraminifera: Marine Micropaleontology, v. 3, p. 175–196.

Haq, B. U., Hardenbol, J., and Vail, P., 1987, Chronology of fluctuating sea levels since the Triassic: Science, v. 235, p. 1156–1166.

Haq, B. U., Hardenbol, J., and Vail, P., 1988, Mesozoic and Cenozoic chronostratigraphy and cycles of sea-level change, *in* Wilgus, C. K., Hastings, B. S., Kendall, C. G., Posamentier, H. W., Ros, C., and Van Wagoner, J. C., eds., Sea-level changes: An integrated approach: Society of Economic Paleontologists and Mineralogists Special Publication 42, p. 71–108.

Hollis, C. J., 1993, Latest Cretaceous to late Paleocene radiolarian biostratigraphy: A new zonation from the New Zealand region: Marine Micropaleontology, v. 21, p. 295–327.

Kaiho, K., 1992, A low extinction rate of intermediate-water benthic foraminifera at the Cretaceous/Tertiary boundary: Marine Micropaleontology, v. 18, p. 229–259.

Keller, G., 1988, Extinction, survivorship and evolution of planktic foraminifera across the Cretaceous-Tertiary boundary at El Kef, Tunisia: Marine Micropaleontology, v. 13, p. 239–263.

Keller, G., 1989, Extended Cretaceous/Tertiary boundary extinctions and delayed population change in planktonic foraminifera from Brazos River, Texas: Paleoceanography, v. 4, p. 287–332.

Keller, G., 1992, Paleoecologic response of Tethyan benthic foraminifera to the Cretaceous-Tertiary boundary transition, *in* Takayanagi, Y., and Saito, T., eds., Studies in benthic foraminifera: Sendai, Tokai University Press, p. 77–91.

Keller, G., 1993, The Cretaceous-Tertiary boundary transition in the Antarctic Ocean and its global implications: Marine Micropaleontology, v. 21, p. 1–45.

Keller, G., and Benjamini, C., 1991, Paleoenvironment of the eastern Tethys in the early Paleocene: Palaios, v. 6, p. 439–464.

Keller, G., Barrera, E., Schmitz, B., and Matison E., 1993, Gradual mass extinction, species survivorship, and long-term environmental changes across the Cretaceous-Tertiary boundary in high latitudes: Geological Society of America Bulletin, v. 35, p. 979–997.

Keller, G., Stinnesbeck, W., Adatte, T., MacLeod, N., and Lowe, D., 1994, Field guide to Cretaceous/Tertiary boundary sections in northeastern Mexico: in Lunar and Planetary Institute Contribution 827, 110 p.

Kennedy, W. J., and Henderson, R. A., 1992, Non-heteromorph ammonites from the Upper Maastrichtian of Pondicherry, South India: Palaeontology, v. 35, p. 381–442.

Koutsoukos, E. A. M., 1994, Event stratigraphy and paleoceanography across the Cretaceous-Tertiary boundary in Pernambuco, northeastern Brazil: 12th Colloque Africain de Micropaleontologie and 2nd Colloque de Stratigraphie et Paleogeographie de l'Atlantique Sud, p. 81–82.

Koutsoukos, E. A. M., 1995, The Cretaceous-Tertiary boundary in southern low latitude regions: Preliminary study in Pernambuco, northeastern Brazil: Reply: Terra Nova (in press).

Krassilov, V. A., 1981, Changes of Mesozoic vegetation and the extinction of dinosaurs: Palaeogeography, Palaeoclimatology, Palaeoecology, v. 34, p. 207–224.

Mabesoone, J. M., and Oliveira, P. E., 1993, Paleontología estratigrafica, *in* Revisão geológica da faixa sedimentar costeira de Pernambuco, Paraíba e Rio Grande do Norte: Instituto de Geociencias, Universidade Federal de Pernambuco, Departamento Geologia, Recife, Estudos Pesquisas, v. 10, p. 105–110.

Macellari, C., 1986, Late Campanian–Maastrichtian ammonite fauna from Seymour Island (Antarctic Peninsula): Paleontological Society Memoir 18, 55 p.

MacLeod, N., and Keller, G., 1991a, Hiatus distribution and mass extinctions at the Cretaceous/Tertiary boundary: Geology, v. 19, p. 497–501.

MacLeod, N., and Keller, G., 1991b, How complete are Cretaceous/Tertiary boundary sections? A chronostratigraphic estimate based on graphic correlation: Geological Society of America Bulletin, v. 103, p. 1439–1457.

Maury, C. J., 1930, O Cretaçeo da Parahyba do Norte: Monografia Servico Geologico e Mineralogico do Brasil, v. 8, p. 1–305.

Méon, H., 1990, Palynologic studies of the Cretaceous/Tertiary boundary interval at El Kef outcrop, northwest Tunisia: Paleogeographic implications: Review of Paleobotany and Palynology, v. 65, p. 85–94.

Neufville, H. M. E., 1979, Paleogene ostracods from northeastern Brazil: Geological Institute of the University of Uppsala Bulletin, v. 8, p. 135–172.

Oliveira, P. E., 1953, Invertebrados fosseis da Formação Maria Farinha I—Cephalopoda: Division Geologica y Minerva Boletin, v. 146, 33 p.

Oliveira, P. E., 1957, Invertebrados Cretacicos do fosfato de Pernambuco: Division Geologica y Minerva Boletim, v. 172, 29 p.

Ruthbun, R., 1875, Preliminary report on the Cretaceous lamellibranchs collected in the vicinity of Pernambuco: Boston, Society of Natural History Proceedings, v. 17, p. 241–256.

Schmitz, B., Keller, G., and Stenvall, O., 1992, Stable isotope and foraminiferal changes across the Cretaceous/Tertiary boundary at Stevns Klint, Denmark: Arguments for long-term oceanic instability before and after bolide impact: Palaeogeography, Palaeoclimatology, Palaeoecology, v. 96, p. 233–260.

Stinnesbeck, W., 1986, Zu den faunistischen und palökologischen Verhältnissen in der Quiriquina Formation Zentral-Chiles: Palaeontographica, v. 194, p. 99–237.

Stinnesbeck, W., 1989, Fauna y microflora en el limite Cretácio-Terciario en el estado de Pernambuco, Noreste de Brasil: Contribucionas a los Simposios sobre el Cretacio de America Latina, Parte A: Buenos Aires, Argentina, Eventos y Registro Sedimentario, p. 215–230.

Stinnesbeck, W., and Keller, G., 1995, The Cretaceous-Tertiary boundary in southern low latitude regions: Preliminary study in Pernambuco, northeastern Brazil: Comment: Terra Nova, v. 7, p. 375–378.

Stinnesbeck, W., and Keller, G., 1996, Near-K/T age of clastic deposits from Texas to Brazil: Impact, volcanism and/or sea-level lowstand?: Terra Nova (in press).

Stinnesbeck, W., and Reyment, R. A., 1988, Note on a further occurrence of *Soudanella laciniosa* Apostolescu in northeastern Brasil: Journal of African Earth Sciences, v. 7, p. 779–781.

Stinnesbeck, W., Ashraf, A. R., and von Salis Perch-Nielsen, K., 1993a, Estudos paleontológicos no limite Cretáceo-Terciario no Estado de Pernambuco, *in* Revisão geológica da faixa sedimentar costeira de Pernambuco, Paraíba e Rio Grande do Norte: Institit
uto de Geociencias, Universidade Federal de Pernambuco, Departamento Geologia, Recife, Estudos Pesquisas, v. 10, p. 141–156.

Stinnesbeck, W., Barbarin, J. M., Keller, G., Lopez-Oliva, J. G., Pivnik, D., Lyons, J., Officer, C., Adatte, T., Graup, G., Rocchia, R., and Robin, E., 1993b, Deposition of channel deposits near the Cretaceous-Tertiary boundary in northeastern Mexico: Catastrophic or "normal" sedimentary deposits?: Geology, v. 21, p. 797–800.

Stott, L. D., and Kennett, J. P., 1990, The paleooceanographic and climatic signature of the Cretaceous/Paleogene boundary in the Antarctic: Stable isotopic results from ODP Leg 113, *in* Proceedings of the Ocean Drilling Program, Scientific results, Volume 113: College Station, Texas, Ocean Drilling Program, p. 829–848.

Tschudy, R. H., and Tschudy, B. D., 1986, Extinction and survival of plant following the Cretaceous-Tertiary boundary event: Geology, v. 14, p. 667–670.

Tschudy, R. H., Pillmore, C. L., Orth, C. J., Gillmore, J. S., and Knight, J. D., 1984, Disruption of the terrestrial plant ecosystem at the Cretaceous-Tertiary boundary: Science, v. 225, p. 1030–1032.

Ward, P., and Kennedy, W. J., 1993, Maastrichtian ammonites from the Biscay region (France, Spain): Paleontological Society Memoir 34, 58 p.

White, C. A., 1987, Contribuiçoes a Paleontología do Brasil: Archivos de Museo Nacional de Rio de Janeiro, v. 7, 273 p.

Wiedmann, J., 1979, Die Ammoniten der NW-deutschen, Regensburger und Ostalpinen Kreide im Vergleich mit den Oberkreidefaunen des westlichen Mediterrangebietes, *in* Wiedmann, J., ed., Aspekte der Kreide Europas: Stuttgart, International Union of Geological Societies Special Paper, ser. A, v. 6, p. 335–350.

Wright, C. W., 1957, Cretaceous Ammonoidea, *in* Moore R. C., ed., Treatise on invertebrate paleontology, Part L, Mollusca 4, Cephalopoda Ammonoidea: Boulder, Colorado, Geological Society of America (and University of Kansas Press), p. L1–L490.

19

Cretaceous-Tertiary Boundary Clastic Deposits in Northeastern Mexico: Impact Tsunami or Sea-Level Lowstand?

Wolfgang Stinnesbeck, *Facultad de Ciencias de la Tierra, Universidad Autónoma de Nuevo León, Linares, Mexico*, Gerta Keller, *Department of Geological and Geophysical Sciences, Princeton University, Princeton, New Jersey*, T. Adatte, *University of Neuchâtel, Institut de Géologie, Neuchâtel, Switzerland*, J. G. Lopez-Oliva, *Department of Geological and Geophysical Sciences, Princeton University, Princeton, New Jersey*, and Norman MacLeod, *Department of Palaeontology, The Natural History Museum, London, United Kingdom*

INTRODUCTION

The recent suggestion that the 180-km-diameter buried circular gravity anomaly near Chicxulub on the north coast of Yucatan represents the Cretaceous-Tertiary (K-T) boundary bolide impact site (Hildebrand and others, 1991; Pope and others, 1991; Sharpton and others, 1992, 1993) has focused renewed attention on sedimentary deposits in the Caribbean, Gulf of Mexico, and Mexico as possibly recording deposition by an impact-generated tsunami megawave (Hildebrand and Boynton, 1990; Alvarez and others, 1992; Smit and others, 1992, 1994a, 1994b). Although in theory testing this hypothesis should be easy, this has not been the case in practice because no consensus exists regarding the identifying attributes of an impact-generated tsunami. Should the deposition pattern be chaotic or stratified? How do such deposits differ from submarine gravity slides, portions of turbidite flow deposits, or channelized sea-level lowstand deposits?

Bourgeois and others (1988) originally proposed a K-T boundary impact-tsunami origin for a thin (0.2–1.0 m) clastic deposit along the Brazos River. The same origin was later proposed by Hildebrand and Boynton (1990) and Smit and others (1994a) for similar deposits in Alabama and Georgia. These very shallow water (inner to middle neritic) clastic units have been generally interpreted as incised valley deposits associated with sea-level lowstands (Donovan and others, 1988; Baum and Vail, 1988; Keller, 1989a; Mancini and others, 1989; Habib and others, 1992; Moshkovitz and Habib, 1993; Savrda, 1993). Especially problematic for the impact-tsunami interpretation of these clastic deposits are their differing stratigraphic ages, which include latest Maastrichtian, early Danian, and late Danian (Mancini and others, 1989; Savrda, 1993; Keller and Stinnesbeck, this volume). Clastic deposits of ages predating and postdating the K-T boundary cannot be deposited by a single event, let alone the K-T boundary bolide impact.

Impact-tsunami deposition has also been proposed for a big boulder bed in Cuba by Bohor and Seitz (1990), although this boulder bed was generated by exfoliation weathering of a coarse calcarenaceous part of a turbidite of middle Maastrichtian age (Brönnimann and Rigassi, 1963; Dietz and McHone, 1990; Iturralde-Vincent, 1992). Ocampo and Pope (1994) interpreted a limestone breccia of unknown age in Belize as a K-T impact-tsunami deposit. Montanari and others (1994) reported an impact-tsunami limestone breccia 1 m below the K-T boundary in the Bochil section of Chiapas in southern Mexico, although a dozen similar limestone breccias occur throughout the Campanian-Maastrichtian sequence (Stinnesbeck and others, 1994a; Keller and Stinnesbeck, this volume) as a result of repeated flysch deposition (Quezada Muñeton, 1990; Michaud and Fourcade, 1989). Impact-tsunami deposits have also been reported from sections that contain major hiatuses spanning from lower or middle Maastrichtian to the lower-upper Paleocene (zone P2) such as the Chilil section in Chiapas (Stinnesbeck and others, 1994a) and Deep Sea Drilling project (DSDP) Sites 536 and 540 in the Gulf of Mexico (Keller and others, 1993a). Alvarez and others (1992) reported the presence of a volcaniclastic tsunami deposit at the K-T boundary in the Gulf of Mexico in DSDP Sites 536 and 540. Reexamination of these cores indicates, however, that the volcaniclastic deposit predates the K-T boundary and is of middle Maastrichtian age (Keller and others, 1993a), and a hiatus is present between the middle Maastrichtian and upper part of the lower Danian zone P1a. From these studies, it appears that impact-tsunami interpretations have often been made hastily and with little consideration of the age and stratigraphic position of the clastic deposits, or deposition of sediments below and above that would indicate sea-level changes.

These same problems are inherent in the impact-tsunami interpretation proposed by Smit and others (1992, 1994a, 1994b) for the clastic deposits at Mimbral and other localities in northeastern Mexico. These sections are generally more complex, with near K-T boundary clastic deposits that exhibit three distinct lithological units representing different depositional regimes (Stinnesbeck and others, 1993; Keller and others, 1994a). However, the only unusual lithological feature of these clastic deposits is a basal spherule-rich layer that contains rare glass fragments with geochemical compositions similar to those of Beloc, Haiti (Smit and others, 1992; Stinnesbeck and others, 1993; Bohor and Betterton, 1993). Thus, a common origin for the Beloc and Mimbral glass is likely, although whether they are of impact or volcanic origin is still in dispute (Jéhanno and others, 1992; Lyons and Officer, 1992; Sigurdsson and others, 1991; Blum and Chamberlain, 1992; Swisher and others, 1992; Koeberl, 1994; Robin and others, 1994).

The northeastern Mexico K-T boundary sections are thus of critical importance in evaluating the impact-tsunami deposition hypothesis. This is the only region where near K-T boundary clastic deposits crop out over an area spanning more than 300 km and where sedimentary deposits can be studied in their regional context. We have studied seven K-T outcrop localities based on 5, 10, and 20 cm sample intervals to determine the age and depositional environment. In this study we report on the stratigraphic, sedimentologic, and mineralogic characteristics of these sections in northeastern Mexico (El Mimbral, Lajilla, El Peñon, El Mulato, La Parida, La Sierrita, Los Ramones) and propose a depositional model consistent with our observations.

MAASTRICHTIAN TO PALEOGENE PALEOGEOGRAPHY

During the Maastrichtian and Paleogene, northeastern Mexico was part of a shallow to moderately deep water shelf-slope region that received a steady influx of terrigenous clastic sediments from the rising Laramide orogeny and uplift of the Sierra Madre Oriental to the west (Sohl and others, 1991; Galloway and others, 1991; Figs. 1 and 2). Along the western Gulf of Mexico,

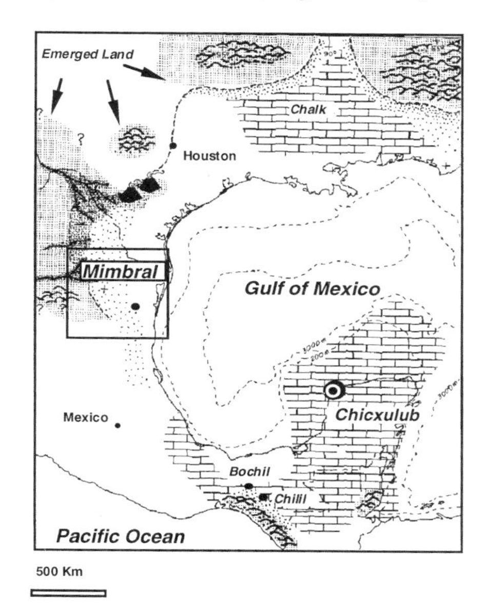

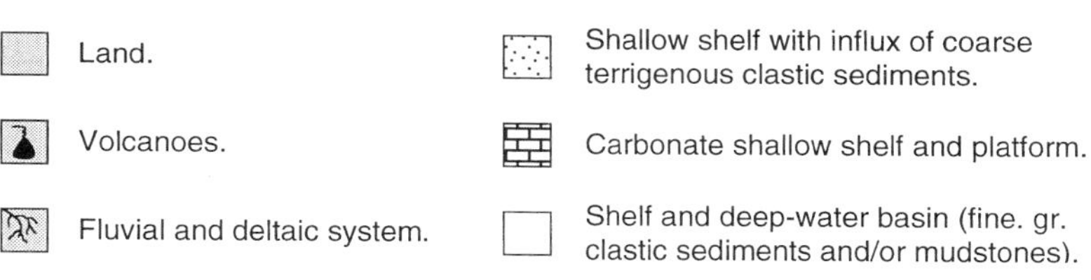

Figure 1. Paleogeography of the western Gulf of Mexico basin during the Maastrichtian (modified from Galloway and others, 1991; Sohl and others, 1991).

a series of basins formed in the present states of Veracruz, Tamaulipas, and in the southern and eastern parts of Nuevo León. In these basins, rhythmically bedded marls and shales and thin layers of ash were deposited in outer neritic to upper bathyal environments. These strata are correlated with the Campanian to Maastrichtian Mendez Formation and the Paleocene Velasco Formation. The source area of the fine-grained terrigenous sediments of these formations is believed to be the Sierra Madre Oriental and the deltaic complex of the Difunta Group in the area of Saltillo-Monterrey to the west and northwest (Weidie and others, 1972).

In the region north and west of Monterrey, terrigenous shallow-water to paralic sediments were deposited during the Maastrichtian and are represented by the Olmos and Escondido formations and the Difunta Group, which locally extends into the Eocene. Sediment influx was high; for example, in the Saltillo-Monterrey region the prodeltaic-deltaic wedge of the Difunta Group is several thousand meters thick and thins out south and east (Weidie and others, 1972).

In the western Gulf of Mexico, several large submarine canyon systems formed

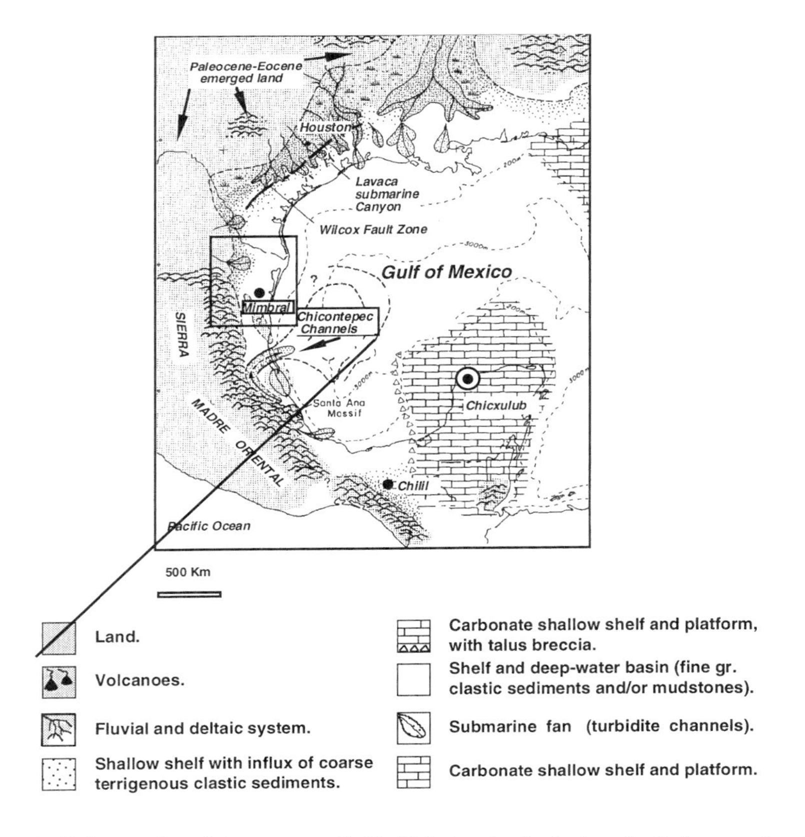

Figure 2. Paleography of the western Gulf of Mexico basin during the Paleocene (modified from Galloway and others, 1991; Sohl and others, 1991).

along the prograding unstable continental margin (Fig. 3). The best known of these submarine canyon systems are the Lavaca and Yoakum paleocanyons in Texas and the Chicontepec paleocanyon system in central-eastern Mexico which are of early Tertiary age (Figs. 1 and 2; Galloway and others, 1991).

The K-T boundary clastic units of northeastern Mexico were deposited in the uppermost Maastrichtian Mendez Formation. Although recent publications have concentrated on the outcrop at El Mimbral (Smit and others, 1992; Stinnesbeck and others, 1993; Keller and others, 1994b), similar lithological sequences crop out over more than 300 km in a north-northwest–south-southeast-trending direction 40 to 80 km east of the front range of the Sierra Madre Oriental (Fig. 3; Keller and others, 1994a). This alignment is not accidental, but is likely determined by the proximity of terrigenous shallow-water sediments to the west and north, and by funneling of sediments in a depression between two north-northwest–south-southeast-trending paleohighs, the Sierra Madre Oriental orogenic belt to the west and the Tamaulipas arch to the east. No sediment source areas are known to the east in the Gulf of Mexico (Sohl and others, 1991; Keller and others, 1994a).

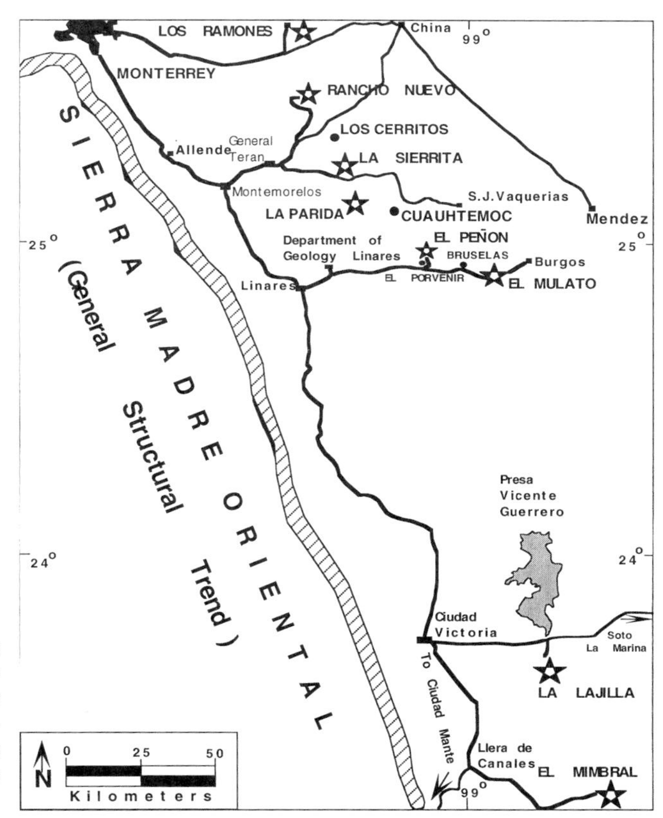

Figure 3. Location map of K-T boundary sections with clastic deposits in northeastern Mexico. Stars mark sections discussed in this report.

BIOSTRATIGRAPHY

The biostratigraphy of the northeastern Mexico K-T boundary sections is based on planktonic foraminifera and the biozonation of Keller (1988, 1993). Keller's zonal scheme is illustrated in Figure 4 along with other commonly used zonal schemes (Smit, 1982; Smit and others, 1992; Berggren and Miller, 1988; Longoria and Gamper, 1992). Index species that define zonal boundaries are marked in bold type. Early Danian zones P0 and P1a correspond to the part of paleomagnetic anomaly C29R above the K-T boundary that spans 230 k.y.; zone P0 is estimated to span 40 to 50 k.y. and zone P1a spans 180 k.y. (Berggren and others, 1985; MacLeod and Keller, 1991a, 1991b; Herbert and D'Hondt, 1990). The *Plummerita hantkeninoides* zone is estimated to span the last 200 k.y. of the

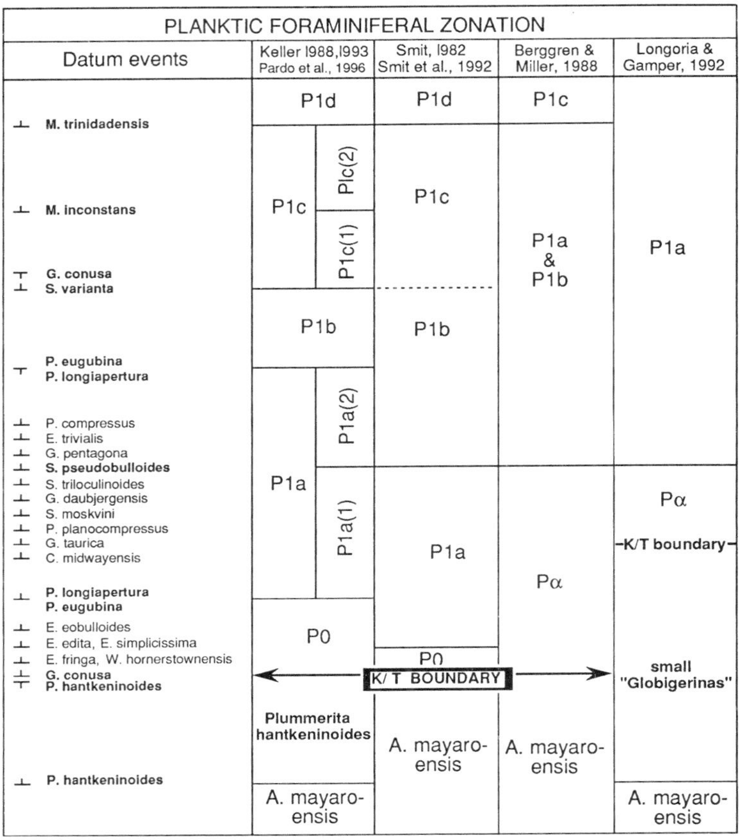

PLANKTIC FORAMINIFERAL ZONATION

Datum events | Keller 1988,1993 Pardo et al., 1996 | Smit, 1982 Smit et al., 1992 | Berggren & Miller, 1988 | Longoria & Gamper, 1992

M. trinidadensis
M. inconstans
G. conusa
S. varianta
P. eugubina
P. longiapertura
P. compressus
E. trivialis
G. pentagona
S. pseudobulloides
S. triloculinoides
G. daubjergensis
S. moskvini
P. planocompressus
G. taurica
C. midwayensis
P. longiapertura
P. eugubina
E. eobulloides
E. edita, E. simplicissima
E. fringa, W. hornerstownensis
G. conusa
P. hantkeninoides
P. hantkeninoides

Keller: P1d; P1c (P1c(2), P1c(1)); P1b; P1a (P1a(2), P1a(1)); P0; **Plummerita hantkeninoides**; A. mayaro-ensis
Smit: P1d; P1c; P1b; P1a; P0; A. mayaro-ensis
Berggren & Miller: P1c; P1a & P1b; Pα; A. mayaro-ensis
Longoria & Gamper: P1a; Pα; **K/T boundary**; **small "Globigerinas"**; A. mayaro-ensis

K/ T BOUNDARY

Figure 4. Correlation of commonly used planktonic foraminiferal schemes for the K-T transition. Note new uppermost Maastrichtian Zone P. Hantkenoides *from Pardo and others (this volume).*

Maaastrichtian, on the basis of paleomagnetic correlations at Agost, Spain (Pardo and others, this volume).

Identification of the K-T boundary is based on biostratigraphic, geochemical, and mineralogical marker horizons that are present at the El Kef stratotype and in complete K-T sequences worldwide (MacLeod and Keller, 1991a, 1991b; Keller, 1993). These include (1) a lithologic break from chalk or marl deposition of the Cretaceous to a thin layer of dark organic-rich and $CaCO_3$-poor clay, known as the boundary clay. At El Kef, this layer is 55 cm thick and represents the most expanded boundary clay observed to date in any K-T boundary sections. More frequently, the boundary clay is only 4–6 cm thick (e.g., Agost, Caravaca, Stevns Klint, Nye Kløv, El Mimbral). (2) A 2–3 mm oxidized red layer at the base of the boundary clay is present at El Kef as well as in all complete K-T sections. (3) Maximum iridium concentrations generally occur in the red layer and boundary clay, although these may trail several tens of centimeters above or below the boundary clay. (4) Ni-rich spinels are usually concentrated in the red layer at the base of the boundary clay. (5) A negative $\delta^{13}C$ shift occurs in marine plankton of low and middle latitudes. (6) The first appearance of Tertiary planktonic foraminifera is at the base or within a few centimeters of the boundary clay, red layer, iridium anomaly, and Ni-rich spinels. At El Kef, the first Tertiary species (*Globconusa conusa*) appears at the base of the boundary clay 1–2 cm above the red layer, and three other new species (*Eoglobigerina fringa*, *E. edita*, and *Woodringina hornerstownensis*) appear within the basal 15 cm of the boundary clay (Keller, 1988; Ben Abdelkader, 1992). (7) The disappearance of Cretaceous tropical taxa is also at or near the K-T boundary.

In the stratotype section at El Kef, all of these criteria are met and most of them are present in all the best and most-complete K-T boundary sequences (e.g., Agost, Caravaca, Nye Kløv, Brazos River, Ocean Drilling Project [ODP] Site 738; MacLeod and Keller, 1991a, 1991b). The coincidence of these lithological, geochemical, and paleontological criteria is unique in the geological record and virtually ensures that the placement of the K-T boundary is uniform and coeval in marine sequences across latitudes. Any of these criteria used in isolation, however, diminishes the chronostratigraphic resolution of the K-T boundary.

In the northeastern Mexico sections, planktonic foraminifera are abundant in Maastrichtian and Tertiary sediments, although they are generally recrystallized and poorly preserved. Nevertheless, typical faunal assemblages with abundant juvenile and adult specimens are always present, except in the clastic deposits, and provide excellent biostratigraphic control. These clastic deposits often contain few foraminifera, except in the spherule-rich basal layer and in mud clasts of the underlying Mendez Formation, and many of these have been reworked. In addition, thin marl layers in the upper part of the clastic deposit contain typical *P. hantkeninoides* zone assemblages.

Upper Maastrichtian

In northeastern Mexico, upper Maastrichtian sediments of the Mendez Formation generally consist of marls in the uppermost several meters and marls with rhythmically interlayered limestone beds below. In all sections examined, the exposed Mendez Formation is of *Abathomphalus mayaroensis* zone age, and the top of the formation is marked by the *Plummerita hantkeninoides* zone, which spans the last 200 k.y. of the Maastrichtian. The clastic sediments and basal unconformity are stratigraphically within the *P. hantkeninoides* zone (Lopez-Oliva and Keller, 1996). This indicates that erosion at the base of the clastic deposit was minimal. For example, at the La Parida section where the clastic deposit is only 10 to 20 cm thick, *P. hantkeninoides* is present in the 20 cm below the unconformity, and suggests erosion of several meters of Mendez marl, assuming hemipelagic to pelagic sedimentation rates similar to those of Agost and Caravaca. *Plummerita hantkeninoides* has been observed in all outcrops, except Mulato, either within, below, or above the clastic deposit, but always below the K-T boundary.

The top 25 cm of the clastic deposit generally consists of one or more discrete sandy limestone layers separated by thin marly interlayers. Abundant burrows of *Chondrites*, *Thalassinoides*, *Zoophycos*, and *Ophiomorpha* are present in these rippled sandy limestone layers. Planktonic foraminiferal assemblages (including the index taxon *A. mayaroensis* and *P. hantkeninoides*) indicate that deposition also occurred during the late Maastrichtian. No Tertiary foraminiferal species are observed in the burrow infillings, suggesting that burrowing occurred during the late Maastrichtian. Recent investigations by Ekdale and Stinnesbeck (1994, in prep.) revealed several horizons of bioturbation within unit 3 and one horizon at the base of unit 2. Their discovery confirms that deposition occurred within the Maastrichtian and over an extended time interval that allowed periods of habitation of burrowing invertebrates.

In three out of four sections where Tertiary sediments are present (Lajilla, El Mulato, La Parida), the clastic deposit is topped by a 5–10-cm-thick marl layer or marly limestone containing the typical *P. hantkeninoides* zone faunal assemblage, including the index taxon. This marl layer indicates that normal hemipelagic deposition of the Mendez Formation resumed after deposition of the clastic units. It also indicates that deposition of the clastic member predates the K-T boundary. Smit and others (1994b) suggested that this Maastrichtian marl layer represents settling through the water column after the tsunami event. However, there is no grain-size grading apparent, or size sorting of foraminifera. Moreover, two K-T boundary sections (Mirador near La Ceiba and Bochil in Chiapas; Macias Pérez, 1988; Montanari and others, 1994) contain 100 cm of Mendez marls overlying the clastic unit, clearly marking this deposit as of Maastrichtian age.

K-T Boundary and Lower Danian

Only the biostratigraphy of El Mimbral has been published to date (Smit and others, 1992; Stinnesbeck and others, 1993; Keller and others, 1994a, 1994b). In this section, the top of the clastic deposit, at the right edge of the channel, is overlain by a 4-cm-thick clay layer with a 4 mm red layer at its base. Stratigraphically, this clay layer (including the red layer) is equivalent to zone P0 in similar K-T boundary lithologies in Spain, Tunisia, and Denmark. It is therefore possible that the Mimbral clay layer also represents zone P0. This interpretation is consistent with the absence of large tropical taxa at the top or below the sandy rippled limestone layer that marks the top of the clastic deposit (though this is negative evidence), and with the first appearance of *Parvularugoglobigerina eugubina*, the index taxon for zone P1a, immediately above the clay layer. It is apparent, however, that the K-T boundary interval is very condensed or even discontinuous, as suggested by the relatively thin boundary clay and its absence in the channel where the clastic deposit is thickest (Smit and others, 1992; Keller and others, 1994b). A disconformity is also suggested by the simultaneous appearance of six Tertiary species (*Eoglobigerina fringa*, *E. edita*, *P. eugubina*, *P. longiapertura*, *Globoconusa daubjergensis*, and *W. hornerstownensis*) that normally evolve in zones P0 and the lower zone P1a, and by their great abundance (Keller and others, 1994b). Moreover, maximum levels of iridium at El Mimbral have been observed near the base of zone P1a 8 cm above the K-T boundary. No elevated levels of iridium are observed within the clastic deposit (Keller and others, 1994a). In the center of the channel fill where the clay layer is absent, maximum iridium concentrations are also found at the base of zone P1a, which at this location directly overlies the clastic deposit (Smit and others, 1992; Keller and others, 1994a). These data suggest that a short hiatus is present between zones P0 and P1a. The absence of Mendez marls above the clastic deposit also suggests that sediment deposition across the K-T boundary is incomplete.

Detailed biostratigraphic analyses of La Lajilla, El Mulato, and La Parida all indicate a short hiatus at the K-T boundary. In all of these sections, Mendez marl of Maastrichtian age tops the clastic deposit (Keller and others, 1994a; Lopez-Oliva and Keller, 1995). Immediately above, this Mendez marl layer is overlain by Tertiary Velasco shales of zone P1a age. No clay layer or evidence of zone P0 is present. Instead, the first Tertiary sediments are characterized by the simultaneous first appearance of six to eight species characteristic of zone P1a (Lopez-Oliva and Keller, 1994, 1996). The simultaneous first appearance and the high abundance of these species mark a K-T boundary hiatus, with at least zone P0 and the lower part of zone P1a missing. A short hiatus at the zone P0-P1a interval is commonly present in most K-T boundary sections worldwide and represents a short sea-level lowstand (MacLeod and Keller, 1991a, 1991b; Keller, 1989a, 1989b, 1993; Keller and others, 1993b; Canudo and others, 1991; Keller and Benjamini, 1991; Keller and Stinnesbeck, this volume).

Thus, the biostratigraphy of all sections examined reveals that deposition of the clastic deposit occurred during the last 200 k.y. of the Maastrichtian and at least several thousand years prior to the K-T boundary event. The K-T boundary is generally characterized by a short hiatus, with the earliest Tertiary zone P0 and the lower part of zone P1a missing. Thereafter, deposition of Tertiary shales of the Velasco Formation appears relatively continuous through zones P1a and P1b.

GENERAL CHARACTERISTICS OF K-T CLASTIC DEPOSITS

The clastic deposits present in K-T transitions spanning a distance of more than 300 km in northeastern Mexico are similar in their lithology, mineralogy, petrology, and stratigraphy. They can be subdivided into three units as characterized below.

Unit 1: Spherule-Rich Layer

The basal unit 1 of the clastic deposit disconformably overlies the Mendez Formation and consists of a soft weathered sediment that contains abundant spherules, foraminiferal tests, clasts of the underlying Mendez Formation, mud clasts containing Turonian age foraminifers, and minor quartz. Unit 1 is either massive or crudely stratified and locally shows large-scale trough cross-stratification. The spherules range from 1 to 5 mm in diameter and are generally infilled with blocky calcite or greenish microcrystalline phyllosilicates, including smectite and glauconite. Many large spherules have several small spherules inside. Some spherules formed around foraminifers or sediment grains, some contain rutile(?) crystals, and many are enclosed by a thin organic wall suggesting that they are infilled algal resting cysts. No glassy spherules were observed by us, although rare glass fragments are present (Stinnesbeck and others, 1993, 1994b; Keller and others, 1994a).

The spherule layer contains a 20–25-cm-thick well-cemented sandy limestone layer which occurs either within or at the top of unit 1 and is present in sections spanning over 300 km. This layer contains few graded spherules at the base and top, and convolute bedding at the top that is draped by the overlying spherule-rich sediments. Microlayering of fine- and coarser-grained sediments can be observed in thin sections within this sandy limestone layer (W. Ward, 1994, personal commun.).

Unit 2: Laminated Sandstone

Unit 2 consists of horizontally weakly laminated sandstone with mud clasts at its base and, at El Mimbral, discrete layers of plant debris. Only rare charcoaled plant debris is generally present in unit 2. Sand beds may grade upward to slightly finer sands containing mica and a clay-rich matrix. Unit 2 disconformably overlies unit 1. Near the base of unit 2, Ekdale and Stinnesbeck (1994, in prep.) observed burrows infilled with spherule-rich sediments of the underlying unit 1 and truncated at the top. This suggests habitation of invertebrates after deposition of unit 1 followed by erosion prior to deposition of the major part of the sandstones of unit 2.

Unit 3: Interlayered Sand-Silt Beds

Unit 3 consists of interlayered sand, shale, and silt beds with diverse sedimentological features such as horizontal laminations, ripple marks, small-scale cross-bedding, flaser bedding, and convolute lamination. Thin, fine-grained layers with typical Maastrichtian faunas indicating normal hemipelagic sedimentation and distinct layers enriched in zeolites are present within this unit (Keller and others, 1994a; Adatte and others, 1994, 1996). Very small charcoaled plant fragments are sometimes present. Where unit 3 directly overlies unit 1 (e.g., El Peñon, La Lajilla, La Sierrita), mud clasts of the Mendez Formation are commonly found near the base. Unit 3 is topped by one or more resistant rippled sandy limestone layers that are always heavily burrowed, most commonly by *Chondrites*, *Zoophycos*, *Thalassinoides*, and *Ophiomorpha* (Stinnesbeck and others, 1993, 1994b; Keller and others, 1994a; Ekdale and Stinnesbeck, 1994, in prep.). These discrete burrowed horizons indicate repeated recolonization by invertebrates and thus show that deposition of unit 3 also occurred over an extended time interval.

OUTCROP DESCRIPTIONS, OBSERVATIONS, AND INTERPRETATIONS

We have examined seven K-T boundary outcrops in northeastern Mexico, spanning over 300 km in a northwest-southeast direction (Fig. 3). All of these outcrops contain clastic deposits at or near the K-T boundary. However, in one section, La Parida, the clastic deposit is very thin and locally disappears, revealing a Mendez Formation to Velasco Formation transition devoid of significant lithological changes. In sections where thick clastic deposits are present, all three lithological units are usually present. Where outcrop exposures span one or more hundreds of meters, unit 2 tends to disappear laterally first, followed by unit 1. The topmost unit 3 is laterally most continuous. This depositional pattern suggests that the clastic member of each outcrop represents an individual channel-fill or fan deposit and does not blanket the region. These channelized deposits are thickest (with all three lithological units present) in the channel center, and thin toward the edges, where unit 2 is generally absent and unit 1 thins out and disappears, as illustrated in Figure 5.

Studies of northeastern Mexico K-T clastic deposits have generally concentrated on the El Mimbral outcrop (Smit and others, 1992; Stinnesbeck and others, 1993; Bohor and Betterton, 1993; Keller and others, 1994b) and little is known of the many other sections. Hypotheses explaining emplacement of the clastic deposit, such as impact-generated tsunami (Smit and others, 1992), turbidite flow (Bohor and Betterton, 1993), or gravity flows and sea-level lowstand deposits (Stinnesbeck and others, 1993; Keller and others, 1994a) are all based on El Mimbral with some later reference to La Lajilla and El Peñon (Stinnesbeck and others, 1993; Smit and others, 1994a, 1994b; Adatte and others, 1994; Keller and others, 1994a). For this reason, we use El Mimbral as the "type section" for describing the K-T boundary transition and clastic deposit. The seven other outcrop localities are described more briefly, concentrating on the manner in which they differ from the succession at El Mimbral.

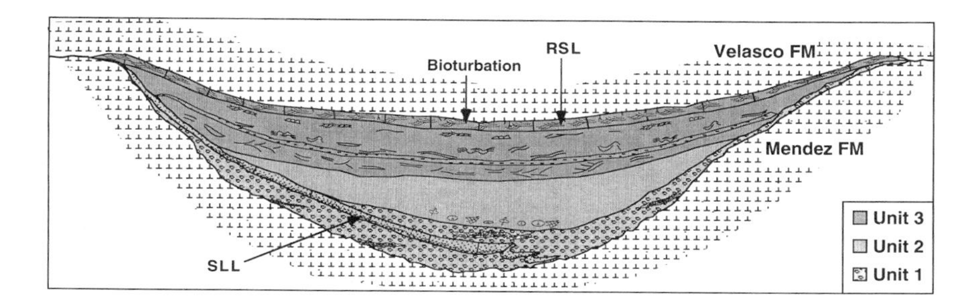

Figure 5. Sketch of idealized channel deposit showing thinning out of units 1, 2, and 3 toward the edge of the channel. S.L.L refers to sandy limestone layer within unit 1. R.S.L. refers to rippled sandy limestone layer that caps unit 3.

El Mimbral

The El Mimbral outcrop is located on the southwestern bank of the Mimbral creek ~10 km east of the main road from Ciudad Victoria to Tampico (23°13'N, 98°40'W). The outcrop was first described by Muir (1936) as marking a hiatus between the Maastrichtian Mendez and Paleocene Velasco formations. Hay (1960) inferred that the clastic deposit marked an angular unconformity. Smit and others (1992, 1994a, 1994b) interpreted the same deposit as an impact-generated tsunami or megawave deposit related to the hypothesized K-T impact crater near Chicxulub on the north coast of Yucatan. Stinnesbeck and others (1993, 1994b), Adatte and others (1996), and Keller and others (1994a, 1994b) disagreed with the tsunami interpretation and proposed deposition over a longer time period related to the latest Maastrichtian sea-level lowstand.

The Mimbral outcrop is ~152 m long and from 1 to 36 m high. The main part of the clastic member is 60 m long and varies between 0.2 and 3.0 m high, as illustrated in Figure 6. In cross-section, the clastic member represents a channel-fill deposit that thins at the channel edge to 20–25 cm at 152 m along the outcrop. Laterally, unit 2, the massive sandstone, disappears first, followed by unit 1, the spherule-rich layer. At the channel edge (152 m along the outcrop), only the topmost 20–25-cm-thick bioturbated limestone layer is present, with thin (1–2 cm) lenses of the spherule-rich sediments of unit 1 at its base. Benthic foraminifera indicate that the depositional environment shallowed from upper bathyal (400–500 m) in the late Maastrichtian Mendez Formation to outer neritic (200–300 m) depth during deposition of the early Tertiary Velasco Formation (Keller and others, 1994b).

Two transects were studied, one at the outer channel at 28 m along the outcrop across the thickest part of the clastic member as shown in Figure 7, and the second at the channel edge (152 m along the outcrop) along the thinnest part of the clastic member as shown in Figure 8. The stratigraphically most continuous K-T transition is found at the channel edge, where only the topmost 20–25 cm bioturbated rippled sandy limestone layer is present and thin (1–2 cm) lenses of unit 1. At this location, the K-T boundary is at the base of a 4-cm-thick clay layer with a 2–4-mm-thick red oxidized layer at its base that overlies the rippled sandy lime-

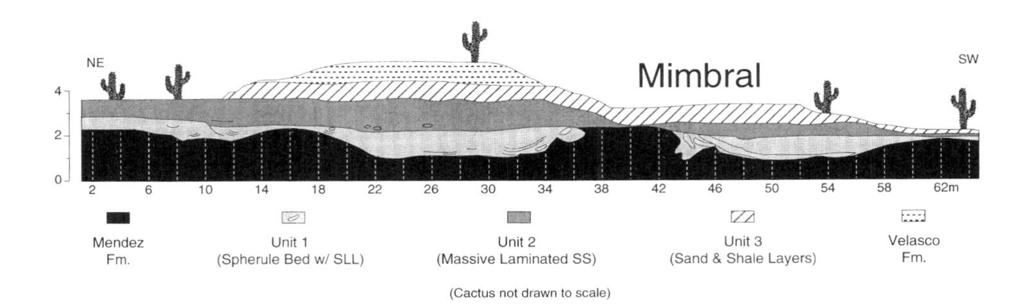

Figure 6. Sketch of El Mimbral outcrop based on measured section at 2 m intervals. S.L.L. refers to sandy limestone layer, S.S. is sandstone.

Figure 7. Outcrop at El Mimbral I showing central channel clastic deposit.

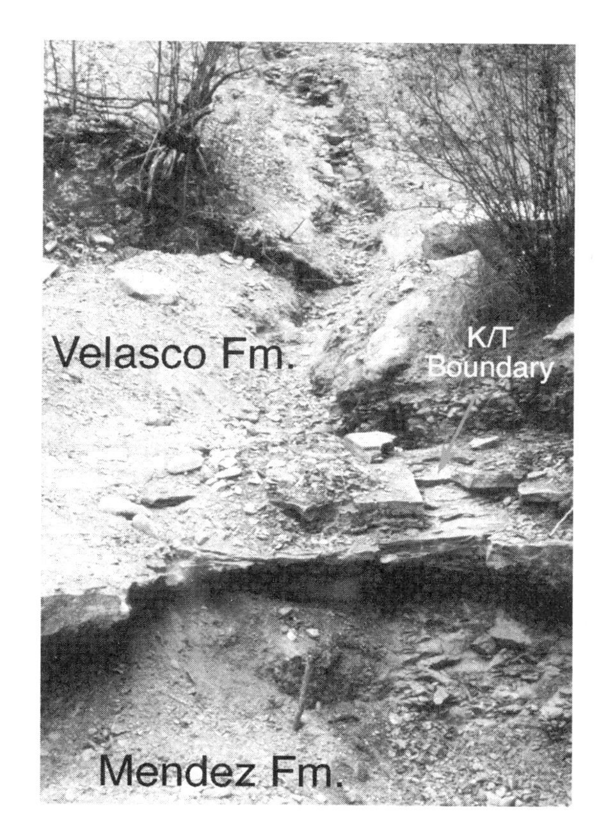

Figure 8. Outcrop at El Mimbral II showing K-T transect at right edge of channel.

stone layer at the top of unit 3 (Fig. 9). Rare shocked quartz grains have been found in the red layer (see Fig. 9 inset; Keller and others, 1994a). Shales of the Paleocene Velasco Formation (zone P1a) overlie the clay layer and K-T boundary.

Figure 10 illustrates the lithostratigraphy and biostratigraphy of the two

Figure 9 (top). Closeup of K-T boundary and red layer at El Mimbral II with insert of shocked quartz grain from the red layer.

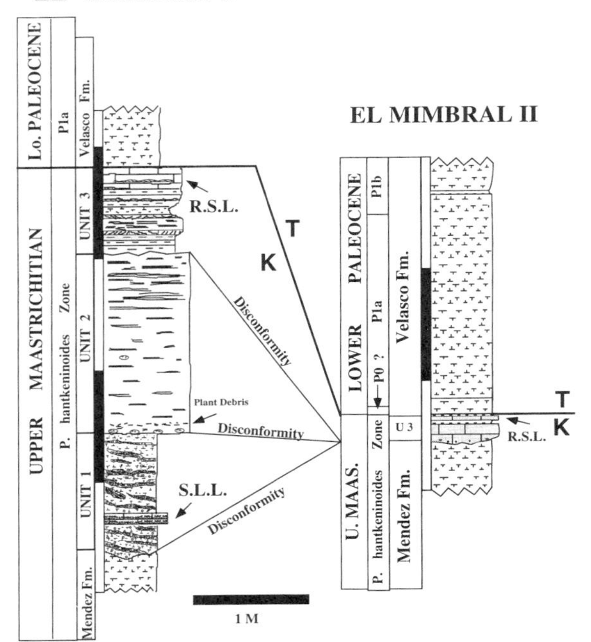

Figure 10 (left). Stratigraphic and lithologic correlation of two K-T boundary transects across the clastic deposit at the center and at the right edge of the channel. Note that at the edge of the channel only the topmost rippled sand limestone (R.S.L.) of the clastic deposit is present and the stratigraphic sequence between Mendez and Velasco formations is more continuous. S.L.L. refers to sandy limestone layer within unit 1.

Mimbral transects. Note that erosional disconformities are present at the base and top of the clastic member as well as between units 1, 2, and 3. These disconformities are marked by undulating surfaces, rip-up clasts, convolute surfaces, and/or sharp lithological contacts. Characteristic features of unit 1 of the clastic member are illustrated in Figure 11, showing the undulating erosional contact between the Mendez Formation and the spherule-rich layer, the presence of the resistant sandy limestone layer, and the overlying spherule-rich sediment. Unit 2 disconformably rests upon unit 1 (Fig. 11, 34 m mark along the outcrop). The sandy limestone layer has few spherules at the base and its top, where the overlying spherule-rich sediments drape over the convolute bedding surface (Fig. 12). Microlayering of thin and coarse sediments is apparent in thin sections of the sandy limestone layer, suggesting variable flow energies (W. Ward, 1994, personal commun.). At 36 m along the outcrop, the sandy limestone layer abuts the overlying sandstone of unit 2, which drapes around it (Fig. 13). This suggests that deposition of unit 2 occurred after the sandy limestone layer was already semi-lithified, and formed a resistant and deformed layer over which renewed sand deposition draped. Alternatively, this deformation could have occurred by sedi-

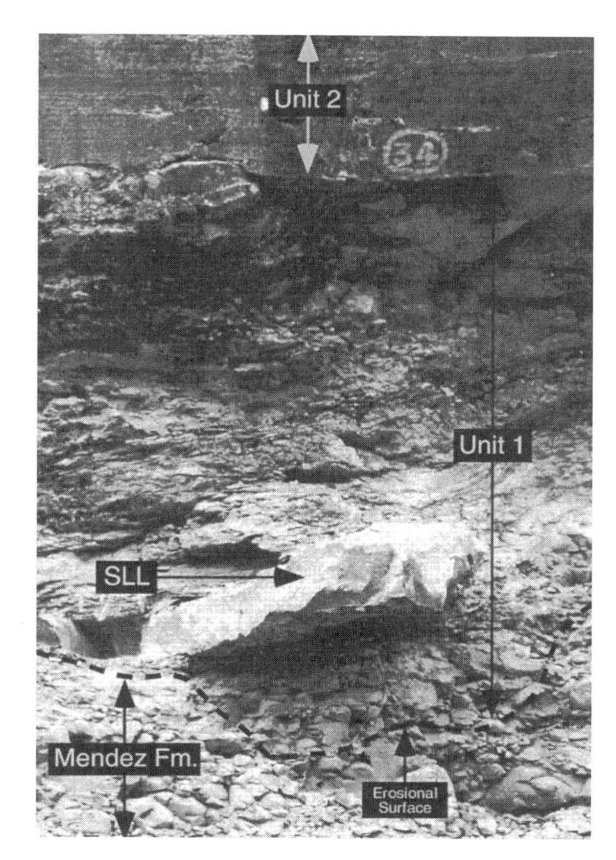

Figure 11. El Mimbral outcrop showing erosional disconformity between unit 1 and the Mendez Formation and the sandy limestone layer (S.L.L.) in unit 1.

Figure 12. Sandy limestone layer (S.L.L.) of unit 1 at El Mimbral showing spherules draped over convolute upper surface.

Figure 13. Sandy limestone layer (S.L.L.) at El Mimbral abuts overlying sandstone of unit 2 which drapes over it, indicating partial lithification of S.L.L. prior to deposition of sandstone of unit 2.

ment loading after deposition and semilithification of the clastic member. Whatever the age of deformation, the resistant sandy limestone layer represents a depositional regime nearly devoid of the underlying and overlying spherules and records a unique depositional event that can be correlated between outcrops spanning over 200 km (Keller and others, 1994a).

Smit and others (1992, 1994a, 1994b) interpreted the spherule-rich layer of unit 1 (including the sandy limestone layer) as the proximal impact ejecta that was deposited within hours after the bolide impact on Chicxulub and prior to the backwash of the first tsunami wave and deposition of the sandstone of unit 2. Our data suggest that unit 1 was deposited over a longer time period by at least three different events with variable detrital influx that resulted first in deposition of the spherule-rich sediments, followed by deposition of the sandy limestone layer, and then by spherule-rich sediments. We know of no depositional process that can deposit these three different layers by a single event over a few hours.

The most unusual features of unit 1 are the abundant spherules and the rare glass shards, which are absent in units 2 and 3. The spherule-rich sediments are loosely cemented and weathered, and spherules range from 1 to 5 mm in diameter. Large spherules may contain smaller spherules, as shown in Figure 14. Most spherules are infilled with blocky calcite and surrounded or partly filled by mixed layers (illite/smectite, chlorite, or mica; Stinnesbeck and others, 1993; Adatte and others, 1994, 1996). Some spherules are replaced by smectite, kaolinite, and glauconite; others contain apatite concretions, rutile crystals (Fig. 15), rounded clasts of limestone, and foraminiferal tests (Fig. 16). Commonly, spherules

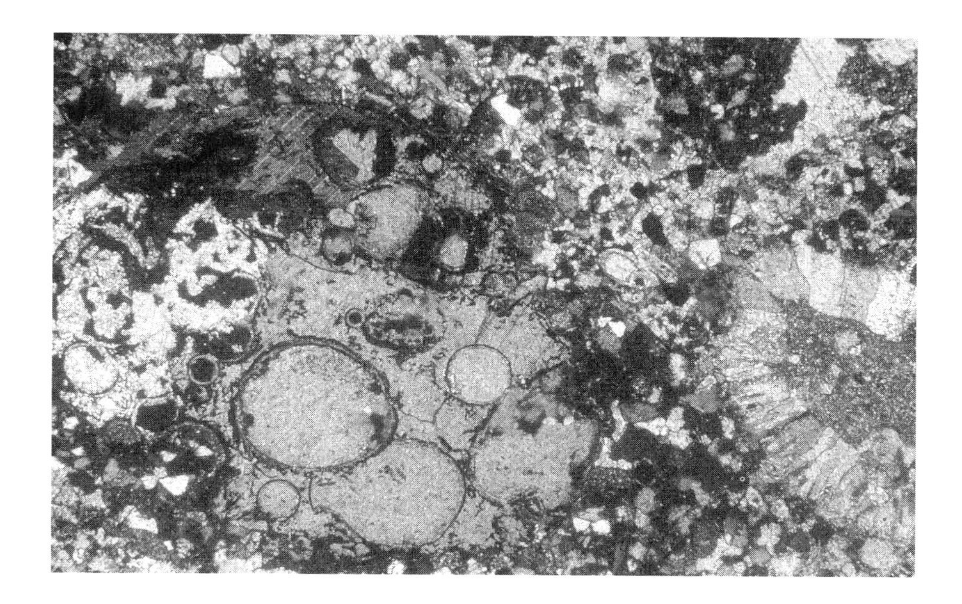

Figure 14. El Mimbral thin section of spherule-rich layer in unit 1 showing large composite spherule with several smaller spherules enclosed.

Figure 15. El Mimbral thin section of spherule-rich layer of unit 1 showing spherule containing opaque mineral (rutile?).

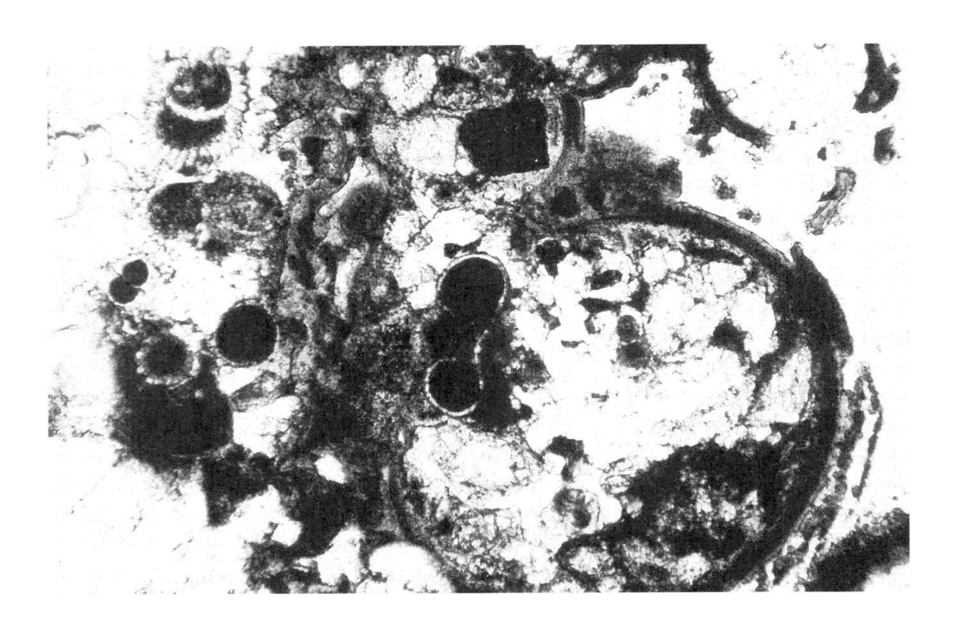

Figure 16. El Mimbral spherule containing foraminifera with glauconite infilling.

infilled with blocky calcite are surrounded by a thin, tan organic layer that remains intact after acid treatment. These hollow spheres are probably algal resting cysts, which have been observed in K-T boundary deposits at Stevns Klint (Hansen and others, 1986). Moreover, there is a significant organic content in unit 1 relative to units 2 and 3 (Stinnesbeck and others, 1993). Acid-treated residues commonly contain green chlorite spherules with radiating structures and brown palagonite of radiating smectite. These two minerals and radiating structures suggest a volcanic affinity (Lyons, 1993, personal commun.). Additional evidence of volcanic input is found in the presence of aluminum chromite with almost no nickel in all three units, as well as above and below the clastic deposit and in two discrete zeolite-enriched layers within unit 3 (Stinnesbeck and others, 1993; Adatte and others, 1994, 1996).

Our observations indicate that the spherules of unit 1 are of multiple origins, including oolites, oncolites, algal resting cysts, and volcanic spherules. We suggest that they originally formed and accumulated in a shallow-water neritic environment, then transported to upper bathyal or outer neritic depths during the latest Maastrichtian sea-level lowstand (see Keller and Stinnesbeck, this volume). In contrast, Smit and others (1992, 1994a, 1994b) interpreted all unit 1 spherules as originally glassy, hollow, impact-produced microtektites that have been subsequently filled and replaced by calcite and phyllosilicates. Although it is possible and even likely that some spherules were originally glassy (impact or volcanic), such an origin cannot be distinguished from calcite, glauconite, or phyllosilicate replacements. Moreover, spherules that contain foraminifers, rock fragments, or rutile crystals, or are surrounded by an organic shell, cannot be of impact origin. Thus, even if some glassy spherules are found, the depositional mechanism must explain these multiple origins from primarily shallow water environments.

Glass shards are extremely rare and restricted to the spherule-rich unit 1 (Smit and others, 1992; Stinnesbeck and others, 1993; Keller and others, 1994a). These vary in size from less than 100 μm to greater than 400 μm, are irregular in outline, and are commonly vesicular. Colors vary from yellow-amber, green, brown, and red to rarely black. Black glass shards have total oxides close to 100%, whereas colored glasses average 87%. These hydrated glasses are probably weathered. The average glass composition is similar to that of glass from the Beloc section in Haiti (Smit and others, 1992; Bohor and Betterton, 1993; Stinnesbeck and others, 1993). Jéhanno and others (1992) and Lyons and Officer (1992) interpreted Beloc glass as of volcanic origin because of its high Fe oxidation state and the absence of lechatelierite, whereas Izett (1990), Blum and Chamberlain (1992), Sigurdsson and others (1991), and Koeberl and Sigurdsson (1992) interpreted Beloc glass as of impact origin. For a discussion of this controversy, see Koeberl (1994) and Robin and others (1994). If the glass shards present in the spherule-rich sediments of unit 1 at Mimbral and other sections are determined to be of impact origin, they stratigraphically predate the K-T boundary and hence represent an earlier event.

Unit 2 consists of a weakly laminated sandstone, in some sections with burrows near the base, which are infilled with the underlying spherule-rich sediment of unit 1 (Ekdale and Stinnesbeck, 1994, in prep.). These burrows indicate habitation of invertebrates during deposition and thus rule out tsunami deposition. The only unusual feature of unit 2 at El Mimbral is the presence of discrete layers of wood and plant debris at its base (Fig. 17). This coarse plant debris has not been observed elsewhere, although rare small charcoaled wood fragments are present throughout units 2 and 3 in all sections examined. Smit and others (1992) interpreted the large plant debris as transported by a tsunami wave from the coastal region, and the small fragments throughout units 2 and 3 as evidence of combustion by an impact-generated wildfire. Because plant debris generally floats and settles at a slower rate than sand grains, they argued that it must have been waterlogged in coastal deposits prior to transport by a tsunami wave. Smit and others (1992) did not explain how small charcoal fragments from an impact-generated wildfire could have settled through 500 m of water within a few hours or even a few days. We suggest that ordinary depositional processes can account for the local large plant debris layers at Mimbral. Moreover, rare charcoal fragments produced by natural fires can be found in terrigenous clastic deposits of any age. No impact-produced wildfires need be invoked.

Unit 3 is lithologically, mineralogically, and petrographically most variable. This interval contains two distinct layers enriched in zeolites and thin layers rich in planktonic foraminifera, indicating periods of volcanogenic influx and normal hemipelagic sedimentation, respectively. Sedimentologic features are diverse and include laminated intervals, ripple marks, small-scale cross-bedding, and coarse-grained layers separated by thin muddy layers. The top of unit 3 consists of a 20–25-cm-thick rippled sandy limestone layer that is burrowed by *Chondrites*, *Zoophycos*, and *Thalassinoides*-like structures. There is no evidence of Tertiary

Figure 17. Plant debris in discrete layers near base of unit 2.

microfossils in the burrows, indicating that burrowing did not originate in the overlying Tertiary shales. Moreover, Ekdale and Stinnesbeck (1994) reported several horizons of bioturbation, indicating habitation at different times. Smit and others (1992, 1994a, 1994b) interpreted unit 3 as the back and forth of weakening tsunami or seiche waves across the Gulf of Mexico that deposited alternating coarse- and fine-grained layers over a period of hours to a few days. In support of this interpretation, they cited upward fining of sediments and the presence of bidirectional currents within individual sediment layers, all pointing in the direction of Chicxulub. However, no upward fining in the alternating coarse- and fine-grained sediment layers of unit 3 could be observed (Stinnesbeck and others, 1994b; Adatte and others, 1996). Moreover, no bidirectional current directions could be confirmed by us or sedimentologists participating in a February 1994 field trip to these sections. Current directions were observed to be unidirectional within each sediment layer, with occasional directional change from one layer to the next at La Lajilla. No uniform current direction pointing to Chicxulub can be observed, as claimed by Smit and others (1994a, 1994b), and the directional change between sediment layers at La Lajilla is in the opposite direction. This is consistent with normal current transport of sediments, rather than seiche waves. Moreover, sediment deposition of these fine- and coarse-grained layers, including several discrete layers of hemipelagic sedimentation and volcanogenic influx that can be correlated over 300 km, and several discrete horizons of bioturbation, are inconsistent with tsunami deposition over a period of hours to days. These sediment layers and horizons of bioturbation suggest that periods of normal sediment deposition alternated with periods of terrigenous influx over an extended time period.

Smit and others (1994a, 1994b) also claimed elevated levels of iridium in fine-grained intervals of unit 3, as well as in the sandy rippled limestone layer that caps the clastic deposit. The only data they published, however, show 0.8 ppb of iridium in the basal shale of the Tertiary Velasco Formation (zone P1a) that disconformably overlies the rippled sandy limestone layer at the top of the clastic deposit. Iridium values at the top of the underlying sandy limestone layer range between 0.01 and 0.4 ppb. Smit and others (1992, 1994a, 1994b) interpret these values as evidence of settling of iridium through the water column and hence support for an impact-induced tsunami deposit. However, these authors neglected to mention that such tailing of iridium values is usually related to leaching, transport, and precipitation at redox boundaries (Sawlowicz, 1993, and references therein). Moreover, iridium measurements by Rocchia and Robin (*in* Stinnesbeck and others, 1993; Keller and others, 1994a) at El Mimbral confirm high values of 0.5 to 0.8 ppb in zone P0 and at the base of zone P1a of the Velasco Formation, and low values of 0.1 to 0.3 ppb in the top of the sandy rippled limestone layer below. No anomalous iridium concentrations were observed in units 1, 2, and 3 of the clastic deposit. Thus, the Tertiary zone P0 and P1a iridium anomaly is consistent with the K-T boundary enrichment observed in K-T sections worldwide, but lends no support to the impact-tsunami interpretation of the clastic deposit.

El Peñon

The El Peñon K-T boundary outcrop is located 40 km east of Linares on the unpaved road leading to Burgos (Fig. 3). The outcrop is ~2 km northwest of the

Porvenir lake (25°58'N, 99°12.5'W). The main El Peñon outcrop (Peñon I) has a 7-m-thick clastic deposit that is exposed for ~100 m and dips 8° to the northeast. A second outcrop (Peñon II) is located 200 m to the north and has a 3-m-thick clastic deposit. Lithological columns of the two outcrops are shown in Figure 18. The two sections differ primarily in that the massive and partly laminated sandstone of unit 2 is absent at the Peñon II outcrop, but present as a 3.7-m-thick member in Peñon I. This suggests a channelized deposit, with Peñon I repre-

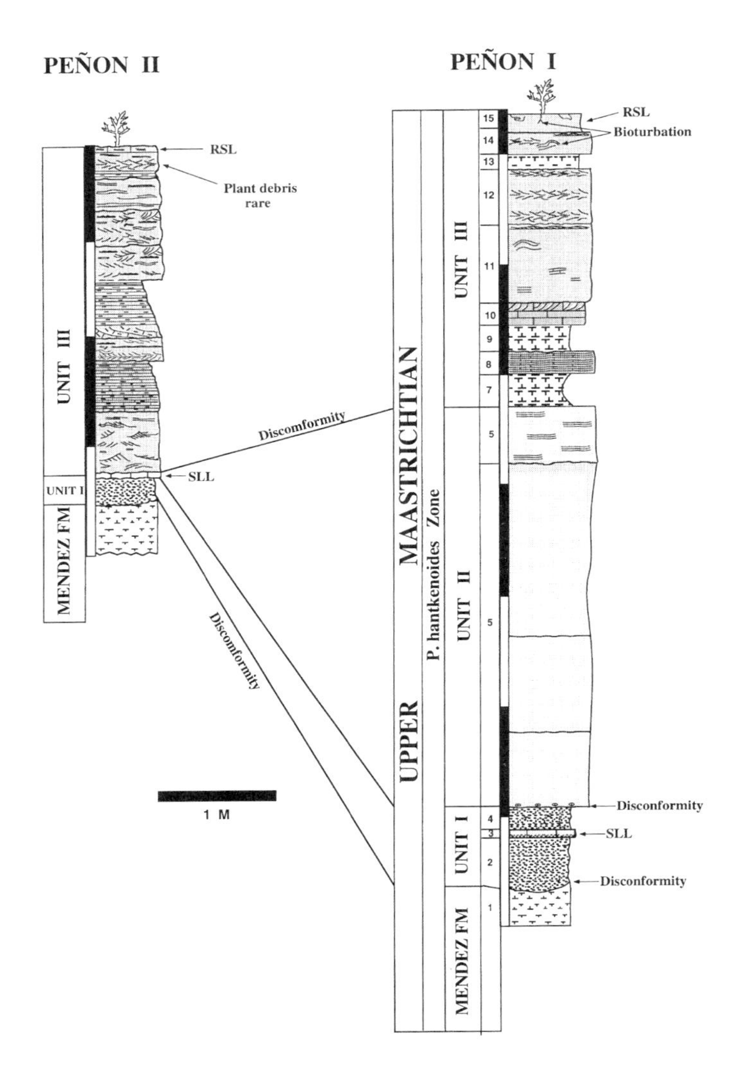

Figure 18. Stratigraphic and lithologic correlation of two outcrops at El Peñon I and II. R.S.L. refers to rippled sandy limestone; S.L.L. refers to sandy limestone layer.

senting the thick central portion and Peñon II the channel edge where unit 2 disappears, as also observed at El Mimbral.

The clastic deposit is similar to that at El Mimbral. Unit 1 disconformably overlies an undulating erosional surface of the Maastrichtian Mendez Formation and contains a 5–8-cm-thick sandy limestone layer with few spherules. At El Peñon I, this resistant layer lies within the loosely cemented spherule-rich sediments (Fig. 19), similar to El Mimbral, whereas at Peñon II the sandy limestone layer caps unit 1 at an erosional disconformity, where the upper spherule-rich sediments and unit 2 are missing (Fig. 18). The 3.7-m-thick laminated sandstone of unit 2 at Peñon I disconformably overlies unit 1 and has mud clasts at its base. No discrete layers of plant debris are present, but bioturbation is evident. Weak upward fining can be observed in unit 2. Unit 3, the interlayered sand, shale, and siltstone beds topped by the bioturbated sandy limestone layers, are ~2.7 m thick at both Peñon outcrops (Fig. 20). Sedimentary features in unit 3 include laminations, convolute bedding, ripple marks, flaser bedding, thin fine-grained muddy layers, and two layers enriched in zeolites (Adatte and others, 1996). The rippled sandy limestone layer that caps unit 3 contains abundant *Chondrites*, *Thalassinoides*, and *Zoophycos* burrows (Fig. 21), and at least two other discrete bioturbated horizons are present below (Ekdale and Stinnesbeck, in prep.). No sediments are present above this bioturbated layer.

Figure 19. El Peñon I, lower part of clastic deposit, showing disconformity between unit 1 and the Mendez Formation, and the sandy limestone layer (S.L.L.) within unit 1.

Figure 20. El Peñon I showing alternating beds of sand and siltstone of unit 3.

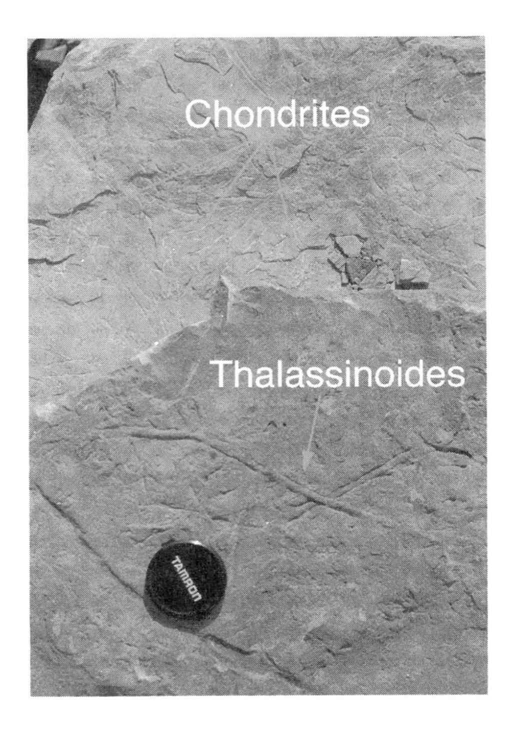

Figure 21. El Peñon showing different bioturbated sediment layers with Chondrites *and* Thalassinoides *burrows.*

La Lajilla

The La Lajilla K-T boundary section is located 40 km east of Ciudad Victoria and 8 km south of the village of Casas (Fig. 3). Outcrops are located on north and south sides along the outlet of the La Lajilla lake dam, ~200 m north of the village of La Lajilla (23°40'N, 98°43'W).

Two sections were analyzed and their lithostratigraphies and biostratigraphies are illustrated in Figure 22. The clastic member is a maximum of 1.3 m thick and underlies 2 m of gray shales of the Paleocene Velasco Formation. Approximately 75 cm of Maastrichtian Mendez marls are exposed below the clastic member. An erosional contact separates the Mendez Formation from the overlying spherule-rich sediments of unit 1, which are variable in thickness and reach a maximum of 35 cm. An 8-cm-thick sandy limestone layer caps unit 1 and is disconformably

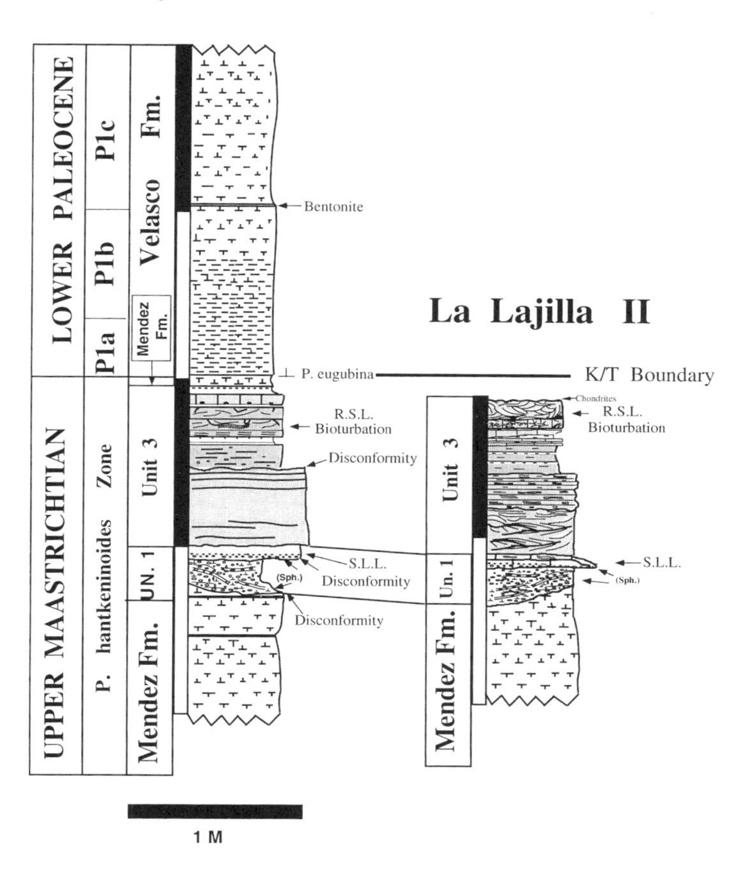

Figure 22. Stratigraphic and lithologic correlation of two outcrops at La Lajilla I and II. R.S.L. refers to rippled sandy limestone. S.L.L. refers to sandy limestone layer. Note the thin layer of Mendez marl with P. hantkeninoides *zone foraminifera on top of unit 3 at La Lajilla I, indicating that the clastic deposit predates the K-T boundary.*

overlain by unit 3 (Fig. 23). Unit 2 is missing at La Lajilla, similar to El Peñon II.

Unit 3 consists of 1 m of interlayered sand, shale, and siltstone beds topped by rippled sandy limestone layers (Fig. 24). Two fine-grained interbeds enriched in zeolites can be correlated with the El Mimbral, El Peñon, and El Mulato sections (Adatte and others, 1994, 1996). Unit 3 at La Lajilla is capped by two to three resistant rippled sandy limestone layers that are separated by thin interlayers of siltstone (Fig. 25). Bioturbation by *Chondrites*, *Thalassinoides*, and *Zoophycos* is present in all sandy limestone layers. Burrows are infilled with Maastrichtian sediments.

Above the rippled sandy limestone layers is a thin (5–10 cm) Mendez marl layer containing a Maastrichtian *P. hantkeninoides* zone foraminiferal assemblage, including the index taxon and no evidence of Tertiary planktonic foraminifera (Lopez-Oliva and Keller, 1996). This marly layer indicates that normal Maastrichtian sedimentation resumed after deposition of the clastic deposit. Immediately above this marly layer, a well-developed zone P1a fauna is present, including *P. eugubina*, *P. longiapertura*, *E. fringa*, *E. edita*, *Planorotalites compressus*, *Chiloguembelina waiparaensis*, *Woodringina hornerstownensis*, and *Globanomalina pentagona*. This assemblage is characteristic of the upper part of zone P1a (subzone P1a2) and indicates a K-T boundary hiatus, with the lower Danian, zone P0, and part of P1a missing (Lopez-Oliva and Keller, 1996).

Figure 23. La Lajilla I showing disconformities above and below the sandy limestone layer (S.L.L.), and overlaying unit 3.

Figure 24. La Lajilla II showing Mendez Formation overlain by a thin spherule-rich unit 1 topped by sand-silt alternations of unit 3.

Figure 25. La Lajilla II showing several beds of the bioturbated sandy limestone separated by thin beds of siltstone.

El Mulato

The village of El Mulato is located on the unpaved road from Linares to Burgos ~60 km east of Linares (Fig. 3, 24°54'N, 98°57'W). The K-T boundary outcrop spans 50 m across a hillside to the north of the village. The outcrop consists of ~20 m of gray marls of the Maastrichtian Mendez Formation followed by a 2-m-thick clastic deposit overlain by several tens of meters of gray shales of the Tertiary Velasco Formation (Fig. 26).

The clastic deposit contains all three lithological units similar to El Mimbral I and El Peñon I (Figs. 27 and 28). The spherule-rich unit 1, however, reaches only a maximum of 10 cm and lacks the prominent resistant sandy limestone layer characteristic of El Mimbral, El Peñon, and La Lajilla. Unit 1 disconformably overlies Mendez marls and is disconformably overlain by unit 2. The laminated sandstone of unit 2 is 1.4 m thick and shows weak upward fining and convolute bedding in the upper part. No plant debris layers are present. Unit 3 disconformably overlies unit 2 and consists of 77 cm of interlayered sand and siltstone beds topped by two layers of the bioturbated rippled sandy limestone. Similar to El Mimbral, La Lajilla, and El Peñon, some layers of unit 3 are enriched in zeolites (Adatte and others, 1994, 1996), and some layers contain typical *P. hantkeninoides* zone assemblages, suggesting intervals of normal hemipelagic sedimentation. Sedimentological features of unit 3 include laminated intervals and rippled, wavy, oblique, and convolute bedding similar to unit 3 at all other outcrops.

Figure 26. El Mulato showing clastic deposit with Mendez and Velasco formations.

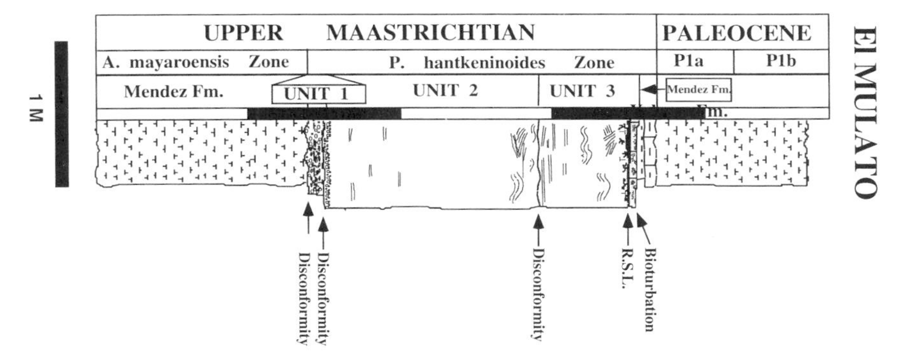

Figure 27. Stratigraphic and lithologic column of the El Mulato K-T transition. Note that the thin layer of Mendez marl with P. hantkeninoides *zone foraminifera on top of unit 3 indicates that the clastic deposit predates the K-T boundary. R.S.L. refers to rippled sandy limestone.*

Figure 28. El Mulato showing units 1, 2, and 3 of the clastic deposit and disconformity with the Mendez Formation.

Above the rippled sandy limestone layer of unit 3 is a 5–10-cm-thick marly limestone layer containing abundant planktonic foraminifera of the *P. hantkeninoides* zone, similar to La Lajilla. This indicates that in both El Mulato and La Lajilla sections (as well as La Parida), deposition of the clastic deposit preceded the K-T boundary by a short time interval. Immediately above the marly limestone layer, a well-developed Danian fauna of zone P1a is present. Similar to the other outcrops, the basal Tertiary zone P0 and the lower part of zone P1a are absent and mark a short K-T boundary hiatus.

La Parida

The hamlet of La Parida is located 40 km northeast of Linares (Fig. 3). The K-T boundary section is found 3 km east of La Parida, and ~100 m north of La Parida creek. At this location, the clastic deposit can be traced over a distance of 50 m. It reaches a maximum of 80 cm in the east and thins westward to 10 cm and disappears. Where the clastic deposit is absent, marls of the Maastrichtian Mendez Formation conformably underlie marls of the Tertiary Velasco Formation. Figure 29 illustrates three transects that depict the thinning out and disappearance of the clastic deposit at La Parida. In transect 1, the clastic deposit is 10 cm thick (Fig. 30), thins to 3–5 cm over a distance of 5 m (Fig. 31), and disappears over a dis-

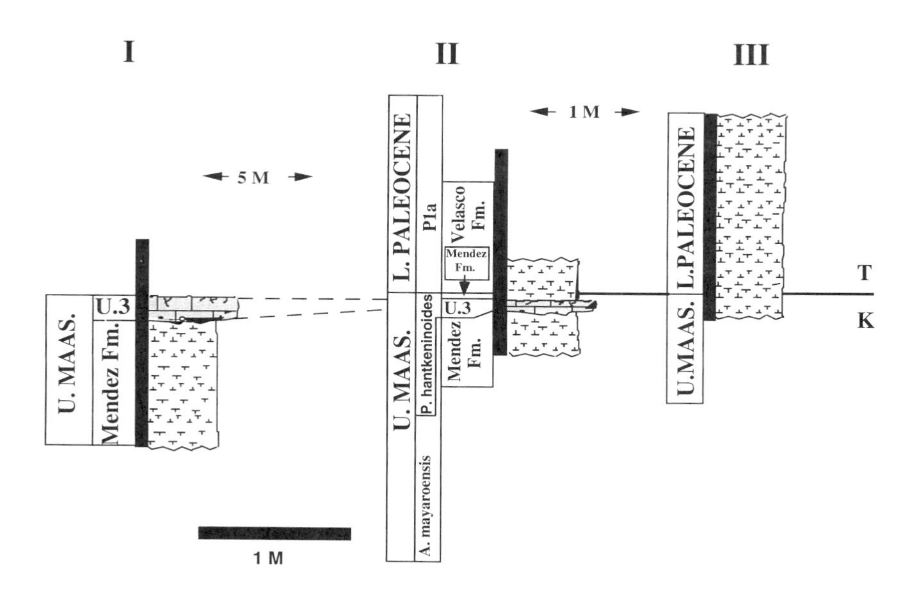

Figure 29. Stratigraphic and lithologic column of three La Parida transects showing lateral thinning out of the clastic deposit. Note that the thin layer of Mendez marl with P. hantkeninoides *zone foraminifera above the clastic deposit indicates that the latter predates the K-T boundary. Note also that only the bioturbated rippled sandy limestone (R.S.L.) layer from the top of unit 3 is present.*

Figure 30. La Parida showing outcrop at transect 1.

Figure 31. La Parida showing outcrop at transect 2.

tance of 1 m in transect 3 (Fig. 32). The clastic deposit in the La Parida transects consists of the bioturbated rippled sandy limestone layer that caps unit 3 in all other sections (e.g., El Peñon, El Mimbral, El Mulato, La Lajilla, La Sierrita). *Chondrites* and *Thalassinoides* burrows are common and infilled with Maastrichtian sediments.

At La Parida, similar to La Lajilla and El Mulato, a 5–10-cm-thick marl layer of latest Maastrichtian *P. hantkeninoides* zone age overlies the clastic deposit (Lopez-Oliva and Keller, 1996). Immediately above this layer, a well-developed early Tertiary planktonic foraminiferal assemblage of zone P1a is present. This indicates a short K-T boundary hiatus with zone P0 and part of zone P1a missing, similar to La Lajilla, El Mulato, and El Mimbral. Thus, three out of four sections that contain Tertiary sediments (all except Mimbral) reveal the presence of latest Maastrichtian sediments overlying the clastic member.

La Sierrita

The K-T boundary outcrops are located near the hamlet of La Sierrita on the road from Montemorelos to Vaquería (25°12.3'N, 99°31'W, Fig. 3). The La Sierrita K-T outcrops (also known as Loma Las Rusias) are found over several kilometers along hilltops and valleys trending in a north-south direction. Three transects are illustrated in Figure 33 to show the lateral variability of the clastic deposit. In some outcrops only unit 3, or part of unit 3, is exposed (La Sierrita II and III), whereas in others units 1 and 2 are present. In all sections, the clastic deposit dis-

Figure 32. La Parida showing outcrop at transect 3 with no clastic sediments present.

conformably overlies marl of the Mendez Formation. The spherule-rich layer of unit 1 varies between 0 and 20 cm and no resistant sandy limestone layer has been observed. At la Sierrita I ~20 cm of Maastrichtian marls overlie unit 1. The unusual presence of this marl layer, however, may be due to local tectonic disturbance. Units 2 and 3 are similar in sedimentological features in all other clastic outcrops.

Los Ramones

The village of Los Ramones is located ~40 km east of the city of Monterrey (25°42' N, 99°35.5'W, Fig. 3). The K-T clastic deposit is exposed 500 m east of the village on the eastern bank of the Pesqueria River. The outcrop spans ~50 m and has a 4-m-thick clastic deposit that dips northeast and disappears into the riverbed (Fig. 34). The clastic deposit disconformably rests on Maastrichtian marls of the Mendez Formation. The presence of the late Maastrichtian index species *P. hantkeninoides* indicates that clastic deposition occurred during the last 200 k.y. of the Maastrichtian. In contrast to all other K-T clastic deposits known to us, the Los Ramones deposit does not permit positive identification of any of the three lithological units observed at other locations. For instance, no spherule-rich layer of unit 1 is present, and we can only tentatively identify undifferentiated units 2 and 3, which consist of generally structureless sandstones with some upward fining (Fig. 35). Within these sandstones are numerous zones with many mud clasts of Mendez marls that

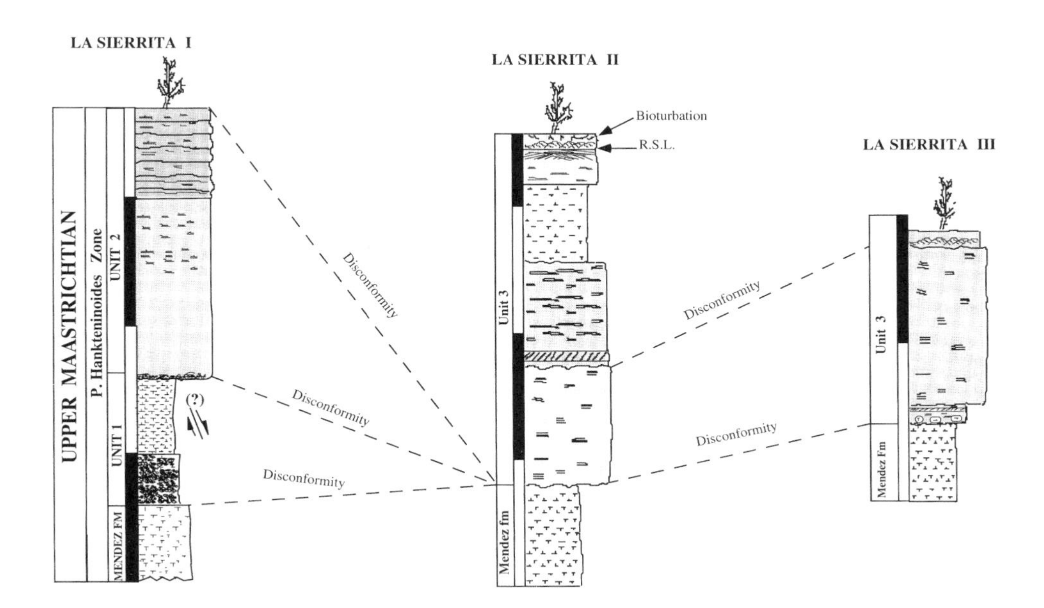

Figure 33. Stratigraphic and lithologic column of three La Sierrita outcrops showing variable thicknesses of the clastic deposit and lateral variations in units 1, 2, and 3. R.S.L. refers to rippled sandy limestone.

Figure 34 (top). Los Ramones outcrop with view of Mendez Formation and overlying clastic deposit.

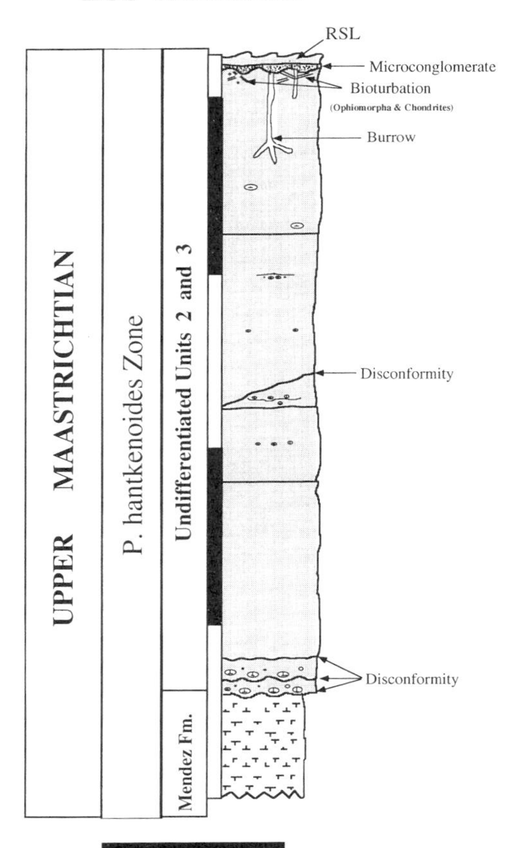

Figure 35 (left). Stratigraphic and lithologic column of the Los Ramones outcrop. Note that in this shallow-water section the spherule-rich unit 1 is absent and units 2 and 3 are not easily differentiated. R.S.L. is rippled sandy limestone.

suggest disconformities. The top of the clastic deposit is bioturbated by shallow-water crabs (*Ophiomorpha*) and underlies a thin microconglomeratic bed. A sandstone layer with unidirectional ripples caps the clastic deposit (Fig. 35) and appears to be equivalent to the rippled sandy limestone observed in all other outcrops. Trace fossils and lithological characteristics suggest that deposition of this deposit occurred in a shallower (neritic) environment than at other locations to the south.

LITHOSTRATIGRAPHIC AND MINERALOGIC CORRELATIONS

Near K-T boundary channelized clastic deposits in northeastern Mexico consist primarily of transported terrigenous and shallow-water neritic debris. Nevertheless, the depositional sequence is surprisingly constant, allowing recognition of three well-defined units based on lithologic and sedimentologic characteristics that can be traced over a distance of more than 300 km. Even mineralogic variations are relatively constant within the three units (Keller and others, 1994b; Adatte and others, 1994, 1996). Specifically, two distinct layers rich in zeolites (clinoptilolite-heulandite) have been recognized near the base and top of unit 3, below the rippled sandy limestone layer that can be correlated in all sections examined. These zeolite-enriched layers indicate a significant volcaniclastic influx during these times. Periodic volcanism occurred throughout the late Maastrichtian and early Tertiary in northeastern Mexico, as indicated by thin bentonite layers in the Mendez and Velasco formations. These zeolite-enriched layers may either represent periods of volcanic eruptions or reworking and transport of volcaniclastic deposits.

The lithostratigraphic correlation of nine K-T outcrops is illustrated in Figure 36: correlation line A at or just above the top of the clastic deposit marks the K-T boundary. Tertiary sediments are absent at Los Ramones, La Sierrita, and El Peñon. Correlation line B marks the base of unit 3 and the top of unit 2. Note that part or all of unit 3 is generally present in all outcrops with clastic deposits. Correlation line C marks the base of unit 2 and the top of unit 1. Note that unit 2 is frequently absent (e.g., Mimbral II, Peñon II, La Lajilla, La Sierrita II, La Parida). The absence of unit 2 appears to be directly related to its distance from the maximum thickness of the clastic deposit. This suggests a channel-fill where unit 2 thins toward the channel edges (see Fig. 5). Correlation line D marks the top of the Mendez Formation and base of unit 1. At least part of unit 1 is present at El Mimbral, La Lajilla, El Peñon, and El Mulato. Within unit 1 at La Lajilla, El Peñon, and El Mimbral is a distinct subunit consisting of a resistant 10–20-cm-thick sandy limestone layer. Depending upon erosion at the disconformity between units 1 and 2 and the distance from the channel center, this sandy limestone layer may be within the spherule layer as at El Mimbral I and El Peñon I, or cap the spherule layer as at La Lajilla, El Mimbral II, and El Peñon II (Fig. 36).

Correlation of zeolite-enriched layers near the top and base of unit 3 is illustrated in Figure 37, along with the lithostratigraphy. The solid line marks the K-T boundary at or just above unit 3 and the top of the clastic deposit, and the dashed

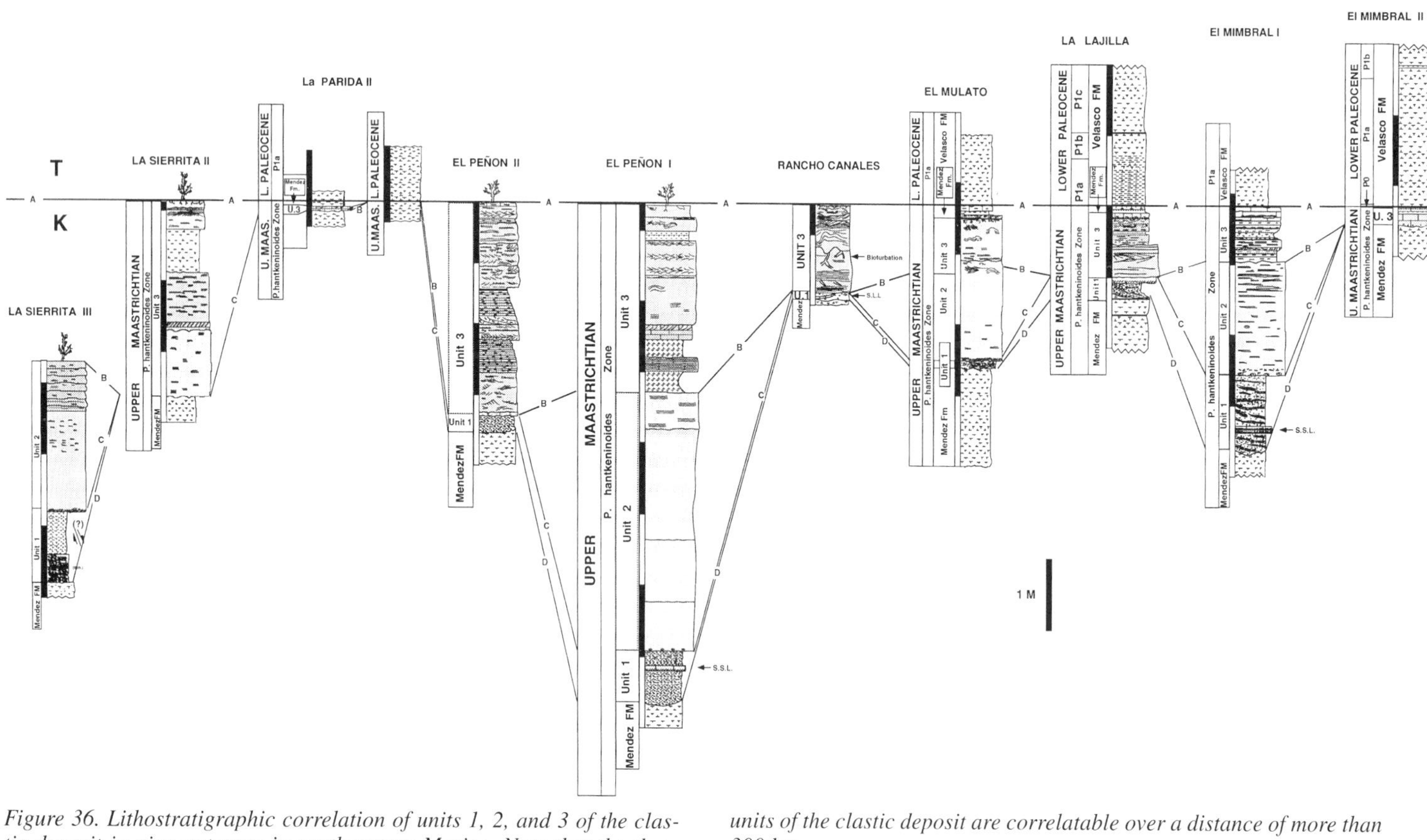

Figure 36. Lithostratigraphic correlation of units 1, 2, and 3 of the clastic deposit in nine outcrops in northeastern Mexico. Note that the three units of the clastic deposit are correlatable over a distance of more than 300 km.

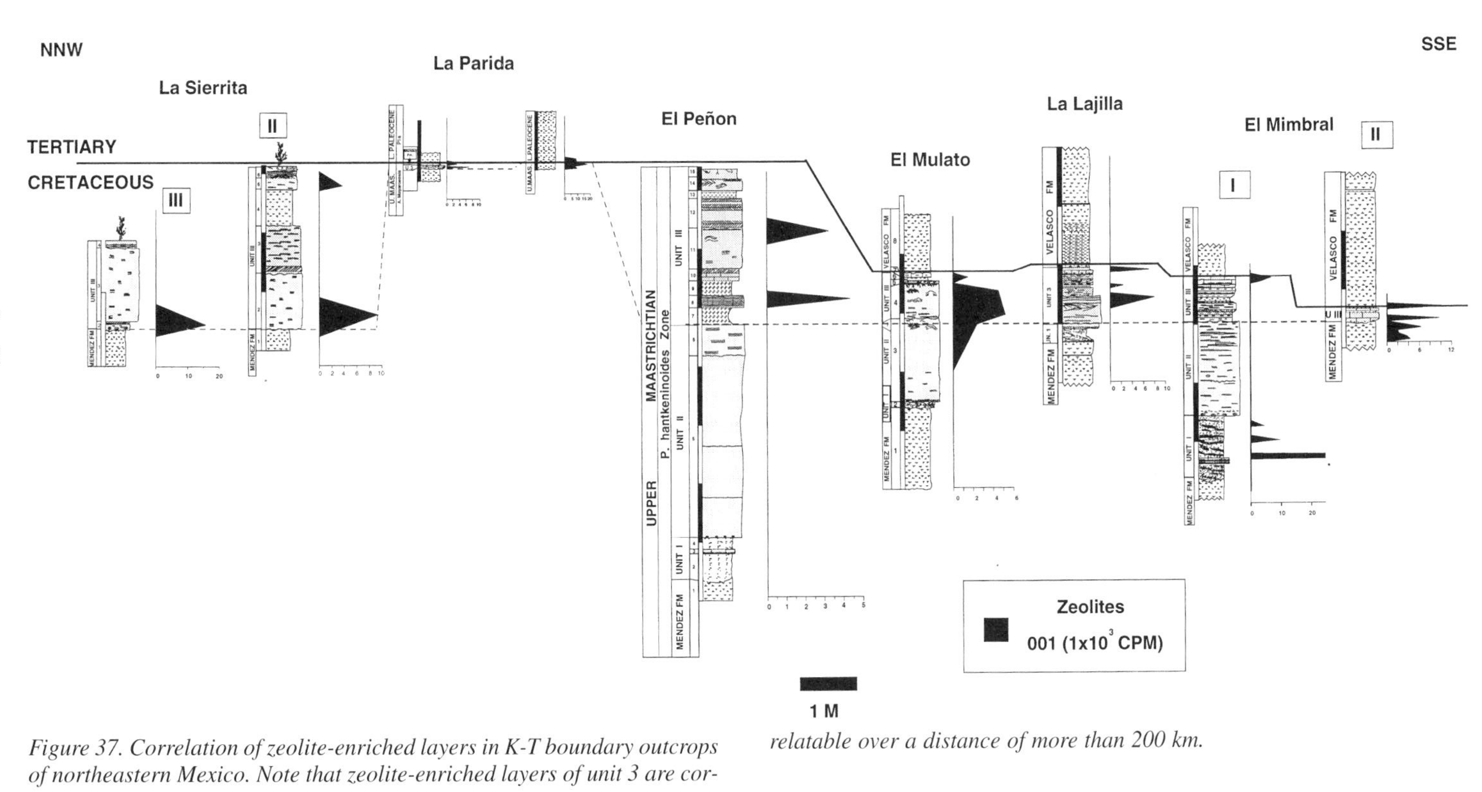

Figure 37. Correlation of zeolite-enriched layers in K-T boundary outcrops of northeastern Mexico. Note that zeolite-enriched layers of unit 3 are correlatable over a distance of more than 200 km.

line marks the base of unit 3. Note that, although zeolite-enriched layers are most prominent in unit 3, they have also been observed in the K-T boundary clay at Mimbral II, in the spherule-rich unit 1 at Mimbral I, and in the Mendez marls (Adatte and others, 1994, 1996).

Lithologically distinct clastic units and mineralogically distinct (zeolite-enriched) layers correlatable over 200–300 km are probably not the result of an impact-generated tsunami megawave over a period of hours to days, as suggested by Smit and others (1992, 1994a, 1994b). Rather, explanation of this stratigraphic sequence requires multievent deposition over an extended time period with systematic pulsed input of nearshore and neritic sediments over an extended time period. A depositional model for these clastic sediments is proposed below.

DISCUSSION AND DEPOSITIONAL MODEL

Sedimentologists who visited the near K-T boundary clastic deposits in northeastern Mexico in conjunction with the 1994 Snowbird III meeting generally agreed that these sediments do not represent classic turbidite sequences or any single event deposit. They also agreed that no obvious depositional mechanism is apparent. Most puzzling is the presence of the spherule-rich unit 1 sediments at the base of the clastic deposit, which has now been found in several northeastern Mexico sections and one section, Tlaxcalantonga, near La Ceiba in east-central Mexico. How was such a unique sediment layer deposited in the same stratigraphic position over a distance of 500 km? If this unit represents a shallow-water deposit consisting primarily of oolites, oncolites, glauconite, and algal resting cysts, then how was it transported to the upper bathyal to outer neritic depths of the K-T boundary sections? We suggest that these sediments were transported during a sea-level lowstand, as suggested by the presence of shallow-water benthic foraminifera (Keller and others, 1994b). This interpretation, however, does not necessarily explain the presence of volcanic debris and rare glass shards in unit 1: the shards have geochemical characteristics similar to those of glass from the Chicxulub cores on Yucatan and the Beloc, Haiti, K-T boundary section. Some workers believe that the presence of this glass proves an impact origin, a Chicxulub impact, and an impact-generated tsunami deposit (Alvarez and others, 1992; Smit and others, 1992, 1994a, 1994b). Unfortunately, this view ignores numerous inconsistencies, such as the difficulties in depositing numerous fine-grained and coarse-grained sediment layers with very different lithologic and mineralogic characteristics over a period of only hours to days; the absence of iridium and, except for extremely rare fragments, shocked quartz; the deposition of well-stratified (rather than chaotic) units correlatable over more than 300 km; multiple and discrete bioturbation horizons in unit 3 of the clastic deposit, suggesting the sequential establishment of burrowing communities during deposition; the presence of burrowing at the base of unit 2 suggesting a time of normal sedimentation; and the presence of Maastrichtian marl on top of the clastic member that indicates that deposition occurred before the K-T boundary. We suggest that the presence of the glass fragments in unit 1 may be coincidental and not related to the depositional events of units 1, 2, or 3. It could represent a concurrent or earlier volcanic event and subsequent transport along with the multiorigin shallow-water spherules. If the glass proves to be of impact origin, then the bolide impact would have to be of pre-K-T boundary age and deposition could have been via airborne fallout. This scenario, however, also fails to explain why the spherule-rich sediments have only been found in channelized deposits, rather than blanketing the region, as would be expected from airborne fallout material. Moreover, to date there is no independent evidence of a pre-K-T boundary bolide impact in biotic, stratigraphic, or geochemical records.

Perhaps the major problem for a K-T boundary impact scenario for these clastic sediments is the evidence that they were deposited before the K-T boundary and that deposition was interrupted by multiple horizons of established burrowing communities. In northeastern and east-central Mexico this is indicated by the 5 cm to 1-m-thick Maastrichtian age marls or marly limestone layers above the clastic deposits at La Lajilla, El Mulato, and La Parida, as well as in central and southern Mexico sections (Mirador and Bochil, Macias Pérez, 1988; Montanari and others, 1994), and the burrowing horizons observed in units 2 and 3. A pre-K-T boundary age was also suggested by Jéhanno and others (1992) for the Beloc spherule layer, on the basis of its stratigraphic position 30 cm below the K-T boundary, iridium anomaly, and Ni-rich spinels. Moreover, Stinnesbeck and Keller (this volume) observed late Maastrichtian limestone and marl layers overlying the breccia deposits in Brazil (Poty Quarry section). This strongly indicates that deposition of the spherule-rich layer and the sandstone and sand-silt layers above are not related to the K-T boundary event. Accordingly, alternative scenarios must be considered. One such scenario involves deposition of the clastic sediments during the latest Maastrichtian sea-level lowstand. Examination of benthic foraminifera across the clastic deposits indicates that sea level dropped by ~70 to 100 m from an uppermost bathyal to outer neritic depth (Keller and others, 1994b). We suggest that deposition of the clastic deposit is a direct result of this sea-level lowstand.

A sea-level lowstand during the latest Maastrichtian is well documented worldwide in deep-sea and continental shelf sections (Haq and others, 1987; Donovan and others, 1988; MacLeod and Keller, 1991a, 1991b; Keller, 1992) and is reviewed in Keller and Stinnesbeck (this volume) with respect to near K-T boundary clastic deposits. We illustrate this sea-level lowstand and systems tract deposition for El Mimbral in Figure 38. In such a scenario, deposition of the entire clastic deposit occurred during low sea levels. Marls of the Mendez Formation were deposited during the late Maastrichtian sea-level highstand, as illustrated in Figure 39. No transport of shallow-water sediments into deeper waters occurred at this time, although deltaic sediments accumulated in nearshore regions.

The unconformity at the base of the clastic deposit represents a type 1 sequence boundary (Fig. 38) associated with a falling sea level. Deltaic sediments are exposed at this time, incised valleys and canyons form, and deltaic sediments from nearshore areas (including the multiorigin spherules) are transported into deeper waters, depositing unit 1 (Fig. 39). Continued lowering of the sea

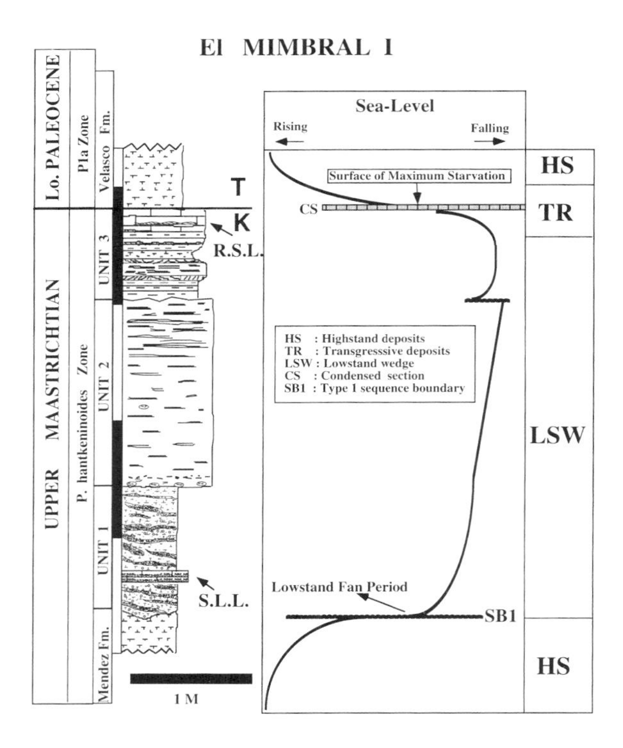

Figure 38 (top). Stratigraphic and lithologic column of El Mimbral I with sea-level curve interpreted from benthic foraminifera and sequence stratigraphic units. R.S.L. is rippled sandy limestone, S.L.L. is sandy limestone layer.

level results in erosion and bypass of inner shelf sediments and deposition of massive sand lenses in submarine canyons and fans (Fig. 39), thus depositing unit 2. Deposition of unit 3 occurred during a stable low sea level. At this time, erosion of shelfal sediments decreased, resulting in deposition of distal turbidites (coarse-grained layers) alternating with normal hemipelagic sedimentation (Fig. 39). The latter is indicated by thin marly interlayers with typical *P. hantkeninoides* zone faunal assemblages.

Sea level began to rise prior to the K-T boundary globally (Schmitz and others, 1992; Keller and others, 1993b; Askin and Jacobson, this volume; Keller and Stinnesbeck, this volume). In northeastern Mexico and east-central Mexico the rising sea level during the latest Maastrichtian deposited hemipelagic marls (La Lajilla, El Mulato, La Parida, Bochil, Mirador). Continued sea-level rise and transgression across the K-T boundary resulted in migration of the depocenters landward and sediment starvation in deeper waters (Fig. 39). At El Mimbral, the maximum sea-level rise is marked by the condensed interval (maximum flooding surface) represented by the boundary clay and a short interval of nondeposition or hiatus at the P0-P1a boundary in all sections examined (Fig. 38). Hemipelagic sedimentation resumed during the sea-level highstand of the later Danian (zones P1a to P1c, Fig. 39).

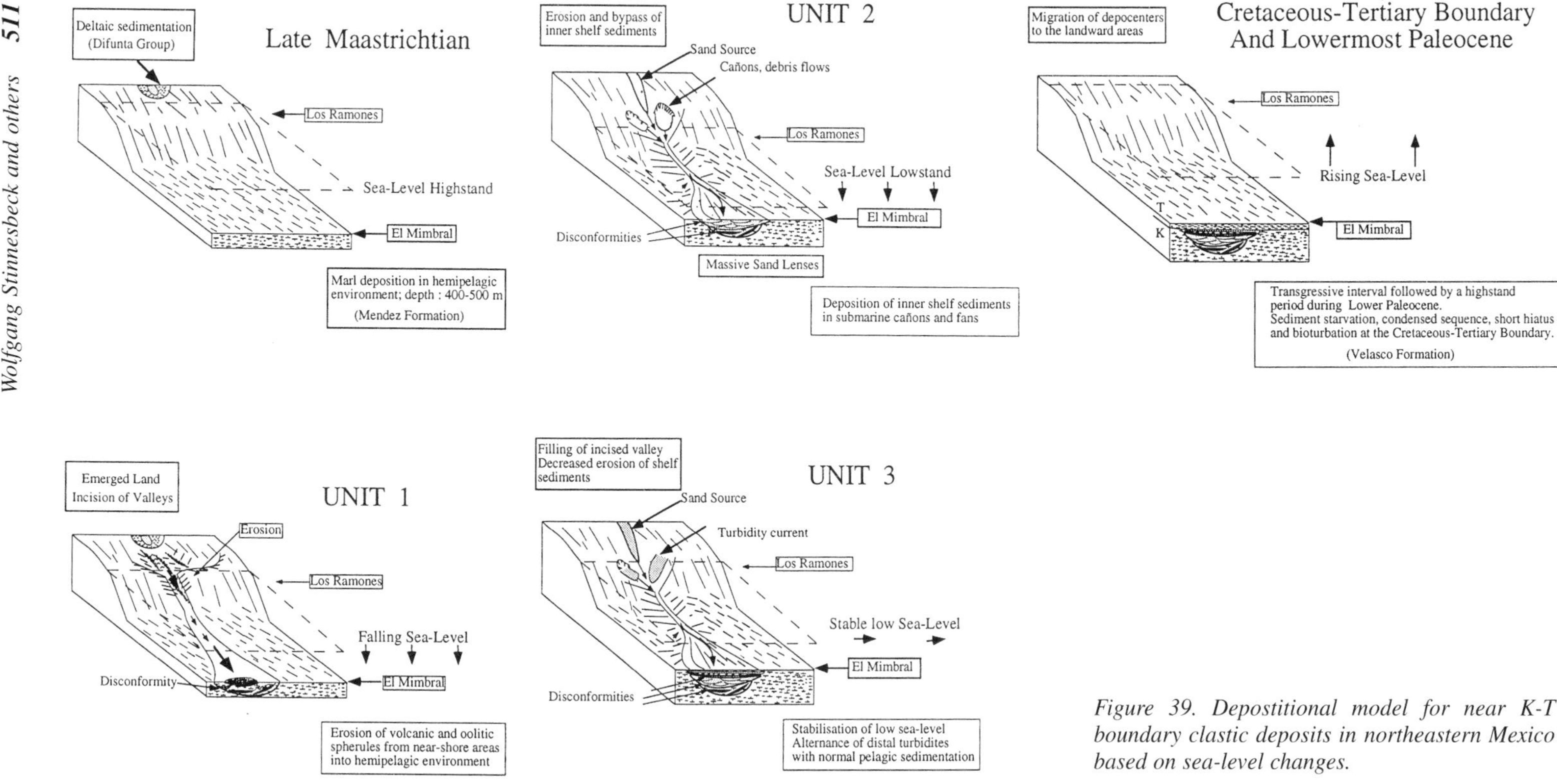

Figure 39. Depostitional model for near K-T boundary clastic deposits in northeastern Mexico based on sea-level changes.

CONCLUSIONS

1. The clastic deposits in northeastern Mexico and east-central Mexico sections are of latest Maastrichtian *P. hantkeninoides* age. This is indicated by the presence above the clastic deposit of marls or marly limestone layers with a typical *P. hantkeninoides* zone fauna. With the exception of Mimbral, these late Maastrichtian age sediments are present in five out of six sections that contain sediments above the clastic deposit (e.g., La Lajilla, El Mulato, La Parida, Bochil, Mirador).

2. Deposition of the clastic sediments occurred during the last 100–200 k.y. of the Maastrichtian. This is indicated by the presence of *Plummerita hantkeninoides*, which spans the last 200 k.y. of the Maastrichtian and is present below, within, and above the clastic deposits.

3. Clastic deposits at each outcrop represent individual channel-fill or fan deposits indicating channelized transport. There is no evidence that these deposits blanket the region outside channel systems.

4. Clastic deposits consist of three sedimentologically, lithologically, and mineralogically well defined units that are correlatable over more than 300 km in a northwest-southeast–trending direction parallel to and 40–80 km distant from the Sierra Madre Oriental. This proximity suggests transport of clastic sediments from this region.

5. The basal spherule-rich unit 1 of the clastic deposit contains abundant spherules of multiple origins, including infilled algal cysts, glauconite spherules, oolites, oncolites, and volcanic spherules. They formed in shallow-water deltaic environments and were subsequently transported into deeper water via submarine channels during the latest Maastrichtian sea-level lowstand.

6. Glass shards present in the spherule-rich layer are geochemically similar to Beloc (Haiti) and Chicxulub glass and may be of volcanic or impact origin. Because this deposit predates the K-T boundary, the event that produced them, whether bolide or volcano, also predates the K-T boundary.

7. A resistant 10–20-cm-thick sandy limestone layer with rare spherules is stratigraphically within the spherule-rich layer of unit 1. Microlayering of coarser grained and fine-grained sediments observed in thin sections indicates different (and slower) current conditions than those of the coarsely stratified spherule-rich sediments above and below. Thus, various different flow regimes are responsible for deposition of unit 1, which could not have happened within the few hours proposed by the tsunami model.

8. The alternating sand and silt layers of unit 3 contain two zeolite-enriched layers correlatable over 300 km. They indicate discrete influxes of volcanogenic sediments.

9. Thin marl and silt interlayers showing no grain-size grading are present in unit 3 with typical *P. hantkeninoides* zone foraminiferal assemblages. These layers indicate that periods of normal hemipelagic sedimentation alternated with periods of increased terrigenous influx during deposition of unit 3.

10. Abundant burrows of *Chondrites*, *Thalassinoides*, *Ophiomorpha*, and *Zoophycos* are present in two to three distinct layers of unit 3 and one layer at the base of unit 2, indicating that these organisms resided at these localities during

the latest Maastrichtian, and that deposition occurred over an extended time interval, rather than over a period of hours or days.

11. We found no evidence of unequivocal impact origin within the clastic deposit, such as common shocked quartz, Ni-rich spinels, iridium concentrations, or microtektites. We found no evidence of a tsunami such as bidirectional currents, or current directions pointing toward Chicxulub, as suggested by Smit and others (1994a, 1994b).

12. We propose that deposition of the clastic deposit occurred during the last 200 k.y. of the Maastrichtian during a sea-level regression. The three lithologically different units represent erosion from deltaic to inner neritic sediments and transport via submarine channels into deeper waters during successively lower sea levels (units 1 and 2), followed by intermittent transport during a stabilized sea-level lowstand (unit 3). Deposition of Maastrichtian marls above the clastic deposit indicates a rising sea level followed by condensed sedimentation (boundary clay) as the depocenters moved shoreward.

ACKNOWLEDGMENTS

We gratefully acknowledge reviews, comments, and discussions by J. Cooper, R. W. Scott, W. C. Ward, and D. R. Lowe. This research was supported by National Science Foundation grants INT-9314080 and OCE-9021338, the Petroleum Research Fund of the American Geochemical Society grant no. 26780-AC8, National Geographic Society grant no. 4620-91, Conacyt Grant no. L120-36-36, and Swiss National Fund grant no. 8220-028367. Technical support was provided by the Laboratoire de Geochemie et Petrographique de l'Institut de Géologie de Neuchâtel and by B. Kübler and C. Beck.

REFERENCES CITED

Adatte, T., Stinnesbeck, W., and Keller, G., 1994, Mineralogical correlations of near-K/T boundary deposits in northeastern Mexico: Evidence for long-term deposition and volcanoclastic influence, *in* New developments regarding the K/T event and other catastrophes in Earth history: Houston, Texas, Lunar and Planetary Institute Contribution 825, p. 1.

Adatte, T., Stinnesbeck, W., and Keller, G., 1996, Lithostratigraphic and mineralogical correlations of near K/T boundary clastic sediments in NE Mexico: Implications for origin and nature of deposition, *in* Ryder, G., Gartner, S., and Fastovsky, D., eds., New developments regarding the K/T event and other catastrophes in Earth history: Geological Society of America Special Paper (in press).

Alvarez, W., Smit, J., Lowrie, W., Asaro, F., Margolis, S. V., Claeys, P., Kastner, M., and Hildebrand, A. R., 1992, Proximal impact deposits at the Cretaceous-Tertiary boundary in the Gulf of Mexico: A restudy of DSDP Leg 77 Sites 536 and 540: Geology, v. 20, p. 697–700.

Baum, G. R., and Vail, P. R., 1988, Sequence stratigraphic concepts applied to Paleogene outcrops, Gulf and Atlantic basins, *in* Wilgus, C. K., Hastings, B. S., Kendall, C. G., Posamentier, H. W., Ros, C., and Van Wagoner, J. C., eds., Sea-level changes: An integrated approach: Society of Economic Paleontologists and Mineralogists Special Publication 42, p. 309–328.

Ben Abdelkader, O. B., 1992, Planktic foraminiferal content of El Kef Cretaceous-Tertiary (K/T) boundary type section (Tunisia) [abs.]: International Workshop on Cretaceous-Tertiary Transitions (El Kef Section), Part I: Tunis, Geological Survey of Tunisia, p. 9.

Berggren, W. A., and Miller, K. G., 1988, Paleogene tropical planktic foraminiferal biostratigraphy and magnetobiochronology: Micropaleontology, v. 34, p. 362–380.

Berggren, W. A., Kent, D. V., and Flynn, J. J., 1985, Paleogene geochronology and chronostratigraphy, *in* Snelling N. J., ed., The chronology of the geological record: Geological Society of London Memoir 10, p. 141–195.

Blum, J. D., and Chamberlain, C. P., 1992, Oxygen isotope constraints on the origin of impact glasses from the Cretaceous-Tertiary boundary: Science, v. 257, p. 1104–1107.

Bohor, B. F., and Betterton, W. J., 1993, Arroyo el Mimbral, Mexico, K/T unit: Origin as debris flow/turbidite, not a tsunami deposit [abs.]: Lunar and Planetary Science, v. 24, p. 143–144.

Bohor, B. F., and Seitz, R., 1990, Cuban K/T catastrophe: Nature, v. 344, p. 593.

Bourgeois, I., Hansen, T. A., Wiberg, P. L., and Kauffman, E. G., 1988, A tsunami deposit at the Cretaceous-Tertiary boundary in Texas: Science, v. 241, p. 567–570.

Brönnimann, P., and Rigassi, D., 1963, Contribution to the geology and paleontology of the area of the city of La Habana, Cuba, and its surroundings: Ecologae Geologicae Helvetiae, v. 56, p. 194.

Canudo, I., Keller, G., and Molina, E., 1991, K/T boundary extinction pattern and faunal turnover at Agost and Caravaca, SE Spain: Marine Micropaleontology, v. 17, p. 319–341.

Dietz, R. S., and McHone, J., 1990, Isle of Pines (Cuba), apparently not K/T boundary impact site: Geological Society of America Abstracts with Programs, v. 22, no. 7, p. A79.

Donovan, A. D., Baum, G. R., Blechschmidt, G. L., Loutit, T. S., Pflum, C. E., and Vail, P. R., 1988, Sequence stratigraphic setting of the Cretaceous/Tertiary boundary in central Alabama, *in* Wilgus, C. K., Hastings, B. S., Kendall, C. G., Posamentier, H. W., Ros, C., and Van Wagoner, J. C., eds., Sea-level changes: An integrated approach: Society of Economic Paleontologists and Mineralogists Special Publication 42, p. 300–307.

Ekdale, A. A., and Stinnesbeck, W., 1994, Sedimentologic significance of trace fossils in the KT "Mimbral Beds" of northeastern Mexico: Geological Society of America Abstracts with Programs, v. 26, no. 7, p. A395.

Galloway, W. E., Bebout, D. G., Fisher, W. L., Dunlap, J. B., Jr., Cabrera-Castro, R., Lugo-Rivera, J. E., and Scott, T. M., 1991, Cenozoic, *in* Salvador, A., ed., The Gulf of Mexico basin: Boulder, Colorado, Geological Society of America, Geology of North America, v. J, p. 245–324.

Habib, D., Moshkowitz, S., and Kramer, C., 1992, Dinoflagellate and calcareous nannofossil response to sea-level change in Cretaceous-Tertiary boundary sections: Geology, v. 20, p. 165–168.

Hansen, H. J., Gwozdz, R., Bromley, R. G., Rasmusseu, K. L., Vogensen, E. W., and Pedersen, K. R., 1986, Cretaceous-Tertiary boundary spherules from Denmark, New Zealand, and Spain: Geological Society of Denmark Bulletin, v. 35, p. 75–82.

Haq, B. U., Hardenbol, J., and Vail, P. R., 1987, Chronology of fluctuating sea levels since the Triassic: Science, v. 235, p. 1156–1166.

Hay, W. W., 1960, The Cretaceous-Tertiary boundary in the Tampico Embayment, Mexico: International Geological Congress, 21st, Copenhagen, Proceedings, Part 5, p. 70–77.

Herbert, T. D., and D'Hondt, S., 1990, Environmental dynamics across the Cretaceous-Tertiary extinction horizon measured 21 thousand year climate cycles in sediments: Earth and Planetary Science Letters, v. 99, p. 263–275.

Hildebrand, A. R., and Boynton, W. V., 1990, Proximal Cretaceous/Tertiary boundary impact deposits in the Caribbean: Science, v. 248, p. 843–847.

Hildebrand, A. R., Penfield, G. T., Kring, D. A., Pilkington, M., Camargo, Z. A., Jacobsen, S. B., and Boynton, W. V., 1991, Chicxulub crater: A possible Cretaceous/Tertiary boundary impact crater on the Yucatan Peninsula: Geology, v. 19, p. 867–869.

Iturralde-Vincent, M. A., 1992, A short note on the Cuban late Maastrichtian megaturbidite (an impact derived deposit?): Earth and Planetary Science Letters, v. 109, p. 225–229.

Izett, G. A., 1990, Tektites in Cretaceous-Tertiary boundary rocks on Haiti and their bearing on the Alvarez extinction hypothesis: Journal of Geophysical Research, v. 96, p. 20,879–201,905.

Jéhanno, C., Boclet, D., Froget, L., Lambert, B., Robin, E., Rocchia, R., and Turpin, L., 1992, The Cretaceous-Tertiary boundary at Beloc, Haiti: No evidence for an impact in the Caribbean area: Earth and Planetary Science Letters, v. 109, p. 229–241.

Keller, G., 1988, Extinction, survivorship and evolution of planktic foraminifera across the Cretaceous/Tertiary boundary at El Kef, Tunisia: Marine Micropaleontology, v. 13, p. 239–263.

Keller, G., 1989a, Extended Cretaceous/Tertiary boundary extinctions and delayed population changes in planktonic foraminifera from Brazos River, Texas: Paleoceanography, v. 4, p. 287–332.

Keller, G., 1989b, Extended period of extinctions across the Cretaceous/Tertiary boundary in planktonic foraminifera of continental shelf sections: Implications for impact and volcanism theories: Geological Society of America Bulletin, v. 101, p. 1408–1419.

Keller, G., 1992, Paleoecologic response of Tethyan benthic foraminifera to the Cretaceous-Tertiary boundary transition, *in* Takayanagi, Y., and Saito, T., eds., Studies in benthic foraminifera: Sendai, Tokai University Press, p. 77–91.

Keller, G., 1993, The Cretaceous-Tertiary boundary transition in the Antarctic Ocean and its global implications: Marine Micropaleontology, v. 21, p. 1–45.

Keller, G., and Benjamini, C., 1991, Paleoenvironment of the eastern Tethys in the early Paleocene: Palaios, v. 6, p. 439–464.

Keller, G., MacLeod, N., Lyons, J. B., and Officer, C. B., 1993a, Is there evidence for Cretaceous-Tertiary boundary–age deep-water deposits in the Caribbean and Gulf of Mexico?: Geology, v. 21, p. 776–780.

Keller, G., Barrera, E., Schmitz, B., and Mattson, E., 1993b, Gradual mass extinction, species survivorship, and long-term environmental changes across the Cretaceous/Tertiary boundary in high latitudes: Geological Society of America Bulletin, v. 105, p. 979–997.

Keller, G., Stinnesbeck, W., Adatte, T., MacLeod, N., and Lowe, D. R., 1994a, Field guide to Cretaceous-Tertiary boundary sections in northeastern Mexico: Houston, Texas, Lunar and Planetary Institute Contribution no. 827, 110 p.

Keller, G., Stinnesbeck, W., and Lopez-Oliva, J. G., 1994b, Age, deposition and biotic effects of the Cretaceous/Tertiary boundary event at Mimbral, NE Mexico: Palaios, v. 9, p. 144–157.

Koeberl, C., 1994, Deposition of channel deposits near the Cretaceous-Tertiary boundary in northeastern Mexico: Catastrophic or "normal" sedimentary deposits?: Comment: Geology, v. 22, p. 957.

Koeberl, C., and Sigurdsson, H., 1992, Geochemistry of impact glasses from the K/T boundary in Haiti: Relation to smectites and a new type of glass: Geochimica et Cosmochimica Acta, v. 56, p. 2113–2129.

Longoria, J. F., and Gamper, M. A., 1992, Planktonic foraminiferal biochronology across the KT boundary from the Gulf coastal plain of Mexico: Implications for timing the extraterrestrial bolide impact in the Yucatan: Boletín de la Asociación Mexicana de Géologos Petroleros, v. 42, p. 19–40.

Lopez-Oliva, G. J., and Keller, G., 1994, Biotic effects of the K/T boundary event in northeastern Mexico, *in* New developments regarding the K/T event and other catastrophes in

Earth history: Houston, Texas, Lunar and Planetary Institute Contribution 825, p. 72–73.
Lopez-Oliva, G. J., and Keller, G., 1996, Age and stratigraphy of near-K/T boundary clastic deposits in NE Mexico, *in* Ryder, G., Gartner, S, and Fastovsky, D., eds., New developments regarding the K/T event and other catastrophes in Earth history: Geological Society of America Special Paper (in press).
Lyons, J. B., and Officer, C. B., 1992, Mineralogy and petrology of the Haiti Cretaceous/Tertiary section: Earth and Planetary Science Letters, v. 109, p. 205–224.
Macias Pérez, F. J., 1988, Estratigrafía detallada del límite Cretácico-Terciario en Xicotepec, Puebla: una alternativa al catastrofismo [Ph.D. thesis]: Mexico City, Institute Politécnico Nacional, 108 p.
MacLeod, N., and Keller, G., 1991a, Hiatus distribution and mass extinction at the Cretaceous/Tertiary boundary: Geology, v. 19, p. 497–501.
MacLeod, N., and Keller, G., 1991b, How complete are Cretaceous/Tertiary boundary sections? A chronostratigraphic estimate based on graphic correlation: Geological Society of America Bulletin, v. 103, p. 1439–1457.
Mancini, E. A., Tew, B., and Smith, C. C., 1989, Cretaceous-Tertiary contact, Mississippi and Alabama: Journal of Foraminiferal Research, v. 19, p. 93–104.
Michaud, F., and Fourcade, E., 1989, Stratigraphie et paléogéographie du Jurassique et du Crétacé du Chiapas (Sud-Est du Mexique): Bulletin de la Société Géologique de France, new ser., v. 8, p. 639–650.
Montanari, A., Claeys, P., Asaro, F., Bermudez, J., and Smit., J., 1994, Preliminary stratigraphy and iridium and other geochemical anomalies across the K/T boundary in the Bochil section (Chiapas, southeastern Mexico), *in* New Developments regarding the K/T event and other catastrophes in Earth history: Houston, Texas, Lunar and Planetary Institute Contribution 825, p. 84.
Moshkovitz, S., and Habib, D., 1993, Calcareous nanofossil and dinoflagellate stratigraphy of the Cretaceous-Tertiary boundary, Alabama and Georgia: Micropaleontology, v. 39, p. 167–191.
Muir, J. M., 1936, Geology of the Tampico region, Mexico: Tulsa, Oklahoma, American Association of Petroleum Geologists, 280 p.
Ocampo, A. C., and Pope, K. O., 1994, A K/T boundary section from northern Belize, *in* New developments regarding the K/T event and other catastrophes in Earth history: Houston, Texas, Lunar and Planetary Institute Contribution 825, p. 86.
Pope, K. O., Ocampo, A. C., and Duller, C. E., 1991, Mexican site for K/T impact crater: Nature, v. 351, p. 105–108.
Quezada Muñeton, J. M., 1990, El Cretácio Medio-Superior, y el Límite Cretácio Superior-Terciario Inferior en la Sierra de Chiapas: Boletín de la Asociación Mexicana de Géologos Petroleros, v. 39, p. 3–98.
Robin, E., Rocchia, R., Lyons, J. B., and Officer, C. B., 1994, Deposition of channel deposits near the Cretaceous-Tertiary boundary in northeastern Mexico: Catastrophic or "normal" sedimentary deposits?: Reply: Geology, v. 22, p. 958.
Sawlowicz, Z., 1993, Iridium and other platinum elements as geochemical markers in sedimentary environments: Palaeogeography, Palaeoclimatology, Palaeoecology, v. 104, p. 253–270.
Savrda, C. E., 1993, Ichnosedimentologic evidence for a noncatastrophic origin of Cretaceous-Tertiary boundary sands in Alabama: Geology, v. 21, p. 1075–1078.
Schmitz, B., Keller, G., and Stenvall, O., 1992, Stable isotope and foraminiferal changes across the Cretaceous/Tertiary boundary at Stevns Klint, Denmark: Arguments for long-term oceanic instability before and after bolide impact: Palaeogeography, Palaeoclimatology, Palaeoecology, v. 96, p. 233–260.
Sharpton, V. L., Dalrymple, G. B., Marin, L. E., Ryder, G., Schuraytz, B. C., and Urrutia-Fucugauchi, J., 1992, New links between the Chicxulub impact structure and the Cretaceous/Tertiary boundary: Science, v. 359, p. 819–821.

Sharpton, V. L., Burke, K., Camargo-Zanoguera, A., Hall, S. A., Lee, S., Marin, L., Suárez-Reynoso, G., Quezada-Muñeton, J. M., Spudis, P. D., and Urrutia-Fucugauchi, J., 1993, Chicxulub multitiring impact basin: Size and other characteristics derived from gravity analysis: Science, v. 261, p. 1564–1567.

Sigurdsson, H., D'Hondt, S., Arthur, M. A., Bralower, T. J., Zachos, J. C., Fossen, M., and Channell, J. E. T., 1991, Glass from the Cretaceous/Tertiary boundary in Haiti: Nature, v. 349, p. 482–486.

Smit, J., 1982, Extinction and evolution of planktonic foraminifera after a major impact at the Cretaceous/Tertiary boundary, *in* Silver, L. T., and Schultz, P. H., eds., Geological implications of impacts of large asteroids and comets on Earth: Geological Society of America Special Paper 190, p. 329–352.

Smit, J., Montanari, A., Swinburne, N. H. M., Alvarez, W., Hildebrand, A., Margolis, S. V., Claeys, P., Lowrie, W., and Asaro, F., 1992, Tektite bearing deep-water clastic unit at the Cretaceous-Tertiary boundary in northeastern Mexico: Geology, v. 20, p. 99–103.

Smit, J., Roep, T. B., Alvarez, W., Claeys, P, and Montanari, A., 1994a, Is there evidence for Cretaceous-Tertiary boundary–age deep-water deposits in the Caribbean and Gulf of Mexico?: Comment: Geology, v. 22, p. 953–954.

Smit, J., Roep, T. B, Alvarez, W., Montanari, A., and Claeys, P., 1994b, Stratigraphy and sedimentology of K/T clastic beds in the Moscow Landing (Alabama) outcrop: Evidence for impact related earthquakes and tsunamis, *in* New developments regarding the K/T event and other catastrophes in Earth history: Houston, Texas, Lunar and Planetary Institute Contribution 825, p. 119.

Sohl, N. F., Martinez, R. E., Salmerón-Ureña, P., and Soto-Jaramillo, F., 1991, Upper Cretaceous, *in* Salvador, A., ed., The Gulf of Mexico basin: Boulder, Colorado, Geological Society of America, Geology of North America, v. J, p. 205–244.

Stinnesbeck, W., Barbarin, J. M., Keller, G., Lopez-Oliva, J. G., Pivnik, D. A., Lyons, J. B., Officer, C. B., Adatte, T., Graup, G., Rocchia, R., and Robin, E., 1993, Deposition of channel deposits near the Cretaceous-Tertiary boundary in northeastern Mexico: Catastrophic or "normal" sedimentary deposits?: Geology, v. 21, p. 797–800.

Stinnesbeck, W., Keller, G., and Adatte, T., 1994a, K/T boundary sections in southern Mexico (Chiapas): Implications for the proposed Chicxulub impact site, *in* New developments regarding the K/T event and other catastrophes in Earth history: Houston, Texas, Lunar and Planetary Institute Contribution 825, p. 121–122.

Stinnesbeck, W., Keller, G., Adatte, T., and MacLeod, N., 1994b, Is there evidence for Cretaceous-Tertiary boundary–age deep-water deposits in the Carribean and Gulf of Mexico?: Reply: Geology, v. 22, p. 955–956.

Swisher, C. C., Grajales Nishimura, J. M., Montanari, A., Margolis, S. V., Claeys, P., Alvarez, W., Renne, P., Cedillo-Pardo, E., Maurasse, F. J-M., Curtis, G. H., Smit, J., and McWilliams, M. O., 1992, Coeval $^{40}Ar/^{39}Ar$ ages of 65.0 million years ago from Chicxulub crater melt rock and Cretaceous-Tertiary boundary tektites: Science, v. 257, p. 954–958.

Weidie, A. E., Wolleben, J. A., and McBride, E. F., 1972, Late Cretaceous depositional systems in northeastern Mexico: Gulf Coast Association of Geological Societies Transactions, v. 22, p. 323–329.

20

Pele Hypothesis: Ancient Atmospheres and Geologic-Geochemical Controls on Evolution, Survival, and Extinction

Gary P. Landis, *U.S. Geological Survey, Denver Federal Center, Denver, Colorado,* J. Keith Rigby, Jr., *Department of Civil Engineering and Geological Sciences, University of Notre Dame, Notre Dame, Indiana,* Robert E. Sloan, *Department of Geology and Geophysics, University of Minnesota, Minneapolis, Minnesota,* Richard Hengst, *Department of Biology, Purdue University, Westville, Indiana,* and Larry W. Snee, *U.S. Geological Survey, Denver Federal Center, Denver, Colorado*

INTRODUCTION

The Pele hypothesis is an attempt to understand three originally independent lines of research (dinosaur and mammalian evolution near Cretaceous-Tertiary [K-T] boundary time, amber and paleoatmospheres, and dinosaur physiology) that converged in our study of the Hell Creek and Tullock geologic sections exposed on the eastern margin of the Fort Peck Reservoir in east-central Montana (Fig. 1). The Pele hypothesis (Landis and others, 1993) is unique in that it proposes a unifying explanation for geological and biological events previously viewed as disparate. Full development of the Pele hypothesis transcends Montana or even the K-T boundary, to encompass the interconnected nature of biotic evolution and continuing Earth processes. We present first a description of the Russell basin amber locality of eastern Montana, then discuss the analysis and interpretation of the amber data, and conclude with our scenario for understanding of the atmospheric oxygen across the K-T boundary by formally proposing the Pele hypothesis.

Perhaps no other continental section anywhere in the world has played a greater role in the discussion of the continental K-T vertebrate extinction than

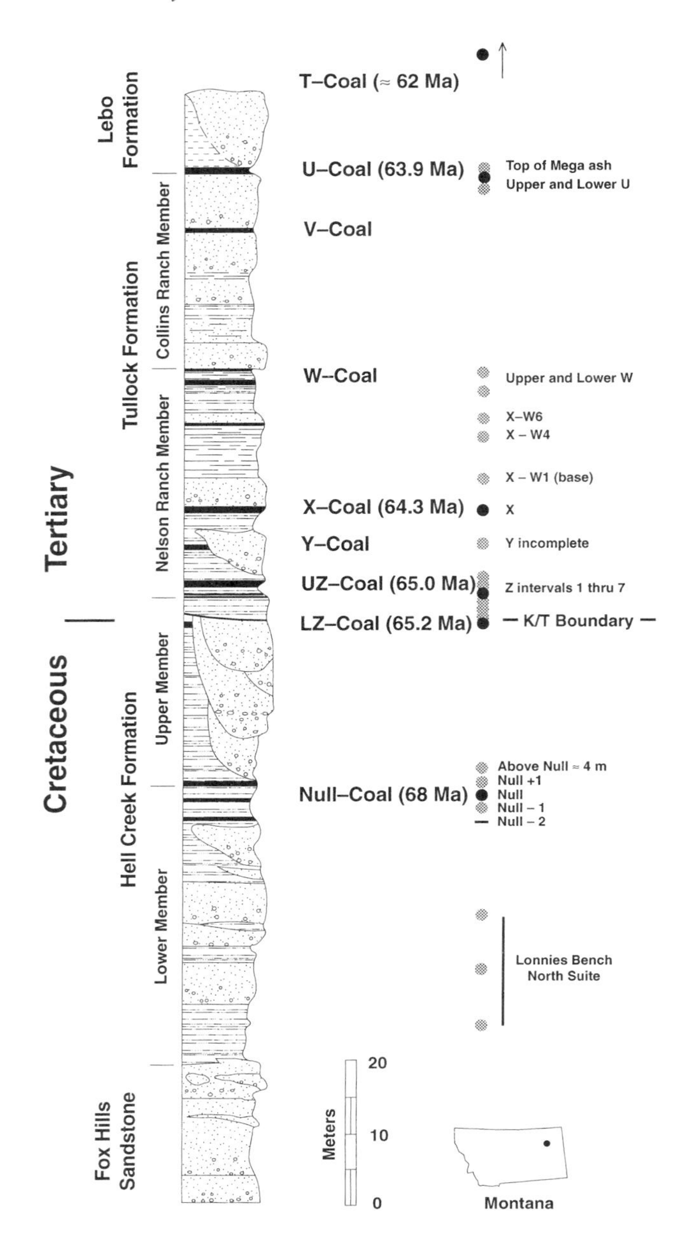

Figure 1. Stratigraphic column of Upper Cretaceous Hell Creek Formation and Tertiary Tullock and Lebo formations in Russell Basin of eastern Montana (location map, inset). Column modified from Rigby and Rigby (1990). Dots to right of column identify location of amber samples collected. Black dots are samples analyzed in this study and reported in Table 1a. K-T boundary is located at ~5 mm above lower Z coal interval between Z4 and Z5 and is dated at ~65.0 Ma. Upper Z coal is basal Tertiary above the regression unconformity at Z5, ~100,000 yr after the K-T boundary. Amber occurs in each of 35 depositional cycles at top of coal interval as indicated in Figure 2.

that of eastern Montana. This succession has long been thought to be one of the most complete continental sections across the K-T boundary and has played a pivotal role in discussions of mode, tempo, selectivity, and size of the terminal Cretaceous event(s), particularly as they affect vertebrates. Mammalian biostratigraphy has traditionally provided the necessary time resolution for the K-T transition interval in terrestrial settings. The mammalian framework is corroborated by recent paleomagnetic and absolute age data (Swisher and others, 1993). Large vertebrate collections made by wet screening (McKenna, 1965) and numerous dinosaur excavations that have been made more or less continuously since the turn of the century have provided critical baseline data on the nature, rate, and magnitude of the K-T extinction event (Archibald and others, 1982; Sloan and others, 1986; Rigby, 1987; Smit and others, 1987; Archibald and Bryant, 1990; Lofgren, 1990; Rigby and Rigby, 1990; Sheehan and others, 1991; Archibald, 1993). Recent studies have tried to determine if the rock record is complete enough to preserve evidence of catastrophic or gradual extinction of fossil vertebrates.

Dinosaur remains have been found in Tertiary sediments (Sloan and Rigby, 1986; Sloan and others, 1986; Rigby, 1987; 1989; Rigby and others, 1987), but the exact age of the fossils has been questioned because of the potential for reworking has not been satisfactorily evaluated (Fastovsky, 1990; Lofgren, 1990; Lofgren and others, 1990). The bulk of fossil vertebrates recovered to date has been gathered by large-scale, screen concentrations of lag deposits from numerous nested channel deposits. Vertebrate fossils were collected from natural concentrates when transport was arrested in ancient streams. As such, they may plausibly represent either primary deposition, or may contain a substantial fraction of reworked material from older sediment. Significant amounts of reworked material in any paleontological sample would reduce the time resolution that is critical for detailed study of rapid geologic events.

The study of fossil materials recovered from overbank deposits avoids many of the reworking problems and is more likely to produce materials of primary deposition. Fossil vertebrates recovered from overbank deposits are rare, but other fossils might yield information about environmental conditions during latest Cretaceous and earliest Tertiary time. Berner and Landis (1988) and Landis and Berner (1988) reported on gas concentrations obtained from the analysis of gas bubbles located within ancient plant resins—amber. Continuing studies of the gas contained in amber provide compelling evidence that primary gas bubbles in amber contain minute samples of ancient atmosphere. These results encouraged us to attempt analyses of amber from the organic-rich coal and/or shale units in the Hell Creek, Tullock, and Lebo formations that span the K-T boundary. Knowing the amount and variation in atmospheric oxygen across the K-T boundary commands attention, considering the magnitude of various proposed K-T boundary events, along with possible major implications of a high O_2 in Late Cretaceous atmosphere and its subsequent drop during the early Tertiary.

Mesozoic life likely existed under environmental conditions of elevated oxygen and carbon dioxide (Berner and Landis, 1988; Berner and Canfield, 1989; Berner, 1990; 1991; 1994; Caldeira and Rampino, 1991; Cerling, 1991). The elevated presence of these gases in Mesozoic atmosphere must have provided powerful selection pressures for life, because they are integral components in the metabolic processes of both plants and animals. The number of large North American dinosaur genera declined from 34 to 12 during the last 10 m.y. of the Cretaceous (Sloan and Rigby, 1986; Rigby, 1987; Weishampel, 1990), overlapping in time with the drop in atmospheric oxygen from 35% to 31% in the last ~1.5 m.y. preceding the K-T boundary. McAlester (1970, 1971; Schopf and others, 1971) noted that high extinction rates in a large number of families at the end of all major geologic periods correlated with high oxygen consumption rates. Atmospheric composition clearly would affect physiological adaptations of evolving organisms, and lung ventilation is a major factor limiting sustained performance (Raven, 1991; Spotila and others, 1991; Hengst and others, 1993). Analysis of respiratory mechanisms and metabolic scope of an *Apatosaurus* (Hengst and others, 1993, this volume) suggests that, assuming present-day oxygen levels, this animal utilized inefficient rib breathing with thorasic and ventilation volumes, and blood chemistry adequate to sustain only a very minimal basal metabolism. Breathing mechanisms do not adapt well to hypoxic conditions, thus leading to lowered overall fitness. Dinosaur extinction may have resulted from a combination of increased environmental stresses, including cooler temperatures, changing food web, forced migration, and, ultimately, decreased oxygen (Hengst and others, 1993; Landis and others, 1993). Thus, analysis of dinosaur respiratory physiology further compelled us to seek a better understanding of the changing atmospheric O_2 concentrations across the K-T boundary.

AMBER OCCURRENCE, COLLECTION, AND DEPOSITIONAL ENVIRONMENT

Amber used in this study was obtained from thin (<0.5 m thick) subituminous coal, lignites, or carbonaceous shale units scattered stratigraphically throughout the Hell Creek, Tullock, and Lebo formations. "Coal" seam designation follows that of Rigby and Rigby (1990). Figure 1 shows the stratigraphic placement of sampled and analyzed amber. In some cases more than one sample suite has been analyzed from a specific coal horizon.

Amber was recovered throughout most of the thin coals sampled but it was most commonly observed to be locally concentrated in 0.5–1.0-cm-thick intervals with a lateral extent of up to 2.0 m. (Fig. 2). Within these bedding-plane concentration intervals, rounded to teardrop-shaped amber blebs occurred. These blebs varied in size (1.0 mm–1.5 cm) and in color (pale, frothy yellow to "crystalline" red and in some cases dark to black-red). Some blebs were collected from the outer margins of relict logs within the coals that were associated with abundant charcoal (fusinite) fragments (particularly in the Null coal). The darkest shades of red to red-black amber were associated with carbonized logs that appeared to be charcoal remains (fusinite), and were not analyzed because of potential alteration due to fire. Other than the charred and thermally brittled red to black color, we do not find that amber color correlates in any way with gas analytical data. The same color variation occurred in amber from all stratigraphic intervals, and commonly the range in color is observed as flow banding within a single larger piece.

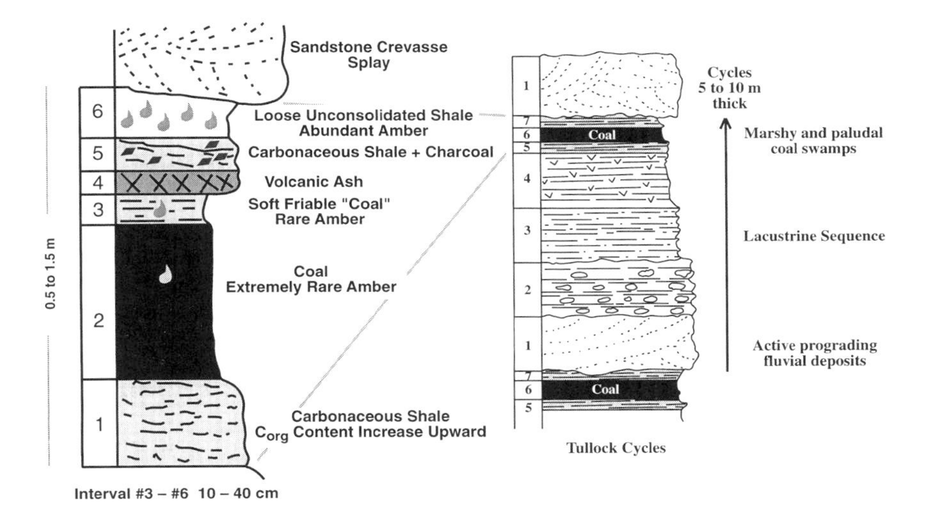

Figure 2. Typical (schematic) depositional cycle as reported in Rigby and Rigby (1990). Amber occurs at top of coal–carbonaceous shale sequence predominantly as "forest litter" droppings of discrete pea-sized blebs deposited in situ. Rare logs and wood fragments are preserved in the upper unconsolidated shale units, some with amber (resin) still in place. Each of 35 documented depositional cycles contains a volcanic air-fall ash, and is terminated by an overbank crevasse splay sandstone unit. Most of analyzed amber is derived from interval #6, illustrated in the left expanded view.

Amber was found most commonly as blebs of widely differing size in what we interpret as syndepositional forest litter dropping deposits in the uppermost part of each cyclic coal sequence. A number of these carbonaceous units were interpreted by Retallack and other (1987) and Fastovsky and McSweeney (1987) as paleosols. These concentrations were clearly stratigraphically above "charcoal" zones suspected to be fire events; we have tentatively interpreted the amber occurrences as representing depositional intervals of increased resin flow in response to burning. Multiple bedding planes of concentrated amber were superimposed two to three high within the coals thicker than 20 cm. Samples were taken from pits scattered over a 20–60 m lateral interval along the exposure face with at least a 3 m spacing between pits. Although amber abundance varied considerably, it generally required a crew of ten up to one day to collect enough amber for multiple analyses. Each analysis requires 25 to 50 mg of clear, bubble-containing amber.

We have documented more than 35 coal–crevasse-splay depositional cycles in the Hell Creek, Tullock, and Lebo formations of Russell basin that all exhibit the same general sequence (schematically illustrated in Fig. 2). The cause of the cyclic sedimentation is not known, but the cyclic pattern is most evident in the Tullock Formation (Rigby and Rigby, 1990). The same general pattern is also present in the upper part of the Hell Creek Formation and that part of the Lebo Formation investigated to date. Each cycle has one or more volcanic ashes preserved within the cycle, generally most visible within the most organic-rich part of the section. The major crevasse-splay sandstones are tuffaceous, suggesting that ash fall may have played some part in causing the channel breach. These host sedimentary rock cycles represent the same reoccurring depositional environment and contain syndepositional accumulations of amber from the same type of amber-producing vegetation over ~6 m.y. Swisher and others (1993) presented details of geochronology and geology of the section, and placement of the K-T boundary. These dates are included in Figure 1 and Table 1a for reference. Their data constrain deposition of the Tullock Formation to 1.1 m.y., or from 65.0 to 63.9 Ma, with recognition that the U-coal is the basal Lebo Formation and not the upper Tullock Formation. The Fox Hills sandstone is 69 Ma on the basis of ammonite zonation. By linear interpolation of Tullock Formation sedimentation rates, the T-coal of the Lebo Formation is estimated as ~62 Ma. These geochronologic dates compare favorably with approximate ages determined by mammalian assemblages and magnetochronology.

The dominant preserved wood throughout the entire section is *Taxodiaceae*. Because its occurrence is so pervasive in the section, we assume that it was one of the largest contributors to the amber populations we have analyzed. We recognize that other types of wood perhaps were at some preservational disadvantage, and therefore possibly are underrepresented in the paleobotanical assemblages. However, we have no data to suggest that these non-*Taxodiaceae* forms were major contributors to the recovered resins. Variation in measured oxygen in the amber bubbles does not appear to be attributable to different depositional environments, degree of diagenesis, depth of burial, temperature, differing host tree species for the resin, variable degrees of weathering, humidity, or length of exposure to the elements, or to general climatic conditions under which the tree originally secreted the resin. We assume that all of these intrinsic variables are the same for each amber accumulation interval.

AMBER ANALYSIS

On the basis of work by Berner and Landis (1988) and ongoing petrographic and gas studies, we recognize four gas sites in amber, and six most reasonable sources of gas expected to be in amber. These sites include bubbles of (1) atmospheric gas (primary gas bubbles) formed near the interface between successive flows of resin (Plate 1, A and B) as it exudes from the resin ducts (Hillis, 1987); (2) matrix gas dissolved in the amber (Landis and Berner, 1988); (3) gas trapped along fractures; and (4) intrinsic gases dissolved in the resin (which were exsolved from the resin) and distributed throughout the amber. Gas extracted from amber include components from atmospheric, metabolic (tree), curing (relased during polymer formation), burial, biologic, and exchange and/or reaction sources. Only primary gas bubbles may host microscopic samples of ancient air, as discussed by Berner and Landis (1988).

Before analysis, all amber samples were examined in thick doubly-polished sections or in transparent fractured pieces by optical microscopy to ascertain the presence of suitable bubbles, the degree of fracturing and chemical and/or thermal alteration, and the gas-bubble petrography. Deeply reddened or blackened amber,

TABLE 1A
Summary Russell Basin Amber Data

Source Unit	Formation	Age (Ma)	O_2 Calc. (Vol. %)	O_2 Avg. 3 High	N_2/Ar
T-coal	Lebo	≈62	31	28.6	41.4
U-coal	Lebo	63.9	30	27.5	40.1
X-coal	Tullock	64.3	33	25.6	39.2
Upper Z-coal	Tullock	65.0	29	22.8	46.5
Lower Z-coal	Hell Creek	65.2	31	29.6	41.7
Null-coal	Hell Creek	≈68	35	33.5	32.2

Based on n = 96 quantitative analyses. Age data from Swisher and other (1993) or estimated where indicated approximate. Heavy line in table indicates Cretaceous-Tertiary boundary. For each sample/locality, calculated oxygen is from end-member mixing line calculations, second oxygen value reported is average of three highest determinations in data set.

TABLE 1B
Summary World Data

Location	Age Estimate (Ma)	O_2 Calc. (Vol. %)	O_2 Avg. 3 High	N_2/Ar
Tertiary/Holocene				
New Zealand (Kauri)	< 0.1	21	22.4	12.3
Philippines	0.1	21	7.0	5.0
Dominican Republic	20	14	3.6	31.2
Chiapas (Mexico)	24	20	16.8	28.0
Baltic Ukraine	45	22	20.5	39.0
Lake Hazen (Ellesmere)	50	16	15.4	32.9
Ellesmere Island	52	20	19.1	25.0
Cretaceous				
North Dakota	70	35	19.5	21.8
Grassy Lakes	74	35	26.4	20.3
Cedar Lakes	75	35	31.0	36.0
Dinosaur Park	78	33	43.5	26.5
Magothy Formation	88	33	22.0	29.2
Raritan Formation	95	34	29.0	32.2
Saudi Arabia	115	29	14.4	19.2
Mt. Hermon	130	28	33.5	35.0

Based on n = 514 quantitative analyses. Age estimates based upon published reports of amber occurrences and other available stratigraphic information. Data from published and unpublished studies of Landis and Berner. Detailed analyses reported in USGS Open File Report (Landis and Berner, in prep).

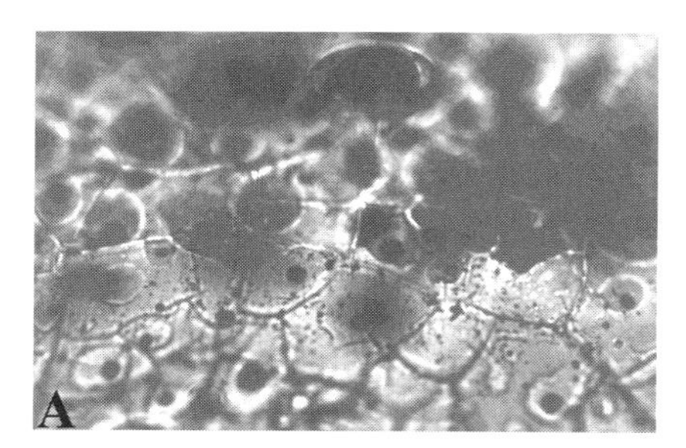

Plate 1. A: Photomicrograph of a planar view of a dessication surface, or schlauben, buried within amber (Cedar Lakes) with numerous primary gas bubbles. View is ~4 mm across. Gas bubbles appear black because internal reflections block light transmission, whereas the anular meniscus of water or immiscible organic fluid surrounding the gas are visible in high contrast. B: Photomicrograph of primary gas bubble (~350 μm across) in cross-sectional view above the dessication interface between older and younger resin flows.

highly fractured amber, amber intermixed with fusinite, and milky or "bone" amber were avoided, because these contain either gas from sites other than primary bubbles, or gas compositions possibly altered by a thermal event. Only amber pieces hosting identified or presumed primary gas bubbles along flow interfaces were selected for analysis. Several 1–4 mm pieces from the same collection site were place in the microcrushing apparatus, which was then mounted on the quadrupole mass spectrometer (QMS) inlet. Next, the chamber was evacuated, and pumped to a low vacuum of between 10^{-7} to 10^{-8} Torr. Stepwise crushing and fracturing of the specimens released amber gas into the QMS as the instrument continuously analyzed the gas-stream mass spectrum in real time.

Details of amber selection and intrumental analysis of microscopic gas inclusions are summarized elsewhere (Berner and Landis, 1988; Landis and Berner, 1988; Landis and Hofstra, 1991). The instrumental configuration and specific procedures for data collection and reduction were designed to maximize the reliability of data, given several unique requirements for analysis of amber gases. (1) The procedure must enable selective opening of specific gas sites in amber and recognition of different compositions obtained from these sites, with specific attention to gases released by primary gas bubbles. (2) Gas released from amber must be analyzed within tens of milliseconds to avoid reaction with newly created amber surfaces. (3) Thermal diffusion and adsorption/desorption (A/D) kinetics in the crushing apparatus and vacuum system must be minimized and the results corrected (use of cryogenic traps for separation of specific gases or static bulk analysis of released gas will produce poor analyses because of thermal diffusion and physico-chemical sorption). (4) Instrumental errors (typically 3%–5% relative) by standard error propagation must be less than the resolution required by the geologic problem from instrumentation capable of measuring small quantities (10^{-12}–10^{-15} mol) of gas mixtures.

Conductance-limited tubulation (~ 0.75 l/s) connected the crushing device to the ion source of the QMS with a rate of gas injection optimized to the vacuum pumping speed (170 l/s), the computer-driven measurement rate (25 μs/12-bit A/D), the rate of gas adsorption-desorption in the system (~3 s/monolayer coverage), and the rate of reaction of oxygen released from bubbles with the newly created surfaces on crushed amber ($t_{1/2}$ ~1 day [t = time]). The QMS was calibrated for ion fragmentation and ionization sensitivity for each gas of interest. These included H_2, CH_4, H_2O, N_2, O_2, Ar, CO_2, various light chain hydrocarbons, and terpenoid and isoprene ion radicals and fragments up to a mass/charge (m/c) of 80. Scans to mass (m/e) =160 verified that no important ion intensities were present above the routine m/e 80 upper limit. Mass intensity data for each release of gas by crushing were reduced by general matrix least squares and iterative convergence methods to yield quantitative analyses in partial pressure or mole fraction for each gas. Error analysis and verification with gas standards and simulated capillary flow injections of gas mixtures define typical uncertainty values of 3%–5%. A second calibrated quadrupole mass spectrometer mounted in the same vacuum chamber within 25 cm of the first, but with an intervening liquid nitrogen optically-dense cold trap, qualitatively distinguished ethanol from oxygen at m/e = 32 and organic fragments from argon at m/e = 40 by comparing its mass spectrum response with that of the main quadrupole instrument.

Larger gas bubbles in amber were opened to the analysis system during the first few crushes, and repeated crushing of the sample opened smaller bubbles, and eventually released only gas from the amber matrix. As each crushing step released gas from several possible sites in unpredictable proportions, end-member gas compositions were determined by simple mixing calculations (Landis and Berner, 1988).

AMBER DATA EVALUATION

In addition to direct analysis of gas bubbles in amber, we conducted numerous experiments to ascertain adsorption and/or desorption and reaction kinetics, gas diffusion in amber, thermal breakdown and gas release, and thermal degradation of amber and modified gas uptake. The details of these studies are beyond the scope of this presentation and will be discussed in a separate manuscript. However, several important observations are summarized here. (1) Gas compositions released from bubbles near both reddened and clear outer margins and from deeper in the interior of the same piece of amber are indistinguishable. (2) Gas of primary bubbles in highly fractured amber is no different than similar bubble gas in unfractured amber of the same locality. (3) Swelling and reddening of amber occurs visibly above ~90–100 °C, but sensitive mass-spectrometer detection of gas release by the onset of polymer breakdown occurs at ≥ 65–75 °C. We conclude that amber subjected to burial or fire temperatures ≥ 70 °C cannot be relied upon to preserve air in primary bubbles. (4) From exceptionally clear amber, free from imperfections, and in the absence of bubbles, only matrix gas is detected; primary bubbles must be present to yield "air" gases. Furthermore, gas from primary bubbles is released within the first several "crushes." (5) Gas release over four days (measured in gm O_2/100 gm amber) from amber held in a vacuum a 1 x 10^{-8} Torr is not a function of time$^{1/2}$, but rather a function of time. Compositions are normal for gases released from an unbaked high vacuum system and not amber-containing gases. These data indicate surface desorption and not bulk diffusion, according to Fick's Law. We previously discussed (Landis and Berner, 1988) difficulties with the propane "diffusion" experiments of Hopfenberg and other (1988). None of our comments (Landis and Berner, 1988) were addressed in a reported duplicate experiment (Miranda and others, 1991). Plate 1C is a photomicrograph of a 1-µm-thick microtone peel of Grassy Lakes amber similar to those used by Hopfenber in their "diffusion" experiements. The mechanically deformed structure and increased surface area is clear. This being the case, these propane uptake experiments do not provide volume diffusion information. (6) We have considered at length how to perform direct diffusion experiments using a thin prepared amber membrane with isotopically labeled gases held on either side, but rejected such attempts because of the inherent difficulties of preparing and sealing such a membrane. Gas diffusion experiments using argon isotopes is described below as an attempt to quantify gas diffusion rates in amber. (7) We cannot detect oxygen reaction with amber at ambient temperature unless the amber has been freshly broken and fractured. For mechanically crushed amber, oxygen reaction rates are similar to those reported by Horibe and Craig (1988). (8) N_2/Ar ratios of the matrix gas (Table 1) vary from 5 to 46 with a mean of 30. N_2/Ar does not represent the amber matrix water equivalent solubility of air gases as proposed by Cerling (1989), but rather a complex relation of gas uptake and osmotic exchange with tree resin while still in the resin-producing ducts (Jane, 1955; Record, 1934). We further note that Cerling failed to analyze oxygen in amber bubbles for two reasons: his use of a liquid N_2 cold trap in front of his mass spectrometer inlet (we have verified that thermal diffusion and selective cryosorption of gas in such low vacuum conditions change the detected gas chemistry by tens of percent), and his use of "bone" amber, which contains intrinsic bubbles of dissolved suboxic to anoxic gases in the resin and not primary gas bubbles.

We are currently engaged in studies to determine if there are unknown surface reactions that might cause anomalous O_2 abundance. We, like the critics of amber

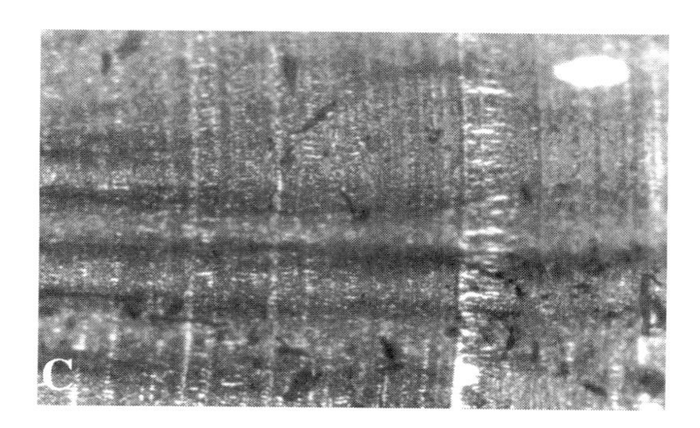

Plate 1. C: Photomicrograph of 1 µm microtomed peel of Grassy Lakes amber. Width of view is ~600 µm across. Rough and crenulate texture reflects the mechanical deformation and increased adsorption and condensation sites of the amber surface created by microtoming the amber.

data, want to know if there are any other explanations for the high O_2 values. For example, a high apparent oxygen abundance actually may be a nitrogen loss from the primary gas bubbles in amber through reaction with resin terpenoids or diffusion. However, no gaseous nitrogen reaction products were detected in any of the analyses. Furthermore, several preliminary $\delta^{15}N$ analyses of matrix nitrogen (unpublished data) indicate substantial ^{15}N enrichment relative to that of atmospheric nitrogen, suggesting denitrification reactions prior to inclusion into tree resin. If nitrogen diffusion from amber were occurring, older amber would be more enriched in ^{15}N by fractional loss of ^{14}N. We observe the opposite: the older amber specimens are less enriched in ^{15}N. We are continuing diffusion/adsorption experiments with isotopically labeled gas to quantify air-gas behavior in amber.

Gas Diffusion in Amber

Landis and Snee (1991) attempted to quantify argon diffusion rates in amber by irradiating amber in the U.S. Geological Survey instrumental nuclear reactor and then measuring the isotopes of argon released both by crushing and stepwise heating. We report here the results from 12 such analyses held for 43 to 1196 days to determine a maximum diffusion coefficient of argon in amber (Table 2). Neutron-induced ^{39}Ar is created in the amber matrix from dispersed trace potassium, and then allowed to diffuse out of the amber while held for a period of time in vacuum-sealed quartz glass ampules. Plate 1D is a four-quadrant element map of Na-K-Ca-Cl distribution in Grassy Lakes amber that illustrates the typical uniform distribution of potassium throughout the 300 μm field of element maps prepared for all samples. Samples were selected to avoid solid inclusions of possibly potassium-bearing minerals, even though inclusions as large as 30–40 μm will lose ^{39}Ar to the amber matrix by recoil in the reactor and therefore do not alter our diffusion arguments. The argon diffusion coefficient calculated is a maximum value because of the minimum detection level of 1×10^{-17} mol ^{39}Ar. Our data indicate that ^{39}Ar induced by irradiation did not excape during the ~2.2 Mev neutron recoil, did not diffuse from amber at ambient temperature during the isolation time (as long as 3 yr), and was released from the amber only upon heating to ~135 °C. Diffusion coefficients calculated by a spherical model range

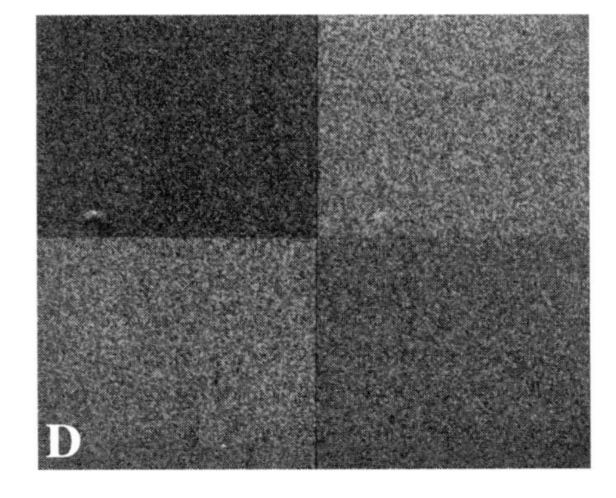

Plate 1. D: Scanning electron microscope energy-dispersive X-ray element map in four quadrants of Grassy Lakes amber illustrating the uniform distribution of potassium throughout the amber matrix. Mapped elements are Na, K, Ca, and Cl. Detection limit is ~500 ppm. No indication of microscopic solid K-bearing mineral inclusions are noted in this 300 μm field of view.

TABLE 2

Summary Argon Data and Diffusion Coefficients for Irradiated Amber

All samples were irradiated in one of five packages in the USGS TRIGA reactor at Denver, Colorado, and the argon isotopes were measured on a MAP215 mass spectrometer in the Branch of Isotope Geology in Denver. All irradiations were at 1 megawatt (MW) and total length of the irradiation for each sample is given in megawatt-hours (MWH). Temperature during irradiation did not exceed 65 °C (likely 45–55 °C). Each amber sample was irradiated within a quartz glass capsule (breakseal) that was evacuated to 10^{-8} Torr by pumping for 12 hours with a mercury diffusion pump before glassblowing closed under vacuum. After irradiation and isolation time, each breakseal was glassblown onto our extraction line for argon analysis. Breakseals were broken directly into the extraction system and released gases purified with charcoal chilled to -50 °C, molecular sieve desiccant, titanium foil heated to 380 °C, Fe-V-Al alloy getter at room temperature, and Zr-Al alloy getter heated to 250 °C. After analyzing the argon within each evacuated capsule, the amber (still under vacuum) was heated in temperature steps beginning at about 65 °C up to about 325 °C or higher, and after purification of the argon released from each temperature interval, the argon isotope abundances were measured by direct release into the mass spectrometer. Isotopic measurements were made as abundances in volts of each isotope of argon on a Faraday cup; sensitivity of the mass spectrometer and J-value were determined by analyzing irradiated MMhb-1 hornblende—the included standard for each irradiation package. Our accepted ages for MMhb-1 hornblende is that of Samson and Alexander (1987), i.e., K-Ar age = 520.4 Ma, percent k = 1.555, $^{40}Ar_R = 1.624 \times 10^{-9}$ mole/gm. Decay constants are those of Steiger and Jäger (1977). The detection limit for ^{39}Ar and ^{37}Ar is 1.0×10^{-17} moles. $(^{40}Ar/^{36}Ar)_{md}$ is the value for atmospheric argon measured by our mass spectrometer on purified atmospheric argon and is used for correction of mass discrimination assuming atmospheric $(^{40}Ar/^{36}Ar) = 295.5$. Abundance of potassium for each amber sample was calculated by comparing the yield of ^{39}Ar/mg amber with that for MMhb-1 hornblende used as the standard in the associated irradiation package; error associated with the potassium abundance calculation is difficult to quantify but is probably less than 10 percent. The apparent age for each amber was calculated from the sum of all argon isotopic abundances for all temperature steps excluding the initial argon measured in the breakseal volume. The ^{37}Ar-decay correction was made only for Cedar Lakes, Dominican Republic, and North Dakota amber samples. All errors reported are 1 σ. Diffusion coefficients calculated are maximum possible based upon detection limit for ^{39}Ar, the isolation time between irradiation (creation of ^{39}Ar) and analysis, the total ^{39}Ar yield, and approximate spherical geometry of the amber pieces, using a spherical bulk diffusion model. Diffusion distance is the maximum linear diffusion distance possible over the stratigraphic age of the amber using the calculated diffusion coefficient. All stratigraphic ages are approximate.

Cedar Lakes Amber

32.0 mg; irradiated 2/21/89 @ 10 MWH; analyzed 4/5/89; Isolation Time = 43 days
$(^{40}Ar/^{36}Ar)_{md} = 298.9$; sensitivity = 3.01×10^{-12} moles/volt; J = 0.00245 ± 0.5%;
340 ppm K; Apparent age = 482 ± 10 Ma; Stratigraphic age ≈ 75 Ma
Diffusion coefficient ≤ 7.3×10^{-7} cm^2-sec^{-1}; Diffusion distance ≤ 0.004 mm

Temp (°C)	**^{40}Ar** (moles x 10^{-12})	**^{39}Ar** (moles x 10^{-15})	**^{37}Ar** (moles x 10^{-15})	**(^{40}Ar/^{36}Ar)** corrected
Breakseal	463.2 ± 0.3	nd	nd	296.7 ± 0.5
55	0.4746 ± 0.0003	nd	nd	303.2 ± 30
70	0.73856 ± 0.00009	nd	nd	299.2 ± 22
90	0.64231 ± 0.00036	nd	0.36 ± 0.18	313.8 ± 14
140	2.4432 ± 0.0004	nd	0.09 ± 0.18	294.0 ± 6
180	0.43209 ± 0.0002	0.99 ± 0.06	0.12 ± 0.12	299.1 ± 20
250	4.1322 ± 0.0011	0.12 ± 0.03	0.24 ± 0.03	305.8 ± 2
520	4.5740 ± 0.0008	2.05 ± 0.18	nd	308.9 ± 3
750	3.2733 ± 0.0003	0.51 ± 0.09	0.24 ± 0.12	302.1 ± 3
Total (w/o breakseal)	16.710	3.67	1.05	303.9 ± 3.0

Dominican Republic Amber

17.5 mg; irradiated 3/6/90 @ 10 MWH; analyzed 5/9/90; Isolation Time = 64 days
$(^{40}Ar/^{36}Ar)_{md}$ = 299.5; sensitivity = 9.713 x 10^{-13} moles/volt; J = 0.002584 ± 0.5%;
275 ppm K; Apparent age = 119 ± 20 Ma; Stratigraphic age ≈ 20 Ma
Diffusion coefficient ≤ 1.3 x 10^{-16} cm^2-sec^{-1}; Diffusion distance ≤ 0.003 mm

Temp (°C)	^{40}**Ar** (moles x 10^{-12})	^{39}**Ar** (moles x 10^{-15})	^{37}**Ar** (moles x 10^{-15})	**(^{40}Ar/^{36}Ar)** corrected
Breakseal	3087.3 ± 1.2	nd	nd	297.4 ± 1.0
63	2.4944 ± 0.0010	nd	nd	292.5 ± 0.9
115	3.2732 ± 0.0010	nd	nd	294.2 ± 0.9
180	2.9188 ± 0.0007	nd	0.09 ± 0.04	298.4 ± 1.0
260	1.1513 ± 0.0002	0.311 ± 0.01	0.456 ± 0.003	327.4 ± 0.1
400	0.3627 ± 0.0034	1.807 ± 0.006	0.29 ± 0.08	291.7 ± 2.9
475	0.2733 ± 0.0044	3.68 ± 0.26	0.68 ± 0.08	290.0 ± 5.8
Total (w/o breakseal)	10.474	5.80	1.52	299.9 ± 1.0

North Dakota Amber

19.7 mg; irradiated 3/6/90 @ 10 MWH; analyzed 6/1/90; Isolation Time = 87 days
$(^{40}Ar/^{36}Ar)_{md}$ = 299.5; sensitivity = 1.825 x 10^{-12} moles/volt; J = 0.002584 ± 0.5%;
660 ppm K; Apparent age = 98 ± 13 Ma; Stratigraphic age ≈ 70 Ma
Diffusion coefficient ≤ 6.3 x 10^{-18} cm^2-sec^{-1}; Diffusion distance ≤ 0.001 mm

Temp (°C)	^{40}**Ar** (moles x 10^{-12})	^{39}**Ar** (moles x 10^{-15})	^{37}**Ar** (moles x 10^{-15})	**(^{40}Ar/^{36}Ar)** corrected
Breakseal	1256.9 ± 0.9	nd	nd	298.9 ± 0.2
67	15.360 ± 0.002	nd	nd	299.1 ± 0.5
145	5.4913 ± 0.0012	1.113 ± 0.07	2.04 ± 0.18	308.9 ± 0.6
223	0.9877 ± 0.0005	2.72 ± 0.22	3.36 ± 0.55	314.6 ± 3.0
320	1.094 ± 0.034	24.62 ± 0.42	38.03 ± 1.08	306.0 ± 5.0
450	0.821 ± 0.029	6.72 ± 1.95	49.86 ± 1.72	310.0 ± 6.0
Total (w/o breakseal)	23.753	35.168	93.294	305.0 ± 4.0

Grassy Lakes Amber

25.2 mg; irradiated 3/6/90 @ 10 MWH; analyzed 1/15/91; Isolation Time = 315 days
$(^{40}Ar/^{36}Ar)_{md}$ = 299.0; sensitivity = 2.078 x 10^{-12} moles/volt; J = 0.002584 ± 0.5%;
637 ppm K; Apparent age = 11 ± 1 Ma; Stratigraphic age ≈ 74 Ma
Diffusion coefficient ≤ 3.4 x 10^{-17} cm^2-sec^{-1}; Diffusion distance ≤ 0.003 mm

Temp (°C)	^{40}**Ar** (moles x 10^{-12})	^{39}**Ar** (moles x 10^{-15})	^{37}**Ar** (moles x 10^{-15})	**(^{40}Ar/^{36}Ar)** corrected
Breakseal	2495.7 ± 1.4	nd	nd	295.8 ± 0.4
64	3.5812 ± 0.0005	0.42 ± 0.19	0.08 ± 0.06	291.6 ± 0.1
110	3.8638 ± 0.0019	0.18 ± 0.08	0.04 ± 0.08	295.4 ± 0.1
200	0.9604 ± 0.0003	1.75 ± 0.12	1.87 ± 0.08	308.1 ± 2.3
280	0.1662 ± 0.0043	6.71 ± 0.17	0.64 ± 0.02	307.5 ± 7.7
325	0.12148 ± 0.00006	0.10 ± 0.04	0.06 ± 0.02	307.7 ± 15.0
Total (w/o breakseal)	8.6931	9.16	2.70	297.3 ± 3.0

Kauri Resin

33.7 mg; irradiated 3/6/90 @ 10 MWH; analyzed 1/15/91; Isolation Time = 315 days
$(^{40}Ar/^{36}Ar)_{md}$ = 299.0; sensitivity = 2.078 x 10^{-12} moles/volt; J = 0.002584 ± 0.5%;
1738 ppm K; Apparent age = -0.5 ± 1.2 Ma; Stratigraphic age ≈ 0.1 Ma [modern]
Diffusion coefficient ≤ 6.7 x 10^{-18} cm^2-sec^{-1}; Diffusion distance ≤ 0.00005 mm

Temp (°C)	^{40}**Ar** (moles x 10^{-12})	^{39}**Ar** (moles x 10^{-15})	^{37}**Ar** (moles x 10^{-15})	**(^{40}Ar/^{36}Ar)** corrected
Breakseal	568.5 ± 0.4	nd	nd	295.8 ± 0.3
52	3.9233 ± 0.0006	1.413 ± 0.008	0.14 ± 0.06	303.5 ± 0.3
198	0.1907 ± 0.0016	3.16 ± 0.08	0.39 ± 0.19	340.0 ± 15.0
435	0.4750 ± 0.0124	29.74 ± 1.41	3.37 ± 1.42	293.1 ± 5.9
Total (w/o breakseal)	4.5889	34.31	3.91	294.1 ± 2.0

Baltic Amber

16.6 mg; irradiated 3/6/90 @ 10 MWH; analyzed 6/14/93; Isolation Time = 1196 days
$(^{40}Ar/^{36}Ar)_{md}$ = 299.7; sensitivity = 1.252 x 10^{-12} moles/volt; J = 0.002584 ± 0.5%;
158 ppm K; Apparent age = 144 ± 10 Ma; Stratigraphic age ≈ 45 Ma
Diffusion coefficient ≤ 5.2 x 10^{-17} cm^2-sec^{-1}; Diffusion distance ≤ 0.003 mm

Temp (°C)	^{40}**Ar** (moles x 10^{-12})	^{39}**Ar** (moles x 10^{-15})	^{37}**Ar** (moles x 10^{-15})	**(^{40}Ar/^{36}Ar)** corrected
Breakseal	327.57 ± 0.14	nd	nd	298.0 ± 0.4
100	1.9844 ± 0.0014	nd	nd	294.3 ± 0.6
130	1.6830 ± 0.0022	0.901 ± 0.088	0.14 ± 0.14	302.8 ± 3.0
141	0.10152 ± 0.00003	0.288 ± 0.050	0.04 ± 0.01	312.0 ± 4.0
240	0.09936 ± 0.00005	0.764 ± 0.005	0.01 ± 0.04	305.2 ± 0.2
300	0.18711 ± 0.00014	0.138 ± 0.25	0.02 ± 0.02	392.0 ± 0.3
Total (w/o breakseal)	4.0554	2.091	0.21	300.5 ± 0.5

Mt. Hermon Amber

60.4 mg; irradiated 6/16/92 @ 33 MWH; analyzed 7/7/93; Isolation Time = 386 days
$(^{40}Ar/^{36}Ar)_{md}$ = 299.0; sensitivity = 1.252 x 10^{-12} moles/volt; J = 0.00775 ± 0.5%;
1156 ppm K; Apparent age = 1500 ± 100 Ma; Stratigraphic age ≈ 130 Ma
Diffusion coefficient ≤ 1.8 x 10^{-18} cm^2-sec^{-1}; Diffusion distance ≤ 0.0009 mm

Temp (°C)	^{40}**Ar** (moles x 10^{-12})	^{39}**Ar** (moles x 10^{-15})	^{37}**Ar** (moles x 10^{-15})	**(^{40}Ar/^{36}Ar)** corrected
Breakseal	3947.5 ± 1.2	nd	nd	312.0 ± 1.0
65	1931.3 ± 0.4	nd	1.39 ± 0.88	298.8 ± 0.6
140	144.42 ± 0.10	nd	nd	299.1 ± 0.4
205	111.376 ± 0.006	nd	nd	301.1 ± 0.9
310	47.654 ± 0.029	2.28 ± 1.20	0.388 ± 0.22	303.4 ± 0.5
400	18.04 ± 0.34	185.9 ± 41.7	17.0 ± 1.93	298.5 ± 6.0
Total (w/o breakseal)	934.59	188.1	18.83	306.4 ± 1.0

Saudi Arabian Amber

37.0 mg; irradiated 6/16/92 @ 33 MWH; analyzed 7/7/93; Isolation Time = 386 days
$(^{40}Ar/^{36}Ar)_{md}$ = 299.0; sensitivity = 1.252 x 10^{-12} moles/volt; J = 0.00775 ± 0.5%;
37 ppm K; Apparent age = 2500 ± 150 Ma; Stratigraphic age ≈ 115 Ma
Diffusion coefficient ≤ 5.7 x 10^{-17} cm^2-sec^{-1}; Diffusion distance ≤ 0.005 mm

Temp (°C)	**^{40}Ar** (moles x 10^{-12})	**^{39}Ar** (moles x 10^{-15})	**^{37}Ar** (moles x 10^{-15})	**$(^{40}Ar/^{36}Ar)$** corrected
Breakseal	1511.5 ± 0.6	nd	nd	296.7 ± 0.7
80	13.473 ± 0.009	nd	nd	301.0 ± 0.3
140	20.42 ± 0.02	nd	nd	304.9 ± 0.6
213	15.74 ± 0.09	0.95 ± 0.83	0.18 ± 0.01	302.5 ± 0.2
305	8.206 ± 0.005	1.39 ± 0.15	0.42 ± 0.06	301.9 ± 0.1
410	2.959 ± 0.001	1.31 ± 0.25	0.20 ± 0.02	297.3 ± 0.1
Total (w/o breakseal)	60.80	3.66	0.80	302.6 ± 0.1

Upper Z Coal

73.1 mg; irradiated 3/23/93 @ 20 MWH; analyzed 10/28/93; Isolation Time = 219 days
$(^{40}Ar/^{36}Ar)_{md}$ = 298.9; sensitivity = 1.252 x 10^{-12} moles/volt; J = 0.00490 ± 0.5%;
34 ppm K; Apparent age = 6.6 ± 10 Ma; Stratigraphic age ≈ 65.0 Ma
Diffusion coefficient ≤ 9.1 x 10^{-16} cm^2-sec^{-1}; Diffusion distance ≤ 0.013 mm

Temp (°C)	**^{40}Ar** (moles x 10^{-12})	**^{39}Ar** (moles x 10^{-15})	**^{37}Ar** (moles x 10^{-15})	**$(^{40}Ar/^{36}Ar)$** corrected
Breakseal	2227.2 ± 2.6	nd	nd	306.4 ± 0.8
70	17.26 ± 0.10	0.60 ± 0.32	0.06 ± 0.06	296.7 ± 1.5
145	21.293 ± 0.019	nd	0.21 ± 0.06	297.8 ± 0.3
200	10.972 ± 0.004	nd	nd	298.6 ± 0.1
300	4.253 ± 0.004	0.81 ± 0.32	0.09 ± 0.07	301.7 ± 0.1
400	2.5327 ± 0.001	2.58 ± 0.56	0.30 ± 0.08	299.3 ± 0.2
Total (w/o breakseal)	56.309	3.99	0.66	300.5 ± 0.5

Lower Z Coal

21.4 mg; irradiated 3/17/93 @ 20 MWH; analyzed 10/29/93; Isolation Time = 226 days
$(^{40}Ar/^{36}Ar)_{md}$ = 298.9; sensitivity = 1.252 x 10^{-12} moles/volt; J = 0.00535 ± 0.5%;
51 ppm K; Apparent age = 561 ± 42 Ma; Stratigraphic age ≈ 65.2 Ma
Diffusion coefficient ≤ 4.5 x 10^{-16} cm^2-sec^{-1}; Diffusion distance ≤ 0.0096 mm

Temp (°C)	**^{40}Ar** (moles x 10^{-12})	**^{39}Ar** (moles x 10^{-15})	**^{37}Ar** (moles x 10^{-15})	**$(^{40}Ar/^{36}Ar)$** corrected
Breakseal	[Broken seal — no analysis]			
155	0.5486 ± 0.0004	0.26 ± 0.04	0.14 ± 0.20	302.2 ± 2.1
162	0.3479 ± 0.0903	0.43 ± 0.19	0.14 ± 0.11	323.1 ± 0.3
220	0.31395 ± 0.00006	0.25 ± 0.06	0.08 ± 0.02	334.3 ± 0.2
320	0.2191 ± 0.0004	0.60 ± 0.12	0.06 ± 0.08	324.1 ± 0.6
400	0.4250 ± 0.0003	0.30 ± 0.06	0.06 ± 0.06	317.2 ± 0.3
Total (w/o breakseal)	1.8546	1.778	0.48	317.2 ± 0.3

Zambian Amber [outer rim]				
38.8 mg; irradiated 3/17/93 @ 20 MWH; analyzed 11/1/93; Isolation Time = 229 days $(^{40}Ar/^{36}Ar)_{md}$ = 298.9; sensitivity = 1.252 x 10^{-12} moles/volt; J = 0.00535 ± 0.5%; 39 ppm K; Apparent age = 560 ± 100 Ma; Stratigraphic age ≈ 3–5 Ma Diffusion coefficient ≤ 8.7 x 10^{-16} cm^2-sec^{-1}; Diffusion distance ≤ 0.004 mm				
Temp (°C)	^{40}**Ar** (moles x 10^{-12})	^{39}**Ar** (moles x 10^{-15})	^{37}**Ar** (moles x 10^{-15})	**(^{40}Ar/^{36}Ar)** corrected
Breakseal	[Broken seal — no analysis]			
65	2.6962 ± 0.0011	0.28 ± 0.30	0.06 ± 0.05	299.1 ± 0.1
150	1.9250 ± 0.0012	0.05 ± 0.01	nd	300.3 ± 0.2
220	0.0126 ± 0.0004	0.45 ± 0.12	nd	326.9 ± 6.5
310	0.3988 ± 0.0010	1.46 ± 0.24	nd	297.7 ± 0.2
402	0.4250 ± 0.0001	0.24 ± 0.14	nd	305.8 ± 3.1
Total (w/o breakseal)	6.3575	2.47	0.06	299.9 ± 0.1

Zambian Amber [interior]				
78.1 mg; irradiated 3/17/93 @ 20 MWH; analyzed 11/2/93; Isolation Time = 230 days $(^{40}Ar/^{36}Ar)_{md}$ = 298.9; sensitivity = 1.252 x 10^{-12} moles/volt; J = 0.00535 ± 0.5%; 18.5 ppm K; Apparent age = 1522 ± 500 Ma; Stratigraphic age ≈ 3–5 Ma Diffusion coefficient ≤ 2.6 x 10^{-15} cm^2-sec^{-1}; Diffusion distance ≤ 0.006 mm				
Temp (°C)	^{40}**Ar** (moles x 10^{-12})	^{39}**Ar** (moles x 10^{-15})	^{37}**Ar** (moles x 10^{-15})	**(^{40}Ar/^{36}Ar)** corrected
Breakseal	[Broken seal — no analysis]			
65	0.6163 ± 0.0004	0.34 ± 0.15	0.12 ± 0.14	303.8 ± 1.5
145	1.1412 ± 0.0009	0.19 ± 0.15	nd	314.3 ± 0.9
208	1.8605 ± 0.0007	0.50 ± 0.15	0.09 ± 0.09	321.0 ± 0.6
310	2.5575 ± 0.0022	0.83 ± 0.21	0.14 ± 0.08	321.7 ± 0.6
405	1.6918 ± 0.0008	0.50 ± 0.38	nd	322.5 ± 0.8
Total (w/o breakseal)	7.8673	2.35	0.35	319.1 ± 0.6

from $D \leq 2.6 \cdot 10^{-15}$ cm^2-s^{-1} to $D \leq 1.8 \cdot 10^{-18}$ cm^2-s^{-1}. Amber from 12 different localities and of different geologic ages all exhibited the same "closed" behavior to argon. Assuming that diffusion of other "air" gases in amber is the same or less than for argon, these gases could not diffuse more than several microns over the geologic age of the amber.

Furthermore, if oxygen diffusion through amber is high, O_2 values should be uniform throughout the sample and very near modern abundance. All Cretaceous amber samples indicate elevated oxygen. Low values of nearly 14% are indicated for Miocene and late Eocene amber, and yet modern Kauri resin yields 21% oxygen (Landis and Berner, 1988, and Table 1b). Kauri resin from New Zealand was buried in a peat bog for ≥ 35,000 yr and only briefly exhumed and exposed to modern atmosphere before analysis, yet primary gas bubbles yield modern air composition and not anoxic sediment gas of the burial environment. If gas diffusion through amber were an issue, all values should have equilibrated to modern oxygen abundance.

Amber preserves organic substances that most certainly would be oxidized (i.e., DNA, insects, small animals, botanical material) if O_2 diffused either as a gas or was released from organic compounds (Cano and others, 1992a, 1993; DeSalle and others, 1992, 1993; Morell, 1993; Pääbo, 1993; Poinar, 1993; Poinar and others, 1993a, 1993b). We know of no "oxygen pump" to selectively enrich O_2 sites within amber. Because the high O_2 values in each primary bubble obviously are not uniform throughout the amber, we view several criticisms of amber gas data as being severely flawed (Beck, 1988; Hopfenberg and others, 1988; Horibe and Craig, 1988; Cerling, 1989).

Conclusions

Only primary gas bubbles yield high O_2 values, whereas other gas sites in amber release anoxic and metabolic gases as expected. We know of no other means to obtain the high oxygen values we report for primary gas bubbles in amber than to conclude that primary bubbles actually trapped ancient air in the original amber resins when it exuded from the tree.

Gas in amber must be geologically old for the following combined reasons. (1) The minimum O_2 in bubbles, normalized to the sum of N_2, O_2, Ar, and CO_2, is higher in Cretaceous than in Tertiary and recent ambers (Berner and Landis, 1988; Landis and Berner, 1988); modern Kauri resin contains "air" gases at present atmospheric levels, whereas low values of nearly 15% are indicated for Miocene and late Eocene with near modern values throughout most of the Tertiary. (2) The variability in gas chemistry with each crush points to mixtures of end-member gas compositions that can be resolved by a series of linear mixing calculations. (3) Diffusion of gas into and out of amber has been addressed using argon isotopes (Landis and Snee, 1991), and from the indicated diffusion coefficient, $D \leq 1.5 \cdot 10^{-17}$ cm^2-sec^{-1}, gas will remain unexchanged in amber for geologically long periods of time. (4) Elevated oxygen in amber is independent of degree of reddened rind, fracturing, length of exposure to modern atmosphere, time of burial in sediments, and size of gas bubbles. (5) Paleoatmospheric oxygen levels through time are both predictable from observable geologic and biologic records and are consistent with those records. The Hell Creek data we report here are based upon six stratigraphic intervals spanning several million years from the same locality, with amber probably produced from the same type of vegetation, deposited in the same depositional environment, subjected to the same burial and diagenesis history, and exposed and weathered to the same degree. Oxygen variation in gas bubbles of these ambers can best be explained only by differing levels of oxygen in the Late Cretaceous–early Tertiary atmosphere.

The atmospheric oxygen concentrations determined from amber that bracket the K-T boundary (Table 1a) show the same elevated values for Late Cretaceous as reported by Berner and Landis (1988). Age control of the samples is based upon recent correlations of argon isotope dates with magnetostratigraphy and biostratigraphy (Swisher and others, 1993). Oxygen varied from nearly 35% ~1.5–3.0 m.y. before the K-T boundary, to 31% just below the Ir anomaly and pollen break, to 29% about 0.2 m.y. after the boundary. Within ~0.5–0.7 m.y. oxygen in the atmosphere climbed to 33% from its previous low of 29%. Atmospheric levels then dropped to 30%–31% over the next 2 m.y. and presumably dropped to 22% by the time (~45 m.y.) atmosphere was sampled in the early Eocene Baltic amber of Berner and Landis (1988). This pattern of variation is illustrated in Figure 3. Two striking freatures of this trend in atmospheric oxygen across the K-T boundary are the gradual oxygen decrease, initiated ≥ 1.5 m.y. before the K-T boundary, and the rate change (measured in several 10^5 yr) at and immediately following the K-T boundary horizon.

PELE HYPOTHESIS

Elevated O_2 in Cretaceous atmosphere as indicated by amber samples has enormous biological and geochemical implications. Because free oxygen can be produced only via net photosynthesis (Berner and Canfield, 1989) with sequestering of unoxidized organic material, there must be a close correlation between plant productivity and atmospheric CO_2 (Graybill and Idso, 1993; Idso and Kimball, 1993). Many factors contribute to increased plant productivity, including temperature, humidity, precipitation, solar light level, available nutrients (Broecker and Peng, 1982; Berner and others, 1993), and carbon dioxide. We suggest that atmospheric CO_2 is the dominant factor in determining rates of photosynthesis and productivity, because it has an impact on weathering rates and release of

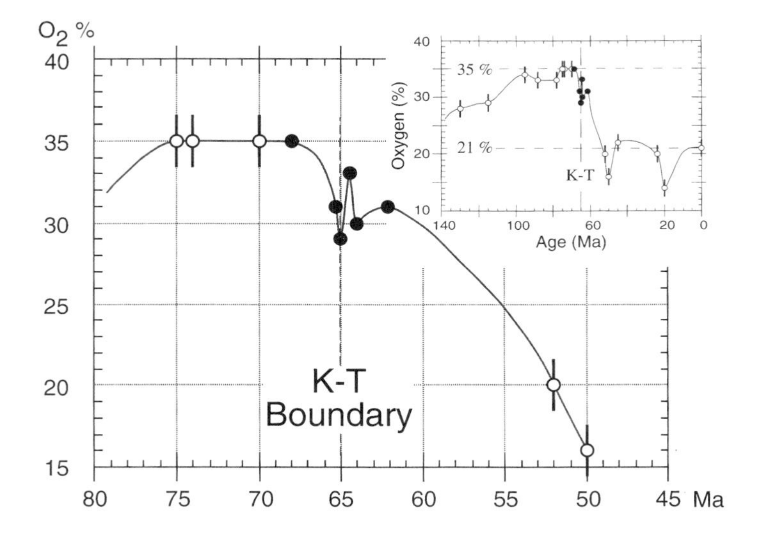

Figure 3. Variation in atmospheric oxygen through time across the K-T boundary. Inset illustrates the oxygen variation in volume percent based upon earlier work on 15 noted world localities reported by Berner and Landis (1988) and Landis and Berner (1988). Open circles are earlier data and solid circles represent new data reported in Table 1 for Russell basin amber. Vertical bars are approximate combined analytical error and uncertainty in mixing calculations. With the exception of precise age for Russell basin amber, the ages of other amber are approximate and rely upon stratigraphic and paleontologic controls. Cretaceous-Tertiary boundary data are based upon 96 separate crush analyses from 6 stratigraphic horizons (Fig. 1) with duplicate splits for each. World data represent more than 500 analyses.

nutrients (nitrates and phosphates), as well as being an important factor in shaping global temperature and precipitation. High levels of O_2 in the K-T boundary atmosphere therefore imply similar large amounts of CO_2 that must have preceded or coincided with observed O_2 abundance (Berner, 1991; Cerling, 1991).

We argue that large volumes of new CO_2 in the atmosphere can only be derived from mantle degassing (McLean, 1978, 1985a, 1985b, 1985c; Caldeira and Rampino, 1991; Larson, 1991a, 1991b) which is in turn related directly to mantle superplumes during times of increased spreading rates and new crust production at ridge crests (Larson and Olson, 1991; Larson, 1991a, 1991b), increased intraplate volcanism, and formation of large igneous provinces (LIPs) (Coffin and Eldholm, 1992, 1993). The correlation between superplume activity, the "quiet" times in Earth's magnetic field reversal rate, and eustatic sea-level changes has been demonstrated previously (Gaffin, 1987; Larson and Olson, 1991; Marzocchi and other, 1992). It seems likely that atmospheric gas concentrations, sea levels, global mean temperatures, and many additional environmental constraints on the biosphere are all ultimately linked to mantle processes (Fyfe, 1990). We consider constraints on the biosphere to include limits on survival, extinction, and diversity.

Pele Hypothesis Definition

We propose the Pele hypothesis (after the Hawaiian volcano goddess) to identify, but not limit, our concept of the relation and interdependence between episodic (possibly periodic) changes in the rate of mantle degassing and other Earth and biologic processes. Major geologic events can have important consequences for evolution and extinction of life on Earth. The Pele hypothesis is the relation between (1) episodic increases in deep mantle CO_2 degassing to the atmosphere from superplumes and large igneous provinces (LIPs), (2) accelerated net photosynthesis from an expanded biosphere and sequestering of large amounts of organic carbon, and (3) the increase in photosynthetically-generated atmospheric O_2.

The Pele hypothesis represents a framework within which to understand both primary and secondary extinction pressures and stimuli for biodiversity. A middle Cretaceous episode of increased mantle CO_2 flux to the atmosphere stimulated plant productivity and yielded vast deposits of organic material in shallow continental and marine shelf anoxic basins that formed or were expanded by the eustatic rise in sea level accompanying accelerated sea-floor spreading and intraplate superplume volcanism. Increased productivity within the terrestrial and marine biosphere causes generation of important Cretaceous oil, gas, coal, chalk, and carbonate deposits on a global scale. Deep-ocean anoxic events occur, the pH of mixed-layer ocean water decreases, the oceanic carbonate compensation depth rises, and the mechanics of ocean chemistry and mixing change. Carbon cycle models must be modified to account for independent mantle CO_2 fluxes to the atmosphere as a function of new ridge crest crust production and LIP events, and to account for increased organic material in anoxic basins. In the Cretaceous, high atmospheric CO_2 and increased moisture caused greenhouse warming, created a more equable global climate lacking seasonality even at high latitudes, and caused O_2 to climb to ~35%.

This major episode of superplume activity and mantle CO_2 degassing ended in the Late Cretaceous, causing world ocean basins to deepen, global temperatures to drop, and atmospheric CO_2/O_2 ratios to change. Eustatically controlled marine regression exposed organic-rich anoxic basin sediments to oxidation, which caused the observed ~12% reduction in atmospheric O_2. Spikes of $\delta^{13}C_{microfossils}$, $^{87}Sr/^{86}Sr_{marine}$, and elemental carbon (soot) in marine K-T boundary sediment cores reflect this selectively accelerated weathering. Productivity decreased with more drastic effects taking place at lower latitudes resulting in many species being thrown into ecologic crisis situations. The primary extinction factor of changing atmospheric CO_2/O_2 modified productivity, altered global temperatures, caused respiratory and metabolic stress, and shifted ecological niches. Secondary extinction factors included disruption of food webs that forced migration, and shifts in anoxia, precipitation and weathering rates, marine water chemistry, and global humidity. Plant response to increased CO_2 is largely nutrient controlled (nitrates and phosphates), with CO_2 becoming toxic above plant productivity "sink" limits. Weathering (nutrient supply) and sediment transport (organic burial) rates limit the atmospheric O_2 maximum to the observed ~35%.

Atmospheric CO_2-O_2 budgets must be considered major survival constraints based upon the complex interdependence of geologic processes. That essentially all major extinction events occur at the nodes of marine transgressions or regressions is not by chance—these are times that major, atmosphere-driven primary and secondary extinction pressures are greatest. The gradual extinction of select species (dinosaurs, certain marine organisms, and others of the terminal Cretaceous extinction) is best explained by the interdependence of terrestrial-based processes of the Pele model.

Possible Consequences of the Pele Hypothesis

We argue that a major period of deep mantle degassing, with CO_2 addition to the atmosphere, occurred between 122 and 83 Ma during the superplume described by Larson (1991a, 1991b). Ridge crest spreading rates increased from 1 cm to 2 cm per year to more than 10 cm per year. Because observed elevated atmospheric O_2, inferred CO_2/O_2, and changing atmospheric compositions through time have such far-reaching implications, we outline below important required and permissive consequences. Figure 4 combines a geologic and absolute time scale and magnetochronographic time scale from the Early Cretaceous to the present, with reconstructed changes in sea level, new crust production, and large igneous provinces (LIPs), estimated global temperatures, changes in atmospheric oxygen and carbon dioxide, genera extinction rates, and number of dinosaur genera. These data indicate a clear interplay of geologic and biologic processes of the Pele hypothesis. Features of these dependencies are discussed below.

The following items illustrated some possible effects of the Pele hypothesis and are openly speculative in nature. When atmospheric gas concentrations shift as rapidly and as much as indicated in primary amber bubbles and when other lines of independent geologic evidence (e.g., organic sequestering or burying of organic matter as indicated by vast coal, oil, and gas deposits, $\delta^{13}C$ shifts of marine tests, and changes in productivity) virtually require these or similarly elevated O_2 values, many past assumptions regarding relations of the biosphere, lithosphere, hydrosphere, and atmosphere may be invalid. We make no claim for comprehensive insight but hope the following will provoke thought and inquiry.

1. The well-known Cretaceous Aptain to Maastrichtian greenhouse episode (Savin, 1977; Barron, 1983; Barron and Washington, 1984; Sloan and Barron,

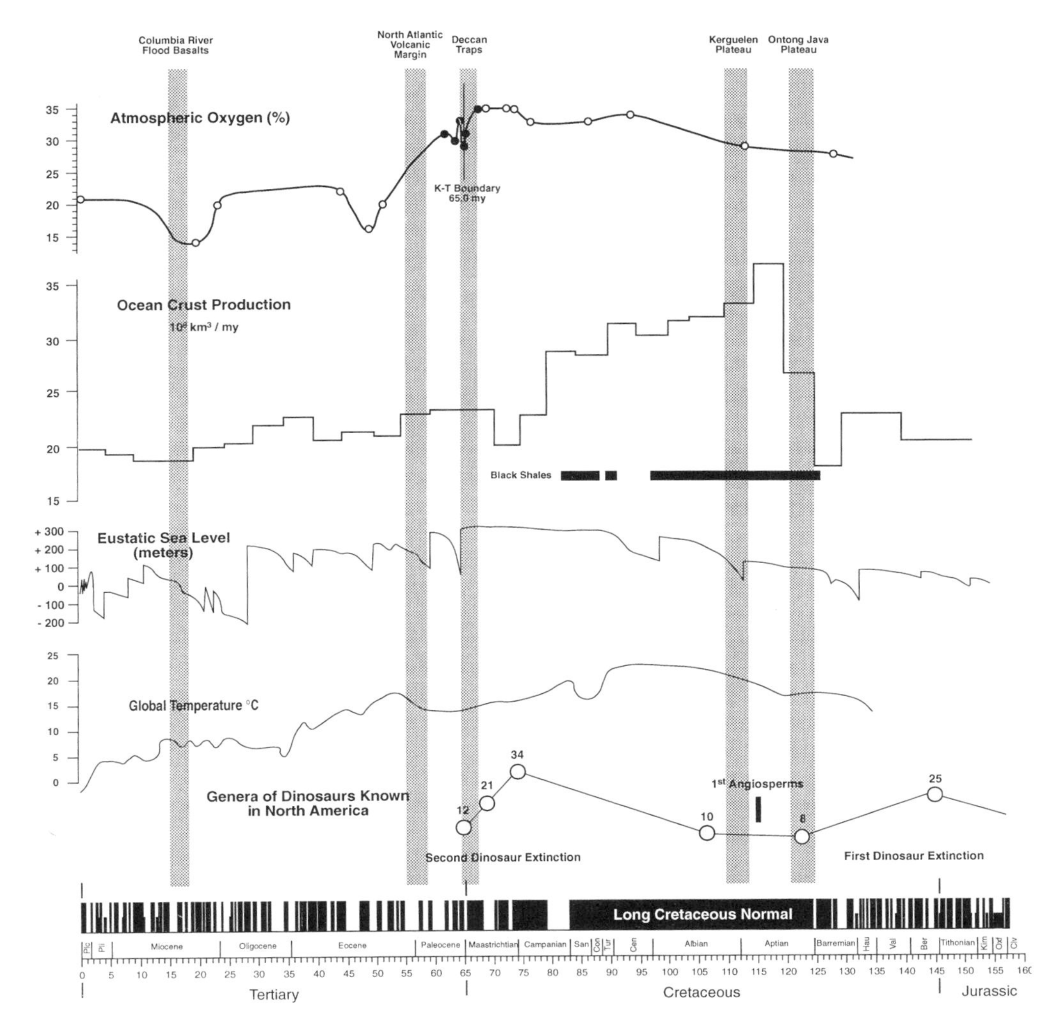

Figure 4. Pele hypothesis parameters, variations in atmospheric oxygen, ocean crust production, eustatic sea level, global temperature, and dinosaur extinction through time. Also located are several major superplume events (large igneous provinces, LIPs) and black shale depositional intervals. Data plotted against magnetochron and absolute time scale of Harland and others (1989). Oxygen data from this paper and Berner and Landis (1988); ocean crust production and black shales from Larson (1991a, 1991b); sea-level data from Vail and others (1977) and Haq and others (1987); global temperature estimates from Barron (1983), Barron and Washington (1984), and Savin (1977); dinosaur extinction by number of genera in North America from Weishampel (1990) and Sloan and others (1986a, 1986b); and LIPs from Larson and Olson (1991) and Coffin and Eldholm (1992, 1993). Minor misalignment of curves is the result of attempting to cast all data to the same time base.

1990; Caldeira and Rampino, 1991) resulted from both paleogeographic plate positions and large amounts of CO_2 in the atmosphere (Arthur and others, 1985, 1987, 1988; Arthur and Dean, 1986; Berner, 1990, 1991; Cerling, 1991; Larson and Olson, 1991; Larson, 1991a, 1991b). Atmospheric CO_2 levels, as much as 10 to 12 times present levels, stimulated increased plant productivity that generated large amounts of photosynthetic O_2 and vast organic deposits that became isolated from the carbon cycle, thus allowing for a large net atmospheric O_2 buildup (Berner, 1989; Berner and Canfield, 1989; Bertrand and Lallier-Vergès, 1992; Diaz and others, 1993; Graybill and Idso, 1993; Hollander and others, 1993; Idso and Kimball, 1993; Riebesell and others, 1993; Riebesell, 1993; Turpin, 1993). It appears from our studies that Cretaceous atmospheric abundance of O_2 reached and maintained an upper limit of ~35%, which closely corresponds to metabolic toxicity, and placed an upper limit on O_2 production via negative feedback into photosynthesis (Raven, 1991). Our data, and those of Berner and Landis (1988), indicate that O_2 levels in the atmosphere change with a rate not much slower than for CO_2 (several hundred thousand years), not several million years (see Holland, 1984). This observation is not surprising given that atmospheric CO_2 is cycled through the surface oceans about once every 20 yr (de Baar and Suess, 1993), the residence time for CO_2 in the oceans is ~100,000 yr, that for nutrients such as phosphate is nearly 180,000 yr, and ocean deep-water overturn is every 2000 yr (Broecker and Peng, 1982).

2. We are reasonably certan that weathering was more severe in the Cretaceous (D'Argenio and Mindszenty, 1986; D'Argenio and Mindszenty, 1989). Weathering rates are expected to increase radically with more acid and oxidizing environments produced by high levels of O_2 and/or CO_2 (Berner, 1989a, 1989b, 1991; Velbel, 1993) in the atmosphere. The exact nature of the nutrient flux (phosphates and nitrates) to marine environments resulting from accelerated weathering is unclear (Berner and others, 1993), as is the effect of accelerated nutrient fluxes on oceanic anoxia and placement of the carbonate compensation depth (or CCD) (Arthur and others, 1987). We do observe a plateau of O_2 values at about 35% during a large part of the Cretaceous. This would imply that some set of controls prevents O_2 from rising above those values. We suspect that weathering rates and subsequent mutrient fluxes, even in the presence of high CO_2, may be primary factors in O_2 control by limiting organic productivity to the available nutrient supply.

3. Anomalous Cretaceous hydrocarbon reserves such as petroleum, natural gas, coal, as well as massive amounts of chalk and other carbonate materials suggest an enormous organic productivity. This productivity may have been responsible for food-web expansion and the subsequent explosion in animal diversity (a 5–20 fold species richness increase) experienced by virtually all major marine invertebrate groups and many terrestrial groups during the Cretaceous. One must conclude from the enormous primary productivity and from the sequestering of organic material that oxygen in the Cretaceous atmosphere was significantly above that of today (Berner and Canfield, 1989). This conclusion is indicated irrespective of the amber data.

4. We suspect that the radical shifts in Cretaceous atmospheric gases controlled the shift from gymnosperm to angiosperm dominated floras nearly 115 m.y. ago

(Retallack and Dilcher, 1986; Wing and others, 1993). Elevated O_2 would accelerate oxidation (possibly burning) of plant reproductive bodies—unprotected seeds would be at a decided disadvantage for survival. Increased fire damage would preferentially favor angiosperms because of a more efficient and rapid reproductive cycle and would also explain observed specialized fire defense structures seen in many Cretaceous floras (Robinson, 1989). The speculated increase in fire damage is less than one might expect because concurrent high levels of CO_2 and atmospheric moisture would have retarded fire activity. Our preliminary combustion experiments in controlled Cretaceous atmospheres support this conclusion.

5. Relative proportions of O_2 and CO_2 may also have played a direct role in biotic extinctions. A sudden rise in atmospheric CO_2 will bring on increased respiratory stress because CO_2 is preferentially absorbed into plant and animal metabolic systems at a rate nearly ten times that of oxygen, causing reduced viability or respiratory failure. Decreasing O_2 levels in the presence of high CO_2 abundance would also produce respiratory stress. In animals with high metabolic needs and inefficient respiratory systems, decreased levels of oxygen may prove inadequate to support basal metabolic rates and may exceed metabolic scope. Both types of respiratory extinctions are exhibited during the Cretaceous. (1) A high CO_2-generated extinction at the end of the Jurassic and beginning of the Cretaceous with large amounts of CO_2 released from the mantle may have promoted the onset of the Cretaceous greenhouse. (2) Falling O_2 levels in the presence of elevated CO_2 at the end of the Cretaceous may have caused respiratory stress in the exact time interval when metabolic needs were increasing because of falling global temperatures. Both periods of dramatic change in atmospheric gases mark major extinction events for dinosaurs (Sloan and Rigby, 1986; Sloan and others, 1986; Weishampel, 1990; Hengst and others, 1993). Declining oxygen concentrations, along with cooler global temperatures, a eustatic sea-level drop, with attendant reduction of shallow-marine environments, changing food webs, changing ocean chemistry, accelerated weathering, and other factors conceptually linked to the Pele hypotheis were likely responsible for most of the mass extinctions at the end of the Cretaceous (Landis and others, 1993).

6. The reversal rate of the Earth's magnetic field is inversely related to sea-level change, with an ~6–9 m.y. lag in eustatic sea-level response (Gaffin, 1987; Larson and Olson, 1991; Marzocchi and others, 1992). If mantle superplume activity originates at the boundary between the outer core and the innermost mantle (D-layer) and its increase decreases the rate of paleomagnetic field reversal (e.g., maximum superplume activity occurs during paleomagnetic "quiet" times such as the Cretaceous long normal period), then reduced mantle superplume activity would slow spreading rates, deepen ocean basins, and cause marine regressions worldwide. The terminal Cretaceous decline in atmospheric O_2 abundance is associated with a drop in eustatic sea level of more than 200 m (Vail and others, 1977; Haq and others, 1987; Hallam, 1992; Gurnis, 1993), which exposed buried organics to oxidation in sites that previously served as organic sinks. Exposure of these sites to oxidation resulting from marine regression would be expected to reduce the net O_2 photosynthetically generated to the atmosphere and increase O_2 consumption by weathering. Thus, relatively high

CO_2 abundances could be maintained while substantially reducing O_2 levels. Rapid recycling and oxidation of estimated exposed epicontinental seaways and shelf anoxic and organic-rich sediments at the end of the Cretaceous (Kauffman, 1977; Brass and others, 1982; Arthur and others, 1987) is more than adequate to account for the decrease in atmospheric oxygen we observe (29%–35%).

7. Eustatic adjustments have worldwide correlation potential and must be controlled by changes in ocean-basin volume or shifts in water volumes within those basins. Deep mantle convection, elevating and lowering ocean-basin floors, and accelerating spreading rates at ridge crests are commonly believed to control marine transgressions and regressions. We assume that oceanic plate buoyancy is the major driving factor in eustatic adjustments: however, thermal volume changes in the oceans and hydration and/or dehydration of the upper asthenosphere as isotherms shift in response to superplume activity over vast ridge crests and interplate areas should also play a part in determining ocean basin and water volumes. The inferred relation between paleomagnetic field reversal rates and sea-level changes also may be linked to the quantity of new crust production (Larson and Olson, 1991; Larson 1991a, 1991b) and mantle degassing of CO_2 and other gases. Coincidence of marine extinctions with eustatic sea-level adjustments alone seems improbable. Introduction of mantle-derived CO_2 to the atmosphere, associated with a sea-level rise, might produce a CO_2 "poisoning" of shelf and epicontinental seaway environments by increasing productivity, expanding anoxic bottom-water conditions for depths $\geq$100 m, and decreasing mixed surface-water pH ($\leq$100 m). Linkage between atmospheric CO_2 and O_2 budgets must now be considered major limiting factors. Because virtually all major extinction events occur at the nodes of either marine transgressions or regressions, it seems reasonable to conclude that these intervals represent times at which atmospheric extinction pressures were at their greatest (Holser and others, 1989).

8. For some time the selective extinction of terrestrial vertebrate lineages at the K-T boundary has posed a problem for those seeking a catastrophic end to dinosaurs and other vertebrate groups (Sheehan and others, 1991). How could known cold-blooded organisms, like lizards, turtles, crocodiles, and high-metabolism animals like birds survive the proposed "nuclear winter" when large numbers of species are supposed to be dying of cold and lack of food in an extended period of darkness. Instead, these very temperature-sensitive groups come through the K-T biotic crisis more or less intact! Dinosaurs suffer irreparable damage and ultimately go extinct at the same time that mammalian groups are radiating. Recent descriptions of Arctic terminal Cretaceous cool-climate dinosaur fauna (Brouwers and others, 1987; Clemens and Nelms, 1993, 1994; Buffetaut 1994) indicate that cold temperatures, seasonal darkness, and fluctuations in food chain structure do not provide adequate extinction pressures. Northern high-latitude marine and nonmarine habitats were not markedly affected by terminal Cretaceous events (Brouwers and DeDeckker, 1993). During periods of decreasing atmospheric O_2 and high CO_2, animals with large metabolic needs and inefficient respiratory systems (e.g., nonavian dinosaurs) would be predicted to suffer respiratory stress and decreased viability while those with high metabolic needs but efficient respiratory mechanisms (birds and mammals) would be predicted to survive (Hengst and others, 1993). It is well documented

that the decline in Cretaceous dinosaur populations predates the terminal Cretaceous crisis (Van Valen and Sloan, 1977; Sloan and Rigby, 1986; Sloan and others, 1986; Rigby, 1987; Weishampel, 1990). These are not catastrophic reductions in the sense that reductions occur over a substantial period of time, not suddenly. The Cretaceous dinosaur decline is more likely a response to gradually induced extinction pressures created by increased metabolic needs in the presence of falling temperatures and onset of seasonality, changing food supplies, loss of the equable Cretaceous greenhouse climate, and high CO_2 in the presence of declining O_2. Whether temperature alone or temperature acting in consort with declining O_2 values and other factors caused virtually two-thirds of the dinosaur extinctions prior to the terminal Maastrichtian is not clear. What is certain is that dinosaur populations and diversity had been declining for more than 30 m.y. and, prior to the terminal event, two-thirds of the ultimate dinosaur extinction had been completed.

9. We note that many changes in sea level throughout the Mesozoic have also been associated with major increases in stagnation intensity of deep oceans, often referred to as oceanic anoxic events (or OAEs) (Jenkyns, 1980; Arthur and others, 1985, 1987). Eustatic sea-level rise may relate to a generally more equable climate resulting from influx of more mantle-derived CO_2 (greenhouse effect), increased crust production, a lack of oxygenated deep-ocean circulation from polar to equatorial regions in the young and much smaller Atlantic Ocean (Barron and others, 1981; Brass and others, 1982), and development of anoxic conditions in the deep ocean basins. The encroachment of seas onto the continents extends shelf and epicontinental marine environments, many with anoxic bottom conditions. Kump and others (1993) described the Cretaceous Western Interior seaway of North America as 200–300 m deep, thousands of meters wide, and significantly longer. This seaway exhibited no mixing except by shallow wind-driven overturn, and was likely to have been thermohaline stratified with a freshwater cap. Increased marine productivity in this environment would have created vast bottom-water anoxia and sequestering of organic material. These anoxic events are recorded in the increased amount of marine versus terrigenous organic matter and in shifts in carbon isotopic compositions from Cretaceous pelagic limestone. High $\delta^{13}C$ values reported for pelagic carbonates reflect partitioning of stable carbon isotopes between carbonate and $\delta^{13}C$-depleted organic matter. Higher rates of biologic productivity stimulated by the increased CO_2 selectively remove ^{12}C in the coupled ocean-sediment package. As a consequence, globally elevated $\delta^{13}C$ values of pelagic carbonates are associated with the major OAEs and $\delta^{13}C$-enriched HCO_{-3} in the oceans. Decreased circulation of oxygenated surface waters into the deep ocean, along with increased levels of ocean-bottom anoxia, can result in the transgression of anoxia onto the continental shelves of the world oceans even during times of increased O_2 in the atmosphere. Furthermore, increased marine productivity and stable thermoclines in shelf and shallow continental seaways might be accompanied by an increased "rain" of organic material that rapidly consumes oxygen in the water column.

10. In support of the importance of ocean anoxia, sulfur isotopes of seawater sulfate (Claypool and others, 1980) at the terminal Cretaceous show a small but significant increase that reflects microbial sulfate reduction in the water column

under anoxic conditions (Arthur and others, 1987; Saelen and others, 1993). A $\delta^{34}S_{SO^4}$ increase has also been proposed for the Cenomanian-Turonian boundary, another major Cretaceous anoxic event. Spirakis (1989) noted that increasing available ferrous iron, from volcanic ash (glass), rapid weathering and transport fluxes of organic sediments, or submarine volcanism, could increase the amount of FeS_2 precipitation by reduction of seawater sulfate. Because the present sulfate reservoir of the ocean is slightly larger than the present oxygen reservoir of the atmosphere (Kump and Garrels, 1986), the stochiometry of reactions suggests that a 15% increase in FeS_2 precipitation by increased flux of available iron to the oceans (especially epicontinental anoxic seaways) will decrease the ocean sulfate by ~30%. By this process alone the atmospheric oxygen concentrations would be expected to increase by more than 50%.

11. At the end of the Cretaceous there was a decline in the equable greenhouse climate and a return to increased seasonal extremes (Barron, 1983). Thermal regulation of dinosaur's body temperature via extensive and elegant "radiators" became widespread during the Upper Cretaceous. Those animals unable to adapt to increased daily and seasonal temperature extremes may have experienced loss of core body temperature, possible sterility, increased metabolic stress, and eventual extinction.

12. Geologic and paleontologic observations argue against extreme combustion of biomass in a high O_2 atmosphere (Robinson, 1989). It has been argued that high oxygen concentrations in the atmosphere would cause a mass conflagration, and therefore is not possible. Certainly there were numerous forest fires in the Cretaceous. Each of 35 documented sedimentary coal sequences of the Russell basin is terminated by an interval of volcanic air-fall ash and abundant charcoal. Uncontrolled burning cannot occur in a forest (vegetation) fire, which experts (e.g., Hottel, 1959) characterize as a diffusion-type fire (Tewarson and others, 1981; Tewarson 1988, 1994). Such fires depend upon many factors in addition to availability of oxygen, such as temperature of pyrolysis and combustion, rates of convective and radiative heat transfer, and amount of soot produced. In all diffusion-type fires, heat fluxes asymptotically approach maximum values defining a fuel-specific upper combustion limit. Combustion experiments performed in oxygen levels up to 60% (Tewarson, 1994) show a logarithmic increase in soot production with increasing oxygen, but no significant accelerated rate of combustion. High CO_2 and increased humidity further moderate combustibility. High oxygen in the Cretaceous atmosphere is indicated from our work. Combustion arguments can place no upper limit on oxygen. A Cretaceous atmosphere of 35% oxygen simply could not foster massive rapidly-burning wildfires. Instead, increased soot production at rates of combustion comparable to forest fires today would occur. One effect of the impact scenario would be combustion of terrestrial vegetation burned first from major fires triggered by the impact, followed by burning of dried vegetation (~25% of terrestrial biomass; Wolbach and others, 1988; Ivany and Salawitch, 1993). Fire science data indicate that the enormity of this suggested burning simply is not possible.

The total carbon (soot from biomass burning) estimated to be at the K-T boundary is ~2×10^{14} gm C (Ivany and Salawitch, 1993; Wolbach and others, 1988). We note that the magnitudes of carbon soot fluxes from historic single-season forest fires measured in arctic glacier ice cores (Chylek and others, 1995) integrate in about 53 yr to the soot content reported for the K-T boundary. A maximum "black" soot content of the GISP2 core at year A.D. 324 (405.4 m depth) is 6.8 μg C/l with a measured accumulation rate (ice equivalent) of 108 cm/yr. This yields a total global flux of 3.8 x 10^{12}gm C/yr from a Northern Hemisphere fire that occurred within historic time. Similar carbon flux is known from the Mt. Logan ice (G. Holdsworth, 1994, personal commun.) with 3–100 yr integration time from a 1896 fire soot concentration level. Assuming hemispheric differences in tropospheric soot transport and uneven deposition, several hundred years of very ordinary sustained forest fires could yield K-T boundary amounts of soot. Given the logarithmic increase in soot with increase in oxygen level during fire combustion, it is clear that a catastrophic worldwide conflagration burning of ~25% of the biomass is not necessary to produce the K-T boundary carbon spike (assuming that the K-T carbon is all carbon black or soot).

13. The breakdown in $\delta^{13}C$ surface to deep gradient above the K-T boundary and the temporary "spike" in more negative $\delta^{13}C$ values for pelagic relative to benthic forams has been argued to reflect massive (~25%) global biomass burning following the impact event (Wolbach and others, 1988; Ivany and Salawitch, 1993). No geologic evidence for such massive burning exists in the form of a fossil charcoal record (Jones and Chaloner, 1991). Elemental carbon and especially carbon interpreted as soot both rise sharply at the boundary. A small but very sharp increase in the $^{87}Sr/^{86}Sr$ ratio of seawater is also reported from K-T boundary sediments (Martin and Macdougall, 1991). The apparent $\delta^{13}C$ shift appears instantaneous and may be simply an artifact of a K-T boundary hiatus (MacLeod and Keller, 1991). With a 50,000 to 100,000 yr time interval indicated, and a "Strangelove Ocean" across the K-T boundary (Hollander and others, 1993), these features are easily understood as expressions of increased riverine discharge to the oceans and decreased primary marine productivity (Zachos and others, 1989) brought on by rapid shifts in mixed-layer ocean chemistry. Decreased atmospheric O_2 and massive CO_2 generated both by accelerated oxidation of rapidly exposed anoxic basin sediments and by CO_2 influx from the Deccan traps (McLean 1985a, 1985b, 1985c) would virtually stop marine productivity. Surface-ocean pH would drop as waters reach CO_2 saturation. Marine waters would receive very negative $\delta^{13}C$ carbon from weathered anoxic organic sediments, both by direct riverine flux and via oxidation to CO_2 and atmospheric transport. Rapid transport of $^{87}Sr/^{86}Sr$ from the continental weathering products originally in the anoxic sediments would then account for the marine strontium spike.

Though assumed to be soot from burning, the morphology of the K-T carbon may also be plausibly interpreted as colloid residues. Organic colloid formation is triggered by increased solutes to the oceans (especially Ca^{++}) brought on by greater riverine flux. Organic colloids will scavenge most metals from the ocean water column as they rapidly settle out. Because sedimentation rates in the deep oceans are extremely slow, most colloidal organics are resorbed and fail to become concentrated in the marine section. However, metals, including iridium, are concentrated. Marine iridium anomalies produced by this mechanism are not accompanied by other elements in chondritic ratios. Iridium anomalies of smaller magnitude than the K-T boundary anomaly are present at most other major mass-extinction boundaries (Holser and others, 1989). In addition to extraterrestrial sources of iridium (Kyte and Wasson, 1986), anomalies may result from several syndepositional and postdepositional processes that include precipitation from seawater under low sedimentation rates and anoxic conditions, microbial concentration, hydrothermal exhalative processes from ridge and intraplate volcanic vents, and leaching, transport, and precipitation at redox boundaries (Colodner and others, 1992; Sawlowicz, 1993). Chronologically inconsistent Ir anomalies found globally, but at horizons other than the K-T boundary, argue strongly for terrestrial iridium enrichment processes (Kyte and others, 1980; Montanari and others, 1993). Iridium enrichments at points of rapid marine regression and shifts in atmosphere-ocean chemistry are expected consequences of the Pele model.

14. The Earth's mantle appears largely undegassed on the basis of rare gas systematics (Allègre and others, 1983). Internal Earth processes that determine mantle convection operate independently of the atmosphere, hydrosphere, biosphere, and surficial processes. Episodic, and possibly periodic, superplume activity is coupled with avalanche downwelling of cold aesthenosphere-lithosphere material, catastrophic breakthrough of the 670 km phase boundary to the lower mantle, and episodic creation of whole-mantle convection (Morgan and Shearer, 1993; Tackley and others, 1993). Varying rates of cylindrical convection transfer across the phase boundary in both directions provides deep-mantle degassing conduits, variably "chills" the core-mantle boundary, temporarily stops the magnetic field reversals, and diminishes the magnetic field strength during "quiet" intervals to about 25%–50% that of today (Hollerbach and Jones, 1993; Pick and Tauxe, 1993). The more common layered mantle convection maintains a base-level CO_2 flux to the atmosphere from the mantle above the 670 km phase boundary. However, major Pele excursions seem to occur at periods of huge catastrophic breakthroughs in the phase boundary and "flushing" of upper mantle material to the core-mantle boundary (Larson and Olson, 1991; Tackley and others, 1993). If these deep mantle processes contribute to the periodicity of plume emplacement, periodic extinctions may best be explained by mantle processes.

15. Features of the K-T boundary transition predicted by the Pele hypothesis can be inferred at other geologic period boundaries associated with major extinctions events. The Permian-Triassic (P-T) boundary extinctions are associated with major anoxic sedimentation events (Holser and others, 1989; Erwin, 1994), oceanic $\delta^{18}O$ and $\delta^{13}C$ shifts reflecting extensive weathering of organic matter and an inferred lowering of atmospheric oxygen (Berner, 1989b; Gruszczynski and others, 1989, 1992), possibly contributing to the mass extinction at the P-T boundary. High CO_2 in Devonian time (Berner and Canfield, 1989; Berner, 1991) led to enormous plant productivity in emerging communities, accelerated weathering by vascular plant root penetration, enhanced nutrient flux to shallow-marine waters with resultant eutrophication and widespread anoxia, and ultimately the major Frasnian-Famennian extinction (Wang and others, 1991; Algeo, 1993; Joachimski and Boggisch, 1993). Sudden changes in paleoredox conditions are indicated for the Devonian-Carboniferous boundary, an interval that exhibits geographically widespread black shales and iridium anomalies, but no shocked quartz and microtektites (Wang and others, 1993). Changes in carbon isotopes, sea level,

and conodont biozones at the Cambrian-Ordovician boundary indicate Pele processes (Ripperdan and others, 1992). Towe argued (1990) that early in Archean time, aerobic respiration must have developed in order to maintain a low but important atmospheric oxygen level (0.2%–0.4%) that otherwise would have evolved to much higher levels based upon the amount of sequestered organic matter and realistic deposition of iron and oxidation of methane. Significant atmospheric free oxygen is indicated from 3.4 Ga biogenic pyrite from Barberton, South Africa (Ohmoto and others, 1993). Mantle degassing and increased photosynthetic productivity likely was at play, even in the earliest Archean. Paleomagnetic reversal "quiet" intervals other than the Cretaceous "long normal" are known. Do these and other events mark major changes in Pele processes?

CONCLUSIONS

Major geologic–geochemical events can have important consequences for evolution. We have defined the Pele hypothesis to encompass a relation among (1) episodic increases in deep-mantle CO_2 release to the atmosphere from superplumes, large igneous provinces (LIPs), and oceanic ridges during times of increased spreading rates at plate margins; (2) accelerated net photosynthesis from an expanded biosphere and sequestering of large amounts of organic carbon predominantly in expanded shallow-marine environments; and (3) increases in atmospheric O_2. Reduction in superplume activity and leads to a decrease in CO_2 degassing of the mantle are predicted to decrease rates of new crust production and ridge-crest spreading rates, resulting in a eustatic sea-level drop, diminished plant productivity, less burial of organic matter, and a gradual drop in atmospheric oxygen. At times of rapid change, large marine regression can produce major atmospheric O_2 decrease by oxidation and recycling of anoxic bottom sediments of the exposed shelf and epicontinental sea. Larger and more rapid changes in mantle CO_2 flux to the atmosphere through geologic time, both increasing and decreasing, create significant primary and secondary environmental pressures that have significant effects on evolution and induce biotic crisis. The Pele hypothesis invokes these processes as a general framework in which to understand extinction pressures and the limiting factors effecting biodiversity.

Atmospheric abundance of O_2 and CO_2 must be closely related because net photosynthesis is the only significant means to increase atmospheric O_2. Increased plant productivity at times of high atmospheric CO_2 would yield vast deposits of organic material in shallow continental and marine shelf anoxic basins that form or are expanded by the eustatic rise in sea level accompanying accelerated sea-floor spreading and intraplate superplume volcanism. This would also result in the generation of economically important oil, gas, coal, chalk, and carbonate deposits on a global scale, along with an increase in atmospheric oxygen. Other results include the occurrence of deep-ocean anoxic events, decrease in pH of mixed-layer ocean water, rising of the oceanic carbonate compensation depth, and the mechanics of ocean chemistry and mixing. Current carbon cycle models should be modified to account for independent mantle CO_2 fluxes to the atmosphere as are recorded by proxy in variable ridge crest spreading rates, production of new ridge-crest crust and LIP events, in paleomagnetic "quiet" intervals, in sea-level change, and in increased burial of organic material in expanded anoxic basins.

An increased rate of crust production in the Cretaceous contributed to marine transgression and high atmospheric CO_2 and humidity that created a greenhouse warming and more equable global climate lacking seasonality even at high latitudes, and that caused O_2 to climb to ~35%. In contrast, superplume activity and mantle CO_2 degassing decreased in the Late Cretaceous, causing world ocean basins to deepen, global temperatures to drop, and atmospheric CO_2/O_2 ratios to change. At the same time, a brief but major eustatic marine regression exposed organic-rich anoxic basin sediments to oxidation, which caused an ~12% reduction in atmospheric O_2. The well-documented δ13C, $^{87}Sr/^{86}Sr$, Ir, and elemental carbon (soot) spikes in marine K-T boundary sediment cores reflect this selective accelerated weathering and sudden changes in marine redox conditions. Productivity decreased (more at lower latitudes), and many species were thrown into ecologic crisis. Fundamental changes in mantle CO_2 fluxes beyond the capacity of the carbon cycle to compensate are mimicked by biosphere productivity and changes in atmosphere, quantities of black shale, coal, and oil and gas deposits, limestone and chalk sedimentation, mammalian radiation, dinosaur diversity and ultimate extinction, marine transgressions and regressions, global temperatures and seasonality, and shifts in terrestrial and marine ecology (including changes in marine mixed-surface layer pH, O_2 and CO_2 content, nutrient levels, carbonate compensation depth, and deep ocean mixing rates and anoxia).

Changing atmospheric CO_2/O_2 alters productivity and global temperatures, selectively imposing respiratory and metabolic stress on more vulnerable organisms. Secondary factors related to biotic extinction include disruption of food webs, forced migration, shifts in anoxia and other aspects of marine water chemistry, and changes precipitation and weathering rates. Plant response to increased CO_2 would largely be nutrient controlled (nitrates and phosphates), with CO_2 becoming toxic to plants. Weathering rates (nutirent supply) and sedimentation rates (organic burial) limit atmospheric O_2 to a maximum of ~35%. Atmospheric CO_2–O_2 budgets must be considered major survival constraints based upon the complex interdependence of geologic processes. That virtually all major extinction events occur at the nodes of marine transgressions or regressions is not by chance—these are times that major, atmosphere-driven primary and secondary extinction pressures are greatest. The gradual extinction of select species (dinosaurs, certain marine organisms, and others of the terminal Cretaceous extinction) is best explained by the interdependence of these terrestrial-based processes. We have proposed that major periods of biodiversity and biotic crisis in Earth history ultimately are mantle controlled.

Many of our ideas are obviously speculative. However, we feel that weight of evidence for our speculations is compelling. The implications of the Pele hypothesis are complex, yet in most cases they are subject to empirical verification. This we encourage. Are there other reasonable explanations for the high oxygen values in primary gas bubbles of Cretaceous amber? We will continue to examine this question. How do plants and animals respond to variable atmospheric O_2 and

CO_2 and what are their tolerance limits? Is the apparent link between sea-level adjustment and atmospheric oxygen valid? Are rates of selective organic weathering adequate to change the atmosphere? Can evidence from the Montana K-T boundary section be replicated at other K-T boundary intervals and at other major regressions? We will be sampling amber across other K-T boundary intervals. What are geologic–geochemical processes that might produce atmospheric oxygen levels below that of today, as indicated for mid-Eocene and Oligocene–Miocene times (Fig. 3)? Many predictions of the Pele hypothesis have been known for all major mass-extinction events that occur at period boundaries. We urge use of the conceptual framework provided by the Pele hypothesis in the ongoing examination of these biotic crises in Earth history.

ACKNOWLEDGMENTS

We wish to thank the large number of Earthwatch volunteers and University of Minnesota students who assisted in sample collection. Special thanks to A. Tewarson for quantification of fire science issues. The Earthwatch Center for Field Research has supported the study of Cretaceous-Tertiary vertebrate extinction for many years and is gratefully acknowledged. Amber investigations were supported by the U.S. Geological Survey National Mineral Resource Assessment Program as an extension of fluid-inclusion ore-fluid gas chemistry studies.

REFERENCES CITED

Algeo, T. J., 1993, The Late Devonian increase in vascular plant "rootedness": Source of coeval shifts in seawater chemistry: Geological Society of America Abstracts with Programs, v. 25, no. 6, p. A83.

Allègre, C. J., Staudacher, T., Sarda, P., and Kurz, M., 1983, Constraints on evolution of earth's mantle from rare gas systematics: Nature, v. 303, p. 762–766.

Archibald, J. D., 1993, Testing K-T extinction senarios with nonmarine vertebrate data: Geological Society of America Abstracts with Programs, v. 25, no. 6, p. A297.

Archibald, J. D., and Bryant, L. J., 1990, Differential Cretaceous/Tertiary extinctions of nonmarine vertebrates, *in* Sharpton, V. L., and Ward, P., eds., Global catastrophes in Earth history: An interdisciplinary conference on impacts, volcanism, and mass mortality: Geological Society of America Special Paper 247, p. 549–562.

Archibald, J. D., Butler, R. F., Lindsay, E. H., Clemens, W. A., and Dingus, L., 1982, Upper Cretaceous–Paleocene biostratigraphy and magnetostratigraphy, Hell Creek and Tullock formations, northeastern Montana: Geology, v. 10, p. 153–159.

Arthur, M. A., and Dean W. E., 1986, Cretaceous paleoceanography of the western North Atlantic Ocean: Boulder, Colorado, Geological Society of America, Geology of North America, v. M, p. 617–630.

Arthur, M. A., Dean, W. E., and Schlanger, S. O., 1985, Variations in the global carbon cycle during the Cretaceous related to climate, volcanism, and changes in atmospheric CO_2, *in* Sundquist, E. T., and Broecker, W. S., eds., The carbon cycle and atmospheric CO_2: Natural variations Archean to present: Washington, D.C., American Geophysical Union, p. 504–529.

Arthur, M. A., Schlanger, S. O., and Jenkyns, H. C., 1987, The Cenomanian-Turonian oceanic anoxic event, II. Palaeoceanographic controls on organic-matter production and

preservation, *in* Brooks, J., and Fleet, A. J., eds., Marine petroleum source rocks: Geological Society of America Special Paper 26, p. 401–420.
Arthur, M. A., Dean, W. E., and others, 1988, Geochemical and climatic effects of increased marine organic carbon burial at the Cenomanian/Turonian boundary: Nature, v. 335, p. 714–717.
Barron, E. J., 1983, A warm, equable Cretaceous: The nature of the problem: Earth Science Reviews, v. 19, p. 305–338.
Barron, E. J., and Washington, W. M., 1984, The role of geographic variables in explaining paleoclimates: Results from Cretaceous climate model sensitivity studies: Journal of Geophysical Reseach, v. 89, p. 1267–1279.
Barron, E. J., Harrison, C. G. A., Sloan, J. L., and Haq, W. W., 1981, Paleogeography. 180 million years ago to the present: Ecologae Geologicae Helvetiae, v. 74, p. 443–469.
Beck, C. W., 1988, Is the air in amber ancient?: Science, v. 241, p. 719.
Berner, R. A., 1989a, Biogeochemical cycles of carbon and sulfur and their effect on atmospheric oxygen over Phanerozoic time: Palaeogeography, Palaeoclimatology, Palaeoecology, v. 75, p. 97–122.
Berner, R. A., 1989b, Drying, O_2 and mass extinction: Nature, v. 340, p. 603–604.
Berner, R. A., 1990, Atmospheric carbon dioxide levels over Phanerozoic time: Science, v. 249, p. 1382–1386.
Berner, R. A., 1991, A model for atmospheric CO_2 over Phanerozoic time: American Journal of Science, v. 291, p. 339–376.
Berner, R. A., 1994, GEOCARB II: A revised model of atmospheric CO_2 over Phanerozoic time: American Journal of Science, v. 294, p. 56–91.
Berner, R. A., and Canfield, D. E., 1989, A new model for atmospheric oxygen over Phanerozoic time: American Journal of Science, v. 289, p. 333–361.
Berner, R. A., and Landis, G. P., 1988, Gas bubbles in fossil amber as possible indicators of the major gas composition of ancient air: Science, v. 239, p. 1406–1409.
Berner, R. A., Ruttenberg, K. C., Ingell, E. D., and Rao, J.-L., 1993, The nature of phosphorus burial in modern marine sediments, *in* Wollast, R., Mackenzie, F. T., and Chou, L., eds., Interactions of C, N, P and S biogeochemical cycles and global change: Berlin, Springer-Verlag, p. 365–378.
Bertrand, P., and Lallier-Vergès, E., 1993, Past sedimentary organic matter accumulation and degradation controlled by productivity: Nature, v. 364, p. 786–788.
Betts, J. N., and Holland, H. D., 1991, The oxygen content of ocean bottom waters, the burial efficiency of organic carbon, and the regulation of atmospheric oxygen: Palaeogeography, Palaeoclimatology, Palaeoecology, v. 97, p. 5–18.
Brass, G. W., Southam, J. R., and Peterson, W. H., 1982, Warm saline bottom water in the ancient ocean: Nature, v. 296, p. 620–623.
Broecker, W. S., and Peng, T.-H., 1982, Tracers in the Sea: Lamont-Doherty Geological Observatory Press, Palisades, NY.
Brouwers, E. M., Clemens, W. A., Spicer, R. A., Ager, T. A., Carter, L. D., and Sliter, W. V., 1987, Dinosaurs on the North Slope, Alaska: high latitude Late Cretaceous environments: Science, v. 237, p. 1608–1610.
Brouwers, E. M., and Deckker, P. D., 1993, Late Maastrichtian and Danian ostracode faunas from Northern Alaska: reconstructions of environment and paleogeography: Palaios, v. 8, p. 140–154.
Buffetaut, E., 1994, Paleoecological implications of Alaskan terrestrial vertebrate fauna in latest Cretaceous time at high paleolatitudes: Comment: Geology, v. 22, p. 191–192.
Caldeira, K., and Rampino, M. R., 1991, The mid-Cretaceous super plume, carbon dioxide, and global warming: Geophysical Research letters, v. 18, p. 987–990.
Cano, R. J., Poinar, H. N., Pieniazek, M.-J., Acra, A., and Poinar, G. O., 1993, Amplification and sequencing of DNA from a 120–135 million year old weevil: Nature, v. 363, p. 536–538.

Cerling, T. E., 1989, Does the gas content of amber reveal the composition of palaeoatmospheres?: Nature, v. 339, p. 695–696.
Cerling, T. E., 1991, Carbon dioxide in the atmosphere: Evidence from Cenozoic and Mesozoic paleosols: American Journal of Science, v. 291, p. 377–400.
Chylek, P., Johnson, B., Damiano, P. A., Taylor, K. C., and Clement, P., 1995, Biomass burning record and black carbon in the GISP2 ice core: Geophysical Research Letters, v. 22, no. 2, p. 89–92.
Claypool, G. E., Holser, W. T., Kaplan, I. R., Sakai, H., and Zak, I., 1980, The age curves of sulfur and oxygen isotopes in marine sulfate and their mutual interpretation: Chemical Geology, v. 28, p. 199–260.
Clemens, W. A., and Nelms, L. G., 1993, Paleoecological implications of Alaskan terrestrial vertebrate fauna in latest Cretaceous time at high paleolatitudes: Geology, v. 21, p. 503–506.
Clemens, W. A., and Nelms, L. G., 1994, Paleoecological implications of Alaskan terrestrial vertebrate fauna in latest Cretaceous time at high paleolatitudes: Reply: Geology, v. 22, p. 191–192.
Coffin, M. F., and Eldholm, O., 1992, Volcanism and continental break-up: A global compilation of large igeneous provinces, *in* Storey, B. C., Alabaster, T., and Pankhurst, R. S., ed., Magmatism and the causes of continental break-up: Geological Society Special Paper 68, p. 17–30.
Coffin, M. F., and Eldholm, O., 1993, Large igneous provinces: Scientific American, v. 269, p. 42–49.
Colodner, D. C., Boyle, E. A., Edmond, J. M., and Thompson, J., 1992, Post-depositional mobility of platinum, iridium and rhenium in marine sediments: Nature, v. 358, p. 402–404.
D'Argenio, B., and Mindszenty, A., 1986, Cretaceous bauxites in the tectonic framework of the Mediterranean: Rendiconti della Società Geologica Italiana, v. 9, p. 257–262.
D'Argenio, B., and Mindszenty, A., 1989, Cretaceous–early Tertiary bauxites of Mediterranean region: A tectonic approach [abs.]: International Geological Congress, 28th, Abstracts, v. 1, p. 1357–1358.
de Baar, H. J. W., and Suess, E., 1993, Ocean carbon cycle and climate change—An introduction to the interdisciplinary union symposium: Global and Planetary Change, v. 8, p. VII—XI.
DeSalle, R., Gatesy, J., Wheeler, W. C., and Grimaldi, D. A., 1992, DNA sequences from a fossil termite in Oligo-Miocene amber and their phylogenetic implications: Science, v. 257, p. 1933–1936.
DeSalle, R., Gatesy, J., Wheeler, W. C., and Grimaldi, D. A., 1993, Working with fossil DNA from amber: Discovery, v. 24, p. 19–24.
Diaz, S., Grime, J. P., and McPherson, E., 1993, Evidence of a feedback mechanism limiting plant response to elevated carbon dioxide: Nature, v. 364, p. 616–617.
Erwin, D. H., 1994, The Permo-Triassic extinction: Nature, v. 367, p. 231–236.
Fastovsky, D. E., 1990, Rocks, resolution and the record; a review of depositional constraints on fossil vertebrate assemblages at the terrestrial Cretaceous/Paleogene boundary, eastern Montana and western North Dakota, *in* Sharpton, V. L., and Ward, P. D., eds., Global catastrophes in Earth history: An interdisciplinary conference on impacts, volcanism, and mass mortality: Geological Society of America Special Paper 247, p. 541–548.
Fastovsky, D. E., and McSweeney, K., 1987, Paleosols spanning the Cretaceous–Paleogene transition, eastern Montana and western North Dakota: Geological Society of America Bulletin, v. 99, p. 66–77.
Fyfe, W. S., 1990, Geosphere forcing: Plate tectonics and the biosphere: Palaeogeography, Palaeoclimatology, Palaeoecology, v. 89, p. 185–191.
Gaffin, S., 1987, Phase differences between sea level and magnetic reversal rate: Nature, v. 329, p. 816–819.

Graybill, D. A., and Idso, S. B., 1993, Detecting the aerial fertilization effect of atmospheric CO_2 enrichment in tree-ring chronologies: Global Biogeochemical Cycles, v. 7, p. 81–95.

Gruszczynski, M., Halas, S., Hoffman, A., and Malkowski, K., 1989, A brachiopod calcite record of the oceanic carbon and oxygen isotope shifts at the Permian/Triassic transition: Nature, v. 337, p. 64–68.

Gruszczynski, M., Hoffman, A., Malkowski, K., and Veizer, J., 1992, Seawater strontium isotopic perturbation at the Permian-Triassic boundary, West Spitsbergen, and its implications for the interpretation of strontium isotopic data: Geology, v. 20, p. 779–782.

Gurnis, M., 1993, Phanerozoic marine inundation of continents driven by dynamic topography above subducting slabs: Nature, v. 364, p. 589–593.

Hallam, A., 1992, Phanerozoic sea-level changes: New York, Columbia University Press, 266 p.

Haq, B. U., Hardenbol, J., and Vail, P. R., 1987, Chronology of fluctuating sea levels since the Triassic: Science, v. 235, p. 1156–1167.

Harland, W. B., Armstrong, R. L., Cox, A. V., Smith, A. G., and Smith, D. G., 1989, A geologic time scale: London, Cambridge University Press, 263 p.

Hengst, R. A., Rigby, J. K., Jr., Landis, G. P., and Sloan, R. E., 1993, Biological consequences of Mesozoic atmospheric gases: Geological Society of America Abstracts with Programs, v. 25, no. 6, p. A297.

Hillis, W. E., 1987, Heartwood and tree exudates: London, Springer–Verlag, 237 p.

Holland, H. D., 1984, The chemical Evolution of the Atmosphere and Oceans: Princeton, Princeton University Press, 582 p.

Hollander, D. J., McKenzie, J. A., and Hsü, K. J., 1993, Carbon isotope evidence for unusual plankton blooms and fluctuations of surface water in CO_2 in "Strangelove Ocean" after terminal Cretaceous event: Palaeogeography, Palaeoclimatology, Palaeoecology, v. 104, p. 229–237.

Hollerbach, R., and Jones, C. A., 1993, Influence of the earth's inner core on geomagnetic fluctuations and reversals: Nature, v. 365, p. 541–543.

Holser, W. T., and 14 others, 1989, A unique geochemical record at the Permian/Triassic boundary: Nature, v. 337, p. 39–44.

Hopfenberg, H. B., Witchey, L. C., and Poinar, G. O., 1988, Is the air in amber ancient?: Science, v. 241, p. 717–718.

Horibe, Y., and Craig, H., 1988, Is air in amber ancient?: Science, v. 241, p. 720–721.

Hottel, H. C., 1959, Review: Certain laws governing the diffusive burning of liquids by Blinov and Khudlakov (1957), Nauk SSSR, v133, 1094: Fire Research Abstract and Reviews, v. 1, p. 41–45.

Idso, S. B., and Kimball, B. A., 1993, Tree growth in carbon dioxide enriched air and its implications for global carbon cycling and maximum levels of atmospheric CO_2: Global Biogeochemical Cycles, v. 7, p. 537–555.

Ivany, L. C., and Salawitch, R. J., 1993, Carbon isotopic evidence for biomass burning at the K-T boundary: Reply: Geology, v. 21, p. 1150–1151.

Jane, F. W., 1955, The Structure of Wood: New York, The Macmillan Co., 274 p.

Jenkyns, H. C., 1980, Cretaceous anoxic events: From continents to oceans: Geological Society of London Journal, v. 137, p. 171–188.

Jaochimski, M. M., and Boggisch, W., 1993, Anoxic events in the late Frasnian—Causes of the Frasnian-Famennian faunal crisis?: Geology, v. 21, p. 675–678.

Jones, T. P., and Chaloner, W. G., 1991, Fossil charcoal, its recognition and palaeoatmospheric significance: Palaeogeography, Palaeoclimatology, Palaeoecology, v. 97, p. 39–50.

Kauffman, E. G., 1977, Geological and biological overview: Western interior Cretaceous basin: Mountain Geologist, v. 14, p. 75–99.

Keller, G., and MacLeod, N., 1993, Carbon isotopic evidence for biomass burning at the K-T boundary: Comment: Geology, v. 21, p. 1149–1150.
Kump, L. R., and Garrels, R. M., 1986, Modeling atmospheric oxygen in the global sedimentary redox cycle: American Journal of Science, v. 286, p. 337–360.
Kump, L. R., Slingerland, R. L., Arthur, M. A., and Barron, E. J., 1993, Modelling anoxia in the Cretaceous western interior seaway: Geological Society of America Abstracts with Programs, v. 25, no. 6, p. A385.
Kyte, F. T., and Wasson, J. T., 1986, Accretion rate of extraterrestrial matter: Iridium deposited 33 to 67 million years ago: Science, v. 232, p. 1225–1229.
Kyte, F. T., Zhou, A., and Watson, J. T., 1980, Siderophile-enriched sediments at the Cretaceous-Tertiary boundary: Nature, v. 288, p. 651–656.
Landis, G. P., and Berner, R. A., 1988, Is air in amber ancient?—Reply: Science, v. 241, p. 721–724.
Landis, G. P., and Hofstra, A. H., 1991, Fluid inclusion gas chemistry as a potential minerals exploration tool: Case studies from Creede, CO, Jerritt Canyon, NV, Coeur d'Alene district, ID and MT, southern Alaska mesothermal veins, and mid-continent MVT's: Journal of Geochemical Exploration, v. 42, p. 25–59.
Landis, G. P., and Snee, L. W., 1991, $^{40}Ar/^{39}Ar$ systematics and argon diffusion in amber: Implications for ancient earth atmospheres: Palaeogeography, Palaeoclimatology, Palaeoecology, v. 97, p. 63–68.
Landis, G. P., Rigby, J. K., Jr., Sloan, R. E., and Hengst, R. A., 1993, Pele hypothesis: A unified model for ancient atmosphere and biotic crisis: Geological Society of America Abstracts with Programs, v. 25, no. 6, p. A362.
Larson, R. L., 1991a, Geological consequences of superplumes: Geology, v. 19, p. 963–966.
Larson, R. L., 1991b, Latest pulse of earth: Evidence for a mid-Cretaceous superplume: Geology, v. 19, p. 547–550.
Larson, R. L., and Olson, P., 1991, Mantle plumes control magnetic reversal frequency: Earth and Planetary Science Letters, v. 107, p. 437–447.
Lofgren, D. L., 1990, The Bug Creek problem and the Cretaceous–Tertiary transition at McGuire Creek, Montana [Ph.D. thesis]: Berkeley, University of California, 461 p.
Lofgren, D. L., Hotton, C., Runkel, A., 1990, Reworking of Cretaceous dinosaurs into Paleocene channel deposits, upper Hell Creek Formation, Montana: Geology, v. 18, p. 874–877.
Martin, E. E., and Macdougall, J. D., 1991, Seawater Sr isotopes at the Cretaceous/Tertiary boundary: Earth and Planetary Science Letters, v. 104, p. 166–180.
Marzocchi, W., Mulargia, F., and Parulo, P., 1992, The correlation of geomagnetic reversals and mean sea level in the last 150 m.y.: Earth and Planetary Science Letters, v. 111, p. 383–393.
McAlester, A. L., 1970, Animal extinctions, oxygen consumption, and atmospheric history: Journal of Paleontology, v. 44, p. 405–409.
McAlester, A. L., 1971, Oxygen consumption rates and their paleontologic significance: Reply: Journal of Paleontology, v. 45, p. 917.
McKenna, M. C., 1965, Collecting microvertebrate fossils by washing and screening, *in* Kummel, B., and Raup, D. M., eds, Handbook of paleontological techniques: San Francisco, California, W. H. Freeman, p. 193–203.
McLean, D. M., 1978, A terminal Mesozoic "Greenhouse": Lessons from the past: Science, v. 201, p. 401–406.
McLean, D. M., 1985a, Deccan traps mantle degassing in the terminal Cretaceous marine extinctions: Cretaceous Research, v. 6, p. 235–259.
McLean, D. M., 1985b, Mantle degassing induced dead ocean in the Cretaceous-Tertiary transition, *in* Sundquist, E. T., and Broecker, W. S., eds., The carbon cycle and atmos-

pheric CO_2: Natural variations Archean to present: Washington, D. C., American Geophysical Union, p. 493–503.
McLean, D. M., 1985c, Mantle degassing unification of the Trans-K-T geobiological record: Evolutionary Biology, v. 20, p. 287–313.
Miranda, N. R., Freeman, B. D., and Hopfenberg, H. B., 1991, The relative contribution of adsorption to the overall sorption and transport of small molecules in amber: Journal of Membrane Science, v. 60, p. 147–155.
Montanari, A., Asaro, F., Michel, H., and Kennett, J. P., 1993, Iridium anomalies of late Eocene age at Massignano (Italy), and ODP site 689B (Maud Rise, Antarctica): Palaios, v. 8, p. 420–437.
Morell, V., 1993, Dino DNA: The hunt and the hype: Science, v. 261, p. 160–162.
Morgan, J. P., and Shearer, P. M., 1993, Seismic constraints on mantle flow and topography of the 660-km discontinuity: Evidence for whole-mantle convection: Nature, v. 365, p. 506–511.
Ohmoto, H., Kakegawa, T., and Lowe, D. R., 1993, 3.4 billion-year-old biogenic pyrites from Barbarton, South Africa: Sulfur isotope evidence: Science, v. 262, p. 555–557.
Pääbo, S., 1993, Ancient DNA: Scientific American, v. 269, p. 86–92.
Pick, T., and Tauxe, L., 1993, Geomagnetic palaeointensities during the Cretaceous normal superchron measured using submarine basaltic glass: Nature, v. 366, p. 238–242.
Poinar, G. O., Jr., 1993, Recovery of antediluvian DNA: Reply: Nature, v. 365, p. 700.
Poinar, G. O., Jr., Waggoner, B. M., and Bauer, U.-C., 1993a, Terrestrial soft-bodied protists and other microorganisms in Triassic amber: Science, v. 259, p. 222–224.
Poinar, H. N., Cano, R. J., and Poinar, G. O., 1993b, DNA from an extinct plant: Nature, v. 363, p. 677.
Raven, J. A., 1991, Plant responses to high O_2 concentrations: Relevance to previous high O_2 episodes: Palaeogeography, Palaeoclimatology, Palaeoecology, v. 97, p. 19–38.
Record, S. J., 1934, Identification of the timbers of temperate North America: New York, John Wiley & Sons, Inc., 312 p.
Retallack, G. J., and Dilcher, D. L., 1986, Cretaceous angiosperm invasion of North America: Cretaceous Research, v. 7, p. 227–252.
Retallack, G. J., Leahy, G. D., and Spoon, M. D., 1987, Evidence from paleosols for ecosystem changes across the Cretaceous/Tertiary boundary in eastern Montana: Geology, v. 15, p. 1090–1093.
Riebesell, U., 1993, Phytoplankton growth and CO_2: Reply: Nature, v. 363, p. 678–679.
Riebesell, U., Wolf-Gladrow, D. A., and Smetacek, V., 1993, Carbon dioxide limitation of marine phytoplankton growth rates: Nature, v. 361, p. 249–251.
Rigby, J. K., Jr., 1987, The last of the North America dinosaurs, *in* Czerkas, S. J., and Olson, E. C., eds, Dinosaurs past and present: Los Angeles, California, Natural History Museum of Los Angeles, p. 119–135.
Rigby, J. K., Jr., 1989, The Cretaceous-Tertiary boundary of the Bug Creek drainage: Hell Creek and Tullock formations, McCone and Garfield counties, Montana, *in* Flynn, J. J., and McKenna, M. C., eds., Mesozoic/Cenozoic vertebrate paleontology classic localities, contemporary approaches: Washington, D.C., American Geophysical Union, p. 67–73.
Rigby, J. K., and Rigby, J. K., Jr., 1990, Geology of the Sand Arroyo and Bug Creek quandrangles, McCone County, Montana: Brigham Young University Geology Studies, v. 36, p. 69–134.
Rigby, J. K., Jr., Newman, K. R., Smit, J., Van der Kaars, W. A., Sloane, R. E., and Rigby, J. K., 1987, Dinosaurs from the Paleocene part of the Hell Creek Formation, McCone County, Montana: Palaios, v. 2, p. 296–302.
Ripperdan, R. L., Magaritz, M., Nicoll, R., and Shergold, J., 1992, Simultaneous changes in carbon isotopes, sea level, and conodont biozones within the Cambrian-Ordovician boundary interval at Black Mountain, Australia: Geology, v. 20, p. 1039–1042.

Robinson, J. M., 1989, Phanerozoic O_2 variation, fire, and terrestrial ecology: Palaeogeography, Palaeoclimatology, Palaeoecology, v. 75, p. 223–240.

Saelen, G., Raiswell, R., Talbot, M., Skei, J., and Bottrell, S., 1993, Heavy sedimentary sulfur isotopes as indicators of super-anoxic bottom-water conditions: Geology, v. 21, p. 1091–1094.

Samson, S. D., and Alexander, E. C., Jr., 1987, Calibration of the interlaboratory 40Ar-39Ar dating standard, MMhb-1: Chemical Geology, Isotope Geoscience Section, v. 66, p. 27–34.

Savin, S. M., 1977, The history of the Earth's surface temperature during the past 100 million years: Annual Review of Earth and Planetary Sciences, v. 5, p. 319–355.

Sawlowicz, Z., 1993, Iridium and other platinum-group elements as geochemical markers in sedimentary environments: Palaeogeography, Palaeoclimatology, Palaeoecology, v. 104, p. 253–270.

Schopf, T. J. M., Farmanfarmaian, A., and Gooch, J. L., 1971, Oxygen consumption rates and their paleontologic significance: Journal of Paleontology, v. 45, p. 247–252.

Sheehan, P. M., Fastovsky, D. E., Hoffman, R. G., Berghaus, C. B., and Gabriel, D. L., 1991, Sudden extinction of the dinosaurs: Latest Cretaceous, Upper Great Plains, USA: Science, v. 254, p. 835–839.

Sloan, C. L., and Barron, E. J., 1990, "Equable" climates during Earth history?: Geology, v. 18, p. 489–492.

Sloan, R. E., and Rigby, J. K., Jr., 1986, Cretaceous-Tertiary dinosaur extinction: Response to Letters to the Editor: Science, v. 234, p. 1170–1174.

Sloan, R. E., Rigby, J. K., Jr., Van Valen, L. M., and Gabriel, D., 1986, Gradual dinosaur extinction and simultaneous ungulate radiation in the Hell Creek Formation: Science, v. 232, p. 629–633.

Smit, J., VanDer Kaars, W. A., and Rigby, J. K., Jr., 1987, Stratigraphic aspects of the Cretaceous-Tertiary boundary in the Bug Creek area of eastern Montana, USA: Société Géologique de France, Memoires, v. 150, p. 53–73.

Spirakis, C. S., 1989, Possible effect of readily available iron in volcanic ash on the carbon to sulfur ratio in lower Paleozoic normal marine sediments and implications for atmospheric oxygen: Geology, v. 17, p. 599–601.

Spotila, J. R., O'Connor, M. P., Dodson, P., and Paladino, F. V., 1991, Hot and cold running dinosaurs: Body size, metabolism and migration: Modern Geology, v. 16, p. 203–227.

Steiger, R. H., and Jager, E., 1977, Subcommission on geochronology: convention on the use of decay constants in geo- and cosmo-chronology: Earth and Planetary Science Letters, v. 36, p. 359–362.

Swisher, C. C., III, Dingus, L., and Butler, R. F., 1993, $^{40}Ar/^{39}Ar$ dating and magnetostratigraphic correlation of the terrestrial Cretaceous-Paleogene boundary and Puercan mammal age, Hell Creek–Tullock formations, eastern Montana: Canadian Journal of Earth Sciences, v. 30, p. 1981–1996.

Tackely, P. J., Stevenson, D. J., Glatzmaier, G. A., and Schubert, G., 1993, Effects of an endothermic phase transition at 670 km depth in a spherical model of convection in the earth's mantle: Nature, v. 361, p. 699–704.

Tewarson, A., 1988, Generation of heat and chemical compounds in fires: The SFPE Handbook of Fire Protection Engineering, v. 1, p. 179–199.

Tewarson, A., 1994, Flammability parameters of materials: Ignition, combustion, and fire propagation: Journal of Fire Sciences, v. 12, p. 329–356.

Tewarson, A., Lee, J. L., and Pion, R. F., 1981, The influence of oxygen concentration on fuel parameters for fire modeling, *in* Proceedings, International Symposium on Combustion, 18th: Pittsburgh, Pennsylvania, The Combustion Institute, p. 563–570.

Towe, K. M., 1990, Aerobic respiration in the Archean?: Nature, v. 348, p. 54–56.

Turpin, D. H., 1993, Phytoplankton growth and CO_2: Nature, v. 363, p. 678.

Vail, P. R., Mitchum, R. M., Todd, R. G., Widmier, J. M., Thompson, S., Songree, J. B., Rubb, J. N., and Hatlelid, W. G., 1977, Seismic stratigraphy and global changes of sea level: American Association of Petroleum Geologists Memoir 26, p. 49–212.

Van Valen, L., and Sloan, R. E., 1977, Ecology and the extinction of the dinosaurs: Evolutionary Theory, v. 2, p. 37–64.

Velbel, M. A., 1993, Temperature dependence of silicate weathering in nature: How strong a negative feedback on long-term accumulation of atmospheric CO_2 and global greenhouse warming?: Geology, v. 21, p. 1059–1062.

Wang, K., Orth, C. J., Attrep, M., Chatterton, B., Hou, H., and Geldsetzer, H., 1991, Geochemical evidence for a catastrophic biotic event at the Frasnian/Famennian boundary in south China: Geology, v. 19, p. 776–779.

Wang, K., Attrep, M., Jr., and Orth, C., 1993, Global iridium anomaly, mass extinction, and redox change at the Devonian-Carboniferous boundary: Geology, v. 21, p. 1071–1074.

Weishampel, D. B., 1990, Dinosaurian distribution, *in* Weishampel, D. B., Dodson, P., and Osmolska, H., eds., The dinosauria: Berkeley, University of California Press, p. 63–139.

Wing, S. L., Hickey, L. J., and Swisher, C. C., 1993, Implications of an exceptional fossil flora for Late Cretaceous vegetation: Nature, v. 363, p. 342–344.

Wolbach, W. S., Gilmour, I., Anders, E., Orth, C. J., and Brooks, R. R., 1988, Global fire at the Cretaceous-Tertiary boundary: Nature, v. 334, p. 665–669.

Zachos, J. C., Arthur, M. A., and Dean, W. E., 1989, Geochemical evidence for suppression of pelagic marine productivity at the Cretaceous/Tertiary boundary: Nature, v. 337, p. 61–64.

Species Index

Page numbers in italics represent illustrations.

Subject Index

Page numbers in italics represent illustrations.